Principles *of* Genetics

Principles *of* Genetics

Fourth Edition

Robert H. Tamarin

Boston University

 **Wm. C. Brown Publishers**

Dubuque, Iowa•Melbourne, Australia•Oxford, England

Book Team

Editor *Kevin Kane*
Developmental Editor *Margaret J. Kemp*
Production Editor *Marc Morehouse*
Designer *Elise A. Lansdon*
Photo Editor *Carrie Burger*
Permissions Editor *Karen L. Storlie*
Art Editor/Processor *Jodi Wagner*
Visuals/Design Developmental Consultant *Donna Slade*

WCB

Wm. C. Brown Publishers
A Division of Wm. C. Brown Communications, Inc.

Vice President and General Manager *Beverly Kolz*
National Sales Manager *Vincent R. Di Blasi*
Assistant Vice President, Editor-in-Chief *Edward G. Jaffe*
Director of Marketing *John W. Calhoun*
Marketing Manager *Carol J. Mills*
Advertising Manager *Amy Schmitz*
Director of Production *Colleen A. Yonda*
Manager of Visuals and Design *Faye M. Schilling*
Design Manager *Jac Tilton*
Art Manager *Janice Roerig*
Publishing Services Manager *Karen J. Slaght*
Permissions/Records Manager *Connie Allendorf*

WCB

Wm. C. Brown Communications, Inc.

Chairman Emeritus *Wm. C. Brown*
Chairman and Chief Executive Officer *Mark C. Falb*
President and Chief Operating Officer *G. Franklin Lewis*
Corporate Vice President, President of WCB Manufacturing *Roger Meyer*

Cover photo provided by Michael Pique and Peter E. Wright, Dept. of Molecular Biology, The Scripps Research Institute, LaJolla, CA, using software from Ray-Tracing Corporation and Sun Microsystems. From *Science* 245: 635, Aug. 11, 1989. Copyright 1989 by the AAAS.

Photo researcher *Kathy Husemann*

Copyedited by *Julie Bach*

Cover and interior design by *Deborah Schneck*

Library of Congress Catalog Card Number: 92–70225

ISBN 0–697–16658–9 (paper)
 0–697–13723–6 (case)

Printed in the United States of America by Wm. C. Brown Communications, Inc., 2460 Kerper Boulevard, Dubuque, IA 52001

10 9 8 7 6 5 4

For Ginger, David, and Bonnie

Contents

17 Non-Mendelian Inheritance 481

PART IV

Quantitative and Evolutionary Genetics 499

18 Quantitative Inheritance 501

19 Population Genetics: The Hardy-Weinberg Equilibrium and Mating Systems 522

20 Population Genetics: Processes that Change Allelic Frequencies 540

21 Genetics of the Evolutionary Process 557

Preface

The science of genetics includes the rules of inheritance in cells, individuals, and populations and the molecular mechanisms by which genes control the growth, development, and appearance of an organism. No area of biology can be truly appreciated or understood without an understanding of genetics because genes not only control cellular processes, they also determine the course of evolution. Genetics is an exciting basic science whose concepts provide the framework for the study of modern biology.

This text provides a balanced treatment of the major areas of genetics in order to prepare students for upper-level courses and to help them share in the excitement of research. Most readers of this text will have taken a general biology course and will have had some background in cell biology and organic chemistry. However, for an understanding of the concepts in this text, the motivated student will need to have completed only an introductory biology course and have had some chemistry and algebra in high school.

Genetics is commonly divided into three areas: classical, molecular, and population. Many genetics teachers feel that a historical approach provides a sound introduction to the field and that a thorough grounding in Mendelian genetics is necessary for an understanding of molecular and population genetics—an approach this book follows. Other teachers, however, may prefer to begin with molecular genetics. For this reason, the chapters have been written as units that allow for flexibility in their use. A comprehensive glossary and separate subject and author indexes will help maintain continuity if the order of the chapters is changed from the original.

An understanding of genetics is crucial to advancements in medicine, agriculture, and animal breeding; and genetic controversies—such as the potential harm of recombinant DNA and the pros and cons of the human genome project—have captured the interest of the general public. Throughout this book, the implications for human health and welfare of the research conducted in laboratories and universities around the world are pointed out. Digressions, in the form of boxed material, give insights into genetic techniques, controversies, and breakthroughs.

Because genetics is the first analytical biology course for many students, they may have difficulty with its quantitative aspects. There is no substitute for work with pad and pencil. This text provides a large number of problems to help the reader learn and retain the material. All problems, whether within the body of the text or at the end of the chapters, should be worked through as they are encountered. After students have worked out the problems, they may want to refer to the answer section at the back of the book. We provide solved problems at the end of each chapter to help.

For those who wish to pursue particular topics, each chapter provides references to review articles, more advanced volumes, and articles in the original literature. Although some of these articles might be difficult for the beginner to follow, each is either a landmark paper, a comprehensive summary, or a paper with some valuable aspect to it. Some papers may contain an insightful photograph or diagram. Some magazines and journals are especially recommended for the student to look at periodically, including *Scientific American, Science,* and *Nature* because they contain nontechnical summaries as well as material at the cutting edge of genetics. Some articles are included to help the teacher with supplementary material or material from which concepts have been developed. Photographs of selected geneticists are also included. After all, geneticists are people who occasionally make mistakes and often disagree with each other. Perhaps the glimpse of a face from time to time will help add a human touch to this science.

NEW TO THIS EDITION

Since the last edition of this book, many exciting discoveries have been made in genetics. All chapters have been updated to reflect those discoveries. For example, we have added material on the human genome project, sex determination in people, the way in which proteins fold, tumor suppressor genes, AIDS, molecular imprinting, and the practical benefits of genetic engineering. We have also added new boxed material on many topics, including prions, polymerase collisions, guide RNA, observing transcription in real time, and molecular chaperones. Each chapter now begins with a list of objectives and ends with several new problems and worked-out solutions. The end-of-chapter questions have been doubled in number to give students a wider variety and level of difficulty. Answers are provided in the appendix. For chapters containing a large amount of questions, only the even-numbered answers are given for the higher-numbered questions. Odd-numbered answers can be found in the Instructor's Manual.

ANCILLARIES

Available upon request to all adopters is an **Instructor's Manual/Solutions Manual** written by William Wellnitz of Augusta College. It contains additional problems and worked-out solutions as well as a **Test Item File** of thirty-five to fifty objective questions for each chapter. Instructors may also request a set of forty-five overhead **transparencies** that will be useful in explaining difficult concepts.

For students who want additional help in mastering problems, a **Student Solutions Manual** is available. Written by Ken Zwicker, the manual provides over 300 problems and worked-out solutions.

Also available upon request, adopters of *Principles of Genetics* may request **wcb Testpak,** a free computerized testing service that includes two options:

• A software package in IBM® 3.5 and 5.25, Apple®, and Mac programs that will enable adopters to print test masters and will allow students to take exams on the computer and teachers to use the program to grade and store exam results.

• Our call-in/mail-in service that will enable teachers to use the test item file found in the back of the instructor's manual to create exams. Choose the questions you want printed and, within two working days of your request, we will put a test master, a student answer sheet, and an answer key in the mail to you. Call-in hours are 8:30–5:00 CST, Monday through Friday.

A Mac-Hypercard program titled *GenPak: A Computer-Assisted Guide to Genetics,* is also available. This program features numerous interactive/tutorial (problem-solving) exercises in Mendelian, molecular, and population genetics at the introductory level.

ACKNOWLEDGMENTS

I would like to thank many people for their encouragement and assistance in the production of this fourth edition. At Wm. C. Brown Publishers, Kevin Kane enthusiastically supported the undertaking of this edition, and Marge Kemp worked on the details of development. It was also a pleasure to work with many other people during the production of this book, especially Marc Morehouse, Julie Bach, Carrie Burger, and Jodi Wagner. Many reviewers greatly helped improve the quality of this edition. I specifically wish to thank Charles H. Green, Rowan College of New Jersey; LaJoyce H. Debro, Jacksonville State University; Larry R. Eckroat, Penn State Erie–The Behrend College; Wade N. Hazel, DePauw University; Carol Ann Rush, La Roche College; John W. Schiefelbein, University of Michigan; David E. Sheppard, University of Delaware; William R. Wellnitz, Augusta College; W. Keith Hartberg, Baylor University; Austin L. Hughes, Penn State University; R. L. Bernstein, San Francisco State University; Susan Lovett, Brandeis University; and Ammini S. Moorthy, Wagner College.

Lastly, thanks are due to the many students, particularly those in my Introductory Genetics, Population Biology, and Graduate Seminar courses, who have helped clarify points, find errors, and discover new and interesting ways of looking at the many topics collectively called genetics.

Robert H. Tamarin
Boston

Part I

Genetics and the Scientific Method

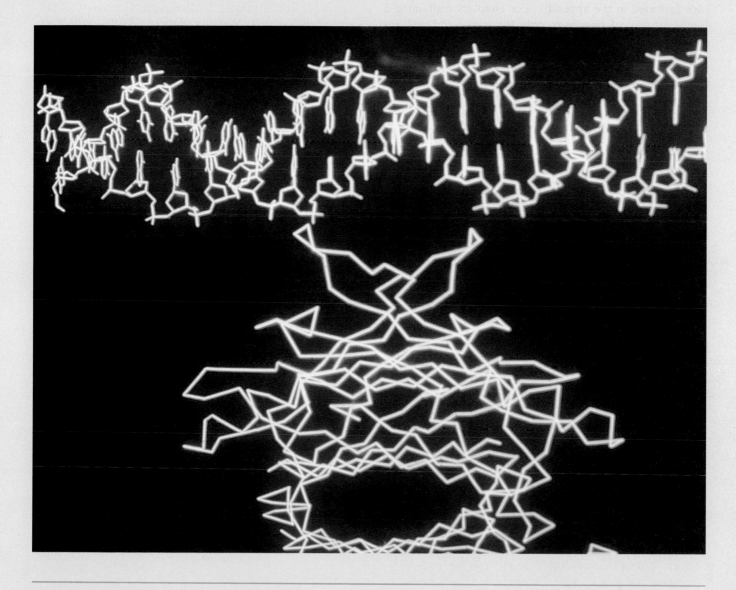

A molecular model of DNA displayed on a cathode ray tube.
© *John Yates/Superstock*

1

Introduction

OBJECTIVES

♦ To gain an overview of the topics included in the subject of genetics
♦ To analyze the scientific method
♦ To look at why certain organisms and techniques have been used preferentially in genetics research

A field of daisies (*Chrysanthemum* sp.) highlights the role of genetics in the development, functioning, and evolution of organisms in nature.
© Bruce Iverson

Genetics is the study of inheritance in all of its manifestations, from the distribution of human traits in a family pedigree to the biochemistry of the genetic material, deoxyribonucleic acid, DNA (or, in some viruses, ribonucleic acid, RNA). It is our purpose in this book to introduce and describe the processes and patterns of inheritance. In this chapter we present a broad outline of the topics to be covered.

THE THREE GENERAL AREAS OF GENETICS

Historically, geneticists have worked in three different areas, each with its own particular problems, terminology, tools, and organisms. These areas are classical genetics, molecular genetics, and evolutionary genetics. In *classical genetics* we are concerned with the chromosomal theory of inheritance, the concept that genes are located in a linear fashion on chromosomes; the relative positions of genes can be determined by the frequencies of offspring in controlled matings. *Molecular genetics* is the study of the genetic material: its structure, replication, and expression. Here we also examine the information revolution emanating from the discoveries of recombinant DNA techniques (genetic engineering). *Evolutionary genetics* is the study of the mechanisms of evolutionary change, the changes of gene frequencies in populations. Darwin's concept of evolution by natural selection is given a firm genetic footing in this area of the study of inheritance (table 1.1).

Today these areas are less clearly defined because of advances made in molecular genetics. Information from molecular genetics is allowing us to understand better the structure and functioning of chromosomes on the one hand and the nature of natural selection on the other. In this book we hope to bring together this information from a historical perspective. From Mendel's work in discovering the rules of inheritance (chapter 2) to genetic engineering (chapter 12) to molecular evolution (chapter 21), we hope to present a balanced view of the various topics that make up genetics.

HOW DO WE KNOW?

Genetics is an empirical science, which means that it works according to the *scientific method:* Our information comes from observations of the real world. The scientific method is a collection of rules on how to obtain an understanding of nature. At its heart is the experiment, in which a guess about how something works, called a hypothesis, is tested. In a good experiment, there are only two outcomes possible: One will support the hypothesis and the other will refute it (fig. 1.1). Scientists refer to this process as *strong inference.*

Figure 1.1

A schematic of the scientific method. An observation leads to a hypothesis from which predictions are made that are tested by an experiment. The results of the experiment either support or refute the hypothesis. If the hypothesis is refuted, a new hypothesis must be developed. If the hypothesis is supported, further experiments should be developed to try to disprove it.

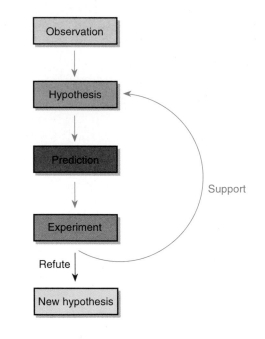

TABLE 1.1   The Three Major Areas of Genetics—Classical, Molecular, and Evolutionary— and the Topics They Cover

Classical Genetics	Molecular Genetics	Evolutionary Genetics
Mendel's principles	Structure of DNA	Quantitative genetics
Meiosis and mitosis	Chemistry of DNA	Hardy-Weinberg equilibrium
Sex determination	Transcription	Assumptions of equilibrium
Sex linkage	Translation	Evolution
Chromosomal mapping	DNA cloning	Speciation
Cytogenetics (chromosomal changes)	Control of gene expression	
	DNA mutation and repair	
	Extrachromosomal inheritance	

For example, you might have the idea that acquired characteristics can be inherited, an idea put forth by Jean-Baptiste Lamarck (1744–1829), a French biologist. Lamarck suggested that giraffes that reached higher into trees to get at leaves to eat developed longer necks. They passed on these longer necks to their offspring (in only small increments each generation), leading to the long-necked giraffes of today. An alternative view, *evolution by natural selection,* was put forward in 1859 by Charles Darwin. According to the Darwinian view, giraffes normally varied in the lengths of their necks, and this variation was inherited. Giraffes with slightly longer necks would be at an advantage in getting leaves to eat from trees. In other words, over time, the longer-necked giraffes would survive and reproduce better than the shorter-necked ones. Thus longer necks would come to predominate because of the loss of shorter-necked giraffes. Any genetic *mutations* (changes) that introduced greater neck length would be favored.

To test Lamarck's view, you would begin by choosing a suitable organism. Giraffes are too difficult to get or to do breeding studies with. You might settle on using lab mice, which are relatively inexpensive to obtain and keep. Instead of looking at neck length, you might simply cut off half their tails, assuming that this experiment is appropriate to test the hypothesis and is not needlessly painful to the animals. You could then mate these short-tailed mice and see if their offspring had shorter tails. If not, you could conclude that shortened tails, an acquired characteristic, is not inherited. If, however, the next generation of mice were born with shorter tails than their parents were born with, you would conclude that acquired characteristics can be inherited.

The point of doing the experiment, as trivial as it might seem, is to actually determine the answer to a question using data based on what really happens. You have not appealed to the authority of someone you respect or asked for divine enlightenment. If your experiment were designed correctly and carried out without error, you should have confidence in your results. If your results are negative, as ours would be here, then you would reject your hypothesis. Testing hypotheses with the understanding that they will be rejected if refuted is the essence of the scientific method.

Nothing in this book is inconsistent with that method. Every fact has been gained by experiment or observation of the real world. If you do not accept something said herein, you can go back to the *original literature,* the publication of original experiments in scientific journals (as cited at the end of each chapter) and read the work yourself. If you still don't believe a conclusion, it is the nature of the scientific method that you can repeat the work in question to either verify or challenge it.

As previously mentioned, the results of scientific studies are usually published in scientific journals. For example, the common journals read by many geneticists include *Genetics, Proceedings of the National Academy of Sciences, Science, Nature, Evolution, Journal of Experimental Zoology, American Journal of Human Genetics, Journal of Molecular Biology,* and hundreds more. The reported research has usually undergone a process called *peer review* in which an article submitted for publication is reviewed by other scientists before it is published to ensure both accuracy and relevance. Scientific articles usually include a detailed justification for the work, an outline of the methods—sufficient to allow another scientist to repeat the work—results, and a discussion of the significance of the work.

At the end of each chapter in this book we cite journal articles describing research that has contributed to that chapter. (In chapter 9 we reprint a research article by J. Watson and F. Crick.) We also cite secondary sources, journals and books that publish syntheses of the literature rather than original contributions. These include *Scientific American, Annual Review of Biochemistry, Annual Review of Genetics, American Scientist,* and others. We also cite upper-level books that have been important in defining a particular area of study. Students are encouraged to look at all of these sources in their efforts to both improve their grasp of genetics and understand how science progresses.

WHY FRUIT FLIES AND COLON BACTERIA?

In reading this book, you will see that certain organisms are used over and over again in experiments. If the goal of science is to uncover generalities about the living world, why do geneticists persist in using the same few organisms in their work? The answer is probably obvious: The organisms that are used for any particular type of study have certain attributes that make them desirable for that research.

In the early stages of genetic research, at the turn of the century, techniques had not been developed to do genetic work with bacteria and viruses. The organism that took center stage then was the fruit fly, *Drosophila melanogaster,* which had been used by developmental biologists (fig. 1.2). It has a relatively short generation time of about two weeks, survives and breeds well in the lab, has very large chromosomes in some of its cells, and has many aspects of its *phenotype* (appearance) genetically controlled. For example, it was easy to see mutations of genes that control eye color, bristle number and type, and wing characteristics such as shape or vein pattern.

At the middle of this century, when techniques were developed for genetic work on bacteria, the common colon bacterium, *Escherichia coli,* became a favorite organism of genetic researchers (fig. 1.3). Because it had a generation time of only twenty minutes, a single copy of each gene, and only a small amount of genetic material, it was

Figure 1.2
Adult female fruit fly, *Drosophila melanogaster*. Mutations of eye color, bristle type and number, and wing characteristics are easily visible when they occur.

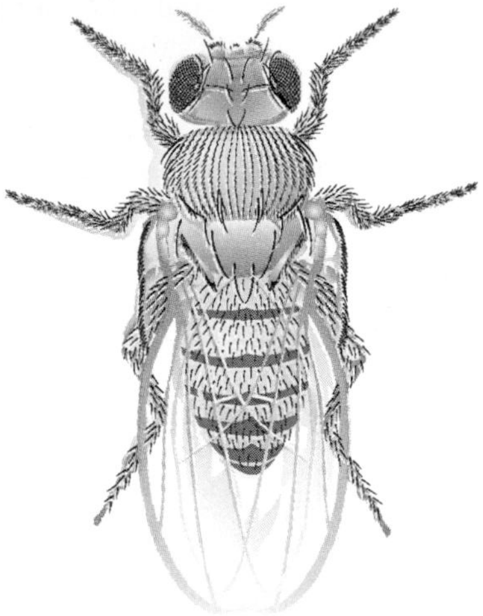

Figure 1.4
Mendel worked with garden pea plants. He observed seven traits of the plant, each with two discrete forms, attributes of the seed, the pod, and the stem. For example, all plants had either round or wrinkled seeds, full or constricted pods, or yellow or green pods.

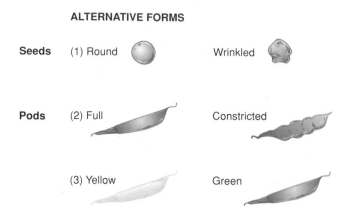

ALTERNATIVE FORMS

Seeds	(1) Round		Wrinkled
Pods	(2) Full		Constricted
	(3) Yellow		Green

Figure 1.3
Electron micrograph of an *Escherichia coli* bacterium. It is a bacillus, a rod-shaped bacterium. The bacterium is magnified about 3,700×.
Courtesy of Wayne Rosenkrans and Dr. Sonia Guterman.

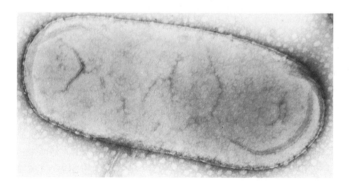

used by many research groups. Still later, bacterial viruses, called *bacteriophages,* became very popular in genetics labs. The viruses are constructed of only a few types of protein molecules and a very small chromosome. Some can replicate a hundredfold in an hour. Our point is not to list the major organisms used by geneticists but to suggest why some are used commonly. Comparative studies are usually done to determine which generalities discovered in the elite genetic organisms are really scientifically universal.

TECHNIQUES OF STUDY

Each area of genetics has its own particular techniques of study. Often the development of a new technique, or an improvement in a technique, has opened up major new avenues of research. As our technology has improved over the years, problems are explored at lower and lower levels

of biological organization. Gregor Mendel, the father of genetics, did simple breeding studies of plants in a garden at his monastery in Brünn, Austria, in the middle of the nineteenth century. Today, with modern biochemical and biophysical techniques, it has become routine to determine the sequence of *nucleotides* (molecular subunits of DNA and RNA) that make up any particular gene. In fact, one of the most ambitious projects being carried out currently in genetics is the mapping of the human genome, all 3.3 billion nucleotides that make up all of our genes. Only recently has the technology been developed to begin a project of this magnitude, which may take a decade or more and cost billions of dollars to accomplish. Many scientists, however, feel that it will be worth the expense.

CLASSICAL, MOLECULAR, AND EVOLUTIONARY GENETICS

In the next three sections we briefly outline the general subject areas covered in the book: classical, molecular, and evolutionary genetics.

Classical Genetics

Gregor Mendel discovered the rules of inheritance in 1859 by doing carefully controlled breeding experiments with the garden pea plant, *Pisum sativum.* He found that traits, such as flower color, were controlled by genetic elements that we now call genes (fig. 1.4). Adult organisms had two copies of each gene (*diploid* state); gametes received just one of these copies (*haploid* state). In other words, one of the two copies was segregated into any given gamete. Upon fertilization, the zygote would get one copy from each gamete, reconstituting the diploid number (fig. 1.5). When Mendel looked at the inheritance of several traits at the same time, he found that they were inherited independently of each other. His work has been distilled into two

Figure 1.5

Mendel crossed tall and dwarf pea plants, demonstrating the rule of segregation. A diploid individual with two copies of the gene for tallness (T) per cell forms gametes that have the T gene. Similarly, an individual that has two copies of the gene for shortness (t) will have gametes that have the t gene. Fertilization will result in zygotes that have both the T and t genes. When both forms are present (T, t), the plant is tall, indicating that the T gene is *dominant* to the t gene, which is *recessive*.

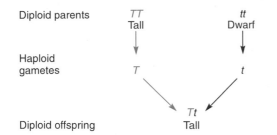

Figure 1.6

Genes are located in linear order on chromosomes, as seen in a diagram of chromosome number 2 of *Drosophila melanogaster,* the common fruit fly. The centromere is a constriction in the chromosome. The numbers are map units.

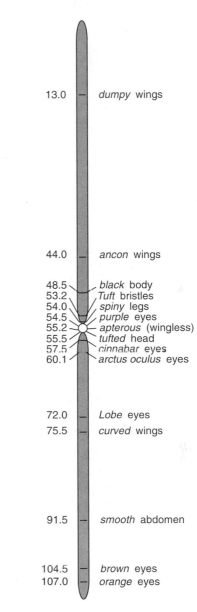

rules, referred to as *segregation* and *independent assortment*. Not until the analysis of the behavior of chromosomes during the latter half of the nineteenth century was Mendel's work accepted. At that time, in the year 1900, the science of genetics was born.

During much of the early part of this century, geneticists discovered many genes by looking for changed organisms, called *mutants*. Crosses were made to determine the genetic control of mutant traits. From this research evolved chromosome mapping, the ability to locate the relative position of genes on chromosomes by doing appropriate crosses. The proportion of recombinant offspring, those with new combinations of parental genes, gives a measure of separation between genes on the same chromosomes in distances called *map units*. From this work arose the chromosomal theory of inheritance: Genes are located at fixed positions on chromosomes in a linear order. This is known as the "beads on a string" model of gene arrangement (fig. 1.6) and was not modified to any great extent until the middle of this century, after the structure of DNA was worked out.

In general, genes function by controlling the synthesis of proteins (called *enzymes*) that act as biological catalysts in cellular pathways (fig. 1.7). G. Beadle and E. Tatum suggested that one gene controls the formation of one enzyme. Although we now know that many proteins are made up of subunits, the products of several genes, and that some genes code for proteins that are not enzymes and others do not code for proteins, the *one-gene-one-enzyme* rule of thumb serves as a general guideline to gene action.

Molecular Genetics

With the exception of some viruses, the genetic material of most organisms is DNA, a double helical molecule shaped like a twisted ladder. The backbones of the helices are repeating units of sugars (deoxyribose) and phosphate

groups. The rungs of the ladder are base pairs, one base from each side (fig. 1.8). There are only four bases found normally in DNA: adenine, thymine, guanine, and cytosine, abbreviated A, T, G, and C, respectively. There is no restriction on the order of bases on one strand. However, there is a relationship between bases forming a rung, called *complementarity*. If one base of the pair is adenine, the other must be thymine; if one base is guanine, the other must be cytosine. James Watson and Francis Crick deduced this structure in 1953, ushering in the modern era of molecular genetics.

From the complementary nature of the base pairs of DNA, the mode of replication was obvious to Watson and

Figure 1.7

Biochemical pathways show the sequential changes in compounds as they are modified by cellular reactions. In this case, we show the first few steps in the glycolytic pathway in which glucose is converted to energy. For example, glucose + ATP is converted to glucose-6-phosphate + ADP with the aid of the enzyme hexokinase. The enzymes are the products of genes.

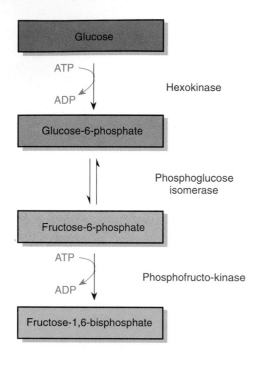

Figure 1.8

A closer look at a DNA double helix, showing the backbone of sugar-phosphate units and the rungs as base pairs. We abbreviate a phosphate group as a "P" within a circle; the pentagonal ring containing an oxygen atom is the sugar deoxyribose. Bases are either adenine, thymine, cytosine, or guanine (A, T, C, G).

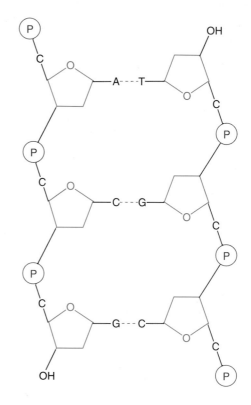

Crick: The double helix would "unzip" and each strand would act as a template for a new strand, resulting in two double helices, exactly like the first (fig. 1.9). Mutation, a change in one of the bases, could result from either an error in complementarity during replication or some damage to the DNA that was not repaired by the time of the next cycle of DNA replication.

Information is encoded in DNA in the sequence of bases on one strand of the double helix. During gene expression, that information is *transcribed* into RNA, the other form of nucleic acid, which actually takes part in protein synthesis. RNA differs from DNA in several respects: It has the sugar ribose in place of deoxyribose; it has the base uracil (U) in place of thymine (T); and it usually occurs in a single-stranded form. RNA is transcribed from DNA by the enzyme *RNA polymerase,* using DNA–RNA rules of complementarity: G, C, A, and T in DNA pair with C, G, U, and A, respectively, in RNA (fig. 1.10). The DNA information that is transcribed into RNA is for the amino acid sequence of proteins. Three nucleotide bases form a *codon* that specifies one of the twenty naturally occurring amino acids used in protein synthesis. The sequences of bases making up the codons are referred to as the genetic code (table 1.2).

The process of *translation,* the converting of nucleotide sequences into amino acid sequences, takes place at the ribosome, a structure found in all cells, made up of RNA and proteins (fig. 1.11). As the RNA moves along the ribosome, one codon at a time, one amino acid is added to the growing protein for each codon.

The major control mechanisms of gene expression usually act at the transcriptional level. For transcription to take place, the RNA polymerase enzyme must be able to pass along the DNA; if it is prevented from this movement, transcription will be stopped. Various proteins can bind to the DNA, thus preventing the RNA polymerase from continuing, providing a mechanism of control of transcription. One particular mechanism is known as the *operon model* and provides the basis for a wide range of control mechanisms, primarily in prokaryotes and viruses.

In eukaryotes there are no operons and, although we know quite a bit about some systems of control of gene expression, the general rules are not as clear. There is an indication that the level of methylation (the addition of methyl groups to DNA bases) may be important in controlling gene expression as well as some odd structural changes in the DNA, causing it to form a zigzag or Z-form.

Figure 1.9

The DNA double helix unwinds during replication. Each half acts as a template for a new double helix. Because of the rules of complementarity, each new double helix will be identical to the original one. In addition, the two new double helices will be identical to each other. Thus, an AT base pair in the original DNA double helix will result in two AT base pairs, one each in the daughter strands.

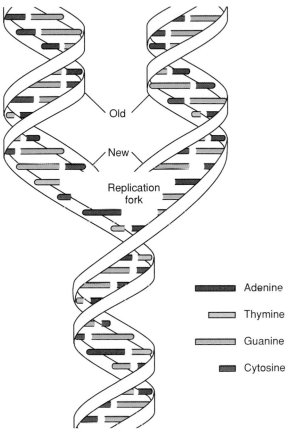

Old

New

Replication fork

Adenine

Thymine

Guanine

Cytosine

Figure 1.10

Transcription is the process whereby RNA is synthesized using the complementarity of DNA bases as a template. The enzyme responsible is RNA polymerase. The DNA double helix is partially unwound during this process, allowing the bases of one strand to be a template in the synthesis of RNA, using the rules of DNA-RNA complementarity: A, T, G, and C of DNA pair with U, A, C, and G, respectively, in RNA. The RNA base sequence is identical to the sequence that would be formed if the DNA were replicating instead, with the exception that RNA has uracil in place of thymine.

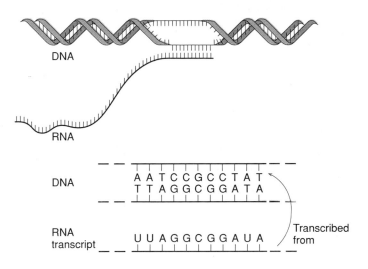

DNA

RNA

DNA   A A T C C G C C T A T
       T T A G G C G G A T A

RNA
transcript   U U A G G C G G A U A

Transcribed from

TABLE 1.2   The Genetic Code Dictionary of RNA

Codon	Amino Acid	Codon	Amino Acid	Codon	Amino Acid	Codon	Amino Acid
UUU	Phe	UCU	Ser	UAU	Tyr	UGU	Cys
UUC	Phe	UCC	Ser	UAC	Tyr	UGC	Cys
UUA	Leu	UCA	Ser	UAA	STOP	UGA	STOP
UUG	Leu	UCG	Ser	UAG	STOP	UGG	Trp
CUU	Leu	CCU	Pro	CAU	His	CGU	Arg
CUC	Leu	CCC	Pro	CAC	His	CGC	Arg
CUA	Leu	CCA	Pro	CAA	Gln	CGA	Arg
CUG	Leu	CCG	Pro	CAG	Gln	CGG	Arg
AUU	Ile	ACU	Thr	AAU	Asn	AGU	Ser
AUC	Ile	ACC	Thr	AAC	Asn	AGC	Ser
AUA	Ile	ACA	Thr	AAA	Lys	AGA	Arg
AUG	Met (START)	ACG	Thr	AAG	Lys	AGG	Arg
GUU	Val	GCU	Ala	GAU	Asp	GGU	Gly
GUC	Val	GCC	Ala	GAC	Asp	GGC	Gly
GUA	Val	GCA	Ala	GAA	Glu	GGA	Gly
GUG	Val	GCG	Ala	GAG	Glu	GGG	Gly

A codon, specifying one amino acid, is three bases long (read in RNA bases in which U replaces the T of DNA). There are 64 different codons, specifying 20 naturally occurring amino acids (abbreviated by three letters: e.g., Phe is phenylalanine—see fig 11.1 for the names and structures of the amino acids). Also present is stop (UAA, UAG, UGA) and start (AUG) information.

Figure 1.11

In prokaryotes, translation of RNA begins shortly after the RNA is synthesized. A ribosome attaches to the RNA and begins reading the codons of the RNA. As the ribosome moves along the RNA, amino acids are added to the growing protein. When the process is finished, the completed protein is released from the ribosome and the ribosome detaches from the RNA. As the first ribosome moves along, a second ribosome can attach at the beginning of the RNA, and so on, creating RNAs with many ribosomes attached.

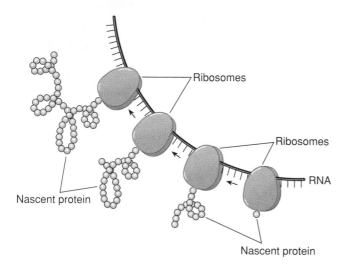

Figure 1.12

Hybrid DNA molecules can be constructed of a plasmid and a piece of foreign DNA. The ends are made compatible by having both DNAs cut with the same restriction endonuclease that leaves complementary ends. These ends will re-form double helices to form intact hybrid plasmids when the two types of DNA are mixed. A repair enzyme, DNA ligase, finishes the patching of the hybrid DNA within the plasmid. The hybrid plasmid is then reinjected into a bacterium, to be grown up into billions of copies that can later be isolated and studied (sequenced), or the hybrid plasmid can express the foreign DNA from within the host bacterium.

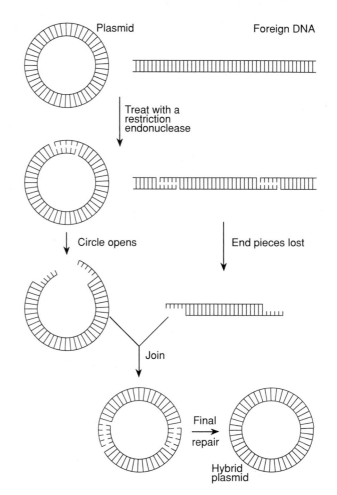

In recent years there has been an explosion of information resulting from *recombinant DNA techniques.* This revolution began with the discovery of *restriction endonucleases,* enzymes that cut DNA at specific sequences. Many of these enzymes leave single-stranded ends on the cut DNA. If a *plasmid,* a small circular chromosome found in some bacteria, and another piece of DNA (called foreign DNA) are both acted on by the same restriction enzyme, they will be left with identical single-stranded free ends (fig. 1.12). If the cut plasmid and cut foreign DNA are mixed together, the free ends can re-form double helices, and thus a plasmid with a single piece of foreign DNA within it can form (fig. 1.12). Final repair processes will create a completely closed circle of DNA. The hybrid plasmid is then reinserted into the bacterium. When the bacterium grows, it replicates the plasmid DNA, producing many copies of the foreign DNA. From that point, the foreign DNA can be isolated and sequenced, a process in which the exact order of bases making up the foreign DNA can be determined. That sequence itself can tell us much about how the gene works. In addition, the gene can function within the bacterium, resulting in bacteria, for example, producing human growth hormone.

This technology has tremendous implications in medicine, agriculture, and industry. The opportunity to locate and study disease-causing genes, such as the genes for cystic fibrosis and muscular dystrophy, is being made available by this technology, as well as potential treatments. Crop plants and farm animals are being modified for better productivity by improving growth and disease resistance. And industries are flourishing that are bringing to fruition the concepts of genetic engineers.

One area of great interest has been in cancer research. We have discovered that cancer can be caused by a single gene that has lost its normal control mechanisms (an *oncogene*). These oncogenes exist normally in noncancerous cells and are also carried by viruses. Cancer-causing viruses are especially interesting because most of them are of the RNA type. The disease AIDS is caused by one of these RNA viruses that attacks one of the cells in the immune system. Discovering the mechanism by which our immune system can produce millions of different protective proteins (*antibodies*) has been another success of modern molecular genetics.

Evolutionary Genetics

From a genetic standpoint, evolution is the change in gene frequencies in a population over time. Charles Darwin described evolution as the result of natural selection. In the 1920s and 1930s, geneticists, primarily Sewall Wright, R. A. Fisher, and J. B. S. Haldane, provided the algebraic models of evolutionary processes. The marriage of Darwinian theory and population genetics has been termed neo-Darwinism.

In 1908, G. H. Hardy and W. Weinberg discovered a simple genetic equilibrium that occurs in a population if the population is large, has random mating, and has negligible effects of mutation, migration, and natural selection. This equilibrium gives population geneticists a baseline from which populations can be compared to see if any evolutionary processes are occurring. The equilibrium condition can be formulated as a statement: If the assumptions are met, the population will not experience changes in gene frequencies, and the frequencies of *genotypes* (gene combinations in individuals, e.g., *AA, Aa,* or *aa*) are predicted by the gene frequencies.

Recently several areas of evolutionary genetics have become active and controversial. It has been discovered by electrophoresis (a method of separation of proteins and other molecules), and then DNA sequencing, that there is much more *polymorphism* (variation) within natural populations than older mathematical models could account for. One of the more interesting explanations for this variability is that it is neutral. That is, natural selection, the guiding force of evolution, does not act differentially on the genetic differences found so commonly in nature. At first, this theory was quite controversial, with few followers. Now it seems to be the view of the majority about the abundance of variation found in natural populations.

Another controversial theory concerns the rate of evolutionary change. It is suggested that most evolutionary change is not gradual, as the fossil record seems to indicate, but occurs in short, rapid bursts, followed by long periods of very little change. This theory is called *punctuated equilibrium.*

A final area of evolutionary biology that has generated much controversy is the theory of *sociobiology,* in which it is suggested that social behavior is under genetic control and is acted upon by natural selection just as any morphological or physiological trait of an organism. It is controversial mainly in the way in which it is applied to human beings; it calls altruism into question and suggests that to some extent we are programmed genetically to act in certain ways. People have criticized the theory because they feel it justifies racism and sexism.

SUMMARY

The purpose of this chapter has been to provide a brief overview of what is presented in the following twenty chapters. We hope it serves to introduce the material and to provide a basis for early synthesis of some of the material that, of necessity, is presented in discrete units called chapters. This chapter is also different from all the other chapters in that it lacks the end materials that should be of value to you as you proceed: a summary, solved problems, exercises and problems, and suggestions for further reading. These parts are left off this chapter; they will be presented chapter by chapter throughout the book at places in which much more detail is presented on each topic. (For students wishing to read further on the philosophy of science, we recommend *Philosophy of Biology* by M. Ruse, ed. [1989], Macmillan, New York; *How We Know* by M. Goldstein and I. F. Goldstein [1978], Plenum, New York; *The Structure of Scientific Revolutions* by T. S. Kuhn [1962], University of Chicago Press; *Conjectures and Refutations: The Growth of Scientific Knowledge* by K. R. Popper [1962], Basic Books, New York; *The Philosophy of Science: An Introduction* by S. Toulmin [1953], Harper Torch Books, Harper & Row, New York.) At the end of the book, before the index, we provide answers to exercises and problems and a glossary of all words first seen in boldface throughout the book.

Part II

Mendelism and the Chromosomal Theory

Metaphase of *Allium* (onion) root tip. (500×).
© *Edwin A. Reschke/Peter Arnold, Inc.*

2

Mendel's Principles

OBJECTIVES

◆ To understand that genes are discrete units that control the appearance of an organism

◆ To understand Mendel's rules of inheritance: segregation and independent assortment

◆ To understand that dominance is a function of the interaction of alleles. Similarly, an interaction of nonallelic genes is termed epistasis

◆ To define how genes generally control the production of enzymes and thus the fate of biochemical pathways

A fruit fly (*Drosophila melanogaster*), one of the most common organisms used in genetic research.

© *Runk/Schoenberger/Grant Heilman Photography, Inc.*

15

Genetics is concerned with the transmission, expression, and evolution of genes, the molecules that control the function, development, and ultimate appearance of individuals. In this section of the book we will look at the rules of transmission of genes, their passage from one generation to the next. Gregor Mendel discovered these rules of inheritance; we derive and expand upon his rules in this chapter (fig. 2.1).

In 1900, three botanists, Carl Correns of Germany, Erich von Tschermak of Austria, and Hugo de Vries of Holland, reported the rules governing the transmission of traits from parent to offspring (see Box 2.1). These rules had been published previously in 1866 by an obscure Austrian monk named Gregor Mendel. Although his work was widely available after 1866, not until the turn of the century was the scientific community ready to appreciate Mendel's great contribution. There are at least four reasons for this lapse of thirty-four years.

First, before Mendel's experiments, biologists were primarily concerned with explaining the transmission of characteristics that could be measured on a continuous scale such as height, cranium size, and longevity. They were looking for rules of inheritance that would explain such **continuous variations,** especially after Darwin's theory of evolution was put forth in 1859 (see chapter 21). Mendel, however, suggested that inherited characteristics were discrete and constant (**discontinuous**): Peas were either yellow or green. Evolutionists were looking for small changes in traits with continuous variation, whereas Mendel presented them with rules for discontinuous variation. His principles did not seem to cover the type of variation that biologists thought prevailed. Second, there was no physical element with which Mendel's inherited particles could be identified. One could not say, upon reading Mendel's work, that a certain subunit of the cell followed Mendel's rules. Third, Mendel worked with large numbers of offspring and converted these numbers to ratios. Biologists, practitioners of a very descriptive science at the time, were not well trained in mathematical tools. And last, Mendel was not well known and did not persevere in his attempts to convince the academic community.

Between 1866 and 1900, two major changes took place in biological science. First, by the turn of the century, not only had scientists discovered chromosomes, but they also had learned to understand chromosomal movement during cell division. Second, biologists were better prepared to handle mathematics by the turn of the century than they were during Mendel's time.

MENDEL'S EXPERIMENTS

Gregor Mendel was an Austrian monk (of Brünn, Austria, which is now Brno, Czechoslovakia). The essence of his experiments was to **crossbreed** plants that had discrete, nonoverlapping characteristics and then to observe the

distribution of these characteristics in the next several generations. Mendel worked with the common garden pea plant, *Pisum sativum.* (See the translation of part of Mendel's original paper in this chapter, Box 2.2.) He chose the pea plant for at least three reasons. (1) The garden pea was easy to cultivate and had a relatively short life cycle. (2) The plant had discontinuous characteristics such as flower color and pea texture. (3) In part because of its anatomy, pollination of the plant could be controlled easily. Foreign pollen could be kept out, and **cross-fertilization** could be accomplished artificially.

Figure 2.2 shows a cross section of the pea flower and indicates the keel in which the anthers and stigma develop. Normally, **self-fertilization** occurs when pollen falls onto the stigma before the bud opens. Mendel cross-fertilized the plants by opening the keel of a flower before the anthers matured and placed pollen from another plant on the stigma. In more than ten thousand plants examined by Mendel, only a few were fertilized other than the way he had intended them to be (either self- or cross-pollinated).

BOX 2.1

WHO REALLY REDISCOVERED MENDEL?

Historically, three botanists—Carl Correns, Erich von Tschermak, and Hugo de Vries—have been given credit for independently rediscovering Mendel's rule of segregation. In recent years, critics have questioned the contributions of all three.

To rediscover prior work means, in this context, to make an independent discovery and then to learn of, and give appropriate credit to, the person who had done the work earlier. Thus implicit in calling the three botanists rediscoverers of Mendel's work is the fact that they independently discovered the rules of inheritance. Later they found out that Mendel had discovered these rules thirty-four years earlier. Correns, Tschermak, and de Vries have been criticized for taking credit for rediscovering Mendel's rules without actually having rediscovered them. That is, their research, which was leading them toward the rules of inheritance, led them instead to Mendel's published work. Presumably, they then took credit for independently discovering the rules when in fact they had only read them in Mendel's paper.

Curt Stern, in the foreword of his 1966 book edited with Eva Sherwood (see Suggestions for Further Reading), cast doubt on Tschermak's contribution by pointing out that Tschermak's papers written in 1900 (three in all) lacked a fundamental understanding of Mendel's rules. In several places Tschermak failed to make appropriate analyses that would be consistent with segregation. For example, when F_1 hybrids are testcrossed to the recessive parent, a 1:1 ratio should result (see fig. 2.8). Tschermak wrote, "The influence of the trait 'yellow' in the seeds of the hybrid was reduced by 57 per cent, that of the trait 'smooth' by 43.5 per cent." In other words, he did not see the 1:1 ratio that was there and a part of the rule of segregation. Later Tschermak wrote that he purposefully avoided putting the results in theoretical terms. However, as Stern points out, it is precisely those theoretical terms that define the rule of segregation.

Corcos and Monaghan, in a 1985 paper, describe the long-standing controversy regarding de Vries's claim to rediscovery of Mendel's work. According to de Vries's critics, his published work did not reveal the essence of the rule of segregation until after he read Mendel's paper. Nor did he give credit to Mendel until after Carl Correns in 1900 published his first paper acknowledging Mendel's contribution. De Vries was also vague about when he first read Mendel's paper, giving different accounts at different times.

In a 1987 paper, Corcos and Monaghan suggested that Correns did not have a flash of independent discovery, as Correns had written, but instead had a flash of understanding. After analyzing Correns's papers and writings on the subject of Mendel's rediscovery, Corcos and Monaghan point out that before Correns wrote his first paper describing the rules of inheritance, he had already read Mendel's paper. In addition, like de Vries and Tschermak, Correns was somewhat vague about the details of events that took place in the fall of 1899, just before he published his rediscovery paper.

To keep the record straight, it should be pointed out that Correns, Tschermak, and de Vries made important discoveries that are not in doubt and that all three men have supporters who believe that they independently rediscovered the rule of segregation. If the critics are correct, all three men might be guilty of self-deception rather than outright deception of others. They had been working toward discovering the rule of segregation. After reading Mendel's "obscure" paper, it is easy to see how they might have convinced themselves that they had thought of the monumental ideas independently.

Mendel used plants that had been obtained from suppliers and grew them for two years to ascertain that they were homogeneous, or true-breeding, for the particular characteristic under study. He chose for study the seven characteristics shown in figure 2.3. Take as an example the characteristic of plant height. Although height is often continuously distributed, Mendel used plants that showed only two alternatives: very tall or dwarf. He made the crosses shown in figure 2.4. In the parental, or P_1, generation, tall plants were pollinated by dwarf plants and,

in a **reciprocal cross,** dwarf plants were pollinated by tall ones to determine whether the results were independent of the parents' sex. As we will see later on, some traits have inheritance patterns related to the sex of the parents carrying the traits. In those cases, reciprocal crosses give different results; with Mendel's tall and dwarf pea plants, the results were the same.

Offspring of the cross of P_1 individuals are referred to as the first **filial generation,** or F_1. Mendel also referred to them as **hybrids** because they were the offspring of

BOX 2.2
SOME EXCERPTS FROM MENDEL'S ORIGINAL PAPER

In February and March of 1865, Mendel delivered two lectures to the Natural History Society of Brünn. These were published as a single, forty-eight-page article handwritten in German. The article appeared in the 1865 *Proceedings of the Society,* which came out in 1866. The paper was entitled "Versuche über Pflanzen-Hybriden," which means "experiments in plant hybridization." Following are some paragraphs from the English translation of the article to give some of the sense of the original.

In his introductory remarks Mendel writes:

> That, so far, no generally applicable law governing the formation and development of hybrids has been successfully formulated can hardly be wondered at by anyone who is acquainted with the extent of the task, and can appreciate the difficulties with which experiments of this class have to contend. A final decision can only be arrived at when we shall have before us the results of detailed experiments made on plants belonging to the most diverse orders.
>
> Those who survey the work done in this department will arrive at the conviction that among all the numerous experiments made, not one has been carried out to such an extent and in such a way as to make it possible to determine the number of different forms under which the offspring of hybrids appear, or to arrange these forms with certainty according to their separate generations, or definitely to ascertain their statistical relations.
>
> . . .
>
> The paper now presented records the results of such a detailed experiment. This experiment was practically confined to a small plant group, and is now, after eight years' pursuit, concluded in all essentials. Whether the plan upon which the separate experiments were conducted and carried out was the best suited to attain the desired end is left to the friendly decision of the reader.

After discussing the origin of his seeds and the nature of the experiments, Mendel then proceeds to discuss the F_1, or hybrids, resulting:

> This is precisely the case with the Pea hybrids. In the case of each of the seven crosses the hybrid-character resembles that of one of the parental forms so closely that the other either escapes observation completely or cannot be detected with certainty. This circumstance is of great importance in the determination and classification of the forms under which the offspring of the hybrids appear. Henceforth in this paper those characters which are transmitted entire, or almost unchanged in the hybridization, and therefore in themselves constitute the characters of the hybrid, are termed the *dominant,* and those which become latent in the process, *recessive.* The expression "recessive" has been chosen because the characters thereby designated withdraw or entirely disappear in the hybrids, but nevertheless reappear unchanged in their progeny, as will be demonstrated later on.

He then writes about the F_2 generation:

> In this generation there reappear, together with the dominant characters, also the recessive ones with their peculiarities fully developed, and this occurs in the definitely expressed average proportion of three to one, so that among each four plants of this generation three display the dominant character and one the recessive. This relates without exception to all the characters which were investigated in the experiments. The angular wrinkled form of the seed, the green colour of the albumen, the white colour of the seed-coats and the flowers, the constrictions of the pods, the yellow colour of the unripe pod, of the stalk, of the calyx, and of the leaf venation, the umbel-like form of the inflorescence, and the dwarfed stem, all reappear in the numerical proportion given, without any essential alteration. *Transitional forms were not observed in any experiment.*
>
> . . .
>
> Expt. 1. Form of seed.—From 253 hybrids 7,324 seeds were obtained in the second trial year. Among them were 5,474 round or roundish ones and 1,850 angular wrinkled ones. Therefrom the ratio 2.96 to 1 is deduced.
>
> If A be taken as denoting one of the two constant characters, for instance the dominant, a the recessive, and Aa the hybrid form in which both are conjoined, the expression
>
> $$A + 2Aa + a$$
>
> shows the terms in the series for the progeny of the hybrids of two differentiating characters.

Mendel also discusses the dihybrids. He mentions the genotypic ratio of $1:2:1:2:4:2:1:2:1$ and the principle of independent assortment.

> The fertilized seeds appeared round and yellow like those of the seed parents. The plants raised therefrom yielded seeds of four sorts, which frequently presented themselves in one pod. In all, 556 seeds were yielded by 15 plants, and of those there were
>
> 315 round and yellow
> 101 wrinkled and yellow
> 108 round and green
> 32 wrinkled and green
>
> Consequently the offspring of the hybrids, if two kinds of differentiating characters are combined therein, are represented by the expression
>
> $$AB + Ab + aB + ab + 2ABb + 2aBb + 2AaB + 2Aab + 4AaBb$$

This expression is indisputably a combination series in which the two expressions for the characters A and a, B and b are combined. We arrive at the full number of the classes of the series by the combination of the expressions

$A + 2Aa + a$
$B + 2Bb + b$

There is therefore no doubt that for the whole of the characters involved in the experiments the principle applies that *the offspring of the hybrids in which several essentially different characters are combined exhibit the terms of a series of combinations, in which the developmental series for each pair of differentiating characters are united.* It is demonstrated at the same time that *the relation of each pair of different characters in hybrid union is independent of the other differences in the two original parental stocks.*

Source: Gregor Mendel, "Versuche über Pflanzen-Hybriden," paper presented to the Natural History Society of Brünn, 1866. Translation by the Royal Horticultural Society, London.

Figure 2.2
Anatomy of the garden pea plant flower

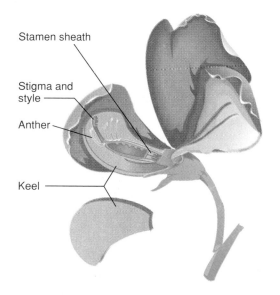

Stamen sheath

Stigma and style

Anther

Keel

unlike parents (tall and dwarf). We will specifically refer to the offspring of tall and dwarf peas as **monohybrids** because they are hybrid for only one characteristic (height). Since all the F_1 offspring plants were tall, Mendel referred to tallness as the **dominant** trait. The alternative, dwarfness, he referred to as **recessive.** The terms dominant and recessive are used to describe both the genes and the traits they control. Thus, we say that the gene for tallness and the trait, tall, are dominant. Dominance applies to the appearance of the trait in the heterozygous condition. It does not imply that the dominant trait is better, more abundant, or will increase over time in a population.

When the F_1 offspring of figure 2.4 were self-fertilized to form the F_2 generation, both tall and dwarf offspring occurred; the dwarf characteristic reappeared.

Among the F_2 offspring, Mendel observed 787 tall and 277 dwarf plants for a ratio of 2.84 to 1. It is an indication of Mendel's insight that he recognized in these numbers an approximation to a 3:1 ratio, a ratio that suggested to him the mechanism of the inheritance of height.

SEGREGATION

Rule of Segregation

Mendel assumed that each plant contained two determinants (which we now call **genes**) for the characteristic of height. Different forms of a gene exist within a population and are termed **alleles.** For example, a hybrid F_1 pea plant possesses the dominant allele for tallness and the recessive allele for dwarfness for the gene that determines plant height. A pair of alleles for dwarfness is required to develop the recessive phenotype. Only one of these alleles is passed into a single **gamete,** and the union of two gametes to form a zygote restores the double complement of alleles. The fact that the recessive trait reappears in the F_2 generation shows that the allele controlling it was unaffected by being hidden in the F_1 individual. This explanation of the passage of these discrete trait determinants, the genes, is referred to as Mendel's first principle, the **rule of segregation.** The rule of segregation can be summarized as follows: A gamete receives only one allele from the pair of alleles possessed by an organism; fertilization (the union of two gametes) reestablishes the double number. We can visualize this process by redrawing figure 2.4 using letters to denote the alleles. Mendel used capital letters to denote alleles that controlled dominant traits and lowercase letters for alleles that controlled recessive traits. Following this notation, T will be used for the allele controlling tallness and t will be used for the allele controlling shortness (dwarfed stature). From figure 2.5 we can see

Figure 2.3

Seven characteristics observed by Mendel in peas. Traits in the left column are dominant.

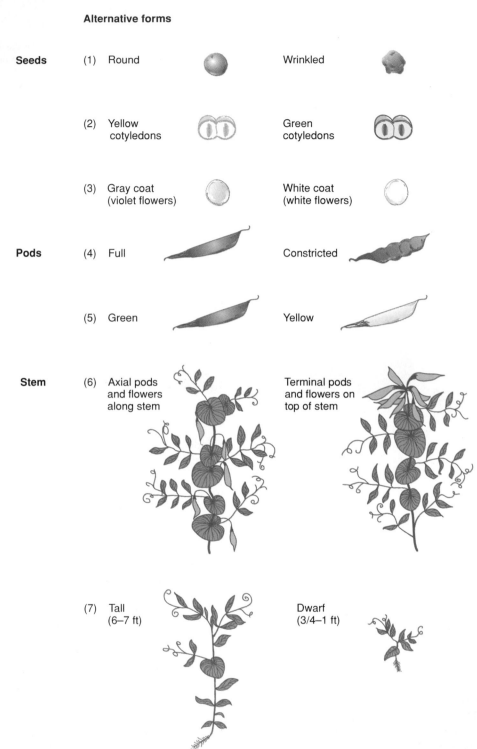

Alternative forms

Seeds	(1)	Round	Wrinkled
	(2)	Yellow cotyledons	Green cotyledons
	(3)	Gray coat (violet flowers)	White coat (white flowers)
Pods	(4)	Full	Constricted
	(5)	Green	Yellow
Stem	(6)	Axial pods and flowers along stem	Terminal pods and flowers on top of stem
	(7)	Tall (6–7 ft)	Dwarf (3/4–1 ft)

that Mendel's rule of segregation explains the homogeneity of the F_1 generation (all tall) and the 3:1 ratio of tall to dwarf offspring in the F_2 generation.

Let us define some terms. The **genotype** of an organism refers to the genes it possesses. In figure 2.5 the genotype of the parental tall plant is TT; that of the F_1 tall plant is Tt. **Phenotype** refers to the observable attributes of an organism. Plants with either of the above two genotypes, TT or Tt, are phenotypically tall. Genotypes come in two general classes: **homozygotes**, in which both alleles are the same, as in TT or tt, and **heterozygotes**, in which the two alleles are different, as in Tt. These last two

Figure 2.4
First two offspring generations from the cross of tall plants with dwarf plants

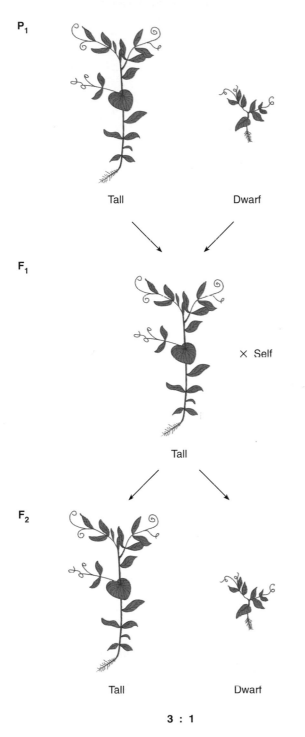

P₁

Tall   Dwarf

F₁

× Self

Tall

F₂

Tall   Dwarf

3 : 1

Figure 2.5
Assigning of genotypes to the cross in figure 2.4

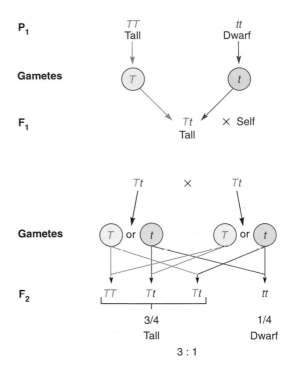

P₁   *TT*   *tt*
            Tall   Dwarf

Gametes   *T*   *t*

F₁   *Tt*   × Self
                  Tall

            Tt   ×   *Tt*

Gametes   *T* or *t*   *T* or *t*

F₂   *TT*   *Tt*   *Tt*   *tt*

              3/4   1/4
              Tall   Dwarf

              3 : 1

dividuals are uniformly heterozygous *Tt,* and each F₁ individual can produce two kinds of gametes in equal frequencies, *T*- or *t*-bearing. In producing the F₂ generation, these two types of gametes randomly pair during the process of fertilization. Figure 2.6 shows three ways of picturing this process.

Testing the Rule of Segregation

We can see from figure 2.6 that the F₂ generation has a phenotypic ratio of 3:1, the classic Mendelian ratio. However, there is also a genotypic ratio of 1:2:1 for dominant homozygote:heterozygote:recessive homozygote. Demonstrating this genotypic ratio provides a good test of Mendel's hypothesis of segregation.

The simplest way to test the hypothesis is by **progeny testing,** that is, by self-fertilizing F₂ individuals to produce an F₃ generation, which Mendel did. From his hypothesis it is possible to predict the frequencies of the phenotypic classes that would result. The dwarf F₂ plants were recessive homozygotes, and so, when **selfed** (self-fertilized), they should have produced only *t*-bearing gametes and had only dwarf offspring in the F₃ generation. The tall F₂ plants, however, were a heterogeneous group of which one-third should have been homozygous *TT* and two-thirds should have been heterozygous *Tt* (fig. 2.7). The tall homozygotes, when selfed, should have produced only tall F₃ offspring (genotypically *TT*). However, the F₂ heterozygotes when selfed should have produced tall and dwarf offspring in a ratio identical to that produced by

terms were coined by William Bateson in 1901. The word *gene* was first used by the Danish botanist Wilhelm Johannsen in 1909.

If we look at figure 2.5, we can see that the *TT* homozygote can produce only one type of gamete, the *T*-bearing kind, and in a similar manner, the *tt* homozygote can produce only *t*-bearing gametes. Thus the F₁ in-

Figure 2.6

Methods of determining F₂ genotypic combinations in a self-fertilized monohybrid. The Punnett square diagram is named after the geneticist Reginald C. Punnett.

Schematic

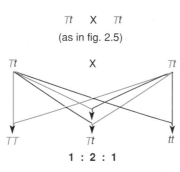

Tt X Tt

(as in fig. 2.5)

1 : 2 : 1

Diagrammatic

(Punnett square)

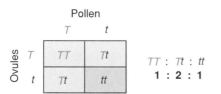

TT : Tt : tt

1 : 2 : 1

Probabilistic

(Multiply; see rule 2, chapter 4.)

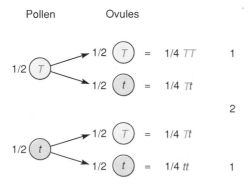

the selfed F₁ plants, three tall to one dwarf. Mendel found that all the dwarf (homozygous) F₂ bred true as predicted. Among the tall, 28% (28/100) bred true and 72% (72/100) produced tall and dwarf offspring. The prediction was one-third (33.3%) and two-thirds (66.7%), respectively. Mendel's observed values are very close to predicted values. We thus conclude that Mendel's progeny-testing experiment confirmed his hypothesis of segregation. In fact, a statistical test—developed in chapter 4—would support this conclusion.

Figure 2.7

Mendel self-fertilized F₂ tall and dwarf plants. He found that all the dwarf plants produced only dwarf progeny. Among the tall plants, 72 percent produced both tall and dwarf progeny in a 3:1 ratio.

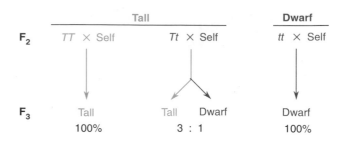

Another way to test the segregation hypothesis is with the extremely useful method of the **testcross,** that is, a cross of any organism with a recessive homozygote. (Another type of cross, a **backcross,** is the cross of a progeny with an individual that has a parental genotype. Hence a testcross can often be a backcross.) Since the gametes of the recessive homozygote contain only recessive alleles, the alleles carried by the gametes of the other parent will determine the phenotypes of the offspring. If a gamete from the organism being tested contains a recessive allele, the resulting F₁ organism will have a recessive phenotype; if it contains a dominant allele, the F₁ organism will have a dominant phenotype. Thus, in using a testcross, the genotypes of the gametes from the organism tested determine the phenotypes of the offspring (fig. 2.8). A testcross of the tall F₂ in figure 2.5 would produce the results shown in figure 2.9. These results are a further confirmation of Mendel's rule of segregation.

DOMINANCE IS NOT UNIVERSAL

If dominance were universal, the heterozygote would always have the same phenotype as the dominant homozygote and we would always see the 3:1 ratio when heterozygotes are crossed. If however, the heterozygote were distinctly different from both homozygotes, we would see a 1:2:1 ratio of phenotypes when heterozygotes are crossed. We refer to **partial dominance (incomplete dominance)** as those cases in which the phenotype of the heterozygote falls on some scale between the two homozygotes. An example is in flower petal color in some plants.

In four o'clocks (*Mirabilis jalapa*), we can cross a plant with red flower petals with another with white flower petals; the offspring have pink flower petals. If these pink-flowered F₁ plants are crossed, the F₂ plants appear in a ratio of 1:2:1, having red, pink, or white flower petals, respectively (fig. 2.10). The pink-flowered plants are heterozygotes in which the color is intermediate between the red and white colors of the homozygotes. In this case, there is an allele for red pigment color (R_1) and another allele

Figure 2.8

Testcross. In a testcross, the phenotype of an offspring is determined by the allele from the parent with the genotype being tested.

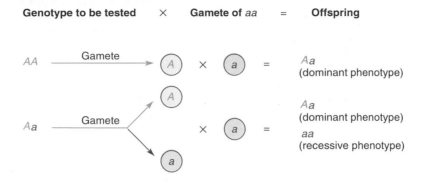

Figure 2.9

Testcrossing the dominant phenotype of the F₂ generation from figure 2.5

Tall (2 classes)

TT X *tt* = all *Tt*

Tt X *tt* = *Tt* : *tt*

1 : 1

that results in no color (*R₂*; the flower petals have a white background color). Flowers in heterozygotes (*R₁R₂*) have about half the red pigment of the flowers in red homozygotes (*R₁R₁*) because the heterozygotes have only one copy of the allele producing color whereas homozygotes have two copies of it.

As technology has improved, more and more cases have been found in which we can differentiate the heterozygote. It is now clear that dominance and recessiveness are phenomena dependent on which alleles are interacting and at what phenotypic level we are looking. For example, in Tay-Sachs disease, homozygous recessive children usually die before the age of three after suffering severe nervous degeneration; heterozygotes seem to be normal. With the discovery of the way in which the disease works, detection of the heterozygotes has now become possible.

As with many genetic diseases, the culprit is a defective **enzyme** (protein catalyst). Afflicted homozygotes have no enzyme activity, heterozygotes have about half the normal level of activity, and, of course, homozygous normal individuals have the full level. In the case of Tay-Sachs disease, the defective enzyme is hexoseaminidase-A, needed for the proper metabolism of lipids. With modern techniques the blood of normal persons can be assayed for this enzyme, and heterozygotes can be identified by their intermediate level of enzyme activity. Two heterozygotes will know that there is a 25% chance that any child they bear will have the disease. They can make an educated decision as to whether or not to have children.

Figure 2.10

Flower color inheritance in four o'clocks. This is an example of partial or incomplete dominance.

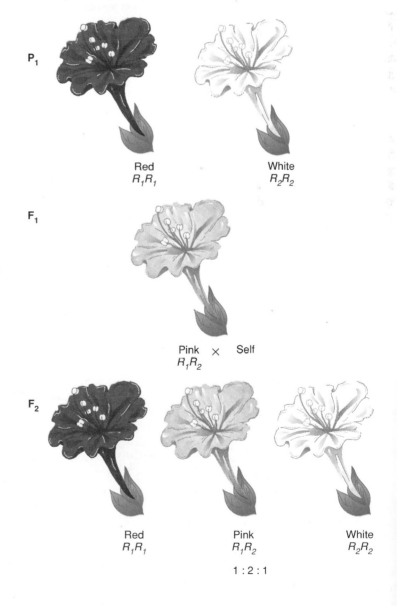

Figure 2.11
Wild-type fruit fly, *Drosophila melanogaster*

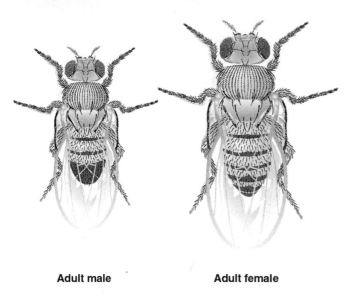

Adult male   **Adult female**

The other category in which the heterozygote is discernible occurs when the phenotype of the heterozygote is not on a scale somewhere between the two homozygotes but actually expresses both phenotypes simultaneously. We refer to this situation as **codominance.** For example, people with blood type AB are heterozygotes who are expressing both the *A* and *B* alleles for blood type (see below under Multiple Alleles for more information about blood types). The technique of electrophoresis (described in chapter 5) lets us see proteins directly and also gives us many examples of codominance when we can see the protein products of both alleles.

NOMENCLATURE

Throughout the last century, botanists, zoologists, and microbiologists have adopted different methods of naming alleles. Botanists tend to prefer the capital-lowercase scheme. Zoologists and microbiologists have adopted schemes that relate to the **wild-type.** The wild-type is the phenotype of the organism in nature. Different people collecting the same organism may come up with organisms with different phenotypes from the wild. However, there is usually an agreed-upon phenotype that is referred to as

Figure 2.12
Wing mutants of *Drosophila melanogaster. Cy,* curly; *sd,* scalloped; *ap,* apterous; *vg,* vestigial; *dp,* dumpy; *D,* Dichaete; *c,* curved.

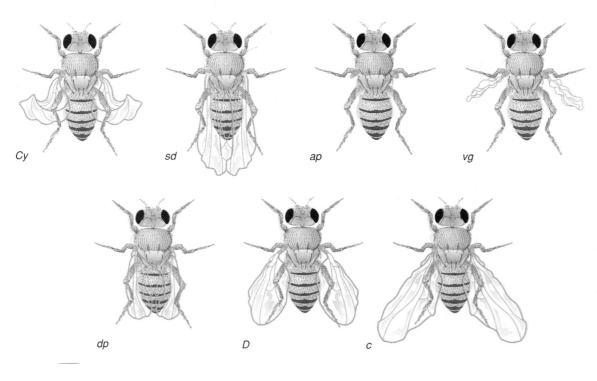

Cy   *sd*   *ap*   *vg*

dp   *D*   *c*

the wild-type. For fruit flies (*Drosophila*), organisms commonly used in genetic studies, the wild-type has red eyes and round wings (fig. 2.11). Alternatives to the wild-type are referred to as **mutants** (fig. 2.12). Thus red eyes are wild-type and white eyes are mutant. Fruit fly genes are named after the mutant, with a capital letter if the mutation is dominant and a lowercase letter if it is recessive. Table 2.1 gives some examples. The wild-type allele is given the symbol of the mutant with a + added as a superscript; by definition, every mutant has a wild-type allele as an alternative. For example, *w* stands for the white-eye allele, a recessive mutation. The wild-type (red eyes) is given the symbol w^+. Hairless is a dominant allele with the symbol *H*. Its wild-type allele is denoted as H^+. Sometimes geneticists will use the + symbol alone for the wild-type. This is done only when there will be no confusion by its use. If we are discussing eye color only, then + is clearly the same as w^+. Both mean red eyes. However, if we are discussing both eye color and bristle morphology, the + alone refers to two aspects of the phenotype and should be avoided if confusing.

TABLE 2.1   Some Mutants of *Drosophila*

Mutant Designation	Description	Dominance Relationship to Wild-Type
abrupt (*ab*)	Shortened, longitudinal, median wing vein	Recessive
amber (*amb*)	Pale yellow body	Recessive
black (*b*)	Black body	Recessive
Bar (*B*)	Narrow, vertical eye	Dominant
dumpy (*dp*)	Reduced wings	Recessive
Hairless (*H*)	Various bristles absent	Dominant
white (*w*)	White eye	Recessive
white-apricot (*wᵃ*)	Apricot-colored eye (allele of white eye)	Recessive

MULTIPLE ALLELES

A given gene can have more than two alleles. Although any particular individual can have only two, there may be many alleles of a given gene in a population. The classic example of multiple alleles in humans is in the ABO blood group, discovered by Landsteiner in 1900. This is the best known of all the red-cell antigen systems primarily because of its importance in blood transfusions. There are four phenotypes produced by three alleles (table 2.2). The *A* and *B* alleles are responsible for the production of the A and B antigens found on the surface of the erythrocytes (red blood cells). Antigens are substances, normally foreign to the body, that induce the immune system of the body to produce antibodies (proteins that bind to antigens). The ABO system is unusual because antibodies can be present (e.g., anti-B in a type A person) without prior exposure to the antigen. Thus people with a particular ABO antigen on their red cells will have in their serum the antibody against the other antigen: Type A persons have A antigen on their red cells and anti-B antibody in their serum; type B persons have B antigen on their red cells and anti-A antibody in their serum; type O persons do not have either antigen but have both antibodies in their serum; and type AB persons have both A and B antigens and do not form anti-A or anti-B antibodies in their serum. The *A* and *B* alleles each cause a different modification to the terminal sugars of a mucopolysaccharide (H structure) found on the surface of red cells. They are codominant because both modifications (antibodies) are present in a heterozygote. In other words, whichever enzyme (product of the *A* or *B* allele) reaches the H structure first will modify it. Once modified, the H structure cannot be acted upon by the other enzyme. Therefore, both A and B antigens will be produced in the heterozygote, in roughly equal proportions. Because the *O* allele causes no change to the H structure, it and its phenotype are recessive; the presence of the *A* or *B* alleles, or both, will eliminate the H product, thus masking the fact that it was ever there (fig. 2.13).

TABLE 2.2   ABO Blood Types with Immunity Reactions

Blood Type Corresponding to Antigens on Red Cells	Antibodies in Serum	Genotype	Reaction to Anti-A	Reaction to Anti-B
O	Anti-A and anti-B	*OO*	−	−
A	Anti-B	*AA* or *AO*	+	−
B	Anti-A	*BB* or *BO*	−	+
AB	None	*AB*	+	+

Figure 2.13

Function of the *A, B,* and *O* alleles of the ABO gene. The gene products of the *A* and *B* alleles of the ABO gene affect the terminal sugars of a mucopolysaccharide (H structure) found on red blood cells. The gene products of the *A* and *B* alleles are the enzymes alpha-3-N-acetyl-D-galactosaminyltransferase and alpha-3-D-galactosyltransferase, respectively.

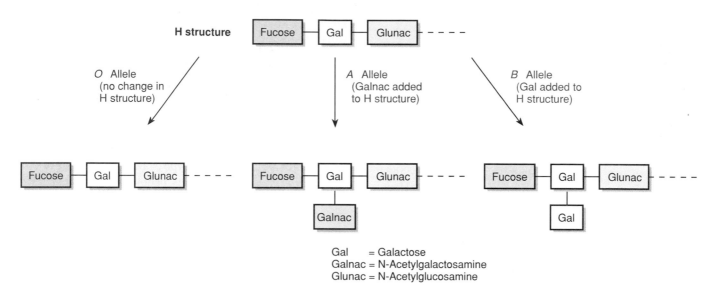

Gal = Galactose
Galnac = N-Acetylgalactosamine
Glunac = N-Acetylglucosamine

Adverse reactions to blood transfusions are primarily a result of the antibody in the serum of the recipient reacting with the antigen on the red cells of the donors. Thus type A persons cannot donate blood to type B persons. Type B persons have anti-A antibody, which reacts with the A antigen on the donor red cells to cause clumping of the cells.

Since both *A* and *B* are dominant to the *O* allele, this system not only shows multiple allelism, it also shows both codominance and simple dominance. (As with virtually any system, intense study yields more information, and subgroups of type A are known. We will not, however, deal with that complexity here.) According to the American Red Cross, of one hundred blood donors in the United States, forty-six are type O, forty are type A, ten are type B, and four are type AB.

Many other genes with multiple alleles are known. In some plants, such as red clover, there are genes with several hundred alleles that prevent self-fertilization. In *Drosophila* there are numerous alleles of the white-eye gene, and in people there are numerous hemoglobin alleles. In fact, multiple alleles are the rule rather than the exception.

INDEPENDENT ASSORTMENT

Mendel also analyzed the inheritance pattern of two traits at the same time. He looked, for instance, at plants that differed in the form and color of the peas: He crossed true-breeding (homozygous) plants that had seeds that were

round and yellow with plants that had seeds that were wrinkled and green. His results are shown in figure 2.14. The F_1 plants all had round, yellow seeds, which demonstrated that round was dominant to wrinkled and yellow was dominant to green. When these F_1 plants were self-fertilized, they produced an F_2 generation that had all four possible combinations of the two seed characteristics: round, yellow seeds; round, green seeds; wrinkled, yellow seeds; and wrinkled, green seeds. The numbers Mendel reported in these categories were 315, 108, 101, and 32, respectively. Dividing each class by 32 gives a 9.84:3.38:3.16:1.00 ratio, which is very close to a 9:3:3:1 ratio, one that, as you will see, we would expect if the genes governing these two traits were behaving independently of each other.

In figure 2.14 the letter *R* has been assigned to the dominant allele, round, and *r* has been assigned to the recessive allele, wrinkled; *Y* and *y* have been used for yellow and green color, respectively. In figure 2.15 we have rediagrammed the cross in figure 2.14. The P_1 plants in this cross produce only one type of gamete each, *RY* for the parent with the dominant traits and *ry* for the parent with the recessive traits. The resulting F_1 plants are heterozygous for both genes (**dihybrid**). Self-fertilizing the dihybrid produces the F_2 generation.

In the construction of the **Punnett square** to form the F_2 generation, a critical assumption is being made: The four types of gametes from each parent will be produced in equal numbers, and hence every offspring category or "box" in the square is equally likely. Thus, because there

Figure 2.14
Independent assortment in garden peas

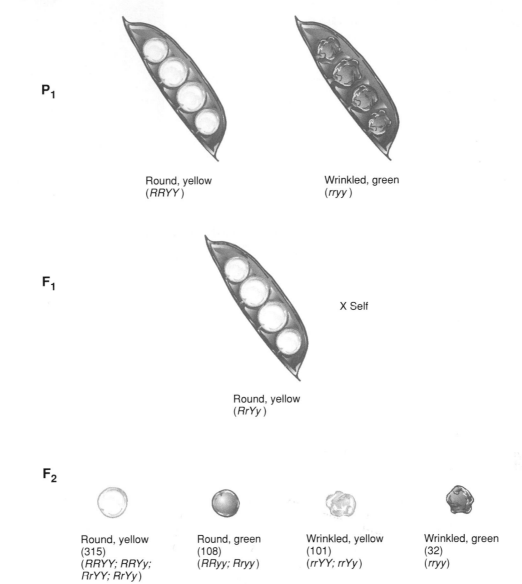

P₁

Round, yellow
(*RRYY*)

Wrinkled, green
(*rryy*)

F₁

X Self

Round, yellow
(*RrYy*)

F₂

Round, yellow
(315)
(*RRYY; RRYy;
RrYY; RrYy*)

Round, green
(108)
(*RRyy; Rryy*)

Wrinkled, yellow
(101)
(*rrYY; rrYy*)

Wrinkled, green
(32)
(*rryy*)

are sixteen boxes in the Punnett square, the ratio of F₂ offspring should be in sixteenths. Grouping the F₂ offspring by phenotype, we find there are 9/16 that have round, yellow seeds; 3/16 that have round, green seeds; 3/16 that have wrinkled, yellow seeds; and 1/16 that have wrinkled, green seeds. This is the origin of the expected 9:3:3:1 F₂ ratio.

Rule of Independent Assortment

This ratio comes about because the two characteristics behave independently of each other. There are four types of gametes from the F₁ plants (check fig. 2.15): *RY, Ry, rY,* and *ry*. These gametes occur in equal frequencies. Regardless of which allele for seed shape a gamete ends up

with, it has a 50:50 chance of getting either of the alleles for color—the two genes are segregating, or assorting, independently. This is the essence of Mendel's second rule, the **rule of independent assortment,** which states that alleles of one gene can segregate independently of alleles of other genes. Are the alleles for the two characteristics of color and form segregating properly according to Mendel's first principle?

If we look only at seed shape (see fig. 2.14), we find that a homozygote with round seeds was crossed with a homozygote with wrinkled seeds in the P₁ generation (*RR* × *rr*) to give only heterozygous plants with round seeds (*Rr*) in the F₁ generation. When these F₁ plants are self-fertilized, the result is 315 + 108 round seeds (*RR* or *Rr*) and 101 + 32 wrinkled seeds (*rr*) in the F₂ generation.

Figure 2.15
Assigning genotypes in the cross of figure 2.14

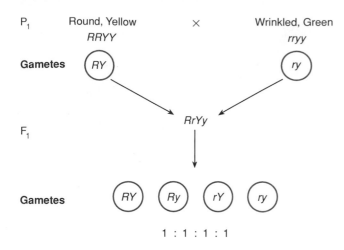

P_1 Round, Yellow   ×   Wrinkled, Green
     RRYY   rryy

Gametes   RY   ry

F_1 RrYy

Gametes   RY   Ry   rY   ry

1 : 1 : 1 : 1

Pollen

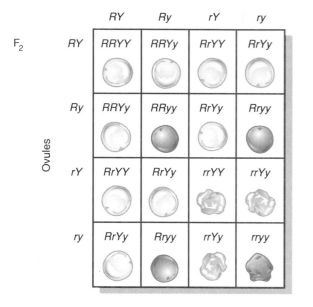

F_2

	RY	Ry	rY	ry
RY	RRYY	RRYy	RrYY	RrYy
Ry	RRYy	RRyy	RrYy	Rryy
rY	RrYY	RrYy	rrYY	rrYy
ry	RrYy	Rryy	rrYy	rryy

Ovules

This is 423:133 or a 3.18:1.00 phenotypic ratio—very close to the expected 3:1 ratio. So the gene for seed shape is segregating normally. In a similar manner, if we look only at the gene for color, we see that the F_2 ratio of yellow to green seeds is 416:140 or 2.97:1.00—again, very close to a 3:1 ratio. Thus, when two genes are segregating normally according to the rule of segregation, their independent behavior will give us the rule of independent assortment.

From the Punnett square of figure 2.15, you can see that because of dominance, all phenotypic classes except the homozygous recessive one—wrinkled, green seeds—are actually genetically heterogeneous with phenotypes that are made up of several genotypes. For example, the dominant class, with round, yellow seeds, is made up of

Reginald C. Punnett (1875–1967)
Genetics, *58 (1968): frontispiece.*

four genotypes: *RRYY, RRYy, RrYY,* and *RrYy.* Grouping all the genotypes by phenotype shows the ratio in figure 2.16. Thus, with complete dominance, a self-fertilized dihybrid gives a 9:3:3:1 phenotypic ratio in its offspring (F_2). With the absence of dominance, a 1:2:1:2:4:2:1:2:1 phenotypic ratio results in the F_2 generation. What ratio would be obtained if one gene exhibited dominance and the other did not? An example of this case is given in figure 2.17.

Testcrossing Multihybrids

A simple test of Mendel's rule of independent assortment is the testcrossing of the dihybrid plant. We would predict, for example, that if an *RrYy* F_1 individual were crossed with an *rryy* individual, the results would be all four phenotypes in a 1:1:1:1 ratio as shown in figure 2.18. Mendel's data verified this prediction. We will proceed to look at a **trihybrid** cross in order to develop general rules for **multihybrids.**

The trihybrid Punnett square is shown in figure 2.19. From this we can see that from a cross of a homozygous dominant with a homozygous recessive individual in the P_1 generation, plants in the resulting generation (F_1) are capable of producing eight gamete types. When these F_1 individuals are selfed, they in turn produce offspring of twenty-seven different genotypes in a ratio of sixty-fourths in the F_2 generation. By extrapolating from the monohybrid through the trihybrid, or simply by the rules of probability, we can construct table 2.3, which contains the rules for F_1 gamete production and F_2 zygote formation in a multihybrid cross. For example, from this table we can figure out the F_2 offspring when a dodecahybrid (twelve segregating genes: *AABBCC . . . LL* × *aabbcc . . . ll*) is selfed. The F_1 organisms in that cross will produce gametes with 2^{12} or 4,096 different genotypes. The proportion of homozygous recessive offspring in the F_2 generation is $1/(2^n)^2$ where $n = 12$, or 1 in 16,777,216. With complete dominance there will be 4,096 different phenotypes in the F_2 generation. If dominance is absent, there can be 3^{12}, or 531,441 different phenotypes present in the F_2 generation.

Figure 2.16
Dihybrid F$_2$ genotypic ratio

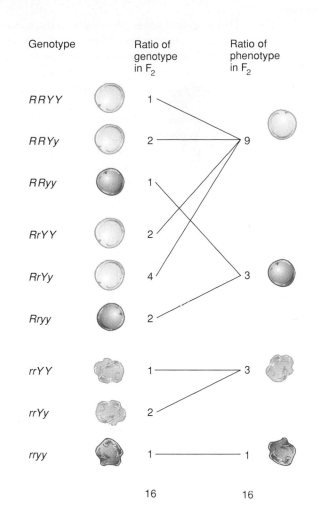

Genotype	Ratio of genotype in F$_2$	Ratio of phenotype in F$_2$
RRYY	1	
RRYy	2	9
RRyy	1	
RrYY	2	
RrYy	4	3
Rryy	2	
rrYY	1	3
rrYy	2	
rryy	1	1
	16	16

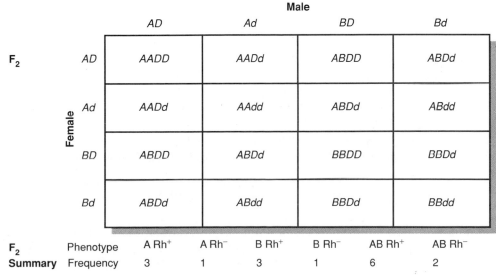

P$_1$ AADD X BBdd

F$_1$ ABDd
(F$_1$ X F$_1$)

Male

	AD	Ad	BD	Bd
AD	AADD	AADd	ABDD	ABDd
Ad	AADd	AAdd	ABDd	ABdd
BD	ABDD	ABDd	BBDD	BBDd
Bd	ABDd	ABdd	BBDd	BBdd

Female

F$_2$ Summary	Phenotype	A Rh$^+$	A Rh$^-$	B Rh$^+$	B Rh$^-$	AB Rh$^+$	AB Rh$^-$
	Frequency	3	1	3	1	6	2

Figure 2.17
Independent assortment of two blood systems in humans. In the ABO system, only the *A* and *B* alleles are segregating (present) here; they are codominant. In a simplified view of the Rhesus system, the Rh$^+$ phenotype (*D* allele) is dominant to the Rh$^-$ phenotype (*d* allele).

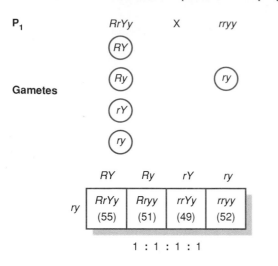

Figure 2.18
Testcross of a dihybrid. A 1:1:1:1 ratio is expected in the offspring.

P₁   RrYy   X   rryy

Gametes

	RY	Ry	rY	ry
ry	RrYy (55)	Rryy (51)	rrYy (49)	rryy (52)

1 : 1 : 1 : 1

GENOTYPIC INTERACTIONS

Often several genes contribute to the same phenotype. An example is the combs of fowl (fig. 2.20). If we cross a rose-combed hen with a pea-combed rooster (or vice versa), all the F_1 offspring are walnut combed. If we cross together the hens and roosters of this heterozygous F_1 group, we will get, in the F_2 generation, walnut-, rose-, pea-, and single-combed fowl in a ratio of 9:3:3:1. Can you figure out the genotypes of this F_2 population before reading further? An immediate indication that two allelic pairs are involved is the fact that the 9:3:3:1 ratio appeared in the F_2 generation. As we have seen, this ratio comes about when we cross dihybrids in which both genes have alleles that control traits with complete dominance.

P₁   AABBCC × aabbcc

F₁   AaBbCc × Self

Figure 2.19
Trihybrid cross

	ABC	ABc	AbC	Abc	aBC	aBc	abC	abc
ABC	AABBCC	AABBCc	AABbCC	AABbCc	AaBBCC	AaBBCc	AaBbCC	AaBbCc
ABc	AABBCc	AABBcc	AABbCc	AABbcc	AaBBCc	AaBBcc	AaBbCc	AaBbcc
AbC	AABbCC	AABbCc	AAbbCC	AAbbCc	AaBbCC	AaBbCc	AabbCC	AabbCc
Abc	AABbCc	AABbcc	AAbbCc	AAbbcc	AaBbCc	AaBbcc	AabbCc	Aabbcc
aBC	AaBBCC	AaBBCc	AaBbCC	AaBbCc	aaBBCC	aaBBCc	aaBbCC	aaBbCc
aBc	AaBBCc	AaBBcc	AaBbCc	AaBbcc	aaBBCc	aaBBcc	aaBbCc	aaBbcc
abC	AaBbCC	AaBbCc	AabbCC	AabbCc	aaBbCC	aaBbCc	aabbCC	aabbCc
abc	AaBbCc	AaBbcc	AabbCc	Aabbcc	aaBbCc	aaBbcc	aabbCc	aabbcc

TABLE 2.3 Multihybrid Self-Fertilization, Where n Equals Number of Genes Segregating Two Alleles Each

	Monohybrid $n = 1$	Dihybrid $n = 2$	Trihybrid $n = 3$	General Rule
F_1 Gamete Genotypes	2	4	8	2^n
Proportion of Homozygous Recessives Among the F_2 Individuals	1/4	1/16	1/64	$1/(2^n)^2$
Number of Different F_2 Phenotypes Given Complete Dominance	2	4	8	2^n
Number of Different Genotypes (or Phenotypes If There Is No Dominance)	3	9	27	3^n

BOX 2.3

DID MENDEL CHEAT?

Overwhelming evidence gathered this century has proven the correctness of Mendel's conclusions. However, close scrutiny of Mendel's paper has led to some suggestions that (1) Mendel failed to report the inheritance of traits that did not show independent assortment and (2) Mendel fabricated numbers. Both these claims are on the surface difficult to ignore; both have been reasonably countered.

The first claim—that Mendel failed to report on crosses involving traits not showing independent assortment—arises from the observation that all seven traits that Mendel studied do show independent assortment and the pea plant has precisely seven pairs of chromosomes. For Mendel to have chosen seven genes, one located on each of the seven chromosomes, by chance alone seems extremely unlikely. In fact the probability would be

$$7/7 \times 6/7 \times 5/7 \times 4/7 \times 3/7 \times 2/7 \times 1/7 = 0.006$$

That is, Mendel had less than one chance in one hundred of randomly picking seven traits on the seven different chromosomes. However, Douglas and Novitski in 1977 analyzed Mendel's data in a different way. To understand their analysis you have to know that two genes sufficiently far apart on the same chromosome will appear to assort independently (discussed in chapter 6). Thus Mendel's choice of characters showing independent assortment has to be viewed in light of the length of all the chromosomes. That is, Mendel could have chosen two genes on the same chromosome that would still show independent assortment. In fact he did. For example, stem length and pod texture (wrinkled or smooth) are on the fourth chromosome pair in peas. In their analysis, Douglas and Novitski report that the probability of randomly choosing seven characteristics that appear to assort independently is actually between one in four and one in three. So it seems that Mendel did not have to manipulate his choice of characters in order to hide the failure of independent assortment. He had a one in three chance of naively choosing the seven characters that he did so that no deviation from independent assortment would be uncovered.

The second claim—that Mendel fabricated data—comes from a careful analysis of Mendel's paper by R. A. Fisher, a brilliant English statistician and population geneticist. In a paper in 1936, Fisher pointed out two problems of Mendel's work. First, all of Mendel's published data taken together fit their expected ratios better than predicted by chance alone. Second, there were cases in which Mendel's data fit incorrect expected ratios. This second "error" on Mendel's part came about as follows. Mendel determined whether a dominant phenotype in the F_2 generation was a homozygote or a heterozygote

by self-fertilizing it and examining ten offspring. In an F_2 generation composed of $1AA:2Aa:1aa$, he expected a 2:1 ratio of heterozygotes to homozygotes within the dominant phenotypic class. In fact this ratio is not precisely correct because of the problem of misclassification of heterozygotes. There is a probability that some heterozygotes will be classified as homozygotes because all their offspring will be of the dominant phenotype. The probability that one offspring from a selfed Aa individual has the dominant phenotype is 3/4, or 0.75: The probability that ten offspring will be of the dominant phenotype is $(0.75)^{10}$ or 0.056. Thus Mendel misclassified heterozygotes as dominant homozygotes 5.6% of the time. He should have expected a 1.89:1.11 ratio instead of a 2:1 ratio to demonstrate segregation. Mendel classified six hundred plants this way in one cross and got a ratio of 201 homozygous to 399 heterozygous offspring. This is an almost perfect fit to the presumed 2:1 ratio and thus a poorer fit to the real 1.89:1.11 ratio. This bias is consistent and repeated in Mendel's trihybrid analysis.

Fisher, believing in Mendel's basic honesty, suggested that Mendel's data do not represent an experiment exactly, but more of a demonstration. In 1971 Weiling published a more convincing case in Mendel's defense. Pointing out that the data of Mendel's rediscoverers are also suspect for the same reason, he suggested that the problem lies with the process of pollen formation in plants, not with the experimenters. In an Aa heterozygote, two A and two a cells develop from a pollen mother cell. These cells tend to stay together on the anther. Thus pollen cells do not fertilize in a strictly random fashion. A bee is more likely to take equal numbers of A and a pollen than would be expected by chance alone. The result is that the statistics used by Fisher are not applicable. By using a different statistic, Weiling showed that, in fact, no manipulations need to have been done by Mendel or his rediscoverers in order to get data that fit the expected ratios so well. By the same reasoning there would have been very little misclassification of heterozygotes.

More recently, Weiling and others have made several points. First, in order for Mendel to be sure of ten offspring, he probably examined more than ten and thus his misclassification was probably lower than 5.6%. Second, despite Fisher's brilliance as a statistician, several compelling arguments have been made that Fisher's statistical analyses were incorrect. In other words, for subtle statistical reasons, many of his analyses involved methods and conclusions that were in error.

We conclude that there is no compelling evidence to suggest that Mendel in any way manipulated his data to demonstrate his rules. In fact, taking into account what is known about him personally, it is much more logical that he did not "cheat."

Figure 2.20
Four types of combs in fowl

Rose

Pea

Walnut

Single

Figure 2.21
Independent assortment in determination of comb type in fowl

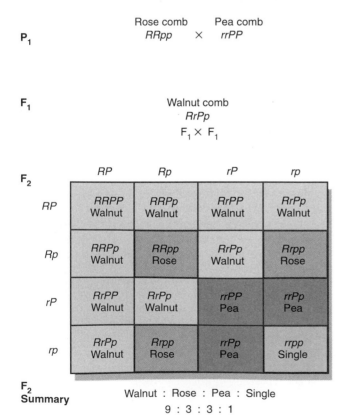

	Rose comb	Pea comb		
P_1	$RRpp$	$\times$	$rrPP$	

F_1 Walnut comb
$RrPp$
$F_1 \times F_1$

F_2	RP	Rp	rP	rp
RP	$RRPP$ Walnut	$RRPp$ Walnut	$RrPP$ Walnut	$RrPp$ Walnut
Rp	$RRPp$ Walnut	$RRpp$ Rose	$RrPp$ Walnut	$Rrpp$ Rose
rP	$RrPP$ Walnut	$RrPp$ Walnut	$rrPP$ Pea	$rrPp$ Pea
rp	$RrPp$ Walnut	$Rrpp$ Rose	$rrPp$ Pea	$rrpp$ Single

F_2
Summary   Walnut : Rose : Pea : Single
9 : 3 : 3 : 1

The analysis of this cross is given in figure 2.21. When dominant alleles of both genes are present in an individual, the walnut comb appears. A dominant allele of the rose gene (R-) with recessive alleles of the pea gene (pp) gives a rose comb. A dominant allele of the pea gene (P-) with recessive alleles at the rose gene (rr) gives pea-combed fowl. When both genes are homozygous for the recessive alleles, the fowl are single combed. Thus a $9:3:3:1$ F_2 ratio arises from crossing dihybrid individuals even though different expressions of the same phenotypic characteristic, the comb, are involved. In our previous $9:3:3:1$ example (fig. 2.15), we dealt with two separate characteristics of peas: shape and color.

In corn, several different field varieties produce white kernels on the ears. In certain crosses, two white varieties will result in an F_1 generation with all the kernels being purple. If plants grown from these purple kernels are selfed, the F_2 individuals have both purple and white kernels in a ratio of $9:7$. How can this be explained? We must be dealing with two genes, each segregating two alleles, because the ratio is in sixteenths. Therefore, the F_1 individuals must have been dihybrid. Furthermore, we can see that the F_2 $9:7$ ratio is a variation of the $9:3:3:1$ ratio. The 3, 3, and 1 categories here are producing the same phenotype and thus make up 7/16 of the F_2 offspring. The cross is outlined in figure 2.22. We can see from this figure that the purple color appears only when dominant alleles of both genes are present. When one or both genes have only recessive alleles, the kernels will be white.

Epistasis

The color of corn kernels illustrates the concept of **epistasis,** the interaction of nonallelic genes in the formation of the phenotype, a process that is analogous to dominance among alleles of one gene. For example, in the preceding corn case, an aa genotype produces white kernels regardless of the alleles of the B gene. Similarly, the bb genotype masks whatever A alleles are present. We say that the epistatic gene masks the expression of the **hypostatic gene.** As another example, the recessive apterous (wingless) gene in fruit flies is epistatic to any gene that controls wing characteristics: Hairy wings is hypostatic to apterous. Without wings, no wing characteristics can be expressed. Note that the genetic control of comb type in fowl does not involve epistasis. There is no masking of genotypes at one locus by allelic combinations at the other locus: the $9:3:3:1$ ratio is not an indication of epistasis. To illustrate further the principle of epistasis, we can look at the control of the coat color of mice.

Figure 2.22

Epistasis in color production in corn

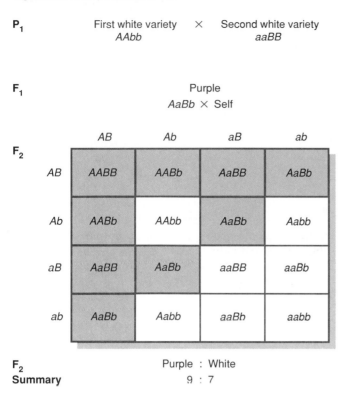

P₁   First white variety   ×   Second white variety
AAbb   *aaBB*

F₁   Purple
AaBb × Self

	AB	Ab	aB	ab
AB	AABB	AABb	AaBB	AaBb
Ab	AABb	AAbb	AaBb	Aabb
aB	AaBB	AaBb	aaBB	aaBb
ab	AaBb	Aabb	aaBb	aabb

F₂ Summary   Purple : White
              9 : 7

If a pure-breeding black mouse is crossed with a pure-breeding albino mouse (pure white because all pigment is lacking), all of the offspring are agouti (the typical brownish-gray mouse color). When the F₁ agouti mice are crossed to each other, agouti, black, and albino offspring appear in the F₂ generation in a ratio of 9:3:4. What are the genotypes in this cross? The answer is given in figure 2.23. By now it should be apparent that the F₂ ratio of 9:3:4 is also a variant of the 9:3:3:1 ratio; it indicates epistasis in a dihybrid cross. What is the mechanism producing this 9:3:4 ratio? Of a potential 9:3:3:1 ratio, one of the 3/16 classes and the 1/16 class are combined to give a 4/16 class. Any genotype that includes $c^a c^a$ will be albino, but as long as at least one dominant C allele is present, the A gene can express itself. Mice with dominant alleles of both genes (*A-C-*) will have the agouti color, whereas mice that are homozygous recessive at the A gene (*aaC-*) will be black. So, at the A gene, *A* for agouti is dominant to *a* for black. The albino gene (c^a) is epistatic to the A gene, which is hypostatic to the albino gene.

Mechanism of Epistasis

The physiological mechanism of epistasis is known in this case. The pigment melanin is present in both the black and agouti phenotypes. The agouti is a modified black hair in which yellowish stripes have been added. Thus with melanin present agouti is dominant. Without melanin we get an albino regardless of the genotype of the black-agouti gene because both agouti and black depend on melanin. Albinism is the result of one of several defects in the enzymatic pathway for the synthesis of melanin (fig. 2.24).

Knowing that epistatic modifications of the 9:3:3:1 ratio come about through interaction at the biochemical level, we can look for a biochemical explanation for the 9:7 ratio in corn kernel color (fig. 2.22). Two possible mechanisms for a 9:7 ratio are shown in figure 2.25. Either there is a two-step process that takes a precursor molecule and turns it into purple pigment, or there are two precursors that need to be converted to final products that then combine to produce purple pigment. The dominants from the two genes control the two steps in the process. Recessive alleles are ineffective. Thus dominants for both steps are necessary to complete the pathways for a purple pigment. Stopping the process at any point will prevent the production of purple color.

Another example of epistasis occurs in the snapdragon (*Antirrhinum majus*). There, a gene called *nivea* has alleles that determine whether any pigment will be produced; the *nn* genotype prevents pigment production whereas the *NN* or *Nn* genotypes permit pigment color genes to express themselves. The *eosinea* gene controls the production of a red anthocyanin pigment. In the presence of the *N* allele of the *nivea* gene, the genotypes *EE* or *Ee* of the *eosinea* gene result in red flowers; the *ee* genotype results in pink flowers. When dihybrids are self-fertilized, red- pink- and white-flowered plants are produced in a ratio of 9:3:4 (fig. 2.26). The epistatic interaction is the *nn* genotype masking the expression of alleles at the *eosinea* gene. In other words, regardless of the genotypes of the *eosinea* gene (*EE, Ee,* or *ee*), the flowers will be white if the *nivea* gene has the *nn* combination of alleles. Thus *nivea* is epistatic to *eosinea,* which is hypostatic to *nivea.* (We should add that at least seven major colors occur in snapdragons along with subtle shade differences, all genetically controlled by the interaction of at least seven genes.)

Other types of epistatic interactions are known. In table 2.4 several are listed. We do not know the exact physiological mechanisms in many cases, especially when developmental processes are involved (e.g., size and shape). However, from an analysis of crosses, we can know the number of genes involved and the general nature of their interactions.

Figure 2.23
Epistasis in coat color in rodents

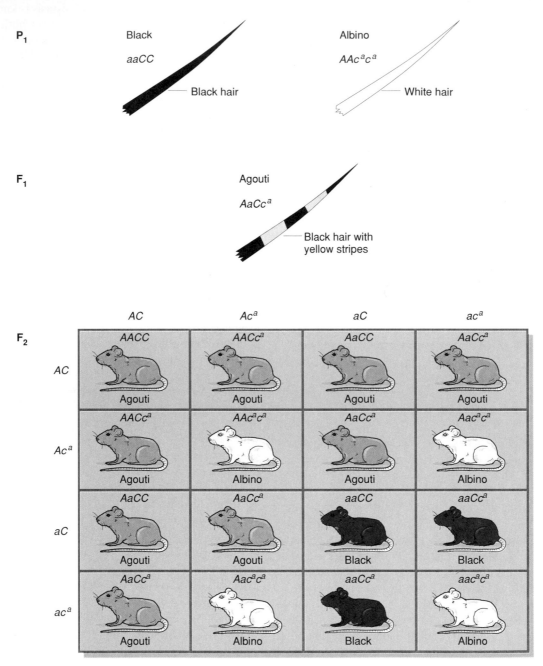

P_1

Black

$aaCC$

— Black hair

Albino

AAc^ac^a

— White hair

F_1

Agouti

$AaCc^a$

— Black hair with yellow stripes

F_2

	AC	Ac^a	aC	ac^a
AC	$AACC$ Agouti	$AACc^a$ Agouti	$AaCC$ Agouti	$AaCc^a$ Agouti
Ac^a	$AACc^a$ Agouti	AAc^ac^a Albino	$AaCc^a$ Agouti	Aac^ac^a Albino
aC	$AaCC$ Agouti	$AaCc^a$ Agouti	$aaCC$ Black	$aaCc^a$ Black
ac^a	$AaCc^a$ Agouti	Aac^ac^a Albino	$aaCc^a$ Black	aac^ac^a Albino

Agouti : Black : Albino
9 : 3 : 4

BIOCHEMICAL GENETICS

Inborn Errors of Metabolism

The examples of mouse coat color, corn kernel color, and snapdragon flower petal color demonstrate that genes control the formation of enzymes, proteins that control steps in biochemical pathways. For the most part, dominant alleles control functioning enzymes that catalyze biochemical steps. Recessive alleles often produce non-functional enzymes that cannot catalyze specific steps. Often a heterozygote is normal because one allele is producing a functional enzyme; usually only half the enzyme quantity of the dominant homozygote is enough. The study of the relationship between genes and enzymes is generally called **biochemical genetics** because it involves the genetic control of biochemical pathways. A. E. Garrod, a British physician, pointed out this general concept of gene action in people in *Inborn Errors of Metabolism,*

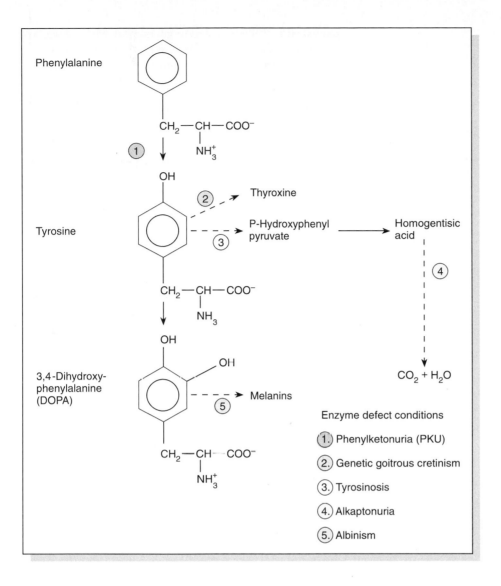

Figure 2.24
Part of the tyrosine (an amino acid) metabolic pathway in human beings with the associated diseases caused by homozygous recessive conditions. The broken arrows indicate that there is more than one step in the pathways.

Phenylalanine

Tyrosine

Thyroxine

P-Hydroxyphenyl pyruvate

Homogentisic acid

$CO_2 + H_2O$

3,4-Dihydroxy-phenylalanine (DOPA)

Melanins

Enzyme defect conditions

1. Phenylketonuria (PKU)

2. Genetic goitrous cretinism

3. Tyrosinosis

4. Alkaptonuria

5. Albinism

Figure 2.25
Possible metabolic pathways of color production yielding 9:7 ratios in the F_2 generation of a self-fertilized dihybrid

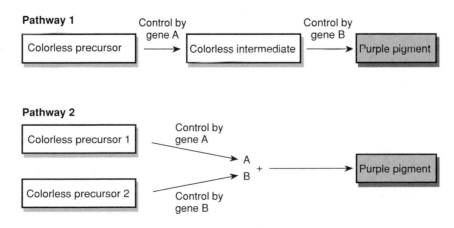

Pathway 1

Colorless precursor → Control by gene A → Colorless intermediate → Control by gene B → Purple pigment

Pathway 2

Colorless precursor 1 → Control by gene A → A
Colorless precursor 2 → Control by gene B → B
A + B → Purple pigment

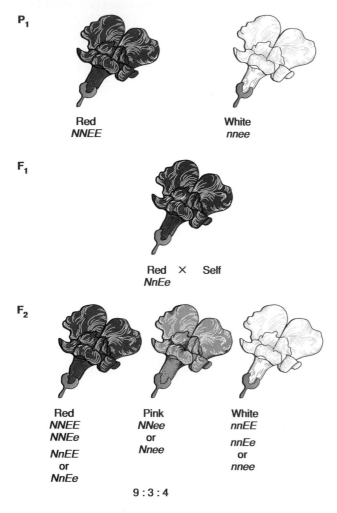

Figure 2.26
Flower color inheritance in snapdragons. This is an example of epistasis: An *nn* genotype masks the expression of alleles at the eosinea gene.

P₁

Red
NNEE

White
nnee

F₁

Red × Self
NnEe

F₂

Red
NNEE
NNEe

NnEE
or
NnEe

Pink
NNee
or
Nnee

White
nnEE

nnEe
or
nnee

9 : 3 : 4

published in 1909. Only nine years after Mendel was re-discovered, Garrod described many human conditions, such as albinism and alkaptonuria, that are the result of homozygosity of recessive alleles (fig. 2.24).

For example, normal people degrade homogentisic acid (alkapton) into maleylacetoacetic acid. Persons with the disease alkaptonuria are homozygous for a nonfunctional form of the enzyme essential to the process. Absence of the appropriate enzyme blocks the degradation reaction so that there is a buildup of homogentisic acid. This acid darkens upon oxidation. Thus, affected persons can be identified by the black color of their urine after its exposure to air. The eventual effects of alkaptonuria are problems of the joints and a darkening of cartilage that is visible in the ears and eye sclera.

One-Gene-One-Enzyme Hypothesis

Pioneering work in the concept that genes control the production of enzymes that control the steps in biochemical pathways was done by George Beadle and Edward Tatum, who eventually shared the Nobel Prize for their work. Not only did they put forth the **one-gene-one-enzyme hypothesis** but they used mutants to work out the details of biochemical pathways. In the early 1940s they united the fields of biochemistry and genetics by using strains of a bread mold that had nutritional requirements to discover the steps in biochemical pathways in that organism.

Through this century, the study of mutations has been the driving force in genetics. The process of mutation results in alternative alleles to a wild-type and shows us that a particular aspect of the phenotype is under genetic control. Beadle and Tatum used mutants to work out steps in the pathway of the biosynthesis of niacin (vitamin B₃) in the pink bread mold, *Neurospora crassa*.

TABLE 2.4 Some Examples of Epistatic Interactions among Alleles of Two Genes

Characteristic	Phenotype of F₁ Dihybrid (*AaBb*)	Phenotypic F₂ Ratio
Corn and Sweet Pea Color	Purple	Purple:white 9:7
Mouse Coat Color	Agouti	Agouti:black:albino 9:3:4
Shepherd's Purse Seed Capsule Shape	Triangular	Triangular:oval 15:1
Summer Squash Shape	Disk	Disk:sphere:elongate 9:6:1
Fowl Color	White	White:colored 13:3

Edward L. Tatum (1909–1975)
Courtesy of National Academy of Sciences.

George W. Beadle (1903–1989)
Courtesy of Muriel B. Beadle.

Normally, *Neurospora* synthesizes niacin via the pathway shown in figure 2.27. Beadle and Tatum isolated mutants that could not grow unless niacin was provided in the culture medium: These mutants had enzyme deficiencies in the synthesizing pathway that ends with niacin. Thus, although wild-type *Neurospora* could grow on medium without additives, the mutants could not. Beadle and Tatum had a general idea, based on the structure of niacin, as to what substances were in the pathway of niacin biosynthesis. They could thus make educated guesses as to what substances might be added to the culture medium to enable the mutants to grow. Mutant B (table 2.5), for example, could grow if given niacin or, alternatively, 3-hydroxyanthranilic acid. It could not grow if given only kynurenine. Thus Beadle and Tatum knew that the B mutation affected the pathway between kynurenine and 3-hydroxyanthranilic acid. Similarly, mutant A could grow

Figure 2.27
Pathway of niacin synthesis in *Neurospora*. Each *arrow* represents an enzyme-mediated step. Each *question mark* represents a presumed, but unknown, compound.

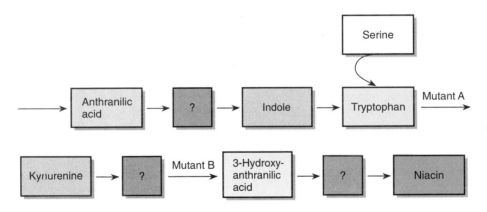

TABLE 2.5 Growth Performance of *Neurospora* Mutants (Plus Sign Indicates Growth; Minus Sign Indicates No Growth)

	Additive			
	Tryptophan	*Kynurenine*	*3-Hydroxyanthranilic Acid*	*Niacin*
Wild-type	+	+	+	+
Mutant A	−	+	+	+
Mutant B	−	−	+	+

if given 3-hydroxyanthranilic acid or kynurenine instead of niacin. Therefore, these two products must be in the pathway after the step interrupted in mutant A. Since these mutant organisms could not grow when given only tryptophan, Beadle and Tatum knew that tryptophan occurred in the pathway before the step with the deficient enzyme. By this type of analysis, they discovered the steps in several biochemical pathways of *Neurospora*. Many biochemical pathways are similar in a huge range of organisms and thus the work of Beadle and Tatum was of general importance. (We will spend more time studying *Neurospora* in chapter 6.)

Beadle and Tatum could further verify their work by observing which substances accumulated in the mutant organisms. If a biochemical pathway is blocked at a certain point, then the substrate at that point cannot be converted into the next product and will build up in the cell. For example, in the niacin pathway (fig. 2.27), if a block occurred just after 3-hydroxyanthranilic acid, then that substance would build up in the cell because it could not be converted into the next substance on the way to niacin.

This analysis could be misleading, however, in the cases where the substance being built up was "siphoned off" into other biochemical pathways in the cell. Also, the cell might attempt to break down or sequester substances that were toxic. Here, too, there might not be an obvious buildup of the substance just before the blocked step.

Beadle and Tatum concluded from their studies that one gene controls the production of one enzyme. The one-gene-one-enzyme hypothesis is an oversimplification that will be clarified later in the book. As a rule of thumb, however, the hypothesis is valid and has served to direct attention to the functional relationship between genes and enzymes in biochemical pathways.

Although a change in a single enzyme usually disrupts a single biochemical pathway, it frequently has more than one effect on the phenotype of an organism. These multiple effects are referred to as **pleiotropy**. A well-known example is sickle-cell anemia, which is caused by a mutation in the gene for the β chain of the hemoglobin molecule. As a homozygote, this mutation causes a sickling of red blood cells (fig. 2.28). The sickling of these cells has two major ramifications.

First, the sickled cells are destroyed by the liver, causing anemia. The phenotypic effects of this anemia include physical weakness, below-average development, and hypertrophy of the bone marrow resulting in the "tower skull" development seen in some of those afflicted with the disease. The second major effect of sickle-cell anemia is the interference of capillary blood flow by clumping of the odd-shaped cells, resulting in damage to every major organ. The individual can suffer pain, heart failure, rheumatism, and other ill effects. Hence a single mutation shows itself in many aspects of the phenotype.

Figure 2.28
Sickle-shaped red blood cells from a person with sickle-cell anemia. Normal red blood cell is about 7 to 8 μm in diameter.
Courtesy of Dr. Patricia N. Farnsworth.

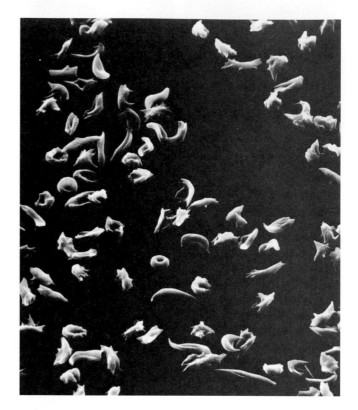

SUMMARY

Genes control the phenotype. They are inherited as discrete units. Higher organisms contain two alleles of each gene, but only one allele enters each gamete. Zygote formation restores the double number of alleles in the cell. This is Mendel's rule of segregation. Alleles of different genes segregate independently of each other. Mendel was the first to recognize the 3:1 phenotypic ratio as a pattern of inheritance in hybridization. Mendel was successful in his endeavor because he performed careful experiments using discrete characteristics, large numbers of offspring, and an organism (the pea plant) amenable to controlled fertilizations.

There can be many alleles for one gene, although each individual organism has only two. A phenotype is dominant if it is expressed with one or two copies of its allele present (heterozygote or homozygote). Dominance depends on at what level of the phenotype one looks. Genes usually control the production of enzymes, which control steps in metabolic pathways. Many metabolic diseases in humans are due to homozygosity of an allele that produces a nonfunctioning enzyme.

Nonallelic genes can interact in the formation of a phenotype so that alleles of one gene alter or mask the expression of alleles of another gene. This process is termed epistasis and results in alterations of expected ratios. Beadle and Tatum used mutants with mutations in the pathway for the biosynthesis of niacin to work out the steps in the biochemical pathway. A single mutation can have many phenotypic effects (pleiotropy).

1. In corn the single-flowered phenotype is dominant to the double flower, starchy endosperm is dominant to waxy endosperm, and serrate (notched or toothed) leaves are dominant to entire leaves. What are the results of the testcrossing of a trihybrid?

ANSWER: The trihybrid has the genotype *SisiStstNn,* naming the alleles for the single flower (*Si*), starchy endosperm (*St*), and notched leaf (*N*), respectively. This parent is capable of producing eight different gamete types in equal frequencies, all combinations of one allele from each gene (*SiStN, SiStn, SistN, Sistn, siStN, siStn, sistN,* and *sistn*). In a testcross, the other parent is a recessive homozygote with the genotype of *sisiststnn,* capable of producing only one type of gamete, *sistn.* Thus zygotes of eight different genotypes (and phenotypes) can be produced in this cross, one for each of the gamete types of the trihybrid parent: *SisiStstNn* (single flower, starchy endosperm, notched leaf), *SisiStstnn* (single, starchy, entire), *SisiststNn* (single, waxy, notched), *Sisiststnn* (single, waxy, entire), *sisiStstNn* (double, starchy, notched), *sisiStstnn* (double, starchy, entire), *sisiststNn* (double, waxy, notched), and *sisiststnn* (double, waxy, entire). These should each make up one eighth of the total number of offspring.

2. Summer squash can be found in three shapes: disk, spherical, and elongate. In one experiment, two squash plants with disk-shaped fruits were crossed. The first 160 seeds planted from this cross produced plants with fruit shapes as follows: 89 disk, 61 sphere, and 10 elongate. What is the mode of inheritance of fruit shape in summer squash?

ANSWER: The numbers are very close to a ratio of 90:60:10, or 9:6:1, an epistatic variant of the 9:3:3:1, in which the two 3/16ths categories are combined. If this is the case, then the parent plants with disk-shaped fruits were dihybrids (*AaBb*). In the offspring, 9/16ths had disk-shaped fruit indicating that it takes at least one dominant allele of each gene to produce disk-shaped fruits (*A-B-: AABB, AaBB, AABb,* or *AaBb*). The 1/16th category of plants with elongate fruits indicates that this fruit shape occurs in homozygous recessive plants (*aabb*). The plants with spherical fruit are thus plants with a dominant allele of one gene but a homozygous recessive combination at the other gene (*AAbb, Aabb, aaBB,* or *aaBb*). In summary then, two genes combine to control fruit shape in summer squash and there are epistatic interactions between the two genes giving a 9:3:4 ratio of offspring phenotypes when dihybrids are crossed.

3. A geneticist studying the pathway of synthesis of phenylalanine in *Neurospora* isolated several mutants that required phenylalanine to grow. She tested whether each mutant would grow when provided several additives, each believed to be in the pathway of the synthesis of phenylalanine (below); a plus indicates growth and minus indicates the lack of growth in the three mutants tested:

	Additive			
	Phenylpyruvate	*Prephenate*	*Chorismate*	*Phenylalanine*
Wild-type	+	+	+	+
Mutant 1	−	−	−	+
Mutant 2	+	+	−	+
Mutant 3	+	−	−	+

Where in the pathway to phenylalanine synthesis does each of the additives belong, if at all?

ANSWER: The wild-type grows in the presence of all additives, a fact that is not surprising since the wild-type can grow, by definition, in the absence of all the additives because it can synthesize phenylalanine *de novo.* Mutant 1 cannot grow in the presence of any additive except phenylalanine, indicating that its mutation is just before the end of the pathway at phenylalanine. In other words, each of the other additives occurs in the phenylalanine pathway before the point of the mutation in mutant 1. Mutant 2 can grow if given all additives but chorismate, indicating that chorismate is at the beginning of the pathway and the mutation occurred just after that substance. Finally, mutant 3 can grow if given phenylpyruvate or phenylalanine, indicating that its mutation occurred before phenylpyruvate and phenylalanine but after the earlier part of the pathway. Putting all of this information together indicates that the pathway to phenylalanine, with mutants indicated, is:

$$\text{chorismate} \xrightarrow{2} \text{prephenate} \xrightarrow{3} \text{phenylpyruvate} \xrightarrow{1} \text{phenylalanine.}$$

1. Mendel crossed tall pea plants with dwarf ones. The F_1 plants were all tall. When these F_1 plants were selfed to produce the F_2 generation, he got a 3:1, tall to dwarf ratio of offspring. If the F_2 generation had been selfed, give the genotypes and phenotypes and relative proportions of the F_3 generation produced.

2. Mendel self-fertilized dihybrid plants ($RrYy$) with round and yellow seeds and got a 9:3:3:1 ratio in the F_2 generation. As a test of Mendel's hypothesis of independent assortment, predict the kinds and numbers of progeny produced in testcrosses of these F_2 offspring.

3. In a variety of onions three bulb colors segregate: red, yellow, and white. A plant with a red bulb is crossed to a plant with a white bulb and all the offspring have red bulbs. When these are selfed, the following plants are obtained:

   Red-bulbed   119
   Yellow-bulbed   32
   White-bulbed   9

   What is the mode of inheritance of bulb color and how do you account for the ratio?

4. Four o'clock plants have a gene for color and a gene for height with the following phenotypes:

RR: red flower	TT: tall plant
Rr: pink flower	Tt: medium height plant
rr: white flower	tt: dwarf plant

   If a dihybrid plant is self-fertilized, give the proportions of genotypes and phenotypes produced.

5. A particular variety of corn has a gene for kernel color and a gene for height with the following phenotypes:

CC, Cc: purple kernels	TT: tall stem
cc: white kernels	Tt: medium height stem
	tt: dwarf stem

   If a dihybrid plant is selfed, give the resulting proportions of genotypes and phenotypes produced.

6. When studying an inherited phenomenon, a geneticist discovers a phenotypic ratio of 9:6:1 among offspring of a given mating. Give a simple, plausible explanation for this result. How would you test this hypothesis?

7. In order to determine the genotypes of the offspring of a cross in which a corn trihybrid ($AaBbCc$) was selfed, a geneticist has three choices. He or she can take a sample of the progeny and (a) self-fertilize the individual plants, (b) testcross the plants, or (c) cross the individuals with a trihybrid (backcross). Which method is preferred?

8. Explain how Tay-Sachs disease can be both a recessive and an incomplete dominant trait. What are the differences between incomplete dominance and codominance?

9. In the ABO blood-group system in human beings, alleles A and B are codominant and both are dominant to the O allele. In a paternity dispute, a type AB woman claimed that one of four men was the father of her type A child. Which of the following men (a–d) could be the father of the child on the basis of the evidence given?
   a. Type A
   b. Type B
   c. Type O
   d. Type AB

10. Under what circumstances can the phenotypes of the ABO system be used to refute paternity?

11. In blood transfusions, one blood type is called the "universal donor" and one the "universal recipient" because of their ABO compatibilities. Which is which?

12. In figure 2.17, the F_2 phenotypic ratio is 3:1:3:1:6:2. What are the phenotypic segregation ratios for each blood system (AB, Rh) separately? Are they segregating properly? What phenotypic ratio in the F_2 generation would indicate the lack of independent assortment?

13. Assume that Mendel looked simultaneously at four traits of his pea and that each trait exhibited dominance. If he crossed a homozygous dominant plant with a homozygous recessive plant, all the F_1 offspring would have been of the dominant phenotype. If he then selfed the F_1 plants, how many different types of gametes would these F_1 plants have produced? How many different phenotypes would have appeared in the F_2 generation? How many different genotypes would have appeared? What proportion of the F_2 offspring would have been of the fourfold recessive phenotype?

14. A geneticist crossed two corn plants, creating an F_1 decahybrid (ten segregating loci). This decahybrid was then self-fertilized. How many different kinds of gametes were produced by the F_1 plant? What proportion of the F_2 offspring were recessive homozygotes? How many different kinds of genotypes and phenotypes were generated in the F_2 offspring? What would your answer be if the decahybrid were testcrossed instead?

15. What properties of fruit flies and corn made them the organisms of choice for geneticists during most of the first half of this century? (Molecular geneticists have made great strides working with bacteria and viruses. You could begin thinking at this point about the properties that have made these organisms so valuable to geneticists.)

16. In fruit flies a new dominant trait, wingless, was discovered. Describe different ways of naming the alleles of the wingless gene.

17. State precisely the rules of segregation and independent assortment.

18. You notice a rooster with a pea comb and a hen with a rose comb in your chicken coop. Outline how you would determine the nature of the genetic control of comb type. How would you proceed if both your rooster and hen had rose combs?

19. How does the biochemical pathway of figure 2.13 explain how alleles A and B are codominant yet both are dominant to allele O?

20. The following is a list of ten genes in fruit flies, each with one of its alleles given. Are the alleles dominant or recessive? Are they mutant or wild-type? What is the designation of the alternative allele of each? Is the alternative allele dominant or recessive in each case?

Name of gene	Allele
yellow	y^+
Hairy wing	Hw
Abruptex	Ax^+
Confluens	Co
raven	rv^+
downy	dow
Minute(2)e	$M(2)e^+$
Jammed	J
tufted	tuf^+
burgundy	bur

21. Among the loci having the greatest number of alleles are those involved in self-incompatibility in plants. In some cases hundreds of alleles are known. What types of constraints might exist to set a limit on the number of alleles that a gene can have?

22. Suggest possible mechanisms for the epistatic ratios given in table 2.4. Can you add any further ratios?

23. What are the differences among dominance, epistasis, and pleiotropy? How can you determine that pleiotropic effects, such as those seen in sickle-cell anemia, are not due to different genes?

24. The following is a pathway from substance Q to substance U with each step numbered:

$$\begin{array}{ccccccccc} & 1 & & 2 & & 3 & & 4 & \\ Q & \to & R & \to & S & \to & T & \to & U \end{array}$$

Which product should build up in the cell and which products should never appear if the pathway is blocked at point 1? At 2? At 3? At 4?

25. Table 2.6 shows the growth ($+$) or lack of growth ($-$) of four mutant strains of *Neurospora* with various additives. The additives are in the pathway of niacin synthesis. Diagram the pathway and show which steps the various mutants block. Which compound would each mutant accumulate? When you complete this problem, compare your results with figure 2.27. What effect on growth would be observed following a mutation in the pathway of serine biosynthesis?

26. Table 2.7 shows the growth ($+$) or the lack of growth ($-$) of various mutants in another pathway of biosynthesis. Determine this pathway, the point of blockage of each mutant, and the substrate accumulated by each mutant.

27. Maple sugar urine disease is a rare inborn error of human metabolism. The urine of affected individuals smells like maple sugar.
    a. If two normal individuals have an affected child, what is the mode of inheritance of the disease?
    b. What is the chance that the second child will be normal?

28. In *Drosophila*, a cross between a dark bodied fly and tan bodied fly yields 76 tan and 80 dark flies. Diagram the cross.

29. If two black mice are crossed, 10 black and 3 white mice result.
    a. Which allele is dominant?
    b. Which allele is recessive?
    c. What is the genotype of the parents?

30. In the human ABO blood system, the alleles A and B are dominant over O. What possible phenotypic ratios do you expect from a mating between a type A individual and a type B individual?

TABLE 2.6

	Additive					
Mutant	Nothing	Niacin	Tryptophan	Kynurenine	3-Hydroxyanthranilic Acid	Indole
1	−	+	+	+	+	−
2	−	+	+	+	+	+
3	−	+	−	−	+	−
4	−	+	−	−	−	−

TABLE 2.7

		Additive				
Mutant	Nothing	A	B	C	D	E
1	−	−	−	−	−	+
2	−	+	+	+	−	+
3	−	+	−	−	−	+
4	−	+	+	+	+	+
5	−	+	−	+	−	+

31. Two short-eared pigs are mated. In the progeny, three have no ears, seven have short ears, and four have long ears. Explain these results by diagramming the cross.

32. In *Drosophila,* two red-eyed flies mate and yield 110 red-eyed and 35 brown-eyed offspring. Diagram the cross and determine which allele is dominant.

33. A plant with red flowers is crossed to a plant with white flowers. All the progeny are pink. When the pink flowers are crossed, the progeny are 11 red, 23 pink, and 12 white. What is the mode of inheritance of color?

34. Consider the following crosses in pea plants and determine the genotypes of the parents in each cross. Yellow and green refer to seed color; tall and short refer to plant height.

Cross	Progeny			
	Yellow, Tall	Yellow, Short	Green, Tall	Green, Short
1) Yellow, tall × yellow, tall	89	31	33	10
2) Yellow, short × yellow, short	0	42	0	15
3) Green, tall × yellow, short	21	20	24	22

35. In screech owls, crosses between red and silver individuals sometimes yield all red; sometimes 1/2 red : 1/2 silver; and sometimes 1/2 red : 1/4 white : 1/4 silver offspring. Crosses between two red owls yield either all red, 3/4 red : 1/4 silver, or 3/4 red : 1/4 white offspring. What is the mode of inheritance?

36. A brown-eyed, long-winged fly is mated to a red-eyed, long-winged fly. The progeny are

51 long, red
53 long, brown
18 short, red
16 short, brown

What are the genotypes of the parents?

37. True breeding flies that have long wings and dark bodies are mated to true breeding flies with short wings and tan bodies. All the F_1 progeny have long wings and tan bodies. The F_1 progeny are allowed to mate and produce:

44 tan, long
16 dark, long
14 tan, short
6 short, dark

What is the mode of inheritance?

38. In peas, tall (T) is dominant to short (t), yellow (Y) is dominant to green (y), and smooth (S) is dominant to wrinkled (s). From a cross of two triple heterozygotes, what is the chance of getting a plant that is:
   a. tall, yellow, smooth?
   b. short, green, wrinkled?
   c. short, green, smooth?

39. In corn, the genotype *A-C-R-* is colored. Recessive individuals homozygous for at least one gene are colorless. Consider the following crosses involving colored plants all with the same genotype. Based on the results, deduce the genotype of the colored plant.

colored × *aaccRR* → 1/2 colored; 1/2 colorless
colored × *aaCCrr* → 1/4 colored; 3/4 colorless
colored × *AAccrr* → 1/2 colored; 1/2 colorless

40. Consider the following crosses in *Drosophila.* Based on the results, deduce which alleles are dominant and the genotypes of the parents. Orange and red are eye colors; crossveins occur on the wings.

Parents	Progeny			
	Orange, crossveins	Orange, crossveinless	Red, crossveins	Red, crossveinless
1) Orange, crossveins × orange, crossveins	83	26	0	0
2) Red, crossveins × red, crossveinless	20	18	65	63
3) Red, crossveinless × red, crossveins	0	0	74	81
4) Red, crossveins × red, crossveins	28	11	93	34

41. In *Drosophila melanogaster*, a recessive autosomal gene, *ebony*, produces a dark body color when homozygous, and an independently assorting autosomal gene, *black*, has a similar effect. If homozygous *ebony* flies are crossed with homozygous *black* flies,
  a. What will be the phenotype of the F_1 flies?
  b. What phenotypes and what proportions would occur in the F_2 generation?
  c. What phenotypic ratios would you expect to find in the progeny of the backcrosses of
      1) $F_1 \times$ *ebony*?
      2) $F_1 \times$ *black*?

42. *A, B,* and *C* are independently assorting Mendelian factors controlling the production of black pigment; alleles of these genes are indicated *a, b,* and *c* respectively. Assume that *A, B,* and *C* act in a pathway as follows:

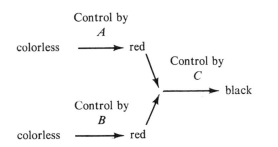

A black *AABBCC* individual is crossed with a colorless *aabbcc* to give black F_1 individuals. The F_1 individuals are selfed to give F_2 progeny.
  a. What proportion of the F_2 generation is colorless?
  b. What proportion of the F_2 generation is red?

43. In *Drosophila*, there are four strains differing in eye color: wild-type, orange-1, orange-2, and pink. The following matings involving true breeding individuals were performed.

Cross	F_1
wild-type $\times$ orange-1	all wild-type
wild-type $\times$ orange-2	all wild-type
orange-1 $\times$ orange-2	all wild-type
orange-2 $\times$ pink	all orange-2
F_1 (orange-1 $\times$ orange-2) $\times$ pink	1/4 orange-2 : 1/4 pink : 1/4 orange-1 : 1/4 wild-type

What F_2 ratio do you expect if the F_1 progeny from orange-1 $\times$ orange-2 are selfed?

44. A pre-med student, Steve, plans to marry the daughter of the dean of nursing. Steve's father, a doctor, puts pressure on Steve to marry someone else because the girl has a sister with PKU (phenylketonuria) and a brother with albinism. The dean's husband was sterile and the three children were all the result of artificial insemination from three different donors. Having served as an anonymous sperm donor, the doctor is concerned that Steve and his fiancé may be half brother and sister. Given the following information, deduce whether Steve and his fiancé are related. The MN and Ss systems are two independent, codominant blood genetic systems.

	Blood type
dean	A, MN, Ss
her daughter	O, M, S
Steve's father	A, MN, Ss
Steve	O, N, s
Steve's mother	B, N, s

45. You are working with the exotic organism, *Phobia laboris*, and are interested in obtaining mutants that work hard. Normal phobes are lazy. Perseverance finally pays off, and you successfully isolate a true-breeding line of workers. You begin a detailed genetic analysis of this trait. To date you have obtained the following results:

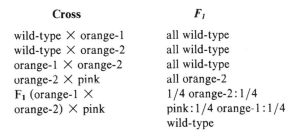

From these results, predict the expected phenotypic ratio from crossing two F_1 nonworkers.

Suggested Readings for chapter 2 are on page 641.

3

Mitosis and Meiosis

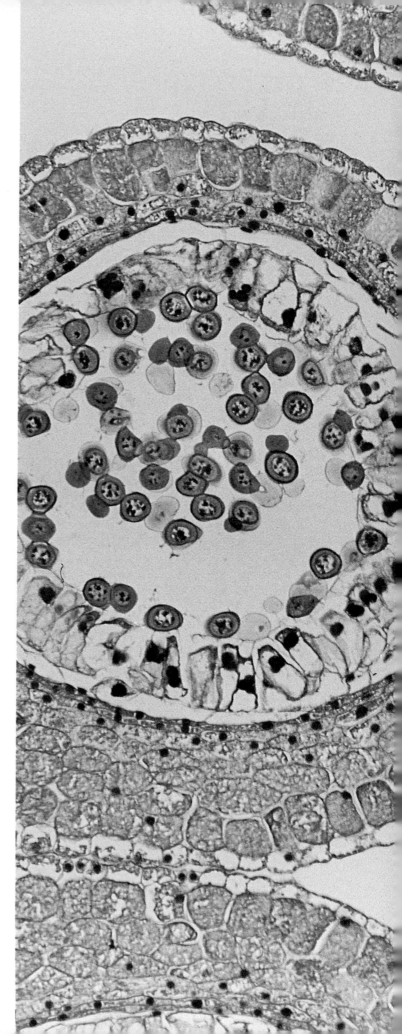

Lily (*Lilium michiganense*) meiosis showing late prophase of the anther in the first meiotic division. Magnification 100×.
© *Bruce Iverson*

The zygote, or fertilized egg, of higher organisms is the starting point of most life cycles. This zygote then divides many times to produce an adult organism. In animals, the adult then produces gametes that combine to start the cycle again. In plants, the adult is a sporophyte that produces spores by genetic reduction. These spores develop into gametophytes, which may or may not be independent. Gametophytes produce gametes that fuse to form the zygote (fig. 3.1). The process of cell division is composed of a nuclear and a cytoplasmic component. Nuclear division (**karyokinesis**) has two forms, a nonreductional **mitosis** in which the mother and daughter cells have exactly the same genetic complement, and a reductional **meiosis** in which the products, gametes in animals and spores in higher plants, have approximately half the genetic material as the parent cell, ensuring that the amount of genetic material of a zygote will be the same from generation to generation. The division of the cytoplasm, resulting in two cells from one original cell, is termed **cytokinesis.** In this chapter we examine the processes of mitosis and meiosis, which allow chromosomes, the gene vehicles, to be properly apportioned among daughter cells. Keep in mind the engineering difficulties posed by these processes and the relationship of meiosis to Mendel's rules.

Mendel's work was rediscovered at the turn of the century after being ignored for thirty-four years. One of the major reasons that it could be appreciated in 1900 was that many of the processes that chromosomes undergo had been described. With those discoveries, a physical basis had been found for genes. That is, the way in which chromosomes behave during gamete formation is precisely the way in which Mendel predicted that genes would behave during gamete formation. In this chapter we look at the morphology of chromosomes and their behavior during somatic-cell division and gamete and spore formation.

Modern biologists classify organisms into two major categories: **eukaryotes,** organisms that have true nuclei, and **prokaryotes,** organisms that lack true nuclei (table 3.1). Bacteria and blue-green algae are prokaryotes. All higher organisms are eukaryotes. In the prokaryotes the genetic material is a simple circle of DNA (deoxyribonucleic acid), although ancillary circles of DNA, called plasmids, are found frequently (see chapters 12 and 17). In eukaryotes the genetic material, located in the nucleus (fig. 3.2), is DNA highly complexed with protein (nucleoprotein). In this chapter we concentrate on the nuclear division processes of eukaryotes.

CHROMOSOMES

Chromosomes were discovered by C. Nägeli in 1842. The term **chromosome,** coined by W. Waldeyer in 1888, means "colored body" and is used because chromosomes stain a bright color with certain techniques of histology. The nucleoprotein material of the chromosomes is referred to as **chromatin.**

Figure 3.1
Generalized life cycle of (*a*) animals and (*b*) plants

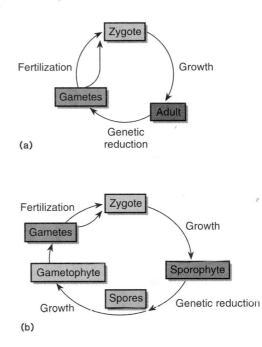

Although all eukaryotes have chromosomes, in the **interphase** between divisions they are spread out or diffused throughout the nucleus and are usually not identifiable. Each chromosome, with very few exceptions, has a distinct attachment point for fibers that makes up the mitotic and meiotic spindle apparatus. (The exceptions are in organisms with a diffuse arrangement of attachment points all along the chromosomes.) The attachment points are called **kinetochores,** and the constrictions in the chromosomes where these occur are called **centromeres.** Kinetochores seem to be composed of protein and RNA (ribonucleic acid; see chapter 9). Chromosomes can be classified according to whether the centromere is in the middle of the chromosome (**metacentric**), at the end of the chromosome (**telocentric**), very near the end of the chromosome (**acrocentric**), or somewhere in between (**subtelocentric** or **submetacentric;** figs. 3.3 and 3.4). For any particular chromosome the position of the centromere is fixed.

Most eukaryotic cells are **diploid** before nuclear division takes place; that is, all their chromosomes occur in pairs. One member of each pair came from each parent. **Haploid** cells, which include the reproductive cells (gametes), have only one copy of each chromosome. In the diploid state, members of the same chromosome pair are referred to as **homologous chromosomes** (homologues); the two make up a homologous pair.

The total chromosome complement of a cell, the **karyotype,** can be photographed during mitosis and rearranged in pairs to make a picture referred to as an **idiogram** (fig. 3.5). From the idiogram it is possible to see

TABLE 3.1 Differences between Prokaryotic and Eukaryotic Cells

	Prokaryotic Cells	Eukaryotic Cells
Taxonomic Groups	Bacteria and blue-green algae (monera)	All plants, fungi, animals, protists (according to Whittaker's five-kingdom classification scheme)
Size	Less than 5 μm in greatest dimension	Greater than 5 μm in smallest dimension
Nucleus	No true nucleus, no nuclear membrane	Nuclear membrane
Genetic Material	One circular molecule of DNA, little protein	Linear histone-containing nucleoproteins
Mitosis and Meiosis	Absent	Present

Figure 3.2

Mouse lung cell. Magnification 4,270×.

Courtesy of Wayne Rosenkrans.

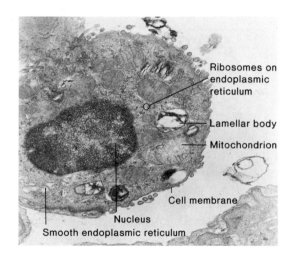

Figure 3.3

(*a*) Submetacentric chromosome. (*b*) Submetacentric chromosome in early mitosis. The chromosome is best seen after it has been duplicated but before separation of the identical halves (sister chromatids).

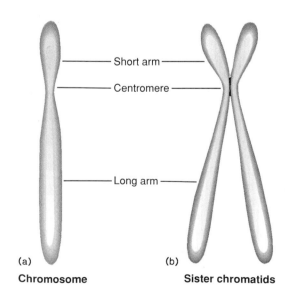

(a) Chromosome

(b) Sister chromatids

Figure 3.4

(*a*) Metacentric, (*b*) submetacentric, and (*c*) acrocentric chromosomes in human beings. With the exception of telocentric chromosomes, the centromere divides the chromosome into two arms.

Source: Reproduced courtesy of Dr. Thomas G. Brewster, Foundation for Blood Research, Scarborough, Maine.

whether there are abnormal numbers of chromosomes and to identify the sex of the organism. As you can see from figure 3.5, all of the homologous pairs are made up of identical partners, referred to as **homomorphic chromosomes.** A potential exception is the sex chromosomes, which in some species are of unequal size and are therefore called **heteromorphic chromosomes.**

The number of chromosomes possessed by individuals of a particular species is constant. Some species exist mostly in the haploid state or have long intervals in their life cycle that are haploid. For example, the pink bread mold *Neurospora crassa,* a fungus, in the haploid state has a chromosome number of seven ($n = 7$). The diploid number is, of course, fourteen ($2n = 14$). The diploid chromosome number of several species is shown in table 3.2.

Figure 3.5

Idiogram or karyotype of a human female (two X chromosomes, no Y chromosome). A male would have one X and one Y chromosome. The chromosomes are grouped into categories (A-G, X, Y) by length and position of their centromeres.

Source: Reproduced courtesy of Dr. Thomas G. Brewster, Foundation for Blood Research, Scarborough, Maine.

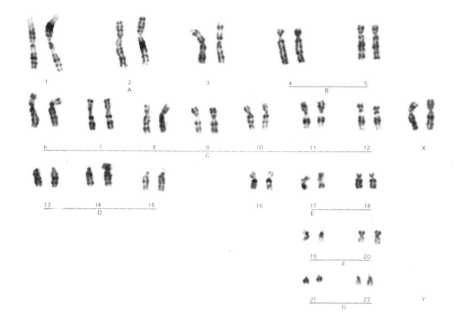

TABLE 3.2 Chromosome Number for Selected Species ($2n$ Is the Diploid Complement)

Species	$2n$
Human being (*Homo sapiens*)	46
Garden pea (*Pisum sativum*)	14
Fruit fly (*Drosophila melanogaster*)	8
House mouse (*Mus musculus*)	40
Roundworm (*Ascaris sp.*)	2
Pigeon (*Columba livia*)	80
Boa constrictor (*Constrictor constrictor*)	36
Cricket (*Gryllus domesticus*)	22
Lily (*Lilium longiflorum*)	24
Indian fern (*Ophioglossum reticulatum*)	1,260

Note: The fern has the highest known diploid chromosome number.

In eukaryotes there are two processes whereby the genetic material is partitioned into offspring, or daughter, cells. One is the simple division of one cell into two. In this process the two daughter cells must each receive an exact copy of the genetic material of the parent cell. The cellular process is simple cell division and the nuclear process accompanying it is mitosis. In the other partitioning process, the genetic material must be precisely halved so that the diploid complement is re-formed by fertilization. The cellular process is gamete formation in animals and spore formation in higher plants and the nuclear process is meiosis. The term *mitosis,* coined by Flemming in the 1880s, is from the Greek word for "a thread," referring to a chromosome. The term *meiosis* is from the Greek word "to lessen."

Chromosomes are separated in both processes of nuclear division. The division of the cytoplasm of the cell, cytokinesis, is much less organized. In animals, there is a constriction of the cell membrane that more or less randomly distributes the cytoplasm. In plants, the growth of a cell wall accomplishes the same purpose. Let us first examine the process of mitosis, in which our emphasis will be on the behavior of the chromosomes because it is their behavior that has genetic implications. In this book we are less concerned with cellular details.

MITOSIS

Consider the engineering problem that must be solved by mitosis. Identical **chromatids,** or **sister chromatids,** the results of chromosome replication, must be separated in such a way that each goes into a different daughter cell (see fig. 3.3). These chromatids are the visible manifestation of the chromosome replication that has taken place previously in the S-phase of the cell cycle. The chromatids

Figure 3.6

Cell cycle in the broad bean, *Vicia faba*. Total time is under twenty hours. The DNA content of the cell is doubled during the S-phase and then reduced back to its original value by mitosis.

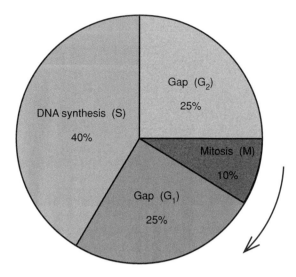

are held together in the region of the centromere and each will be called a chromosome when it separates and becomes independent. This separation must occur for each chromosome. Each of the two daughter cells will then end up with a chromosome complement identical to that of the parent cell. Mitosis is nature's elegant process to achieve that end—surely an engineering marvel.

Mitosis is a continuous process. However, for descriptive purposes it is broken into four stages: **prophase, metaphase, anaphase,** and **telophase** (Greek: *pro-,* before; *meta-,* mid; *ana-,* back; *telo-,* end). Replication (duplication) of the genetic material occurs during the S-phase of the **cell cycle** (fig. 3.6). The timing of the four stages varies from species to species, from organ to organ within a species, and even from cell to cell within a given cell type.

In general, however, a cell usually divides when it has doubled its volume. Although the control seems to involve many genes and differs among many organisms, a common thread of genetic control of the cell cycle in eukaryotes has emerged. Apparently, cell division is under the general control of two proteins, one that remains constant throughout the cell cycle (called cdc2 for cell-division cycle protein number two) and another, which oscillates during the cell cycle, called cyclin. Cyclin's oscillation is caused by the fact that it is degraded during part of the cell cycle. When combined and modified, these two proteins initiate cell division. The cdc2 protein is a kinase, an enzyme that transfers a phosphate group from ATP to another protein. A common way that cells regulate enzyme activity is by controlling the level of phosphorylation of the enzymes. Hence it is not surprising that a kinase takes part in regulating the cell cycle.

Prophase

This stage of mitosis is characterized by a shortening and thickening of the chromosomes so that individual chromosomes become distinct. (Details of the molecular structure of the eukaryotic chromosome and the processes of coiling and shortening are given in chapter 14.) At this time also, the nuclear envelope (membrane) disintegrates, the **nucleolus** disappears, the **centrioles,** when present, duplicate and migrate to opposite poles of the cell, and the **spindle** apparatus forms (fig. 3.7). The nucleolus is a darkly stained body in the nucleus. It is involved in ribosome construction and forms around a **nucleolar organizer** on one of the chromosomes. The number of nucleoli varies in different species, but in the simplest case there will be two nucleolar organizers per nucleus, one each on the two members of a homologous pair of chromosomes. Nucleoli are re-formed after mitosis.

Centrioles are cylindrical organelles found in virtually every eukaryotic group except the higher plants (fig. 3.8). They organize the spindle, a conglomerate of **microtubules**—hollow cylinders made up of the protein tubulin—that attach to the chromosomes at the kinetochores and pull the chromosomes toward a pole of the spindle.

As prophase progresses, each chromosome can be seen to be composed of two identical (sister) chromatids (see fig. 3.3). Spindle fibers can be seen to attach to the individual chromosomes at their kinetochores (fig. 3.9), two disk-shaped objects at each centromere. The kinetochores are on opposite sides of the centromere, and there is one associated with each sister chromatid. This geometry assures that the chromatids are separated from each other during the next stage of mitosis. The number of microtubules attaching to each kinetochore differs in different species. Four to seven microtubules attach per kinetochore in the cells of the rat fetus, whereas 70 to 150 attach in the plant *Haemanthus* (fig. 3.9).

Metaphase

With the attachment of the spindle fibers and the completion of the spindle itself, the chromosomes are jockeyed into position in the plane of the equator of the spindle, called the **metaphase plate.** Alignment of the chromosomes on this plate marks the end of metaphase (fig. 3.10).

Anaphase

This stage is initiated by the separation of sister chromatids (fig. 3.11). Almost simultaneously, all of the chromatids in the cell separate, with sisters being pulled to opposite poles of the cell. The dragging occurs at a uniform rate of speed, and the chromosomes appear to be pulled at the centromere by the action of the spindle fibers. Thus metacentric chromosomes appear V-shaped, subtelocentrics, J-shaped, and telocentrics, rod-shaped.

Figure 3.7

Interphase and prophase of mitosis. In this cell, $2n = 4$, consisting of a pair of long and a pair of short metacentric chromosomes. Maternal chromosomes are *red*, paternal chromosomes are *blue*.

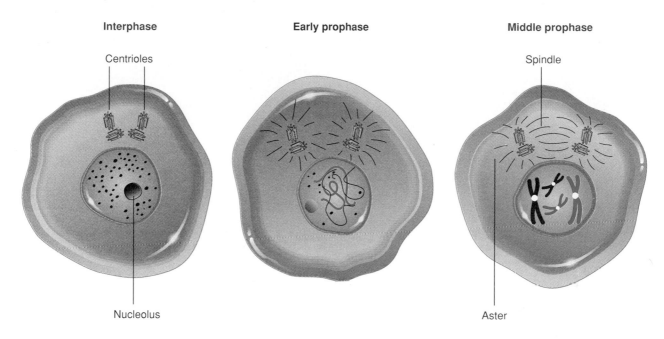

Interphase   **Early prophase**   **Middle prophase**

Centrioles   Spindle

Nucleolus   Aster

Figure 3.8

Centriole. Magnification 111,800✕.

Reproduced from the Journal of Cell Biology *37 (1968): 381. F. R. Turner, "An Ultrastructural Study of Plant Spermatogenesis: Spermatogenesis in Nitella." By copyright permission of the Rockefeller University Press.*

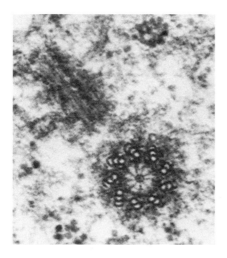

Figure 3.9

Scanning electron micrograph of the centromeric region of a metaphase chromosome from the plant *Haemanthus katherinae.* Spindle fiber bundles on either side of the centromere are directed in opposite directions. A fiber not connected to the kinetochore can be seen in the center. Fibers are 60 to 70 nm in diameter.

Source: Waheeb K. Heneen, "The centromeric region in the scanning electron microscope," Hereditas *97 (1982):311–314. Reproduced by permission.*

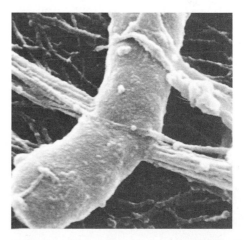

Telophase

At the end of anaphase (fig. 3.12), the separated sister chromatids (now full-fledged chromosomes) have been pulled to opposite poles of the cell. The cell now reverses the steps of prophase to return to the active interphase state (fig. 3.13). The chromosomes uncoil and begin to carry out their physiological functions (directing protein synthesis, replication, etc.). A nuclear envelope re-forms about each set of chromosomes, nucleoli form, and cyto-

kinesis takes place. The cell has now entered the G_1 phase of the cell cycle (fig. 3.6). A summary of mitosis is shown in figure 3.14.

Significance of Mitosis

Cytokinesis and mitosis result in two daughter cells, each with genetic material identical to that of the one parent cell. There is an exact distribution of the genetic material, in the form of chromosomes, to the daughter cells.

Figure 3.10

Metaphase, mitosis. $2n = 4$. Maternal chromosomes are *red,* paternal chromosomes are *blue.*

Metaphase plate

Figure 3.11

(*a*) Mitotic spindle of anaphase. $2n = 4$. (*b*) Fluorescent microscope image of a cultured cell in anaphase. Microtubules are *red* and chromosomes (DNA) are stained *yellow.*

(a) From E. J. DuPraw, DNA and Chromosomes. Copyright © 1970 Holt, Rinehart & Winston, Inc., Orlando, FL. (b) John M. Murray, Department of Anatomy, University of Pennsylvania. Cover of BioTechniques, volume 7, number 3, March 1989. Reproduced with permission.

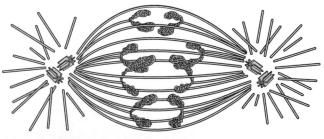

(a)

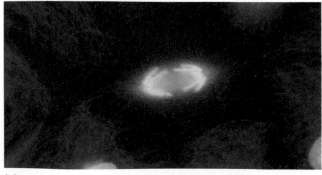

(b)

Figure 3.12

Late anaphase, mitosis. $2n = 4$. Karyokinesis has occurred but not cytokinesis. Maternal chromosomes are *red,* paternal chromosomes are *blue.*

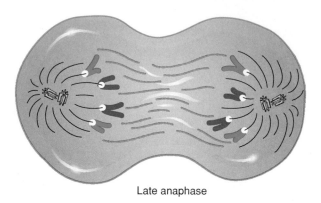

Late anaphase

Figure 3.13

Telophase and interphase, mitosis. $2n = 4$. Maternal chromosomes are *red,* paternal chromosomes are *blue.*

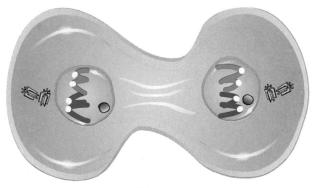

Telophase

Interphase

MEIOSIS

Gamete formation is an entirely new engineering problem to be solved. To form gametes in animals (and, for the most part, to form spores in plants), a diploid organism with its two copies of each chromosome must form daughter cells that have only one copy of each chromosome. In other words, the genetic material must be reduced to half so that when gametes recombine to form zygotes, the original number of chromosomes is restored, not doubled.

Figure 3.14
Cells in various stages of mitosis in the onion root tip. The average cell is about 50 μm long.
Source: Carolina Biological Supply Company.

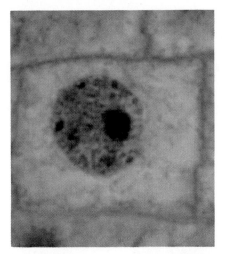

Interphase

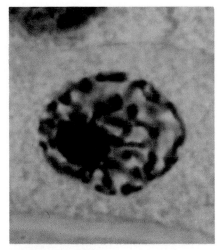

Early prophase

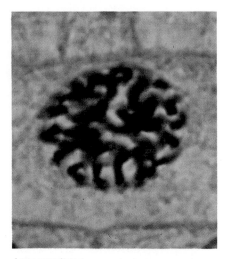

Late prophase

Early metaphase

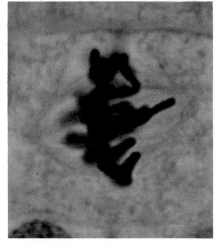

Late metaphase

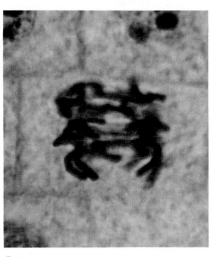

Early anaphase

Late anaphase

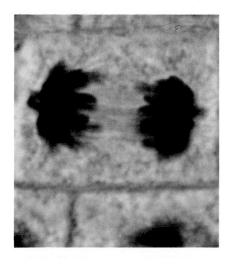

Early telophase

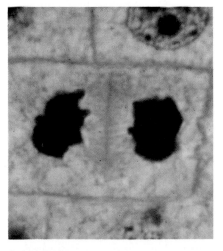

Late telophase

Figure 3.15

Synaptonemal complex. (*a*) In the electron micrograph, M is the central element, La are lateral elements, and F are chromosome fibers. Magnification 400,000×. (*b*) Diagram of the structure.

Source: (a) R. Wettstein and J. R. Sotelo, "The molecular architecture of synaptonemal complexes," in E. J. DuPraw, ed., Advances in Cell and Molecular Biology, Vol. 1 (New York: Academic Press, 1971), p. 118. Reproduced by permission. (b) From B. John and K. R. Lewis, Chromosome Hierarchy. Copyright © 1975 Oxford University Press, London, England. Reprinted by permission of the Oxford University Press.

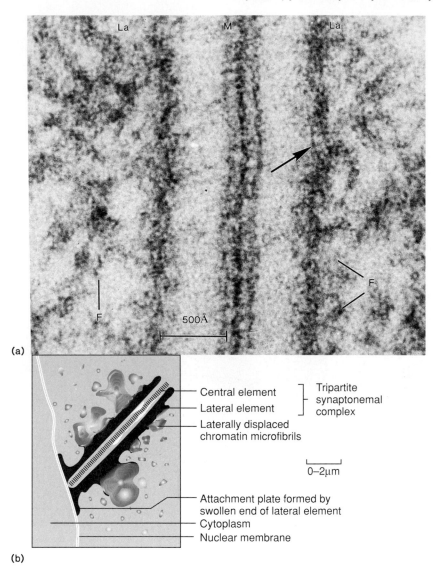

If we were to try to engineer this task, we would have to be able to recognize homologous chromosomes. We could then push one member into one daughter cell and the other member into the other daughter cell. If we were unable to recognize homologues, we would not be able to ensure that each daughter cell received one and only one member of each pair. In accomplishing this task, the cell solves the problem by having homologous chromosomes pair up during an extended prophase. The spindle apparatus then separates members of the homologous chromosome pairs. There is one complication. As in mitosis, cells entering meiosis have already replicated their chromosomes. Therefore, two nuclear divisions without an intervening chromosome replication are necessary in meiosis to produce haploid gametes or spores. Meiosis is, then, a two-division process that results in four cells from each original parent cell. The two divisions are known as meiosis I and meiosis II.

Unlike mitosis, meiosis occurs only in certain kinds of cells. In animals, meiosis takes place in the primary and secondary gametocytes; in higher plants, which show an alternation of generations, the process takes place only in the spore-mother cells of the sporophyte generation (see fig. 3.1). At the end of this chapter, we review the processes of gamete and spore formation in animals and plants, respectively.

Figure 3.16

Prophase I of meiosis (nuclei shown). Maternal chromosomes are *red,* paternal chromosomes are *blue.* Note that crossing over is evident at diakinesis. $2n = 4$.

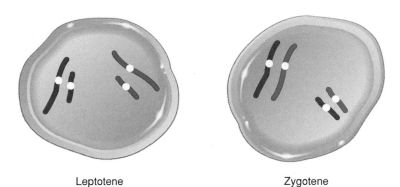

Leptotene

Zygotene
(synapsis)

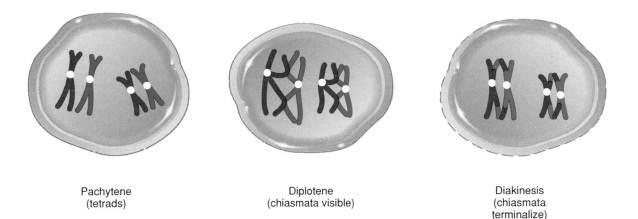

Pachytene
(tetrads)

Diplotene
(chiasmata visible)

Diakinesis
(chiasmata
terminalize)

Prophase I

Cytogeneticists have divided the prophase of meiosis I into five stages: **leptonema, zygonema, pachynema, diplonema,** and **diakinesis** (Greek: *lepto-,* thin; *zygo-,* yoke-shaped; *pachy-,* thick; *diplo-,* double; *dia-,* across). Since we are primarily interested in the genetic consequences of this process, we will not concern ourselves with the cytological details of these substages. A cell entering prophase I (leptotene stage) behaves similarly to one entering prophase of mitosis, with the centriole, spindle, nuclear envelope, and nucleolus behaving the same way. As coiling down in size takes place, homologous chromosomes pair point-for-point along their lengths. This process in the zygotene stage is referred to as **synapsis.** Synapsis is mediated, in an unknown way, by a proteinaceous complex appearing between the homologous chromosomes, referred to as a **synaptonemal complex** (fig. 3.15). At this point, the chromosome figures are referred to as **bivalents,** one bivalent per homologous pair (fig. 3.16). As the chromosomes continue to shorten and thicken in pachynema, each chromosome can be seen to be made of two sister chromatids. Now the chromosome figures are referred to as **tetrads** because they are made up of four chromatids (fig. 3.16). At

about this time the synaptonemal complex disintegrates in most species. Further on in prophase I (diplonema), the chromosomes, while still shortening and thickening, appear to repel each other along most of their length. At this point one can see X-shaped configurations along the tetrads (fig. 3.17).

These X-shaped configurations are called **chiasmata** (singular: chiasma) and are of enormous significance because they indicate **crossing over,** a process whereby homologous chromosomes exchange parts. When two chromatids come to lie in close proximity, enzymes can break both chromatid strands and reattach them in an alternative way (fig. 3.18). Thus, although genes have a fixed position on a chromosome, alleles that started out attached to a paternal centromere can end up attached to a maternal centromere. (We examine the molecular mechanism of this process in chapter 16.) Crossing over can greatly increase the genetic variability in gametes by associating alleles that were not previously joined. As prophase I moves into diakinesis, spindle-fiber attachment takes place, and chiasmata terminalize: They slip down the length of the chromosome until they reach the ends, freeing the chromosomes from each other along their lengths but not at the tips.

Figure 3.17
A tetrad from the grasshopper, *Chorthippus parallelus,* at diplotene with five chiasmata
Source: Courtesy of Bernard John.

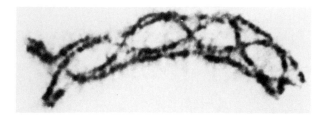

Figure 3.19
Metaphase and anaphase of meiosis I. Maternal chromosomes are *red,* paternal chromosomes are *blue.* 2n = 4.

Metaphase I

Figure 3.18
Crossing over in a tetrad during prophase, meiosis I. *Circles* represent centromeres, one paternal, one maternal.

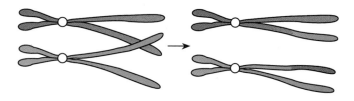

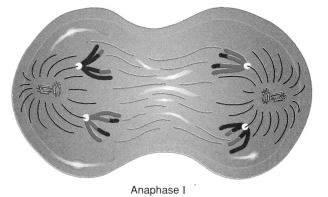

Anaphase I

Metaphase I and Anaphase I

In metaphase I, tetrads are pulled to a metaphase plate by the spindle fibers. In anaphase I, homologous centromeres, each with its two chromatids attached, are pulled apart (fig. 3.19)—the centromeres do not divide as they do in mitosis. This meiotic division is therefore called a **reductional division** because it reduces the number of chromosomes and centromeres to half the diploid number in each daughter cell. For every tetrad there is now one chromosome in the form of a chromatid pair, known as a **dyad,** at each pole of the cell. The initial objective of meiosis, that of separating homologues into different daughter cells, is accomplished. However, since each dyad consists of two sister chromatids, a second division is required to reduce each chromosome to a single chromatid.

Telophase I and Prophase II

Depending on the organism, telophase I may or may not be greatly shortened in time. In some organisms all the expected stages take place; chromosomes enter an interphase configuration as cytokinesis takes place. However, no chromosome duplication (DNA replication) occurs during this abbreviated interphase, termed **interkinesis.** Then prophase II begins and meiosis II proceeds. In other organisms the late anaphase I chromosomes go almost directly into metaphase II, virtually skipping telophase I, interphase, and prophase II.

Meiosis II

In any case, meiosis II is basically a mitotic division in which the chromatids of each chromosome are pulled to opposite poles. For each original cell entering meiosis I, four cells emerge at telophase II. Meiosis II is referred to as an **equational division;** although it reduces the amount of genetic material per cell by half, it does not further reduce the chromosome number per cell (fig. 3.20). (Note that sometimes it is simpler to concentrate on the behavior of centromeres during meiosis rather than on the chromosomes and chromatids. Meiosis I separates maternal from paternal centromeres and meiosis II separates sister centromeres.) Figure 3.21 is a summary of meiosis in corn (*Zea mays*).

Significance of Meiosis

Meiosis is significant for several reasons. First, the diploid number of chromosomes is reduced in such a way that each of four daughter cells has one complete haploid chromosome set. Second, because of crossing over, there is an opportunity for increasing the allelic combinations of a

Figure 3.20

Meiosis II. Maternal chromosomes are *red*, paternal chromosomes are *blue*. $2n = 4$.

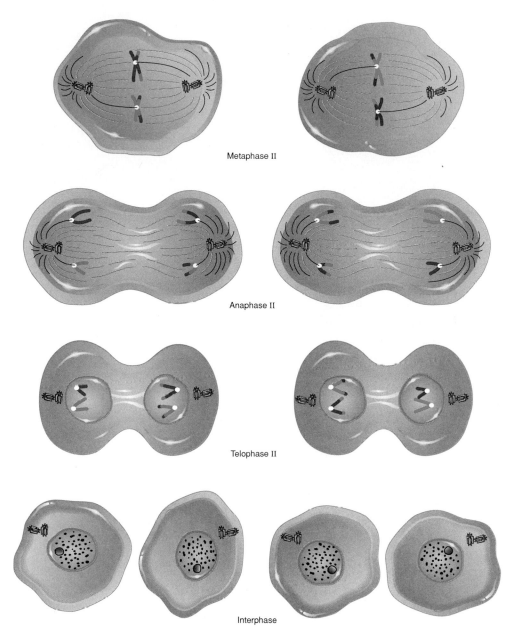

Metaphase II

Anaphase II

Telophase II

Interphase

gamete. Rather than maternal alleles staying together and paternal alleles doing the same, in each generation new combinations of maternal and paternal alleles can form. New combinations are, of course, also introduced by the process of random assortment itself, in which maternal and paternal chromosomes are randomly combined in each gamete. The process of creating new arrangements, either by crossing over or by independent segregation of homologous pairs of chromosomes, is called **recombination.** Assuming ten thousand genes in an organism with two alleles each, $2^{10,000}$ different gametes could arise by meiosis.

The behavior of any tetrad follows the pattern of Mendel's rule of segregation. At spore or gamete formation (meiosis), the diploid number of chromosomes is halved; each gamete receives only one chromosome from a homologous pair. This process, of course, explains Mendel's rule of segregation. Independent assortment is also explained by chromosome behavior at meiosis. In anaphase I the direction of separation is independent in different tetrads. Whereas one pole may get the maternal centromere from chromosome pair number 1, it could get either the maternal or the paternal centromere from chromosome pair number 2, and so on (fig. 3.22). Alleles of

Figure 3.21

Meiosis in corn (*Zea mays*)

Courtesy: Dr. M. M. Rhoades. "Meiosis in Maize," Journal of Heredity *41 (1950):59–67. Reproduced by permission.*

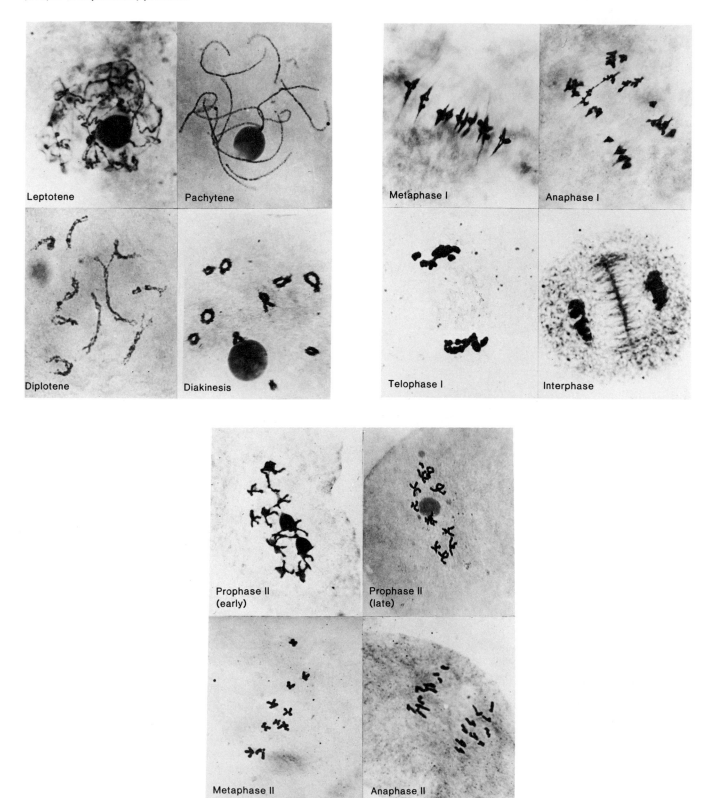

Leptotene

Pachytene

Metaphase I

Anaphase I

Diplotene

Diakinesis

Telophase I

Interphase

Prophase II (early)

Prophase II (late)

Metaphase II

Anaphase II

Figure 3.22
Significance of meiosis to the rule of independent assortment. P and
M refer to paternal and maternal centromeres, which separate
independently in different tetrads.

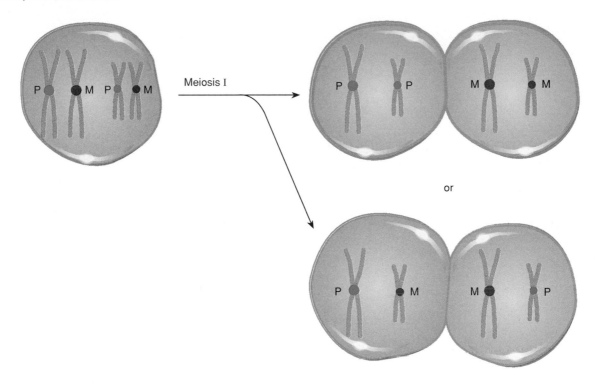

one gene segregate independently of alleles of other genes.
Very shortly after rediscovery of Mendel's principles at
the turn of the century, geneticists were quick to realize
this.

MEIOSIS IN ANIMALS

In male animals, each meiosis produces four equally sized
sperm cells (fig. 3.23) in a process called **spermatogenesis.**
In vertebrates, a cell type in the testes known as a **spermatogonium** produces **primary spermatocytes,** as well as
additional spermatogonia, by mitosis. The primary spermatocytes undergo meiosis. After the first meiotic division, these cells are known as **secondary spermatocytes;**
after the second meiotic division, they are known as **spermatids.** The spermatids mature into spermatozoa by a
process called **spermiogenesis**—four sperm cells resulting
from each primary spermatocyte. In human beings and
other vertebrates without a specific mating season, the
process of spermatogenesis is continuous throughout adult
life. A normal human male may produce several hundred
million sperm cells per day.

During embryonic development in human females,
cells in the ovary known as **oogonia** proliferate by numerous mitotic divisions to form **primary oocytes.** About
one million per ovary are formed. These begin the first
meiotic division and then stop before the birth of the female
in a prolonged diplonema, called the **dictyotene** stage. A
primary oocyte does not resume meiosis until past puberty
when, under hormonal control, ovulation takes place, a
process usually occurring for only one oocyte per month
during the female's reproductive life span (from about
twelve to forty-five years of age). Meiosis then proceeds
in the ovulated oocyte. The two cells formed by meiosis I
are of unequal size. One, termed the **secondary oocyte,**
contains almost all the nutrient-rich cytoplasm; the other,
a **polar body,** receives very little cytoplasm. The second
meiotic division in the larger cell yields another polar body
and an **ovum.** The first polar body may or may not divide
to form two other polar bodies. Thus **oogenesis** (fig. 3.24)
produces cells of unequal size—an ovum and two or three
polar bodies. The polar bodies disintegrate. Cells of
unequal size are produced because the oocyte nucleus
resides very close to the surface of this large cell.

Figure 3.23
Spermatogenesis

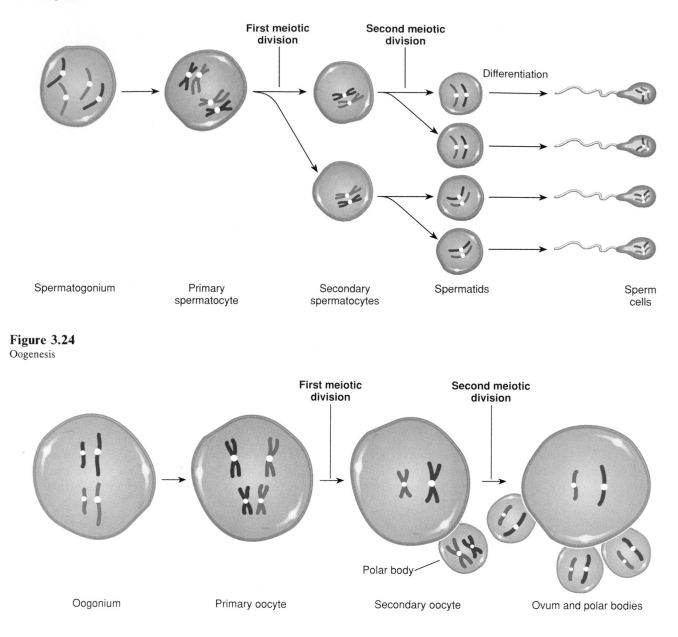

First meiotic division

Second meiotic division

Differentiation

Spermatogonium

Primary spermatocyte

Secondary spermatocytes

Spermatids

Sperm cells

Figure 3.24
Oogenesis

First meiotic division

Second meiotic division

Polar body

Oogonium

Primary oocyte

Secondary oocyte

Ovum and polar bodies

LIFE CYCLES

For eukaryotes, the basic pattern of the life cycle is an alternation between a diploid and a haploid state (fig. 3.1). All life cycles are modifications of this general pattern. Most animals are diploids that form gametes by meiosis. The diploid number is restored by fertilization. Exceptions, however, are numerous. For example, in most bee species males are haploid and produce gametes by mitosis; females are diploid. Some fishes exist by **parthenogenesis,** in which the offspring come from unfertilized eggs. And, in some copepods, sexual and parthenogenetic stages of their life cycles are alternated.

The general pattern of the life cycle of plants is one of an alternation of two distinct generations, each of which,

depending on the species, may exist independently. In lower plants the haploid generation predominates, whereas in higher plants the diploid generation is dominant. In flowering plants (**angiosperms**), the plant that you see is the diploid **sporophyte** (see fig. 3.1). It is referred to as a sporophyte because, through meiosis, it will give rise to spores. The spores germinate into the alternate generation, the haploid **gametophyte,** which produces gametes by mitosis. Fertilization then produces the next generation of diploid sporophytes. In lower plants the gametophyte has an independent existence; in angiosperms this generation is radically reduced. For example, in corn (fig. 3.25), an angiosperm, the mature corn plant that you are undoubtedly familiar with is the sporophyte. In the male

Figure 3.25
Life cycle of the corn plant

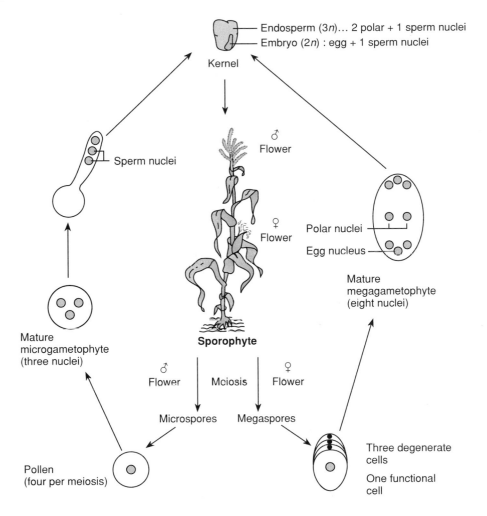

flowers, microspores are produced by meiosis. After mitosis, three nuclei exist in each spore, a structure that we call a **pollen grain,** the male gametophyte. In female flowers meiosis produces megaspores. Mitosis within a megaspore produces an embryo sac of seven cells with eight nuclei. This is the female gametophyte. The egg cell is fertilized by a sperm nucleus. The two polar nuclei of the embryo sac are fertilized by a second sperm nucleus, producing nutritive endosperm tissue that is triploid ($3n$). The sporophyte grows from the diploid fertilized egg.

Many fungi and protista are haploid. Fertilization produces a diploid stage, which almost immediately undergoes meiosis to form haploid cells. These cells, in turn, increase in number by mitosis. More detail will be given when organisms such as *Neurospora,* the pink bread mold, are analyzed genetically (see chapter 6).

Much of our knowledge of genetics derives from the study of specific organisms with unique properties. Mendel found pea plants useful because he could carefully control matings, their generation time was only a year, they were easily grown in his garden, and they had the discrete traits

that he was looking for. Our interest in human beings is obvious. However, we are members of a very difficult species to study experimentally. We have a long generation time and a small number of offspring from matings that cannot be tailored for research purposes. The fruit fly, *Drosophila melanogaster,* is one of the organisms most extensively studied by geneticists. Fruit flies have a short generation time (twelve to fourteen days), which means that many matings can be carried out in a reasonable amount of time. In addition, they do exceptionally well in the laboratory, they have many easily observable mutants, and in several organs they have giant banded chromosomes of great interest to cytogeneticists.

It is interesting to note that species used in food production tend to be intermediate in their life cycles. That is, many crop plants, such as peas and corn, have only one generation interval per year under normal circumstances. (We use generation interval here in the broadest sense of the time for the completion of an entire life cycle; see also chapter 19.) Crop plants are easier to work with from a genetic standpoint than people, but much more difficult

TABLE 3.3	Approximate Generation Intervals of Some Organisms of Genetic Interest	
Organism	**Approximate Generation Interval**
Intestinal bacterium (*Escherichia coli*)	20 minutes
Bacterial virus (*lambda*)	1 hour
Pink bread mold (*Neurospora crassa*)	2 weeks
Fruit fly (*Drosophila melanogaster*)	2 weeks
House mouse (*Mus musculus*)	2 months
Corn (*Zea mays*)	6 months
Sheep (*Ovus aries*)	1 year
Cattle (*Bos taurus*)	2 years
Human being (*Homo sapiens*)	14 years

than, say, *Drosophila,* or bacteria (table 3.3). Because of their relatively long generation interval, crop plants are limited in their utility for studying basic genetic concepts or applying genetic technology to agriculture.

Other organisms will be examined extensively later on. As you make your way through this book and through other readings on genetics, and as you come across studies involving new organisms, you should ask yourself the question, What are the properties of this organism that make it ideal for this type of research?

CHROMOSOMAL THEORY OF HEREDITY

In a paper in 1903, Walter Sutton, a cytologist, firmly stated the concepts we have developed here: The behavior of chromosomes during meiosis explains Mendel's principles. Genes, then, must be located on chromosomes. This idea, also being developed by several other biologists at the time, was immediately accepted; it ushered in the era of the **chromosomal theory of inheritance** wherein intensive effort was devoted to studying the relationships between genes and chromosomes. The major portion of the first section of this book is devoted to classical studies of **linkage** and **mapping.** Linkage deals with the association of genes to each other and to specific chromosomes. Mapping deals with the sequence in which genes appear on a chromosome and their distance apart. This is basic information for a study of the structure and function of genes. Here we introduce a new term for the gene. The term **locus** (plural: *loci*), meaning "place" in Latin, refers to the location of a gene on the chromosome.

SUMMARY

During cell division in eukaryotes, mitosis and meiosis are the processes whereby the chromosomes are apportioned to the daughter cells. Both processes are preceded by chromosome replication during the S-phase of the cell cycle. In mitosis, the two sister chromatids making up each replicated chromosome are separated into two daughter cells. Sex cells—gametes in animals and spores in plants—are produced by the two-stage process of meiosis in which homologous chromosomes are first separated into two daughter cells, and then the sister chromatids making up each chromosome are distributed to two new daughter cells; there are then four cells, each with the haploid chromosomal complement.

The spindle is the apparatus in both processes that separates chromosomes. Mendel's principles, segregation and independent assortment, are explained by the behavior of chromosomes during meiosis.

At the end of this chapter we define the chromosomal theory of inheritance, the concept that shapes the first section of this book. Genes are located on chromosomes; their positions and order on the chromosomes can be discovered by mapping techniques described in later chapters.

SOLVED PROBLEMS

1. What are the differences between chromosomes and chromatids?

   ANSWER: In higher organisms, a chromosome is a linear molecule of DNA complexed with protein and, generally, having a centromere somewhere along its length. During the cell cycle, in the S-phase, the DNA replicates and each chromosome is duplicated. The duplication can be seen in the early stages of mitosis and meiosis when chromosomes shorten. At this point each chromosome has been duplicated and is made up of two chromatids that are then called chromosomes when the centromere divides and each chromatid becomes independent.

2. What are the relationships between mitosis and meiosis and Mendel's rules of segregation and independent assortment?

   ANSWER: The process of mitosis does not relate directly to Mendel's rules. The behavior of chromosomes during meiosis, however, explains both segregation and independent assortment. Segregation is explained by the fact that only one chromosome from each homologous pair goes into a gamete, the same pattern as maternal and paternal alleles of a given gene. Independent assortment is explained by the independent behavior of each tetrad at meiosis. That is, the separation of maternal and paternal alleles in one tetrad is independent of the separation of alleles in any other tetrad.

3. A hypothetical organism has 6 chromosomes ($2n = 6$). How many different combinations of maternal and paternal chromosomes can appear in the gametes?

   ANSWER: You could do this empirically by listing all combinations. For example, let A, B, and C = maternal chromosomes and A′, B′, and C′ = paternal chromosomes. Two combinations in the gametes could be A B C′ and A′ B′ C. It is easier to recall that $2^n =$ number of combinations, where $n =$ the number of chromosome pairs. In this case, $n = 3$, so we expect 8 different combinations.

EXERCISES AND PROBLEMS

1. What are the major differences between prokaryotes and eukaryotes?

2. What is the difference between a centromere and a kinetochore?

3. What is the difference between sister and nonsister chromatids? Between homologous and nonhomologous chromosomes?

4. You are working with a species with $2n = 6$, in which one pair of chromosomes is telocentric, one pair subtelocentric, and one pair metacentric. The A, B, and C loci, each segregating a dominant and recessive allele (*A* and *a, B* and *b, C* and *c*), are each located on a different chromosome. Draw the stages of mitosis.

5. Given the same information as in problem 4, diagram one of the possible meioses. How many different gametes can arise, excluding crossing over? What variation in gamete genotype is introduced by a crossover between the A locus and its centromere?

6. Given the following stages in nuclear division, identify the process, stage, and diploid number (e.g., meiosis I, prophase, $2n = 10$). Keep in mind that one picture could possibly represent more than one process and stage.

(a)

(d)

(b)

(e)

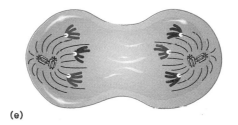

(c)

(f)

7. In human beings $2n = 46$. How many chromosomes would you find in a
   a. brain cell?
   b. red blood cell?
   c. polar body?
   d. sperm cell?
   e. secondary oocyte?

8. In corn (fig. 3.25), the diploid number is twenty. How many chromosomes would you find in a(n)
   a. sporophyte leaf cell?
   b. sporophyte embryo cell?
   c. endosperm cell?
   d. pollen grain?
   e. polar nucleus?

9. How many bivalents, tetrads, and dyads would you find during meiosis in human beings? In fruit flies? In the other species of table 3.2?

10. When during the cell cycle does chromosome replication take place?

11. Can you devise a method of chromosome partitioning during gamete formation that would not involve synapsis—that is, can you reengineer meiosis without passing through a stage of synapsis?

12. What are the differences between a reductional and an equational division? What do the terms refer to?

13. How does the process of meiosis explain Mendel's two rules of inheritance?

14. If a dihybrid corn plant is self-fertilized, what genotypes of the triploid endosperm can result? If you know the endosperm genotype, can you determine the genotype of the sporophyte embryo?

15. How many sperm come from ten primary spermatocytes? How many ova from ten primary oocytes?

16. Take the generalized life cycles of figure 3.1 and change them to describe the life cycle of human beings, peas, and *Neurospora*.

17. If inheritance were controlled primarily by the cytoplasm rather than nuclear genes, what might be the relationship between an organism's phenotype and genotype and its parents' phenotypes and genotypes in
   a. *Drosophila*?
   b. corn?
   c. *Neurospora*?

18. If a drone (male) honeybee is haploid (arising from unfertilized eggs) and a queen (female) is diploid, draw a testcross between a dihybrid queen and a drone. How many different kinds of sons and daughters result from this cross?

19. How do the quantity of genetic material and the ploidy change from stage to stage of spermatogenesis and oogenesis (figs. 3.23 and 3.24)? (Consider the spermatogonium and the oogonium to be diploid with a DNA content arbitrarily set at two.)

20. The plant *Arabidopsis thaliana* has five pairs of chromosomes AA, BB, CC, DD, and EE. If this plant is self-fertilized, what chromosome complements would be found in the roots of the offspring?

   a. A B C D   d. AA BB CC   g. AA BB CC
   b. B C E   e. DD EE   DD EE
   c. B C D   f. AA BB CC EE   h. AAAA BBBB
                                                        CCCC

21. *Drosophila* has four pairs of chromosomes. Let chromosomes from the male parent be A, B, C, and D, and those from the female parent be A′, B′, C′, and D′. What fraction of the gametes from an AA′ BB′ CC′ DD′ individual will be
   a. all of paternal origin?
   b. all of maternal origin?
   c. half of maternal origin and half of paternal origin?

22. Wheat has $2n = 42$ and rye has $2n = 14$ chromosomes. Explain why a wheat-rye hybrid is usually sterile.

23. In wheat the haploid number is 21. How many chromosomes would you expect to find in
   a. the tube nucleus?
   b. a leaf cell?
   c. the endosperm?

24. How many sperm cells will be formed from
   a. 50 primary spermatocytes?
   b. 50 secondary spermatocytes?
   c. 50 spermatids?

25. In human beings, how many eggs will be formed from
   a. 50 primary oocytes?
   b. 50 secondary oocytes?

26. The arctic fox has 50 small chromosomes and the red fox has 38 larger chromosomes. Hybrids of these two species are sterile, but cytological studies during meiosis in these hybrids reveal both paired and unpaired chromosomes.
   a. Account for the sterility of the hybrids.
   b. How can you explain the paired chromosomes?

27. An organism has six pairs of chromosomes. In the absence of crossing over, how many different chromosomal combinations are possible for the gametes?

28. A mature human sperm cell has c amount of DNA. How much DNA (c, $2c$, $4c$, etc.) will a somatic cell have if it is
   a. in G_1?
   b. in G_2?
   c. How much DNA will be in a cell at the end of meiosis I?

29. A hypothetical organism has two distinct chromosomes ($2n = 4$) and 50 known genes, each of which has two different alleles. If an individual is heterozygous at all known loci, how many gametes can be produced if
   a. all genes behave independently?
   b. all genes are completely linked?

Suggested Readings for chapter 3 are on page 642.

4

Probability and Statistics

OBJECTIVES

♦ To understand the rules of probability and how they apply to genetics

♦ To understand the use of the chi-square statistical test in genetics

The results of a genetic cross are influenced by probability.
© Spencer Grant/Photo Researchers, Inc.

In an experimental science, such as genetics, decisions about hypotheses are made on the basis of data gathered during experiments. Geneticists must therefore have an understanding of probability theory and statistical tests of hypotheses. Probability theory allows for accurate predictions of what to expect from an experiment. Statistical testing of hypotheses, particularly with the chi-square test, allows geneticists to have confidence in interpretations of the data gathered from experiments.

PROBABILITY

Part of Gregor Mendel's success was his ability to work with simple mathematics. He was capable of turning numbers into ratios and from them deducing the mechanisms of inheritance. Taking numbers that did not exactly fit a ratio and rounding them off to fit was at the heart of Mendel's deductive powers. The underlying rules that make the act of "rounding to a ratio" reasonable are the rules of probability.

In the **scientific method,** predictions are made, experiments are performed, and data are gathered that are then compared with the original predictions (see chapter 1). However, even if the predictions are correct, the data almost never fit exactly to the predicted outcome. The problem is that we live in a world permeated by random, or **stochastic,** events. A bright new penny when flipped in the air twice in a row will not always give one head and one tail. In fact, that penny if flipped one hundred times could conceivably give one hundred heads. In a stochastic world, we can guess how often a coin should land heads up, but we cannot know for certain what the next toss will bring. We can guess how often a pea should be yellow from a given cross, but we cannot know with certainty what the next pod will contain. Thus the need for **probability theory;** it tells us what to expect from data. This chapter closes with some thoughts on statistics, a branch of mathematics that helps us with criteria for supporting or rejecting our hypotheses.

Types of Probabilities

The **probability** (P) that an event will occur is the number of favorable cases (a) divided by the total number of possible cases (n):

$$P = a/n$$

The probability can be determined either by observation or the nature of the event. For example, we observe that about one child in ten thousand is born with phenylketonuria. Therefore, the probability that the next child born has phenylketonuria is 1/10,000. The odds based on the geometry of an event are, for example, like the familiar toss of dice. A die (singular of dice) has six faces. When that die is tossed, there is no reason one face should land

up more often than any other. Thus the probability of any one of the faces being up (for example, a four) is one sixth:

$$P = a/n = 1/6$$

Similarly, the probability of drawing the seven of clubs from a deck of cards is

$$P = 1/52$$

The probability of drawing a spade from a deck of cards is

$$P = 13/52 = 1/4$$

The probability (assuming a 1:1 sex ratio, which actually is about 1.06 males per female at birth in the United States) of having a daughter is

$$P = 1/2$$

And the probability that an offspring from a self-fertilized dihybrid will show the dominant phenotype is

$$P = 9/16$$

From the probability formula, we can say that an event with certainty has a probability of one and an event that is an impossibility has a probability of zero. If an event has the probability of P, all the other alternatives combined will have a probability of $Q = 1 - P$; thus $P + Q = 1$. That is, the probability of the completely dominant phenotype in the F_2 of a selfed dihybrid is 9/16. The probability of any other phenotype is 7/16, which added to 9/16 equals 16/16, or 1.

Combining Probabilities

The basic principle of probability can be stated as follows: If one event has c possible outcomes and a second event has d possible outcomes, then there are cd possible outcomes of the two events. From this principle we obtain three rules that concern us as geneticists.

1. Sum Rule

When the occurrence of one event precludes the occurrence of the other events, that is, when the events are mutually exclusive, the **sum rule** is used: The probability of the occurrence of one of several mutually exclusive events is the sum of the probabilities of the individual events. This is known as the *either/or rule*— for example, what is the probability, when we throw a die, of its showing *either* a four *or* a six? According to the sum rule:

$$P = 1/6 + 1/6 = 2/6 = 1/3$$

2. Product Rule

When the occurrence of one event is independent of the occurrence of other events, the **product rule** is used: The probability of the occurrence of independent events is the product of their separate probabilities. This is known as the *and rule*. For example, the probability of throwing a die two times and getting a four *and* then a six, in that order, is

$$P = 1/6 \times 1/6 = 1/36$$

3. Binomial Theorem

The **binomial theorem** is used for unordered events: The probability of the occurrence of some arrangement in which the final order is not specified is defined by the binomial theorem. For example, what is the probability when tossing two pennies simultaneously of getting a head and a tail?

USE OF RULES

There are several ways to calculate the probability just asked for. To put the problem in the form for rule 3 is the quickest method, but this problem can also be solved by using a combination of rules 1 and 2 in the following manner: For each penny the probability of getting a head (H) *or* a tail (T) is

for H: $P = 1/2$

for T: $Q = 1/2$

Tossing the pennies one at a time, it is possible to get a head *and* a tail in two ways:

first head, then tail (HT)

or

first tail, then head (TH)

Within a sequence (HT or TH) the probabilities are of independent events. Thus the probability for any one of the two ways involves the product rule (rule 2):

1/2 X 1/2 = 1/4 for HT or TH

The two sequences (HT or TH) are mutually exclusive. Thus the probability of getting either of the two sequences is one of a set of mutually exclusive events and involves the sum rule (rule 1):

1/4 + 1/4 = 1/2

Thus, for unordered events, we can obtain the probability by a combination of rules 1 and 2. The binomial theorem (rule 3) provides the shorthand method.

To use rule 3, we must state it as follows: If the probability of an event (X) is p and an alternative (Y) is q, then the probability in n trials that event X will occur s times and Y will occur t times is

$$P = \frac{n!}{s!t!}p^s q^t$$

In the above equation, $s + t = n$, and $p + q = 1$. The symbol (!), as in $n!$, is called **factorial**, as in "n factorial," and is the product of all integers from n down to one. For example, $7! = 7 \times 6 \times 5 \times 4 \times 3 \times 2 \times 1$. Zero factorial equals one, as does anything to the power of zero ($0! = n^0 = 1$).

Now, what is the probability of tossing two pennies and getting one head and one tail? In this case, $n = 2$, s and $t = 1$, and p and $q = 1/2$. Thus,

$$P = \frac{2!}{1!1!}(1/2)^1(1/2)^1 = 2(1/2)^2 = 1/2$$

This is, of course, our original answer. Now on to a few more genetically relevant problems. What is the probability that a family with six children will have precisely five girls and one boy? (We assume that the probability of either a son or a daughter equals $1/2$.) Since the order is not specified, we use rule 3:

$$P = \frac{6!}{5!1!}(1/2)^5(1/2)^1 = 6(1/2)^6 = 6/64 = 3/32$$

What would happen if we asked for a specific family order, in which four girls were born, then one boy, and then one girl? This would entail rule 2; for a sequence of six independent events:

$$P = \frac{1}{2} \times \frac{1}{2} \times \frac{1}{2} \times \frac{1}{2} \times \frac{1}{2} \times \frac{1}{2} = \frac{1}{64}$$

When no order is specified, the probability is six times larger than when the order is specified; the reason is simply that there are six ways of getting five girls and one boy, and the sequence 4–1–1 is only one of them. Rule 3 tells us that there are six ways. These are (letting B stand for boy and G for girl):

Birth Order

1	2	3	4	5	6
B	G	G	G	G	G
G	B	G	G	G	G
G	G	B	G	G	G
G	G	G	B	G	G
G	G	G	G	B	G
G	G	G	G	G	B

If two persons, heterozygous for albinism (a recessive condition), have four children, what is the probability that all four will be normal? The answer is simply $(3/4)^4$ by rule 2. What is the probability that three will be normal and one albino? If we specify which of the four children will be albino (e.g., the fourth), then the probability is $(3/4)^3(1/4)^1$. If, however, we do not specify order:

$$P = \frac{4!}{3!1!}(3/4)^3(1/4)^1 = 4(3/4)^3(1/4)^1$$

This is precisely four times the ordered probability because the albino child could have been born first, second, third, or last.

The formula for rule 3 is the formula for the terms of the **binomial expansion.** That is, if $(p + q)^n$ is expanded, the formula $(n!/s!t!)p^sq^t$ gives the probability for one of these terms, given that $p + q = 1$ and that $s + t = n$. Since there are $(n + 1)$ terms in the binomial, the formula gives the probability for the term numbered $(t + 1)$. Two bits of useful information come from recalling that rule 3 is in reality the rule for the terms of the binomial expansion. First, if you have difficulty calculating the term, you can use **Pascal's triangle** to get the coefficients:

$$
\begin{array}{ccccccccccc}
 & & & & & 1 & & & & & \\
 & & & & 1 & & 1 & & & & \\
 & & & 1 & & 2 & & 1 & & & \\
 & & 1 & & 3 & & 3 & & 1 & & \\
 & 1 & & 4 & & 6 & & 4 & & 1 & \\
1 & & 5 & & 10 & & 10 & & 5 & & 1 \\
\end{array}
$$

Pascal's triangle is a triangular array made up of coefficients in the binomial expansion and is calculated by starting any row with a 1, proceeding by adding two adjacent terms from the row above, and then ending with a 1.

For example, the next row would be:

1, (1 + 5), (5 + 10), (10 + 10), (10 + 5), (5 + 1), 1
or 1, 6, 15, 20, 15, 6, 1

These numbers give us the combinations for any p^sq^t term. That is, in our previous example, $n = 4$; so we use the $(n + 1)$, or fifth, row of Pascal's triangle. (The second number in any row of the triangle gives the power of the expansion or n. Here, 4 is the second number of the row.) We were interested in the case of one albino child in a family of four children, or p^3q^1, where p is the probability of the normal child (3/4) and q is the probability of an albino child (1/4). Hence, we are interested in the $(t + 1)$—that is, the $(1 + 1)$—or the second term of the fifth row of Pascal's triangle, which will tell us the number of ways of getting a four-child family with one albino child. That number is 4. Thus, using Pascal's triangle, we see that the solution to the problem is:

$$4(3/4)^3(1/4)^1$$

This is the same as the answer obtained the conventional way.

The second advantage from knowing that rule 3 is the binomial expansion formula is that we can now generalize to more than two events. The general form for the **multinomial expansion** is $(p + q + r + \ldots)^n$ and the general formula for the probability is:

$$P = \frac{n!}{s!t!u! \ldots}p^sq^tr^u \ldots$$

where $s + t + u \ldots = n$ and $p + q + r + \ldots = 1$. For example, our albino-carrying heterozygous parents may have wanted an answer to the following question. If we have a family of five, what is the probability that we will have two normal sons, two normal daughters, and one albino son? (This family will have no albino daughter.) By rule 2:

probability of a normal son = (3/4)(1/2) = 3/8

probability of a normal daughter = (3/4)(1/2) = 3/8

probability of an albino son = (1/4)(1/2) = 1/8

probability of an albino daughter = (1/4)(1/2) = 1/8

Thus:

$$P = \frac{5!}{2!2!1!0!}(3/8)^2(3/8)^2(1/8)^1(1/8)^0$$

$$= 30(3/8)^4(1/8)^1 = 30(3)^4/(8)^5$$

$$= 2{,}430/32{,}768$$

$$= 0.074$$

STATISTICS

In one of Mendel's experiments, F_1 heterozygous pea plants, all of which were tall, were self-fertilized. In the next generation (F_2), he recorded 787 tall offspring and 277 dwarf offspring for a ratio of 2.84:1. Mendel saw this as a 3:1 ratio, which supported his proposed rule of inheritance. In fact, is 787:277 "roundable" to a 3:1 ratio? From a brief discussion of probability, we expect some deviation from an exact 3:1 ratio (798:266), but how much of a deviation is acceptable? Would 786:278 still support Mendel's rule? Would 785:279 support it? Would 709:355 (a 2:1 ratio) or 532:532 (a 1:1 ratio)? Where do we draw the line? It is at this point that the discipline of statistics provides help.

We can never speak with certainty about stochastic events. For example, take the case of Mendel's cross. Although a ratio of 3:1 is expected on the basis of Mendel's hypothesis, chance could give a 1:1 ratio in the data (532:532), yet the mechanism could be the one that Mendel suggested. We could flip an honest coin and get ten heads in a row. Conversely, Mendel could have gotten exactly a 3:1 ratio (798:266) in his F_2 generation, yet his hypothesis of segregation could have been wrong. The point is that any time we deal with probabilistic events there is some chance that the data will lead us to support a bad hypothesis or reject a good one. Statistics quantifies these chances. We cannot say with certainty that a 2.84:1 ratio represents a 3:1 ratio; we can say, however, that we have a certain degree of confidence in the ratio. It is statistics that helps us ascertain these **confidence limits.**

Statistics is a branch of probability theory that helps the experimental geneticists in three ways. First, part of statistics is called **experimental design.** A bit of thought before the performance of an experiment may help design the experiment in the most efficient way. Although he did not know statistics, Mendel's experimental design was very good. The second way in which statistics is helpful is the summarization of data. Such familiar terms as *mean* and *standard deviation* are part of the body of descriptive statistics that takes large masses of data and reduces them to one or two meaningful values. We will examine further some of these terms and concepts in the chapter on quantitative inheritance (chapter 18).

Hypothesis Testing

The third way that statistics is valuable to geneticists is in the **testing of hypotheses:** determining whether to support or reject a proposed hypothesis based on how close the data are to the predictions of the hypothesis. This area is the most germane to our current discussion. For example, was the ratio of 787:277 really indicative of a 3:1 ratio? Since we know now that we cannot answer with an absolute "yes," how can we attach a level of support to our answer?

Statisticians would have us proceed as follows. To begin with, we need to establish what kind of variation to expect. This can be determined by calculating a **sampling distribution:** the frequencies with which various possible events could occur in a particular experiment. For example, if we self-fertilized a heterozygous tall plant, we would expect a 3:1 ratio of tall to dwarf plants among the progeny. (The 3:1 ratio is our hypothesis based on the assumption of genetic control of height by one locus with two alleles.) If we looked at the first four offspring, what is the probability of getting three tall and one dwarf plant? The answer is calculated using the formula for the terms of the binomial expansion:

$$P = \frac{4!}{3!1!}(3/4)^3(1/4)^1 = 108/256 = 0.42$$

Similarly, we can calculate the probability of getting all tall ($81/256 = 0.32$), two tall and two dwarf ($54/256 = 0.21$), one tall and three dwarf ($12/256 = 0.05$), and all dwarf ($1/256 = 0.004$). This distribution, as well as the distribution for samples of eight and forty progeny, is shown in table 4.1. These distributions are graphed in figure 4.1.

As sample sizes increase (from four to eight to forty in fig. 4.1), the sampling distribution takes on the shape of a smooth curve with a peak at the true ratio of 3:1 (75% tall progeny)—that is, there is a high probability of getting very close to the true ratio. However, there is some chance the ratio will be fairly far off, and a very small part of the time our ratio will be very far off. It is important to see that any ratio could arise in a given experiment even though the true ratio is 3:1. So where do we draw the line? At what ratio do we decide that an experimental result is not indicative of a 3:1 ratio?

Statisticians have agreed on a convention. When all the frequencies are plotted, as in figure 4.1, the area under the curve is taken as one unit, and we draw lines to include 95% of this area (fig. 4.2). Any ratios included within the 95% limits are considered supportive of (failing to reject) the hypothesis of a 3:1 ratio. Any ratio in the remaining 5% area is considered unacceptable. (Other conventions also exist, such as rejection within the outer 10% or 1% limits; we consider these at the end of the chapter.) Thus

TABLE 4.1 Sampling Distribution for Sample Sizes of Four, Eight, and Forty, Given a 3:1 Ratio of Tall and Dwarf Plants in This Experiment

n = 4		n = 8		n = 40	
No. Tall Plants	Probability*	No. Tall Plants	Probability*	No. Tall Plants	Probability*
4	$\frac{81}{256} = 0.32$	8	0.10	40	0.00001
3	$\frac{108}{256} = 0.42$	7	0.27	39	0.0001
2	$\frac{54}{256} = 0.21$	6	0.31	38	0.0009
1	$\frac{12}{256} = 0.05$	5	0.21		. . .
0	$\frac{1}{256} = 0.004$	4	0.09	30	0.14
		3	0.02		. . .
		2	0.004	2	0.59×10^{-20}
		1	0.0004	1	0.10×10^{-21}
		0	0.00002	0	0.83×10^{-24}

*Probabilities are calculated from the binomial theorem.
probability $= (n!/s!t!)p^s q^t$
where n = number of progeny observed
  s = number of progeny that are tall
  t = number of progeny that are dwarf
  p = probability of a progeny plant being tall (3/4)
  q = probability of a progeny plant being dwarf (1/4)

Figure 4.1
Sampling distributions from an experiment with an expected ratio of three tall to one dwarf plant. As the sample size, n, gets larger, the distribution becomes smoother. These distributions are the plotted terms of the binomial expansion (table 4.1). Note also that as n gets larger, the peak of the curve gets lower because as more points (possible ratios) are squeezed in along the X axis, the probability of any one ratio decreases.

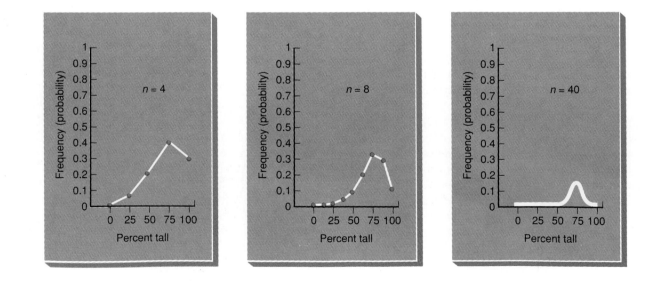

Figure 4.2

Sampling distribution of figure 4.1 ($n = 40$); 5 percent of the area is marked off (2.5 percent at each end).

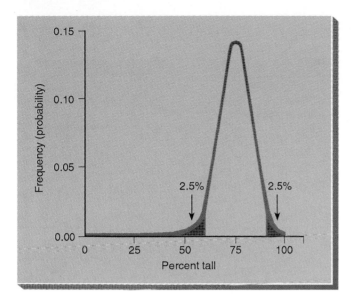

TABLE 4.2 Chi-Square Analysis of One of Mendel's Experiments, Assuming a 3:1 Ratio

	Tall Plants	Dwarf Plants	Total
Observed Numbers (O)	787	277	1,064
Expected Ratio	3/4	1/4	
Expected Numbers (E)	798	266	1,064
$O - E$	-11	11	
$(O - E)^2$	121	121	
$(O - E)^2/E$	0.15	0.45	$0.60 = \chi^2$

TABLE 4.3 Chi-Square Analysis of One of Mendel's Experiments, Assuming a 1:1 Ratio

	Tall Plants	Dwarf Plants	Total
Observed Numbers (O)	787	277	1,064
Expected Ratio	1/2	1/2	
Expected Numbers (E)	532	532	1,064
$O - E$	255	-255	
$(O - E)^2$	65,025	65,025	
$(O - E)^2/E$	122.23	122.23	$244.45 = \chi^2$

it is possible to see if the experimental data support our hypothesis (in this case the hypothesis of 3:1). One in twenty times (5%) we will make a **type I error:** We will reject a true hypothesis. (A **type II error** is that of failing to reject a false hypothesis.)

Is it necessary to calculate a sampling distribution every time we do an experiment? To determine whether or not to reject a hypothesis, a frequency distribution must be derived for each type of experiment. Mendel could have used the distribution shown in figure 4.1 for seed coat or seed color, as long as he was expecting a 3:1 ratio and had a similar sample size. What about independent assortment, where a 9:3:3:1 ratio is expected? A geneticist would have to calculate a new sampling distribution based on a 9:3:3:1 ratio and a particular sample size. Statisticians have devised shortcut methods by using standardized distributions from which to calculate probabilities. Many are in use, such as the *t*-distribution, binomial distribution, and chi-square distribution. Each is useful for particular kinds of data; geneticists usually use the chi-square distribution to test hypotheses regarding breeding data.

Chi-Square

When sample subjects are distributed among discrete categories such as tall and dwarf plants, the **chi-square distribution** is frequently used. The formula for converting categorical experimental data to a chi-square value is

$$\chi^2 = \sum \frac{(O - E)^2}{E}$$

where χ is the Greek letter chi, O is the observed number for a category, E is the expected number for that category, and Σ means to sum the calculations for all categories.

A chi-square (χ^2) value of 0.60 is calculated in table 4.2 for Mendel's data on the basis of a 3:1 ratio. If Mendel had originally expected a 1:1 ratio, he would have calculated a chi-square of 244.45 (table 4.3). However, these χ^2 values have little meaning of themselves: They are not probabilities. They must be converted to probabilities by determining where the chi-square value falls in relation to the area under the chi-square distribution curve. We usually use a chi-square table in which these probabilities have already been calculated (table 4.4). Before we can use this table, we must define the concept of **degrees of freedom.**

Reexamination of the chi-square formula and tables 4.2 and 4.3 reveal that there is a contribution to the total chi-square value from each category, because chi-square is a summed value. We expect the chi-square value to increase as the total number of categories increases. That is, the more categories involved, the larger the chi-square value even if the sample is a relatively good fit to the hypothesized ratio. Hence we need some way of keeping track

TABLE 4.4 Chi-Square Values

Degrees of Freedom	Probabilities						
	0.99	0.95	0.80	0.50	0.20	0.05	0.01
1	0.000	0.004	0.064	0.455	1.642	3.841	6.635
2	0.020	0.103	0.446	1.386	3.219	5.991	9.210
3	0.115	0.352	1.005	2.366	4.642	7.815	11.345
4	0.297	0.711	1.649	3.357	5.989	9.488	13.277
5	0.554	1.145	2.343	4.351	7.289	11.070	15.086
6	0.872	1.635	3.070	5.348	8.558	12.592	16.812
7	1.239	2.167	3.822	6.346	9.803	14.067	18.475
8	1.646	2.733	4.594	7.344	11.030	15.507	20.090
9	2.088	3.325	5.380	8.343	12.242	16.919	21.666
10	2.558	3.940	6.179	9.342	13.442	18.307	23.209
15	5.229	7.261	10.307	14.339	19.311	24.996	30.578
20	8.260	10.851	14.578	19.337	25.038	31.410	37.566
25	11.524	14.611	18.940	24.337	30.675	37.652	44.314
30	14.953	18.493	23.364	29.336	36.250	43.773	50.892

From: C. M. Thompson, *Biometrika*, 32:188–189. Copyright © 1941 Biometrika Trustees. Reprinted by permission.

of categories. This is done with degrees of freedom, which is basically a count of independent categories. With Mendel's data, the total is 1,064, of which 787 had tall stems. Therefore, the short-stem group had to consist of 277 plants and isn't an independent category. For our purposes here, degrees of freedom equal the number of categories minus one. Thus, with two phenotypic categories, there is only one degree of freedom.

Table 4.4, the table of chi-square probabilities, is read as follows. Degrees of freedom are read in the left column. We are interested in the first row where there is one degree of freedom. The numbers across the top of the table are the probabilities. We are interested in the next-to-the-last column, headed by the 0.05. We thus get the following information from the table: The probability is 0.05 of getting a chi-square value of 3.841 or larger by chance alone, given that the hypothesis is correct. If we examine this statement, it is a formalization of what we have been talking about in our discussion of frequency distributions. Hence we are interested in how large a chi-square value will be found in the 5% unacceptable area of the curve. For Mendel's plant experiment, the **critical chi-square** (at $p = 0.05$, one degree of freedom) is 3.841. This is the value to which we compare the calculated χ^2 values (0.60 and 244.45). Since the chi-square value for the 3:1 ratio is 0.60 (table 4.2), which is less than the critical value of 3.841, we fail to reject the hypothesis of a 3:1 ratio. But since χ^2 for the 1:1 ratio (table 4.3) is 244.45, which is greater than the critical value, we reject the hypothesis of a 1:1 ratio. Notice, however, that once we did the chi-square test for the 3:1 ratio and failed to reject the hypothesis, no other statistical tests were needed: Mendel's data are consistent with a 3:1 ratio.

A word of warning when using the chi-square: If the expected number in any category is less than 5, the conclusions are not reliable. In that case the experiment can be repeated to obtain a larger sample size, or categories can be combined. Note also that chi-square tests are always done on whole numbers, not on ratios or percentages.

Failing to Reject Hypotheses

The hypothesis against which the data are tested is referred to as the **null hypothesis.** Hypothesis testing involves testing the assumption that there is no difference between the observed and the expected samples. If the null hypothesis is not rejected, then we say that the data are consistent with it, not that the hypothesis has been proved. (As previously discussed, there are built-in possibilities of not rejecting false hypotheses or rejecting true ones.) If, however, the hypothesis is rejected, as we rejected a 1:1 ratio for Mendel's data (table 4.3), the only other choice is not to reject the alternative hypothesis that there is a difference between the observed and the expected values. The data may then be retested against some other hypothesis. (We don't say "accept the hypothesis" but rather "fail to reject the hypothesis," because supportive numbers could arise for many reasons. Our failure to reject is a tentative acceptance of a hypothesis. However, we are on stronger ground when we reject a hypothesis.)

The use of the 0.05 probability level as a cutoff for rejecting a hypothesis is a convention. It is called the **level of significance.** When a hypothesis is rejected at that level, statisticians say that the data depart *significantly* from the expected ratio. Other levels of significance are also used, such as 0.01. If a calculated chi-square is greater than the critical value in the table at the 0.01 level, we say that the data depart in a *highly significant* manner from the null hypothesis. Since the chi-square value at the 0.01 level is larger than the value at the 0.05 level, it is more difficult to reject a hypothesis at this level and hence more convincing when it is rejected.

Other levels of rejection are also set. In clinical trials of medication, for example, an attempt is made to make it very easy to reject the null hypothesis: A level of significance of 0.10 or higher is set. The rationale is that it is not desirable to discard a drug or treatment that might be beneficial. Since the null hypothesis states that the drug has no effect—that is, the control and drug groups show the same response—clinicians would rather be overly conservative. Not rejecting the hypothesis means concluding that the drug has no effect. Rejecting the hypothesis means that the drug has some effect and it should be tested further. It is much better to have to retest some drugs that are actually worthless than to discard drugs that have potential value.

SUMMARY

We have examined the rules of probability theory relevant to genetic experiments. Probability theory allows us to predict the outcome of experiments. The probability (P) of independent events is calculated by multiplying their separate probabilities. The probability of mutually exclusive events is calculated by adding their individual probabilities. And the probability of unordered events is defined by the polynomial expansion $(p + q + r + \ldots)^n$:

$$P = \frac{n!}{s!t!u! \ldots} p^s q^t r^u \ldots$$

In order to assess whether data gathered during an experiment actually support a particular hypothesis, it is necessary to determine what the probability is of getting a particular data set when the null hypothesis is correct. We have considered the chi-square test:

$$\chi^2 = \sum \frac{(O - E)^2}{E}$$

It is a method of quantifying the confidence we may have in the results obtained from typical genetic experiments. The rules of probability and statistics allow us to devise hypotheses about inheritance and to test these hypotheses with experimental data.

SOLVED PROBLEMS

1. Mendel self-fertilized a dihybrid plant that had round, yellow peas. In the offspring generation, what is the probability that a pea picked at random will be round and yellow? What is the probability that five peas picked at random will be round and yellow? What is the probability that of five peas picked at random, four will be round and yellow and one will be wrinkled and green?

   ANSWER: The offspring peas will be round and yellow, round and green, wrinkled and yellow, and wrinkled and green in a ratio of 9:3:3:1. Thus the probability that a pea picked at random will be one of the above four categories will be 9/16, 3/16, 3/16, and 1/16, respectively. Thus the probability that a pea picked at random will be round and yellow is 9/16, or 0.563. The probability of getting five of these peas in a row is $(9/16)^5$, or 0.056. The probability that of five peas picked at random, four will be round and yellow and one will be wrinkled and green is (substituting into the binomial equation): $(5!/4!1!)(9/16)^4(1/16)^1 = 5(9^4)/(16^5) = 5(0.006) = 0.031$.

2. On a chicken farm, walnut-combed fowl were crossed with each other with the following offspring produced: walnut combed, 87; rose combed, 31; pea combed, 30; and single combed, 12. What is your hypothesis about the control of comb shape in fowl and do the data support that hypothesis?

   ANSWER: The numbers 87, 31, 30, and 12 are very similar to 90, 30, 30, and 10, which would be a perfect fit to a 9:3:3:1 ratio. We might expect that ratio knowing something about how comb type is inherited in fowl (chapter 2). Thus we hypothesize that inheritance of comb type is by two loci in which dominant alleles at both result in walnut combs, a dominant allele at one locus and recessives at the other result in rose or pea combs, and the recessive homozygote has a single comb. Then the results of the cross of dihybrids should produce fowl with the four comb types in a 9:3:3:1 ratio, of walnut-, rose-, pea-, and single-combed fowl, respectively. Therefore our observed numbers are 87, 31, 30, and 10 (sum = 160). Our expected ratio is 9:3:3:1, or 90, 30, 30, and 10 fowl, which are 9/16, 3/16, 3/16, and 1/16, respectively, of the sum of 160. We therefore set up the following chi-square table:

	Comb type				
	Walnut	Rose	Pea	Single	Total
Observed Numbers (O)	87	31	30	12	160
Expected Ratio	9/16	3/16	3/16	1/16	
Expected Numbers (E)	90	30	30	10	160
$O - E$	-3	1	0	2	
$(O - E)^2$	9	1	0	4	
$(O - E)^2/E$	0.1	0.033	0	0.4	$0.533 = \chi^2$

There are three degrees of freedom since there are four categories. The critical chi-square value with three degrees of freedom and probability of 0.05 = 7.815 (table 4.4). Since our calculated chi-square value (0.533) is less than this critical value, we cannot reject our hypothesis. In other words, our data are consistent with the hypothesis of a 9:3:3:1 ratio of phenotypes, indicative of a two-locus genetic model with dominance at each locus.

EXERCISES AND PROBLEMS

1. Assuming a 1:1 sex ratio, what is the probability that a family of five children will consist of
   a. three daughters and two sons?
   b. alternating sexes, starting with a son?
   c. alternating sexes?
   d. all daughters?
   e. all the same sex?
   f. at least four daughters?
   g. a daughter as the eldest child and a son as the youngest?

2. Phenylthiocarbamide (PTC) tasting is dominant (T) to nontasting (t). If a taster woman with a nontaster father married a taster man who, in a previous marriage, had a nontaster daughter, what would be the probability that
   a. their first child would be a nontaster?
   b. their first child would be a nontaster girl?
   c. if they had six children, they would have two nontaster sons, two nontaster daughters, and two taster sons?
   d. their fourth child would be a taster daughter?

3. Albinism is recessive, as are blue eyes. (Albinos have blue eyes.) What is the probability that two brown-eyed persons, heterozygous for both traits, produce (remembering epistasis)
   a. five albino children?
   b. five albino sons?
   c. four blue-eyed daughters and a brown-eyed son?
   d. two sons genotypically like their father and two daughters genotypically like their mother?

4. On the average, about one child in every ten thousand live births in the United States has phenylketonuria (PKU). What is the probability that
   a. the next child born in a Boston hospital will have PKU?

   b. after a PKU child is born, the next child born will have PKU?
   c. two children born in a row will have PKU?

5. In fruit flies, the diploid chromosome number is eight.
   a. What is the probability that a male gamete will contain only paternal centromeres or only maternal centromeres?
   b. What is the probability that a zygote will contain only centromeres from male grandparents? (Disregard the problems that the sex chromosomes may introduce.)

6. The following data are from Mendel's original experiments. Suggest a hypothesis for each set and test this hypothesis with the chi-square test. Do you reach different conclusions with different levels of significance?
   a. Self-fertilization of round-seeded hybrids produced 5,474 round seeds and 1,850 wrinkled ones.
   b. One particular plant from part (a) yielded 45 round seeds and 12 wrinkled ones.
   c. Of the 565 plants raised from F_2 round-seeded plants, 372 gave both round and wrinkled seeds in a 3:1 proportion, whereas 193 yielded only round seeds when self-fertilized.
   d. A violet-flowered, long-stemmed plant was crossed with a white-flowered, short-stemmed plant with the following offspring:

   47 violet, long-stemmed plants
   40 white, long-stemmed plants
   38 violet, short-stemmed plants
   41 white, short-stemmed plants

7. Mendel self-fertilized pea plants with round and yellow peas. In the next generation he recovered the following numbers of peas:

315 round and yellow peas
108 round and green peas
101 wrinkled and yellow peas
 32 wrinkled and green peas

What is your hypothesis about the genetic control of the phenotype? Do the data support this hypothesis?

8. Two agouti mice are crossed and over a period of a year they produce 48 offspring with the following phenotypes:

28 agouti mice
 7 black mice
13 albino mice

What is your hypothesis about the genetic control of coat color in these mice? Do the data support that hypothesis?

9. How many seeds should Mendel have tested to determine with complete certainty that a plant with a dominant phenotype was heterozygous? With 99% certainty? With 95% certainty? With "pretty reliable" certainty?

10. Assuming that a couple wants one son and one daughter, what chance do they have of achieving this goal? If all couples wanted at least one child of each sex, approximately what would be the average family size?

11. PKU and albinism are two autosomal recessive disorders, unlinked in human beings. If two people, each heterozygous for both traits, marry, what is the chance of their having a child
  a. with PKU?
  b. with either PKU or albinism?
  c. with both traits?

12. In human beings, the absence of molars is inherited as a dominant trait. If two heterozygotes have four children, what is the probability that
  a. all will have no molars?
  b. three will have no molars and one will have molars?
  c. the first two will have molars and the second two will have no molars?

13. Galactosemia is inherited as a recessive trait. If two normal heterozygotes marry, what is the chance that
  a. one of four children will be affected?
  b. of three children, they will be in the order: normal boy, affected girl, affected boy?

14. A normal man (A) whose grandfather had galactosemia, marries a normal woman (B) whose mother was galactosemic. What is the probability that their first child will be galactosemic?

15. A city had 900 deaths during the year, and of these 300 were from cancer and 200 from heart disease. What is the probability that the next death will be
  a. from cancer?
  b. from either cancer or heart disease?

16. Two curly-winged flies, when mated, produce 61 curly- and 35 straight-winged progeny. Use a chi-square test to determine whether these numbers fit a 3:1 ratio.

17. A short-winged, dark-bodied fly is crossed to a long-winged, tan-bodied fly. All the F_1 progeny are long-winged and tan-bodied. F_1 flies are selfed to yield 84 long-winged, tan-bodied; 27 long-winged, dark-bodied; 35 short-winged, tan-bodied; and 14 short-winged, dark-bodied flies.
  a. What ratio do you expect in the progeny?
  b. Is the observed ratio within the expected range? Use the chi-square test to evaluate your hypothesis.

18. The ability to taste phenylthiocarbamide is dominant in human beings. If a heterozygous taster marries a nontaster, what is the probability that of their five children, only one will be a taster?

19. In mice, coat color is determined by two independent genes, A and C, as indicated below:
A-C-: agouti; aaC-: black; A-$c^a c^a$: albino; $aac^a c^a$: albino.
If the cross is $AaCc^a \times Aac^a c^a$, what is the probability that among the first six offspring, two will be black, two will be agouti, and two will be albino?

20. An individual that has the genotype *AAbbccDDEE* is mated with an individual who is *aaBBCCddee*. F_1 individuals are selfed to produce an F_2 generation. What is the chance of getting an individual whose genotype is identical to one of the parents?

Suggested Readings for chapter 4 are on page 642.

5

Sex Determination, Sex Linkage, and Pedigree Analysis

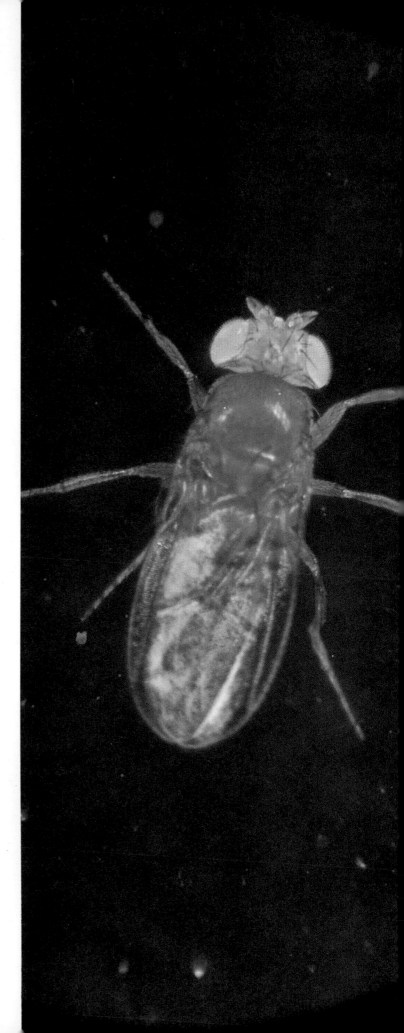

The white eye of the fruit fly (*Drosophila melanogaster*) is controlled by a gene on the X chromosome. Magnification 17×.
© *Grant Heilman/Grant Heilman Photography*

This chapter begins a four-chapter sequence on the analysis of the relationship of genes to chromosomes. We begin with the study of sex determination.

SEX DETERMINATION

Patterns

A heteromorphic pair of chromosomes, termed the **sex chromosomes,** are involved in sex determination in many species. However, they are not the only way in which sex is determined. Sex can be controlled by the ploidy of an individual, as in many hymenoptera (bees, ants, wasps), in which males are haploid and females are diploid; by allelic mechanisms in which sex is determined by a single allele or multiple alleles not associated with heteromorphic chromosomes; or by environmental factors. The sex of some geckos is determined by temperature. At 25° C, females hatch from eggs; at 32° C, males hatch from the eggs. At 28° or 29° C, about half the eggs hatch as males and half as females. In some marine worms and gastropods, the sex of the individual depends on the substrate on which it lands. For example, in the slipper limpet, *Crepidula,* individuals are frequently found stacked up on one another. The reproductive system can develop as either male or female, depending on the sexes of the other organisms in the cluster. If the organism is attached to a female, the reproductive system will develop as male. Isolation or the presence of a large number of males will induce a male to become a female. Once a female, the individual will no longer change.

Sex Chromosomes

Basically four types of chromosomal sex-determining mechanisms exist: the XY, ZW, X0, and compound chromosome mechanisms. In the XY case, the females have a homomorphic pair of chromosomes (XX), as in human beings or fruit flies; males are heteromorphic (XY). In the ZW case, males are homomorphic (ZZ), and females are heteromorphic (ZW). (XY and ZW are chromosome notations and imply nothing about the size or shape of these chromosomes.) In the X0 case, there is only one sex chromosome, as in some grasshoppers and beetles; females are usually XX and males X0. And in the compound chromosome case, several X and Y chromosomes are involved in sex determination, as in bedbugs and some beetles. We need to emphasize that the chromosomes themselves are not determining sex, but the genes they carry are. In general, the genotype determines the type of gonad that then determines the phenotype of the organism through male or female hormone production.

The XY System

The XY situation occurs in human beings, in which females have forty-six chromosomes arranged in twenty-three homologous, homomorphic pairs. Males, with the same number of chromosomes, have twenty-two homomorphic pairs and one heteromorphic pair that is referred to as the XY pair (fig. 5.1). During meiosis, females will produce gametes that contain only the X chromosome, whereas males will produce two kinds of gametes, X- and Y-bearing (fig. 5.2). For this reason females are referred to as **homogametic** and males as **heterogametic.** As you

Figure 5.1

Human karyotype. Note the X and Y chromosomes. A female would have a second X chromosome in place of the Y.

Source: Reproduced courtesy of Dr. Thomas G. Brewster, Foundation for Blood Research, Scarborough, Maine.

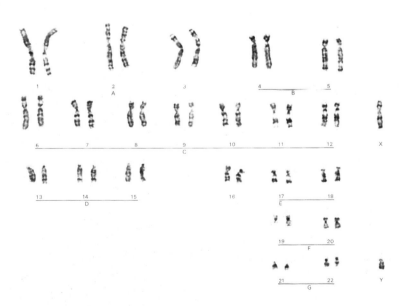

Figure 5.2

Segregation of human sex chromosomes during meiosis, with subsequent zygote formation

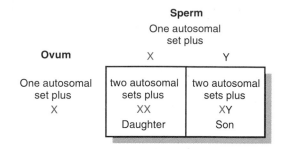

	Sperm One autosomal set plus	
Ovum	X	Y
One autosomal set plus X	two autosomal sets plus XX Daughter	two autosomal sets plus XY Son

Figure 5.3

Chromosomes of *Drosophila melanogaster*

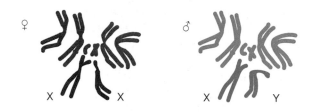

Calvin B. Bridges (1889–1938)
Genetics, 25 (1940): frontispiece.

can see from figure 5.2, in people, fertilization will cause equal numbers of male and female offspring to be formed. In *Drosophila* the system is the same, but, unlike in human beings, the X chromosome is smaller than the Y chromosome (fig. 5.3).

Since both human and *Drosophila* females normally have two X chromosomes and males have an X and a Y chromosome, it is not possible to know from the normal condition whether maleness is determined by the presence of a Y chromosome or the absence of a second X chromosome. One way to resolve this problem would be to isolate individuals with odd numbers of chromosomes. In chapter 8 we examine the causes and outcomes of anomalous chromosome numbers. Here we consider two facts from that chapter. First, in rare instances, individuals are formed, although not necessarily viable, with extra sets of chromosomes. These individuals are referred to as **polyploids** (*triploids* with 3n, *tetraploids* with 4n, etc.). Second, also infrequently, individuals are formed that have more or less than the normal number of any one chromosome. These **aneuploids** usually come about by the failure of a pair of chromosomes to separate properly during meiosis, an occurrence called **nondisjunction.** The existence of polyploid and aneuploid individuals makes it possible to test whether the Y chromosome is male determining. For example, a person or a fruit fly that has all the proper nonsex chromosomes, or **autosomes** (forty-four in human beings, six in *Drosophila*), but only a single X without a Y would answer our question. If the Y were absolutely male determining, then this X0 individual should be female. However, if the sex-determining mechanism is a

result of the number of X chromosomes, this individual should be a male. As it turns out, an X0 individual is a *Drosophila* male and a human female.

Genic Balance in *Drosophila*. When geneticist Calvin Bridges, working with *Drosophila,* crossed a triploid (3n) female with a normal male, he observed many combinations of autosomes and sex chromosomes in the offspring. Bridges suggested in 1922 that sex in *Drosophila* is determined by the balance (ratio) of autosomal alleles that favor maleness and alleles on the X chromosomes that favor femaleness. He calculated a ratio of X chromosomes to autosomal sets in order to see if this ratio would predict the sex of a fly. An **autosomal set** (A) in *Drosophila* consists of three chromosomes. (An autosomal set in human beings consists of twenty-two chromosomes.) Table 5.1, which presents his results, shows that Bridges's **genic balance theory** of sex determination was essentially correct. When the X:A ratio is 1.00, as in a normal female, or greater, the organism is a female. When this ratio is 0.50, as in a normal male, or less, the organism is a male. At 0.67 the organism is an **intersex. (Metamales** [X/A = 0.33] and **metafemales** [X/A = 1.50] are usually very weak and sterile. The metafemales usually do not even emerge from their pupal cases.)

The analysis so far should not give a misleading view of sex determination, which is a very complex, multistage developmental process. In addition to the determinants already discussed, environment and alleles at several other loci can also influence the final sex of the fly. For example, the intersex fly with an X:A ratio of 0.67 can have its development altered by the temperature at which it is raised. At higher temperatures the flies tend toward the female end of the intersex spectrum; at lower temperatures the flies tend toward the male end of this spectrum. Also, there are autosomal loci with alleles that can override the chromosomal constituency. The recessive **doublesex** (*dsx*) allele converts males and females into developmental intersexes.

TABLE 5.1   Data Supporting Bridges's Theory of Sex Determination by Genic Balance in *Drosophila*

Number of X Chromosomes	Number of Autosomal Sets (A)	Total Number of Chromosomes	$\frac{X}{A}$ Ratio	Sex
3	2	9	1.50	Metafemale
4	3	13	1.33	Female
4	4	16	1.00	Female
3	3	12	1.00	Female
2	2	8	1.00	Female
1	1	4	1.00	Female
2	3	11	0.67	Intersex
1	2	7	0.50	Male
1	3	10	0.33	Metamale

Transformer (*tra*) is a recessive allele that converts chromosomal females into sterile males. Thus the framework suggested by Bridges's genic balance theory can be influenced by the environment as well as by specific loci.

A **sex-switch** gene has been discovered that seems to direct female development. This gene, **Sex-lethal** (*Sxl*), is located on the X chromosome. Apparently, it has two states of activity. In the "on" state, it directs female development; in the "off" state maleness ensues. This sex-switch gene is regulated by other genes located on the X chromosome and the autosomes. Genes on the X chromosome that act to regulate *Sxl* into the on state (female development) are called **numerator elements** because they act on the numerator of the X/A genic balance equation. Genes on the autosomes that act to regulate *Sxl* into the off state (male development) are called **denominator elements**. Three numerator elements have been discovered. Research on these has led scientists to predict that the number of numerator elements will be relatively small. Thus only a few genes are acting to regulate the sex switch that controls sex determination in fruit flies. Although the mechanism of action of *Sxl* and the regulator genes is unknown, the X/A genic balance theory of Bridges can be seen to be the result of a few genes on the X chromosome and the autosomes.

Sex Determination in Human Beings.   Since the X0 genotype in human beings is a female (having Turner syndrome), it seems reasonable to conclude that the Y chromosome is male-determining in human beings. This is verified by the fact that persons with Klinefelter syndrome (XXY, XXXY, XXXXY) are all male, and XXX, XXXX, and other multiple-X karyotypes are all female. (More details on these anomalies will be presented in chapter 8.) For a long time there has been a search for a single gene, a **testis-determining factor** (TDF) that acts as a sex switch on the Y chromosome and initiates maleness. Human embryologists had discovered that during the first month of embryonic development, the gonads that develop are neither testes nor ovaries, but instead indeter-

David Page
Courtesy of Dr. David Page.

minate. At about six or seven weeks of development, the indeterminate gonads become either ovaries or testes. In the 1950s, Ernst Eichwald found that males had a protein on their cell surfaces not found in females; he discovered that female mice rejected skin grafts from genetically identical brothers, whereas grafts from sisters were accepted by their brothers. This implies the existence of an antigen on the surface of male cells not found on female cells. This protein was called the *histocompatibility Y-antigen* (H-Y antigen). The gene for this protein was found to be located on the Y chromosome, near the centromere, probably on the long arm of the chromosome. At first it was believed to be the sex switch: If the gene were present, the gonads would begin development as testes. Further development of maleness, such as male secondary sexual characteristics, comes about through the testosterone produced by the functional testes. If the gene were absent, development of the gonads would proceed to form as ovaries. Recently, however, from the study of "sex-reversed" individuals, this theory was shown to be wrong.

Sex-reversed individuals are XX males or XY females. David Page, at the Whitehead Institute for Biomedical Research, found twenty XX males who had a

Peter Goodfellow (1951–)
Courtesy of Peter Goodfellow.

Robin Lovell-Badge (1953–)
Courtesy of Robin Lovell-Badge.

small piece of the short arm of the Y chromosome attached to one of their X chromosomes. He found six XY females in whom the Y chromosome was missing the same small piece at the end of its short arm. This region is now known to carry the testis-determining factor. The first candidate gene believed to code for the testis-determining factor was named the **ZFY gene,** for zinc finger on the Y chromosome. Zinc fingers are protein configurations known to interact with DNA (discussed in detail in chapter 15). Thus, it was believed that the *ZFY* gene, coding for the testis-determining factor, worked by directly interacting with DNA. (Later in the book we will look at the way regulatory genes work that interact with DNA.) However, men lacking the *ZFY* gene have been found, suggesting that the testis-determining factor is very close to but not the *ZFY* gene. From work in mice it has been suggested that the *ZFY* gene controls the initiation of sperm cell development, but not maleness.

Figure 5.4
Normal male mouse (*left*) and female littermate (*right*) given the *SRY* gene. Both mice are indistinguishably male.
Courtesy of Robin Lovell-Badge.

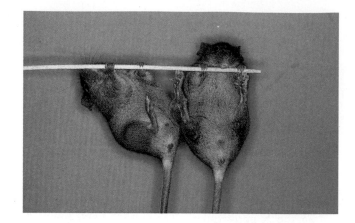

In 1991, Robin Lovell-Badge and Peter Goodfellow and their colleagues in England isolated a gene called **Sex-determining region Y (SRY)**—*Sry* in mice—adjacent to the *ZFY* gene. *Sry* has been positively identified as the testis-determining factor because, when injected into normal (XX) female mice, it caused them to develop as males (fig. 5.4). Although these XX males are sterile, they appear as normal males in every other way. (We discuss in chapter 12 how the procedures of getting new genes into an organism are carried out.) Note also that the mouse and human systems are very similar genetically and the homologous genes have been isolated from both. However, at present, the human *SRY* gene does not convert XX female mice into males. Like the *ZFY* gene product, *SRY* protein also binds to DNA.

We should point out that the determination of maleness or femaleness (testis or ovary development) is not a single-gene phenomenon but rather a developmental sequence involving numerous genes. The sex switch initiates this process. Eva Eicher and Linda Washburn, of the Jackson Laboratory, have presented a model in which two pathways of coordinated gene action are involved in sex determination, one for each sex. The first gene in the ovary-determining pathway is termed *ovary determining* (*Od*). The first gene in the testis-determining pathway must function in time before the *Od* gene, so as to allow XY individuals to develop as males. Once the steps of a pathway are initiated, the other pathway is inhibited (fig. 5.5).

The X0 System

This system is also sometimes referred to as an X0-XX system. It occurs in many species of insects in which the situation is as described for the XY chromosomal mechanism, except that instead of a Y chromosome the heterogametic sex (male) has only one X chromosome. Males

Figure 5.5
A model for the initiation of gonad determination in mammals

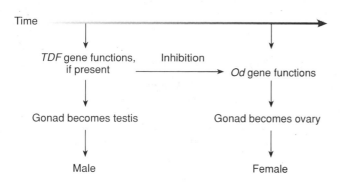

Figure 5.6
The human Y chromosome. The X chromosome also contains the *MIC2* gene and a TDF-like gene.

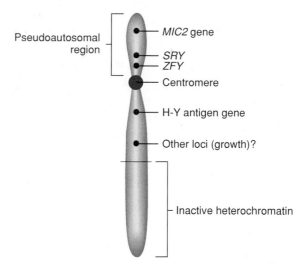

produce gametes that contain either an X chromosome or no sex chromosome, whereas all the gametes from a female contain the X chromosome. The result of this arrangement is that females have an even number of chromosomes (all in homomorphic pairs) and males have an odd number of chromosomes.

The ZW System
The ZW system is identical to the XY system except that males are homogametic and females are heterogametic. This situation occurs in birds, some fishes, and moths.

Compound Chromosome Systems
The compound chromosome systems tend to be complex. For example, in *Ascaris incurva,* a nematode, there are eight X chromosomes and one Y. The species has twenty-six autosomes. Males have thirty-five chromosomes (26A + 8X + Y), and females have forty-two chromosomes (26A + 16X). During meiosis the X chromosomes unite end-to-end and so behave as one unit.

SEX LINKAGE

In an XY chromosomal system of sex-determination, the pattern of inheritance for loci on the heteromorphic sex chromosomes differs from the pattern for loci on the homomorphic autosomal chromosomes, because sex-chromosome alleles are inherited in association with the sex of the offspring. Alleles on a male's X chromosome go to his daughters but not to his sons, because the presence of his X chromosome normally determines that his offspring is a daughter. For example, the inheritance pattern of hemophilia (failure of blood to clot), in which the common form is caused by an allele located on the X chromosome, has been known since the end of the eighteenth century. It was known that mostly men had the disease, whereas women could pass on the disease without actually having it. (In fact the general nature of the inheritance of this trait was known in biblical times. The Talmud—the Jewish book of laws and traditions—specified exemptions to circumcision on the basis of hemophilia among relatives consistent with an understanding of who was at risk.)

Before we continue, there is a small distinction to be made. Since both X and Y are sex chromosomes, three different patterns of inheritance are possible, all sex linked (for loci found only on the X chromosome, only on the Y chromosome, or on both). However, the term **sex linked** usually refers to loci found only on the X chromosome; the term **Y linked** is used to refer to loci found only on the Y chromosome, which control **holandric traits** (traits found only in males). Loci found on both the X and Y chromosomes are called **pseudoautosomal.** In humans there are at least three hundred loci known to be on the X chromosome; there are only a few known to be on the Y chromosome. We already mentioned the H-Y antigen and the *ZFY* genes (fig. 5.6). In addition, on the short arm of the Y chromosome is a region of homology and pairing of the two sex chromosomes that is referred to as the pseudoautosomal region of Y. This region contains an antibody gene, *MIC2,* a pseudoautosomal gene found on both the X and Y chromosomes. A *ZFY*-like gene on the X chromosome also occurs in this region of homology. There are several nonfunctioning genes on the Y chromosome, including the gene for steroid sulphatase and the gene for Kallmann syndrome, a disease of gonadal and olfactory functions. There are also some genes that are probably located on the Y chromosome but have not been absolutely established. Candidates include height genes (on average, XYY men are taller than XY men who are taller than X0 women), a gene for tooth growth, a gene for speed of maturation, and two nonfunctional genes similar to the genes for the protein actin and the enzyme arginosuccinate.

X-Linkage in *Drosophila*

T. H. Morgan, a 1933 Nobel laureate, demonstrated the **X-linked** pattern of inheritance in *Drosophila* in 1910 when a white-eyed male appeared in a culture of wild-type (red-eyed) flies (fig. 5.7). This male was crossed with

Thomas Hunt Morgan (1866–1945)
Genetics, 32 (1947): frontispiece.

Figure 5.7
(*a*) Wild-type and (*b*) white-eyed fruit flies
Carolina Biological Supply Company.

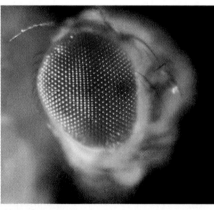

(a)

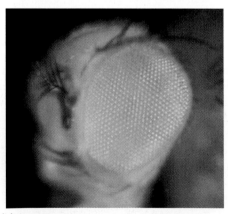

(b)

Figure 5.8
Pattern of inheritance of the white-eye trait in *Drosophila*

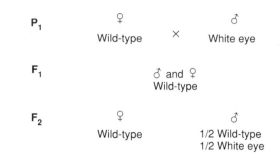

a wild-type female. All of the offspring were wild-type. When these F_1 individuals were crossed with each other, their offspring fell into two categories (fig. 5.8). All the females and half the males were wild-type, whereas the remaining half of the males were white eyed. Morgan interpreted this to mean that the white-eye locus was on the X chromosome. (Historically, this was a very important finding because it helped strengthen the chromosomal theory of inheritance by indicating that a particular gene was on a specific chromosome.) We can redraw figure 5.8 to include the sex chromosomes of Morgan's flies (fig. 5.9). We denote the X chromosome with the white-eye allele as X^w. Similarly X^+ is the X chromosome with the wild-type allele, and Y is the Y chromosome, which has no allele at this locus.

Another property of sex linkage is seen in figure 5.9. Since females have two X chromosomes, they can have normal homozygous and heterozygous gene combinations. But males, with only one copy of the X chromosome, can be neither homozygous nor heterozygous. Instead, the term **hemizygous** is used for X-linked genes in males. Since only one allele is present, a single copy of a recessive allele will determine the phenotype, a phenomenon called **pseudodominance.** Thus a male with one *w* allele is white eyed, the allele acting in a dominant fashion. This is the same way that one copy of an autosomal dominant allele would determine the phenotype of a normal diploid organism. Hence the term pseudodominance.

Nonreciprocity

The X-linked pattern has long been known as the **criss-cross pattern of inheritance** because the father passes a trait to his daughters, who pass it to their sons. That this analysis is correct and that the inheritance pattern is not reciprocal are shown in figure 5.10 in which a white-eyed female is crossed with a wild-type male. Here the F_1 males are white eyed, the F_1 females are wild-type, and 50% of each sex in the F_2 are white eyed. Such nonreciprocity and different ratios in the two sexes suggest sex linkage, which the criss-cross pattern confirms.

Figure 5.9

Crosses of figure 5.8 redrawn to include the sex chromosomes

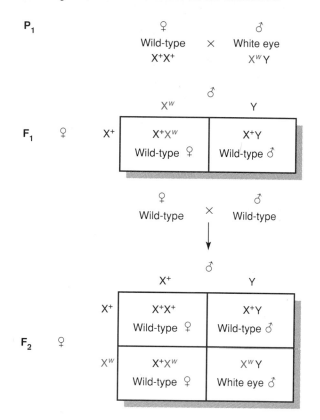

Figure 5.10

Reciprocal cross to that in figure 5.9

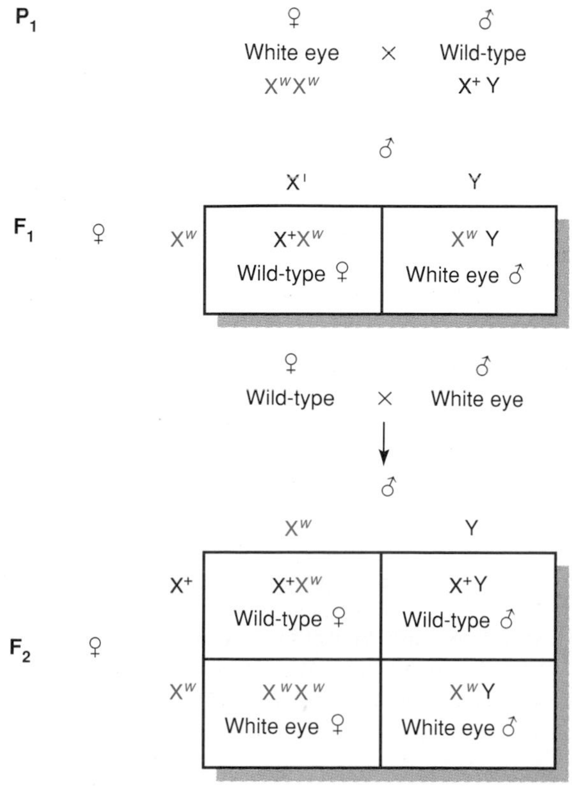

Figure 5.11

Inheritance pattern of barred plumage in chickens

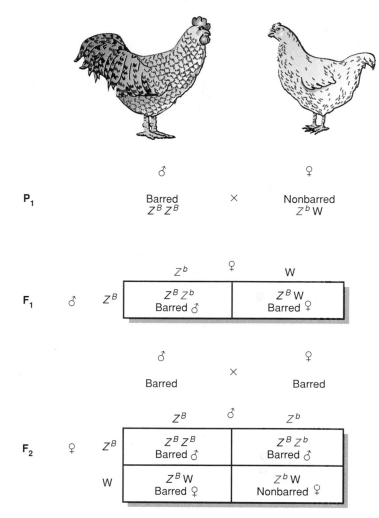

Figure 5.11 shows the inheritance pattern of a sex-linked trait in chickens, in which the male is the homogametic sex. The gene for barred plumage is Z linked and dominant to the gene for nonbarred plumage. If we substitute white eyed for nonbarred and male for female, we get the same pattern as in fruit flies (fig. 5.9) in which, of course, females are homogametic.

The Y chromosome in fruit flies is known to carry six loci that are required for male fertility; they act only in primary spermatocytes. In addition, it carries the pseudoautosomal bobbed locus (bb). In the homozygous recessive state it causes bristles to be shortened. The locus occurs on both the X and Y chromosomes: it is the nucleolar organizer (see chapter 3). Figures 5.12 and 5.13 show the results of reciprocal crosses involving bobbed. In both cases, one quarter of the F_2 are bobbed. In one cross it is males and in the other it is females.

Figure 5.12

Inheritance pattern of the bobbed locus in *Drosophila*

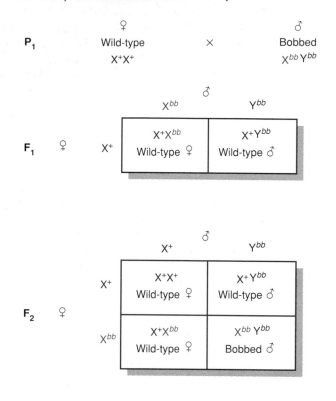

Figure 5.13

Reciprocal cross to that in figure 5.12

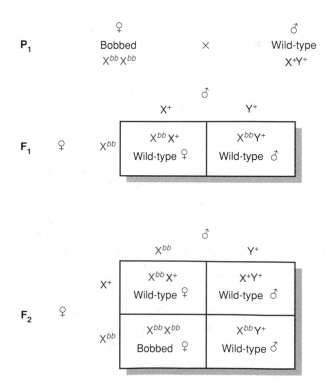

Sex-Limited and Sex-Influenced Traits

Aside from X-linked, holandric, and pseudoautosomal inheritance, two inheritance patterns show nonreciprocity without necessarily being controlled by loci on the sex chromosomes. **Sex-limited traits** are traits expressed in only one sex, although the genes are present in both sexes. In women, breast and ovary formation are sex-limited traits as are facial hair distribution and sperm production in men. Nonhuman examples are plumage patterns in birds—in many species the male is brightly colored—and horns found only in males of certain sheep species. Milk yield in mammals is expressed phenotypically only in females. **Sex-influenced, or sex-controlled, traits** appear in both sexes but occur in one sex more than the other. Pattern, or premature, baldness in human beings is an example of a sex-influenced trait. It is controlled by an allele that appears to be dominant in men but recessive in women; in the latter it is usually expressed as a thinning of hair rather than a balding. Apparently testosterone, the male hormone, is required for the full expression of the allele.

PEDIGREE ANALYSIS

Inheritance patterns in many organisms are relatively easy to determine, because crucial crosses can be made to test hypotheses about the genetic control of a particular trait. Many of these same organisms produce an abundance of offspring so that numbers large enough to compute ratios can be gathered. Recall Mendel's work with garden peas, in which his 3:1 ratio in the F_2 generation led him to suggest the rule of segregation. If Mendel's sample sizes had been smaller, he might not have seen the ratio. Think of what Mendel's difficulty would have been had he decided to work with human beings instead of pea plants. The same problems are faced by human geneticists today. The occurrence of a trait in one of four children is not necessarily indicative of a true 3:1 ratio.

To determine the inheritance pattern of many human traits, human geneticists often have little more to go on than a **pedigree** that many times does not include critical mating combinations. Frequently there are uncertainties and ambiguities in pedigree analysis, a procedure whereby conclusions are often arrived at by a process of elimination. Other difficulties encountered by human geneticists are the lack of **penetrance** and different degrees of **expressivity** in many traits. Both are aspects of the expression of a phenotype.

Penetrance and Expressivity

Penetrance refers to the appearance in the phenotype of traits determined by the genotype. Unfortunately for geneticists, not all genotypes "penetrate" into the phenotype. For example, a person could have the genotype that specifies vitamin-D-resistant rickets and yet not have rickets. This disease is caused by a sex-linked dominant gene and is distinguished from normal vitamin D deficiency by its failure to respond to low levels of vitamin D. It does, however, respond to very high levels of vitamin D and is thus treatable. In any case, family trees are known in which affected children are born from normal parents. This would violate the rules of dominant inheritance because one of the parents must have had the allele yet did not express it. That the parent actually had the allele is demonstrated by the occurrence of low levels of phosphorus in the blood, a pleiotropic effect of the same allele. The low-phosphorus aspect of the phenotype is always fully penetrant.

Thus certain genotypes, especially those for developmental traits, are not always fully penetrant. Most genotypes, however, are fully penetrant. For example, there are no known cases of individuals homozygous for albinism who do not lack pigment. Vitamin-D-resistant rickets illustrates another phenomenon in which a phenotype that is not genetically determined mimics a phenotype that is genetically controlled. This **phenocopy** is the result of dietary deficiency or environmental trauma. A dietary deficiency of vitamin D, for example, will produce rickets that is virtually indistinguishable from genetically caused rickets.

Many developmental traits not only fail to penetrate sometimes but also show a variable pattern of expression, from very mild to very extreme, when they do penetrate. For example, cleft palate is a trait that shows both variable penetrance and variable expressivity. Once the genotype penetrates, the severity of the impairment varies considerably, from a very mild external cleft to a very severe clefting of the hard and soft palates. Failure to penetrate and variable expressivity are not unique to human traits but are symptomatic of developmental traits in many organisms.

Family Tree

One way to examine a pattern of inheritance is to draw a family tree. The symbols used in constructing a family tree, or pedigree, are defined in figure 5.14; a pedigree is shown in figure 5.15. The circles represent females and the squares represent males. Symbols that are filled in represent individuals who have the trait under study; the individuals are said to be **affected.** The open symbols represent those who do not have the trait. The direct horizontal lines between two individuals (one male, one female) are called marriage lines. Children are attached to a marriage line by a vertical line. All the brothers and sisters (**siblings** or **sibs**) from the same parents are connected by a horizontal line above their symbols. Siblings are numbered below their symbols according to birth order, and generations are numbered on the right in Roman numerals. When the sex of a child is unknown, the symbol used is diamond shaped (e.g., the children of III-1 and

Figure 5.14
Symbols used in a pedigree

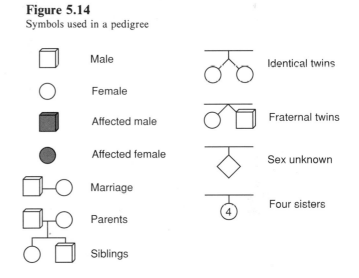

Figure 5.15
Part of a pedigree for polydactyly

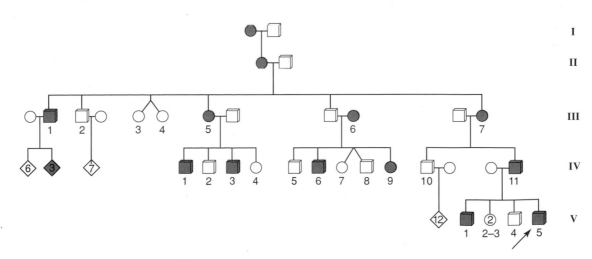

Figure 5.16

Hands of a person with polydactyly. Manifestations of polydactyly range in severity from one extra finger or toe to one or more extra digits on each hand and foot.

© *Lester Bergman and Associates*

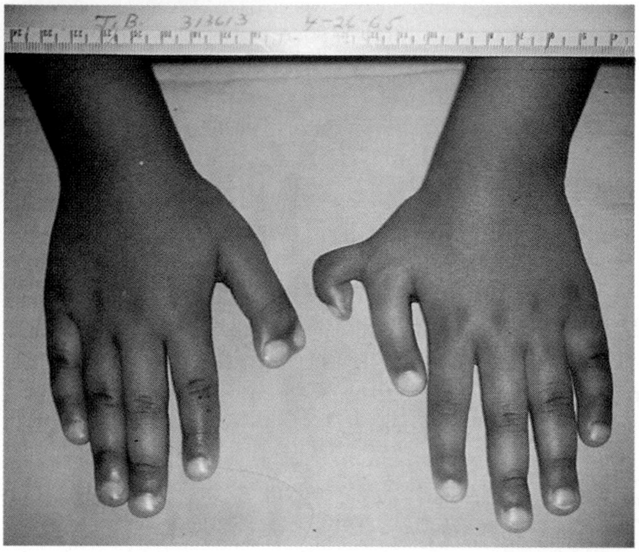

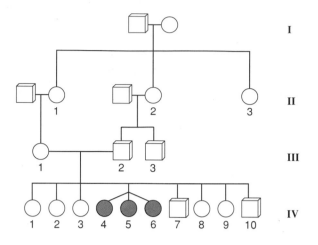

Figure 5.17

Part of a pedigree of hypotrichosis (early hair loss)

III-2 in figure 5.15). A number within a symbol represents the number of siblings not separately listed. Individuals IV-7 and IV-8 of figure 5.15 are fraternal (nonidentical) twins: They originate from the same point. Individuals III-3 and III-4 are identical twins: They originate from a short vertical line.

When other symbols occur in a pedigree, they are usually defined in the legend. Individual V-5 is called a **proband** or **propositus** (female, **proposita**). The arrow pointing to individual V-5 in figure 5.15 indicates that it was through this individual that the pedigree was ascertained, usually by a physician or clinical investigator.

On the basis of the information in a pedigree, geneticists attempt to determine the mode of inheritance of a trait. There are two types of questions the pedigree might be used to answer. First, are there patterns within the pedigree that are consistent with a particular mode of inheritance? Second, are there patterns within the pedigree that are inconsistent with a particular mode of inheritance? Often, however, it is not possible to determine the mode of inheritance of a particular trait with certainty. McKusick has reported that, as of 1990, the mode of inheritance of about 5,000 loci in human beings was known with some confidence, including autosomal dominant, autosomal recessive, and sex-linked genes.

Dominant Inheritance

In looking again at the pedigree of figure 5.15, several points emerge. First, polydactyly (fig. 5.16) occurs in every generation. That is, every affected child has an affected parent—no generations are skipped. This suggests domi-

nant inheritance. Second, the trait occurs about equally in both sexes. There are six affected males and six affected females in the pedigree. This indicates autosomal rather than sex-linked inheritance. Thus so far we would categorize the trait as an autosomal dominant. Note also that individual IV-11, a male, passed on the trait to two of his three sons. This would rule out sex linkage. Remember that a male gives his X chromosome to all of his daughters but none of his sons. His sons receive his Y chromosome. By consistency in many such pedigrees, it has been confirmed that polydactyly is caused by an autosomal dominant gene.

Polydactyly shows variable penetrance and expressivity. The most extreme manifestation of the trait is the occurrence of an extra digit on each hand (fig. 5.16) and one or two extra toes on each foot. However, some individuals have only extra toes, some have extra fingers, and some have an asymmetrical distribution of digits such as six toes on one foot and seven on the other.

Recessive Inheritance

Figure 5.17 is a pedigree with a different pattern of inheritance. Here affected individuals are not found in each generation. The affected daughters, identical triplets, come from normal parents. They are, in fact, the first appearance of the trait in the pedigree. A telling point here is that the parents of the triplets are first cousins. A marriage between relatives is referred to as **consanguineous.** If the degree of relatedness is closer than law permits, the union is called **incestuous.** In all states, brother–sister and mother–son marriages are forbidden; and in all states except Georgia, father–daughter marriages are forbidden. Georgia did not intend to permit father–daughter marriages. However, the law was drafted using biblical terminology that inadvertently did not prohibit a man from marrying his daughter or his grandmother. Thirty states prohibit the marriage of first cousins.

Consanguineous marriages often produce offspring that have rare recessive, and often deleterious, traits. The reason is that through common ancestry (e.g., first cousins have a pair of grandparents in common), an allele found in an ancestor can be inherited on both sides of the pedigree and become homozygous in the child. The occurrence of a trait in a pedigree with common ancestry is often good evidence for an autosomal recessive mode of inheritance. Consanguinity by itself does not guarantee that the trait being examined has an autosomal recessive mode of inheritance; all modes of inheritance are found in consanguineous pedigrees. Conversely, recessive inheritance is not confined to consanguineous pedigrees. Hundreds of recessive traits are known from pedigrees lacking consanguinity.

Sex-Linked Inheritance

Figure 5.18 is the pedigree of Queen Victoria of England. Through her children, hemophilia was passed on to many of the royal houses of Europe. Several interesting aspects of this pedigree help to confirm the method of inheritance. First, generations are skipped. Although Alexis (1904–1918) was a hemophiliac, neither his parents nor his grandparents were. This pattern occurs in several other places in the pedigree and indicates a recessive mode of inheritance. From other pedigrees and from the biochemical nature of the defect, it has been determined that hemophilia is a recessive trait.

Further inspection of the pedigree in figure 5.18 will reveal that all the affected individuals are sons, strongly suggestive of sex linkage. Since males are hemizygous for the X chromosome, more males than females should have the phenotype of a sex-linked recessive trait because males do not have a second X chromosome that might carry the normal allele. If this mode is correct, we can make several predictions. First, since all males get their X chromosomes from their mothers, affected males should be offspring of carrier (heterozygous) females. A female is a carrier if her father had the disease. She has a 50 percent chance of being a carrier if her brother, but not her father, has the disease. In that case her mother was a carrier. The pedigree is consistent with these predictions.

In no place in the pedigree is the trait passed on from father to son. This would defy the route of an affected X chromosome. We can conclude from the pedigree that hemophilia is a sex-linked recessive trait. (There are several different inherited forms of hemophilia known, each deficient in one of the steps in the pathway of the formation of fibrinogen, the blood clot protein. Two of these forms, "classic" hemophilia A and hemophilia B, which is called Christmas disease, are sex linked. Other hemophilias are autosomal.)

One other interesting point about this pedigree is that there was no evidence of the disease in Queen Victoria's ancestors, yet she was obviously a heterozygote, having one affected son and two daughters that were known carriers. Thus, from what appears to be a homozygous normal mother and a hemizygous normal father, one of Queen Victoria's X chromosomes had the hemophilia allele. This could have happened if there had been a change (mutation) in one of the gametes that formed Queen Victoria. (We will explore the mechanisms of mutation in chapter 16.)

Figure 5.19 is another pedigree in which dominant inheritance is suspected because no generations are skipped. The pedigree shows the distribution of low blood-phosphorus levels, the fully penetrant aspect of vitamin-D-resistant rickets, among the sexes. Affected males pass on the trait to their daughters, but not their sons. This pattern is the pattern of the X chromosome: A male passes it on to all of his daughters but to none of his sons. Although this pedigree is in good agreement with a sex-linked dominant mode of inheritance, the sample is too small to rule out autosomal inheritance. The pedigree shown is a small part of one involving hundreds of people, all with phenotypes consistent with the hypothesis of sex-linked dominant inheritance.

In figure 5.19 there is the slight possibility that the trait is recessive. This could be true if the husband in generation I and the wives of II-5 and II-7 were all heterozygotes. Since this is a rare trait, the possibility of all these events is small. For example, if one person in fifty (0.02) is a heterozygote, then the probability of three heterozygotes marrying into the same pedigree is $(0.02)^3$, or eight in one million. The rareness of this event further supports the hypothesis of dominant inheritance. The expected patterns for the various types of inheritance in pedigrees can be summarized in the following four categories.

Autosomal Recessive Inheritance

1. Trait often skips generations.

2. There should be an almost equal distribution of affected individuals among sexes.

3. Traits are often found in pedigrees with consanguineous marriages.

4. If both parents are affected, all children should be affected.

5. In most cases of normal people married to affected individuals, all children produced are normal. When at least one child is affected (indicating that the normal parent is heterozygous), then approximately half the children should be affected.

6. Most affected individuals have normal parents.

Figure 5.18

Hemophilia in the pedigree of Queen Victoria of England. In this photograph of the Queen and some of her descendants, three carriers—Queen Victoria, Princess Irene of Prussia (*right*), and Princess Alix (Alexandra) of Hesse (*left*)—are indicated.

Photo © Mary Evans Picture Library/Photo Researchers, Inc.

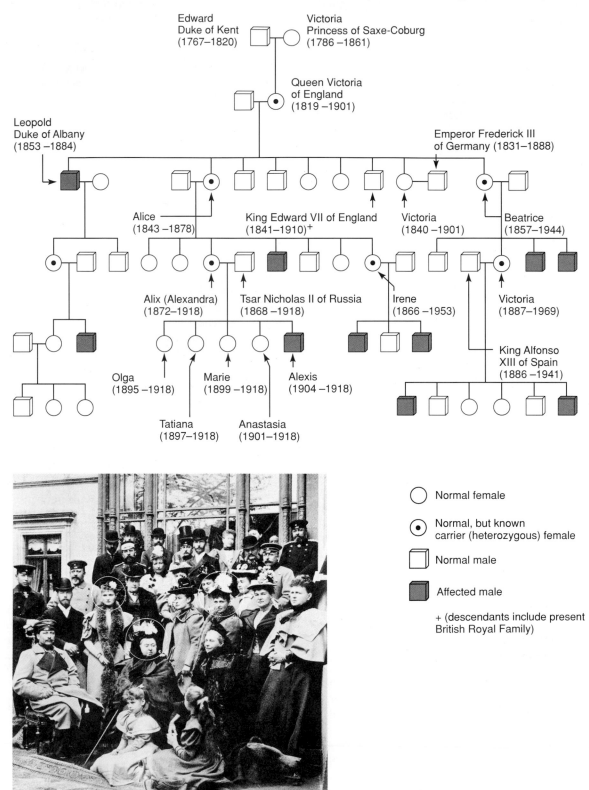

Edward
Duke of Kent
(1767–1820)

Victoria
Princess of Saxe-Coburg
(1786–1861)

Queen Victoria
of England
(1819–1901)

Leopold
Duke of Albany
(1853–1884)

Emperor Frederick III
of Germany (1831–1888)

Alice
(1843–1878)

King Edward VII of England
(1841–1910)+

Victoria
(1840–1901)

Beatrice
(1857–1944)

Alix (Alexandra)
(1872–1918)

Tsar Nicholas II of Russia
(1868–1918)

Irene
(1866–1953)

Victoria
(1887–1969)

Olga
(1895–1918)

Marie
(1899–1918)

Alexis
(1904–1918)

King Alfonso
XIII of Spain
(1886–1941)

Tatiana
(1897–1918)

Anastasia
(1901–1918)

○ Normal female

⊙ Normal, but known
carrier (heterozygous) female

▢ Normal male

▤ Affected male

+ (descendants include present
British Royal Family)

Figure 5.19

Part of a pedigree of vitamin-D-resistant rickets. Affected individuals have low blood-phosphorus levels. Although the sample is too small for certainty, dominance is indicated because no generations were skipped and sex linkage is suggested by the distribution of affected individuals.

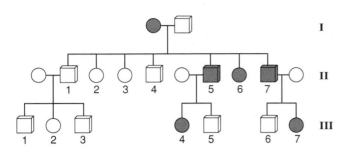

Mary F. Lyon (1925–   )
Courtesy of Dr. Mary F. Lyon.

Autosomal Dominant Inheritance

1. Trait should not skip generations (unless penetrance is reduced).

2. An affected person married to a normal person should produce approximately 50% affected offspring (indicating also that the affected individual is heterozygous).

3. Distribution of the trait among sexes should be almost equal.

Sex-Linked Recessive Inheritance

1. Most affected individuals are male.

2. Affected males result from mothers who are affected or who are known to be carriers (heterozygotes) by having affected brothers, fathers, or maternal uncles.

3. Affected females come from affected fathers and affected or carrier mothers.

4. The sons of affected females must be affected.

5. Approximately half the sons of carrier females should be affected.

Sex-Linked Dominant Inheritance

1. The trait does not skip generations.

2. Affected males must come from affected mothers.

3. Approximately half the children of an affected female are affected.

4. Affected females come from affected mothers or fathers.

5. All the daughters, but none of the sons, of an affected father are affected.

DOSAGE COMPENSATION

Recall that in the XY chromosome system of sex determination, males have only one X chromosome whereas females have two. Thus males have half the number of X-linked genes as females. The question arises as to how the organism compensates for this dosage difference between the sexes, given that there is a potential for serious abnormality. In general, an incorrect number of autosomes is usually highly deleterious to an organism (see chapter 8). In humans and other mammals, the necessary **dosage compensation** is accomplished by inactivation of one of the X chromosomes in females so that both males and females have only one functional X chromosome per cell.

A condensed body in the nucleus that was not the nucleolus was first observed by M. Barr and E. Bertram. Noting that normal female cats show a single condensed body whereas males show none, these researchers referred to the body as sex chromatin, which has since been referred to as a **Barr body.** Mary Lyon then suggested that this Barr body represented an inactive X chromosome, which in females becomes tightly coiled into **heterochromatin,** a condensed, and therefore visible, form of chromatin (fig. 5.20). Recent microscopic work indicates that the tips of the Barr body are in close proximity, forming a ring.

Various lines of evidence support the **Lyon hypothesis.** First, XXY males have a Barr body whereas X0 females have none. Second, persons with abnormal numbers of X chromosomes have one fewer Barr body than they have X chromosomes per cell: XXX females have two Barr bodies and XXXX females have three.

Proof of the Lyon Hypothesis

Direct proof of the Lyon hypothesis came when cytologists identified the Barr body in normal females as an X chromosome. Genetic evidence also supports the Lyon hypothesis: Females heterozygous for a locus on the X chromosome show a unique pattern of phenotypic expression.

Figure 5.20

Barr body (*arrow*) in the nucleus of a cheek mucosal cell of a normal female. This visible mass of heterochromatin has been identified as an inactivated X chromosome.

Source: Thomas G. Brewster and Park S. Gerald, "Chromosome Disorders Associated with Mental Retardation," Pediatric Annals 7, No. 2 (1978). Reproduced courtesy of Dr. Thomas G. Brewster, Foundation for Blood Research, Scarborough, Maine.

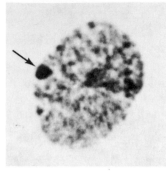

Figure 5.21

Electrophoretic gel stained for glucose-6-phosphate dehydrogenase. Lanes 1–3 contain blood from (1) *AA* homozygote, (2) *BB* homozygote, (3) *AB* heterozygote. Lanes 4–10 contain homogenates of individual cells of an *AB* heterozygote.

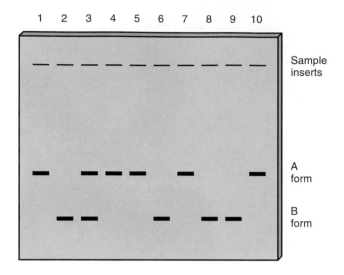

It is now known that in humans the X chromosome is inactivated in each cell at about the twelfth day of embryonic life and further that it is randomly determined which X is inactivated in a given cell. From that point on, the same X remains a Barr body for future cell generations. Thus heterozygous females show a **mosaicism** at the cellular level for X-linked traits. Instead of being typically heterozygous, they are effectively hemizygous for one or the other of the X-chromosome alleles in each cell.

Glucose-6-phosphate dehydrogenase (G-6-PD) is an enzyme controlled by a locus on the X chromosome. The enzyme occurs in several different allelic forms that differ by single amino acids. Thus both forms (A and B) will dehydrogenate glucose-6-phosphate—both are fully functional enzymes—but because they differ by an amino acid, they can be detected by their rate of migration in an electrical field (one form moves faster than another). This electrical separation, termed **electrophoresis,** is carried out by placing samples of the enzymes being tested in a supporting gel, usually starch, polyacrylamide, or cellulose acetate (fig. 5.21; see also the material on electrophoresis in Box 5.1). After the electric current is applied for several hours, the gel is stained in a way that shows the enzymes in the gel as bands, revealing the distance traveled by each enzyme. Since blood serum is a conglomerate of proteins from many cells, the serum of the heterozygote (fig. 5.21, slot 3) has both A and B forms (bands), whereas any single cell (slots 4–10) has only one or the other band. Since the gene for G-6-PD is carried on the X chromosome, this electrophoretic display indicates that only one X is active in any particular cell.

Another aspect of the G-6-PD system provides further proof of the Lyon Hypothesis. If a cell has both alleles functioning, there should be both A and B proteins present. Since the functioning G-6-PD enzyme is a dimer (made up of two protein subunits), 50% of the enzymes should be heterodimers (AB). These would form a third, intermediate band between the A form (AA dimer) and the B form (BB dimer; fig. 5.22). The lack of heterodimers in the blood of heterozygotes is further proof that both alleles of G-6-PD are not active within the same cells.

The Lyon hypothesis has been demonstrated with many X-linked loci, but the most striking examples are those for the color phenotypes in some mammals. For example, the calico pattern of cats (fig. 5.23) is due to the inactivation of X chromosomes. Calico cats are normally females heterozygous for the yellow and black alleles of the X-linked color locus. They exhibit patches of these two colors, thereby indicating that at a certain stage in development one or the other of the X chromosomes was inactivated and that all of the ensuing daughter cells in that line kept the same X chromosome inactive. The result is broad patches of coat color rather than the microscopic color pattern to be expected if every new cell produced had one of its X chromosomes randomly inactivated.

The X chromosome is inactivated starting at a point called the **X-inactivation center (XIC).** That region contains a gene called *XIST* (for X inactive-specific transcripts, referring to the transcriptional activity of this gene in the inactivated X chromosome). The *XIST* gene has been assigned putatively as the gene that initiates the inactivation of the X chromosome. This gene is known to be active only in the inactive X chromosome in a normal XX female. Another aspect of "Lyonization" is that several other loci are known to be active on the inactivated X chromosome; they are active in both X chromosomes even though one is heterochromatic (inactived). Although several of these loci are in the pseudoautosomal region of the

BOX 5.1
ELECTROPHORESIS

Electrophoresis, a technique for separating relatively similar types of molecules (proteins, nucleotide segments), has opened up new and exciting areas of research in population, biochemical, and molecular genetics. It has allowed us to see variations in large numbers of loci, previously difficult or impossible to sample. In population genetics (see chapter 21), electrophoresis has made it possible to estimate the amount of variability occurring in natural populations. The resulting discovery that a great deal of heterozygosity occurs in nature was a revelation and has opened up whole new areas of theoretical as well as empirical study. In biochemical genetics, electrophoretic techniques can be used to study enzyme pathways. In molecular genetics, electrophoresis is used to sequence nucleotides (see chapter 12) and to assign various loci to particular chromosomes.

In the following we will discuss protein electrophoresis, a process that entails placing a sample—often blood serum or a cell homogenate—at the top of a gel prepared from a suitable substrate (e.g., hydrolyzed starch, polyacrylamide, cellulose acetate) and a suitable buffer. An electrical current is passed through the gel (Box fig. 1), and the gel is then treated with a dye that will stain the protein. In the simplest case, if a protein is homogeneous (usually the product of a homozygote), it will form a single band on the gel. If it is heterogeneous (usually the product of a heterozygote), it will form two bands when the two allelic protein products differ by an amino acid in such a way that they have different electrical charges and therefore travel through the gel at different rates (fig. 5.21). The term **allozyme** is used for different electrophoretic forms of an enzyme controlled by alleles at the same locus.

Samples of mouse blood serum that have been stained for protein are shown in Box figure 2. Most of the staining is accounted for by albumins and β-globulins (transferrin). Because they are present in very small concentrations, many enzymes present are not visible, but a stain that is specific for a particular enzyme can make that enzyme

Box figure 1

Vertical starch gel apparatus. Current flows from the upper buffer chamber to the lower one by way of the paper wicks and the starch gel. Cooling water flows around the system.

Source: R. P. Canham, "Serum Protein Variations and Selection in Fluctuating Populations of Cricetid Rodents," Ph.D. thesis, University of Alberta, 1969. Reproduced by permission.

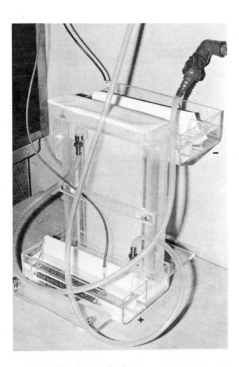

Box figure 2

Ten samples of deer mouse (*Peromyscus*) blood studied for general protein. Al is albumin and Tf is transferrin, the two most abundant proteins in mammalian blood. Note the transferrin allozyme variation in the different samples.

Source: R. P. Canham, "Serum Protein Variations and Selection in Fluctuating Populations of Cricetid Rodents," Ph.D. thesis, University of Alberta, 1969. Reproduced by permission.

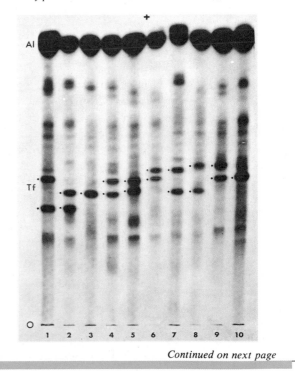

Continued on next page

BOX 5.1—*Continued*

visible on the gel. For example, lactate dehydrogenase (LDH) can be stained for specifically because it catalyzes the reaction:

$$\text{lactic acid} + \text{NAD} \underset{}{\overset{\text{LDH}}{\rightleftarrows}} \text{pyruvic acid} + \text{NADH}$$

Thus we can specifically stain for the LDH enzyme by adding the substrates of the enzyme (lactic acid and NAD) and a suitable stain system specific for a product of the enzyme reaction (pyruvic acid or NADH). Thus, if lactic acid and NAD are added to the gel, only LDH will convert them to pyruvic acid and NADH. We can then test for the presence of NADH by having it reduce the dye, nitro blue tetrazolium, to the blue precipitate, formazan, an electron carrier. We then add all the preceding reagents and look for the blue bands on the gel (Box fig. 3).

In addition to its uses in population genetics and chromosome mapping, electrophoresis has been extremely useful in determining the structure of many proteins and for studying developmental pathways. As we can see from the LDH gel in Box figure 3, five bands can occur. In some tissues of a homozygote these bands occur roughly in a ratio of 1:4:6:4:1. This can come about if the enzyme is a tetramer whose four subunits are random mixtures of two gene products (from the *A* and *B* loci). Thus we would get

AAAA (1/16)

AAAB (4/16)

AABB (6/16)

ABBB (4/16)

BBBB (1/16)

(Note that the ratio 1:4:6:4:1 is the expansion of $[A + B]^4$ and the relative "intensity" of each band—number of protein doses—is calculated from the rule of unordered events described in chapter 4.)

This tetramer model has been verified by protein chemists. In this way electrophoresis has helped us determine the structure of several enzymes. (The term **isozymes** is used for multiple electrophoretic forms of an enzyme caused by subunit interaction rather than allelic differences.) It has also been discovered that during development the five forms differ in their various concentrations in different tissues of the body (Box fig. 4). This has led to various hypotheses as to how the production of enzymes is controlled developmentally.

Box figure 3

LDH isozyme patterns in pigeons. Note the five bands for some individual samples. Lanes I, II, and III under each tissue type indicate the range of individual variation.

Source: W. H. Zinkham et al., "A Variant of Lactate Dehydrogenase in Somatic Tissues of Pigeons: Physicochemical Properties and Genetic Control," Journal of Experimental Zoology 162, no. 1 (June 1966): 45–46. Reproduced by permission.

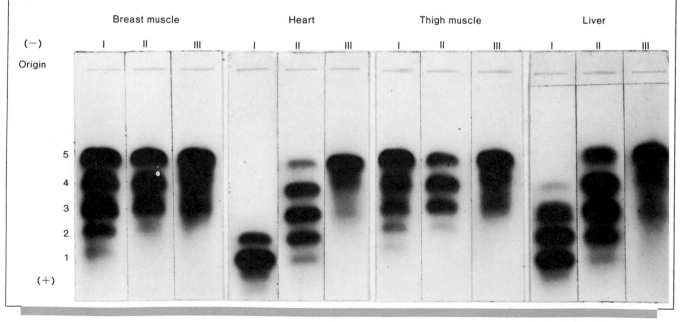

BOX 5.1—*Continued*

Electrophoresis is of clinical diagnostic value. In various diseases there is cell destruction that causes proteins to be released into the bloodstream. Thus the LDH pattern of certain cell types is found in the blood in certain disease states (Box fig. 5). Hence examination of the blood LDH is often a diagnostic test used to pick up early signs of heart and liver diseases (among others).

Box figure 4
LDH patterns found in different tissues in human beings

Box figure 5
LDH pattern of normal human serum and various disease states

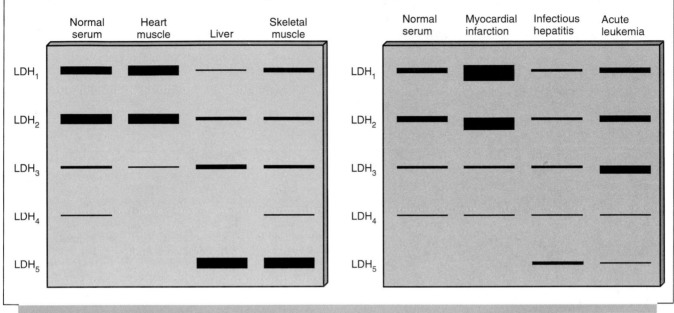

Figure 5.22
Electrophoretic gel stained for glucose-6-phosphate dehydrogenase. Lanes 1 and 2 contain blood serum from *AA* and *BB* individuals, respectively, and lane 3 contains serum from an *AB* heterozygote. Lane 4 shows the pattern expected if both the *A* and *B* alleles were active within the same cell.

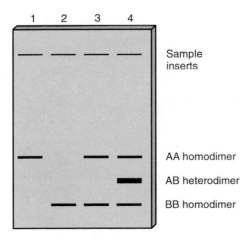

short arm of the X chromosome, several other of the six or more genes known to be active are on other places on the mammalian X chromosome, including at least two on the long arm (the pseudoautosomal region is on the short arm). Active genes on the inactive X include the gene for the enzyme steroid sulphatase, the red-cell antigen Xg^a, *MIC2* (see fig. 5.6), a *ZFY*-like gene termed *ZFX*, the gene for Kallmann syndrome, and several others. The actual mechanism of X-chromosome inactivation is unknown at the present time, although several models have been suggested that involve the role of methylation of the DNA base cytosine (see chapter 15).

Dosage Compensation for *Drosophila*

Dosage compensation occurs in fruit flies in which it also appears that gene activity of X-chromosome loci is about equal in males and females. The mechanism is different from that in mammals since no Barr bodies are found in fruit flies. The male's single X seems to be hyperactive,

Figure 5.23
Calico cat. A female heterozygous for the X-linked *yellow* and *black* alleles. The *white* underside is due to a dominant piebald spotting allele at another locus.

© *Danny Brass/Photo Researchers, Inc.*

approaching the level of activity (transcription—see chapter 10) of both of the female's X chromosomes combined. In recent genetic analysis, at least four autosomal loci have been located whose wild-type alleles seem to cause the increased gene transcription of X chromosome loci in males. Alleles of these loci that disrupt dosage compensation in males are lethal. However, they appear to have no effect in females. We thus conclude that in *Drosophila melanogaster,* dosage compensation is achieved in males by hyperactivity of the X chromosome under the control of autosomal loci. The exact mechanism by which this occurs is still unknown.

SUMMARY

Sex determination in animals is often based on chromosomal differences. In human beings and fruit flies, females are homogametic (XX) and males are heterogametic (XY). In human beings a locus on the Y chromosome, near the *ZFY* gene, determines maleness; in *Drosophila,* sex is determined by the balance between genes on the X chromosome and genes on the autosomes that regulate the state of the sex-switch gene, *Sxl.*

Since different chromosomes are normally associated with each sex, inheritance of loci located on these chromosomes shows specific, nonreciprocal patterns. The white-eye locus in *Drosophila* was the first case of a locus assigned to the X chromosome. About three hundred sex-linked loci are now known in humans.

Pedigree analysis is used in human genetic studies to determine inheritance patterns because it is impossible to carry out large-scale, controlled crosses. However, not all traits determined by the genotype are apparent in the phenotype, and this lack of penetrance can create problems in genetic analysis.

Problems of dosage compensation for loci on the X chromosome are solved in different ways in different organisms. In human beings, one of the female X chromosomes is Lyonized, or inactivated. Lyonization in women leads to cellular mosaicism for most loci on the X chromosome. In *Drosophila,* the male X chromosome is hyperactive.

SOLVED PROBLEMS

1. A yellow-bodied female fruit fly is discovered in a wild-type culture. The female is crossed with a wild-type male. In the F_1 generation, the males are yellow-bodied and the females are wild-type. When these flies are crossed among themselves, the F_2 produced are both yellow-bodied and wild-type, equally split among males and females. Explain the genetic control of this trait.

ANSWER: Since the results in the F_1 are different between the two sexes, we suspect that a sex-linked locus is responsible for the control of body color. If we assume that it is a recessive trait, then the female parent must have been a recessive homozygote and the male must have been a wild-type hemizygote. If we assign the wild-type allele as X^+, the yellow-body allele as X^y, and the Y chromosome as Y, then the figure at right, showing the crosses into the F_2 generation, is consistent with the data. Thus yellow-body in fruit flies is controlled by a recessive X-linked locus.

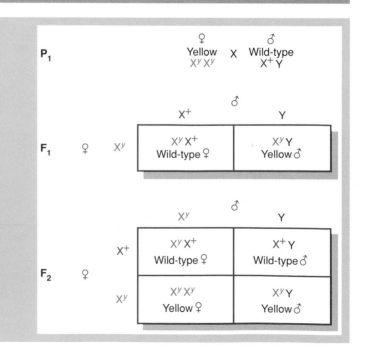

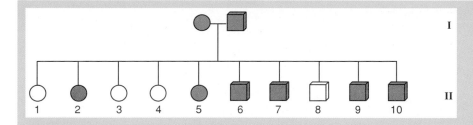

2. The affected individuals in the pedigree above are chronic alcoholics (data from the National Institute of Alcohol Abuse and Alcoholism). What can you say about the inheritance of this trait?

   ANSWER: We begin by assuming 100% penetrance. If that is the case, then we can rule out either sex-linked or autosomal recessive inheritance because both parents had the trait yet they produced some normal children. The mode of inheritance cannot be by a sex-linked dominant allele because if that were the case, then an affected male must have only affected daughters, since his daughters get copies of his single X chromosome. We are thus left with autosomal dominance as the mode of inheritance. If that is the case, then both parents must be heterozygotes, otherwise all the children would be affected. If both parents are heterozygotes, we expect a 3:1 ratio of affected to normal offspring (a cross of $Aa \times Aa$ produces offspring of A-:aa in a 3:1 ratio); here the ratio is 6:4. If we did a chi-square test, the expected numbers would be 7.5:2.5 (3/4 and 1/4, respectively, of 10). Although the expected value of 2.5 makes it inappropriate to do a chi-square test (too small), we can see that the observed and expected numbers are very close. Thus, from the pedigree we would conclude that chronic alcoholism is controlled by an autosomal dominant allele. (Although the analysis is consistent, we actually cannot draw that conclusion about alcoholism because other pedigrees are not consistent with 100% penetrance, a one-gene model, or the lack of environmental influences. In fact, it is currently being debated whether alcoholism is inherited at all. These types of problems related to complex human traits will be discussed in chapter 18).

3. A female fly with orange eyes is crossed to a male fly with short wings. The F_1 females have wild-type (red) eyes and long wings; the F_1 males have orange eyes and long wings. The F_1 flies are crossed to yield

47 long, red

45 long, orange

17 short, red

14 short, orange

with no differences between sexes. What is the genetic basis of each trait?

ANSWER: In the F_1 flies we see a difference in eye color between the sexes, indicating some type of sex linkage. Since the females are wild-type, wild-type must be dominant to orange. We can thus diagram the cross for eye color as:

X^0X^0	X	X^+Y	P_1
	$\downarrow$		
X^+X^0		X^0Y	F_1
red		orange	
	$\downarrow$		
X^+X^0 X^0X^0	X^+Y	X^0Y	F_2
red orange	red	orange	

We expect to see equal numbers of red and orange males and females, which is observed. Now look at long versus short wings. If wing length were sex linked, the F_1 males should be X^LY, and then all F_2 females should have long wings. Since this result is not seen, wing length must be autosomal. If this is true, we expect a 3 long:1 short ratio in both sexes of the F_2, and this is what is seen. Orange eye is an X-linked recessive trait and short wings is an autosomal recessive trait. A 3:3:1:1 ratio may suggest two genes, one autosomal and one X-linked.

1. In *Drosophila,* the lozenge phenotype, caused by a sex-linked recessive allele (*lz*), is of narrow eyes. Diagram, to the F_2 generation, a cross of a lozenge male and a homozygous normal female. Diagram the reciprocal cross.

2. What is the difference between an X and a Z chromosome?

3. Sex linkage was originally detected in 1906 in moths with a ZW sex-determining mechanism. In the currant moth, a pale color (*p*) is recessive to the wild-type and located on the Z chromosome. Diagram reciprocal crosses to the F_2 generation in these moths.

4. Under what circumstances of hemophilia in family members would you exempt a newborn male baby from circumcision?

5. The electrophoretic gel below shows activity for a particular enzyme. Slot 1 is a sample from a "fast" homozygote. Slot 2 is a sample from a "slow" homozygote. In slot 3 the blood from the first two was mixed. Slot 4 comes from one of the children of the two homozygotes.

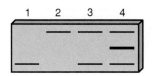

Can you guess the structure of the enzyme? If this were an X-chromosome trait, what pattern would you expect from a heterozygous female's
   a. whole blood?
   b. individual cells?

6. How many different zones of activity (bands) would you see on a gel stained for lactate dehydrogenase (LDH) activity from a person homozygous for the gene for the A protein but heterozygous for the B protein gene? Are the bands due to the activity of allozymes or isozymes?

7. What is the difference between penetrance and expressivity?

8. How many Barr bodies would you see in the nuclei of persons with the following sex chromosomes?

   a. X0   e. XXX
   b. XX   f. XXXXX
   c. XY   g. XX/XY mosaic
   d. XXY

   What would the sexes of these persons be? If these were the sex chromosomes of individual *Drosophila* that were diploid for all other chromosomes, what would their sexes be?

9. What are the possible modes of inheritance in the pedigrees below (*a–c*)? What modes of inheritance are not possible for a given pedigree?

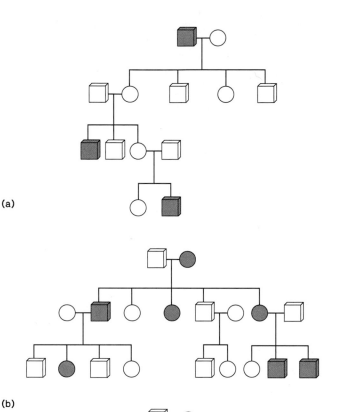

(a)

(b)

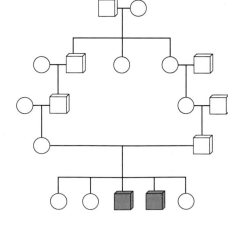

(c)

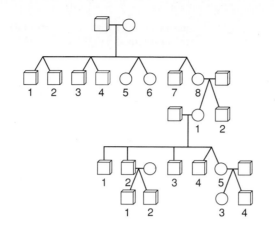

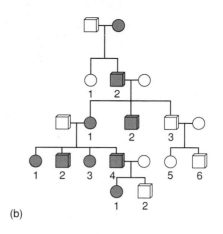

(a)

(b)

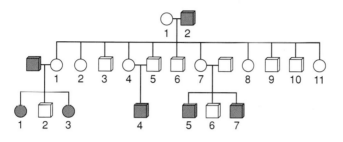

(c)

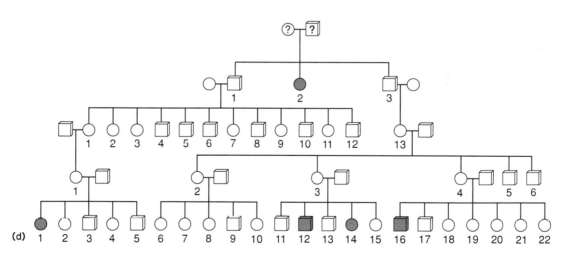

(d)

10. In the pedigrees of the rare human traits (a–d, above), including twin production, determine which modes of inheritance are most probable, possible, or impossible.

11. What is the difference between pseudodominance and phenocopy?

12. Hairy ears, a human trait expressed as excessive hair on the rims of ears in men, shows reduced penetrance (less than 100% penetrant). Mechanisms proposed include Y-linkage, autosomal dominance, and autosomal recessiveness. Construct a pedigree consistent with each of these mechanisms.

13. A female *Drosophila* heterozygous for the transformer allele (*tra*) is mated to a normal male homozygous for transformer. What is the sex ratio of their offspring? What is the sex ratio of their offspring's offspring?

14. Two fruit flies, both heterozygous for the *doublesex* (*dsx*) allele, are mated. What are the sexes of their offspring?

15. What is a sex switch? What genes serve as sex switches in human beings and *Drosophila*?

16. What does the fact that calico cats have large patches of colored fur indicate about the age of onset of Lyonization (early or late)? Tortoiseshell cats have very small color patches. Explain the difference between the two phenotypes.

17. Construct pedigrees for traits that *could not be*

    a. autosomal recessive.
    b. autosomal dominant.
    c. sex-linked recessive.
    d. sex-linked dominant.

18. In *Drosophila,* cut wings are controlled by a recessive sex-linked allele (ct), and fuzzy body is controlled by a recessive autosomal allele (fy). A fuzzy female is mated with a cut male and all the F_1 are wild-type. What are the proportions of F_2 phenotypes, by sex?

19. Consider the following crosses in canaries:

Parents	Progeny
1. pink-eyed female × pink-eyed male	all pink-eyed
2. pink-eyed female × black-eyed male	all black-eyed
3. black-eyed female × pink-eyed male	all females pink-eyed all males black-eyed

    Explain these results by determining which allele is dominant and how eye color is inherited.

20. Consider the following crosses involving yellow and gray true-breeding *Drosophila*:

Cross	F_1	F_2
1. gray female × yellow male	all males gray all females gray	97 gray females 42 yellow males 48 gray males
2. yellow female × gray male	all females gray all males yellow	?

    a. Is color autosomal or X-linked? Explain.
    b. Which allele, gray or yellow, is dominant? Explain.
    c. Assume 100 F_2 offspring are produced in cross 2. What kinds and what numbers of progeny do you expect? List males and females separately.

21. You discover a man with brown teeth. He marries a woman with white teeth. They have four daughters, all with brown teeth, and three sons, all with white teeth. The sons all marry women with white teeth, and all their children have white teeth. One of the daughters (A) marries a man with white teeth (B), and they have two brown-toothed daughters, one white-toothed daughter, one brown-toothed son, and one white-toothed son.

    a. Explain these observations.
    b. Based on your answer to (a), what is the chance that the next child of the A-B couple will have brown teeth?

22. In human beings, red-green color blindness is inherited as an X-linked recessive trait. A woman with normal vision, whose father was color-blind, marries a man with normal vision, whose father was also color-blind. This couple has a color-blind daughter with a normal complement of chromosomes. Is infidelity suspected? Explain.

23. A white-eyed male fly is mated with a pink-eyed female. All the F_1 offspring have wild-type red eyes. F_1 individuals are selfed to yield:

Females		Males	
red-eyed	450	red-eyed	231
pink-eyed	155	white-eyed	301
		pink-eyed	70

    Provide a genetic explanation for the results.

24. In *Drosophila, white eye* is an X-linked recessive trait, and *ebony* body is an autosomal recessive trait. A homozygous white-eyed female is crossed with a homozygous ebony male.
    a. What phenotypic ratio do you expect in the F_1 generation?
    b. What phenotypic ratio do you expect in the F_2 generation?
    c. Suppose the initial cross was reversed: ebony female × white-eyed male. What phenotypic ratio would you expect in the F_2 generation?

25. In *Drosophila,* abnormal eyes can result from mutations in many different genes. A true-breeding wild-type male is mated with three different females, each with abnormal eyes. The results of these crosses are given below.

	Females	Males
male × abnormal-1	all normal	all normal
male × abnormal-2	1/2 normal, 1/2 abnormal	1/2 normal, 1/2 abnormal
male × abnormal-3	all abnormal	all abnormal

    Explain the results by providing the mode of inheritance for each abnormal trait.

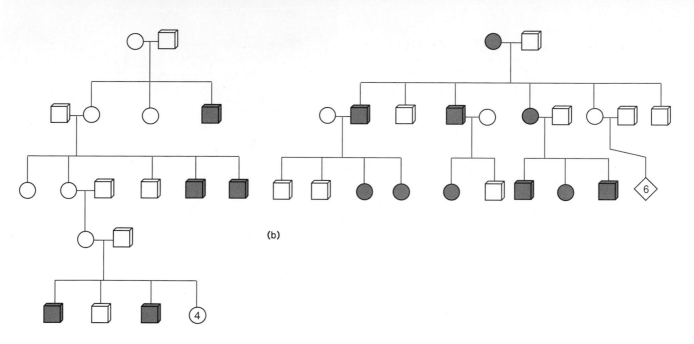

(b)

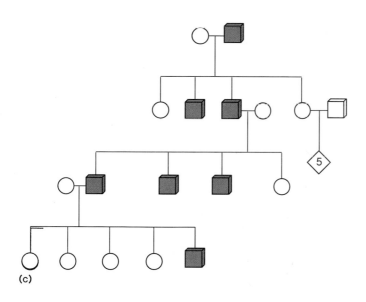

(a)

(c)

26. For the pedigrees (a–c), determine the possible modes of inheritance for each trait.

27. A black and orange female cat is crossed with a black male, and the progeny arc:

females: 2 black, 3 orange and black
males: 2 black, 2 orange

Explain the results.

28. Based on the following *Drosophila* crosses, explain the genetic basis for each trait and determine the genotypes of all individuals:

white-eyed, dark-bodied female × red-eyed, tan-bodied male

F_1: females: all red, tan
 males: all white, tan
F_2: 27 red, tan
 24 white, tan
 9 red, dark
 7 white, dark

(No differences between males and females in the F_2 generation.)

Suggested Readings for chapter 5 are on page 642.

6

Linkage and Mapping in Eukaryotes

Common baker's yeast, *Saccharomyces cerevisiae* (scanning electron micrograph, 430×).

© *Dr. Jeremy Burgess/Science Photo Library/Science Source/Photo Researchers, Inc.*

Each chromosome of an organism contains many genes. If loci were locked together permanently on a chromosome, allelic combinations would always segregate together. However, at meiosis, crossing over between loci allows the alleles of these loci to show some measure of independence. Crossing over between loci can be used by the geneticist as a tool to determine how close one locus actually is to another on a chromosome, and thus to map an entire chromosome and eventually the entire **genome** (genetic complement) of an organism.

Loci carried on the same chromosome are said to be linked to each other. There are as many **linkage groups** as there are chromosomes in the haploid set. *Drosophila* has four linkage groups ($2n = 8$, $n = 4$), whereas human beings have twenty-four linkage groups ($2n = 46$, $n = 22 + X + Y$). Prokaryotes and viruses, which usually have a single chromosome, will be treated in chapter 7.

DIPLOID MAPPING

Two-Point Cross

In *Drosophila,* the recessive band gene (*bn*) causes a dark transverse band to be on the thorax, and the detached gene (*det*) causes the crossveins of the wings to be either detached or absent (fig. 6.1). A banded fly was crossed with a detached fly to produce wild-type, dihybrid offspring in the F_1 generation. F_1 females were then testcrossed to banded, detached males (fig. 6.2). (There is no crossing over in male fruit flies; in experiments designed to detect linkage, heterozygous females—in which crossing over will occur—are usually crossed with homozygous recessive males.) If the loci were assorting independently, a 1:1:1:1 ratio of the four possible phenotypes would be expected. However, of the first one thousand offspring examined, a ratio of 2:483:512:3 was recorded.

Figure 6.1
Wild-type (+) and detached (*det*) crossveins in *Drosophila*

Wild-type   Detached

Figure 6.2
Testcrossing a dihybrid *Drosophila*

	Banded	×	Detached
P_1	*bn bn det$^+$ det$^+$* (homozygous for the *bn* allele and the *dct$^+$* allele)		*bn$^+$bn$^+$ det det* (homozygous for the *bn$^+$* allele and the *det* allele)

Testcross

Wild-type		Banded, detached
bn$^+$bn det$^+$det	×	*bn bn det det*
♀		♂

♀

		bn det	*bn det$^+$*	*bn$^+$det*	*bn$^+$det$^+$*
♂	*bn det*	*bn bn det det*	*bn bn det$^+$det*	*bn$^+$bn det det*	*bn$^+$bn det$^+$det*
Phenotype		Banded, detached	Banded	Detached	Wild-type
Number		2	483	512	3

Figure 6.3

Chromosomal arrangement of the two loci in the crosses of figure 6.2. A line arbitrarily represents the chromosomes on which these loci are actually situated.

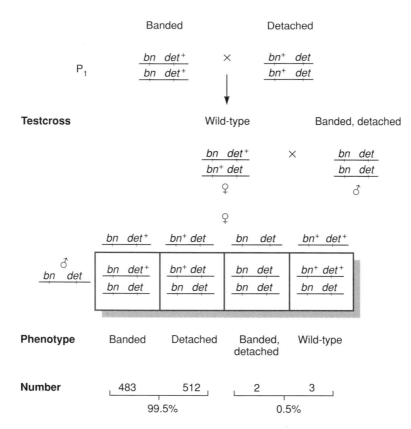

Several points emerge from the data in figure 6.2. First, no simple ratio is apparent. If we divide through by 2, we get a ratio of 1:241:256:1.5. Although the first and last categories seem to be about equal, as do the middle two categories, there appears to be no simple numerical relation between the middle and end categories. Second, the two categories in very high frequency have the same phenotypes as the original parents in the cross (P_1 of fig. 6.2). That is, banded flies and detached flies were the original parents as well as the testcross offspring in very high frequency. We call these phenotypic categories **parentals,** or **nonrecombinants.** The testcross offspring in low frequency combine the phenotypes of the two original parents (P_1). These two categories are referred to as **nonparentals,** or **recombinants.** The simplest explanation for these results is that the banded and detached loci are located near each other on the same chromosome (linkage group), and therefore associated alleles move together during meiosis.

The original cross can be analyzed by drawing the loci as points on a chromosome. This is done in figure 6.3, which shows that 99.5% of the testcross offspring (the nonrecombinants) come about through the simple linkage of the two loci. The remaining 0.5% (the recombinants) must have arisen through a crossover of homologues, from a

chiasma at meiosis, between the two loci (fig. 6.4). Note that since it is not possible to tell from these crosses which chromosome the loci actually are on or where the centromere is in relation to the loci, the centromeres are not included in the figures. The crossover event is viewed as a breakage and reunion of two chromatids lying adjacent to each other during prophase I of meiosis. Later in this chapter we will find cytological proof that this is correct; in chapter 16 we will explore the molecular mechanisms of this breakage and reunion phenomenon.

From the testcross of figure 6.2, we see that 99.5% of the gametes produced by the dihybrid are nonrecombinant, whereas only 0.5% are recombinant. This very small frequency of recombinant offspring indicates that the two loci lie very close to each other on their particular chromosome. In fact, we can use recombination percentage as an estimate of distance between loci on a chromosome: 1% recombinant offspring is referred to as one **map unit** (or one **centimorgan** in honor of the geneticist T. H. Morgan, the first geneticist to win the Nobel Prize). Although a map unit is not exactly a physical distance along a chromosome, it does provide a relative distance and thereby makes it possible to know the order and relative separation of loci on a chromosome. In this case the two loci are 0.5 map units apart.

Figure 6.4

Crossover of homologues during meiosis between the *bn* and *det* loci in the tetrad of the dihybrid female

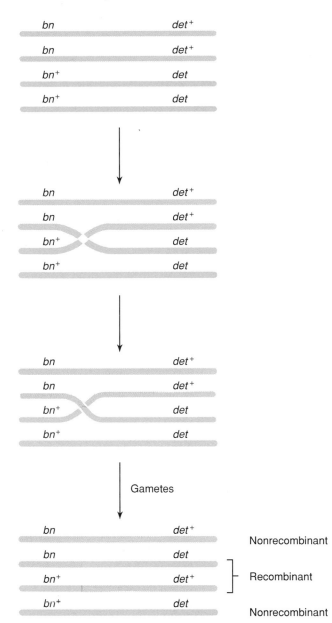

Figure 6.5

Trans (repulsion) and *cis* (coupling) arrangements of dihybrid chromosomes

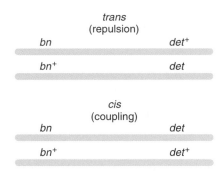

more importantly, analyze the effects of multiple crossovers on map distances, which cannot be detected in a two-point cross. That is, double crossovers between two loci will appear as if no crossovers took place, causing us to underestimate map distances. Thus we need a third locus, between the first two, to detect these events.

Three-Point Cross

Analysis of three loci, each segregating two alleles, is referred to as a **three-point cross.** We will examine wing morphology, body color, and eye color in *Drosophila*. Black body (*b*), purple eyes (*pr*), and curved wings (*c*) are all recessive traits. Since the most efficient way of studying linkage is with the testcross of a multihybrid, we will study these three loci by means of the crosses shown in figure 6.6. A point should be clarified in this figure. Since the organisms are diploid, they have two alleles at each locus. Various ways are used by geneticists to present this situation. For example, the recessive homozygote can be pictured as

(a) *bb prpr cc*

(b) *b/b pr/pr c/c*   or $\dfrac{b}{b}\ \dfrac{pr}{pr}\ \dfrac{c}{c}$

(c) *b pr c/b pr c*   or $\dfrac{b\ pr\ c}{b\ pr\ c}$

A slash or a rule line is used to separate alleles on homologous chromosomes. Thus (a) is used tentatively when we do not know the linkage arrangement of the loci, (b) would be used to indicate that the three loci are on different chromosomes, and (c) would indicate that all three loci are on the same chromosome. We will start by using the notation in (a) in this example and switch to the appropriate notation as we gain more information.

In figure 6.6, the trihybrid organism is testcrossed. If there were independent assortment, the eight types of gametes produced by the trihybrid should appear with equal frequencies, and thus the eight phenotypic classes would

The arrangement of the *bn* and *det* alleles in the dihybrid of figure 6.3 is termed the **trans** configuration, meaning "across," because the two mutants are across from each other as are the two wild-type alleles. The alternative arrangement, in which one chromosome carries both mutants and the other chromosome carries both wild-type alleles (fig. 6.5), is referred to as the **cis** configuration. (Two other terms, **repulsion** and **coupling,** have the same meanings as *trans* and *cis,* respectively.)

A cross involving two loci is usually referred to as a **two-point cross;** it is a powerful tool for dissecting the makeup of a chromosome. The next step in our analysis is to look at three loci simultaneously so that we can determine relative order of the loci on the chromosome and,

Figure 6.6
Alternative results in the testcross progeny of the *b–pr–c* trihybrid

P₁

Black, purple, curved $\times$ Wild-type
b b prpr c c   *b⁺b⁺ pr⁺pr⁺ c⁺c⁺*

Testcross the trihybrid   Wild-type (trihybrid) $\times$ Black, purple, curved
b⁺b pr⁺pr c⁺c   *b b prpr c c*

If unlinked
1/8 *b b prpr c c*
1/8 *b b prpr c⁺c*
1/8 *b b pr⁺pr c c*
1/8 *b b pr⁺pr c⁺c*
1/8 *b⁺b prpr c c*
1/8 *b⁺b prpr c⁺c*
1/8 *b⁺b pr⁺pr c c*
1/8 *b⁺b pr⁺pr c⁺c*

If completely linked
1/2 *b b prpr c c*
1/2 *b⁺b pr⁺pr c⁺c*

TABLE 6.1   Results of Testcrossing Female *Drosophila* Heterozygous for Black Body Color, Purple Eye Color, and Curved Wings (*b⁺b pr⁺pr c⁺c* $\times$ *bb prpr cc*)

Phenotype	Genotype	Number	Alleles from Trihybrid Female	Number Recombinant between		
				b-pr	*pr-c*	*b-c*
Wild-type	*b⁺b pr ⁺pr c⁺c*	5,701	*b⁺ pr⁺ c⁺*			
Black, purple,	*bb prpr cc*	5,617	*b pr c*			
curved	*b⁺b prpr cc*	388	*b⁺ pr c*	388		388
Purple, curved	*bb pr⁺pr c⁺c*	367	*b pr⁺ c⁺*	367		367
Black	*b⁺b pr⁺pr cc*	1,412	*b⁺ pr ⁺ c*		1,412	1,412
Curved	*bb pr pr c⁺c*	1,383	*b pr c⁺*		1,383	1,383
Black, purple	*b⁺b prpr c⁺c*	60	*b⁺ pr c⁺*	60	60	
Purple	*bb pr⁺pr cc*	72	*b pr ⁺ c*	72	72	
Black, curved		15,000		887	2,927	3,550
Total				5.9	19.5	23.7
Percent						

each be 1/8 of the offspring. However, if there were *complete linkage,* in which the loci are so close together on the same chromosome that virtually no crossing over ever takes place, we would expect the trihybrid to produce only two gamete types in equal frequency and yield two phenotypic classes identical to the original parents. This would occur because under complete linkage the trihybrid would produce only two chromosome types: the *b pr c* type from one parent and the *b⁺ pr⁺ c⁺* type from the other. Crossing over between linked loci would produce eight phenotypic classes in various proportions depending on the distances between loci. The actual data are presented in table 6.1.

The data in the table are arranged in reciprocal classes. Two classes are reciprocal if between them they contain each mutant phenotype just once. Wild-type and black, purple, curved classes are reciprocal as are the purple, curved, and the black classes. Reciprocal classes occur in roughly equal numbers: 5,701–5,617, 388–367, 1,412–1,383, 60–72. As we shall see, a single meiotic recombination event produces reciprocal classes. Wild-type and black, purple, curved are the two nonrecombinant classes. The purple, curved class of 388 is grouped with the black class of 367. These two would be the products of a crossover between the *b* and the *pr* loci if we assume

Figure 6.7

Results of a crossover between the black and purple loci in *Drosophila*

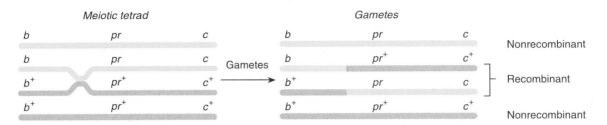

Figure 6.8

Results of a double crossover in the *b–pr–c* region of the *Drosophila* chromosome

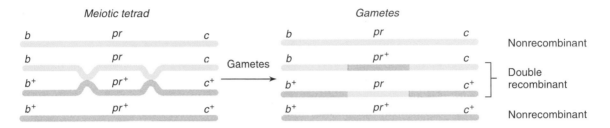

Figure 6.9

Tentative map of the black, purple, and curved chromosome in *Drosophila*. Numbers are map units.

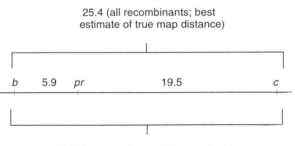

that the three loci are linked and that the gene order is *b pr c* (fig. 6.7). The next two classes, of 1,412 and 1,383 flies, would result from a crossover between *pr* and *c,* and the last set, 60 and 72, would result from two crossovers, one between *b* and *pr,* and the other between *pr* and *c* (fig. 6.8). Groupings according to these recombinant events are shown at the right of table 6.1.

In the final column of table 6.1, recombination between *b* and *c* is scored. Only those recombinant classes that have a new arrangement of *b* and *c* alleles, as compared with the parentals, are counted. This last column shows us what a *b-c,* two-point cross would have revealed had we been unaware of the *pr* locus in the middle.

Map Distances

The totals row in table 6.1 reveals that 5.9% (887/15,000) of the offspring resulted from recombination between *b* and *pr,* 19.5% between *pr* and *c,* and 23.7% between *b* and *c*. These numbers allow us to form a tentative map of the loci (fig. 6.9). There is, however, a discrepancy. The distance between *b* and *c* can be calculated in two ways. By adding the two distances, *b-pr* and *pr-c,* we get 5.9 + 19.5 = 25.4 map units; yet by directly counting the recombinants (the last column of table 6.1), we get a distance of only 23.7 map units. What causes this discrepancy of 1.7 map units?

Returning to the last column of table 6.1, we observe that the double crossovers (60 and 72) are not counted, yet each actually represents two crossovers in this region. The reason they are not counted is simply that if only the end loci of this chromosome segment were observed, the double crossovers would not be detected; the first one of the two crossovers causes a recombination between the two end loci, whereas the second one returns these outer loci to their original configuration (fig. 6.8). If we took the 3,550 recombinants between *b* and *c* and added in twice the total of the double recombinants, 264, we would get a total of 3,814. This is 25.4 map units, which is the more precise figure calculated before. As loci farther and farther apart on a chromosome are observed, more and more double crossovers occur between them. Double crossovers tend to mask recombinants, as in our example, so that distantly linked loci usually appear closer than they really

BOX 6.1
THE NOBEL PRIZE

On December tenth each year, the Nobel Prizes are awarded by the king of Sweden at the Stockholm Concert Hall. The date is the anniversary of Alfred Nobel's death. Awards are given annually in physics, chemistry, medicine and physiology, literature, economics, and peace. In 1989, each award was worth about $469,000, although they are sometimes split among two or three recipients. The prestige is priceless.

Winners of the Nobel Prize are chosen according to the will of Alfred Nobel, a wealthy Swedish inventor and industrialist, who held over three hundred patents when he died in 1896 at the age of sixty-three. Nobel developed a detonator and processes for detonation of nitroglycerine, a substance invented by the Italian chemist Ascanio Sobrero in 1847. In the form developed by Nobel, the explosive was patented as dynamite. Nobel invented several other forms of explosives. He was a benefactor of Sobrero, hiring him as a consultant and paying his wife a pension after Sobrero died.

Nobel believed that dynamite would be so destructive that it would serve as a deterrent to war. Later, realizing that this would not come to pass, he instructed that his fortune be invested and the interest used to fund the awards. The first ones were given in 1901. Each award consists of a diploma, medal, and check.

American, British, German, French, and Swedish citizens have been awarded the most prizes (Box table 1). Highlights of Nobel laureate achievements in genetics are shown in Box table 2.

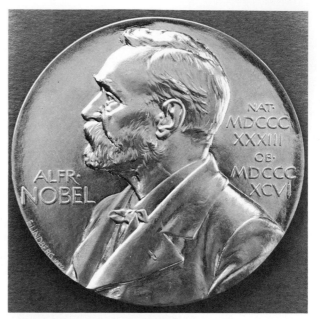

The Nobel medal. The medal is half a pound of 23-karat gold, measures about two and a half inches across, and has Nobel's face and the dates of his birth and death on the front. The diplomas that accompany the awards are individually designed.
Source: Reproduced by permission of the Nobel Foundation.

BOX TABLE 1 Distribution of Nobel Awards According to Country for the Top Five Nations with Recipients (including 1991 winners)

	Physics	Chemistry	Medicine & Physiology	Peace	Literature	Economics	Total
United States	62	35	75	19	8	19	218
Britain	20	22	24	8	8	4	86
Germany	18	26	13	4	6	0	67
France	10	7	7	8	12	1	45
Sweden	4	4	6	5	7	2	28

BOX 6.1—*Continued*

BOX TABLE 2 Some Nobel Laureates in Genetics (Physiology or Medicine; Chemistry)

Name	Year	Nationality	Cited for
Thomas Hunt Morgan	1933	USA	Discovery of the way that chromosomes govern heredity
Hermann J. Muller	1946	USA	X-ray inducement of mutations
George W. Beadle	1958	USA	Genetic regulation of biosynthetic pathways
Edward L. Tatum	1958	USA	
Joshua Lederberg	1958	USA	Bacterial genetics
Severo Ochoa	1959	USA	Discovery of enzymes that synthesize nucleic acids
Arthur Kornberg	1959	USA	
Francis H. C. Crick	1962	British	Discovery of the structure of DNA
James D. Watson	1962	USA	
Maurice Wilkins	1962	British	
François Jacob	1965	French	Regulation of enzyme biosynthesis
Andrè Lwoff	1965	French	
Jacques Monod	1965	French	
Peyton Rous	1966	USA	Tumor viruses
Robert W. Holley	1968	USA	Unraveling of the genetic code
H. Gobind Khorana	1968	USA	
Marshall W. Nirenberg	1968	USA	
Max Delbrück	1969	USA	Viral genetics
Alfred Hershey	1969	USA	
Salvador Luria	1969	USA	
Renato Dulbecco	1975	USA	Tumor viruses
Howard Temin	1975	USA	Discovery of reverse transcriptase
David Baltimore	1975	USA	
Werner Arber	1978	Swiss	Discovery of, sequencing of, and mapping with restriction endonucleases
Hamilton Smith	1978	USA	
Daniel Nathans	1978	USA	
Walter Gilbert	1980	USA	Techniques of sequencing DNA
Frederick Sanger	1980	British	
Paul Berg	1980	USA	Pioneer work in recombinant DNA
Baruj Benacerraf	1980	USA	Work on genetically determined cell surface structures that regulate immune reactions
Jean Dausset	1980	French	
George Snell	1980	USA	
Aaron Klug	1982	British	Crystallographic work on protein-nucleic acid complexes
Barbara McClintock	1983	USA	Transposable genetic elements
Cesar Milstein	1984	British/ Argentine	Immunogenetics
Georges Koehler	1984	German	
Niels K. Jerne	1984	British/ Danish	
Susumu Tonegawa	1987	Japanese	Antibody diversity
J. Michael Bishop	1989	USA	Proto-oncogenes
Harold E. Varmus	1989	USA	
Thomas R. Cech	1989	USA	Enzymatic properties of RNA
Sidney Altman	1989	Canada	

Figure 6.10
Mapping function

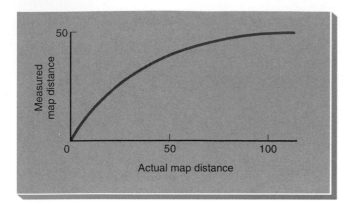

are. The most accurate map distances are those established on very closely linked loci. In other words, summed short distances are more accurate than directly measured larger distances.

The results of the previous experiment show that we can obtain at least two map distances between any two loci: measured and actual. Measured map distance between two loci is the value obtained from a two-point cross. Actual map distance is an idealized, accurate value obtained from summing short distances between many intervening loci. The short distances are obtained in crosses involving other loci between the original two. When measured map distance is plotted against actual map distance, the curve of figure 6.10 is obtained. This curve is called a **mapping function.** This graph is of both practical and theoretical value. Pragmatically, it allows us to convert a measured map distance into a more accurate one. Theoretically, it shows that measured map distance never exceeds 50 map units in any one cross. Multiple crossovers reduce the apparent distance between two loci to a maximum of 50 map units, the value that independent assortment produces (50% parentals, 50% recombinants).

Gene Order

Although the previous analysis was done assuming that *pr* was in the middle, the data of table 6.1 confirm our original assumption that the gene order is *b pr c*. Of the four pairs of reciprocal phenotypic classes in table 6.1, one pair has the highest frequency and one pair has the lowest. The pair with highest frequency is the nonrecombinant group. The one with the lowest frequency is the double recombinant group, the one in which only the middle locus has been changed from the parental arrangement. A comparison of either of the double recombinant classes with either of the nonrecombinant classes will show the gene that is in the middle and therefore the gene order. Since $b^+ pr^+ c^+$ was one of the nonrecombinant gametes and $b^+ pr c^+$ was one of the double recombinant gametes, *pr* stands out as the odd locus, or the one in the middle, since both

end loci show the same pattern as the nonrecombinant. In a similar manner, comparing $b pr^+ c$ with $b pr c$ would also point to *pr* as the inside locus (or **inside marker),** as would comparing $b^+ pr c^+$ with $b pr c$ or $b pr^+ c$ with $b^+ pr c^+$. In each case, the middle locus, *pr,* displays the different pattern whereas the allele arrangements of the outside markers, *b* and *c,* behave in concert.

From the data in table 6.1 we can confirm the association of alleles in the trihybrid parent. That is, since the data came from testcrossing a trihybrid, the allelic configuration in that trihybrid is reflected in the nonrecombinant classes of offspring. In this case, one is the result of a $b^+ pr^+ c^+$ gamete, the other, of a $b pr c$ gamete. Thus the trihybrid had the genotype $b pr c/b^+ pr^+ c^+$: The alleles were in the *cis* configuration.

Coefficient of Coincidence

The next question in our analysis of this three-point cross is, are crossovers occurring independently of each other? That is, are the observed number of double recombinants equal to the expected number? In the example, there were 132/15,000 double crossovers, or 0.88%. The expected number is based on the independent occurrence of crossing over in the two regions measured. That is, 5.9% of the time there is a crossover in the *b-pr* region, which can be expressed as a probability of occurrence of 0.059. Similarly, 19.5% of the time there is a crossover in the *pr-c* region, or a probability of occurrence of 0.195. Then a double crossover should occur as a product of the two probabilities: 0.059 X 0.195 = 0.0115, or 1.15% of the gametes (1.15% of 15,000 = 172.5) should be double recombinants. In our example, the observed number of double recombinant offspring is less than expected (132 observed, 172.5 expected), implying a **positive interference** in which the occurrence of the first crossover reduces the chance of the second. We can express this as a **coefficient of coincidence,** defined as

$$\text{coefficient of coincidence} =$$

$$\frac{\text{observed number of double recombinants}}{\text{expected number of double recombinants}}$$

In the example, the coefficient of coincidence is 132/172.5 = 0.77. In other words, only 77% of the expected double crossovers occurred. Sometimes this reduced quantity of double crossovers is measured as the degree of interference, defined as

$$\text{interference} = 1 - \text{coefficient of coincidence}$$

In our example, the interference is 23%.

It is also possible to have **negative interference,** the situation in which there are more observed double recombinants than expected. In this situation it seems that the occurrence of one crossover enhances the probability of crossovers in adjacent regions.

TABLE 6.2 Offspring from a Trihybrid ($h^+h\ ry^+ry\ th^+th$) Testcross ($x\ hh\ ryry\ thth$) in *Drosophila*

Phenotype	Genotype (order unknown)	Number
Thread	$h^+ry^+th/h\ ry\ th$	359
Rosy, thread	$h^+ry\ th/h\ ry\ th$	47
Hairy, rosy, thread	$h\ ry\ th/h\ ry\ th$	4
Hairy, thread	$h\ ry^+\ th/h\ ry\ th$	98
Rosy	$h^+\ ry\ th^+/h\ ry\ th$	92
Hairy, rosy	$h\ ry\ th^+/h\ ry\ th$	351
Wild-type	$h^+ry^+th^+/h\ ry\ th$	6
Hairy	$h\ ry^+th^+/h\ ry\ th$	43

TABLE 6.3 Data from Table 6.2 Arranged to Show Recombinant Regions

Trihybrid's Gamete	Number	h–th	th–ry	h–ry
$h^+th\ ry^+$	359			
$h\ th^+\ ry$	351			
$h\ th\ ry^+$	98	98		98
h^+th^+ry	92	92		92
$h^+th\ ry$	47		47	47
$h\ th^+ry^+$	43		43	43
$h\ th\ ry$	4	4	4	
$h^+th^+ry^+$	6	6	6	
Total	1,000	200	100	280

Another Example

Let us work out one more three-point cross, in which neither the middle gene nor the *cis-trans* relationship of the alleles in the trihybrid F_1 parent is given. On the third chromosome of *Drosophila*, hairy (h) causes extra hair on the body, thread (th) causes a thread-shaped arista (antenna tip), and rosy (ry) causes the eyes to be reddish brown. All three traits are recessive. Trihybrid females were testcrossed; the phenotypes from one thousand offspring are given in table 6.2. At this point it is possible to determine from the data what the parental genotypes were, what the gene order is, the map distances, and the coefficient of coincidence. The table presents the data in no particular order, as it would have been recorded by an experimenter. Phenotypes are tabulated and, from these, the genotypes can be reconstructed. Notice that the data can be put into the form found in table 6.1; there is a large reciprocal set (359 and 351), a small reciprocal set (4 and 6), and large and small intermediate sets (98 and 92, 47 and 43).

From the data presented, is it obvious that the three loci are linked? The pattern, as just mentioned, is identical to that of the previous example in which the three loci were linked. (What pattern would appear if two of the loci were linked and one assorted independently?) Next, what is the allele arrangement in the trihybrid parent? The offspring with the parental, or nonrecombinant, arrangements are the reciprocal pair in highest frequency. Table 6.2 shows that thread and hairy, rosy offspring are the nonrecombinants. Thus the nonrecombinant gametes of the trihybrid F_1 parent were $h\ ry\ th^+$ and $h^+\ ry^+\ th$, which is the allele arrangement of the trihybrid with the actual order still unknown—$h\ ry\ th^+/h^+\ ry^+\ th$. (What were the genotypes of the parents of this trihybrid, assuming they were homozygotes?) Continuing, which gene is in the middle? From table 6.2 we know that $h\ ry\ th$ and $h^+\ ry^+\ th^+$ are the double recombinant gametes of the trihybrid parent because they occur in such low numbers. Comparison of these chromosomes with either of the nonrecombinant chromosomes ($h^+\ ry^+$

Figure 6.11

Map of the *h–th–ry* region of the *Drosophila* chromosome, with numerical discrepancy in distances. Numbers are map units.

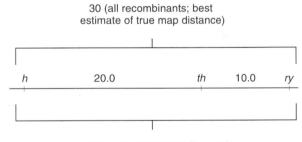

th or $h\ ry\ th^+$) will show that the thread (th) locus is in the middle. We now know that the original trihybrid had the following chromosomal composition: $h\ th^+\ ry/\ h^+\ th\ ry^+$. The h and ry alleles are in the *cis* configuration with th in the *trans* configuration.

We can now compare the chromosome from the trihybrid in each of the eight offspring categories with the parental arrangement and determine the regions that had crossovers. This is done in table 6.3. We can see that the *h-th* distance is 20 map units, the *th-ry* distance is 10 map units, and the apparent *h-ry* distance is 28 map units (fig. 6.11). As in the earlier example, the *h-ry* discrepancy is from not counting the double crossovers twice each: $280 + 2(10) = 300$, which is 30 map units and the more accurate figure. Last, we wish to know what the coefficient of coincidence is. The expected occurrence of double recombinants is $0.200 \times 0.100 = 0.020$, or 2%. Two percent of 1,000 = 20. Thus

coefficient of coincidence =

$$\frac{\text{observed number of double recombinants}}{\text{expected number of double recombinants}}$$

$$= 10/20 = 0.50$$

Only 50% of the expected double crossovers occurred.

Figure 6.12
Giant salivary gland chromosomes of *Drosophila*. X, 2, 3, and 4 are
the four nonhomologous chromosomes. *L* and *R* indicate the *left* and
right arms (in relation to the centromere).
B. P. Kaufman, "Induced Chromosome Rearrangements in Drosophila
melanogaster," Journal of Heredity 30 (1939):178–190. Reproduced by
permission.

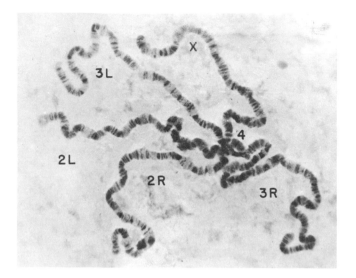

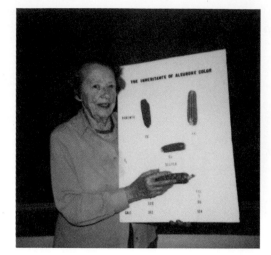

Harriet B. Creighton
Courtesy of Harriet B. Creighton.

From three-point crosses of this type, the chromosomes of many eukaryotic organisms have been mapped—
those of *Drosophila* are probably the most extensively
studied. *Drosophila* and other species of flies have giant
polytene salivary gland chromosomes, which arise as a
result of **endomitosis.** In this process the chromosomes
replicate but the cell does not divide. In the salivary gland
of the fruit fly, homologous chromosomes synapse and then
replicate to about one thousand copies, forming very thick
structures with a distinctive pattern of bands called **chromomeres** (fig. 6.12). In methods to be discussed in chapter
8, many loci have been mapped to particular bands. Current evidence indicates a correspondence between the
number of bands and the number of genes in a region. It
is possible that each chromomere (band) represents one
gene, although controversy certainly exists. The *Drosophila* chromosomal map is presented in figure 6.13.
Locate the loci we have mapped so far to verify the map
distances.

Cytological Demonstration of Crossing Over

If we are correct that a chiasma during meiosis is the visible result of a physical crossover, then we should be able
to demonstrate that genetic crossing over is accompanied
by cytological crossing over. That is, the recombination
event should entail the exchange of physical parts of homologous chromosomes. This can be demonstrated if we
can distinguish between two homologous chromosomes, a
technique first used by Creighton and McClintock in maize
(corn) and by Stern with *Drosophila*. We will look at
Creighton and McClintock's experiment.

Barbara McClintock (1902–1992)
*Courtesy of Cold Spring Harbor Research Library Archives. Photographer,
David Miklos.*

Harriet Creighton and Barbara McClintock worked
with chromosome number 9 in maize (*n = 10*). In one
strain they found a chromosome with abnormal ends. One
end had a knob and the other had an added piece of chromatin from another chromosome (fig. 6.14). This knobbed,
interchange chromosome was thus clearly different from
its normal homologue. It also carried the dominant colored (*C*) allele and the recessive waxy texture (*wx*) allele.

Figure 6.13

Partial map of the chromosomes of *Drosophila*. The centromere is marked by an *open circle*.

Source: Data from C. B. Bridges.

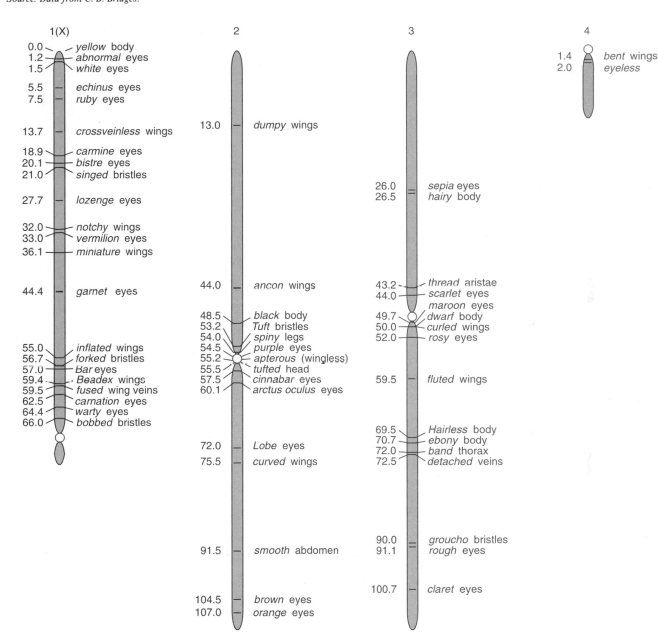

Following mapping studies in which it was established that *C* was very close to the knob and *wx* was close to the added piece of chromatin, Creighton and McClintock made the cross shown in figure 6.14. The dihybrid plant with heteromorphic chromosomes was crossed to the normal homomorphic plant (only normal chromosomes) that had the genotype of *c Wx/c wx* (colorless and nonwaxy phenotype). If a crossover occurred during meiosis in the heteromorph in the region between *C* and *wx*, there should also be a physical crossover, visible cytologically (under the microscope), in which the knob would become associated with an otherwise normal chromosome and the extra

piece of 9 would be associated with a knobless chromosome. Four types of gametes would result (fig. 6.14).

Of twenty-eight offspring examined, all were consistent with the predictions of the Punnett square in figure 6.14. Those of class 8 (lower right box) with the colored, waxy phenotype all had a knobbed interchange chromosome as well as a normal homologue. Those with the colorless, waxy phenotype (class 4) had a knobless interchange chromosome. All of the colored, nonwaxy phenotypes (class 5, 6, and 7) had a knobbed, normal chromosome, which indicated that only classes 5 and 6 were in the sample. Of the two that were tested, both were

BOX 6.2
THE FIRST CHROMOSOME MAP

The first chromosome map ever published included just six loci on the X chromosome of *Drosophila melanogaster*. It was published in 1913 by Alfred H. Sturtevant, who began the work while an undergraduate student at Columbia University, working in the "fly lab" of Thomas Hunt Morgan. The fly lab included H. J. Muller, later to win a Nobel Prize, and Calvin B. Bridges, whose work on sex determination in *Drosophila* we discussed in the last chapter.

Sturtevant worked with six loci: yellow body (y), white (w), eosin (w^e), and vermilion eyes (v), and miniature (m) and rudimentary wings (r). (White and eosin are actually allelic; Sturtevant found no crossing over between the two "loci.") Using crosses, similar to the ones we outline in this chapter, he constructed the map shown in Box figure 1. The map distances that we accept today are very similar to the ones he obtained.

Sturtevant's work was especially important at this point in the development of genetics because his data supported several basic concepts, including the linear arrangement of genes, which itself argued for the placement of genes on chromosomes, the only linear structures in the nucleus. Sturtevant also pointed out crossover interference. His summary is clear and succinct:

> It has been found possible to arrange six sex-linked factors in *Drosophila* in a linear series, using the number of crossovers per one hundred cases as an index of the distance between any two factors. This scheme gives consistent results, in the main.
>
> A source of error in predicting the strength of association between untried factors is found in double crossing over. The occurrence of this phenomenon is demonstrated, and it is shown not to occur as often as would be expected from a purely mathematical point of view, but the conditions governing its frequency are as yet not worked out.
>
> These results . . . form a new argument in favor of the chromosome view of inheritance, since they strongly indicate that the factors investigated are arranged in a linear series, at least mathematically.

Alfred H. Sturtevant (1891–1970)
Courtesy of the Archives, California Institute of Technology.

Box figure 1

The first chromosome linkage map. Five loci in *Drosophila melanogaster* are mapped to the X chromosome. The numbers in parentheses are the more accurately mapped distances recognized today. We also show today's allele designations rather than Sturtevant's original nomenclature.

Source: Data from Sturtevant, "The Linear Arrangement of Six Sex-Linked Factors in Drosophila, *as shown by Their Mode of Association" in* Journal of Experimental Zoology, *14:43–59, 1913.*

	w^e				
y	w		v	m	r
0.0	1.0		30.7	33.7	57.6
(0.0	1.5)		(33.0)	(36.1)	(54.5)

Figure 6.14
Creighton and McClintock experiment in maize

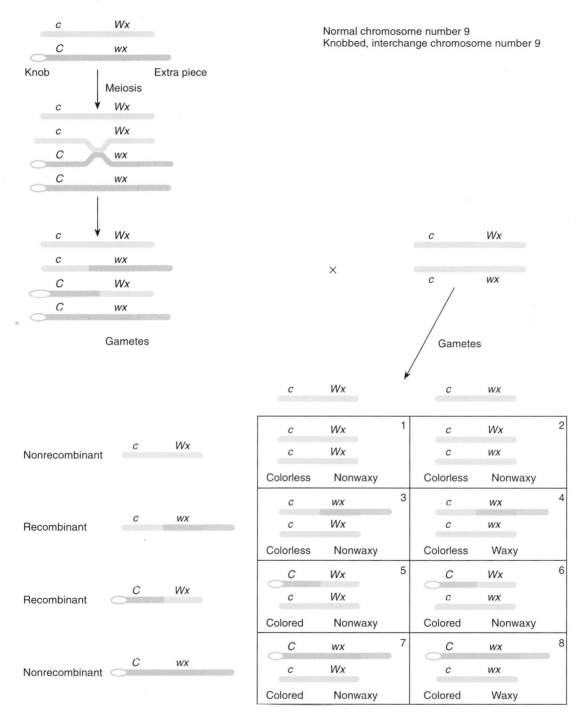

Normal chromosome number 9
Knobbed, interchange chromosome number 9

Wx Wx, indicating that they were of class 5. The remaining classes (1, 2, and 3) were of the colorless, nonwaxy phenotype. All were knobless. Of those that contained only normal chromosomes, some were *Wx Wx* (class 1) and some were heterozygotes (*Wx wx,* class 2). Of those containing interchange chromosomes, two were heterozygous, representing class 3. Two were homozygous, *Wx Wx,* yet interchange-normal heteromorphs. These represent a crossover in the region between the waxy

locus and the extra piece of chromatin. This would give a knobless-*c-Wx*-extra-piece chromosome that when combined with a *c-Wx*-normal chromosome, would give these anomalous genotypes. The sample size was not large enough to pick up the reciprocal event. Creighton and McClintock concluded: "Pairing chromosomes, heteromorphic in two regions, have been shown to exchange parts at the same time they exchange genes assigned to these regions."

Figure 6.15
Life cycle of *Neurospora*. *A* and *a* are mating types; *n* is a haploid stage; 2*n* is diploid.

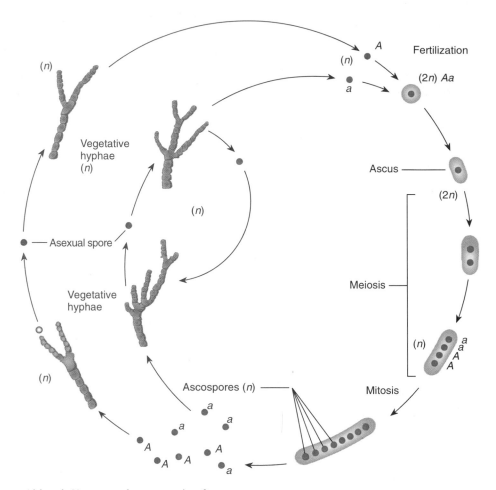

Figure 6.16
Meiosis in *Neurospora*. Although *Neurospora* has seven pairs of chromosomes at meiosis, only one pair is shown. *A* and *a*, the two mating types, represent the two centromeres of the tetrad.

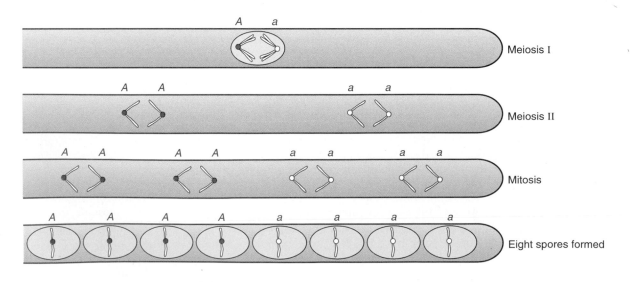

HAPLOID MAPPING (TETRAD ANALYSIS)

For *Drosophila* and other diploid eukaryotes, genetic analysis such as that considered earlier in this chapter is referred to as **random-strand analysis.** Sperm, each of which carry only one chromatid of a meiotic tetrad, unite with eggs that also carry only one chromatid from a tetrad. Thus, zygotes are a result of the random uniting of chromatids.

Fungi of the class Ascomycetes retain the four haploid products of meiosis in a sac called an **ascus.** These organisms, which usually exist as haploids, provide a unique opportunity to look at the total products of meiosis in a tetrad. Different techniques are used for these analyses. We will look at two fungi, the pink bread mold, *Neurospora crassa,* and the common baker's yeast, *Saccharomyces cerevisiae,* both of which retain the products of meiosis as **ascospores.** *Neurospora* has ordered spores whereas yeast does not. *Neurospora's* life cycle is shown in figure 6.15. Fertilization takes place within an immature fruiting body after a spore or filament of one **mating type** contacts a special filament extending from the fruiting body of the opposite mating type. The zygote nucleus undergoes meiosis without any intervening mitosis. (Mating types are generally the result of a one-locus, two-allele genetic system that determines that only opposite types can mate. We discuss this system in more detail in chapter 15.)

Ordered Spores

Since the *Neurospora* ascus is narrow, the meiotic spindle is forced to lie along the cell's long axis. The two nuclei then undergo the second meiotic division, which is also oriented along the long axis of the ascus. The result is that the spores are ordered according to their centromeres (fig. 6.16). That is, if we label one centromere *A* and the other *a* for the two mating types, a tetrad at meiosis I will consist of one *A* and one *a* centromere. At the end of meiosis in *Neurospora,* the four ascospores are in the order *A A a a* or *a a A A* in regard to centromeres. (We talk more simply of centromeres rather than chromosomes or chromatids because of the complications that crossing over adds. A type *A* centromere is always a type *A* centromere whereas, due to crossing over, a chromosome attached to that centromere may be part from the type *A* parent and part from the type *a* parent.)

Before maturation of the ascospores in *Neurospora,* a mitosis takes place in each nucleus so that four pairs of spores (eight) rather than just four spores are formed. With the exception of phenomena such as mutation or gene conversion, to be discussed later in the book, pairs will always be identical (fig. 6.16). As we will see in a moment, by the nature of ordered spores, loci in *Neurospora* can be mapped in relation to their centromeres.

Figure 6.17
Technique of spore isolation in *Neurospora*

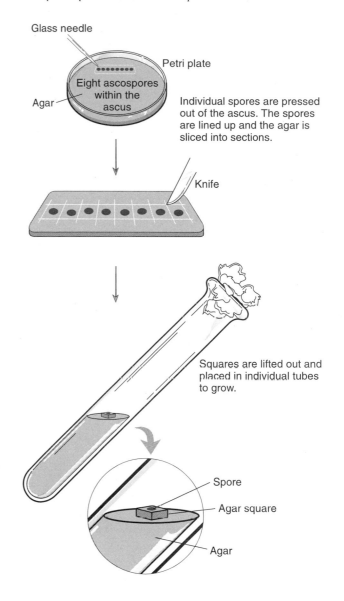

Glass needle

Petri plate

Eight ascospores within the ascus

Agar

Individual spores are pressed out of the ascus. The spores are lined up and the agar is sliced into sections.

Knife

Squares are lifted out and placed in individual tubes to grow.

Spore

Agar square

Agar

Phenotypes of *Neurospora*

At this point you might wonder what phenotypes are expressed by *Neurospora.* In general, microorganisms have phenotypes that fall into three broad categories: colony morphology, drug resistance, and nutritional requirements. Many microorganisms can be cultured in petri plates or test tubes that contain a supporting medium such as agar, to which various components can be added (fig. 6.17). Wild-type *Neurospora* is the familiar pink bread mold. Various mutations exist that change the colony morphology. For example, fluffy (*fl*), tuft (*tu*), dirty (*dir*), and colonial (*col4*) are all mutants of the basic growth form. Also, wild-type *Neurospora* is sensitive to the sulfa drug sulfonamide, whereas one of the mutants (*Sfo*) actually requires sulfonamide in order to survive and grow.

Figure 6.18
Isolation of nutritional-requirement mutants in *Neurospora*.
Mutations are induced in mating type *a* by irradiation.

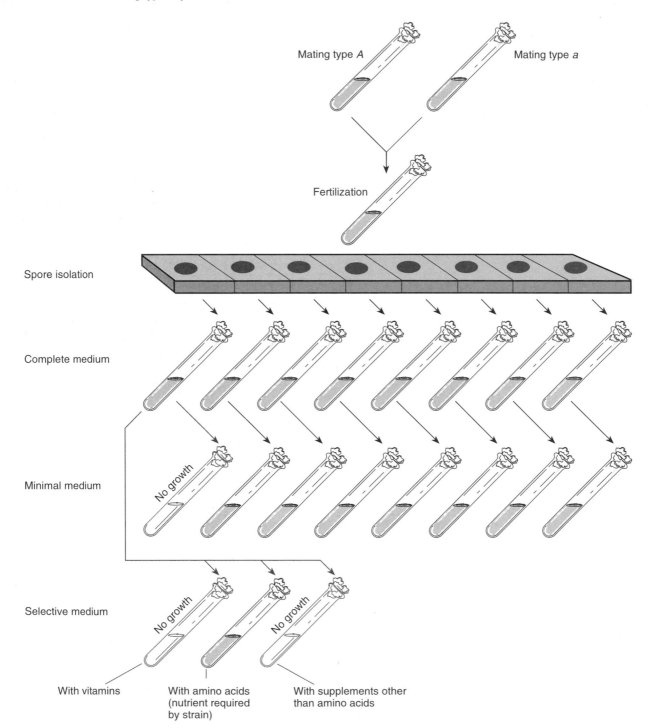

Mating type *A*

Mating type *a*

Fertilization

Spore isolation

Complete medium

Minimal medium

No growth

Selective medium

No growth

No growth

With vitamins

With amino acids
(nutrient required
by strain)

With supplements other
than amino acids

Nutritional-requirement phenotypes have provided great insight not only into genetic analysis but also into biochemical pathways of metabolism as mentioned in chapter 2. Wild-type *Neurospora*—that are haploid, remember—can grow on a medium containing only sugar, a nitrogen source, some organic acids and salts, and the vitamin biotin. This is referred to as **minimal medium.**

However, there are several different mutant types, or mutant strains, of *Neurospora* that cannot grow on this minimal medium until some essential nutrient is added. For example, one mutant strain will not grow on minimal medium but will grow if one of the amino acids, arginine, is added (fig. 6.18). From this we can infer that the wild-type, +, has a normal, functional enzyme in the synthetic

Figure 6.19

Arginine biosynthetic pathway of *Neurospora*

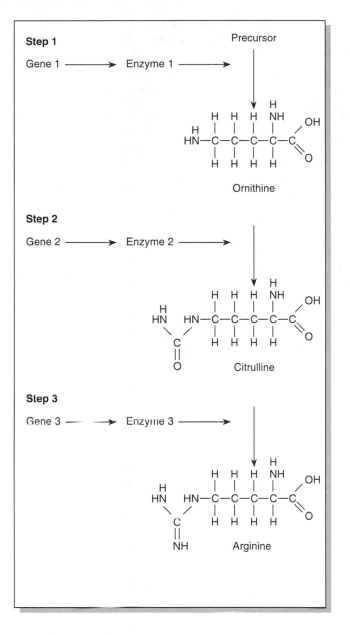

Figure 6.20

The six possible ascospore patterns in *Neurospora*

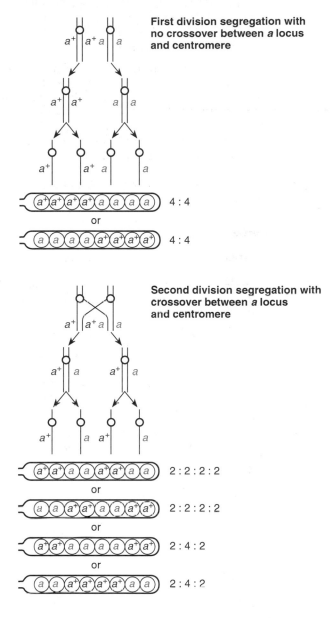

pathway of arginine. The arginine-requiring mutant has an allele that specifies an enzyme that is incapable of converting one of the intermediates in the pathway directly into arginine or into one of the precursors to arginine. We can see that if the synthetic pathway is long, there may be many different loci with alleles that cause the strain to require arginine (fig. 6.19). This, in fact, happens and the different loci are usually named arg_1, arg_2, and so on. There are numerous biosynthetic pathways in *Neurospora,* and mutants exhibit many different nutritional requirements. Mutants can be induced experimentally by irradiation. These, then, are the tools with which we analyze and map chromosomes of *Neurospora*. These techniques will be expanded in the next chapter.

First and Second Division Segregation

Recall that there is a 4:4 segregation of the centromeres in the ascus of *Neurospora*. Two kinds of patterns will be found among the loci on these chromosomes. These patterns depend on whether or not there was a crossover between the locus and its centromere (fig. 6.20). If there was no crossover between the locus and its centromere, the allelic pattern will be the same as the centromeric pattern, which is referred to as **first division segregation** (**FDS**), because alleles are separated from each other at meiosis I. If, however, there has been a crossover between the locus and its centromere, patterns of a different type will emerge (2:4:2 or 2:2:2:2), each of which is referred to as **second division segregation** (**SDS**) pattern. Because the spores are

TABLE 6.4 Genetic Patterns Following Meiosis in an a^+a Heterozygous *Neurospora* (Ten Asci Are Examined)

Spore Number	Ascus Number									
	1	*2*	*3*	*4*	*5*	*6*	*7*	*8*	*9*	*10*
1	a	a	a^+	a	a	a^+	a	a^+	a^+	a^+
2	a	a	a^+	a	a	a^+	a	a^+	a^+	a^+
3	a	a	a^+	a^+	a^+	a^+	a	a	a	a^+
4	a	a	a^+	a^+	a^+	a^+	a	a	a	a^+
5	a^+	a^+	a	a^+	a	a	a^+	a	a^+	a
6	a^+	a^+	a	a^+	a	a	a^+	a	a^+	a
7	a^+	a^+	a	a	a^+	a	a^+	a^+	a	a
8	a^+	a^+	a	a	a^+	a	a^+	a^+	a	a
	FDS	FDS	FDS	SDS	SDS	FDS	FDS	SDS	SDS	FDS

Map distance (*a* locus to centromere) = (1/2)% SDS
= (1/2) 40%
= 20 map units

ordered, the centromeres always follow an FDS pattern. Hence we should be able to map the distance of a locus to its centromere. Under the simplest circumstances (fig. 6.20), every SDS configuration has four recombinant and four nonrecombinant chromatids (spores). Thus half of the chromatids (spores) in an SDS ascus are recombinant. Therefore, remembering that 1% recombinant chromatids equals 1 map unit:

$$\text{map distance} = \frac{(1/2)\text{SDS asci}}{\text{total asci}} \times 100$$

An example using this calculation is given in table 6.4.

Three-point crosses in *Neurospora* can also be done. Let us map two loci and their centromere. For simplicity, we will use the *a* and *b* loci. Dihybrids are formed from fused mycelia ($a\ b \times a^+\ b^+$), which then undergo meiosis. One thousand asci are analyzed, keeping the spore order intact. Before presenting the data, we should consider the way in which they will be grouped. Given that each locus can show six different patterns (fig. 6.20), two loci scored together should give thirty-six possible spore arrangements (6 × 6). Some thought, however, will tell us that many of these patterns are really random variants of each other. The tetrad in meiosis is a three-dimensional entity rather than a flat, four-rod object as it is usually drawn. At the first meiotic division, either centromere can go to the left *or* the right, and when centromeres split at the second meiotic division, movement within the future half-ascus (the four spores to the left or the four spores to the right) is also random. Thus one genetic event can produce up to eight "different" patterns. For example, consider the arrangements shown in figure 6.21 in which a crossover occurs between the *a* and *b* loci. All eight arrangements, producing the asci patterns of table 6.5, are equally likely. The thirty-six possible patterns then reduce to only the seven unique patterns shown in table 6.6.

Figure 6.21
The eight random arrangements possible (table 6.5) when a single crossover occurs between the *a* and *b* loci in *Neurospora*. Circular *arrows* represent rotation of a centromere with respect to the original configuration.

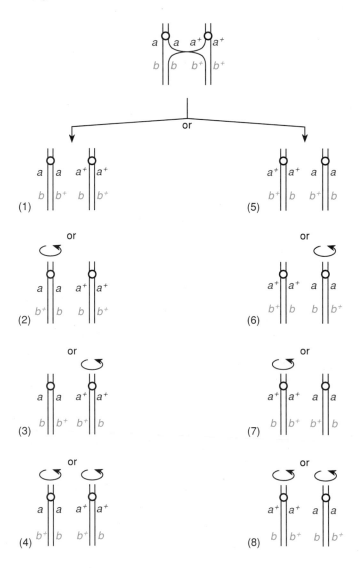

TABLE 6.5 Eight of the Thirty-Six Possible Spore Patterns in *Neurospora* Scored for Two Loci, *a* and *b* (These Are All Random Variants of the Same Genetic Event)

Spore Number	Ascus Number							
	1	*2*	*3*	*4*	*5*	*6*	*7*	*8*
1	a b	a b⁺	a b	a b⁺	a⁺b⁺	a⁺b⁺	a⁺ b	a⁺ b
2	a b	a b⁺	a b	a b⁺	a⁺b⁺	a⁺b⁺	a⁺ b	a⁺ b
3	a b⁺	a b	a b⁺	a b	a⁺ b	a⁺ b	a⁺b⁺	a⁺b⁺
4	a b⁺	a b	a b⁺	a b	a⁺ b	a⁺ b	a⁺b⁺	a⁺b⁺
5	a⁺ b	a⁺ b	a⁺b⁺	a⁺b⁺	a b⁺	a b	a b⁺	a b
6	a⁺ b	a⁺ b	a⁺b⁺	a⁺b⁺	a b⁺	a b	a b⁺	a b
7	a⁺b⁺	a⁺b⁺	a⁺ b	a⁺ b	a b	a b⁺	a b	a b⁺
8	a⁺b⁺	a⁺b⁺	a⁺ b	a⁺ b	a b	a b⁺	a b	a b⁺

TABLE 6.6 The Seven Unique Classes of Asci Resulting from Meiosis in a Dihybrid *Neurospora*, *a b/a⁺b⁺*

Spore Number	Ascus Number						
	1	*2*	*3*	*4*	*5*	*6*	*7*
1	a b	a b⁺	a b	a b	a b	a b⁺	a b
2	a b	a b⁺	a b	a b	a b	a b⁺	a b
3	a b	a b⁺	a b⁺	a⁺ b	a⁺b⁺	a⁺ b	a⁺b⁺
4	a b	a b⁺	a b⁺	a⁺ b	a⁺b⁺	a⁺ b	a⁺b⁺
5	a⁺b⁺	a⁺ b	a⁺b⁺	a⁺b⁺	a⁺b⁺	a⁺ b	a⁺ b
6	a⁺b⁺	a⁺ b	a⁺b⁺	a⁺b⁺	a⁺b⁺	a⁺ b	a⁺ b
7	a⁺b⁺	a⁺ b	a⁺ b	a b⁺	a b	a b⁺	a b⁺
8	a⁺b⁺	a⁺ b	a⁺ b	a b⁺	a b	a b⁺	a b⁺
	729	2	101	9	150	1	8
SDS for *a* locus				9	150	1	8
SDS for *b* locus			101		150	1	8

Figure 6.22

Two possible arrangements of the *a* and *b* loci and their centromere. Distances are in map units.

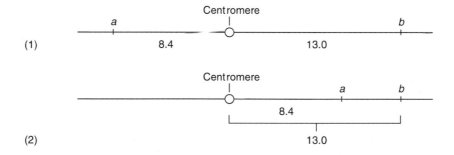

Gene Order

We can now determine the distance from each locus to its centromere and the linkage arrangement of the loci if they are both linked to the same centromere. The procedure is as follows. We can establish by inspection that the two loci are linked to each other and therefore to the same centromere. This is done by examining classes 1 and 2 in table 6.6. If the two loci are unlinked, these two categories would represent two equally likely alternative events when no crossover takes place. Since category 1 is almost 75% of all the asci, we can be sure the two loci are linked. (A convincing way to demonstrate this is to draw a meiosis with two tetrads to show that 1 and 2 should be equally likely.)

To determine the distance of each locus to the centromere, we calculate 1/2 the percentage of SDS patterns for each locus. For the *a* locus, classes 4, 5, 6, and 7 are SDS patterns. For the *b* locus, classes 3, 5, 6, and 7 are SDS patterns. Therefore, the distances to the centromere, in map units, are

for locus *a*: $(1/2) \dfrac{9 + 150 + 1 + 8}{1{,}000} \times 100$

$= 8.4$ centimorgans

for locus *b*: $(1/2) \dfrac{101 + 150 + 1 + 8}{1{,}000} \times 100$

$= 13.0$ centimorgans

(It should now be possible to describe exactly what type of crossover event produced each of the seven classes of table 6.6.)

Unfortunately, these two distances do not provide a unique solution to the gene order. In figure 6.22 we see that two alternatives are possible. How do we determine between these? The simplest way is to find out what happens to the *b* locus when a crossover occurs between the

Figure 6.23

Life cycle of yeast. Mature cells are mating types + or −; *n* is the haploid stage; 2*n* is diploid.

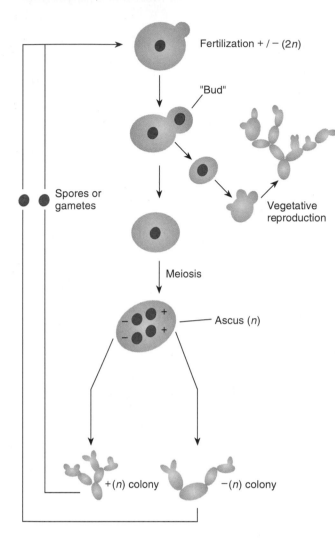

Fertilization + / − (2*n*)

"Bud"

Vegetative reproduction

Spores or gametes

Meiosis

Ascus (*n*)

+(*n*) colony

−(*n*) colony

TABLE 6.7 The Three Ascus Types in Yeast Resulting from Meiosis in a Dihybrid, $aa^+ bb^+$

1 (PD)	2 (NPD)	3 (TT)
a b	*a b⁺*	*a b*
a b	*a b⁺*	*a b⁺*
a⁺b⁺	*a⁺ b*	*a⁺ b*
a⁺b⁺	*a⁺ b*	*a⁺b⁺*
75	5	20

meiosis are contained in the ascus, but not in order. Therefore the order of the spores is not a centromere marker. Let us look at a mapping problem, using the *a* and *b* loci for convenience. Yeast has phenotypes similar to *Neurospora*.

When an *a b* spore (or gamete) fuses with an *a⁺ b⁺* spore (or gamete) and the diploid then undergoes meiosis, the spores can be isolated and grown as haploid colonies, which are then observed for the phenotypes controlled by the two loci. Only three patterns can occur (table 6.7). Class 1 has two types of spores, which are identical to the parental haploid spores. This ascus type is, therefore, referred to as a **parental ditype (PD).** The second class also has only two spore types, but they are of the recombinant type. This ascus type is referred to as a **nonparental ditype (NPD).** The third class has all four possible spore types and is referred to as a **tetratype (TT).**

All three ascus types can be generated whether or not the two loci are linked. As can be seen from figure 6.24, if the loci are linked, PDs come from the lack of a crossover, whereas NPDs come about from four-strand double crossovers (double crossovers involving all four chromatids). We should thus expect PDs to be more numerous than NPDs. However, if the loci are not linked, both PDs and NPDs come about through independent assortment—they should occur in equal frequencies. We can therefore determine whether or not the loci are linked by comparing PDs and NPDs. In table 6.7, the PDs greatly outnumber the NPDs; the two loci are, therefore, linked. What, then, is the map distance between the loci?

A return to figure 6.24 shows that in an NPD, all four chromatids are recombinant, whereas in a TT only half the chromatids are recombinant. Remembering that 1% recombinant offspring equals 1 map unit, we can use the following formula:

$$\text{map units} = \frac{(1/2)\text{TT} + \text{NPD asci}}{\text{total asci}} \times 100$$

Thus for the data of table 6.7:

$$\text{map units} = \frac{10 + 5}{100} \times 100 = 15$$

a locus and its centromere. If order (1) is correct, crossovers between the *a* locus and its centromere should have no effect on the *b* locus; if (2) is correct, most of the crossovers that move the *a* locus in relation to its centromere should also move the *b* locus, because the latter is only a short distance farther down the chromosome.

Asci classes 4, 5, 6, and 7 include all the SDS patterns for the *a* locus. Of 168 asci, 150 of them (class 5) have similar SDS patterns for the *b* locus. Thus 89% of the time a crossover between the *a* locus and its centromere is also a crossover between the *b* locus and its centromere, compelling evidence in favor of alternative (2). (What form would the data take if alternative [1] were correct?)

Unordered Spores

In fungi such as baker's yeast, *Saccharomyces cerevisiae* (fig. 6.23), the ascus shape places no restriction on the arrangement of the spindle in meiosis. All the products of

Figure 6.24
Formation of PD, NPD, and TT at meiosis in a dihybrid yeast through linkage or independent assortment

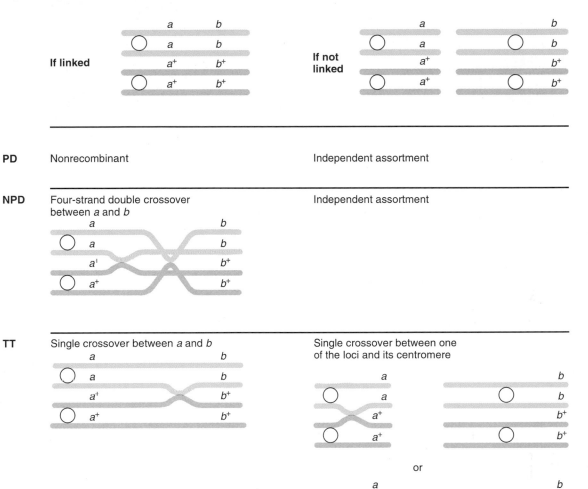

PD	Nonrecombinant	Independent assortment
NPD	Four-strand double crossover between *a* and *b*	Independent assortment
TT	Single crossover between *a* and *b*	Single crossover between one of the loci and its centromere

SOMATIC (MITOTIC) CROSSING OVER

Crossing over is known to occur in somatic cells as well as during meiosis. It apparently occurs by a similar reciprocal mechanism whereby two homologous chromatids come to lie next to each other and breakage and reunion events follow. Unlike meiosis, there is no synaptonemal complex formed. Since mitotic chromosomes do not normally lie side by side, the occurrence of mitotic crossing over is relatively rare. In the fungus *Aspergillus nidulans,* mitotic crossing over occurs about once in every one hundred cell divisions.

Mitotic recombination was discovered in 1936 by Curt Stern, who noticed the occurrence of *twin spots* in fruit flies that were dihybrid for the yellow allele of body color (*y*) and the singed allele (*sn*) for bristle morphology (fig.

6.25). A twin spot could be explained by mitotic crossing over between the *sn* locus and its centromere (fig. 6.26). A crossover in the *sn-y* region would produce only a yellow spot, whereas a double crossover, one between *y* and *sn* and the other in the *sn*-centromere region, would produce only a singed spot. (You should verify this for yourself.) These three phenotypes were found in the relative frequencies expected. That is, given that the gene locations are drawn to scale in figure 6.26, we would expect double spots to be most common, followed by yellow spots, with singed spots rarest of all because they require a double crossover. This in fact occurred and no other explanation was consistent with these facts.

Mitotic crossing over has been used in fungal genetics as a supplemental, or even a primary, method for determining linkage relations. Although gene orders are consistent between mitotic and meiotic mapping, relative

Curt Stern (1902–1981)
Courtesy of the Science Council of Japan.

Figure 6.25
Twin spots on the thorax of a female *Drosophila*

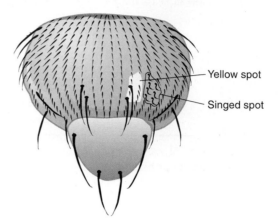

Figure 6.26
Formation of twin spots and single mutant spots through somatic crossing over

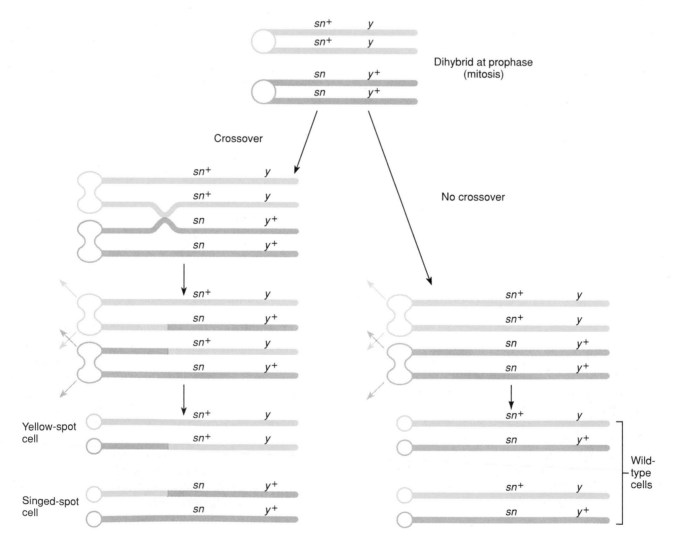

distances are usually not, which is not totally unexpected. We know that neither meiotic nor mitotic crossing over is uniform along a chromosome. Apparently, the factors that cause deviation from uniformity differ in the two processes.

HUMAN CHROMOSOME MAPS

Conceptually, human chromosomes can be mapped like those of any other organism. Realistically, the problems mentioned earlier (the inability to make specific crosses, coupled with the relatively small family sizes in human beings) make human chromosome mapping very difficult. However, there has been some progress based on pedigrees, especially in assigning genes to the X chromosome. As the pedigree analysis in the previous chapter has shown, X-chromosome traits have unique patterns of inheritance and those loci on the X chromosome are easily identified. Currently there are about 350 loci known to be on the X chromosome. It has been estimated, by several different methods, that there are about fifty thousand loci on human chromosomes. In later chapters we will discuss several additional methods of human chromosome mapping using molecular-genetic techniques.

X-Linkage

After determining that a gene is X linked, the next problem is to determine the position of the locus on the X chromosome and to determine map units between loci. This can be done with the proper pedigrees, in which crossing over can be ascertained. An example of what is referred to as the "grandfather method" is shown in figure 6.27. In this example, a grandfather is found who has one of the traits in question (here, color blindness). We then find that he has a grandson who is G-6-PD deficient. From this we can infer that the mother (of the grandson) was dihybrid in the *trans* configuration. That is, she received her colorblindness allele on one of her X chromosomes from her father, and she must have received the G-6-PD deficiency allele on the other X chromosome from her mother (why?). Thus the two sons on the left of figure 6.27 are nonrecombinant and the two on the right are recombinant. Theoretically, we can determine map distance by simply totaling the recombinant grandsons and dividing by the total number of grandsons. Of course, the methodology would be the same if the grandfather were both color-blind and G-6-PD deficient. The mother would then be dihybrid in the *cis* configuration and the sons would be tabulated in the reverse manner. The point is that the grandfather's phenotype gives us information from which to infer that the mother was dihybrid, as well as telling us the *cis-trans* arrangement of her alleles. We can then score her sons as either recombinant or nonrecombinant.

Figure 6.27
"Grandfather method" of determining crossing over between loci on the human X chromosome

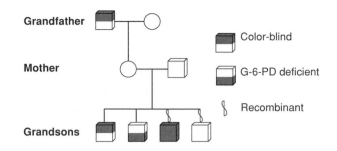

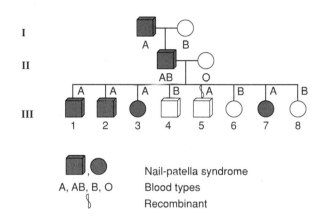

Figure 6.28
Linkage of the nail-patella syndrome locus and the ABO locus

Autosomal Linkage

From this we can see that it is relatively easy to map the X chromosome. The autosomes are another story. Since there are twenty-two autosomal linkage groups (twenty-two pairs of nonsex chromosomes), it is virtually impossible to determine from simple pedigrees which chromosome two loci are on. Pedigrees can tell us if two loci are linked to each other, but not on which chromosome (fig. 6.28). The nail-patella syndrome includes, among other things, abnormal nail growth coupled with the absence or underdevelopment of knee caps. It is a dominant trait. The male in generation II of figure 6.28 is dihybrid with the *A* allele of the ABO system associated with the nail-patella allele (*Nps1*), and the *B* allele with the normal nail-patella allele (*nps1*). Thus only one child in eight (III-5) is recombinant. Actually, the map distance is about 10%. It should be noted that, in general, map distances appear greater in females than in males because there is more crossing over in females.

We now turn our attention to the localization of loci to particular human chromosomes. The first locus that was definitely established to be on a particular autosome was the Duffy blood group on chromosome 1. This was ascertained in 1968 from a family that had a morphologically

BOX 6.3

LOD SCORES

Human population geneticists can increase their accuracy of linkage analysis by using a probability technique, developed by Newton Morton, called the **lod score method** (*Log Odds*). The geneticist is asking what the probability is of getting a particular pedigree assuming a particular recombination frequency (θ), as compared with getting the same pedigree assuming independent assortment ($\theta = 0.50$). In other words, a value is calculated that is the ratio of the probability of genotypes in a family given a certain crossover frequency compared with the probability of those genotypes if the loci are unlinked. Logarithms are used for ease of calculation and the parameter is called *z*, the *lod* score. Using this method, different crossover frequencies can be tried until the one giving the highest *lod* score is found.

For example, take the cross in figure 6.28. The father in generation II can have one of two allele arrangements: *A Nps1/B nps1* or *A/B Nps1/nps1*. The former assumes linkage whereas the latter does not. Our initial estimate of recombination, assuming linkage, was (1/8) $\times$ 100, or 12.5 map units. We now need to calculate the ratio of two probabilities:

$$z = \log \frac{\text{Probability of birth sequence assuming 12.5 map units}}{\text{Probability of birth sequence assuming independent assortment}}$$

Assuming 12.5 map units (or a probability of 0.125 of a crossover; $\theta = 0.125$), the probability of child III-1 is 0.4375. This child would be a nonrecombinant and thus its probability of having the nail-patella syndrome and type A blood is the probability of no crossover during meiosis, or $(1 - 0.125)/2$. We divide by 2 since there are two nonrecombinant types. This is the same probability for all children except III-5, who has a probability of occurrence of $0.125/2 = 0.0625$ since he is a recombinant. Thus the numerator of the previous equation is $(0.4375)^7(0.0625)$.

If the two loci are not linked, then any genotype has a probability of 1/4, or 0.25. Thus the sequence of the eight children has the probability of $(0.25)^8$. This is the denominator of the equation. Thus

$$z = \log \frac{(0.4375)^7(0.0625)}{(0.25)^8}$$

$$z = \log [12.566] = 1.099$$

Newton E. Morton
Courtesy Dr. Newton E. Morton.

BOX TABLE 1 *Lod* Scores for the Cross in Figure 6.28

Recombination Frequency (θ)	Lod Score
0.05	0.951
0.10	1.088
0.125	1.099
0.15	1.090
0.20	1.031
0.25	0.932
0.30	0.801
0.35	0.643
0.40	0.457
0.45	0.244
0.50	0.000

Any *lod* score greater than zero favors recombination. A *lod* score less than zero suggests that θ has been underestimated. A *lod* of 3.0 or greater (10^3 or one thousand times more likely than independent assortment) is considered a strong likelihood of linkage. Thus, in our example, we have an indication of linkage with a recombination frequency of 0.125. Now we can calculate *lod* scores assuming other values of recombination. That is done in Box table 1. You can see that the recombination frequency as calculated, 0.125 (12.5 map units), gives the best *lod* score.

odd, or "uncoiled," chromosome 1. Inheritance in the Duffy blood group system followed the pattern of inheritance of the "uncoiled" chromosome. Real strides have been made since then; about seventeen hundred autosomal loci have been mapped. Two techniques, chromosome banding and somatic-cell hybridization, have been crucial to autosomal mapping.

Chromosome Banding

Techniques were developed around 1970 involving certain histochemical stains that produce repeatable banding patterns on the chromosomes. For example, Giemsa staining is one technique; the resulting bands are called **G-bands.** More detail on these techniques is presented in chapter 14. Before these techniques, human and other mammalian chromosomes were grouped into general size categories because of the difficulty of differentiating many of them. With banding techniques came the ability to identify each human chromosome in a karyotype (see fig. 5.1).

Somatic-Cell Hybridization

The ability to distinguish each human chromosome is required for the technique of somatic-cell hybridization, in which human and mouse (or hamster) cells are fused in culture to form a hybrid. The fusion is usually mediated chemically with polyethylene glycol, which affects cell membranes, or by an inactivated virus, for example the Sendai virus, which has the property of fusing cells by being able to fuse to more than one cell at the same time. The virus does this by having a lipid membrane derived from host cells that easily fuses with new host cells. Because of this property, the virus can fuse to two cells close together, forming a cytoplasmic bridge between them that facilitates their fusion. When two cells fuse, their nuclei are at first separate, forming a **heterokaryon,** a cell with nuclei from different sources. When the nuclei fuse, a hybrid cell is formed, which tends to lose human chromosomes preferentially through succeeding generations. Upon stabilization, the result is a cell with one or more human chromosomes in addition to the original mouse or hamster chromosome complement. Banding techniques allow the human chromosomes to be recognized. A geneticist then looks for specific human phenotypes, such as enzyme products, and can then assign the phenotype to one of the human chromosomes in the cell line.

When cells are mixed together for hybridization, some cells do not hybridize. It is thus necessary to be able to select for study just those cells that are hybrids. One technique, originally devised by J. W. Littlefield in 1964, makes use of genetic differences in the cell lines in regard to DNA synthesis. Normally, in mammalian cells, DNA is synthesized de novo. It can be inhibited by the chemical aminopterin, an enzyme inhibitor. Two enzymes, hypoxanthine phosphoribosyl transferase (HPRT) and thymidine kinase (TK), can bypass aminopterin inhibition by making use of secondary, or salvage, pathways in the cell. If hypoxanthine is provided, it is converted to a purine by HPRT, and if thymidine is provided, it is converted to the nucleotide thymidylate by TK. (Purines are converted to nucleotides and nucleotides are the subunits of DNA—see chapter 9.) Thus normal cells in the absence of aminopterin synthesize DNA even if they lack HPRT activity (HPRT$^-$) or TK activity (TK$^-$). In the presence of aminopterin, HPRT$^-$ or TK$^-$ cells will die. However, in the presence of aminopterin, HPRT$^+$ TK$^+$ cells can synthesize DNA and survive. Using this information, the following selection system was developed.

Mouse cells that have the phenotype of HPRT$^+$ TK$^-$ are mixed with human cells that have the phenotype of HPRT$^-$ TK$^+$ in the presence of Sendai virus or polyethylene glycol. Fusion takes place in some of the cells and the mixture is grown in a medium containing hypoxanthine, aminopterin, and thymidine (called **HAT medium).** In the presence of aminopterin, unfused mouse cells (TK$^-$) and unfused human cells (HPRT$^-$) die. Hybrid cells, however, will survive because they are HPRT$^+$ TK$^+$. Eventually, the hybrid cells will end up with random numbers of human chromosomes. There will be one restriction: All cell lines selected will be TK$^+$. Not only does this HAT method (using the HAT medium) select for hybrid clones, but it localized the TK gene to human chromosome 17, the one human chromosome found in every successful cell line.

After successful cell hybrids are formed, two particular tests are used to map human genes. A **synteny test** (same linkage group) determines whether two loci are in the same linkage group if the phenotypes of the two loci are either always together or always absent in various hybrid cell lines. An **assignment test** determines which chromosome a particular locus is on by the concordant appearance of the phenotype whenever that particular chromosome is in a cell line or by the lack of the particular phenotype when a particular chromosome is absent from a cell line. The first autosomal synteny test was performed in 1970 and demonstrated that the *B* locus of lactate dehydrogenase (LDH$_B$) was linked to the *B* locus of peptidase (PEP$_B$). (Both enzymes are formed from subunits controlled by two loci each. In addition to the *B* locus, each protein has subunits controlled by an *A* locus.) Later, by assignment, these loci were shown to reside on chromosome 12.

TABLE 6.8 Assignment of Blood Coagulating Factor III to Human Chromosome 1 Using Human-Mouse Hybrid Cell Lines

Hybrid Cell Line	Tissue Factor Score	1	2	3	4	5	6	7	8	9	10	11	12	13	14	15	16	17	18	19	20	21	22	X
WIL1	−	−	−	−	−	−	−	−	+	−	−	−	−	−	+	−	−	+	−	−	−	+	−	+
WIL6	−	−	+	−	+	+	+	+	+	−	+	+	−	−	+	−	−	+	−	+	+	+	−	+
WIL7	−	−	+	+	−	+	+	−	+	−	+	+	−	+	+	−	−	+	+	−	−	+	−	+
WIL14	+	+	−	+	−	−	−	+	+	−	+	−	+	−	+	+	−	+	−	−	−	−	−	+
SIR3	+	+	+	+	+	+	+	+	+	−	+	+	+	+	+	−	+	+	+	+	+	+	+	+
SIR8	+	+	+	+	+	+	−	+	+	+	+	+	+	+	+	+	+	+	+	−	−	+	+	+
SIR11	−	−	−	−	−	−	−	−	+	−	−	−	+	−	+	−	−	−	−	−	−	+	+	+
REW7	+	+	+	+	+	+	+	+	+	−	+	+	+	+	+	+	−	+	+	+	+	+	+	+
REW15	+	+	+	+	+	+	+	+	+	−	+	−	+	+	+	+	−	+	+	+	+	+	+	+
DUA1A	−	−	−	−	−	−	−	−	−	−	−	−	−	−	t	−	−	−	−	−	−	−	−	t
DUA1CsAzF	−	−	−	−	−	−	+	−	−	−	−	−	−	−	−	−	−	−	−	−	−	−	−	−
DUA1CsAzH	−	−	−	−	−	−	+	−	−	−	−	−	−	−	−	−	−	−	−	−	−	−	−	−
TSL1	−	−	−	+	+	−	−	−	−	−	+	+	−	+	+	−	+	+	+	−	+	−	−	−
TSL2	−	−	+	t	−	+	+	−	−	−	+	+	−	+	−	−	−	t	+	−	+	+	−	+
TSL2CsBF	−	−	−	−	−	−	+	−	−	−	−	−	−	−	−	−	−	−	−	−	−	−	−	−
XTR1	+	+	−	t	−	+	+	+	+	+	+	+	+	+	+	−	−	+	+	+	+	+	+	+
XTR2	−	−	−	t	−	+	−	−	+	−	+	−	+	+	−	−	−	−	+	−	+	+	−	t
XTR3BsAgE	+	+	−	t	−	+	+	+	+	+	+	−	+	−	+	−	−	+	+	+	−	+	−	t
XTR22	−	−	+	t	+	+	−	−	+	−	+	+	−	−	−	+	−	−	+	+	+	+	+	t
XER9	−	−	+	−	+	−	−	−	+	−	+	t	+	−	+	−	+	+	+	−	+	+	−	t
XER11	+	+	−	+	+	−	+	+	+	−	+	t	+	+	−	+	+	+	+	+	+	+	+	t
REX12	−	−	−	−	+	−	−	+	−	−	−	+	−	−	−	−	−	−	−	−	−	−	+	t
JSR29	+	+	+	+	+	+	+	t	+	t	+	+	+	+	+	+	+	+	+	+	+	+	+	+
JVR22	+	+	+	+	+	+	+	+	+	−	+	+	+	+	+	+	+	+	+	+	+	+	+	+
JWR22H	+	t	t	−	+	−	+	−	−	−	+	+	+	−	+	+	−	+	+	−	+	−	−	−
ALR2	+	+	+	+	+	+	+	+	+	−	+	+	+	+	+	+	+	+	−	+	+	+	+	+
ICL15	−	−	−	−	−	−	−	−	−	−	−	−	+	−	−	−	−	+	−	−	+	+	−	−
ICL15CsBF	−	−	−	−	−	−	−	−	−	−	−	−	+	−	−	−	−	−	−	−	+	+	−	−
MH21	−	−	−	−	−	−	−	−	−	−	−	−	−	−	−	−	−	−	−	−	−	+	−	−
% Discord*		0	32	17	24	31	21	21	31	21	24	30	21	21	28	14	24	21	28	17	34	41	21	27

*Discord refers to cases in which the tissue factor score is plus, whereas the human chromosome is absent, or the score is minus with the chromosome present. A "t" represents a translocation in which only part of a chromosome is present.
From Steven D. Carson, et al., "Tissue Factor Gene Localized to Human Chromosome 1 (1pter–1p21)" in *Science,* 229, September 6, 1985. Copyright 1985 by the AAAS.

In another example, a blood-coagulating glycoprotein (a protein-polysaccharide complex), tissue factor III, was localized by assignment tests to chromosome 1. Table 6.8 shows twenty-nine human-mouse hybrid cell lines, or **clones,** the human chromosomes they contain, and their tissue factor score, the results of an assay for the presence of the coagulating factor. (Clones are cells arising from a single ancestor.) It is obvious from table 6.8 that the gene for tissue factor III is on human chromosome 1. Every time human chromosome 1 is present in a cell line, so is tissue factor III. Every time human chromosome 1 is absent, so is the tissue factor (zero discordance or 100% concordance). No other chromosome showed that pattern.

The human map as we know it now (compiled by V. McKusick at Johns Hopkins University), about eighteen hundred assigned loci of about five thousand known to

Victor A. McKusick (1921–   )
Courtesy of Victor A. McKusick.

Figure 6.29

Human G-banded chromosomes with their accompanying assigned loci; *p* and *q* refer to the short and long arms of the chromosomes, respectively. A key to the loci is given in McKusick, 1990.

McKusick, Victor A., Mendelian Inheritance in Man, 9th edition. The Johns Hopkins University Press, Baltimore/London, 1990, Figure B1.

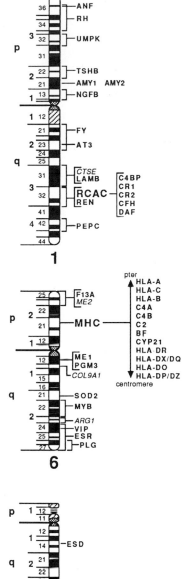

1

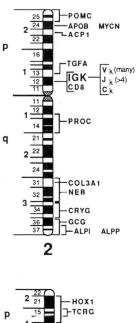

2

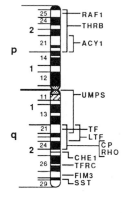

3

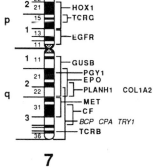

6

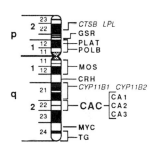

7

8

13

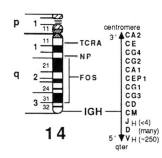

14

15

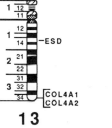

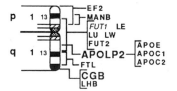

19

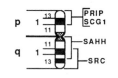

20

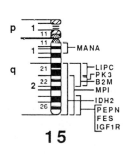
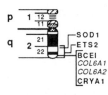

21

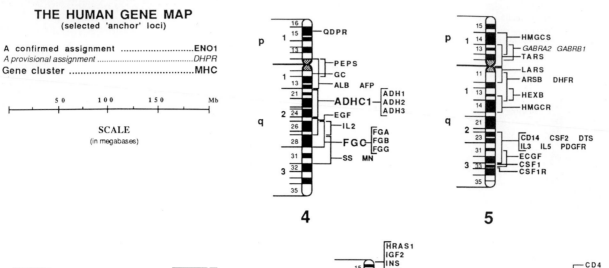

THE HUMAN GENE MAP
(selected 'anchor' loci)

A confirmed assignment ENO1
A provisional assignment .... *DHPR*
Gene cluster .... MHC

50   100   150   Mb

SCALE
(in megabases)

4

5

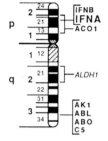

9

10

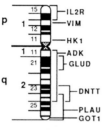

11

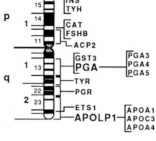

12

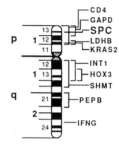

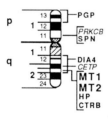

16

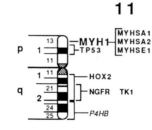

17

18

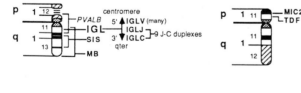

22

Y

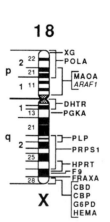

X

Figure 6.29 (continued)

TABLE 6.9   Definition of Selected Loci of the Human Chromosome Map (Figure 6.29)

Locus	Protein Product	Chromosome
ABO	ABO blood group	9
AG	Alpha globin gene family	16
ALB	Albumin	4
AMY1	Amylase, salivary	1
AMY2	Amylase, pancreatic	1
BCS	Breast cancer susceptibility	16
C2	Complement component-2	6
CAT	Catalase	11
CBD	Color blindness, deutan	X
CBP	Color blindness, protan	X
CML	Chronic myeloid leukemia	22
DMD	Duchenne muscular dystrophy	X
FES	Feline sarcoma virus oncogene	15
Fy	Duffy blood group	1
GLB1	Beta-galactosidase-1	3
H1	Histone-1	7
HBB	Hemoglobin beta chain	11
HEMA	Classic hemophilia	X
HEXA	Hexosaminidase A	15
HLA	Human leukocyte antigens	6
HP	Haptoglobin	16
HYA	Y histocompatibility antigen, locus A	Y
IDDM	Insulin-dependent diabetes mellitus	6
IFF	Interferon, fibroblast	9
IGH	Immunoglobulin heavy-chain gene family	14
IGK	Immunoglobulin kappa-chain gene family	2
INS	Insulin	11
LDHA	Lactate dehydrogenase A	11
MDI	Manic depressive illness	6
MHC	Major histocompatibility complex	6
MN	MN blood group	4
MYB	Avian myeloblastosis virus oncogene	6
NHCP1	Nonhistone chromosomal protein-1	7
NPS1	Nail-patella syndrome	9
PEPA	Peptidase A	18
PVS	Polio virus sensitivity	19
Rh	Rhesus blood group	1
RN5S	5S RNA gene(s)	1
RNTMI	Initiator methionine tRNA	6
RWS	Ragweed sensitivity	6
S1	Surface antigen 1	11
SIS	Simian sarcoma virus oncogene	22
STA	Stature	Y
TF	Transferrin	3
Xg	Xg blood group	X
XRS	X-ray sensitivity	13

A more complete list is found in McKusick, 1990.

exist, is shown in figure 6.29 and table 6.9. At the present time, geneticists studying human chromosomes are hampered not by a lack of techniques but by a lack of marker loci. When a new locus is discovered, it is now relatively easy to assign it to its proper chromosome.

The problem still exists of determining exactly where on a chromosome a particular locus belongs. This is being facilitated by particular cell lines with chromosomes that have been broken such that parts are either missing or have been moved to other chromosomes. These processes reveal new linkage arrangements and make it possible to determine on which region of a particular chromosome a particular locus is situated. In chapter 12 we will describe additional techniques used to locate and sequence genes on human chromosomes, including a description of the human genome mapping program, a program to sequence the entire human genome in the next fifteen years in order to locate all of our genes.

SUMMARY

The principle of independent assortment is violated by loci lying near each other on the same chromosome. Recombination between these loci results from the crossing over of chromosomes during meiosis. The amount of recombination provides a measure of the distance between these loci. One map unit (centimorgan) equals 1% recombinant gametes. Map units can be determined by testcrossing a dihybrid and recording the percentage of recombinant offspring. If three loci are used (a three-point cross), double crossovers can be revealed. A coefficient of coincidence, the ratio of observed to expected double crossovers, can be calculated to determine if one crossover changes the probability of a second one nearby.

The chiasma seen during prophase I of meiosis represent both a physical and a genetic crossing over. This can be demonstrated when homologous chromosomes with morphological distinctions are used.

Because of multiple crossovers, the measured percentage recombination underestimates the true map distance, especially for loci relatively far apart: The best map estimates come from using closely linked loci. A mapping function can be used to translate observed map distances into more accurate ones.

Organisms that retain all the products of meiosis are mapped by techniques known as tetrad analysis (haploid mapping). Map units between a locus and its centromere in organisms with ordered spores, such as *Neurospora,* can be calculated as

$$\text{map units} = \frac{(1/2)\text{SDS asci}}{\text{total asci}} \times 100$$

With unordered spores, such as in yeast, we use

$$\text{map units} = \frac{(1/2)\text{TT} + \text{NPD asci}}{\text{total asci}} \times 100$$

Crossing over also occurs during mitosis but at a much reduced rate. Somatic (mitotic) crossing over can be used to map loci.

Human chromosomes can be mapped. Recombination distances can be established by pedigrees, and loci can be attributed to specific chromosomes by synteny and assignment tests in hybrid cell lines.

SOLVED PROBLEMS

1. A homozygous claret (*ca,* ruby eye color), curled (*cu,* upcurved wings), fluted (*fl,* creased wings) fruit fly is crossed to a pure-breeding wild-type fly. The F$_1$ females are testcrossed with the following results:

fluted	4
claret	173
curled	26
fluted, claret	24
fluted, curled	167
claret, curled	6
fluted, claret, curled	298
wild-type	302

    a. Are the loci linked?
    b. If so, give the gene order, map distances, and coefficient of coincidence.

ANSWER: The pattern of numbers among the eight offspring classes is the pattern that we are used to seeing for linkage of three loci. We can tell from the two groups in largest numbers (the nonrecombinants—fluted, claret, curled, and wild-type) that the alleles are in the coupling (*cis*) arrangement. If we compare either of the nonrecombinant classes with either of the double crossover classes (fluted and claret, curled), you should see that the fluted locus is in the center. For example, compare fluted, a double crossover offspring, with the wild-type, a nonrecombinant; clearly fluted has the odd pattern. Thus the trihybrid female parent had the following arrangement of alleles:

$$\frac{ca \quad fl \quad cu}{ca^+ \, fl^+ \, cu^+}$$

A crossover in the *ca-fl* region will produce claret and fluted, curled offspring and a crossover in the *fl-cu* region will produce claret, fluted and curved offspring.

Counting up the crossovers in each region, including the double crossovers in each, and converting to percentages yields a claret to fluted distance of 35.0 map units $(173 + 167 + 10)$ and a fluted to curled distance of 6.0 map units $(26 + 24 + 10)$. We expect $0.35 \times 0.06 \times 1,000 = 21$ double crossovers, but we observed only $6 + 4 = 10$. Thus the coefficient of coincidence is $10/21 = 0.48$.

2. The *ad5* locus in *Neurospora* is in the pathway for synthesis of the amino acid adenine. A wild-type strain (*ad5$^+$*) is crossed with an adenine-requiring strain, *ad5$^-$*. The diploid undergoes meiosis and 100 asci are scored for their segregation patterns with the following results:

ad5$^+$ ad5$^+$ ad5$^+$ ad5$^+$ ad5$^-$ ad5$^-$ ad5$^-$ ad5$^-$ 40

ad5$^-$ ad5$^-$ ad5$^-$ ad5$^-$ ad5$^+$ ad5$^+$ ad5$^+$ ad5$^+$ 46

ad5$^+$ ad5$^+$ ad5$^-$ ad5$^-$ ad5$^-$ ad5$^-$ ad5$^+$ ad5$^+$ 5

ad5$^-$ ad5$^-$ ad5$^+$ ad5$^+$ ad5$^+$ ad5$^+$ ad5$^-$ ad5$^-$ 3

ad5$^-$ ad5$^-$ ad5$^+$ ad5$^+$ ad5$^-$ ad5$^-$ ad5$^+$ ad5$^+$ 4

ad5$^+$ ad5$^+$ ad5$^-$ ad5$^-$ ad5$^+$ ad5$^+$ ad5$^-$ ad5$^-$ 2

What can you say about linkage arrangements of this locus?

ANSWER: You can see that 14 $(5 + 3 + 4 + 2)$ are of the second-division-segregation type (SDS) and 86 $(40 + 46)$ are of the first-division-segregation type (FDS). To map the distance of the locus to its centromere, we merely divide the percentage of SDS types by 2:14/100 = 14%—divided by 2 is 7%. Thus the *ad5* locus is 7 map units from its centromere.

3. In yeast, the *his5* locus is in the pathway for the synthesis of the amino acid histidine and *lys11* locus is in the pathway for the synthesis of the amino acid lysine. A haploid wild-type strain (*his5$^+$ lys11$^+$*) is crossed with the double mutant (*his5$^-$ lys11$^-$*). The diploid is allowed to undergo meiosis and 100 asci are scored with the following results:

his5$^+$lys11$^+$ his5$^+$lys11$^+$ his5$^-$lys11$^-$ his5$^-$lys11$^-$ 62

his5$^+$lys11$^+$ his5$^-$lys11$^-$ his5$^-$lys11$^+$ his5$^+$lys11$^-$ 30

his5$^+$lys11$^-$ his5$^+$lys11$^-$ his5$^-$lys11$^+$ his5$^-$lys11$^+$ 8

What is the linkage arrangement of these loci?

ANSWER: Of the 100 asci analyzed, 62 were parental ditypes (PD), 30 were tetratypes (TT), and 8 were nonparental ditypes (NPD). To map the distance between the two loci, we take the percentage of NPD (8%) plus half the percentage of TT (1/2 of 30 = 15%) = 23% or 23 centimorgans between loci.

4. A given human enzyme is present only in clone B. The human chromosomes present (+) in clones A, B, and C appear below.
Determine the probable chromosomal location of the gene for the enzyme.

Clone	Human Chromosome							
	1	*2*	*3*	*4*	*5*	*6*	*7*	*8*
A	+	+	+	+	−	−	−	−
B	+	+	−	−	+	+	−	−
C	+	−	+	−	+	−	+	−

ANSWER: If a gene is located on a chromosome, the chromosome must be present in the positive clones. Chromosomes 1, 2, 5, 6 are present in B. If the gene in question were located on chromosome 1, the enzyme should have been present in all three clones. The only chromosome that is unique to clone B is 6. Therefore, the gene is located on chromosome 6.

EXERCISES AND PROBLEMS

1. A homozygous groucho fly (*gro* = bristles clumped above the eyes) is crossed with a homozygous rough fly (*ro* = eye abnormality). The F_1 females are testcrossed, with the following offspring produced:

Groucho	518
Rough	471
Groucho, rough	6
Wild-type	5
	1,000

a. What is the linkage arrangement of these loci?
b. What offspring would result if the F_1 dihybrids were crossed among themselves instead of being testcrossed?

2. A female fruit fly with abnormal eyes (*abe*) of a brown color (*bis,* bistre) is crossed with a wild-type male. Her sons have abnormal, brown eyes; her daughters are of the wild-type. When these F_1 flies are crossed among themselves, the following offspring are produced:

	Sons	Daughters
Abnormal, brown	219	197
Abnormal	43	45
Brown	37	35
Wild-type	201	223

What is the linkage arrangement of these loci?

3. In *Drosophila*, the loci inflated (*if*, small, inflated wings) and warty (*wa*, abnormal eyes) are about 10 map units apart on the X chromosome. Construct a data set that would allow you to determine this linkage arrangement. What differences would be involved if the loci were located on an autosome?

4. A geneticist crossed female fruit flies that were heterozygous at three electrophoretic loci, each with fast and slow alleles, with males homozygous for the slow alleles. The three loci were *got1* (glutamate oxaloacetate transaminase-1), *amy* (alpha-amylase), and *sdh* (succinate dehydrogenase). The first 1,000 offspring isolated had the following genotypes:

class 1) $got^s got^s amy^s amy^s sdh^s sdh^s$ . . . 441

class 2) $got^f got^s amy^f amy^f sdh^f sdh^s$ . . . 421

class 3) $got^f got^s amy^s amy^s sdh^s sdh^s$ . . . 11

class 4) $got^s got^s amy^f amy^f sdh^f sdh^s$ . . . 14

class 5) $got^f got^s amy^f amy^s sdh^s sdh^s$ . . . 58

class 6) $got^s got^s amy^s amy^s sdh^f sdh^s$ . . . 53

class 7) $got^f got^s amy^s amy^s sdh^f sdh^s$ . . . 1

class 8) $got^s got^s amy^f amy^s sdh^s sdh^s$ . . . 1

What are the linkage arrangements of these three loci, including map units? If the three loci are linked, what is the coefficient of coincidence?

5. The following three recessive markers are known in lab mice: *h*, hotfoot; *o*, obese; and *wa*, waved. A trihybrid of unknown origin is testcrossed, producing the following offspring:

hotfoot, obese, waved	357
hotfoot, obese	74
waved	66
obese	79
wild-type	343
hotfoot, waved	61
obese, waved	11
hotfoot	9
	1,000

a. If the genes are linked, determine the relative order and the map distance between them.
b. What was the *cis-trans* allele arrangement in the trihybrid parent?
c. Is there any crossover interference? If yes, how much?

6. The following three recessive genes are found in corn: *bt1*, brittle endosperm; *gl17*, glossy leaf; *rgd1*, ragged

seedling. A trihybrid of unknown origin is testcrossed producing the following offspring:

brittle, glossy, ragged	236
brittle, glossy	241
ragged	219
glossy	23
wild-type	224
brittle, ragged	17
glossy, ragged	21
brittle	19
	1,000

a. If the genes are linked, determine the relative order and map distances.
b. Reconstruct the chromosomes of the trihybrid.
c. Is there any crossover interference? If yes, how much?

7. In *Drosophila*, ancon (*an* = legs and wings short), spiny legs (*sple* = irregular leg hairs), and arctus oculus (*at* = small narrow eyes) have the following linkage arrangement on chromosome 3:

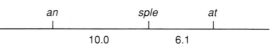

a. Devise a data set with no crossover interference that would yield these map units.
b. What data would yield the same map units but with a coefficient of coincidence of 0.60?

8. Ancon (*an*) and spiny legs (*sple*), of problem 7, are 10 map units apart on chromosome 3. Notchy (*ny* = wing tips nicked) is on the X chromosome (chromosome 1). Create a data set that would result if you were making crosses to determine the linkage arrangement of these three loci. How would you know that the notchy locus is on the X chromosome?

9. The Duffy blood group with alleles Fy^a and Fy^b was localized to chromosome 1 in humans when an "uncoiled" chromosome was associated with it. Construct a pedigree that would verify this.

10. Given the following cross in *Neurospora*: $ab \times a^+b^+$. Give results showing that crossing over occurs in two of the four chromatids of a tetrad at meiosis. What would the results be if crossing over occurred before each chromosome became two chromatids? If each crossover event involved three or four chromatids?

11. A strain of yeast requiring both tyrosine (tyr^-) and arginine (arg^-) is crossed to the wild-type. After meiosis, the following ten asci are dissected. Classify each ascus as

to segregational type (PD, NPD, TT). What is the linkage relationship of these two loci?

(1) arg^-tyr^- arg^+tyr^+ arg^+tyr^+ arg^-tyr^-

(2) arg^+tyr^+ arg^+tyr^+ arg^-tyr^- arg^-tyr^-

(3) arg^-tyr^+ arg^-tyr^+ arg^+tyr^- arg^+tyr^-

(4) arg^-tyr^- arg^-tyr^- arg^+tyr^+ arg^+tyr^+

(5) arg^-tyr^- arg^-tyr^+ arg^+tyr^- arg^+tyr^+

(6) arg^+tyr^+ arg^+tyr^+ arg^-tyr^- arg^-tyr^-

(7) arg^-tyr^- arg^+tyr^+ arg^-tyr^+ arg^+tyr^-

(8) arg^+tyr^+ arg^+tyr^+ arg^-tyr^- arg^-tyr^-

(9) arg^+tyr^+ arg^-tyr^- arg^-tyr^- arg^+tyr^+

(10) arg^-tyr^- arg^+tyr^+ arg^+tyr^+ arg^-tyr^-

12. A certain haploid strain of yeast was deficient for the synthesis of the amino acids tryptophan (try^-) and methionine (met^-). It was crossed to the wild-type and meiosis occurred. One dozen asci were analyzed for their tryptophan and methionine requirements. The following results with the inevitable lost spores were obtained:

(1) try^-met^- ?   ? try^-met^-
(2) ? try^-met^- try^+met^+ try^+met^+
(3) try^-met^+ try^-met^- try^+met^- try^+met^+
(4) try^-met^- try^+met^+ ? try^+met^-
(5) try^-met^+ ?   ? try^+met^-
(6) try^+met^+ try^+met^+ try^-met^- try^-met^-
(7) try^+met^+ try^+met^- ? try^-met^-
(8) try^+met^+ try^-met^- ? try^+met^+
(9) try^-met^+ try^+met^- try^-met^+ try^+met^-
(10) try^-met^- try^+met^+ try^-met^- try^+met^+
(11) try^+met^+ try^+met^+ ?   ?
(12) ? try^+met^- ? try^-met^+

a. Classify each ascus as to segregational type (note that some asci may not be classifiable).
b. Are the genes linked?
c. If so, how far apart are they?

13. In *Neurospora,* a haploid strain requiring arginine (arg^-) is crossed to the wild-type (arg^+). Meiosis occurs and ten asci are dissected with the following results. Map the *arg* locus.

(1) $arg^+arg^+arg^-arg^-arg^+arg^+arg^-arg^-$
(2) $arg^-arg^-arg^+arg^+arg^-arg^-arg^+arg^+$
(3) $arg^+arg^+arg^+arg^+arg^-arg^-arg^-arg^-$
(4) $arg^+arg^+arg^+arg^+arg^-arg^-arg^-arg^-$
(5) $arg^-arg^-arg^-arg^-arg^+arg^+arg^+arg^+$
(6) $arg^+arg^+arg^-arg^-arg^-arg^-arg^+arg^+$
(7) $arg^-arg^-arg^+arg^+arg^+arg^+arg^-arg^-$
(8) $arg^+arg^+arg^+arg^+arg^-arg^-arg^-arg^-$
(9) $arg^-arg^-arg^+arg^+arg^+arg^+arg^-arg^-$
(10) $arg^-arg^-arg^-arg^-arg^+arg^+arg^+arg^+$

14. A haploid strain of *Neurospora* with fuzzy colony morphology (*f*) was crossed to the wild-type (*f⁺*). Twelve asci were scored. The following results with the inevitable lost spores were obtained:

(1) ? f f ? ? $f^+f^+f^+$ (7) f^+f^+f f f f f^+f^+

(2) f f $f^+f^+f^+f^+f$ f (8) f f f ? ? $f^+f^+f^+$

(3) f ? ? ? ? f^+? ? ?   (9) f^+? ? ? ? ? f f ?

(4) f^+? ? ? ? f f f f (10) f f f^+f^+f f f^+f^+

(5) f f ? ? ? f^+? f^+ (11) f f f f $f^+f^+f^+f^+$

(6) ? f f ? ? ? ? ?   (12) f f ? ? ? ? f^+f^+

a. Classify each ascus as to segregational type and note which asci cannot be classified.
b. Draw a map of the chromosome containing the *f* locus and give all the relevant measurements.

15. Draw ten of the remaining twenty-eight ascus patterns not included in table 6.5. To which of the seven major categories of table 6.6 does each belong?

16. In yeast, the *a* and *b* loci are 12 map units apart. Construct a data set to demonstrate this.

17. In *Neurospora,* the *a* locus is 12 map units from its centromere. Construct a data set to show this.

18. An *a b Neurospora* was crossed with an *a⁺b⁺* form. Meiosis occurred and one thousand asci were dissected. Using the classes of table 6.6, the following data resulted:

Class 1	700	Class 5	5
Class 2	0	Class 6	5
Class 3	190	Class 7	10
Class 4	90		

What is the linkage arrangement of these loci?

19. Given the following linkage arrangement in *Neurospora,* construct a data set similar to that in table 6.6 that is consistent with it.

    a   15   ◯   15   *b*

20. Determine crossover events that led to each of the seven classes of table 6.6.

21. What pattern of scores would you expect to get, using the hybrid clones of table 6.8, for a locus on human chromosome 6? 14? X?

22. A man with X-linked color blindness and X-linked Fabry disease (alpha-galactosidase-A deficiency) marries a normal woman and has a normal daughter. She marries a normal man and produces ten sons (as well as eight normal daughters). Of the sons, five were normal, three

were like their grandfather, one was only color-blind and one had Fabry disease. From these data what can you say about the relationship of these two X-linked loci?

23. In people, the ABO system (*A, B, O* alleles) is linked to the aldolase-B locus (*al, al⁺* alleles), a gene that functions in the liver. Deficiency, which is recessive, results in fructose intolerance. A man, with blood type AB, had a fructose intolerant, type B father and a normal, type AB mother. He married a woman with blood type O and fructose intolerance. Together they had ten children, of which five were type A and normal, three were fructose intolerant and type B, and two were type A and intolerant to fructose. Draw a pedigree of this family and determine the map distances involved. (Calculate a *lod* score to determine the most likely recombination frequency between the loci.)

24. In the house mouse, the autosomal alleles Trembling and Rex (short hair) are dominant to normal and long hair, respectively. Heterozygous Trembling, Rex females were crossed to normal, long-haired males and yielded the following offspring.

Trembling, Rex	42
Trembling, long-haired	105
normal, Rex	109
normal, long-haired	44

a. Are the two genes linked? How do you know?
b. In the heterozygous females, were Trembling and Rex in *cis* or *trans* position? Explain.
c. Calculate the percent recombination between the two genes.

25. In corn, a trihybrid tunicate (*T*), glossy (*G*), liguled (*L*) plant was crossed with a nontunicate, nonglossy, liguleless one and produced the following offspring:

Tunicate, liguleless, glossy	58
Tunicate, liguleless, nonglossy	15
Tunicate, liguled, glossy	55
Tunicate, liguled, nonglossy	13
Nontunicate, liguled, glossy	16
Nontunicate, liguled, nonglossy	53
Nontunicate, liguleless, glossy	14
Nontunicate, liguleless, nonglossy	59

a. Determine which genes are linked.
b. Determine the genotypes of the heterozygote; be sure to indicate which alleles are on which chromosome.
c. Calculate the map distance between the linked genes.

26. In *Drosophila,* kidney bean eye (*k*), cardinal eye (*cd*), and ebony body (*e*) are three recessive alleles. If homozygous kidney, cardinal females are crossed to

homozygous ebony males, the F₁ offspring are all wild-type. If heterozygous F₁ females are mated with kidney, cardinal, ebony males, the following two thousand progeny appear:

880 kidney, cardinal

887 ebony

 64 kidney, ebony

 67 cardinal

 49 kidney

 46 ebony, cardinal

  3 kidney, ebony, cardinal

  4 wild-type

a. Determine the chromosomal composition of the F₁ females.
b. Derive a map of the three genes.

27. Below is a partial map of the third chromosome in *Drosophila.* The genes listed are all recessive.

19.2 javelin bristles (*ja*)

43.2 thread arista (*th*)

66.2 delta veins (*d*)

74.7 ebony body (*e*)

a. If flies heterozygous in *cis* position for javelin and ebony are mated among themselves, what phenotypic ratio do you expect in the progeny?
b. A true-breeding thread, ebony fly is crossed with a true-breeding delta fly. An F₁ female is testcrossed to a thread, delta, ebony male. Predict the expected progeny and their frequencies for this cross. Assume no interference.
c. Repeat (b), but assume a coefficient of coincidence of 0.4.

28. In *Neurospora,* a cross is made between *ab⁺* and *a⁺b* individuals. The following one hundred ordered tetrads are obtained.

Spores	I	II	III	IV	V	VI	VII	VIII
1, 2	a^+b	a^+b	a^+b	a^+b^+	a^+b^+	a^+b	a^+b	ab^+
3, 4	a^+b	a^+b^+	a^+b^+	a^+b	a^+b	ab^+	ab^+	a^+b
5, 6	ab^+	ab	ab^+	ab	ab^+	a^+b	ab^+	a^+b
7, 8	ab^+	ab^+	ab	ab^+	ab	ab^+	a^+b	ab^+
	85	2	3	2	3	3	1	1

a. Are genes *a* and *b* linked? How do you know?
b. Calculate the gene-to-centromere distances for *a* and *b*.

29. In *Neurospora*, there are four genes—a, b, c, and d—that control four different phenotypes. Your job is to map these genes by performing pairwise crosses. You obtain the following ordered tetrads.

ab⁺ × a⁺b

Spores	I	II	III
1,2	ab^+	ab	ab^+
3,4	ab^+	ab	a^+b^+
5,6	a^+b	a^+b^+	a^+b
7,8	a^+b	a^+b^+	ab
	45	43	12

bc⁺ × b⁺c

Spores	I	II	III
1,2	bc^+	b^+c^+	b^+c
3,4	bc^+	b^+c^+	b^+c^+
5,6	b^+c	bc	bc
7,8	b^+c	bc	bc^+
	70	4	26

cd⁺ × c⁺d

Spores	I	II	III	IV	V	VI	VII
1, 2	cd^+	cd	cd	cd	cd^+	cd	cd^+
3, 4	cd^+	cd	cd^+	c^+d	c^+d	c^+d^+	c^+d
5, 6	c^+d	c^+d^+	c^+d^+	c^+d^+	c^+d	c^+d^+	c^+d^+
7, 8	c^+d	c^+d^+	c^+d	cd^+	cd^+	cd	cd
	42	2	30	15	5	1	5

a. Calculate the gene-to-centromere distances.
b. Which genes are linked? Explain.
c. Derive a complete map for all four genes.

30. Hemophilia and color blindness are X-linked recessive traits. A normal woman whose mother was color-blind and whose father was a hemophiliac marries a normal man whose father was color-blind. They have the following children.

3 normal daughters

1 color-blind daughter

1 normal son

2 color-blind sons

2 hemophiliac sons

1 color-blind, hemophiliac son

Estimate the distance between the two genes.

31. The results of an analysis of five human-mouse hybrids for five enzymes are given in the table below along with the human chromosomal content of each clone. (+ = enzyme or chromosome present; − = absence). Deduce which chromosome carries which gene.

Human Chromosome

	1	2	3	4	5	6	7	8	9	10	11	12	13	14	15	16	17	18	19	20	21	22
clone A	−	−	−	−	+	+	+	−	+	−	−	−	−	+	+	−	−	−	−	+	+	
clone B	+	+	−	+	−	−	−	+	−	−	+	−	−	+	−	−	+	−	−	+	−	−
clone C	−	−	−	+	−	−	+	−	−	+	−	+	+	+	+	−	+	−	+	−	−	+
clone D	+	−	+	−	+	−	−	−	−	+	−	−	−	+	+	−	−	+	+	+	+	−
clone E	−	−	−	+	−	−	−	−	+	+	+	+	−	+	−	+	−	+	−	+	+	+

Clone

Human enzyme	A	B	C	D	E
glutathione reductase	+	+	−	−	−
malate dehydrogenase	−	+	−	−	−
adenosine deaminase	−	+	−	+	+
galactokinase	−	+	+	−	−
hexosaminidase	+	−	−	+	−

32. You have selected three mouse-human hybrid clones and analyzed them for the presence of human chromosomes. You then analyze each clone for the presence or absence of particular human enzymes (+ = presence of human chromosome or enzyme activity). Based on the results below, indicate the probable chromosomal location for each enzyme.

Human Chromosomes

Clone	3	7	9	11	15	18	20
X	−	+	−	+	+	−	+
Y	+	+	−	+	−	+	−
Z	−	+	+	−	−	+	+

Enzyme

Clone	A	B	C	D	E
X	+	+	−	−	+
Y	+	−	+	+	+
Z	−	−	+	−	+

33. You have isolated a new fungus and have obtained an *arg⁻* and an *ad⁻* auxotroph. You cross these two strains and collect four hundred random spores that you plate on minimal media. If twenty-five spores grow, what is the distance between these two genes?

34. Suppose that you have determined the order of three genes to be a, c, b, and that by doing two-point crosses you have determined distances as a — c = 10 and c — b = 5. If interference is −1.5, and the three-point cross is

$$\frac{A\,C\,B}{a\,c\,b} \times \frac{a\,c\,b}{a\,c\,b}$$

what frequency of double crossovers do you expect?

35. Three distinct traits, pab, pk, and ad, were involved in a cross of *Neurospora*. From the cross *pab pk⁺ ad⁺* × *pab⁺ pk ad,* the following ordered tetrads were recovered:

Spores	I	II	III	IV	V	VI	VII	VIII
1, 2	*pab pk⁺ad⁺*	*pab pk⁺ad⁺*	*pab pk⁺ad⁺*	*pab/ pk⁺ad⁺*	*pab pk⁺ad⁺*	*pab pk⁺ad⁺*	*pab pk⁺ad*	*pab pk⁺ad*
3, 4	*pab pk⁺ad⁺*	*pab⁺pk ad*	*pab pk ad*	*pab pk⁺ad*	*pab⁺pk ad*	*pab⁺pk ad*	*pab⁺pk ad*	*pab⁺pk ad⁺*
5, 6	*pab⁺pk ad*	*pab pk⁺ad⁺*	*pab⁺pk⁺ad⁺*	*pab⁺pk ad⁺*	*pab pk ad*	*pab pk⁺ ad*	*pab⁺ pk ad⁺*	*pab pk⁺ ad⁺*
7, 8	*pab⁺pk ad*	*pab⁺pk ad*	*pab⁺pk ad*	*pab⁺pk ad*	*pab⁺pk⁺ad⁺*	*pab⁺pk ad⁺*	*pab pk⁺ad⁺*	*pab⁺pk ad*
	34	35	9	7	2	2	1	3

Based on the data, construct a map of the three genes. Be sure to indicate centromeres.

36. Three mouse-human cell lines were scored for the presence (+) or absence (−) of human chromosomes, and the results appear below.

Cell Line	Human Chromosome							
	1	2	3	4	5	14	15	18
A	+	+	+	+	−	−	−	−
B	+	+	−	−	+	+	−	−
C	+	−	+	−	+	−	+	−

If a particular gene is located on chromosome 3, which clones should be positive for the enzyme?

Suggested Readings for chapter 6 are on page 643.

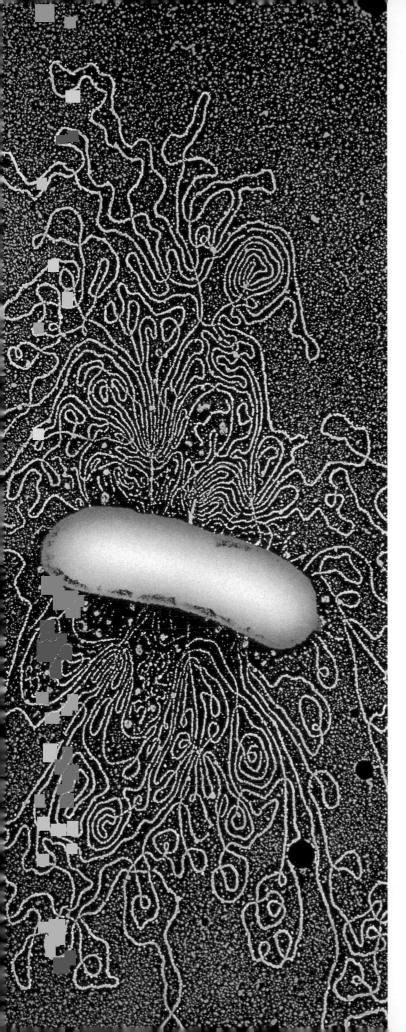

7

Linkage and Mapping in Prokaryotes and Viruses

OBJECTIVES

♦ To learn techniques for cultivation of bacteria and bacterial viruses

♦ To learn what constitutes phenotypes of bacteria and their viruses

♦ To study life cycles and sexual processes in bacteria and their viruses

♦ To make use of bacterial and phage life-cycle processes to map their chromosomes

Transmission electron micrograph of the common intestinal bacterium, *Escherichia coli* (7,857×)

© *Dr. Gopal Murti/Science Photo Library/Photo Researchers, Inc.*

All organisms and viruses have genes located sequentially in the genetic material; and all, with the possible exception of a small group of viruses, can have recombination between homologous (equivalent) pieces of genetic material. Because such recombination can occur, it is possible to map the location and sequence of genes along the chromosomes of all organisms and almost all viruses. The unique properties of the life cycles of bacteria and viruses require special mapping techniques. We will study these techniques in this chapter because of the enormous importance bacteria and viruses have assumed in genetic research in the past four decades. It is through work with bacteria and viruses that we have entered the modern era of molecular genetics, the subject of the second section of this book.

Bacteria (including the cyanobacteria that were formally called blue-green algae) make up the prokaryotes. The true bacteria can be classified according to shape: A spherical bacterium is called a **coccus;** a rod-shaped bacterium is called a **bacillus;** and a spiral bacterium is called a **spirillum.** Prokaryotes do not undergo mitosis or meiosis but simply divide in half after their single chromosome, a circle of DNA (deoxyribonucleic acid), the genetic material, has replicated (see chapter 9). Viruses do not even divide; they are mass-produced within a host cell. Several properties of bacteria and viruses have made them especially suitable for genetic research.

BACTERIA AND VIRUSES IN GENETIC RESEARCH

First, bacteria and viruses generally have a very short generation time. Some viruses replicate a hundredfold in about an hour; an *E. coli* cell (*Escherichia coli,* the common intestinal bacterium, discovered by Theodor Escherich in 1885) doubles every twenty minutes. In contrast, there is a generation time of fourteen days in fruit flies, a year in corn, and twenty years or more in human beings.

Second, bacteria and viruses have much less genetic material than do eukaryotes, and the organization of this material is much simpler. The term prokaryote arises from the fact that these organisms do not have true nuclei (*pro* means before and *karyon* means kernel or nucleus); they have no nuclear membranes (see fig. 3.2) and only a single, relatively "naked," chromosome: they are haploid. Viruses are even simpler. They consist almost entirely of genetic material surrounded by a protein coat. Or, more precisely, the viruses in which we are interested in this chapter, the bacterial viruses or **bacteriophages**—or just **phages** (Greek, eating)—are exclusively genetic material surrounded by a protein coat (fig. 7.1). Some animal viruses are more complex (see Box 7.1). We will look further at animal viruses in chapters 12 and 15.

Figure 7.1

Phage T2 and its chromosome. (*a*) The chromosome, which is about 50 μm long, has burst from the head. (*b*) The intact phage.

(a) A. K. Kleinschmidt, et al. "Darstellung und Langen messungen des gesamten Deoxyribose-nucleinsaüre—Inhaltes von T2-Bacteriophagen." Biochemica et Biophysica Acta 61(1962): 857–864. Reproduced by permission.

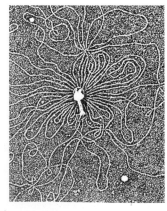

(a)

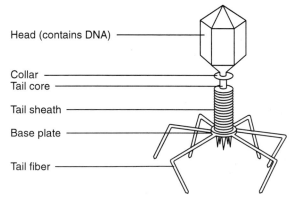

Head (contains DNA)

Collar
Tail core

Tail sheath

Base plate

Tail fiber

(b)

The cause of animal influenza (the flu) is a virion (virus particle) of extreme complexity compared with a phage. It is about 100 nanometers in diameter (Box fig. 1) and covered with spikes of two types (Box fig. 2). The H spikes, so called because of their hemagglutinin ability (they cause red blood cells to clump), allow the virions to attach to host cells. The other type of spike is an N spike, so called because it is the enzyme neuraminidase. Presumably this enzyme allows the virions to get out of host cells. Immunity, primarily to the H spikes, protects a person from being reinfected by the same strain of influenza. The worldwide epidemics of human influenza that occur every ten years or so are the result of a major change in the structure of the H spike, which creates a new strain. These H spike changes are probably due to recombination with animal forms of the virus.

The H and N spikes are embedded in a lipid bilayer membrane that surrounds a protein matrix. This bilayer originates from the infected host cell's own membrane. Within the virion are eight segments of single-stranded RNA. Each segment is capable of directing the synthesis of one or two of the thirteen known proteins produced by the virion.

The exact sequence of events during an infection is not precisely known; however, much is understood, especially regarding the complex structure of the virion. All of the RNAs have been sequenced (see chapter 12) and the three-dimensional structures of the most important proteins have been established. Many interesting questions remain regarding the functioning of this virus.

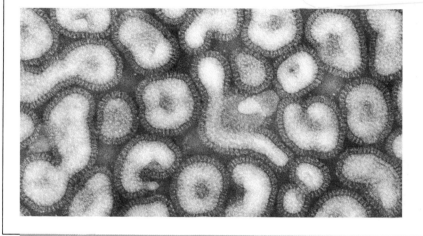

Box figure 1

Influenza virion. Magnification 282,100×.

Source: Courtesy of Dr. K. G. Murti, St. Jude Children's Research Hospital. Reproduced with permission from the Annual Review of Genetics, *Volume 12, copyright 1978 by Annual Reviews, Inc.*

Continued on next page

Viruses are usually classified first by the type of genetic material (nucleic acid) they have (DNA or RNA; see chapter 9), then by structural features of their protein surfaces (**capsids**) such as type or symmetry, presence of an envelope, number of discrete protein subunits (**capsomeres**) in the capsid, and general size. Viruses also can be classified by their host preference (animal, plant, or bacterium; figs. 7.2, 7.3). Viruses are obligate parasites. Outside of a host, they are inert molecules. Once their genetic material penetrates a host cell, they can take over the metabolism of that cell and construct multiple copies of themselves. Details of this and alternative infection pathways are discussed later in the chapter.

A third reason for the use of bacteria and viruses in genetic study is their ease of handling. Millions of bacteria can be handled in a single culture with a minimal amount of work compared with the effort required to grow the same number of eukaryotic organisms such as fruit flies or corn. (Some eukaryotes, such as yeast or *Neurospora,* can, of course, be handled using prokaryotic techniques, as we saw in chapter 6.) The following discussion is an expansion of the techniques introduced in chapter 6 that are used in bacterial and viral studies.

TECHNIQUES OF CULTIVATION

Since different groups of bacteria have diverse nutritional requirements, different media have been developed on which they are grown in the laboratory. All organisms need an energy source, a carbon source, nitrogen, sulphur, phosphorus, several metallic ions, and water. Those that

BOX 7.1—*Continued*

Box figure 2
Structure of the influenza virion
From "The Epidemiology of Influenza" by Martin M. Kaplan and Robert G.
Webster. Copyright © 1977 by Scientific American, Inc. All rights reserved.

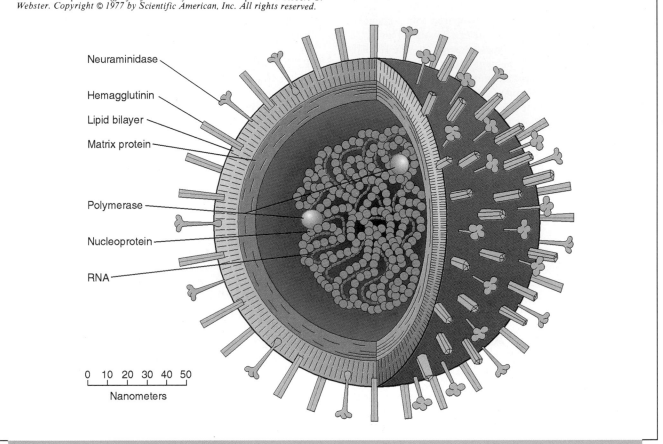

Figure 7.2
Plant and animal viruses. (*a*) Turnip yellow mosaic virus particles, magnification 128,250×, like most plant viruses, contains single-stranded RNA. (*b*) Electron micrograph of an adenovirus particle, a DNA animal virus. (*c*) Ping-pong ball model of an adenovirus particle. Each capsomere (spherical subunit) is about 50 Å in diameter.

Source: (a) is reproduced courtesy of Dr. T. C. Allen, Jr.; (b) and (c) are reproduced with permission from R. W. Horne et al., "The Icosahedral Form of an Adenovirus," Journal of Molecular Biology 1 (1959): 84–86. Copyright by Academic Press, Inc. (London) Ltd.

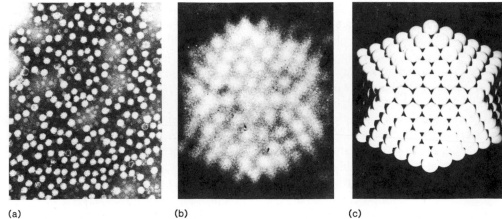

(a)   (b)   (c)

Figure 7.3

Stereo views of the capsid (protein coat) of the poliovirus virion, an animal virus made up of four coat proteins (VP1, *blue;* VP2, *yellow;* VP3, *red;* and VP4, *green*), each in sixty copies. In (*a*) the interior is exposed, showing VP4, which is not on the outer surface, shown in (*b*).

Source: Hogle, Chow, and Filman, 1985. Science 229:1362, Figure 5a, b. *"Three-Dimensional Structure of Poliovirus at 2.9 Å Resolution."* Copyright 1988 by the AAAS.

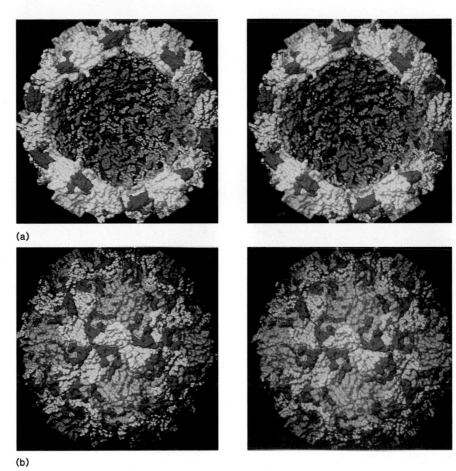

(a)

(b)

require an organic form of carbon are termed **heterotrophs.** Those that can utilize carbon as carbon dioxide are termed **autotrophs.** All bacteria obtain their energy either by photosynthesis or chemical oxidation. Bacteria are usually grown in or on a chemically defined **synthetic medium** either in liquid, or in test tubes or on petri plates using an agar base to supply rigidity. When one cell is placed on the medium in the plate, it will begin to divide. After incubation, often overnight, a colony, or clone, will exist where there was previously only one cell. Overlapping colonies form a confluent growth (fig. 7.4). A culture medium that has only the minimal necessities required by the bacterial species being grown is referred to as minimal medium (table 7.1).

Alternatively, bacteria can be grown on a medium that supplies, in addition to their minimal requirements, the more complex substances that the bacteria normally synthesize, including amino acids, vitamins, and so on. A medium of this kind will allow the growth of strains of bacteria, called **auxotrophs,** that have nutritional requirements. The parent, or wild-type, strain is referred to as a **prototroph.** For example, a strain that has an enzyme defect in the pathway of the production of the amino acid histidine will not grow on a minimal medium because it has no way of obtaining histidine. If, however, histidine were provided in the medium, the organisms could grow. This type of mutant is called a **conditional-lethal mutant.** The organism would normally die, but under appropriate conditions, such as the addition of histidine, the organism can survive.

This histidine-requiring auxotrophic mutant could grow only on an **enriched,** or **complete, medium,** whereas the parent prototroph could grow on a minimal medium. Media are often enriched by adding complex mixtures of organic substances such as blood, beef extract, yeast extract, or peptone, a digestion product. Many media are made up of a minimal medium with the addition of only one other substance, such as an amino acid or a vitamin.

Figure 7.4

Bacterial colonies on a petri plate. Bacteria were streaked on the petri plate with an inoculation loop, a metal wire with a looped end, covered with bacteria. Streaks were begun at the upper right and continued around clockwise. With a heavy inoculation on the needle, bacterial growth is confluent. Eventually, only a few bacteria are left that form single colonies at the upper left.

Photo by Robert Tamarin

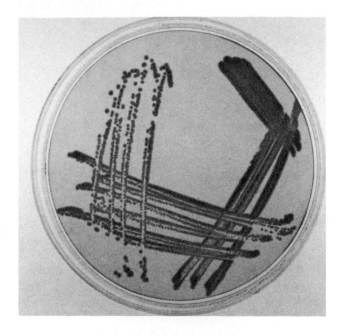

TABLE 7.1 Minimal Synthetic Medium for Growing *E. coli,* a Heterotroph

Component	Quantity
$NH_4H_2PO_4$	1 g
Glucose	5 g
NaCl	5 g
$MgSO_4 \cdot 7H_2O$	0.2 g
K_2HPO_4	1 g
H_2O	1,000 ml

Source: Data from M. Rogosa, et al., *Journal of Bacteriology,* 54:13, 1947.

These are called **selective media;** their uses will be discussed later in the chapter. In addition to minimal, complete, and selective media, other media exist for purposes such as aiding in counting colonies, helping maintain cells in a nongrowth phase, and so on.

The experimental cultivation of viruses is somewhat different. Since viruses are obligate parasites, they can grow only in living cells. Thus, for the cultivation of phages, petri plates of appropriate media are inoculated with enough bacteria to form a continuous cover, or **bacterial lawn.** This bacterial culture serves as a medium for the growth of viruses added to the plate. Since the virus attack eventually results in rupture, or **lysis,** of the bacterial cell, addition of the virus produces clear spots, known as

Figure 7.5

Small viral plaques (phage λ) on a bacterial lawn of *E. coli*
© *Bruce Iverson, BSc*

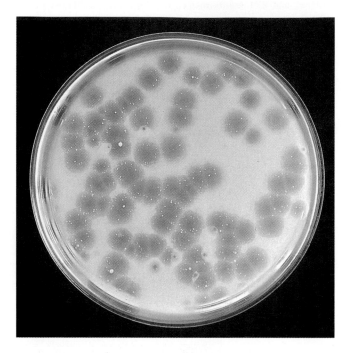

plaques, on the petri plates (fig. 7.5). Different types of bacteria can be used to determine growth characteristics of the various viral strains under study.

BACTERIAL PHENOTYPES

Bacterial phenotypes fall into three general classes: colony morphology, nutritional requirement, and drug or infection resistance.

Colony Morphology

The first of these classes, colony morphology, relates simply to the form, color, and size of the colony that grows from a single cell. A bacterial cell growing on an agar slant or petri plate divides as frequently as once every twenty minutes. Each cell will give rise to a colony, or clone, at the site of its original position. In a relatively short amount of time (e.g., overnight), the colonies will consist of enough cells to be seen with the unaided eye. The different morphologies observed among the colonies are usually under genetic control (fig. 7.6).

Nutritional Requirements

The second basis for classifying bacteria—nutritional requirements—reflects the failure of one or more enzymes in the biosynthetic pathways of the bacteria. If an auxotroph has a requirement for the amino acid cysteine that the parent strain (prototroph) does not have, then that auxotroph most likely has a nonfunctional enzyme in the

Figure 7.6
Various bacterial colonies form on petri plates. (*a*) Smooth, circular, raised surface. (*b*) Granular, circular raised surface. (*c*) Elevated folds on a flat colony with irregular edges. (*d*) Irregular elevations on a raised colony with an undulating edge.

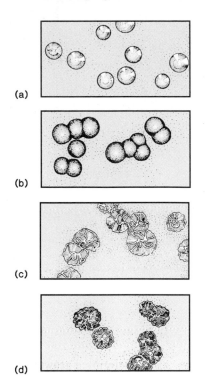

(a)

(b)

(c)

(d)

pathway for the synthesis of cysteine. Figure 7.7 shows five steps in cysteine synthesis; it also shows that each step is controlled by a different enzyme. All enzymes are proteins, and the sequences in the strings of amino acids that make up those proteins are determined by information in one or more genes (chapter 11). A normal or wild-type allele produces a normal, functional enzyme. The alternative allele may produce a nonfunctional enzyme. Recall the one-gene-one-enzyme hypothesis from chapter 2.

A technique known as **replica plating,** devised by Joshua Lederberg, is a rapid **screening technique** that makes it possible to determine quickly whether a given strain of bacteria is auxotrophic for a particular metabolite. In this technique, a petri plate of complete medium is inoculated with bacteria. The resulting growth will have a certain configuration of colonies. This plate of colonies is pressed onto a piece of sterilized velvet. Then any number of petri plates, each containing a medium that lacks some specific metabolite, can be pressed onto this velvet to pick up inocula in the same pattern as the growth on the original plate (fig. 7.8). If a colony grows on the complete medium but does not grow on a plate with a medium in which a metabolite is missing, the inference is that the colony is made of auxotrophic cells that require the metabolite absent from the second plate. Samples of this bacterial strain can be obtained from the colony growing on

complete medium for further study. The nutritional requirement of this strain is its phenotype. The methionine-requiring auxotroph of figure 7.8 would be designated as Met⁻ (methionine-minus or Met-minus).

In terms of energy sources, the notation means a different thing. For example, a strain of bacteria that can utilize the sugar galactose as an energy source would be Gal⁺. If it could not utilize galactose it would be termed Gal⁻. The latter strain will not grow if galactose is its sole carbon source. It will grow if a sugar other than galactose is present. Note that a Met⁻ strain needs methionine to grow, whereas a Gal⁻ strain needs a carbon source other than galactose; it cannot use galactose.

Resistance and Sensitivity

The third class of phenotypes in bacteria involves resistance and sensitivity to drugs, phages, and other environmental insults. For example, penicillin, an antibiotic that prevents the final stage of cell-wall construction in bacteria, will kill growing bacterial cells. Nevertheless, we frequently find a number of cells that do grow in the presence of penicillin. These colonies are resistant to the drug. This resistance is under simple genetic control. The phenotype is penicillin resistant (Pcn^r) as compared with penicillin sensitive (Pen^s), the normal condition, or wild-type. Numerous antibiotics are used in bacterial studies (table 7.2).

Drug sensitivity provides another rapid screening technique for isolating nutritional mutations. For example, if we were looking for mutants that lacked the ability to synthesize a particular amino acid (e.g., methionine), we could grow large quantities of bacteria (prototrophs) and then place them on a medium that lacked methionine but had penicillin. Here any growing cells would be killed. But methionine auxotrophs would not grow and, therefore, they would not be killed. The penicillin could then be washed out and the cells reinoculated onto a complete medium. The only colonies that form should be composed of cells that are methionine auxotrophs (Met⁻).

Screening for resistance to phages is similar to screening for drug resistance. When bacteria are placed in a medium containing phages, only those bacteria that are resistant to the phages will grow and produce colonies. They can thus be isolated easily and studied.

VIRAL PHENOTYPES

In regard to viral phenotypes, we will consider only the phenotypes of bacteriophages, which fall roughly into two categories: plaque morphology and growth characteristics on different bacterial strains. For example, T2, an *E. coli* phage, produces small plaques with fuzzy edges (genotype *r*⁺). Rapid-lysis mutants (genotype *r*) produce large,

Figure 7.7

Five-step conversion of methionine to cysteine. Each of the five steps is controlled by a different enzyme.

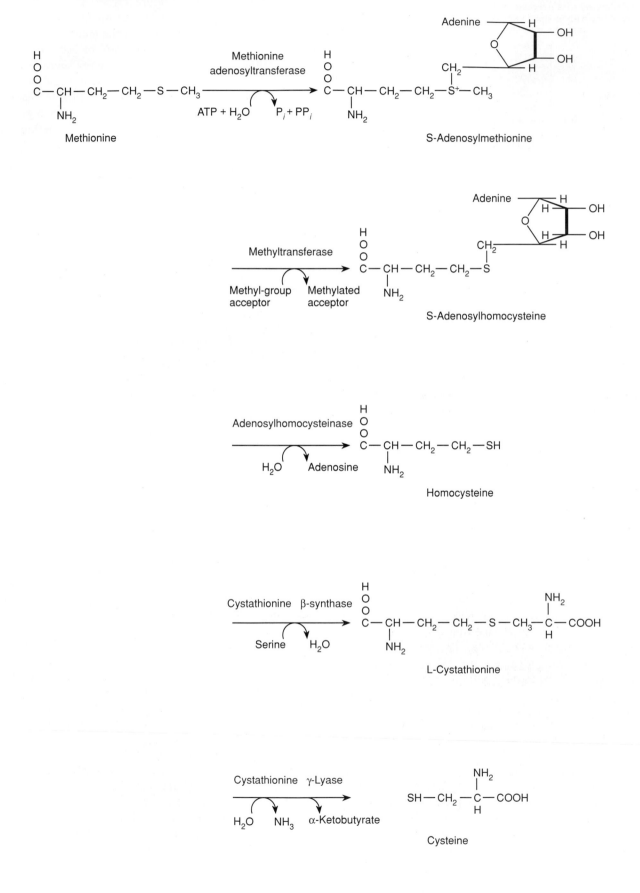

Figure 7.8

Technique of replica plating. (*a*) A pattern of colonies from a plate of complete medium is transferred (*b*) to a second plate of medium that lacks methionine. (*c*) Where colonies fail to grow on the second plate, we can infer that the original colony in that location was a methionine-requiring auxotroph.

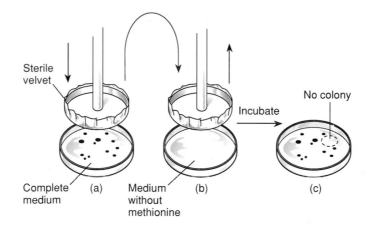

Sterile velvet

No colony

Incubate

Complete medium (a) Medium without methionine (b) (c)

TABLE 7.2   Some Antibiotics and Their Antibacterial Mechanisms

Antibiotic	Microbial Origin	Mode of Action
Penicillin G	*Penicillium chrysogenum*	Blocks cell-wall synthesis
Tetracycline	*Streptomyces aureofaciens*	Blocks protein synthesis
Streptomycin	*Streptomyces griseus*	Interferes with protein synthesis
Terramycin	*Streptomyces rimosus*	Blocks protein synthesis
Erythromycin	*Streptomyces erythraeus*	Blocks protein synthesis
Bacitracin	*Bacillus subtilis*	Blocks cell-wall synthesis

Joshua Lederberg (1925–   )
Courtesy of Dr. Joshua Lederberg.

smooth-edged plaques (fig. 7.9). Similarly, T4, another *E. coli* phage, has rapid-lysis mutants that produce large, smooth-edged plaques on *E. coli* B but will not grow at all on *E. coli* K12, a different strain. Rapid-lysis mutants thus illustrate both colony morphology phenotypes and growth-restriction phenotypes of phages.

SEXUAL PROCESSES IN BACTERIA AND VIRUSES

Although bacteria and viruses are ideal subjects for biochemical analysis, they would not be useful for genetic study if they did not have sexual processes. If we define a sexual process as the combining of genetic material from two individuals, then the life cycles of bacteria and viruses include sexual processes. Although they do not undergo sexual reproduction by means of the fusion of haploid gametes, bacteria and viruses do undergo processes in which genetic material from one cell or virus can be incorporated into another cell or virus, forming recombinants. Actually, bacteria have four different methods to gain access to foreign genetic material: **transformation, conjugation, transduction,** and **sexduction** (fig. 7. 10).

Phages can exchange genetic material when a bacterium is infected by more than one virus particle (**virion**). During the process of viral infection, the genetic material of different phages can exchange parts (recombine; see fig. 7.10). We will examine the exchange processes in bacteria and then in bacteriophages and proceed to the use of these methods for mapping bacterial and viral chromosomes.

Linkage and Mapping in Prokaryotes and Viruses   **143**

Figure 7.9

Normal (r^+) and rapid-lysis (r) mutants of phage T2. Mottled plaques occur when r and r^+ phages grow together.

From Molecular Biology of Bacterial Viruses. *By Gunther S. Stent, copyright © 1963 by W. H. Freeman and Company. Reprinted by permission.*

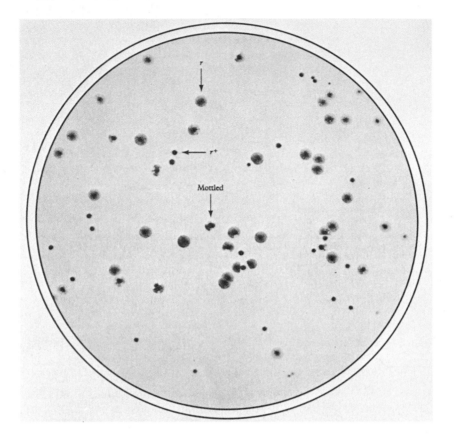

(Chromosome refers to the structural entity in the cell or virus made up of the genetic material. In eukaryotes it is double-stranded DNA complexed with proteins [chapter 14]. Staining of this eukaryotic organelle led to the term chromosome, which means "colored body." In prokaryotes the chromosome is a circle of double-stranded DNA. In viruses it is virtually any combination of linear or circular, single- or double-stranded, RNA or DNA. Sometimes the term **genophore** is used for the prokaryotic and viral genetic material, limiting the word chromosome to the eukaryotic organelle. We will use the term chromosome for the intact genetic material of any organism or virus.)

Transformation

Transformation was first observed in 1928 by F. Griffith and later (1944) examined at the molecular level by O. Avery and his colleagues, who demonstrated that DNA was the genetic material of bacteria. The details of these experiments are presented in chapter 9. In transforma-

tion, a cell takes up extraneous DNA found in the environment and incorporates it into its genome (genetic material) through recombination. Not all cells are competent to be transformed, and not all extracellular DNA is competent to transform. To be competent to transform, the extracellular DNA must be double-stranded and relatively large. To be competent to be transformed, a cell must have the surface protein, **competence factor,** which binds to the extracellular DNA in an energy-requiring reaction.

Mechanisms of Transformation

As the extracellular DNA is brought into the cell, one of the strands is hydrolyzed by an intracellular DNAase (DNA-degrading enzyme) that apparently uses the energy of hydrolysis to pull the remaining strand into the cell. The single strand brought into the cell can then be incorporated into the host genome by a form of crossing over (fig. 7.11), whose molecular mechanisms will be discussed in chapter 16. Note that unlike eukaryotic crossing over,

Figure 7.10
Summary of bacterial and viral sexual processes

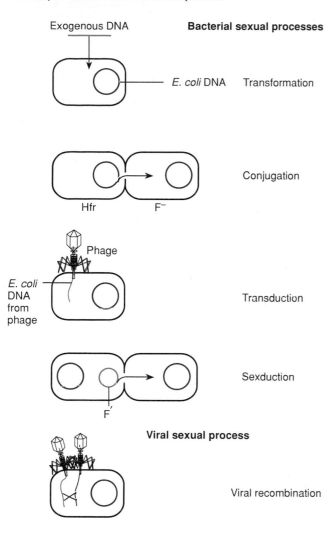

Bacterial sexual processes

Exogenous DNA — *E. coli* DNA — Transformation

Hfr — F⁻ — Conjugation

Phage — *E. coli* DNA from phage — Transduction

F′ — Sexduction

Viral sexual process

Viral recombination

this is not a reciprocal process. The bacterial chromosome is incorporating part of the foreign DNA. The remaining single-stranded DNA will be degraded by host enzymes; linear DNA is degraded rapidly in prokaryotes.

Transformation is a very efficient method of mapping in some bacteria, especially those that are inefficient in other mechanisms of DNA intake, such as transduction. For example, a good deal of the mapping in *Bacillus subtilis* (*B. subtilis*) has been done through the process of transformation; *E. coli,* however, is inefficient in transformation and hence other methods are used to map its chromosome.

Transformation Mapping

The general procedure for transformation mapping is as follows. Two strains of bacteria are selected such that one strain has mutations at two loci. For example, strain B might have wild-type alleles for the synthesis of the amino acids histidine (*his⁺*) and methionine (*met⁺*). Strain A should then be auxotrophic for both of these amino acids (*his⁻* and *met⁻*). The DNA is isolated from strain A and put in the culture medium with strain B under conditions that favor transformation; for example, DNA uptake can be greatly enhanced by the addition of calcium chloride. After a time interval for transformation to take place, the B strain is tested for its methionine and histidine properties (fig. 7.12).

Since we are interested in transformants (auxotrophs), the nontransformed cells can be eliminated by culturing on minimal medium with penicillin. The auxotrophs will not grow and hence will not be killed. The nontransformed prototrophs (*his⁺* and *met⁺*), however, will be killed. After penicillin is washed away, the remaining cells can be plated on complete medium. By replica plating onto media lacking histidine and methionine, it is possible to determine the phenotype of each transformed cell. Any transformation requires two crossovers, of which one must

Figure 7.11
Exogenous linear genetic material can be incorporated into a circular bacterial chromosome by two crossovers.

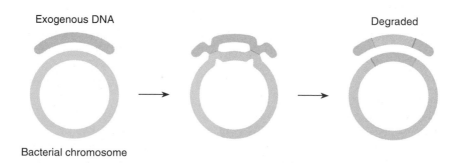

Exogenous DNA

Degraded

Bacterial chromosome

Figure 7.12

Procedure for isolating bacterial transformants. The two strains of bacteria differ in their histidine and methionine requirements; both strains are penicillin sensitive.

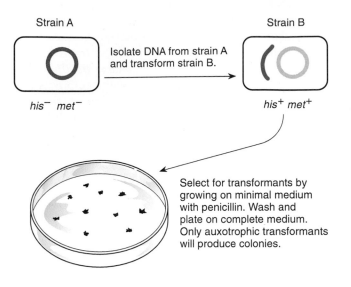

Strain A

his⁻ met⁻

Isolate DNA from strain A and transform strain B.

Strain B

his⁺ met⁺

Select for transformants by growing on minimal medium with penicillin. Wash and plate on complete medium. Only auxotrophic transformants will produce colonies.

Figure 7.13

Three crossover events that result in transformation in the experiment of figure 7.12. In (*a*) and (*b*) only one locus is transformed because a crossover has occurred between the loci. In (*c*) both loci have been transformed because no crossover has occurred between them.

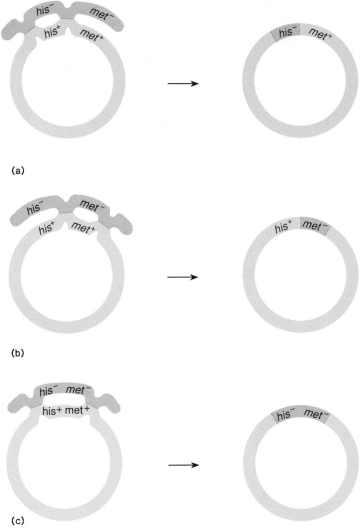

(a)

(b)

(c)

be outside the *his-met* region (fig. 7.13). The second crossover can be either on the other side of the region (double transformant; fig. 7.13*c*) or between the loci (single transformants; fig. 7.13*a, b*). A ratio of single transformants to all transformants will measure the relative occurrence of a crossover between the two loci. Presumably, the higher this value is, the more probable it is that there was a crossover between the two loci and therefore the further the two loci are apart. For example, below are numbers obtained in the experiment of figure 7.12:

34 *his⁻ met⁺* transformants

28 *his⁺ met⁻* transformants

194 *his⁻ met⁻* transformants

Then relative recombination frequency would be

number of single transformants/total number of transformants =

$(34 + 28)/(34 + 28 + 194) = 0.24$

Note again the differences with mapping eukaryotes like *Drosophila*. There every chromatid is in a position to cross over in the tetrad. Here only a few chromosomes will undergo transformation, dependent on the uptake of the foreign DNA, the probability of correct positioning of the DNA for crossing over, and then the actual crossing over. In *Drosophila* we take the percentage of recombinants as a map distance directly. Here we must use the relative occurrence of crossovers between loci among only those cells that were transformed in at least one of the loci. In other words, here we must first select all transformants and then

get a relative occurrence of the crossover between the loci in order to calculate a relative recombination index. Unlike mapping in eukaryotes, in which all offspring are usually either parental or recombinants, prokaryotic genetic work frequently involves selection techniques in which cells having some event are selected first and then among them a relative index is calculated. The selection techniques remove all cells in which no event took place. However, in both systems, the greater the distance between two loci, the greater the probability that there will be a crossover between them.

By systematically examining many loci, relative order can be obtained. For example, if locus *A* is closely linked to locus *B* and *B* to *C*, we can establish the order *A-B-C*. It is not possible by this method to determine exact order for very closely linked genes. For this information we need to rely upon transduction, which will be considered shortly.

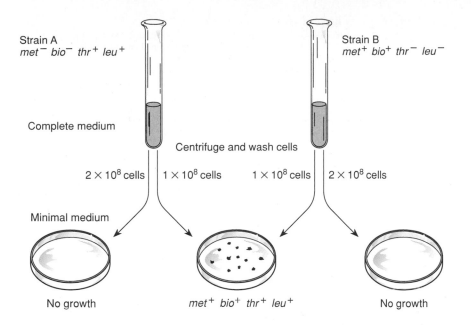

Strain A
met⁻ bio⁻ thr⁺ leu⁺

Complete medium

Centrifuge and wash cells

Strain B
met⁺ bio⁺ thr⁻ leu⁻

2×10^8 cells 1×10^8 cells 1×10^8 cells 2×10^8 cells

Minimal medium

No growth

met⁺ bio⁺ thr⁺ leu⁺

No growth

Figure 7.14
Lederberg and Tatum's cross showing that
E. coli undergoes genetic recombination

However, transformation has allowed us to determine that the map of *B. subtilis* is circular, a phenomenon found in all prokaryotes and many phages. (The *E. coli* map is shown later in fig. 7.30.)

Conjugation

In 1946, Joshua Lederberg and Edward L. Tatum (later to be Nobel laureates) discovered that *E. coli* cells can exchange genetic material through the process of conjugation. They mixed two auxotrophic strains of *E. coli*. One strain was methionine and biotin requiring (Met⁻ Bio⁻), and the other was threonine and leucine requiring (Thr⁻ Leu⁻). This cross is shown in figure 7.14. Remember that if a strain is Met⁻ Bio⁻, it is, without saying, wild-type for all other loci. Thus a cell with the Met⁻ Bio⁻ phenotype actually has the genotype of *met⁻ bio⁻ thr⁺ leu⁺*. Similarly, the Thr⁻ Leu⁻ strain is actually *met⁺ bio⁺ thr⁻ leu⁻*. (Remember that symbols such as "Thr⁻" represent phenotypes; symbols such as "*thr⁻*" represent genotypes.)

Lederberg and Tatum used multiple auxotrophs in order to rule out spontaneous reversion (mutation). About one in 10^6 Met⁻ cells will spontaneously become prototrophic (Met⁺) every generation. However, with multiple auxotrophs the probability of a spontaneous reversion (e.g., *met⁻ → met⁺*) of several loci simultaneously becomes vanishingly small. (In fact, the control plates in the experiment, illustrated in fig. 7.14, showed no growth for parental double mutants.) After mixing the strains, Lederberg and Tatum found that about one cell in 10^7 was prototrophic (*met⁺ bio⁺ thr⁺ leu⁺*). Transformation was ruled out as an explanation by conducting several types of experiments showing that direct cell-to-cell contact was required for this type of recombination.

In one experiment, one strain was put in each arm of a U-tube at the bottom of which was a sintered glass filter

Figure 7.15
The U-tube experiment of B. Davis

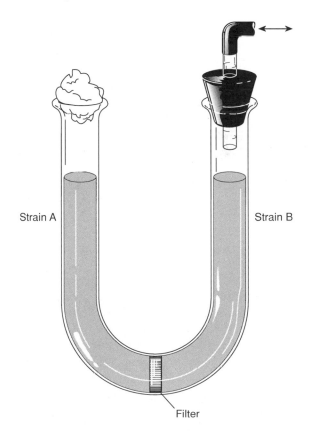

Strain A

Strain B

Filter

(fig. 7.15). The liquid and large molecules, including DNA, were mixed by alternate application of pressure and suction to one arm of the tube; whole cells did not pass through the filter. The result of this mixture was that the fluids surrounding the cells, as well as any large molecules (e.g., DNA), could be freely mixed while the cells were kept separate. After cell growth stopped in the two arms (in complete medium), the contents were plated out on minimal medium. There were no prototrophs in either arm. Therefore, cell-to-cell contact was required for the genetic material of the two cells to come in contact and then recombine.

At first, Lederberg and Tatum interpreted their results in light of conventional sexual processes in which two cells fused forming a diploid zygote that then underwent meiosis. This conventional view of bacterial sexuality was shown to be incorrect. In bacteria, conjugation is a one-way transfer with one strain acting as a donor and the other as a recipient. This was demonstrated by experiments in which one or the other strain was killed just before mixing. When one strain, the recipient, was killed before mixing, the experiment failed (there was no recombination). However, when the other strain, the donor, was killed before mixing, the experiment still worked (recombination occurred). These experiments demonstrated that conjugation was not a symmetrical, reciprocal process, but one in which one strain acted as a donor whereas the other acted as a recipient.

F Factor

It was shown that sometimes, if stored for a long time, donor cells can lose the ability to be donor cells but can regain the donor ability if they are mated with other donor strains. This led to the hypothesis that a **fertility factor, F,** made any strain that carried it a male (donor) strain, termed F^+. The strain that did not have the F factor, referred to as a female, or F^-, strain, served as a recipient for genetic material during conjugation.

The F factor is a **plasmid,** a term originally coined by Lederberg, which refers to independent, self-replicating genetic particles. Plasmids are almost exclusively circles of double-stranded DNA. (Plasmids are at the heart of recombinant DNA technology, which will be discussed in detail in chapter 12.) They are auxiliary circles of DNA that many bacteria carry. They are usually much smaller than the bacteria's own chromosome.

It was found that the transfer of the F factor occurred far more frequently than did the transfer of other genetic material. That is, during conjugation, there was about one recombinant in 10^7 cells, whereas transfer of the F factor occurred at a rate of about one conversion of F^- to F^+ in every five conjugations. An *E. coli* strain was then discovered that transferred its genetic material at a rate about one thousand times that of the normal F^+ strain. This strain was called **Hfr,** for high frequency of recombination. Several other phenomena occurred simultaneously with this high rate of transfer. First, the ability to transfer the F factor itself dropped to almost zero in this strain. Second, not all loci were transferred at the same rate. Some loci were transferred much more frequently than others.

Escherichia coli cells are normally coated with hair-like **pili (fimbriae).** F^+ and Hfr cells have one to three additional pili (singular: pilus) called **F-pili,** or sex pili.

Figure 7.16

Electron micrograph of conjugation between an F^+ (upper right) and F^- (lower left) cell with the F-pilus between them. Magnification 3,700×.

Courtesy of Wayne Rosenkrans and Dr. Sonia Guterman.

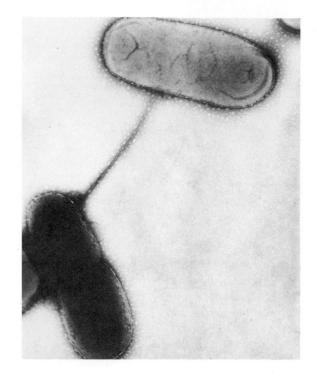

During conjugation these sex pili form a connecting bridge between the F^+ (or Hfr) and F^- cells (fig. 7.16). Once a connection is made, the sex pilus then contracts to bring the two cells into contact. DNA transfer then takes place by a nick in either the plasmid (in F^+ cells) or the bacterial chromosome (in Hfr cells). A single strand of the double-stranded donor DNA is then passed from the F^+ or Hfr cell to the F^- cell across the cell membranes. DNA replication in both the donor and recipient cells reestablishes double-stranded DNA in both cells. It is the F factor itself that has the genes for sex-pilus formation and transfer of DNA to a conjugating F^- cell. At least 22 genes are involved in the transfer process, including genes for the pilus protein, nicking the DNA, and regulation of the process.

In the transfer process of conjugation, the donor cell does not lose its F factor or its chromosome because only a single strand of the double helix of DNA is transferred; the remaining single strand is quickly replicated. (The process of DNA replication is described in chapter 9.) For a short while the F^- cell that has conjugated with an Hfr cell has two copies of whatever chromosomal loci were transferred: one copy of its own and one transferred in. Having these two copies, the cell is a partial diploid, or a **merozygote.** The new foreign DNA (**exogenote**) can be incorporated into the host chromosome (**endogenote**) by an even number of breakages and reunions between the

Figure 7.17

Incorporation of external DNA into the host chromosome. This incorporation is accomplished by an even number of breakage and reunion events after the exogenote lines up (synapses) with the identical (homologous) region of the host chromosome.

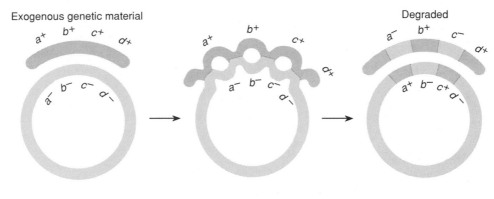

Exogenous genetic material

Degraded

Bacterial chromosome

Elie Wollman (1917–   )
Courtesy of Dr. Elie Wollman and the Pasteur Institute.

TABLE 7.3 Genotypes of Hfr and F⁻ Cells Used in an Interrupted Mating Experiment

Hfr	F⁻
*str*ˢ	*str*ʳ
*azi*ʳ	*azi*ˢ
*tonA*ʳ	*tonA*ˢ
leu⁺	*leu*⁻
galB⁺	*galB*⁻
lac⁺	*lac*⁻

two, just as in transformation. The unincorporated linear DNA is soon degraded by enzymes. This process is diagrammed in figure 7.17.

Interrupted Mating

To demonstrate that the transfer of genetic material from the donor to the recipient cell during conjugation was a linear event, F. Jacob and E. Wollman devised the technique of **interrupted mating.** In this technique, F⁻ and Hfr strains were mixed together in a food blender. After a specific amount of time, the blender was turned on. The spinning separated cells that were conjugating and thereby interrupted mating. Then the F⁻ cells were tested for various alleles originally in the Hfr cell. In an experiment like this, the Hfr strain is usually sensitive to an antibiotic such as streptomycin that will kill all the Hfr cells. Then the genotypes of only the F⁻ cells can be determined by replica plating without fear of contamination by Hfr cells.

The mating outlined in table 7.3 was carried out. In the food blender, an Hfr strain sensitive to streptomycin (*str*ˢ) but resistant to azide (*azi*ʳ), resistant to phage T1 (*tonA*ʳ), and prototrophic for the amino acid leucine (*leu*⁺) and the sugars galactose (*galB*⁺) and lactose (*lac*⁺), was

added to an F⁻ strain that was resistant to streptomycin (*str*ʳ), sensitive to azide (*azi*ˢ), sensitive to T1 (*tonA*ˢ), and auxotrophic for leucine, galactose, and lactose (*leu*⁻, *galB*⁻, and *lac*⁻). After a specific number of minutes (ranging from zero to sixty), the food blender was turned on. To kill all the Hfr cells, the cell suspension was plated on a medium containing streptomycin. The cells remaining were then plated on medium without leucine. The only colonies that resulted were F⁻ recombinants. They must have received the *leu*⁺ allele from the Hfr in order to grow on a medium lacking leucine. Hence all colonies had been selected to be F⁻ recombinants. Then, by replica plating onto specific media, the *azi, tonA, lac,* and *galB* alleles were determined and the percentage of recombinant colonies that had the original Hfr allele (*leu*⁺) was noted. (Note that by trial and error it was determined that leucine should be the locus to use to select for recombinants. As we will see, the leucine locus entered first.)

The graph of figure 7.18 shows that as time of mating increases, two things happen. First, new alleles enter the F⁻ cells from the Hfr cells. The *tonA*ʳ allele first appears among recombinants after ten minutes of mating, whereas

Figure 7.18

Frequency of Hfr genetic characters among recombinants after interrupted mating

From F. Jacob and E. L. Wollman, Sexuality and the Genetics of Bacteria. Copyright © 1961 Academic Press, Orlando, FL. Reprinted by permission.

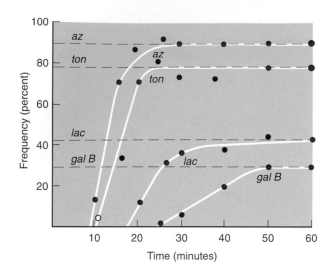

Figure 7.19

Conjugation between an Hfr cell (with the a^+ allele) and an F⁻ cell (with the a^- allele)

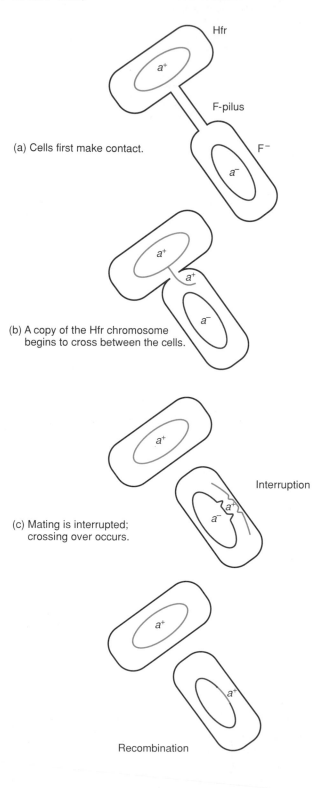

(a) Cells first make contact.

(b) A copy of the Hfr chromosome begins to cross between the cells.

(c) Mating is interrupted; crossing over occurs.

(d) The F⁻ cell now has the a^+ allele.

galB⁺ first enters the F⁻ cells after about twenty-five minutes. This suggests a sequential entry of loci into the F⁻ cells from the Hfr (fig. 7.19). Second, as time proceeds, the percentage of recombinants with a given allele from the Hfr increases. At ten minutes, *tonA*ʳ is first found among recombinants. After fifteen minutes, about 40% of recombinants have the *tonA*ʳ allele from the Hfr; and after about twenty-five minutes, about 80% of the recombinants have the *tonA*ʳ allele. This limiting percentage does not increase with additional time. The limiting percentage is lower for later entering loci, a fact explained by assuming that even without the food blender, mating is usually interrupted before completion by normal agitation alone.

Mapping and Conjugation

Lederberg, using conjugation to map genes, had found that some genes did not fit a linear pattern. In fact, he constructed a map with three branches to accommodate his data. But the work of Jacob and Wollman indicated that the bacterial chromosome was linear. A breakthrough in understanding occurred when interrupted matings were done with several different Hfr strains. These strains were of independent origin. The results were quite striking (table 7.4).

If we ponder this table for a short while, one fact will become obvious. The relative order of the loci is always the same. What differs is the point of origin and the direction of the transfer. These findings led Jacob and Wollman to suggest that the *E. coli* chromosome was circular, which not only fit perfectly with their data, but also solved Lederberg's problem of a nonlinear map.

TABLE 7.4 Gene Order of Various Hfr Strains Determined by Means of Interrupted Mating

Types of Hfr	HfrH	1	2	3	4	5	6	7	AB 311	AB 312	AB 313
	0	0	0	0	0	0	0	0	0	0	0
	T	L	Pro	Ad	B₁	M	Isol	T₁	H	Sm	Mtl
	L	T	T₁	Lac	M	B₁	M	Az	Try	Mal	Xyl
	Az	B₁	Az	Pro	Isol	T	B₁	L	Gal	Xyl	Mal
	T₁	M	L	T₁	Mtl	L	T	T	Ad	Mtl	Sm
	Pro	Isol	T	Az	Xyl	Az	L	B₁	Lac	Isol	S-G
	Lac	Mtl	B₁	L	Mal	T₁	Az	M	Pro	M	H
	Ad	Xyl	M	T	Sm	Pro	T₁	Isol	T₁	B₁	Try
	Gal	Mal	Isol	B₁	S-G	Lac	Pro	Mtl	Az	T	Gal
	Try	Sm	Mtl	M	H	Ad	Lac	Xyl	L	L	Ad
	H	S-G	Xyl	Isol	Try	Gal	Ad	Mal	T	Az	Lac
	S-G	H	Mal	Mtl	Gal	Try	Gal	Sm	B₁	T₁	Pro
	Sm	Try	Sm	Xyl	Ad	H	Try	S-G	M	Pro	T₁
	Mal	Gal	S-G	Mal	Lac	S-G	H	H	Isol	Lac	Az
	Xyl	Ad	H	Sm	Pro	Sm	S-G	Try	Mtl	Ad	L
	Mtl	Lac	Try	S-G	T₁	Mal	Sm	Gal	Xyl	Gal	T
	Isol	Pro	Gal	H	Az	Xyl	Mal	Ad	Mal	Try	B₁
	M	T₁	Ad	Try	L	Mtl	Xyl	Lac	Sm	H	M
	B₁	Az	Lac	Gal	T	Isol	Mtl	Pro	S-G	S-G	Isol

*Order of Transfer of Genetic Characters**

*The 0 refers to origin of transfer.

From F. Jacob and E. L. Wollman, *Sexuality and the Genetics of Bacteria*. Copyright © 1961 Academic Press, Orlando, FL. Reprinted by permission.

Jacob and Wollman proposed that normally the F factor is an independent circular DNA entity in the F⁺ cell and that during conjugation only the F factor is passed to the F⁻ cell. Since it is a small fragment of DNA, it can be passed entirely in a high proportion of conjugations before spontaneous separation of the cells. Every once in a while, however, the F factor becomes integrated into the chromosome of the host, which then becomes an Hfr cell. The point of integration can be different in different strains. However, once the F factor is integrated, it determines the initiation point of transfer of the *E. coli* chromosome, as well as the direction of transfer.

The F factor is the last part of the *E. coli* chromosome to be passed from the Hfr cell. This explains why an Hfr, in contrast to an F⁺, rarely passes the F factor itself. In the original work of Lederberg and Tatum, the one recombinant in 10^7 cells was most likely from a conjugation between an F⁻ cell and an Hfr that had formed spontaneously from an F⁺ cell. Integration of the F factor is diagrammed in figure 7.20. The F factor can also reverse this process and loop out of the *E. coli* chromosome, a process we will examine in detail shortly, under F-duction (sexduction).

We could now diagram the *E. coli* chromosome and show the map location of all the known loci. The map units would be in minutes, having been obtained by interrupted mating. However, at this point the map would not be complete. Interrupted mating is most accurate in giving the relative position of loci that are not very close to each other.

With this method alone there would be a great deal of ambiguity as to the specific order of very close genes on the chromosome. The two remaining sexual processes in bacteria, sexduction and transduction, will provide the details unattainable by interrupted mating or transformation.

Sexduction (F-duction)

In figure 7.20 we saw how the F factor can become integrated into the host genome. It leaves the host genome in the reverse process, one of excision, or looping out. Occasionally, however, the process of looping out is not precise: The F factor takes with it some of the cell's genome (fig. 7.21). This new F factor is referred to as F′ (F-prime), and it endows the bacterial cells with certain interesting characteristics. First, F′ cells transfer genes at a very high rate even though they are not Hfr strains. This makes sense because we know that F⁺ cells transfer their F factor at a high rate during conjugation.

Second, the F′ has a much higher rate of spontaneous integration; and, unlike a normal F factor, it usually integrates at the same point that it originally occupied when in an Hfr cell. This too makes sense since a transferred F′ will have a region of homology with the chromosome in the new *E. coli* cell and will, therefore, tend to pair at that point. A single crossover will then reintegrate it at its original point. The first F′ discovered carried the *lac* (lactose) locus (fig. 7.21). Since then, many F′ factors have been

Figure 7.20

Integration of the F factor by a single crossover. A simultaneous breakage in both the F factor and the *E. coli* chromosome is followed by a reunion of the two broken circles to make one large circle, the Hfr chromosome. In this case, integration is between the *ton* and *lac* loci.

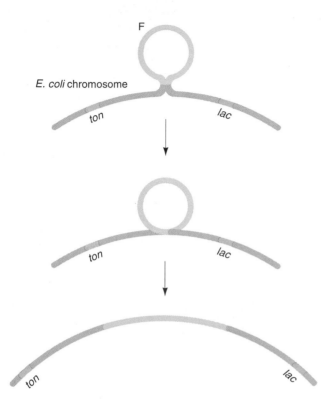

Figure 7.21

Occasional imprecise looping out of the F factor with part of the cell's genome included in the loop. The circular F factor is freed by a single recombination (crossover) at the loop point.

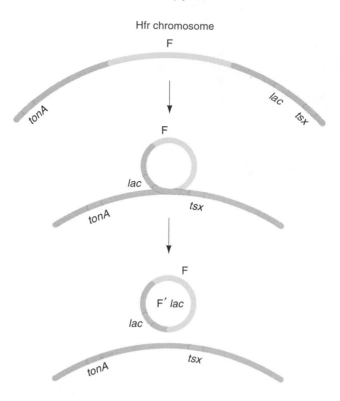

isolated, each with a different *E. coli* region incorporated (fig. 7.22). Thus, although no longer used much, sexduction is an additional tool available for mapping; generally, two loci must be close to each other if they both are on the same F′ factor.

More will be said about the details of mapping in the section on transduction, which is the most common way of determining gene order of very closely linked genes. The same methods used for transductional mapping are used in sexductional mapping and will not be repeated here. Sexduction has other uses besides mapping because partial diploids, or merozygotes, are formed in which a bacterial gene is found both on the bacterial chromosome and on the F′ factor. Their existence allows the study of the interaction of alleles in a normally haploid organism (chapter 13). Transduction, the final method of getting foreign DNA into a bacterial cell, is a phage-mediated process. Discussion of transduction must, therefore, be deferred until after some phage characteristics have been described.

LIFE CYCLES OF BACTERIOPHAGES

Phages are obligate intracellular parasites. Phage genetic material enters the bacterial cell after the phage has adsorbed to the cell surface. Once inside, the viral genetic material takes over the metabolism of the host cell. During the infection process, the cell's genetic material is destroyed while the genetic material of the virus is replicated many times. Viral genetic material controls the mass production of various protein components of the virus. New virus particles are then assembled within the host cell, which bursts open (is lysed) releasing a **lysate** of upwards of several hundred viral particles to infect other bacteria. This life cycle is shown in figure 7.23.

Recombination

The primary genetic work on phages has been done with a group of seven *E. coli* phages called the T series (T-odd: T1, T3, T5, T7; T-even: T2, T4, and T6) and several others including phage λ (lambda; fig. 7.24). The complex structure of T2 was shown in figure 7.1. The phage can undergo recombination processes when a cell is infected with two virions that are genetically distinct. Hence the phage

Figure 7.22

The circular chromosome of *E. coli* with arcs showing the genes carried by various F′ particles. The loci are identified in table 7.8.

From Episomes *by Allan M. Campbell. Copyright © 1969 by Allan M. Campbell. Reprinted by permission of HarperCollins Publishers.*

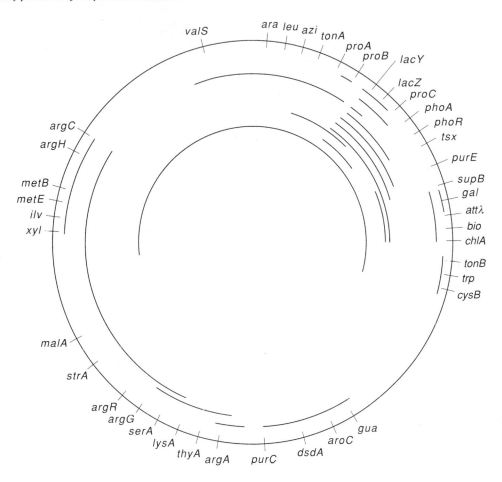

Figure 7.23

Viral life cycle using T4 infection of *E. coli* as an example

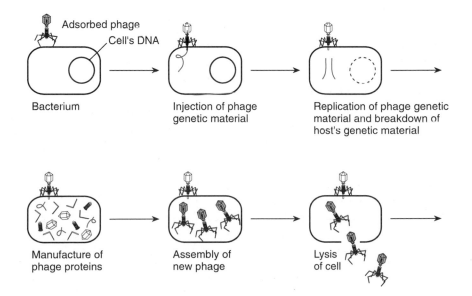

Figure 7.24

Phage λ. Magnification 167,300×.

Gunther S. Stent, Molecular Genetics, *(San Francisco) W. H. Freeman, 1963, p. 421. Reproduced courtesy of Dr. Robley Williams.*

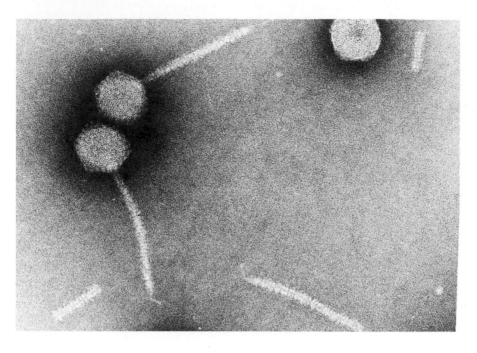

genome can be mapped by recombination. As an example, the host-range and rapid-lysis loci will be looked at here. Rapid-lysis mutants (*r*) of the T-even phages produce large, sharp-edged plaques. The wild-type produces a smaller, more fuzzy-edged plaque (fig. 7.9).

Alternative alleles are known also for host-range loci, which determine the strains of bacteria a phage can infect. For example, T2 can infect *E. coli* cells. These phages can be designated as $T2h^+$ for the normal host range. The *E. coli* would then be Tto^s, referring to their sensitivity to the T2 phage. In the course of evolution, an *E. coli* mutant has arisen that is resistant to the normal phage. This mutant is Tto^r for T2 resistance. In the further course of evolution, the phages have produced mutant forms that can grow on the Tto^r strain of *E. coli*. These phage mutants are designed as *T2h* for host-range mutant. Remember, host range is a mutation in the phage genome, whereas phage resistance is a mutation in the bacterial genome.

In 1945, Max Delbrück (a 1969 Nobel laureate) developed the technique of mixed indicators that can be used to demonstrate four phage phenotypes on the same petri plate (fig. 7.25). A bacterial lawn of mixed Tto^r and Tto^s is grown. On this lawn, the rapid-lysis phage mutants (*r*) produce large plaques, whereas the wild-type (r^+) produce smaller plaques. Phages with host-range mutation

Figure 7.25

Four types of plaques produced on a mixed lawn of *E. coli* by mixed phage T2

From Molecular Biology of Bacterial Viruses *by Gunther S. Stent. W. H. Freeman & Company. Copyright © 1963. Reproduced by permission.*

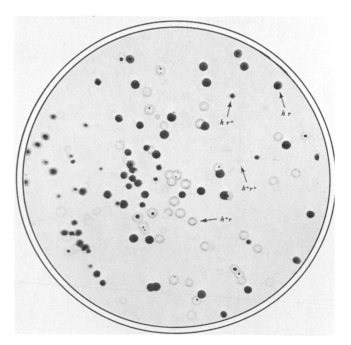

(*h*) lyse both *Tto*^r and *Tto*^s bacteria. They produce plaques that are clear (but appear dark) in figure 7.25. Since phages with the wild-type host-range allele (*h*⁺) can only infect the *Tto*ˢ bacteria, they produce turbid plaques. The *Tto*ʳ bacteria growing in these plaques produce the turbidity. (These plaques appear light-colored in fig. 7.25.)

From the wild stock of phages, we can isolate host-range mutants by looking for plaques on a *Tto*ʳ bacterial lawn. Only *h* mutants will grow. These phages can then be tested for the *r* phenotype. Hence the double mutants can be isolated. Once the two strains (double mutant and wild-type) are available, they can be added in large numbers to sensitive bacteria (fig. 7.26). Large numbers of phage are used in order to ensure that each bacterium is infected by at least one of each phage type that then provides the possibility for recombination within the host bacterium. After a round of phage multiplication, the phages are isolated and plated out on Delbrück's mixed-indicator stock. From this growth the phenotype (and hence genotype) of each phage can be recorded. The percentage of recombinants can be read directly from the plate. For example, on a given petri plate (e.g., fig. 7.25) there might be

h r 46	*h*⁺ *r*⁺ 52
h⁺ *r* 34	*h r*⁺ 26

The first two, *h r* and *h*⁺ *r*⁺, are the original, or parental, phage genotypes. The second two categories result from recombination between the *h* and *r* loci on the phage chromosome. A single crossover in this region will produce the recombinants. The proportion of recombinants is

$$(34 + 26)/(46 + 52 + 34 + 26)$$
$$= 60/158 = 0.38 \text{ or } 38\%$$

This percentage recombination is the map distance, which (as in eukaryotes) is a relative index of distance between loci: The greater the physical distance, the greater the amount of recombination and thus the larger the map distance. One map unit (1 centimorgan) is equal to 1% of recombinant offspring.

Lysogeny

Certain phages are capable of two different life-cycle stages. Some of the time they will replicate in the host cytoplasm and cause destruction of the host cell. At other times these phages are capable of integrating into the host chromosome. The host is then referred to as **lysogenic** and the phage as **temperate.** Lysogeny is the phenomenon of integration of viral and bacterial host genomes. (The term lysogeny means "giving birth to lysis." It refers to the fact that a lysogenic bacterium can be induced to initiate the virulent phase of the phage life cycle.)

Figure 7.26
Crossing *hr* and *h*⁺*r*⁺ phage. Enough of both types are added to sensitive bacterial cells (*Tto*ˢ) to ensure multiple infections. The lysate, consisting of four genotypes, is grown on a mixed-indicator bacterial lawn (*Tto*ˢ and *Tto*ʳ). Plaques of four types appear (fig. 7.25), indicative of genotypes of parental and recombinant phages.

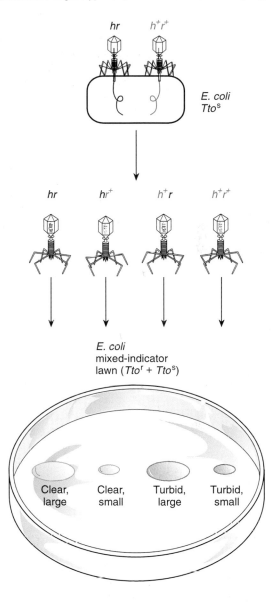

The majority of research on lysogeny has been done on phage λ (fig. 7.24). Phage λ, unlike the F factor, attaches at a specific point, termed *att*λ. This locus can be mapped on the *E. coli* chromosome and lies between the galactose (*gal*) and biotin (*bio*) loci. When the phage is integrated it protects the host from further infection (superinfection) by other λ phages. The integrated phage is termed a **prophage.** Presumably it becomes integrated by a single crossover between itself and the host after apposition of the two at the *att*λ site. (This process resembles the F factor integration shown in fig. 7.20.)

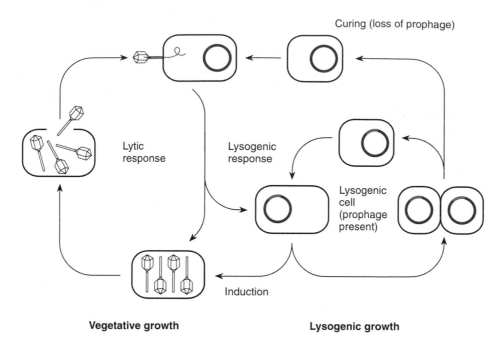

A prophage can become virulent by a process of **induction,** which involves the excision of the prophage followed by the virulent or lytic stage of the viral life cycle. We will consider the control mechanisms in detail in chapter 13. Induction can take place through a variety of mechanisms including UV irradiation and passage of the integrated prophage during conjugation (**zygotic induction**). The complete life cycle of a temperate phage is shown in figure 7.27.

TRANSDUCTION

Before lysis, when phage DNA is being packaged into phage heads, an occasional error occurs whereby bacterial DNA is incorporated into the phage head. When this happens, bacterial genes can be transferred to another bacterium via the phage coat. This process is called transduction and has been of great use in mapping the bacterial chromosome. Transduction occurs in two patterns: specialized (restricted) and generalized.

Specialized Transduction

The process of **specialized**, or **restricted, transduction** was first discovered in phage λ by Lederberg and his students. Specialized transduction is completely analogous to sexduction—it depends upon a mistake made during the looping-out process. In sexduction the error is in the F factor. In specialized transduction the error is in the λ pro-

phage. Figure 7.28 shows the λ prophage looping out incorrectly to create a defective phage carrying the adjacent *gal* locus. Since only loci adjacent to the phage attachment site can be transduced in this process, specialized transduction has not proven very useful for mapping the host chromosome.

Generalized Transduction

Generalized transduction, discovered by Zinder and Lederberg, was the first mode of transduction discovered. The bacterium was *Salmonella typhimurium* and the phage was named P22. Virtually any locus can be transduced by generalized transduction. The mechanism, therefore, does not depend on a faulty excision, but rather on the random inclusion of a piece of the host chromosome within the phage protein coat. A defective phage, one that carries bacterial DNA rather than phage DNA, is called a **transducing particle.** Transduction is complete when the genetic material from the transducing particle is both injected into a new host and enters the new host's chromosome by recombination.

For P22, the rate of transduction is about once for every 10^5 infecting phages. Since a transducing phage can carry only 2 to 2.5% of the host chromosome, only genes very close to each other can be transduced together (**cotransduced**). Cotransduction can thus be used to fill in the details of gene order over short distances after the general pattern has been ascertained by interrupted mating.

Figure 7.28

Excision, or looping out, of the λ prophage resulting in a defective phage carrying the *gal* locus

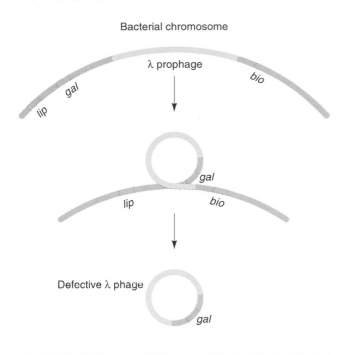

Bacterial chromosome

λ prophage

Mapping with Transduction

Transduction can be used to establish gene order and map distance. Gene order can be established by two-factor transduction. For example, if gene *A* is cotransduced with gene *B* and *B* with gene *C*, but *A* is never cotransduced with *C*, then we have established the order *A-B-C* (table 7.5). This would also apply to quantitative differences in cotransduction. For example, if *E* is often cotransduced with *F* and *F* often with *G*, but *E* is very rarely cotransduced with *G*, then we have established the order *E-F-G*.

However, more valuable is three-factor transduction, in which gene order and relative distance can be established simultaneously. Three-factor transduction is especially valuable when the three loci are so close as to make ordering decisions on the basis of two-factor transduction very difficult. For example, if genes *A*, *B*, and *C* are usually cotransduced, we can find the order and relative distances by taking advantage of the rarity of multiple crossovers. Let us use the prototroph $(A^+B^+C^+)$ to make transducing phages that then infect the $A^-B^-C^-$ stock. Transduced cells are selected for by growth in complete medium minus the nutritional requirements of *A*, *B*, or *C* (parental $A^-B^-C^-$ cells will die). These cultures are then checked for cotransduction by replica plating.

In this example, colonies that grow on complete medium without the requirement of the *A* mutant are replica plated onto complete medium without the requirement of the *B* mutant and then onto complete medium without the requirement of the *C* mutant. In this way, each transductant can be scored for all three loci (table 7.6). Now let us take all those transductants for which the *A* allele was brought in (A^+). These can be of four categories: $A^+B^+C^+$, $A^+B^+C^-$, $A^+B^-C^+$, and $A^+B^-C^-$. We now compare the relative numbers of each of these four categories. The rarest category will be caused by the event that brings in the outer two markers but not the center one because this event requires four crossovers (fig. 7.29).

TABLE 7.5 Gene Order Established by Two-Factor Cotransduction*

Transductants	Number
A^+B^+	30
A^+C^+	0
B^+C^+	25
$A^+B^+C^+$	0

*An $A^+B^+C^+$ strain of bacteria is infected with phage. The lysate is used to infect an $A^-B^-C^-$ strain. The transductants are scored for the wild-type alleles they contain. The data above include only those bacteria transduced for two or more of the loci. Since there are *AB* cotransductants and *BC* cotransductants, but no *AC* types, the order of *A–B–C* is inferred.

TABLE 7.6 Method of Scoring Three-Factor Transductants

Colony Number	Minimal Medium			Genotype
	Without A Requirement	*Without B Requirement*	*Without C Requirement*	
1	+	+	−	$A^+B^+C^-$
2	+	−	−	$A^+B^-C^-$
3	+	−	−	$A^+B^-C^-$
4	−	+	+	$A^-B^+C^+$
5	−	−	+	$A^-B^-C^+$
.	.	.	.	.
.	.	.	.	.
.	.	.	.	.

The plus indicates growth and the minus indicates lack of growth.
An $A^-B^-C^-$ strain was transduced by phage from an $A^+B^+C^+$ strain.

Figure 7.29

The rarest transductant requires four crossovers.

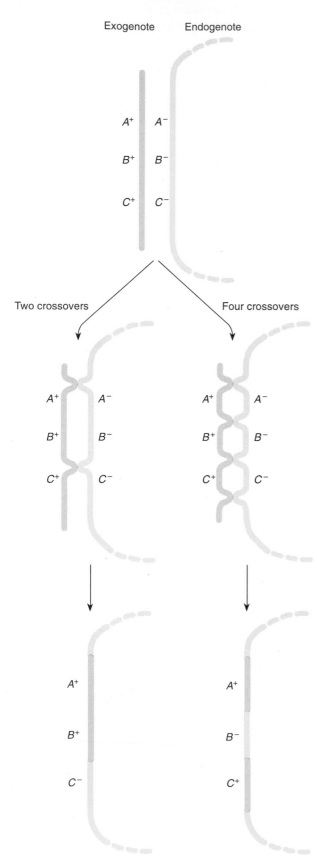

Exogenote Endogenote

Two crossovers   Four crossovers

TABLE 7.7 Numbers of Transductants in the Experiment Used to Determine the $A–B–C$ Gene Order (Table 7.6); Relative Cotransduction Frequencies Are Also Given

Class	Number
$A^+B^+C^+$	50
$A^+B^+C^-$	75
$A^+B^-C^+$	1
$A^+B^-C^-$	300
	426

Relative Cotransductance
$A–B$: $(50 + 75)/426 = 0.29$
$A–C$: $(50 + 1)/426 = 0.12$

Thus by looking at the number of the various categories we can determine the gene order to be A-B-C (table 7.7) since the $A^+B^-C^+$ category is the rarest.

The relative cotransduction frequencies are calculated in table 7.7. Note that in all organisms and viruses, the higher the frequency of co-occurrence of alleles of two loci, the closer those loci are on the chromosome. We usually measure the separation of loci by crossing over between them; the closer together, the lower those crossing-over values are and hence the smaller the measure of map units apart. Here we are measuring the co-occurrence directly and therefore the measure—cotransductance—is the inverse of standard map distance. In other words, the greater the cotransduction rate, the closer the two loci are: The more frequently two loci are transduced together the closer they are and the higher the cotransduction value will be.

The data of table 7.7 cannot be used to calculate the B-C cotransduction rate because the data given are selected values, all of which are A^+; they are not the total data. Recall our discussion under the topic of transformation about the need to select progeny in which some event, in this case transduction, has occurred. In our attempt to study transduction, we must eliminate all the cells that were not transduced ($A^-B^-C^-$). We did that by growing the potentially transduced cells on a medium without one of the growth requirements of the $A^-B^-C^-$ cells. In this case we removed substance A that A^- cells need. Thus only A^+ cells could survive: Some will be B^+, some C^+, some B^-C^-, and some B^+C^+. Parental $A^-B^-C^-$ cells die, but so do $A^-B^+C^+$, $A^-B^-C^+$, and $A^-B^+C^-$ cells. Thus, by the nature of the process, a relatively rare event, we must select for some indication of successful transduction, in this case the transduction of the A^+ allele. Had we removed the requirements for A^-, B^-, and C^- cells simultaneously, we would select only for $A^+B^+C^+$ cells and therefore the experiment would tell us very little.

Figure 7.30

Selected loci on the circular map of *E. coli*. Definitions of loci not found in the text can be found in table 7.8. Units on the map are in minutes. *Arrows* within the circle refer to Hfr-strain transfer starting points, with directions. The two thin regions are the only areas not covered by P1 transducing phages.

From B. J. Bachmann, et al., "Recalibrated Linkage Map of Escherichia coli K–12" in Bacteriological Reviews, 40: 116–117. Copyright © 1976 American Society for Microbiology, Washington, DC. Reprinted by permission.

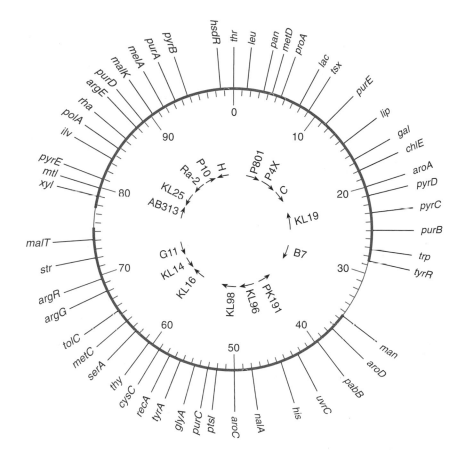

From these sorts of transduction experiments, it is possible to round out the details of map relations for which the overall picture is obtained by interrupted mating. The map of *E. coli* is presented in figure 7.30. Definitions of loci can be found in table 7.8. Unlike eukaryote mapping, map distances of prokaryotes are not generally thought of in map units (centimorgans). Rather, the general distance between loci is determined in minutes with cotransduction values used for loci that are very close to each other. (In chapter 12, we will discuss methods of mapping by direct sequencing of the DNA.)

SUMMARY

Prokaryotes (bacteria) have a single circular chromosome of DNA. The bacterial sexual processes—transformation, conjugation, transduction, and sexduction—can be used to map the bacterial chromosome. A bacteriophage consists of a chromosome wrapped in a protein coat. Its chromosome can be DNA or RNA. The phage chromosome can be mapped by measuring recombination after a bacterium has been simultaneously infected by two strains of the virus carrying different alleles.

Phenotypes of bacteria include colony morphology, nutritional requirements, and drug resistance. Phage phenotypes include plaque morphology and host range. Replica plating is a rapid screening technique for assessing the phenotype of a bacterial clone.

In transformation, a competent bacterium can take up relatively large pieces of double-stranded DNA from the medium. This DNA can be incorporated into the bacterial chromosome if homologous regions come to lie adjacent to each other and an even number of crossovers then take place.

During the process of conjugation, the fertility factor, F, is passed from an F⁺ to an F⁻ cell. If the F factor integrates into the host chromosome, an Hfr cell results that can pass its entire

TABLE 7.8 Symbols Used in the Gene Map of the *E. coli* Chromosome

Genetic Symbols	Mutant Character	Enzyme or Reaction Affected
araD	Cannot use the sugar arabinose as a carbon source	L-Ribulose-5-phosphate-4-epimerase
araA		L-Arabinose isomerase
araB		L-Ribulokinase
araC		
argB	Requires the amino acid arginine for growth	N-Acetylglutamate synthetase
argC		N-Acetyl-γ-glutamokinase
argH		N-Acetylglutamic-γ-semialdehyde dehydrogenase
argG		Acetylornithine-*d*-transaminase
argA		Acetylornithinase
argD		Ornithine transcarbamylase
argE		Argininosuccinic acid synthetase
argF		Argininosuccinase
argR	Arginine operon regulator	
aroA, B, C	Requires several aromatic amino acids and vitamins for growth	Shikimic acid to 3-enolpyruvyl-shikimate-5-phosphate
aroD		Biosynthesis of shikimic acid
azi	Resistant to sodium azide	
bio	Requires the vitamin biotin for growth	
carA	Requires uracil and arginine	Carbamate kinase
carB		
chlA–E	Cannot reduce chlorate	Nitrate-chlorate reductase and hydrogen lysase
cysA	Requires the amino acid cysteine for growth	3-Phosphoadenosine-5-phosphosulfate to sulfide
cysB		Sulfate to sulfide; 4 known enzymes
cysC		
dapA	Requires the cell-wall component diaminopimelic acid	Dihydrodipicolinic acid synthetase
dapB		N-Succinyl-diaminopimelic acid deacylase
dap + hom	Requires the amino acid precursor homoserine and the cell-wall component diaminopilmelic acid for growth	Aspartic semialdehyde dehydrogenase
dnaA–Z	Mutation, DNA replication	DNA biosynthesis
Dsd	Cannot use the amino acid D-serine as a nitrogen source	D-Serine deaminase
fla	Flagella are absent	
galA	Cannot use the sugar galactose as a carbon source	Galactokinase
galB		Galactose-1-phosphate uridyl transferase
galD		Uridine-diphosphogalactose-4-epimerase
glyA	Requires glycine	Serine hydroxymethyl transferase
gua	Requires the purine guanine for growth	
H	The H antigen is present	
his	Requires the amino acid histidine for growth	10 known enzymes*
hsdR	Host restriction	Endonuclease R
ile	Requires the amino acid isoleucine for growth	Threonine deaminase
ilvA	Requires the amino acids isoleucine and valine for growth	α-Hydroxy-β-keto acid rectoisomerase
ilvB		α,β-dihydroxyisovaleric dehydrase*
ilvC		Transaminase B
ind (indole)	Cannot grow on tryptophan as a carbon source	Tryptophanase
λ *(attλ)*	Chromosomal location where prophage λ is normally inserted	
lacI	*Lac* operon regulator	
lacY	Unable to concentrate β-galactosides	Galactoside permease
lacZ	Cannot use the sugar lactose as a carbon source	β-Galactosidase
lacO	Constitutive synthesis of lactose operon proteins	Defective operator
leu	Requires the amino acid leucine for growth	3 known enzymes*
lip	Requires lipoate	
lon (long form)	Filament formation and radiation sensitivity are affected	
lys	Requires the amino acid lysine for growth	Diaminopimelic acid decarboxylase
lys + met	Requires the amino acids lysine and methionine for growth	
λ *rec, malT*	Resistant to phage λ and cannot use the sugar maltose	Regulator for 2 operons
malK	Cannot use the sugar maltose as a carbon source	Amylomaltase (?)
man	Cannot use mannose sugar	Phosphomannose isomerase
melA	Cannot use melibiose sugar	Alpha-galactosidase

Table 7.8 (Continued)

Genetic Symbols	Mutant Character	Enzyme or Reaction Affected
met A–M	Requires the amino acid methionine for growth	10 or more genes
mtl	Cannot use the sugar mannitol as a carbon source	Mannitol dehydrogenase (?)
muc	Forms mucoid colonies	Regulation of capsular polysaccharide synthesis
nalA	Resistance to nalidixic acid	
O	The O antigen is present	
pan	Requires the vitamin pantothenic acid for growth	
pabB	Requires p-aminobenzoate	
phe A, B	Requires the amino acid phenylalanine for growth	
pho	Cannot use phosphate esters	Alkaline phosphatase
pil	Has filaments (pili) attached to the cell wall	
plsB	Deficient phospholipid synthesis	Glycerol 3-phosphate acyltransferase
polA	Repairs deficiencies	DNA polymerase I
proA	Requires the amino acid proline for growth	
proB		
proC		
ptsI	Defective phosphotransferase system	Pts-system enzyme I
purA	Requires certain purines for growth	Adenylosuccinate synthetase
purB		Adenylosuccinase
purC, E		5-Aminoimidazole ribotide (AIR) to 5-aminoimidazole-4-(N-succino carboximide) ribotide
purD		Biosynthesis of AIR
pyrB	Requires the pyrimidine uracil for growth	Aspartate transcarbamylase
pyrC		Dihydroorotase
pyrD		Dihydroorotic acid dehydrogenase
pyrE		Orotidylic acid pyrophosphorylase
pyrF		Orotidylic acid decarboxylase
R gal	Constitutive production of galactose	Repressor for enzymes involved in galactose production
R1 pho, R2 pho	Constitutive synthesis of phosphatase	Alkaline phosphatase repressor
R try	Constitutive synthesis of tryptophan	Repressor for enzymes involved in tryptophan synthesis
RC (RNA control)	Uncontrolled synthesis of RNA	
recA	Cannot repair DNA radiation damage or recombine	
rhaA–D	Cannot use the sugar rhamnose as a carbon source	Isomerase, kinase, aldolase, and regulator
rpoA–D	Problems of transcription	Subunits of RNA polymerase
serA	Requires the amino acid serine for growth	3-Phosphoglycerate dehydrogenase
serB		Phosphoserine phosphatase
str	Resistant to or dependent on streptomycin	
suc	Requires succinic acid	
supB	Suppresses ochre mutations	t-RNA
tonA	Resistant to phages T1 and T5 (mutants called B/1, 5)	T1, T5 receptor sites absent
tonB	Resistant to phage T1 (mutants called B/1)	T1 receptor site absent
T6, colK rec	Resistant to phage T6 and colicine K	T6 and colicine receptor sites absent
T4 rec	Resistant to phage T4 (mutants called B/4)	T4 receptor site absent
tsx	T6 resistance	
thi	Requires the vitamin thiamine for growth	
tolC	Tolerance to colicine E1	
thr	Requires the amino acid threonine for growth	
thy	Requires the pyrimidine thymine for growth	Thymidylate synthetase
trpA	Requires the amino acid tryptophan for growth	Tryptophan synthetase, A protein
trpB		Tryptophan synthetase, B protein
trpC		Indole-3-glycerolphosphate synthetase
trpD		Phosphoribosyl anthranilate transferase
trpE		Anthranilate synthetase
tyrA	Requires the amino acid tyrosine for growth	Chorismate mutase T-prephenate dehydrogenase
tyrR		Regulates 3 genes
uvrA–E	Resistant to ultraviolet radiation	Ultraviolet-induced lesions in DNA are reactivated
valS	Cannot charge Valyl-tRNA	Valyl-tRNA synthetase
xyl	Cannot use the sugar xylose as a carbon source	

*Denotes enzymes controlled by the homologous gene loci of *Salmonella typhimurium*.
From B. J. Bachmann and K. B. Low, "Linkage Map of *Escherichia coli* K-12" in *Microbiological Reviews*, 44:1–56. Copyright © 1980 American Society for Microbiology, Washington, DC. Reprinted by permission.

chromosome into an F⁻ cell. The F factor is the last region to cross into the F⁻ cell. The bacterial chromosome can be mapped by the process of interrupted mating, which determines the times of entry of loci from the Hfr into the F⁻ cell.

The F factor can loop out of the host chromosome. In some cases it takes part of the host chromosome with it. When this defective F, or F′, is passed to an F⁻ cell, recombination of the original host chromosome segment with the F⁻ chromosome can take place. In transduction, a phage protein coat containing some of the host chromosome is passed to a new host bacterium. Again, recombination with this new chromosomal segment can take place.

In *E. coli,* mapping is done most efficiently via interrupted mating and transduction. The former provides information on general gene arrangement and the latter provides finer details.

SOLVED PROBLEMS

1. A wild-type strain of *B. subtilis* is transformed by DNA from a strain that cannot grow on galactose (*gal⁻*) and also needs biotin for growth (*bio⁻*). Transformants are isolated by exposing the transformed cells to minimal medium with penicillin, killing the wild-type cells. After the penicillin is removed, replica plating is used to establish the genotypes of 30 transformants:

class 1 *gal⁻bio⁻*   17

class 2 *gal⁻bio⁺*   4

class 3 *gal⁺bio⁻*   9

What is the relative recombination frequency?

ANSWER: The three classes of colonies above represent the three possible transformant groups. Classes 2 and 3 are single transformants and class 1 is the double transformant. We are interested in the relative occurrence of a crossover between the two loci. Therefore we divide the number of single transformants, which have that crossover, by the total: $(4 + 9)/30 = 13/30 = 0.43$. This is a relative value similar to a map distance; the larger it is, the further apart are the loci.

2. A *gal⁻ bio⁻ attλ⁻* strain of *E. coli* is transduced by P22 phages from a wild-type strain. Transductants are selected for by growing the cells with galactose as the sole energy source. Replica plating and testing for lysogenic ability gives the genotypes of 106 transformants:

class 1 *gal⁺ bio⁻ attλ⁻*   71

class 2 *gal⁺ bio⁺ attλ⁻*   0

class 3 *gal⁺ bio⁻ attλ⁺*   9

class 4 *gal⁺ bio⁺ attλ⁺*   26

What is the gene order and what are the relative cotransduction frequencies?

ANSWER: We have selected all transductants that are *gal⁺*. Class 2 is in the lowest frequency (0) and therefore represents the double crossover between the transducing DNA and the host chromosome. From this we see that *attλ* must be in the middle because this low-probability event is the one that would have switched only the middle locus. In other words, the two end loci would be recombinant and the middle locus would have the host allele. We can only calculate two cotransduction frequencies because these are selected data. Note that in class 1 there is no cotransduction between *gal* and either of the other two loci; class 2 would have been cotransduction of *gal* and *bio*; class 3 represents cotransduction of *gal* and *attλ*; and class 4 represents cotransduction of *gal* and both other loci. Therefore, cotransduction values are:

$$gal\text{-}att\lambda = (9 + 26)/106 = 35/106 = 0.33$$
$$gal\text{-}bio = (0 + 26)/106 = 26/106 = 0.25.$$

EXERCISES AND PROBLEMS

1. What is the nature and substance of the chromosomes of prokaryotes and viruses? Are viruses alive?

2. What genotypic notation indicates a bacterium that
   a. is resistant to penicillin?
   b. is sensitive to azide?
   c. requires histidine for growth?
   d. cannot grow on galactose?
   e. can grow on glucose?
   f. is susceptible to phage T1 infection?

3. What are the differences between a heterotroph and an auxotroph? A minimal and a complete medium? An enriched and a selective medium?

4. What are the differences between a plaque and a colony?

5. What is a plasmid? How does one integrate into a host's chromosome? How does it leave?

6. In conjugation experiments, one Hfr strain should carry a gene for some sort of sensitivity (e.g., *aziˢ* or *strˢ*) so that

the Hfr donors can be eliminated on selective media after conjugation has taken place. Should this locus be near to or far from the origin of transfer point of the Hfr chromosome? What are the consequences of either alternative?

7. How does a geneticist, doing interrupted mating experiments, know that the locus for the drug-sensitivity allele, used to eliminate the Hfr bacteria after conjugation, has crossed into the F⁻ strain?

8. Diagram the step-by-step events required to integrate foreign DNA into a bacterial chromosome in the four processes outlined in the chapter. Do the same for viral recombination.

9. An *E. coli* cell is placed on a petri plate containing λ phages. It produces a colony overnight. By what mechanisms might it have survived?

10. The DNA from a prototrophic strain of *E. coli* is isolated and used to transform an auxotrophic strain deficient in the synthesis of purines ($purB^-$), pyrimidines ($pyrC^-$), and the amino acid tryptophan (trp^-). Tryptophan was used as the marker to determine whether transformation had occurred (the selected marker). What are the gene order and the relative recombination frequencies between loci, given the data below?

$trp^+ \; pyrC^+ \; purB^+$	86
$trp^+ \; pyrC^+ \; purB^-$	4
$trp^+ \; pyrC^- \; purB^+$	67
$trp^+ \; pyrC^- \; purB^-$	14

11. Using the data of figure 7.18, draw a tentative map of the *E. coli* chromosome.

12. Three Hfr strains of *E. coli* (P4X, KL98, and Ra-2) are mated individually with an auxotrophic F⁻ strain, using interrupted mating techniques. Using these data construct a map of the *E. coli* chromosome, including distances in minutes.

Approximate Time of Entry

Donor Loci	Hfr P4X	Hfr KL98	Hfr Ra-2
gal^+	11	67	70
thr^+	94	50	87
xyl^+	73	29	8
lac^+	2	58	79
his^+	38	94	43
ilv^+	77	33	4
$argG^+$	62	18	19

How many different petri plates and selective media are needed?

13. Design an experiment and possible data set that would correctly map five of the loci on the *E. coli* chromosome (fig. 7.30) using interrupted mating.

14. An *E. coli* lawn is formed on a petri plate containing complete medium. Replica plating is used to transfer material to plates containing minimal medium and combinations of the amino acids arginine and histidine. Give the genotype of the original strain as well as the genotypes of the odd colonies found growing on the plates.

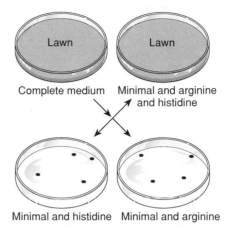

15. A prototrophic Hfr *E. coli* strain, sensitive to penicillin, is mated to an F⁻ strain resistant to penicillin and requiring several amino acids (histidine, arginine, leucine, lysine, and methionine). Recombinants are selected for by plating on a medium with penicillin and amino acids: five colonies are shown on a plate with penicillin and all five amino acids in question. These colonies are replica plated onto minimal medium containing various amino acids. What are the genotypes of each of the five colonies?

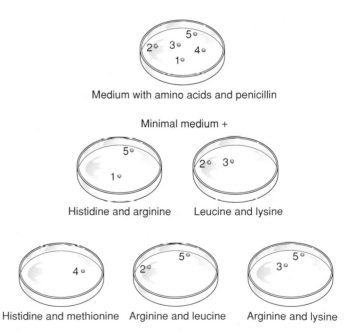

16. A petri plate with complete medium has six colonies growing on it after one of the conjugation experiments described earlier. The colonies are numbered and the plate is used as a master to replicate onto plates of glucose-containing selective (minimal) medium with various combinations of additives. From the following data showing the presence (+) or absence (-) of growth, give your best assessment of the genotypes of the six colonies.

Colony On Minimal Medium Plus	1	2	3	4	5	6
Nothing	−	−	+	−	−	−
Xylose + arginine	+	−	+	+	−	−
Xylose + histidine	−	−	+	−	−	−
Arginine + histidine	−	+	+	+	−	−
Galactose + histidine	−	−	+	−	−	+
Threonine + isoleucine + valine	−	−	+	−	+	−
Threonine + valine + lactose	−	−	+	−	−	−

17. Lederberg and his colleagues (Nester, Schafer, and Lederberg, 1963, *Genetics* 48:529) determined gene order and relative distance using three markers in the bacterium, *Bacillus subtilis*. DNA from a prototrophic strain (trp^+ his^+ tyr^+) was used to transform the auxotroph. The seven classes of transformants, with their numbers, are tabulated as follows:

trp^+	trp^-	trp^-	trp^+	trp^+	trp^-	trp^+
his^-	his^+	his^-	his^+	his^-	his^+	his^+
tyr^-	tyr^-	tyr^+	tyr^-	tyr^+	tyr^+	tyr^+
2,600	418	685	1,180	107	3,660	11,940

Outline the techniques used to obtain these data. Taking the loci in pairs, calculate recombination distances. Construct the most consistent linkage map of these loci.

18. Give possible genotypes of an *E. coli*/phage T1 system in which the phage cannot grow on the bacterium. Give the same in which the phage can grow on the bacterium.

19. Define prophage, lysate, lysogeny, and temperate phage.

20. Define and illustrate specialized and generalized transduction.

21. In *E. coli* the three loci *ara, leu,* and *ilvH* are within 1/2 minute map distance apart. To determine the exact order and relative distance, the prototroph (ara^+ leu^+ $ilvH^+$) was infected with transducing phage P1. The lysate was used to infect the auxotroph (ara^- leu^- $ilvH^-$). The ara^+ classes of transductants were selected to produce the following data:

ara^+	ara^+	ara^+	ara^+
leu^-	leu^+	leu^-	leu^+
$ilvH^-$	$ilvH^-$	$ilvH^+$	$ilvH^+$
32	9	0	340

Outline the specific techniques used to isolate the various transduced classes. What is the gene order and relative distance between genes? Why do some classes occur so infrequently?

22. Outline an experiment to demonstrate that two phages do not undergo recombination until a bacterium is infected simultaneously with both.

23. Doermann (1953, *Cold Spr. Harb. Symp. Quant. Biol.* 18:3) mapped three loci of phage T4: minute, rapid lysis, and turbid. He infected *E. coli* cells with both the triple mutant (m r tu) and the wild-type (m^+ r^+ tu^+) and obtained the following data:

m	m^+	m	m	m^+	m^+	m	m^+
r	r	r^+	r	r^+	r	r^+	r^+
tu	tu	tu	tu^+	tu	tu^+	tu^+	tu^+
3,467	474	162	853	965	172	520	3,729

What is the linkage relationship among these loci? In your answer include gene order, relative distance, and coefficient of coincidence.

24. Consider the following portion of an *E. coli* chromosome:

thr   *ara leu*

Three nonidentical *ara^-* mutants, *ara-1, ara-2,* and *ara-3,* were isolated, and their order with respect to *thr* and *leu* determined by transduction. The donor was always thr^+ leu^+ and the recipient was always thr^- leu^-. Each *ara^-* mutant was used as a donor in one cross and as a recipient in another; ara^+ transductants were selected in each case. The ara^+ transductants were then scored for leu^+ and thr^+. Based on the following results, determine the order of the *ara^-* mutants with respect to *thr* and *leu*.

Cross	Recipient	Donor	$\frac{ara^+ leu^+}{ara^+} \times 100$	$\frac{ara^+ thr^+}{ara^+} \times 100$
1	$ara\text{-}1^-$	$ara\text{-}2^-$	63	1.3
2	$ara\text{-}2^-$	$ara\text{-}1^-$	16.1	7.0
3	$ara\text{-}1^-$	$ara\text{-}3^-$	25.4	6.3
4	$ara\text{-}3^-$	$ara\text{-}1^-$	51.7	2.7
5	$ara\text{-}2^-$	$ara\text{-}3^-$	13.5	9.1
6	$ara\text{-}3^-$	$ara\text{-}2^-$	66.3	2.6

25. An *E. coli* strain that is leu^+ thr^+ azi^r is used as a donor in a transduction of a strain that is leu^- thr^- azi^s. Either leu^+ or thr^+ transductants are selected and then scored for unselected markers. The following results are obtained:

selected marker	unselected markers
leu^+	48% azi^r
leu^+	2% thr^+
thr^+	3% leu^+
thr^+	0% azi^r

What is the order of the three loci?

26. Wild-type phage T4 (r^+) produce small turbid plaques, whereas rII mutants produce large, clear plaques. Four rII mutants (a-d) are crossed and the following percentages of wild-type plaques are obtained:

$a \times b$	0.3
$a \times c$	1.0
$a \times d$	0.4
$b \times c$	0.7
$b \times d$	0.1
$c \times d$	0.6

Deduce a genetic map of these four mutants.

27. A phage cross is performed between a^+ b^+ c^+ and a b c phage. Based on the following results, derive a complete map.

a^+ b^+ c^+	1,801
a^+ b^+ c	954
a^+ b c^+	371
a^+ b c	160
a b^+ c^+	178
a b^+ c	309
a b c^+	879
a b c	1,850
	6,502

28. In a transformation experiment, an a^+ b^+ c^+ strain is used as the donor, and an a^- b^- c^- strain as the recipient. One hundred a^+ transformants are selected, then replica plated to determine whether b^+ and c^+ are present. The genotypes of the transformants appear below. What can you conclude about the relative position of the genes?

a^+ b^- c^-	21
a^+ b^- c^+	69
a^+ b^+ c^-	3
a^+ b^+ c^+	7

29. In a transformation experiment, an a^+ b^+ c^- strain is used as a donor and an a^- b^- c^+ strain as recipient. If you select for a^+ transformants, the least frequent class is a^+ b^+ c^+. What is the order of the genes?

30. A mating between his^+, leu^+, thr^+, pro^+ str sensitive cells (Hfr) and his^-, leu^-, thr^-, pro^-, str resistant cells (F$^-$) is allowed to continue for 25 minutes. The mating is stopped and the genotypes of the recombinants determined. The results appear at the top of the next column. What is the first gene to enter and what is the probable gene order?

Genotype	Number of Colonies
his^+	0
leu^+	12
thr^+	27
pro^+	6

31. a. In a transformation experiment, the donor is trp^+ leu^+ arg^+ and the recipient is trp^- leu^- arg^-. Trp$^+$ transformants are selected and then further tested. Forty percent are trp^+ arg^+; 5 percent were trp^+ leu^+. In what two possible orders are the genes arranged?
    b. You can do only one more transformation to determine gene order. You must use the same donor and recipient, but you can change the selection procedure for the initial transformants. What should you do and what results should you expect for each order you proposed above?

32. DNA from a bacterial strain that is a^+ b^+ c^+ is used to transform a strain that is a^- b^- c^-. The numbers of each transformed genotype appear below. What can be said about the relative position of the genes?

Genotype	Number
a^+ b^- c^-	214
a^- b^+ c^-	231
a^- b^- c^+	206
a^+ b^+ c^-	11
a^+ b^+ c^+	6
a^+ b^- c^+	93
a^- b^+ c^+	14

33. An Hfr strain that is a^+ b^+ c^+ d^+ e^+ is mated to an F$^-$ that is a^- b^- c^- d^- e^-. The mating is interrupted every 5 minutes and the genotypes of the F$^-$ recombinants determined. The results appear below. Draw a map of the chromosome and indicate the position of the F factor, the direction of transfer, and the minutes between genes.

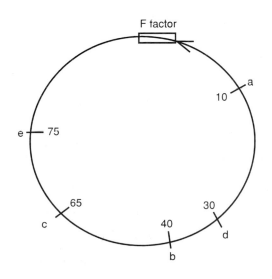

34. A bacterial strain that is *lys+ his+ val+* is used as a donor, and *lys− his− val−* is the recipient. Initial transformants are isolated on minimal medium + histidine + valine.
   a. What are the genotypes that will grow on this medium?
   b. These colonies are replicated to minimal medium + histidine, and 75% of the original colonies grow. What are the genotypes that will grow on this medium?
   c. The colonies are also replicated to minimal medium + valine, and 6% of the colonies grow. What genotypes will grow on this medium?
   d. Finally, the original colonies are replicated to minimal medium. No colonies grow. From this information, what is the genotype that grows on minimal medium + histidine? On minimal medium + valine?
   e. Based on the above information, which gene is closer to *lys?*
   f. The original transformation is repeated, but the original plating is on minimal medium + lysine + histidine. Fifty colonies appear. These colonies are replicated to determine their genotypes and the following results appear:

val+ his+ lys+	0
val+ his− lys+	37
val+ his+ lys−	3

Based on all the results, what is the most likely gene order?

35. The *r*II mutants of T4 phage will grow and produce large plaques on strain B; *r*II mutants will not grow on strain K12, lysogenic for phage λ. The following crosses are performed in strain B. The progeny of the crosses are diluted 10^{-7} and are used to reinfect strain B. The following numbers of plaques are obtained:

1 × 2	250
1 × 3	250
2 × 3	250

The progeny of the above crosses are also diluted 10^{-4} and plated on K12. The following numbers of plaques are seen:

1 × 2	50
1 × 3	25
2 × 3	75

Draw a map of these three mutants (1, 2, and 3) and indicate the distances between them.

Suggested Readings for chapter 7 are on page 644.

8

Cytogenetics

OBJECTIVES

♦ To observe the nature and consequences of chromosomal breakage and reunion

♦ To observe the nature and consequences of the variation in chromosome numbers in human and nonhuman organisms

Normal strawberries (*right*) are smaller than strawberries that have extra sets of chromosomes (polyploid, *left*).
© *William E. Ferguson*

167

Our understanding of the chromosomal theory of genetics was derived primarily through the mapping of loci, using techniques that required alternative allelic forms, or mutations, of these loci. Changes in the genetic material also occur at a much coarser level—changes in large parts of chromosomes or changes in chromosome numbers. In this chapter we investigate how these alterations happen and what their consequences are to the organism.

VARIATION IN CHROMOSOMAL STRUCTURE

In general, chromosomes can break spontaneously or be broken by ionizing radiation, physical stress, or chemical compounds. Chromosomal breaks can occur at either the chromatid or the chromosomal level. When a break in the chromosome occurs before DNA replication, during the S-phase of the cell cycle (see fig. 3.6), the break itself is replicated. In this case we have a chromosome-level break involving both sister chromatids at the same point. After the S-phase, any breaks that occur are in single chromatids—that is, they are chromatid-level breaks.

For every break in a chromatid, two ends are produced. These ends have been described as "sticky," meaning simply that enzymatic processes of the cell tend to reunite them. Broken ends do not attach to the undamaged terminal ends of other chromosomes. (Normal chromosomal ends are capped with structures called *telomeres*—see chapter 14.) If broken ends are not brought together, they can remain broken. But, if broken chromatid ends are brought into apposition, there are several alternative ways in which they can be rejoined. First, the two broken ends of a single chromatid can be reunited. Second, the broken end of one chromatid can be fused with the broken end of another chromatid, resulting in an exchange of chromosomal material and a new combination of alleles. Multiple breaks can lead to a variety of alternative recombinations. These chromosomal aberrations have major genetic, evolutionary, and medical consequences. The types of breaks and reunions discussed in this chapter can be summarized as follows:

I. Noncentromeric breaks
   A. Single breaks (chromatid)
      1. Restitution
      2. Deletion
   B. Single breaks (chromosome): Dicentric bridge
   C. Two breaks (same chromatid)
      1. Deletion
      2. Inversion
   D. Two breaks (nonsister chromatids)
      1. Translocation, reciprocal
      2. Translocation, nonreciprocal
II. Centromeric breaks
   A. Fission
   B. Fusion

Figure 8.1
Chromosomal break with subsequent reunion to form a dicentric chromosome and an acentric fragment

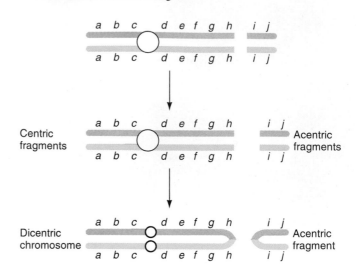

Single Breaks: Chromatid

If the break is of the chromatid type, the broken ends may be rejoined. When the broken ends of a single chromatid are rejoined (restitution), there is no consequence of the break. If they are not rejoined, the result is an **acentric fragment**, without a centromere, and a **centric fragment**, with a centromere. The centric fragment will migrate normally during the division process because it has a centromere. The acentric fragment, however, is soon lost. It is subsequently excluded from the nuclei formed and is eventually degraded. In other words, the viable, centric part of the chromosome has suffered a deletion of the region. After mitosis, the daughter cell that receives the **deletion chromosome** may show several effects.

Pseudodominance may be observed. (This term was used in chapter 5 when we described alleles located on the X chromosome. With only one copy of the locus present, a recessive allele in males shows itself in the phenotype as if it were dominant—hence the term pseudodominance.) The normal chromosome homologous to the deletion chromosome will have loci in the region, and recessive alleles will show pseudodominance. A second effect is that, depending on the length of the deleted segment and the specific loci lost, the imbalance created in the daughter cell by a deletion chromosome may be lethal. If the deletion occurs before or during meiosis, it may be observed under the microscope. This is discussed later in the chapter.

Single Breaks: Chromosomal

A single break can have another effect if it is a chromosomal break. Occasionally the two centric fragments of a single chromosome may join, forming a two-centromere, or **dicentric, chromosome** and leaving the two acentric fragments to join or, alternatively, remain as two fragments (fig. 8.1). The acentric fragments will be lost, as

Figure 8.2

Breakage of a dicentric bridge causes duplications and further deficiencies.

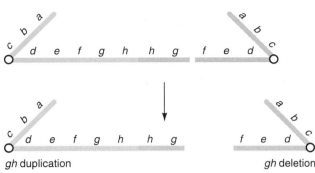

gh duplication   *gh* deletion

Figure 8.4

Bulge in a meiotic tetrad indicates that a deletion has occurred.

Normal chromatids

Deletion chromatids

Figure 8.3

Two possible consequences of a double break in the same chromosome.

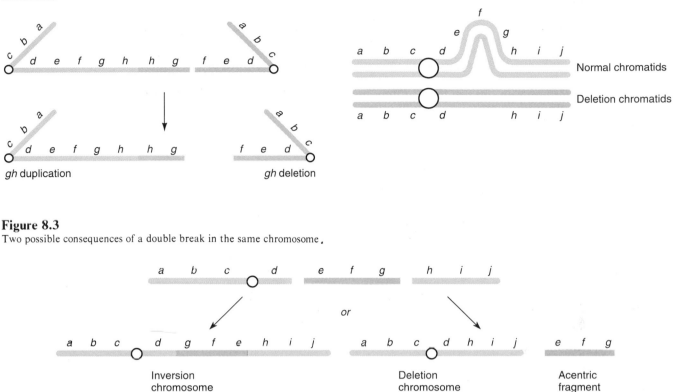

or

Inversion chromosome   Deletion chromosome   Acentric fragment

mentioned before. Because the centromeres are on sister chromatids, the dicentric fragment is pulled to opposite ends of a mitotic cell forming a bridge there, or, if meiosis is occurring, the dicentric fragment is pulled during the second meiotic division. The ultimate fate of this bridge is to be broken by the spindle fibers pulling the centromeres to opposite poles (or possibly to be excluded from a new nucleus if the bridge is not broken).

The dicentric chromosome is not necessarily broken in the middle, and subsequent processes will exacerbate the imbalance created by an off-center break: Duplications will occur on one strand, whereas more deletions will occur on the other (fig. 8.2). In addition, the "sticky" ends produced on both fragments by the break increase the likelihood of repeating this **breakage-fusion-bridge cycle** each generation. The great imbalances resulting from the duplications and deletions usually result in the death of the cell line within several generations.

Two Breaks in the Same Chromatid

Figure 8.3 shows two of the possible results when two breaks occur in the same chromatid. One alternative is a reunion that omits an acentric fragment, which will then be lost. The centric piece, missing the acentric fragment (*e-f-g* in fig. 8.3), is a deletion chromosome. An organism having this chromosome and a normal homologue will

have, during meiosis, a bulge in the tetrad if the deleted section is large enough (fig. 8.4). The bulge also can be seen in *Drosophila* in the giant salivary gland chromosomes, unless, of course, its effects are lethal.

Inversions

Two breaks in the same chromosome can also lead to **inversion,** in which the middle section is reattached but in the inverted configuration (see fig. 8.3). An inversion has several interesting properties. To begin with, fruit flies homozygous for an inversion will show new linkage relations when their chromosomes are mapped. An outcome of this new linkage arrangement is the possibility of a **position effect,** a change in the expression of a gene due to a changed linkage arrangement. Position effects are either stable, as in *Bar* eye of *Drosophila* (to be discussed), or variegated, as with *Drosophila* eye color. A normal female fly that is heterozygous (X^wX^+) has red eyes. If, however, the white locus is moved, through an inversion, so that it comes to lie next to heterochromatin (fig. 8.5), the fly will show a **variegation**—patches of the eye will be white. Whether this is caused by a product of the heterochromatic region or whether the tight coiling in the heterochromatin intermittently "turns off" the allele is not known. It is known, however, that this type of position effect is limited to loci placed in proximity to heterochromatin.

Figure 8.5

Inversion in the X chromosome of *Drosophila* that will produce a variegation in eye color in a female if her other chromosome is normal and carries the white-eye allele (X*)

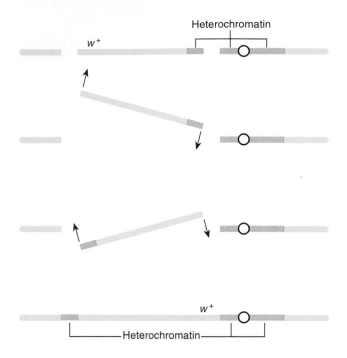

Figure 8.7

Drosophila salivary gland inversion heterozygote chromosome loop. Compare with figure 8.6.

Figure 8.6

Tetrad at meiosis showing the loop characteristic of an inversion heterozygote

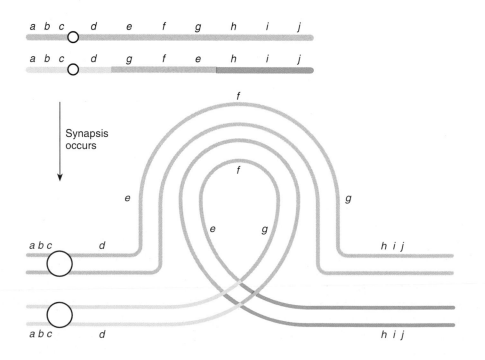

Figure 8.8
Consequences of a crossover in an inversion heterozygote

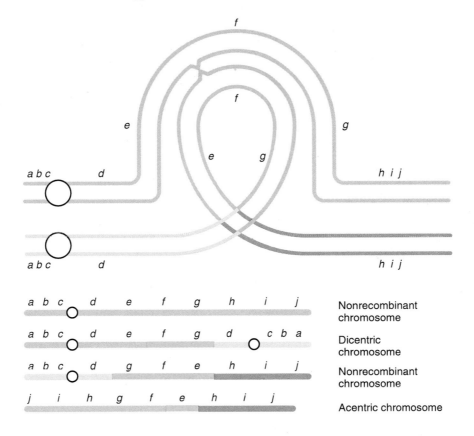

a b c d e f g h i j	Nonrecombinant chromosome
a b c d e f g d c b a	Dicentric chromosome
a b c d g f e h i j	Nonrecombinant chromosome
j i h g f e h i j	Acentric chromosome

When synapsis occurs in an inversion heterozygote, either at meiosis or in the *Drosophila* salivary gland during endomitosis, a loop is often formed to accommodate the point-for-point pairing process (figs. 8.6, 8.7). An outcome of this looping tendency is **crossover suppression.** That is, an inversion heterozygote shows very little recombination for alleles within the inverted region. The reason is not that crossing over is actually suppressed but rather that the products of recombination within a loop are usually lost. A crossover within a loop is shown in figure 8.8. The two nonsister chromatids not involved in a crossover in the loop will result in normal gametes (carrying either the normal chromosome or the intact inverted chromosome). The products of the crossover, rather than being a simple recombination of alleles, are a dicentric and an acentric chromatid. The acentric chromatid is not incorporated into a gamete nucleus, whereas the dicentric chromatid begins a breakage-fusion-bridge cycle that creates a genetic imbalance in the gametes: The gametes carry chromosomes with duplications and deficiencies.

The inversion pictured in figure 8.8 is a **paracentric inversion,** one in which the centromere is outside the in-

version loop. A **pericentric inversion** is one in which the inverted section contains the centromere. It too suppresses crossovers, but for slightly different reasons (fig. 8.9). All four chromatid products of a single crossover within the loop have centromeres and are thus incorporated into the nuclei of gametes. However, the two recombinant chromatids are unbalanced—they both have duplications and deficiencies. One has a duplication for *a-b-c-d* and is deficient for *h-i-j*, whereas the other is the reciprocal of this—deficient for *a-b-c-d* and duplicated for *h-i-j* (in fig. 8.9). These duplication-deletion gametes tend to form inviable zygotes. The result, as with the paracentric inversion, is the apparent suppression of crossing over.

Results of Inversion
Crossing over within inversion loops results in **semisterility.** Almost all gametes that contain dicentric or imbalanced chromosomes form inviable zygotes. Thus a certain proportion of the progeny of inversion heterozygotes are not viable.

There are several evolutionary ramifications of inversions. Those alleles originally together in the noninversion

Figure 8.9

Consequences of a crossover in a pericentric inversion loop

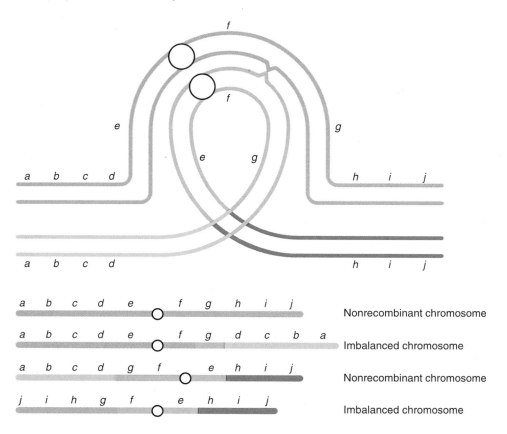

Nonrecombinant chromosome

Imbalanced chromosome

Nonrecombinant chromosome

Imbalanced chromosome

chromosome and those found together within the inversion loop tend to stay together because of the low rate of successful recombination within the inverted region. In the case in which several loci affect the same trait, the alleles are referred to as a **supergene.** Until careful genetic analysis is done, the loci in a supergene could be mistaken for a single locus: They affect the same trait and are inherited apparently as a single unit. Examples include shell color and pattern in land snails and mimicry in butterflies, a topic covered in chapter 21. Supergenes can be beneficial when they involve favorable gene combinations. However, at the same time, their inversion structure prevents the formation of new complexes. Supergenes, therefore, have evolutionary advantages and disadvantages. More on these evolutionary topics will be covered in chapter 21.

Sometimes the inversion process produces a record of the evolutionary history of a group of species. As species evolve, inversions can occur on preexisting inversions. This leads to very complex arrangements of loci as compared with the original arrangement. These patterns are readily studied in Diptera by, for example, noting the changed patterns of bands in salivary gland chromosomes. Since certain arrangements can only come about by a specific sequence of inversions, it is possible to know which species evolved from which other species.

In summary then, inversions result in suppressed crossing over, semisterility, variegation position effects, and new linkage arrangements. They are of evolutionary importance.

Breaks in Nonhomologous Chromosomes

Breaks can occur simultaneously in two nonhomologous chromosomes. There are various ways in which reunion can take place. The most interesting case occurs when the ends of two nonhomologous chromosomes are translocated to each other, which is called a **reciprocal translocation** (fig. 8.10). The organism in which this has happened, a reciprocal translocation heterozygote, has all the genetic material of the normal homozygote. Two outcomes of a reciprocal translocation, like an inversion, are new linkage arrangements in a homozygote—an organism with translocated chromosomes only—and variegation position effects.

During synapsis, either at meiosis or endomitosis, a point-for-point pairing in the translocation heterozygote can be accomplished by the formation of a cross-shaped figure (fig. 8.10). Such a figure is diagnostic of a reciprocal translocation. Unlike the case for an inversion heterozygote, a single crossover in a reciprocal translocation

BOX 8.1

A Case History of the Use of Inversions to Determine Evolutionary Sequence

In 1966, David Futch published a study of the chromosomes of the fruit fly, *Drosophila ananassae,* which is widely distributed throughout the tropical Pacific. The study was originally designed to determine something about the species status of various melanic forms of the fly. In the course of this work, Futch looked at the salivary gland chromosomes of flies from twelve different localities. He discovered twelve paracentric inversions, three pericentric inversions, and one translocation. Because of the precise banding patterns of these chromosomes, it was possible to determine the points of breakage for each inversion.

Observation of several populations, which have had sequential changes in their chromosomes, makes it possible to determine the sequence of events if each successive alteration occupied part of the previous change. By knowing the sequence of changes in different populations of

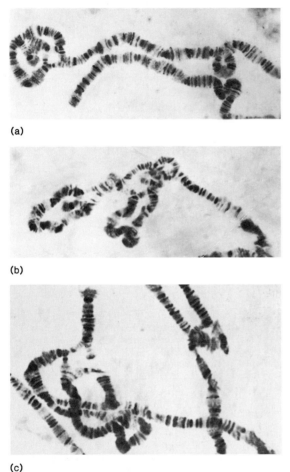

Box figure 1

Photomicrographs of the left arm of chromosome 2 (2L) from larvae heterozygous for various complex gene arrangements. (*a*) Pairing when heterozygous for standard gene sequence and overlapping inversions (2LC; 2LD) and inversion 2LB. (Standard × Tutuila light) (*b*) Pairing when heterozygous for standard gene sequence and single inversion 2LC and overlapping inversions (2LE; 2LB). (Standard × New Guinea) (*c*) Pairing when heterozygous for overlapping inversions (2LD; 2LE; 2LF). (Tutuila light × New Guinea)

David G. Futch, "A Study of Speciation in South Pacific Populations of Drosophila ananassae," *in Marshall R. Wheeler, ed.,* Studies in Genetics, *no. 6615 (Austin: University of Texas Press 1966). Reproduced by permission.*

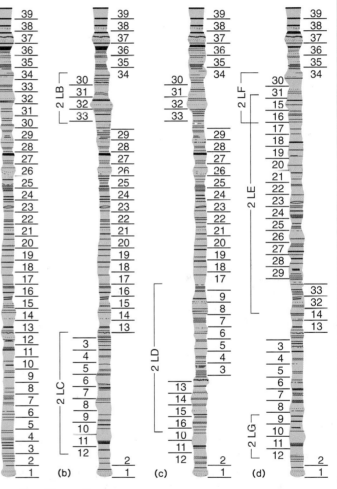

Box figure 2

Chromosomal maps of 2L. (*a*) Standard gene sequence. (*b*) Ponape: Break-points of 2LC and 2LB are indicated and the segments are shown inverted. (*c*) Tutuila light: Break-points of 2LD are indicated. 2LC and 2LB are inverted. 2LD, which overlaps 2LC, is also shown inverted. (*d*) New Guinea: Break-points of 2LE and 2LG are indicated. 2LC, 2LB, and 2LE are shown inverted. Note: Only the break-points of 2LF and LG are shown; neither of these inversions is inverted in the map.

From David G. Futch, "A Study of Speciation in South Pacific Populations of Drosophila ananassae" *in* Studies in Genetics, *no. 6615, 1966, edited by Marshall R. Wheeler. Reprinted by permission of the author.*

Continued on next page

BOX 8.1—*continued*

Drosophila ananassae and knowing the geographic location of the populations, it is possible to determine the history of colonization of these tropical islands by the flies. *D. ananassae* is particularly suited to this type of work because it is believed to be a recent invader to most of the Pacific Islands that it occupies. It is of interest to know something about the spread of this species as an adjunct to studies of human migration in the Pacific Islands, because *D. ananassae* is commensal with people.

Some of Futch's results are shown in the four accompanying figures, which diagram the left and right arms of the fly's second chromosome, as well as the synaptic patterns. We can see vividly the sequence of change in which one inversion occurs after a previous inversion has already taken place. In the figures, the standard (*a*) gave rise to (*b*), which then gave rise independently to (*c*) and (*d*). The standard is from Majuro in the Marshall Islands and is believed to be in the ancestral group of the species. Ponape is the home of (*b*), (*c*) is from Tutuila (eastern Samoa), and (*d*) is from New Guinea. Thus the sequence is Majuro to Ponape and from there the same stock was transferred to Tutuila and New Guinea. This type of analysis has been useful in the *Drosophila* group throughout its range but especially in the Pacific Island populations and in the southwestern United States.

(a)

(b)

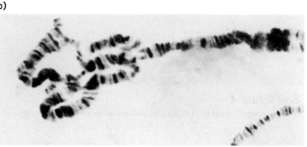

(c)

Box figure 3
Photomicrographs of the right arm of chromosome 2 (2R) from larvae heterozygous for various complex gene arrangements.
(*a*) Pairing when heterozygous for standard gene sequence and overlapping inversions (2RA; 2RB). (Standard × Tutuila light)
(*b*) Pairing when heterozygous for standard gene sequence and overlapping inversions (2RA; 2RC) and inversion 2RD. (Standard × New Guinea) (*c*) Pairing when heterozygous for overlapping inversions 2RB, 2RC, and 2RD. Inversion 2RA is homozygous. (Tutuila light × New Guinea)

David G. Futch, "A Study of Speciation in South Pacific Populations of Drosophila ananassae," in Marshall R. Wheeler, ed., Studies in Genetics, *no. 6615 (Austin: University of Texas Press, 1966). Reproduced by permission.*

Continued on next page

heterozygote will not produce chromatids that are imbalanced. However, nonviable progeny are produced by reciprocal translocation heterozygotes. Problems can arise when centromeres separate at the first meiotic division.

Segregation after Translocation
Since there are two homologous pairs of chromosomes involved, we have to keep track of the independent segregation of the centromeres of the two tetrads. There are two common possibilities (fig. 8.11). The first, called **alternate segregation,** occurs when the first centromere assorts with the fourth centromere, leaving the second and third centromeres to go to the opposite pole. The end result will be balanced gametes, one with normal chromosomes and the other with a reciprocal translocation. Also likely is the **adjacent-1** type of segregation, in which the first and third centromeres segregate together in the opposite direction from the second and fourth centromeres. Here both

BOX 8.1—*continued*

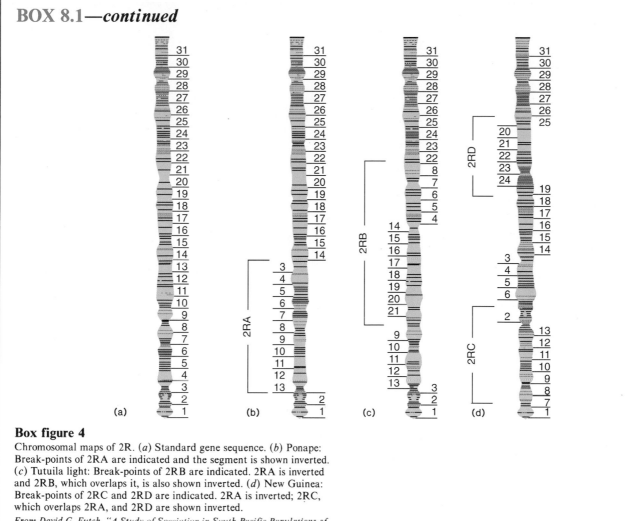

(a) (b) (c) (d)

Box figure 4

Chromosomal maps of 2R. (*a*) Standard gene sequence. (*b*) Ponape:
Break-points of 2RA are indicated and the segment is shown inverted.
(*c*) Tutuila light: Break-points of 2RB are indicated. 2RA is inverted
and 2RB, which overlaps it, is also shown inverted. (*d*) New Guinea:
Break-points of 2RC and 2RD are indicated. 2RA is inverted; 2RC,
which overlaps 2RA, and 2RD are shown inverted.

From David G. Futch, "A Study of Speciation in South Pacific Populations of
Drosophila ananassae*" in* Studies in Genetics, *no. 6615, 1966, edited by*
Marshall R. Wheeler. Reprinted by permission of the author.

types of gametes are unbalanced, carrying duplications and
deficiencies that are usually lethal. Since adjacent-1 segregation occurs at a relatively high frequency, a significant amount of sterility results from the translocation (as
much as 50%).

There is also an **adjacent-2** type of segregation (fig.
8.11) in which homologous centromeres go to the same
pole (first with second, third with fourth). This can result
when the cross-shaped double tetrad opens out into a circle
in late prophase I. In the German cockroach, adjacent-2
patterns have been observed in 10 to 25% of meioses, depending upon which chromosomes were involved.

In summary then, reciprocal translocations result in
new linkage arrangements, variegated position effects, a
cross-shaped figure during synapsis, and semisterility.

Fusions

Another interesting variant of the simple reciprocal translocation occurs when two acrocentric chromosomes become
joined together at or very near their centromeres. The
process is called a **Robertsonian fusion,** after the cytologist W. Robertson, and produces a decrease in the number
of chromosomes, although virtually the same amount of
genetic material is maintained. Often closely related species will have markedly different chromosome numbers
without any significant difference in the quantity of their
genetic material. Such situations could be the result of
Robertsonian fusions. Therefore, cytologists frequently
count the number of chromosomal arms rather than the
number of chromosomes to get a more accurate picture of
species affinities. The number of arms is referred to as the

Figure 8.10
Reciprocal translocation heterozygote

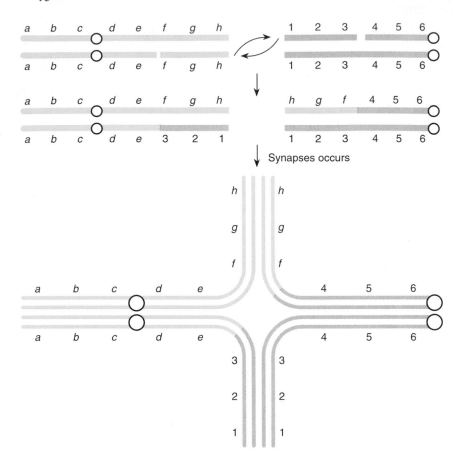

Synapses occurs

fundamental number or **NF** (French: nombre fondamentale). In a similar fashion, **centromeric fission** will increase the chromosome number without changing the NF.

Duplications

Duplications of chromosomal segments can occur, as we have just seen, by the breakage-fusion-bridge cycle or by crossovers within the loop of an inversion. There is another way that duplications arise, specifically, duplications of small adjacent regions of a chromosome. We will illustrate this with a particularly interesting example, the *Bar* eye phenotype in *Drosophila* (fig. 8.12). The wild-type fruit fly has about 800 facets in each eye. The *Bar* (*B*) homozygote has about seventy (a range of 20–120 facets). Another allele, *Doublebar* (*BB:* sometimes referred to as *Ultrabar, B^U*), brings the facet number of the eye down to about twenty-five when homozygous. Around 1920 it was shown that about one progeny in 1,600 from homozygous *Bar* females was *Doublebar*. This is much more frequent than we expect from mutation.

Alfred Sturtevant found that in every *Doublebar* fly there was a crossover between loci on either side of the *Bar* locus. He suggested that the change to *Doublebar* was due to **unequal crossing over** rather than to a simple mutation of one allele to another (fig. 8.13). If the homologous chromosomes do not line up exactly during synapsis, a crossover will produce an unequal distribution of chromosomal material. Later, an analysis of the banding pattern of the salivary glands confirmed Sturtevant's hypothesis. It was found that *Bar* is a duplication of six bands in the 16A region of the X chromosome (fig. 8.14). *Doublebar* is a triplication of the segment.

There is also a position effect in the *Bar* system. A *Bar* homozygote (*B/B*) and a *Doublebar*/wild-type heterozygote (*BB/B⁺*) both have four copies of the 16A region. It would therefore be reasonable to expect that both genotypes would produce the same phenotype. However, the *Bar* homozygote has about seventy facets in each eye, whereas the other has about forty-five. Thus not only the amount of genetic material but also its configuration determines the extent of the phenotype. *Bar* eye was the first position effect discovered.

Figure 8.11
Three possibilities of chromatid separation in the meiosis of a
reciprocal translocation heterozygote

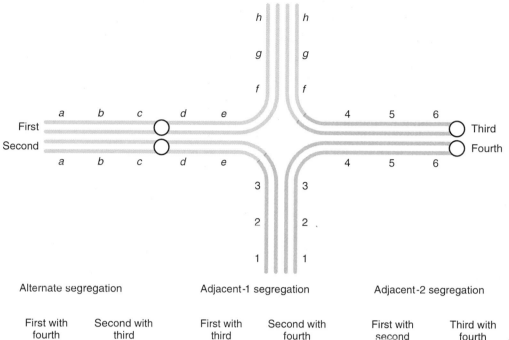

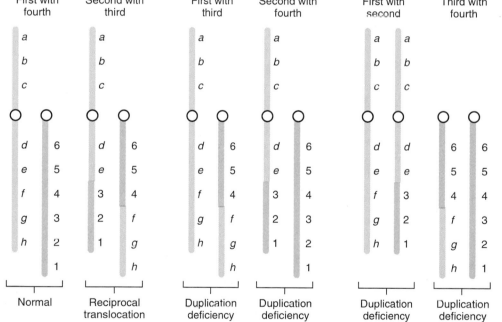

VARIATION IN CHROMOSOME NUMBER

Anomalies of chromosome number occur as two types—
euploidy and **aneuploidy.** Euploidy involves changes in
whole sets of chromosomes; aneuploidy involves changes
in chromosomes in which the change is less than an addition or deletion of a whole set.

Aneuploidy

The terminology of aneuploid change is given in table 8.1.
A diploid cell missing a single chromosome is **monosomic.**
If a cell is missing both copies of that chromosome, it is
nullisomic. A cell missing two nonhomologous chromosomes is called a double monosomic. A similar terminology exists for extra chromosomes. For example, a
diploid cell with an extra chromosome is called **trisomic.**

Figure 8.12

Bar eye in *Drosophila* females

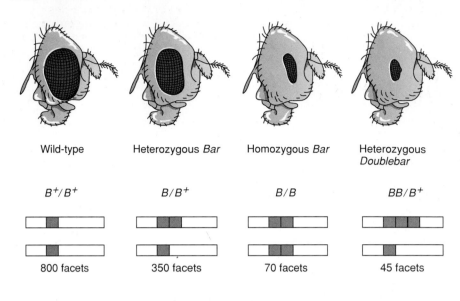

Wild-type	Heterozygous *Bar*	Homozygous *Bar*	Heterozygous *Doublebar*
B^+/B^+	B/B^+	B/B	BB/B^+
800 facets	350 facets	70 facets	45 facets

Figure 8.13

Unequal crossing over in a female *Bar*-eyed *Drosophila* homozygote as a result of improper pairing. The production of a *Doublebar* (and concomitant reversion to wild-type) is accompanied by a crossover between forked (*f*) and fused (*fu*), two flanking loci.

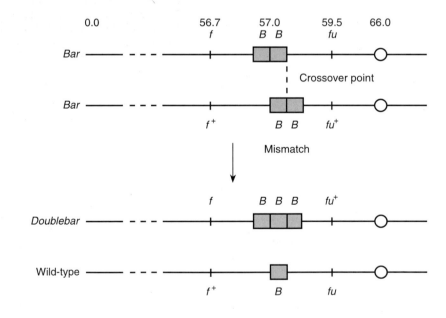

Figure 8.14

Bar region of the X chromosome of *Drosophila*

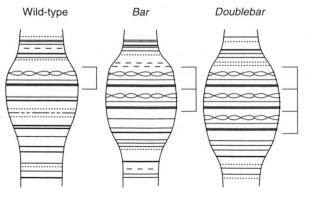

TABLE 8.1 Partial List of Terms to Describe Aneuploidy, Using *Drosophila* as an Example (8 Chromosomes: X, X, 2, 2, 3, 3, 4, 4)

Type	Formula	Number of Chromosomes	Example
Normal	$2n$	8	X, X, 2, 2, 3, 3, 4, 4
Monosomic	$2n - 1$	7	X, X, 2, 2, 3, 4, 4
Nullisomic	$2n - 2$	6	X, X, 2, 2, 4, 4
Double Monosomic	$2n - 1 - 1$	6	X, X, 2, 3, 4, 4
Trisomic	$2n + 1$	9	X, X, 2, 2, 3, 3, 4, 4, 4
Tetrasomic	$2n + 2$	10	X, X, 2, 3, 3, 3, 3, 4, 4
Double Trisomic	$2n + 1 + 1$	10	X, X, 2, 2, 3, 3, 3, 4, 4

Aneuploidy results from nondisjunction in meiosis or by chromosomal lagging whereby one chromosome moves more slowly than the others during anaphase and is thus lost. Here nondisjunction is illustrated using the sex chromosomes in XY organisms such as human beings or fruit flies. Four examples are shown (fig. 8.15): nondisjunction in either the male or female at either the first or second meiotic divisions. The types of zygotes that can result when these nondisjunctional gametes fuse with normal gametes are shown in figure 8.16. All of the offspring produced are chromosomally abnormal. The names and kinds of these imbalances in human beings are detailed later in this chapter.

The occurrence of nondisjunction in *Drosophila* was first shown by Bridges in 1916 with crosses involving the white-eye locus. When a white-eyed female was crossed with a wild-type male, typically the daughters were wild-type and the sons were white eyed. However, occasionally (one or two per thousand), a white-eyed daughter or a wild-type son appeared. This could be explained most easily by a nondisjunctional event in the white-eyed females, where X^wX^w and O eggs (without sex chromosomes) were formed. If an X^wX^w egg were fertilized by a Y-bearing sperm, the offspring would be an X^wX^wY white-eyed daughter. If the egg without sex chromosomes was fertilized by a normal X^+-bearing sperm, the result would be an X^+O wild-type son. Subsequently, these exceptional individuals were found by cytological examination to have precisely the predicted chromosomes (XXY daughters and XO sons). The other types produced by this nondisjunctional event are the XX egg fertilized by an X-bearing sperm and the O egg fertilized by the Y-bearing sperm. The XXX zygotes are genotypically $X^wX^wX^+$, or wild-type daughters (which usually die) and YO flies (which always die).

Mosaicism

Rarely, an individual is made up of several cell lines, each with different chromosome numbers. These individuals are referred to as **mosaics** or **chimeras.** Such conditions can be the result of nondisjunction or chromosomal lagging during mitosis in the zygote or in nuclei in the early embryo. This is demonstrated, again for sex chromosomes, in figure 8.17. A lagging chromosome is shown in figure 8.18, in which the X chromosome is shown to be lost in one of the dividing somatic cells resulting in an XX cell line and an XO cell line. In *Drosophila,* if this chromosomal lagging occurs early in development, an organism that is part male (XO) and part female (XX) develops. Figure 8.19 shows a fruit fly in which chromosomal lagging has occurred at the one-cell stage, causing half the fly to be male and half female. A mosaic of this type, involving male and female phenotypes, has a special name—**gynandromorph.** (A **hermaphrodite** is an individual, not necessarily mosaic, with both male and female reproductive organs.) Many sex-chromosomal mosaics are known in humans, including XX/X, XY/X, XX/XY, and XXX/X. At least one case is known of a human XX/XY chimera that resulted from the fusion of two zygotes, one formed by a sperm fertilizing an ovum and the other formed by a second sperm fertilizing a polar body of that ovum.

Aneuploidy in Human Beings

In human beings, approximately 50% of spontaneous abortions (miscarriages) among women in the United States are found to involve fetuses with some chromosomal abnormality, of which about half are autosomal trisomics. About one in 160 human live births has some sort of chromosomal anomaly, of which most are balanced translocations, autosomal trisomics, or sex-chromosomal aneuploids.

In the standard system of nomenclature, a normal human chromosome complement is 46,XX for a female and 46,XY for a male. The total chromosome number is given first, followed by the description of the sex chromosomes, and, finally, a description of autosomes if some autosomal anomaly is evident. For example, a male with an extra X chromosome would be 47,XXY. A female with a single X chromosome would be 45,X. Since all the autosomes are numbered, we describe their changes by referring to their addition ($+$) or deletion ($-$). For example, a female with trisomy 21 would be 47,XX,$+$21. The short

Figure 8.15
Nondisjunction of the sex chromosomes in *Drosophila* or human beings

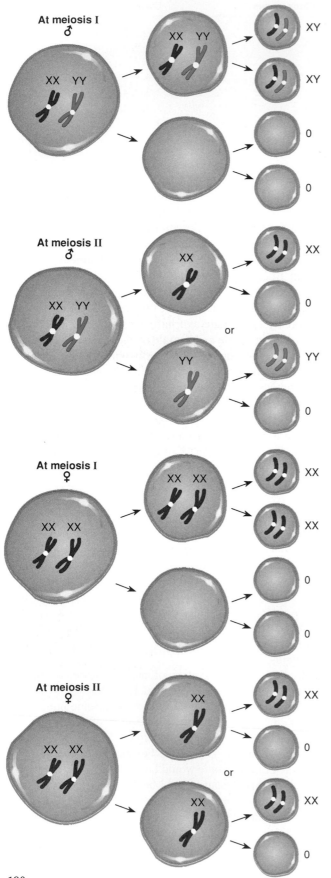

Figure 8.16
Results of fusion of a nondisjunction gamete with a normal gamete

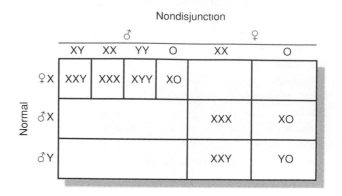

		Nondisjunction				
		♂			♀	
		XY	XX	YY	O	XX	O
Normal	♀X	XXY	XXX	XYY	XO		
	♂X					XXX	XO
	♂Y					XXY	YO

Figure 8.17
Mitotic nondisjunction of the sex chromosomes

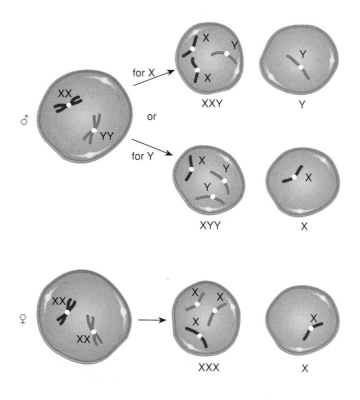

arm of a chromosome is designated p, the longer arm, q. When a change in part of the chromosome occurs, a + after the arm indicates an increase in the length of that arm, whereas a − indicates a decrease in its length. For example, a translocation (t) in which part of the short arm of 9 is transferred to the short arm of 18 would be 46,XX,t(9p−;18p+). The semicolon indicates that both chromosomes kept their centromeres.

Figure 8.18
Chromosomal lagging at mitosis in the X chromosomes of a female *Drosophila*

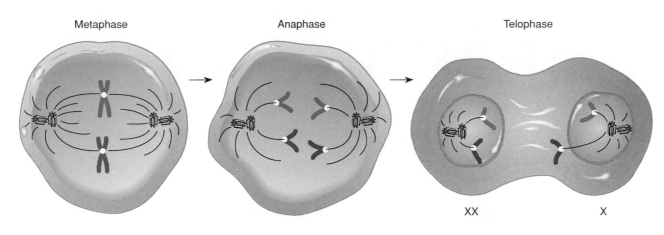

Metaphase   Anaphase   Telophase

XX   X

Figure 8.19
Drosophila gynandromorph. The *left side* is wild-type XX female; the *right side* is XO male, hemizygous for white eye and miniature wing.

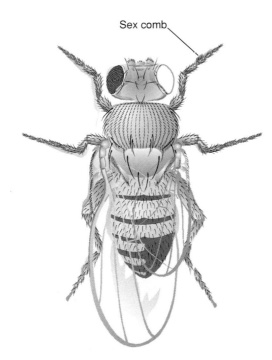

Sex comb

Jérôme Lejeune (1926–   )
Institut de Progenese, Paris.
Courtesy of Dr. Jérôme Lejeune, Institut de Progenese, Paris.

Trisomy 21 (Down Syndrome), 47,XX or XY,+21

Down syndrome (figs. 8.20, 8.21) affects about one in seven hundred live births. Most affected individuals are mildly to moderately mentally retarded and have congenital heart defects and a very high (1/100) risk of acute leukemia. They are usually short and have a broad, short skull; hyperflexibility of joints; and excess skin on the back of the neck. The physician John Langdon Down first described this syndrome in 1866. It was the first human syndrome found to be due to a chromosomal disorder, discovered by Jérôme Lejeune, a physician in Paris who published this finding in 1959. An interesting aspect of this syndrome is the increased incidence among children of older mothers (fig. 8.22), a fact known more than 25 years before the discovery of the cause of the syndrome. Since the future ova are in prophase I of meiosis (dictyotene) since before birth, all ova are the same age as the female. Presumably, older ova have an increased likelihood of nondisjunction of chromosome 21. Despite earlier reports, there appears to be no consistent effect of the father's age on the incidence of Down syndrome.

Trisomy 18 (Edward Syndrome), 47,XX or XY,+18

Edward syndrome affects one in ten thousand live births (fig. 8.23). Most affected individuals are female, with 80 to 90% mortality by two years of age. The affected usually have an elfin appearance with small nose and mouth, a receding lower jaw, abnormal ears, and a lack of distal

Figure 8.20

Karyotype of an individual with trisomy 21

Source: Reproduced courtesy of Dr. Thomas G. Brewster, Foundation for Blood Research, Scarborough, Maine.

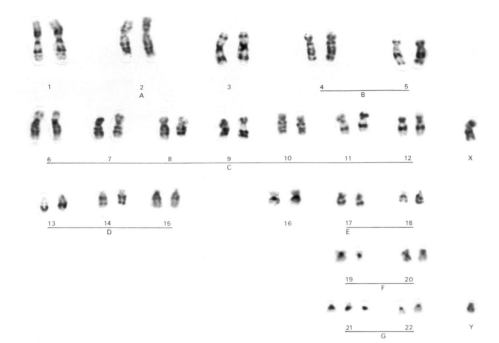

Figure 8.21

Individual with trisomy 21

Source: Reproduced courtesy of Dr. Jérôme Lejeune, Institut de Progenese, Paris.

Figure 8.22

Increased risk of trisomy 21 attributed to the age of the mother

Source: Data from E. Hook, "Estimates of Maternal Age-Specific Risks of a Down-Syndrome Birth in Women Age 34–41" in Lancet, 2: 33–34, 1976.

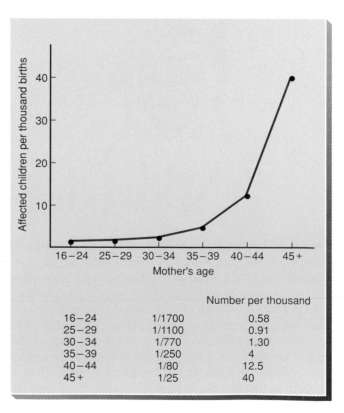

		Number per thousand
16–24	1/1700	0.58
25–29	1/1100	0.91
30–34	1/770	1.30
35–39	1/250	4
40–44	1/80	12.5
45+	1/25	40

Figure 8.23
Child with trisomy 18
Source: Reproduced courtesy of Dr. Jérôme Lejeune, Institut de Progenese, Paris.

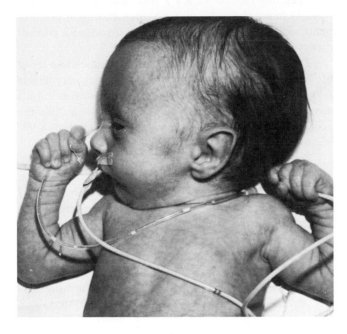

flexion creases on the fingers. There is limited motion of the distal joints and a characteristic posturing of the fingers in which the little and index fingers overlap the middle two. The syndrome is usually accompanied by severe mental retardation.

Trisomy 13 (Patau Syndrome), 47,XX or XY,+13, and Other Trisomic Disorders

Patau syndrome affects one in twenty thousand live births. Diagnostic features are cleft palate, cleft lip, congenital heart defects, polydactyly, and severe mental retardation. Mortality is very high in the first year of life.

Other autosomal trisomics are known but are extremely rare. These include trisomy 8 (47,XX or XY,+8) and cat's eye syndrome, a trisomy of an unknown, small acrocentric chromosome (47,XX or XY,[+acrocentric]). Several aneuploids involving sex chromosomes are known.

Turner Syndrome, 45,X

About one in ten thousand live female births is an infant with Turner syndrome. This and 45,XX or XY,−21 and 45,XX or XY,−22 are the only nonmosaic, viable monosomics recorded in human beings (fig. 8.24), indicating

Figure 8.24
Karyotype of a person with Turner syndrome (XO)
Source: Reproduced courtesy of Dr. Thomas G. Brewster, Foundation for Blood Research, Scarborough, Maine.

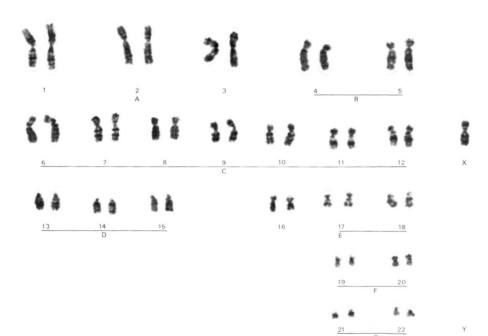

the severe consequences of monosomy of all but the two smallest autosomes and a sex chromosome. However, individuals with Turner syndrome have about normal intelligence but underdeveloped ovaries, abnormal jaws, webbed necks, and shieldlike chests.

The symptoms of Turner syndrome have been logically deduced to be caused by a single dosage of genes that are normally present and active in two dosages. Thus they would be genes that are located on both the X and Y chromosomes (pseudoautosomal) to have two dosages in normal XY males and also active in both X chromosomes in normal XX females. Therefore they should be located on regions of the X chromosome that escape inactivation (see chapter 5). By studies of persons with small X-chromosomal deletions and by molecular analyses of the X and Y chromosomes (outlined in chapter 12), two genes have emerged as candidates: *ZFY* (on the Y chromosome, termed *ZFX* on the X chromosome) and *RPS4Y* (on the Y chromosome, termed *RPS4X* on the X chromosome). You are familiar with *ZFY* (zinc finger on the Y chromosome)—it was once believed to be the male determining gene in mammals. *RPS4Y* encodes a ribosomal protein, one of the many proteins making up the ribosome.

It is interesting to note a dosage-compensation difference in people and mice, which have analogous genes termed *Zfx* and *Rps4x*. In mice, unlike in people, these genes are inactivated in the "Lyonized" X chromosome in females and have restricted activity in the Y chromosome. Hence, mouse cells seem normally to have only one copy of these genes functioning in normal XY males and XX females. Therefore, we would predict that the XO genotype in mice would show few if any negative effects as compared to human XO individuals, since mouse cells of both sexes normally only have one copy of each gene in question functioning. In fact, there is a 99% prenatal mortality of human Turner syndrome fetuses but virtually no prenatal mortality in mouse fetuses with the XO genotype (born of XX mothers), confirming our predictions and pointing to differences between people and mice in dosage-compensation mechanisms for specific genes.

XYY Karyotype, 47,XYY

About one in one thousand live births of males is an individual with an XYY karyotype. (We avoid the term "syndrome" here because there is no clearly defined series of attributes in XYY men.) They are taller than normal. There has been some controversy surrounding this karyotype because it was originally reported that there was an abundance of this karyotype in a group of mentally subnormal males in a prison hospital. Seven XYY males were found among 197 inmates, whereas only one in about two thousand control men were XYY. This study has subsequently been expanded and corroborated. Although it is

now fairly well established that there is about a twenty-fold higher incidence of XYY males in prison than in society at large, the statistic is somewhat misleading: The overwhelming number of XYY men seem to lead normal lives. At most, about 4% of XYY men end up in penal or mental institutions, where they make up about 2 percent of that population.

There is some indication that the XYY men had lower intelligence test levels. Thus criminal tendency may be due to lowered intelligence rather than a predisposition toward criminality caused by an extra Y chromosome. For the most part, expanded studies have indicated that XYY criminals did not commit violent crimes.

Research on this karyotype has produced its own problems. A research project at Harvard University on XYY males came under intense public pressure and was eventually terminated. The project, under the direction of Stanley Walzer (a psychiatrist) and Park Gerald (a geneticist), involved screening all newborn boys at the Boston Hospital for Women and following the development of those with chromosomal anomalies. The criticism of this work centered mainly on the necessity of informing parents that their sons had an XYY karyotype and that there could be behavioral problems. Opponents of this work claimed that telling the parents would constitute a self-fulfilling prophecy. That is, parents who were told that their children were not normal and "might" cause trouble would then behave toward their children in a manner that would increase the probability that their children would "cause trouble." The opponents claimed that the risks of this research outweighed the benefits. The project was terminated in 1975 primarily because of the harassment that Walzer received.

Klinefelter Syndrome, 47,XXY

The incidence of Klinefelter syndrome is about one in one thousand live births. Tall stature and infertility are common. Diagnosis is usually by buccal smear to ascertain the presence of a Barr body in a male, indicative of an XXY karyotype. Some problems with behavior and speech development have been associated with this syndrome.

Triple-X Female, 47,XXX, and Other Aneuploid Disorders of Sex Chromosomes

A triple-X female appears in about one in one thousand female live births. Fertility can be normal, but there is usually a mild mental retardation. Delayed growth, as well as congenital malformations, have been noted. Other sex-chromosomal aneuploids, including XXXX, XXXXX, and XXXXY, are extremely rare. All seem to be characterized by mental retardation and growth deficiencies.

Figure 8.25
Human metaphase chromosomes with the fragile-X site indicated by an arrow

From Ian Craig Nature *(1991) 349: 742. Copyright © 1991 Macmillan Magazines, Ltd.* "Methylation and the Fragile X."

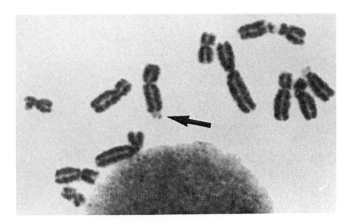

Chromosomal Rearrangements in Human Beings

In addition to the aneuploids mentioned, several human syndromes and abnormalities are the result of chromosomal rearrangements, including deletions and translocations. The most common disorders are described here. Three points should be kept in mind. First, all are extremely rare. Second, the deletion syndromes are often found to be caused by a balanced translocation in one of the parents. And third, about one in five hundred live births contains a balanced rearrangement of some kind, either a reciprocal translocation or inversion.

Fragile-X Syndrome

The most common cause of inherited mental retardation is the **fragile-X syndrome.** It is found in about one in every 1,250 males and about one in every 2,000 females. The symptoms include mental retardation, altered speech patterns, large testes in males, and other physical attributes. It is called the fragile-X syndrome because it is related to a region at the tip of the X chromosome that breaks more frequently than other chromosomal regions. However, the break is not required for the syndrome, and the fragile-X chromosome is usually identified by the lack of chromatin condensation at the site (fig. 8.25). The gene responsible for the syndrome is called *FMR-1,* for fragile-X mental retardation-1.

An interesting property of the syndrome is an unusual pattern of expression of the symptoms. Approximately 20% of males with the fragile-X chromosome do not have symptoms but have grandchildren that do have the symptoms. The daughters of the symptomatic males also don't have symptoms, but they would not be expected to be-cause they have another X chromosome to mask the symptoms. We will discuss further unusual modes of inheritance like this in chapter 17.

Cri-du-Chat Syndrome, 46,XX or XY,5p−

The syndrome known as *cri-du-chat* is so called because of the catlike cry that about half the affected infants show. Microcephaly, congenital heart disease, and severe mental retardation are also common. This disorder arises from a deletion in chromosome 5 (fig. 8.26); most other deletions studied (4p−, 13q−, 18p−, 18q−) also result in microcephaly and severe mental retardation. The rarity of viable deletion heterozygotes is consistent with the fact that viable monosomics are rare. An individual heterozygous for a deletion is, in effect, monosomic for that region of the chromosome that is deleted. Evidently, monosomy or heterozygosity for large deleted regions of a chromosome are generally lethal in human beings.

Familial Down Syndrome

Down syndrome (trisomy 21) was described earlier as the result of a nondisjunctional event during gametogenesis. It is a function of parental age and is not inherited. (Although about half the children of a trisomy 21 person will have trisomy 21 because of aneuploid gamete production, the possibility of an unaffected relative of the trisomy 21 person having abnormal children is no greater than for a person of the same age chosen at random from the general population.) However, about 4% of those with Down syndrome have been found to have a translocation of chromosome 21, usually associated with chromosomes 14, 15, or 22. The translocational and nontranslocational types of Down syndrome have identical symptoms; however, a balanced translocation can be passed on to offspring (see fig. 8.11). Alternate separation of centromeres in the translocation heterozygote will produce either a normal gamete or one carrying the balanced translocation. Adjacent separation will cause partial trisomy for certain chromosomal parts. When this occurs for most of chromosome 21, Down syndrome results.

It is worth mentioning that aside from trisomy and translocation, Down syndrome can come about through mosaicism or a centromeric event. It is found that about 2% of individuals with Down syndrome are mosaic for cells with both two and three copies of chromosome 21. There is some evidence that the original zygotes were trisomic but then a daughter cell lost one of the copies of chromosome 21. The severity of the symptoms of Down syndrome in these individuals is related to the percentage of trisomic cells they possess. Mosaicism increases with maternal age, just as trisomy in general does. In extremely rare cases, Down syndrome has been caused by the occurrence of an abnormal chromosome 21 that has, rather

Figure 8.26

Karyotype of individual with *cri-du-chat* syndrome, due to a partial deletion of the short arm of chromosome 5 (5p—)

Source: Reproduced courtesy of Dr. Thomas G. Brewster, Foundation for Blood Research, Scarborough, Maine.

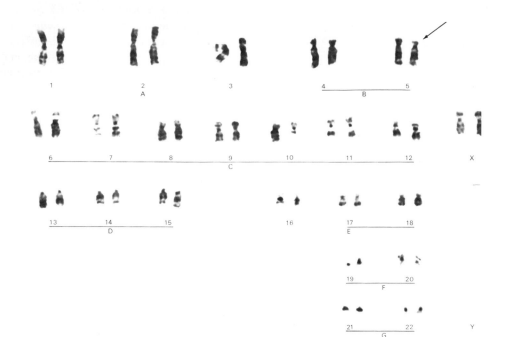

than a short and long arm, two identical long arms attached to the centromere. This type of chromosome is called an **isochromosome** and presumably occurs by an odd centromeric fission (fig. 8.27). Hence a person with a normal chromosome 21 and an isochromosome 21 has three copies of the long arm of the chromosome and therefore shows Down syndrome.

Euploidy

Euploid organisms have varying numbers of complete haploid chromosome sets. We are already familiar with haploids (n) and diploids ($2n$). Organisms with higher numbers of sets, such as **triploids** ($3n$) and **tetraploids** ($4n$), are called polyploids. Three kinds of problems plague polyploids. First, there is potential for a general imbalance in the organism due to the extra genetic material in each cell. For example, a triploid human fetus has about a one in a million chance to survive to birth at which time death usually occurs due to problems in all organ systems. Second, if there is a chromosomal sex-determining mechanism, it may be disrupted by polyploidy. And third, meiosis produces unbalanced gametes in many polyploids.

If the polyploid has an odd number of sets of chromosomes, such as triploid ($3n$), two of the three homologues will tend to pair at prophase I of meiosis, producing a bivalent and a univalent. The bivalent will separate normally, but the third chromosome will go independently to one of the poles. This separation results in a 50% chance of aneuploidy in each of the n different chromosomes, rapidly decreasing the probability of a balanced gamete as n increases. Therefore, as n increases, so does the likelihood of sterility. An alternative to the bivalent-univalent type of synapsis is the formation of trivalents, which have similar problems (fig. 8.28). Even-numbered polyploids, such as tetraploids ($4n$), can do better during meiosis. If the segregation of centromeres is two and two in each of the n meiotic figures, balanced gametes can result. Often, however, the multiple copies of the chromosomes form complex figures during synapsis, including monovalents, bivalents, trivalents, and quadrivalents, tending to result in aneuploid gametes and sterility.

Some groups of organisms, primarily plants, have many polyploid members. For example, the genus of wheat, *Triticum,* has members with fourteen, twenty-eight, and forty-two chromosomes. Because the basic *Triticum* chromosome number is $n = 7$, these forms are $2n$, $4n$, and $6n$ species, respectively. Chrysanthemums have species of eighteen, thirty-six, fifty-four, seventy-two, and ninety chromosomes. With a basic number of $n = 9$, these species represent a $2n$, $4n$, $6n$, $8n$, and $10n$ series. In both these examples, the even-numbered polyploids are viable, but the odd-numbered polyploids are not.

Figure 8.27

The break of the centromere of chromosome 21 perpendicular to the normal division axis can form an isochromosome of the long arms and either an isochromosome of the short arms or two acentric pieces. This can happen during anaphase of mitosis or meiosis II.

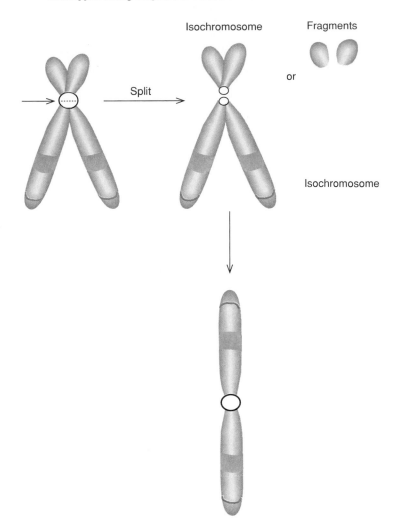

Isochromosome   Fragments

Split

or

Isochromosome

Isochromosome

Autopolyploidy

Polyploidy can come about in two different ways. In **autopolyploidy** all of the chromosomes come from within the same species. In **allopolyploidy** the chromosomes come from the hybridization of two different species. Autopolyploidy occurs in several different ways. The fusion of nonreduced gametes will create polyploidy. For example, if a diploid gamete fertilizes a normal haploid gamete, the result will be a triploid. Similarly, a diploid gamete fertilized by another diploid gamete will produce a tetraploid. The equivalent of a nonreduced gamete will come about in meiosis if the parent cell is polyploid to begin with. For example, if a branch of a diploid plant is tetraploid, its flowers will produce diploid gametes. These gametes are not the result of a failure to reduce chromosome

numbers meiotically but rather the result of successful meiotic reduction in a polyploid flower. The tetraploid tissue of the plant in this example can originate by the process of **somatic doubling** of diploid tissues.

Somatic doubling can come about spontaneously or be caused by anything that disrupts the normal sequence of a nuclear division. For example, colchicine will induce somatic doubling by inhibiting microtubule formation. This prevents the formation of a spindle and thus prevents the movement of the chromosomes during either mitosis or meiosis. The result is a cell with double the chromosome number. Other chemicals, temperature shock, and physical shock can produce the same effect.

Allopolyploidy

Allopolyploidy comes about by cross-fertilization between two species. The resulting offspring have the sum of the reduced chromosome number of each parent species. If the chromosome set of each is distinctly different, the new organisms have difficulty in meiosis because no two chromosomes are sufficiently homologous. Every chromosome will form a univalent (unpaired) figure. They will separate independently during meiosis, producing aneuploid gametes. However, if an organism can survive by vegetative growth until somatic doubling takes place in gamete precursor cells ($2n \rightarrow 4n$), or alternatively, if the zygote was formed by two unreduced gametes ($2n + 2n$), the resulting offspring will be fully fertile because each chromosome has a pairing partner at meiosis. An example can be drawn from the work of a Russian geneticist, G. D. Karpechenko.

In 1928, Karpechenko worked with the radish (*Raphanus sativus,* $2n = 18$, $n = 9$) and cabbage (*Brassica oleracea,* $2n = 18$, $n = 9$). When these two plants are crossed, an F_1 results with $2n = 18$ (9 + 9). This plant has characteristics intermediate between the two parental species (fig. 8.29) and is an allodiploid. If somatic doubling takes place, the chromosome number is doubled to thirty-six, and the plant becomes an allopolyploid (an allotetraploid of $4n$). Since each chromosome has a homologue, this allotetraploid is referred to as an **amphidiploid.** Without knowledge of its past history, this plant would simply be classified as a diploid with $2n = 36$. In this case the new amphidiploid cannot successfully breed with either parent because the offspring are sterile triploids. It is, therefore, a new species and has been named *Raphanobrassica.* As an agricultural experiment, however, it was not a success because it did not combine the best features of the cabbage and radish.

Polyploidy in Plants and Animals

Although polyploids in the animal kingdom are known (in some species of lizards, fish, and invertebrates), polyploidy as a successful evolutionary strategy is primarily a

Figure 8.28

Meiosis in a triploid ($3n = 9$) and one possible resulting arrangemen of gametes. The probability of a "normal" gamete is $(1/2)^n$ where n equals the haploid chromosome number. Here, $n = 3$ and $(1/2)^3 = 1/8$.

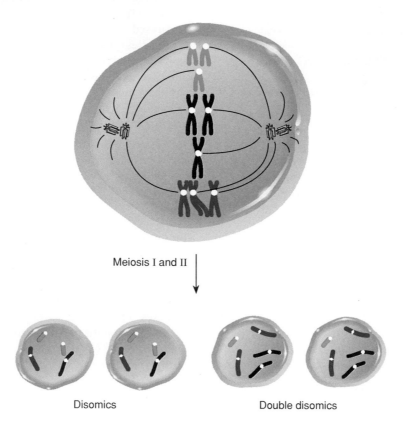

Meiosis I and II

Disomics   Double disomics

plant phenomenon. There are several reasons for this difference between plants and animals. To begin with, many more animals than plants have chromosomal sex-determining mechanisms. These mechanisms are severely disrupted by polyploidy. For example, Bridges discovered a tetraploid female fruit fly. However, it has not been possible to produce a tetraploid male. The tetraploid female's progeny were triploids and intersexes. Second, plants can generally avoid the meiotic problems of polyploidy longer than most animals. Some plants can exist vegetatively, allowing more time and hence more of a probability that the rare somatic doubling event will take place that will produce an amphidiploid; animal life spans are more precisely defined, allowing less time for a somatic doubling. And third, many plants depend on the wind or insect pollinators to fertilize them and thus have more of an opportunity for hybridization. Many animals have relatively elaborate courting rituals that tend to restrict hybridization.

Polyploidy has been used in agriculture for the production of "seedless" as well as "jumbo" varieties of crops. Seedless watermelon, for example, is a triploid. Its seeds are mostly sterile and do not develop. It is produced by growing seeds from the cross between a tetraploid variety and a diploid variety. Jumbo Macintosh apples are tetraploid.

SUMMARY

Variation can occur in the structure and the number of chromosomes an organism can have. When chromosomes break, the ends act in a "sticky" fashion; they tend to reunite with other broken ends. A single break can lead to the possibility of deletions or the formation of acentric or dicentric chromosomes. Dicentrics tend to go through breakage-fusion-bridge cycles, which result in duplications and deficiencies.

Two breaks in the same chromosome can yield deletions and inversions. Variegation position effects as well as new linkage arrangements can result. Inversion heterozygotes produce loop fig-

Figure 8.29

Hybridization of cabbage and radish showing the fruiting structures

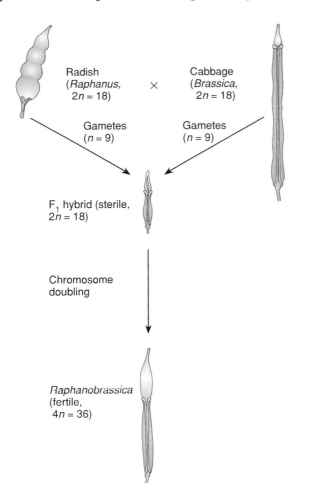

Radish
(*Raphanus*,
2n = 18)

×

Cabbage
(*Brassica*,
2n = 18)

Gametes
(n = 9)

Gametes
(n = 9)

F₁ hybrid (sterile,
2n = 18)

Chromosome
doubling

Raphanobrassica
(fertile,
4n = 36)

ures during synapsis, which can be either at meiosis or in polytene chromosomes. Heterozygosity for an inversion suppresses crossovers, and organisms that are heterozygotes are semisterile.

Reciprocal translocations can result from single breaks in nonhomologous chromosomes. These produce cross-shaped figures at synapsis and result in semisterility. The *Bar* eye phenotype of *Drosophila* is an example of a duplication that causes a position effect.

Changes in chromosome number can involve whole sets (euploidy) or partial sets (aneuploidy) of chromosomes. Aneuploidy usually results from nondisjunction or chromosomal lagging. Several medical syndromes are caused by aneuploidy, such as Down, Turner, and Klinefelter syndromes, and the XYY karyotype.

Polyploidy leads to difficulties in chromosomal sex-determining mechanisms, general chromosomal imbalance, and problems during meiotic segregation. It has been more successful in plants than in animals because plants generally lack chromosomal sex-determining mechanisms. Plants can also avoid meiotic problems by propagating vegetatively. In both animals and plants, even-numbered polyploids do better than odd-numbered polyploids because they have a better chance of producing balanced gametes during meiosis. Somatic doubling provides each chromosome in a hybrid organism with a homologue and thus makes tetrad formation at meiosis possible. New species have arisen by polyploidy.

SOLVED PROBLEMS

1. What are the consequences of an inversion?

   ANSWER: In an inversion homozygote, the consequences are change in linkage arrangements, including new orders and map distances, and the possibility of position effects if a locus is newly placed into heterochromatin. In an inversion heterozygote, the mechanism of crossover suppression causes semisterility because of the loss of zygotes that carry genic imbalances. Inversion heterozygotes can be seen as meiotic loop structures or loops formed in endomitotic chromosomes such as those found in salivary glands of fruit flies. In an evolutionary sense, inversions result in supergenes, the locking together of allelic combinations.

2. What are the consequences of a monosomic chromosome in human beings?

   ANSWER: In human beings, the monosomic state is rare, meaning that with a few exceptions it is deleterious. In fact, monosomics are also rare in spontaneous abortions, indicating that most monosomics lead to loss of the fetus before the woman is aware of the pregnancy. The only monosomics known to be viable in human beings are Turner syndrome (45,X) and monosomics of chromosomes 21 and 22, the two smallest autosomal chromosomes.

3. Ebony body (*e*) in flies is an autosomal recessive trait. A true-breeding ebony female (*ee*) is mated to a true-breeding wild-type male that has been irradiated. Among the wild-type progeny is a single ebony male. Explain this observation.

ANSWER: The cross is $ee \times e^+e^+$, and all F_1s should be e^+e, or wild-type. The use of irradiation alerts us to the possibility of chromosomal breaks, as well as simple mutations. What type of chromosomal aberration would allow a recessive trait to appear unexpectedly? The observation could be explained by a deletion, which creates a situation of pseudodominance because there is no second allele. Thus the male in question could have gotten the ebony allele from its mother and no homologous allele from its father. Alternatively, the wild-type allele from the father could have mutated to an ebony allele.

EXERCISES AND PROBLEMS

1. What kind of figure is observed in meiosis of a reciprocal translocation homozygote?

2. Can a deletion result in the formation of a variegation position effect? If so, how?

3. Does crossover suppression occur in an inversion homozygote? Explain.

4. Which rearrangements of chromosomal structure cause semisterility?

5. What are the consequences of single crossovers during tetrad formation in a reciprocal translocation heterozygote?

6. Is a tetraploid more likely to show irregularities in meiosis or mitosis? Explain. What about these processes in a triploid?

7. How many chromosomes would a human tetraploid have? How many chromosomes would a human monosomic have?

8. Do autopolyploids or allopolyploids experience more difficulties during meiosis? Do amphidiploids have more or less trouble than auto- or allopolyploids?

9. If a diploid species of $2n = 16$ hybridizes with one of $2n = 12$ and the resulting hybrid doubles its chromosome number to produce an allotetraploid (amphidiploid), how many chromosomes will it have? How many chromosomes will an allotetraploid have if both parent species had $2n = 20$?

10. If nondisjunction of the sex chromosomes occurs in a female at the second meiotic division, what type of eggs will arise?

11. In terms of acentrics, dicentrics, duplications, and deficiencies, give the gametic complement when a three-strand, double crossover occurs within a paracentric inversion loop.

12. In studying a new sample of fruit flies, a geneticist noted phenotypic variegation, semisterility, and the nonlinkage of previously linked genes. What probably caused this and what cytological evidence would strengthen your hypothesis?

13. In a second sample of flies, the geneticist found a position effect and semisterility. The linkage groups were correct, but the order was changed and crossing over was suppressed. What probably caused this and what cytological evidence would strengthen your hypothesis?

14. Diagram the results of alternate segregation for a three-strand double crossover between a centromere and the cross center in a reciprocal translocation heterozygote.

15. How might an XO/XYY human mosaic arise? An XX/XXY mosaic? How might a trisomy-21 individual arise?

16. A heterozygous plant $ABCDE/abcde$ is testcrossed to an $abcde/abcde$ plant. The following progeny appear.

$ABCDE/abcde$

$abcde/abcde$

$Abcde/abcde$

$aBCDE/abcde$

$ABCDe/abcde$

$abcdE/abcde$

What is unusual about the results and how can you explain them?

17. White eye color in *Drosophila* is an X-linked recessive trait. A wild-type male is irradiated and mated with a white-eyed female. Among the progeny is a white-eyed female.
   a. Why is this result unexpected, and how could you explain it?
   b. What type of progeny do you expect if this white-eyed female is crossed to a normal, nonirradiated male?

18. You are trying to locate an enzyme-producing gene in *Drosophila,* which you know is located on the third chromosome. You have five strains with deletions for different regions of the third chromosome ("/" indicates deleted region).

```
Normal    0  10  20  30  40  50  60     map units
Strain A  ////_____
Strain B  __///////////// _____
Strain C  _____ ///////////_____
Strain D  _____ /////////___
Strain E  _____ ////////
```

You cross each strain with wild-type flies and measure the amount of enzyme in F_1 progeny. The results appear below. In what region is the gene located?

Strain Crossed	Percent of wild-type enzyme produced in F_1 progeny
A	100
B	45
C	54
D	98
E	101

19. Consider the following crosses in a plant in which number of viable progeny under standard conditions is measured. Provide an explanation for the results.

P_1:	strain A × strain A	strain B × strain B	strain A × strain B
F_1:	765	750	775
F_2:	712	783	416

20. Below is the map position for three X-linked recessive genes in *Drosophila* (v = vermilion eyes, m = miniature wings, and s = sable body).

v	m	s
33.0	36.1	43.0

A wild-type male is X-rayed and mated to a vermilion, miniature, sable female. Among the progeny, a single vermilion-eyed, long-winged, tan-bodied female is recovered. When this female is mated with a $v\ m\ s$ hemizygous male, the progeny are:

Females	Males
87 vermilion, miniature, sable	89 vermilion, miniature, sable
93 vermilion	1 vermilion

Explain these results by drawing a genetic map.

21. Plant species P has $2n = 18$ and species U has $2n = 14$. A fertile hybrid is found. How many chromosomes does it have?

22. A normal-visioned woman whose father was color-blind marries a normal-visioned man. They have a color-blind daughter with Turner syndrome. In which parent did nondisjunction occur?

23. A color-blind man marries a normal woman whose father was color-blind. They have a color-blind son with Klinefelter syndrome. In which parent did nondisjunction occur?

24. Explain what genetic events must happen to produce an XYY man.

25. Chromosomal analysis of a spontaneously aborted fetus revealed that the fetus was 92,XXYY. Propose an explanation to account for this unusual karyotype.

26. In *Drosophila*, recessive genes clot (ct) and black body (b) are located at 16.5 and 48.5 map units, respectively, on the second chromosome. Wild-type females that are $ct^+b^+/ct\ b$ are mated with $ct\ b/ct\ b$ males and produce the following progeny:

wild-type	1,250
clot, black	1,200
black	30
clot	20

What is unusual about the results and how can you explain them?

27. You have four strains of *Drosophila* (1–4) that were isolated from different geographic regions. You compare the banding patterns of the second chromosome and obtain the following results (each letter corresponds to a band):

(1) m n r q p o s t u v

(2) m n o p q r s t u v

(3) m n r q t s u p o v

(4) m n r q t s o p u v

If (3) is the ancestral strain, in what order did the other strains arise?

28. In *Drosophila*, the recessive gene for white eyes is located near the tip of the X chromosome. A wild-type male is irradiated and mated with a white-eyed female. Among the progeny is one red-eyed male. How can you explain the red-eyed male and how could you test your hypothesis?

Suggested Readings for chapter 8 are on page 644.

Part III

Molecular Genetics

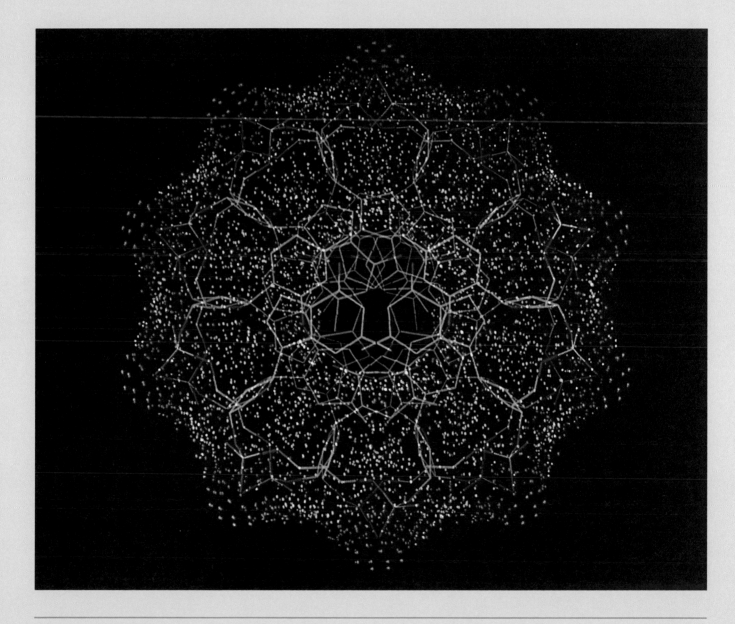

An end view of a computer-graphic model of DNA
© R. Langudge/D. McCoy/Rainbow

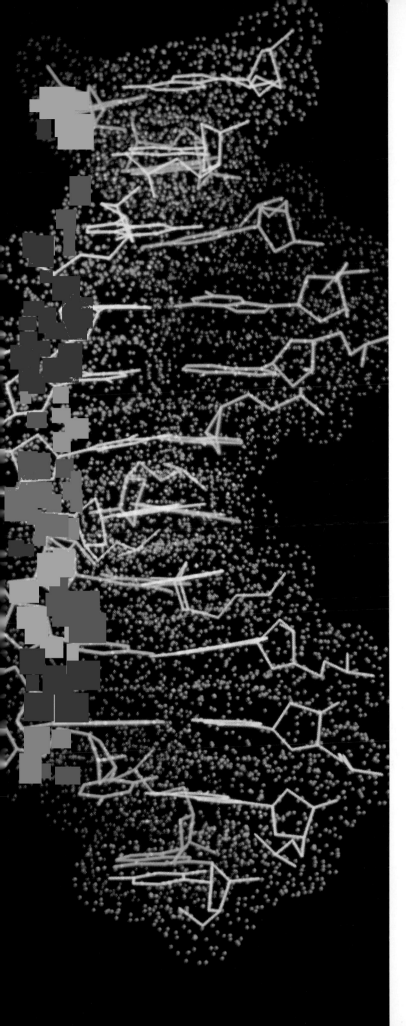

Chemistry of the Gene

A computer-graphic image showing the double helical structure of DNA. DNA consists of two linked strands of nucleotides coiled into a helix.

© Oxford Molecular Biophysics Laboratory/Science Photo Library/Photo Researchers, Inc.

In 1953, James Watson and Francis Crick published a two-page paper in the journal *Nature* entitled "Molecular Structure of Nucleic Acids: A Structure for Deoxyribose Nucleic Acid." It began as follows: "We wish to suggest a structure for the salt of deoxyribose nucleic acid (D.N.A.). This structure has novel features which are of considerable biological interest." This paper, in which the correct model of DNA was first put forth, has become a milestone in the modern era of molecular genetics, compared by some to the work of Mendel and Darwin. (Watson, Crick, and X-ray crystallographer Maurice Wilkins won Nobel Prizes for this work; Rosalind Franklin, also an X-ray crystallographer, was acknowledged, posthumously, to have played a major role in the discovery of the structure of DNA.) Once the structure of the genetic material had been determined, an understanding of its functioning and its method of replication followed quickly.

James D. Watson (1928–   )
Cold Spring Harbor Laboratory Research Library Archives. Margot Bennet, photographer.

Francis Crick (1916–   )
Reproduced by permission of Herb Weitman, Washington University, St. Louis, Missouri.

Maurice H. F. Wilkins (1916–   )
Courtesy of Dr. Maurice H. F. Wilkins and Biophysics Department, King's College, London.

IN SEARCH OF THE GENETIC MATERIAL

Required Properties of a Genetic Material

We will begin with a look at the properties that a genetic material must have and review the evidence that nucleic acids make up the genetic material. To comprise the genes, DNA must carry the information to control the synthesis of the enzymes and proteins within a cell or organism, self-replicate with high fidelity yet show a low level of mutation, and be located in the chromosomes.

Control of the Enzymes

The growth, development, and functioning of a cell are controlled by the proteins within it, primarily its enzymes. Thus the nature of a cell's phenotype is controlled by the protein synthesis within that cell. The genetic material must determine the presence and effective amounts of the enzymes in a cell. For example, given inorganic salts and glucose, an *E. coli* cell can synthesize, through its enzyme-controlled biochemical pathways, all of the compounds it needs for growth, survival, and reproduction.

At this point we need to review some basic information regarding enzymes. An enzyme is a protein that acts as a catalyst of a specific metabolic process without itself being markedly altered by the reaction. Most reactions that are catalyzed by enzymes could occur anyway, but only under conditions too extreme for them to take place within living systems. For example, many oxidations occur naturally at high temperatures. Enzymes allow these reactions to occur within the cell by lowering what is called the **activation energy (ΔG)** of a particular reaction. In other words, an enzyme allows a reaction to take place without needing the boost in energy usually supplied by heat (fig. 9.1).

Figure 9.1
An enzyme lowers the activation energy (ΔG) of a particular reaction.

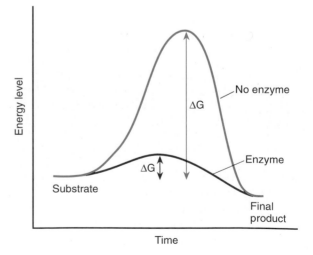

Molecular Structure of Nucleic Acids: A Structure for Deoxyribose Nucleic Acid

We wish to suggest a structure for the salt of deoxyribose nucleic acid (D.N.A.). This structure has novel features which are of considerable biological interest.

A structure for nucleic acid has already been proposed by Pauling and Corey.[1] They kindly made their manuscript available to us in advance of publication. Their model consists of three intertwined chains, with the phosphates near the fibre axis, and the bases on the outside. In our opinion, this structure is unsatisfactory for two reasons: (1) We believe that the material which gives the X-ray diagrams is the salt, not the free acid. Without the acidic hydrogen atoms it is not clear what forces would hold the structure together, especially as the negatively charged phosphates near the axis will repel each other. (2) Some of the van der Waals distances appear to be too small.

Another three-chain structure has also been suggested by Fraser (in the press). In his model the phosphates are on the outside and the bases on the inside, linked together by hydrogen bonds. This structure as described is rather ill-defined, and for this reason we shall not comment on it.

We wish to put forward a radically different structure for the salt of deoxyribose nucleic acid. This structure has two helical chains each coiled round the same axis (see diagram). We have made the usual chemical assumptions, namely, that each chain consists of phosphate diester groups joining β-D-deoxyribofuranose residues with 3', 5' linkages. The two chains (but not their bases) are related by a dyad perpendicular to the fibre axis. Both chains follow right-handed helices, but owing to the dyad the sequences of the atoms in the two chains run in opposite directions. Each chain loosely resembles Furberg's[2] model No. 1; that is, the bases are on the inside of the helix and the phosphates on the outside. The configuration of the sugar and the atoms near it is close to Furberg's "standard configuration," the sugar being roughly perpendicular to the attached base. There is a residue on each chain every 3.4 A in the z-direction. We have assumed an angle of 36° between adjacent residues in the same chain, so that the structure repeats after 10 residues on each chain, that is, after 34 A. The distance of a phosphorus atom from the fibre axis is 10 A. As the phosphates are on the outside, cations have easy access to them.

The structure is an open one, and its water content is rather high. At lower water contents we would expect the bases to tilt so that the structure could become more compact.

The novel feature of the structure is the manner in which the two chains are held together by the purine and pyrimidine bases. The planes of the bases are perpendicular to the fibre axis. They are joined together in pairs, a single base from one chain being hydrogen-bonded to a single base from the other chain, so that the two lie side by side with identical z-co-ordinates. One of the pair must be a purine and the other a pyrimidine for bonding to occur. The hydrogen bonds are made as follows: purine position 1 to pyrimidine position 1; purine position 6 to pyrimidine position 6.

If it is assumed that the bases only occur in the structure in the most plausible tautomeric forms (that is, with the keto rather than the enol configurations) it is found that only specific pairs of bases can bond together. These pairs are: adenine (purine) with thymine (pyrimidine), and guanine (purine) with cytosine (pyrimidine).

In other words, if an adenine forms one member of a pair, on either chain, then on these assumptions the other member must be thymine; similarly for guanine and cytosine. The sequence of bases on a single chain does not appear to be restricted in any way. However, if only specific pairs of bases can be formed, it follows that if the sequence of bases on one chain is given, then the sequence on the other chain is automatically determined.

It has been found experimentally[3,4] that the ratio of the amounts of adenine to thymine, and the ratio of guanine to cytosine, are always very close to unity for deoxyribose nucleic acid.

It is probably impossible to build this structure with a ribose sugar in place of the deoxyribose, as the extra oxygen atom would make too close a van der Waals contact.

The previously published X-ray data[5,6] on deoxyribose nucleic acid are insufficient for a rigorous test of our structure. So far as we can tell, it is roughly compatible with the experimental data, but it must be regarded as unproved until it has been checked against more exact results. Some of these are given in the following communications. We were not aware of the details of the results presented there when we devised our structure, which rests mainly though not entirely on published experimental data and stereo-chemical arguments.

Continued on next page

Most metabolic processes, such as the biosynthesis or degradation of molecules, occur in pathways in which an enzyme facilitates each step in the pathway (see chapter 2). The metabolic pathway for the conversion of threonine into isoleucine (two amino acids) is shown in figure 9.2. Each reaction product in the pathway is altered by an enzyme that converts it to the next product. The enzyme threonine dehydratase, for example, converts threonine into α-ketobutyric acid. Enzymes are composed of folded polymers of amino acids; the sequence of amino acids determines the final structure of an enzyme. The genetic material determines the sequence of the amino acids (see chapter 11).

BOX 9.1—*continued*

Box figure 1

This figure is purely diagrammatic. The two *ribbons* symbolize the two phosphate-sugar chains, and the *horizontal rods,* the pairs of bases holding the chains together. The *vertical line* marks the fiber axis.

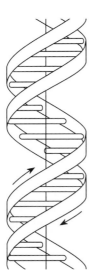

It has not escaped our notice that the specific pairing we have postulated immediately suggests a possible copying mechanism for the genetic material.

Full details of the structure, including the conditions assumed in building it, together with a set of co-ordinates for the atoms, will be published elsewhere.

We are much indebted to Dr. Jerry Donohue for constant advice and criticism, especially on interatomic distances. We have also been stimulated by a knowledge of the general nature of the unpublished experimental results and ideas of Dr. M. H. F. Wilkins, Dr. R. E. Franklin and their coworkers at King's College, London. One of us (J. D. W.) has been aided by a fellowship from the National Foundation for Infantile Paralysis.

J. D. Watson
F. H. C. Crick

Medical Research Council Unit for the Study of the Molecular Structure of Biological Systems, Cavendish Laboratory, Cambridge. April 2.

[1]Pauling, L., and Corey, R. B., *Nature,* 171, 346 (1953); *Proc. U.S. Nat. Acad. Sci.,* 39, 84 (1953).

[2]Furberg, S., *Acta Chem. Scand.,* 6, 634 (1952).

[3]Chargaff, E., for references see Zamenhof, S., Brawerman, G., and Chargaff, E., *Biochim. et Biophys. Acta,* 9, 402 (1952).

[4]Wyatt, G. R., *J. Gen. Physiol.,* 36, 201 (1952).

[5]Astbury, W. T., *Symp. Soc. Exp. Biol. 1, Nucleic Acid* 66 (Camb. Univ. Press, 1947).

[6]Wilkins, M. H. F., and Randall, J. T., *Biochim. et Biophys. Acta,* 10, 192 (1953).

The three-dimensional structure of enzymes permits them to perform their function. An enzyme combines with its substrate or substrates (the molecules it works on) at a part of the enzyme called the **active site** (fig. 9.3). The substrates "fit" into the active site, which has a shape that allows only the specific substrates to enter. This view of the way an enzyme interacts with its substrates is called the *lock-and-key model* of enzyme functioning. When the substrates are in their proper position in the active site of the enzyme, the particular reaction that the enzyme cat- alyzes takes place. The reaction products then separate from the enzyme and leave it free to repeat the process. Enzymes can work at phenomenal speeds. Some can catalyze as many as a million reactions per minute.

Not all of the cell's proteins function as catalysts. Some are structural proteins, such as keratin, the main component of hair. Other proteins are regulatory—they control the rate of production of other enzymes. Still others are involved in different functions; albumins, for example, help regulate the osmotic pressure of blood.

Figure 9.2
Metabolic pathway of conversion of the amino acid threonine into isoleucine

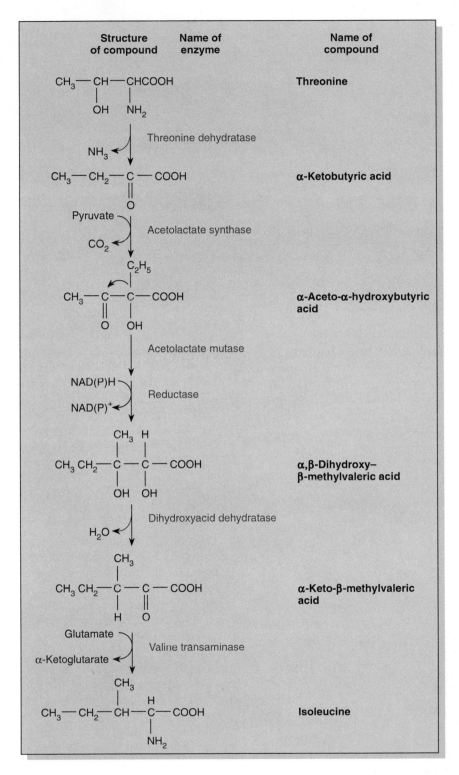

Figure 9.3

Active site of an enzyme recognizing a specific substance. In this case, ATP plus glucose is converted into ADP and glucose-6-phosphate by the enzyme hexokinase. The active site is in *red*.

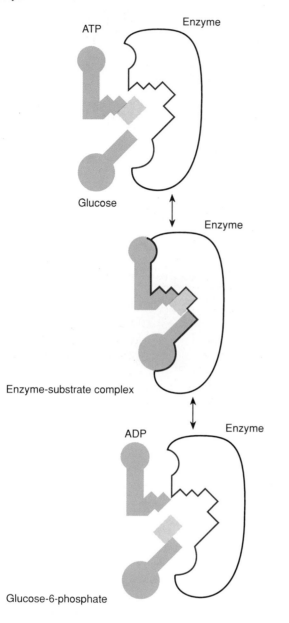

Replication

The genetic material must replicate itself precisely so that every daughter cell receives an exact copy. Some **mutability,** or the ability to change, is also required because we know that the genetic material has changed, or evolved, in the history of life on earth. In their 1953 paper, Watson and Crick had already worked out the replication process based on the structure of DNA. As we shall see, mutability is also a natural derivative of this process.

Oswald T. Avery (1877–1955)
Courtesy of the National Academy of Science.

Location

It has been known since the turn of the century that genes, the discrete functional units of genetic material, are located in chromosomes within the nuclei of eukaryotic cells: The behavior of chromosomes during the cellular division stages of mitosis and meiosis mimics the behavior of genes. Thus the genetic material in eukaryotes must be a part of the chromosomes.

For a long time, proteins were considered the most probable cell components to be the genetic material because they have the necessary molecular complexity. Amino acids can be combined in an almost unlimited variety, creating thousands and thousands of different proteins. The first proof that the genetic material is deoxyribonucleic acid (DNA) was provided in 1944 by Oswald Avery and his colleagues. The Watson and Crick model in 1953 ended a period when DNA was thought to be the genetic material, but its structure was unknown.

Evidence for DNA as the Genetic Material

Transformation

In 1928, F. Griffith reported that heat-killed bacteria of one type could "transform" living bacteria of a different type. Griffith demonstrated this transformation using two strains of the bacterium *Streptococcus pneumoniae*. One strain (S) produced smooth colonies because the cells had polysaccharide capsules. It caused a fatal bacteremia (bacterial infection) in mice. Another strain (R), which lacked polysaccharide capsules, produced rough colonies on petri plates (fig. 9.4); it did not have a pathological effect on mice. Bacteria of the rough strain are engulfed by white blood cells of the mice; the virulent smooth-strain bacteria survive because they are protected by their polysaccharide coating.

Griffith found that neither heat-killed S-type nor live R-type cells, by themselves, caused bacteremia in mice. However, if a mixture of live R-type and heat-killed S-type cells was injected into mice, the mice developed a

Figure 9.4

Petri plate with smooth and rough colonies of *Streptococcus pneumoniae*. R (rough) strain colonies on the *left* and S (smooth) colonies on the *right* on the same agar. Magnification 3.5×.

O. T. Avery, C. M. Macleod, and M. McCarty, "Studies on the Chemical Nature of the Substance Inducing Transformation of Pneumococcal Types." Reproduced from the Journal of Experimental Medicine *79 (1944):137–158, fig. 1 by copyright permission of the Rockefeller University Press. Reproduced by permission. Photograph made by Mr. Joseph B. Haulenbeek.*

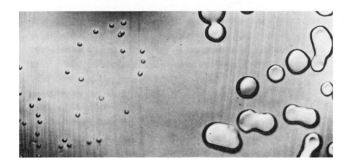

Figure 9.5

Griffith's experiment with *Streptococcus*. S-type cells will kill mice as will heat-killed S-type cells injected with live R-type cells.

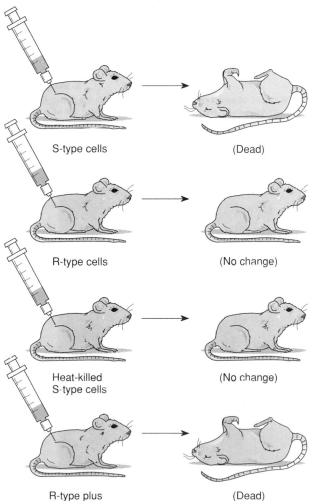

S-type cells — (Dead)

R-type cells — (No change)

Heat-killed S-type cells — (No change)

R-type plus heat-killed S-type cells — (Dead)

A. D. Hershey (1908–   )
Courtesy of Dr. A. D. Hershey.

bacteremia identical to that caused by injection of living S-type cells (fig. 9.5). Thus something in the heat-killed S cells transformed the R-type bacteria into S-type cells.

In 1944, Oswald Avery and two of his associates, C. MacLeod and M. McCarty, reported the nature of the transforming substance. Avery and his colleagues did their work *in vitro* (literally, in glass), using colony morphology on culture media rather than bacteremia in mice as evidence of transformation. They ruled out proteins, carbohydrates, and lipids by their extraction procedure, by the chemical analysis of the transforming material, and by demonstrating that the only enzymes that destroyed the transforming ability were enzymes that destroyed DNA. This study provided the first experimental evidence that DNA was the genetic material: DNA transformed R-type bacteria into S-type bacteria.

Phage Labeling

Valuable information about the nature of the genetic material has also been obtained from viruses. Of particular value are studies of bacterial viruses, the bacteriophages, or phages. Since phages consist of only nucleic acid surrounded by protein, they lend themselves nicely to the problem of determining whether the protein or the nucleic acid is the genetic material.

A. D. Hershey and M. Chase published, in 1952, the results of research that supported the notion that DNA is the genetic material and helped to explain the nature of the viral infection process. Since all nucleic acids contain phosphorus whereas proteins do not, and since most proteins contain sulfur (in the amino acids cysteine and methionine) whereas nucleic acids do not, Hershey and Chase designed an experiment using radioactive isotopes of sulfur and phosphorus to keep track separately of the viral proteins and nucleic acids during the infection process. They used the bacteriophage named T2 and the bacterium *Escherichia coli*. The phages were labeled by growing them on bacteria growing in culture medium containing

Figure 9.6

The Hershey and Chase experiments using ^{35}S-labeled T2 and ^{32}P-labeled bacteriophage. The nucleic acid label (^{32}P) enters the bacterium during infection. The protein label (^{35}S) does not.

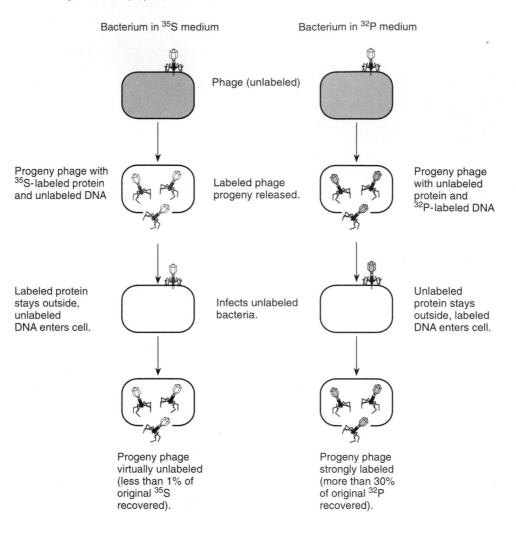

Bacterium in ^{35}S medium

Bacterium in ^{32}P medium

Phage (unlabeled)

Progeny phage with ^{35}S-labeled protein and unlabeled DNA

Labeled phage progeny released.

Progeny phage with unlabeled protein and ^{32}P-labeled DNA

Labeled protein stays outside, unlabeled DNA enters cell.

Infects unlabeled bacteria.

Unlabeled protein stays outside, labeled DNA enters cell.

Progeny phage virtually unlabeled (less than 1% of original ^{35}S recovered).

Progeny phage strongly labeled (more than 30% of original ^{32}P recovered).

the radioactive isotopes ^{35}S or ^{32}P. They then proceeded to identify the material injected into the cell by phages attached to the bacterial wall.

When ^{32}P-labeled phages were mixed with unlabeled *E. coli* cells, Hershey and Chase found that the ^{32}P label entered the bacterial cells and that the next generation of phages that burst from the infected cells carried a significant amount of the ^{32}P label. When ^{35}S-labeled phages were mixed with unlabeled *E. coli,* the researchers found that the ^{35}S label stayed on the outside of the bacteria for the most part. Hershey and Chase thus demonstrated that the outer protein coat of a phage does not enter the bacterium it infects, whereas the phage's inner material, con-

sisting of DNA, does enter the bacterial cell (fig. 9.6). Since the DNA is responsible for the production of the new phages during the infection process, the DNA, not the protein, must be the genetic material.

RNA as Genetic Material

In some viruses, RNA (ribonucleic acid) is the genetic material (see Box 9.2). The tobacco mosaic virus (TMV) that infects tobacco plants consists only of RNA (ribonucleic acid) and protein. The single, long RNA molecule is packaged within a rodlike structure formed by over two thousand copies of a single protein (fig. 9.7). No DNA is present in TMV virions (viral particles). In 1955,

Prions: The Biological Equivalent of Ice-Nine

Without exception, the genetic material is either DNA or RNA; it is RNA only in a few viruses. Since virtually all transmissible diseases are of bacterial or viral origin, this means that all transmissible diseases are also caused by organisms with DNA or RNA as the genetic material. However, there is an interesting situation in which a transmissible disease appears to be caused by an agent without genetic material. Three neurological diseases of human beings and four of animals are caused, we believe, by proteins without DNA or RNA. The diseases in people are kuru, Creutzfeldt-Jakob disease, and Gerstmann-Sträussler-Scheinker disease. The animal diseases are scrapie (sheep and goats), bovine spongiform encephalopathy, transmissible mink encephalopathy, and chronic wasting disease (deer and elk). All of these diseases are extremely slow to develop, all are fatal, and all are believed to be caused by the ingestion of a protein from an infected individual; there are no cures and the mechanism of action is unknown.

The diseases appear to be caused by a protein, similar to one normally produced in the brain of healthy individuals. The term **prion** (short for proteinaceous infectious particle) has been given to these agents by Stanley Prusiner at the University of California in San Francisco. He, along with colleagues, isolated the prion protein (PrP) and most recently located the gene that codes for the protein. In addition to the infective form, there is a familial form of these diseases (inherited) resulting from a mutation of the gene that codes for the prion protein active in normal individuals (probably at least all mammals).

Although there are no cures for these diseases, kuru, at least, seems to be almost eradicated. It was found only among people in part of New Guinea who practiced cannibalism. Once the people stopped this practice, the spread of the disease was stopped; kuru does not seem to be generated to any major extent by mutation in the people of the area. By controlling feeding practices, it is believed that bovine spongiform encephalopathy will also disappear. In the past, cows had been fed protein supplements made from infected animals.

The obvious question to ask is how does a protein that does not appear to contain genetic material result in a transmissible disease when ingested? Unfortunately, at the moment there is no answer. However, Prusiner has suggested several mechanisms by which an infective protein could induce copies of the normal protein to become infective. One of these mechanisms involves a cascade in which an infective PrP binds with a normal PrP resulting in two infective PrPs. From this, one produces two, two produce four, four produce eight, and so on. As Nancy Touchette, writing in *The Journal of NIH Research,* points out, this is the way Kurt Vonnegut described the behavior of the mythical ice-nine in his 1963 book, *Cat's Cradle.* As those of you who read the book may remember, a single seed caused all of the water on earth, by a chain reaction cascade as described above, to form into a novel type of ice. We have not yet resorted to science fiction answers to the mystery of prion function; however, it seems reasonable to guess that an eventual understanding of the mechanism of prion function will provide us with a biological novelty.

H. Fraenkel-Conrat and R. Williams showed that a virus can be separated, *in vitro,* into its component parts and reconstituted as a viable virus. This finding led to experiments by Fraenkel-Conrat and B. Singer, who reconstituted TMV with parts from different strains (fig. 9.7). For example, they combined the RNA from the common TMV with the protein from the masked (M) strain of TMV. They then made the reciprocal combination of common-type protein and M-type RNA. In both cases the TMV produced during the process of infection was of the type associated with the RNA, not with the protein. Thus it was the nucleic acid (RNA in this case) that was shown to be the genetic material. This was confirmed in subse-

quent experiments in which pure TMV RNA was rubbed into plant leaves. Normal infection and a new generation of typical, protein-coated TMV resulted.

CHEMISTRY OF NUCLEIC ACIDS

Having identified the genetic material as the nucleic acid DNA (or RNA), we need to examine the chemical structure of these molecules. Their structure will tell us a good deal about how they function.

Nucleic acids are made by joining **nucleotides** in a repetitive way into long, chainlike polymers. Nucleotides are made of three components: phosphate, sugar, and a nitrogenous base (table 9.1 and fig. 9.8). When incorporated

Figure 9.7

(a) Electron micrograph of TMV (tobacco mosaic virus). Magnification 37,428×. (b) Reconstitution experiment of Fraenkel-Conrat and Singer. Inheritance is controlled by the nucleic acid (RNA), not the protein component of the virus.

(a) © Biology Media/Photo Researchers, Inc.

(a)

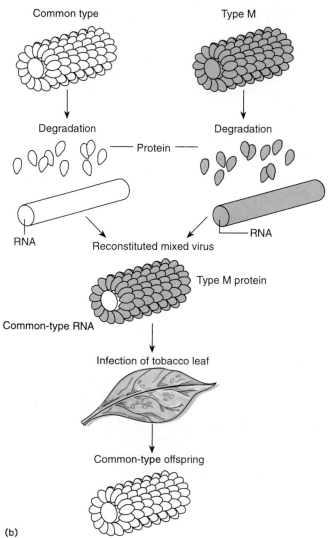

(b)

TABLE 9.1 Components of Nucleic Acids

			Base	
	Phosphate	*Sugar*	*Purines*	*Pyrimidines*
DNA	Present	Deoxyribose	Guanine	Cytosine
			Adenine	Thymine
RNA	Present	Ribose	Guanine	Cytosine
			Adenine	Uracil

Figure 9.8

Components of nucleic acids: phosphate, sugars, and bases. Primes are used in the numbering of the ring positions in the sugars to differentiate them from the ring positions of the bases.

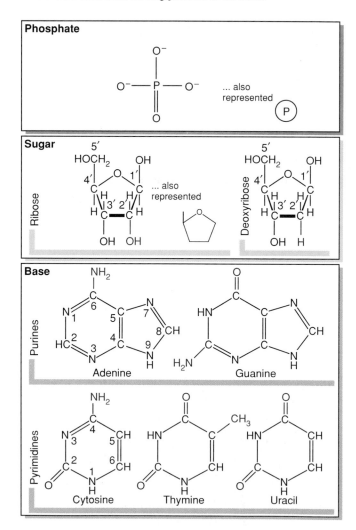

Figure 9.9
The structure of a nucleoside and two nucleotides: a nucleoside monophosphate and a nucleoside triphosphate

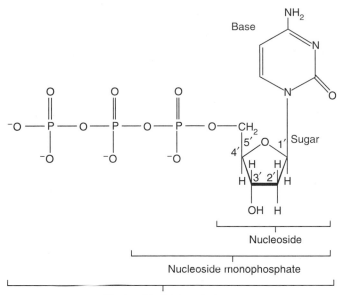

into a nucleic acid, a nucleotide contains one each of the three components. But, when free in the cell pool, nucleotides usually occur as triphosphates. The energy held in the extra phosphates is used, among other purposes, to synthesize the polymer. A **nucleoside** is a sugar-base compound. Nucleotides are therefore nucleoside phosphates (fig. 9.9). (Note that ATP, adenosine triphosphate, the energy currency of the cell, is a nucleoside triphosphate.)

The sugars differ only in the presence (ribose in RNA) or absence (deoxyribose in DNA) of an oxygen in the 2' position. The carbons of the sugars are numbered 1' to 5'. The primes are used to avoid confusion with the numbering system of the bases (fig. 9.8). DNA and RNA both have four types of bases (two **purines** and two **pyrimidines**) in their nucleotide chains. Both molecules have the purines **adenine** and **guanine** and the pyrimidine **cytosine.** DNA has the pyrimidine **thymine;** RNA has the pyrimidine **uracil.** Thus three of the nitrogenous bases are found in both DNA and RNA, whereas thymine is unique to DNA and uracil is unique to RNA.

A nucleotide is formed in the cell by attachment of a base to the 1' carbon of the sugar and attachment of a phosphate to the 5' carbon of the same sugar (fig. 9.10); the nucleotide takes its name from the base (table 9.2). Nucleotides are linked together (**polymerized**) by the formation of a bond between the phosphate of one nucleotide

and the hydroxyl (OH) group at the 3' carbon of an adjacent molecule. Very long strings of nucleotides can be polymerized by this **phosphodiester bonding** (fig. 9.11).

Biologically Active Structure

Although the identity of the nucleotides that polymerized to form a strand of DNA or RNA was known, the actual structure of these nucleic acids when they function as the genetic material remained unknown until 1953. The general feeling was that the biologically active structure of DNA was more complex than a single string of nucleotides linked together by phosphodiester bonds and that several interacting strands were involved. In 1953, Linus Pauling, a Nobel laureate who had discovered the α-helical structure of proteins, was investigating a three-stranded structure for the genetic material, whereas Watson and Crick decided that a two-stranded structure was more consistent with available evidence. Three lines of evidence directed Watson and Crick: the chemical nature of the components of DNA, X-ray crystallography, and Chargaff's ratios.

DNA X-Ray Crystallography
Maurice Wilkins, Rosalind Franklin, and their colleagues were using **X-ray crystallography** to analyze the structure of DNA. The molecules in a crystal are arranged in an orderly fashion, such that when a beam of X rays is passed through the crystal, the beam will be scattered. The pattern of the scatter can be recorded on photographic film. The nature of this pattern depends on the structure of the crystal (fig. 9.12). The cross in the center of the photograph indicates that the molecule is a helix; the dark areas at the top and bottom come from the bases, stacked perpendicularly to the main axis of the molecule.

Chargaff's Ratios
Until Erwin Chargaff's work, scientists had labored under the erroneous **tetranucleotide hypothesis** in which it was believed that DNA was made up of equal quantities of the four bases; therefore, a subunit of this DNA consisted of one copy of each base. Chargaff carefully analyzed the base composition of DNA in various species (table 9.3). He found that although the relative amount of a given nucleotide differs among species, the amount of adenine equaled that of thymine and the amount of guanine equaled that of cytosine. That is, in the DNA of all the organisms studied, there is a 1:1 correspondence between the purine and pyrimidine bases. This is known as **Chargaff's rule.** Chargaff's observations disproved the tetranucleotide hypothesis; the four bases of DNA were not in a 1:1:1:1 ratio. His results were extremely important to Watson and Crick in the development of their model.

Figure 9.10
Structure of the four deoxyribose nucleotides

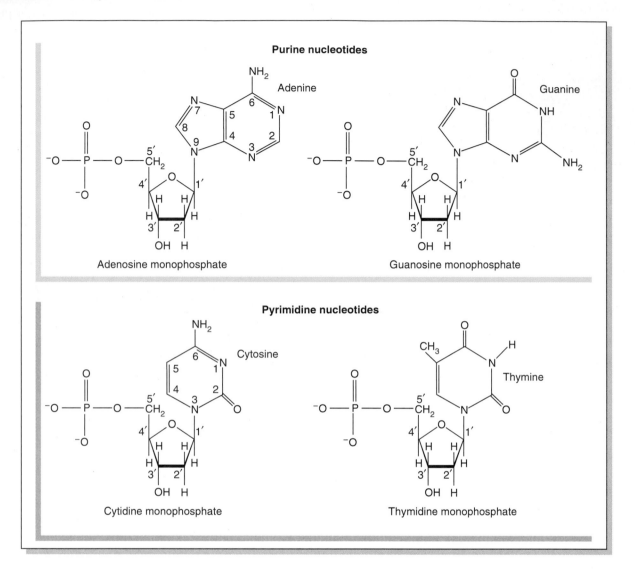

TABLE 9.2 Nucleotide Nomenclature

Base	Nucleotide (Nucleoside monophosphate)	Abbreviation					
		Monophosphate		Diphosphate		Triphosphate	
		Ribose	Deoxyribose	Ribose	Deoxyribose	Ribose	Deoxyribose
Guanine	Guanosine monophosphate	GMP		GDP		GTP	
	Deoxyguanosine monophosphate		dGMP		dGDP		dGTP
Adenine	Adenosine monophosphate	AMP		ADP		ATP	
	Deoxyadenosine monophosphate		dAMP		dADP		dATP
Cytosine	Cytidine monophosphate	CMP		CDP		CTP	
	Deoxycytidine monophosphate		dCMP		dCDP		dCTP
Thymine	Deoxythymidine monophosphate		dTMP		dTDP		dTTP
Uracil	Uridine monophosphate	UMP		UDP		UTP	

Figure 9.11

Polymerization of adjacent nucleotides to form a sugar-phosphate strand. There is no limit to the length that the strand can be, or the type of base attached to each nucleotide residue.

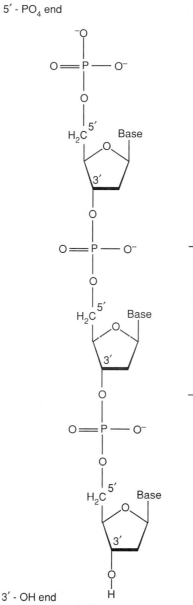

5' - PO₄ end

Nucleotide residue

3' - OH end

Figure 9.12

Scatter pattern of a beam of X rays passed through crystalline DNA

Source: Reprinted by permission from R. E. Franklin and R. Gosling, "Molecular Configuration in Sodium Thymonucleate," Nature 171:740–741. Copyright 1953 by Macmillan Journals Limited.

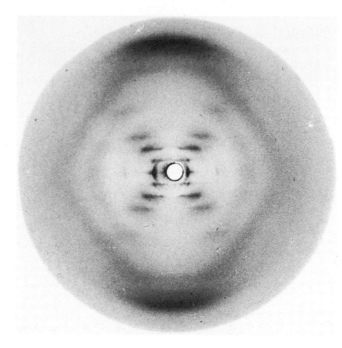

Erwin Chargaff (1905–   )
Courtesy of Dr. Erwin Chargaff.

Rosalind E. Franklin (1920–1958)
Courtesy of Cold Spring Harbor Laboratory.

In DNA, nucleotides are connected across phosphates. Recently (1987, *Science* 235:1173–78), F. H. Westheimer asked the question: "Why did nature choose phosphates?" The answer seems simple enough. The phosphate molecule has several properties that make it ideal for linking subunits into polymers in the biochemical world (fig. 9.8, *top*). First, the phosphate group can form linking bonds—it can thus connect two compounds (fig. 9.11). The linking bonds that are formed from phosphates (phosphodiester bonds) have the additional property of being stable yet are easily undone by enzymatic hydrolysis. In other words, the bonds are stable, but the nucleotide residues can be removed to conserve them during, for example, various replication and repair processes that we will discuss later. When a nucleotide is removed, the nucleotide is not destroyed in the process.

After the phosphodiester bond is formed, one oxygen atom of the phosphate group is still negatively ionized. This property is extremely important, because it makes less likely the event that the phosphodiester bond will be spontaneously broken (a negative charge protects against a nucleophilic attack), and negative ionization keeps the nucleotides and the DNA within membranes, specifically the nuclear membrane of eukaryotes and the cell membrane of prokaryotes. Since negatively charged compounds are extremely insoluble in lipids, phosphates are kept within membranes by this ionization. Westheimer concludes: "All of these conditions are met by phosphoric acid, and no alternative is obvious."

TABLE 9.3   Percentage Base Composition of Some DNAs

Species	Adenine	Thymine	Guanine	Cytosine
Human Being (Liver)	30.3	30.3	19.5	19.9
Mycobacterium tuberculosis	15.1	14.6	34.9	35.4
Sea Urchin	32.8	32.1	17.7	18.4

Source: Data from Chargaff and Davidson, 1955.

The Watson-Crick Model

With the information available, Watson and Crick began making molecular models. They found that a possible structure was one in which two helices coiled around one another (a **double helix**) with the sugar-phosphate backbones on the outside and the bases on the inside. This structure would fit the dimensions established for DNA by X-ray crystallography if the bases from the two strands were opposite each other and formed rungs in a helical ladder (fig. 9.13). The diameter of the helix could only be kept constant (about 20 Å—angstrom units) if there were one purine and one pyrimidine base per rung. Two purines per rung would be too big and two pyrimidines would be too small.

After further experimentation with models of the bases, Watson and Crick found that the hydrogen bonding necessary to form the rungs of their helical ladder could occur readily between certain base pairs, the pairs that Chargaff found in equal frequencies. (Hydrogen bonds are very weak bonds that involve the sharing of a hydrogen between two electronegative atoms, such as O and N.)

Thermodynamically stable hydrogen bonding occurs between thymine and adenine and between cytosine and guanine (fig. 9.14). The relation is called **complementarity.** There are two hydrogen bonds between adenine and thymine and three between cytosine and guanine.

Another point about DNA structure relates to the fact that **polarity** exists in each strand. That is, one end of a DNA strand will have a 5′ phosphate and the other end will have a 3′ hydroxyl group. Watson and Crick found that hydrogen bonding could only occur if the polarity of the two strands ran in opposite directions; that is, the two strands were **antiparallel** (fig. 9.15).

DNA Denaturation

Denaturation studies indicated that the hydrogen bonding in DNA occurs in the way suggested by Watson and Crick. Hydrogen bonds, although individually very weak, give structural stability to a molecule with large numbers of them. However, the hydrogen bonds can be broken and the DNA strands separated by heating the DNA molecule. A point is reached in which the thermal agitation

Figure 9.13
Double helical structure of DNA. (*a*) Component parts. (*b*) Line drawing.

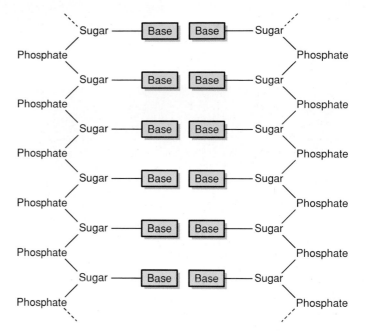

(a)

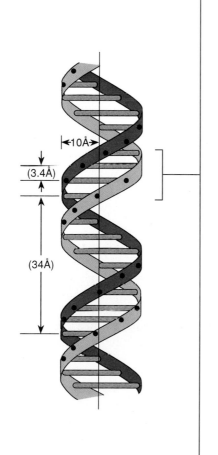

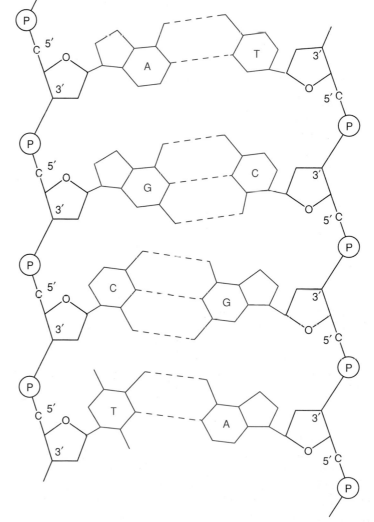

(b)

Figure 9.13 (continued)
(c) DNA magnified 25 million times by scanning tunneling microscopy.
© *John D. Baldeschweiler*

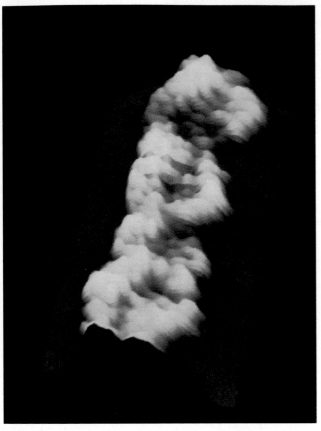

(c)

overcomes the hydrogen bonding and the molecule becomes **denatured** (or "melts"). It is logical that the more hydrogen bonds DNA contains, the higher the temperature needed to denature it. It follows that since a G-C (guanine-cytosine) base pair has three hydrogen bonds to every two in an A-T (adenine-thymine) base pair, the higher the G-C content in a given molecule of DNA, the higher the temperature required to denature that DNA. This relationship has been found to exist (fig. 9.16).

Requirements of Genetic Material

Let us now return briefly to the requirements we have said a genetic material needs to meet: (1) control of protein synthesis, (2) self-replication, and (3) location in the nucleus (in organisms with nuclei). Does DNA (or when DNA is absent, RNA) meet these requirements?

Control of Enzymes

In the next several chapters we will examine the details of protein synthesis. At this point it is necessary only to observe that DNA does possess the complexity required to direct protein synthesis. Although complementarity re-

Figure 9.14
Hydrogen bonding between the nitrogenous bases in DNA

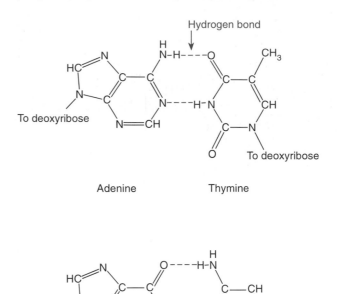

Adenine   Thymine

Guanine   Cytosine

Figure 9.15
Polarity of the DNA strands

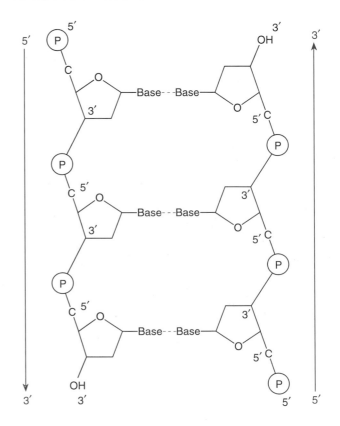

Figure 9.16

Relationship of the number of hydrogen bonds (G-C content) and the thermal stability of DNA from different sources

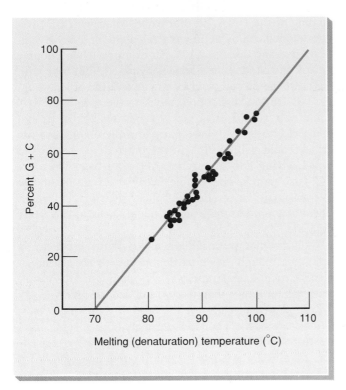

Figure 9.17

Complementarity provides a possible mechanism for the accuracy of DNA replication. The parent duplex opens and each strand becomes a template for a new duplex.

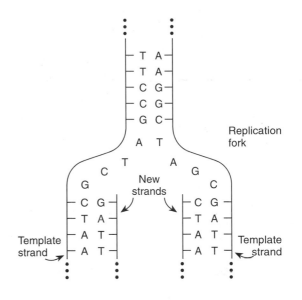

stricts the base opposite a given base in a double helix, there is no restriction to the sequence of bases on a given strand. Later we will show that a sequence of three bases in DNA specifies a particular amino acid during protein synthesis. The **genetic code** gives the relationship of bases in DNA to amino acids in proteins.

Replication

Watson and Crick suggested in their 1953 paper how DNA might replicate. Their observation stemmed from the property of complementarity. Since the base sequence on one strand is complementary to the base sequence on the opposite strand, each strand could act as a template for a new double helix. This could happen if the molecule simply "unzipped," allowing each strand to specify the sequence of bases on a new strand by complementarity (fig. 9.17). Mutability would be due to mispairings or other errors in replication.

Location

The location requirement is that DNA must reside in the nucleus of eukaryotes, where the genes occur on chromosomes, or in the chromosomes of prokaryotes and viruses. In both prokaryotes and eukaryotes the majority of the cell's DNA is in the chromosomes. And all viruses contain either DNA or RNA. Thus DNA potentially fulfills all the requirements of a genetic material. RNA can fulfill the same requirements in RNA viruses and viroids.

Alternative Forms of DNA

The form of DNA we have described so far is called **B DNA.** It is a right-handed helix: It turns in a clockwise manner when viewed down its axis. The bases are stacked almost exactly perpendicular to the main axis with about ten base pairs per turn (34 Å; see fig. 9.13). However, DNA can exist in other forms. If the water content increases to about 75%, the A form of DNA (**A DNA**) will occur. In this form the bases are tilted in regard to the axis and there are more base pairs per turn. However, this and other known forms of DNA are relatively minor variations on the right-handed B form.

In 1979, Alexander Rich and his colleagues at MIT discovered a left-handed helix that they called **Z DNA** because its backbone formed a zigzag structure (fig. 9.18). Z DNA was found by X-ray crystallographic analysis of very small DNA molecules composed of repeating G-C sequences on one strand with the complementary C-G sequences on the other (alternating purines and pyrimidines). The Z DNA looks like B DNA in which each base was rotated 180 degrees, resulting in a zigzag, left-handed structure (fig. 9.19). (The original configuration of the bases is referred to as the *anti* configuration; the rotated configuration is called the *syn* configuration.)

Originally the Z DNA was thought to be a structure that would not prove of interest to biologists because it required very high salt concentrations to become stable. However, recently it has been found that Z DNA can be stabilized in physiologically normal conditions if methyl groups are added to the cytosines. Z DNA may be involved in regulating gene expression in eukaryotes. We will return to this topic in chapter 15.

Figure 9.18

Z (*left*) and B (*right*) DNA. The *dark lines* connect phosphate groups.

From A. Wang, et al., "Left-Handed Double Helical DNA: Variations in the Backbone Conformation" in Science, *81:211. Copyright 1981 by the AAAS.*

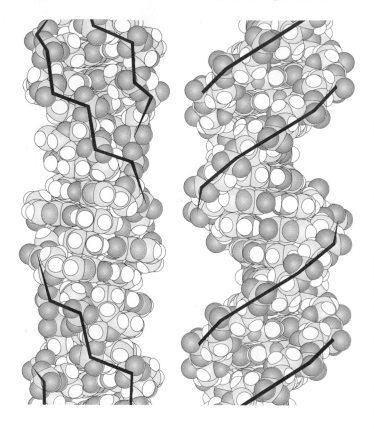

Figure 9.19

B DNA is converted to Z DNA by the rotation of bases as indicated by curved arrows.

Reproduced, with permission, from the Annual Review of Biochemistry, *Volume 53, © 1984 by Annual Reviews, Inc.*

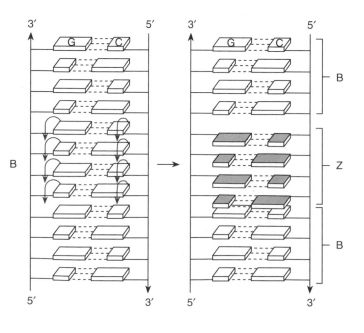

DNA REPLICATION

In their 1953 paper, Watson and Crick suggested that the replication of the double helix could take place by unwinding of the DNA, so that each strand would form a new double helix by acting as a **template** for a newly synthesized strand (see fig. 9.17). For example, when a double helix is unwound at an adenine-thymine (A-T) base pair, one unwound strand would carry A and the other would carry T. During replication, the A in the template DNA would pair with T in a newly replicated DNA strand, giving rise to an A-T base pair. T in the other template strand would pair with A in the other newly replicated strand, giving rise to another A-T base pair. Thus one A-T base pair in one double helix would result in two A-T base pairs in two double helices. This process would repeat at every base pair in the double helix of the DNA molecule.

This mechanism is called **semiconservative** replication because, although the entire double helix is not conserved in replication, each strand is. Every daughter DNA molecule has an intact template strand and an intact newly replicated strand. It is not the only way in which replication could occur. The alternative methods of replication are **conservative** and **dispersive.** In conservative replication, in which the whole original double helix acts as a template for a new one, one daughter molecule would consist of the original parent DNA and the other daughter would be totally new DNA. In dispersive replication, some parts of the original double helix are conserved and some parts are not. Daughter molecules would consist of part template and part newly synthesized DNA. In reality, the dispersive category is the all-inclusive "garbage can" category, including any possibility other than conservative and semiconservative replication.

The Meselson and Stahl Experiment

In 1958, M. Meselson and F. Stahl reported an elegant experiment designed to determine the mode of DNA replication. They grew *E. coli* in a medium containing a heavy isotope of nitrogen, ^{15}N. (The normal form of nitrogen is ^{14}N.) After growing for several generations on the ^{15}N

Matthew Meselson (1930–   )
Courtesy of Dr. Matthew Meselson. Photograph by Bud Gruce.

Franklin W. Stahl (1929--   )
Courtesy of Dr. Franklin W. Stahl.

Figure 9.20

The Meselson and Stahl experiment to determine the mode of replication of DNA. The bands in the centrifuge tube are seen under ultraviolet light. The pattern of bands (*left*) comes about from semiconservative DNA replication (*right*) of ¹⁵N DNA replicating in a ¹⁴N medium.

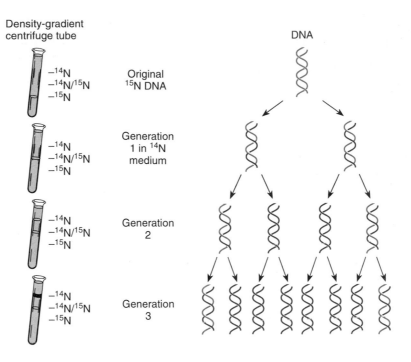

medium, the DNA of *E. coli* was denser. The density of the strands was determined using a technique known as **density-gradient centrifugation.** In this technique a cesium chloride (CsCl) solution is spun in an ultracentrifuge at high speed for several hours. Eventually an equilibrium between centrifugal force and diffusion occurs, such that a density gradient is established in the tube with an increasing concentration of the CsCl from the top to the bottom of the tube. If DNA (or any other substance) is added, it will concentrate and form a band in the tube at the point where its density is the same as that of the CsCl. If there are several types of DNA with different densities, they will form several bands. The bands can be detected by observing the tubes with ultraviolet light at a wavelength of 260 nm (nanometers), in which nucleic acids absorb strongly.

Meselson and Stahl transferred the bacteria with heavy (¹⁵N) DNA to a medium containing only ¹⁴N. The new DNA, replicated in the ¹⁴N medium, was intermediate in density between light (¹⁴N) and heavy (¹⁵N) DNA, because replication was semiconservative (fig. 9.20). If replication had been conservative, there would have been two bands at the first generation of replication—an orig-

inal ¹⁵N DNA and a new ¹⁴N double helix. And, throughout the experiment, if the method of replication had been conservative, the original DNA would have continued to show up as a ¹⁵N band. This of course did not happen. If the method of replication had been dispersive, the result would have been various multiple-banded patterns, depending on the degree of dispersiveness. The results shown in figure 9.20 are consistent only with semiconservative replication.

Autoradiographic Demonstration of DNA Replication

The semiconservative method of replication was verified photographically by J. Cairns in 1963 who used the technique of **autoradiography.** This technique makes use of the fact that radioactive atoms expose photographic film. The visible silver grains on the film can then be counted to provide an estimate of the quantity of radioactive material present. Cairns grew *E. coli* bacteria in a medium containing radioactive thymine, a component of one of the DNA nucleotides. The radioactivity was in tritium (³H). The DNA was then carefully extracted from the bacteria

BOX 9.4

Triple-Stranded DNA

Under natural conditions, single-stranded RNA and double-stranded DNA are the rule. However, it has been found that under laboratory conditions it is possible to induce a third strand of DNA to interdigitate itself into the major groove of the double helix of normal DNA in a sequence-specific fashion. That is, the third strand of DNA will not just interdigitate anywhere, but will form a stable triplex at a specific sequence (Box fig. 1). The rules of binding are a little less precise than normal in that not all sequences are recognized and recognition can depend on surrounding sequences. However, given that, a thymine in the third strand will recognize an adenine in an adenine-thymine base pair (T·A-T) and a cytosine in the third strand will recognize a guanine in a guanine-cytosine base pair (C·G-C).

Triple-stranded nucleotide chains were first created by three scientists at the National Institutes of Health in 1957—Alexander Rich, David Davies, and Gary Felsenfeld—while they were creating artificial nucleic acids. At the time, triple-stranded DNA seemed like a laboratory curiosity. Now it seems to be of interest because it may have valuable uses both experimentally and clinically. (Rich apparently had the same experience in his codiscovery of Z DNA, which at first seemed like an oddity but now is the focus of much attention—see chapter 15.) Recently however, two research groups, one in California and one in Paris, have been able to form triplexes in naturally occurring DNA. Two applications of this technology are being pursued actively.

Both applications arise because a single strand of DNA is capable of recognizing a relatively long sequence of the double-stranded DNA in a chromosome. Thus it is possible to locate selectively a particular genic sequence. Once a particular sequence on a chromosome is located by the third strand, two things can be accomplished. First, due to triplex DNA formation, a particular gene can be prevented from expressing itself. This fact is being investigated in combating AIDS (see chapter 15). By the same technique, triplex DNA can also be an abortifacient, a safe method of preventing implantation of a fetus by preventing the expression of genes affected by the hormone progesterone. The second use of triplex DNA is to cut DNA at a specific place by adding a cleaving compound to both ends of the third strand of DNA. Once the third strand has interdigitated, it can then break the original double helix. For example, Strobel and Dervan at the California Institute of Technology have used a chemical complex containing iron attached to both 3′ and 5′ ends of the third strand of DNA. The cleavage reaction is then initiated by the addition of a third chemical. The cleavage of the original duplex can be of extreme value in modern recombinant DNA technology, such as in the human genome project (see chapter 12).

Whether triplex DNA will ever be of value is not certain at this time. However, its potential to be of therapeutic use and to help in studying and mapping the human genome appears very good.

Continued on next page

---

and autoradiographs were made. Figure 9.21 is an example. Interpretation of this autoradiograph reveals several points. The first, known at the time, is that the *E. coli* DNA is a circle. The second point is that the DNA is replicated while maintaining the integrity of the circle. That is, the circle does not appear to be broken in the process of DNA replication; an intermediate **theta structure** is formed (similar in shape to the Greek letter theta, θ). Third, replication of the DNA seems to be occurring at one or two moving **Y-junctions** in the circle, which further supports the semiconservative mode of replication. The DNA is unwound at a point and replication proceeds at a Y-junction, in a semiconservative manner, in one or both directions (see fig. 9.17).

The way in which the two Y-junctions move along the circle to the final step in which two new circles are formed is diagrammed in figure 9.22. The steps by themselves do not support either a unidirectional or a bidirectional mode of replication. That is, a theta structure will develop if either one or both Y-junctions is active in replication. But with autoradiography it is possible to determine whether new growth is occurring in only one or in both directions.

In some cases, radioactivity had not been applied to the cell until DNA replication had already begun. In these cases, the radioactive label appeared after the theta structure had already begun forming. Figure 9.23 illustrates

BOX 9.4—*continued*

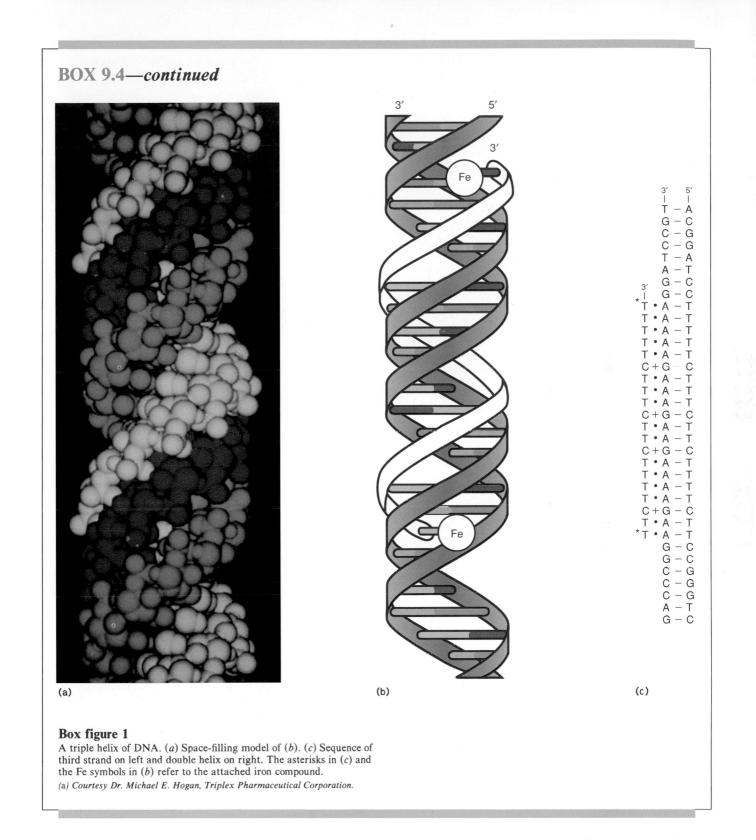

(a) (b) (c)

Box figure 1
A triple helix of DNA. (*a*) Space-filling model of (*b*). (*c*) Sequence of third strand on left and double helix on right. The asterisks in (*c*) and the Fe symbols in (*b*) refer to the attached iron compound.

(*a*) *Courtesy Dr. Michael E. Hogan, Triplex Pharmaceutical Corporation.*

Figure 9.21

(*a*) Autoradiograph of *E. coli* DNA during replication. (*b*) Diagram has labels on the three segments, A, B, C, created by the existence of two forks, X and Y, in the DNA. Forks are created when the circle opens for replication. Length of the chromosome is about 1,300 μm.

(b) *From J. Cairns, "The Chromosome of* E. coli*" in* Cold Spring Harbor Symposia on Quantitative Biology, *28. Copyright © 1963 by Cold Spring Harbor Laboratory Press, Cold Spring Harbor, NY. Reprinted by permission.*

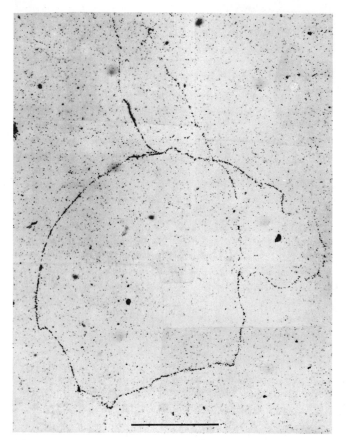

(a)

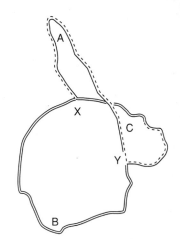

(b)

Figure 9.22

Observable stages in the DNA replication of a circular chromosome, assuming bidirectional DNA synthesis. The intermediate figures are called theta structures.

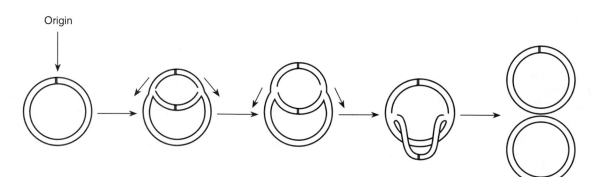

Figure 9.23

Patterns of radioactive label distinguishing unidirectional from bidirectional DNA replication. In these hypothetical experiments, DNA replication was allowed to begin, at which time a radioactive label was added. After a short period of time, the process was stopped and the autoradiographs prepared. In bidirectional replication (the actual case), label is seen at both Y-junctions.

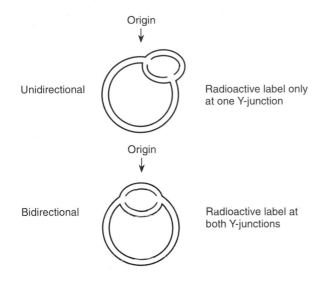

Figure 9.24

Replication bubbles. (*a*) Formation of bubbles (eyes) in eukaryotic DNA because of multiple sites of origin of DNA synthesis. (b) Electron micrograph (and explanatory line drawing) of replicating *Drosophila* DNA showing these bubbles.

(b) H. Kriegstein and D. Hogness, "Mechanism of DNA Replication in Drosophila Chromosomes: Structure of Replication Forks and Evidence for B₁ directionality," Proceeding of the National Academy of Sciences USA, 71 (1974):135–139. Reproduced by permission.

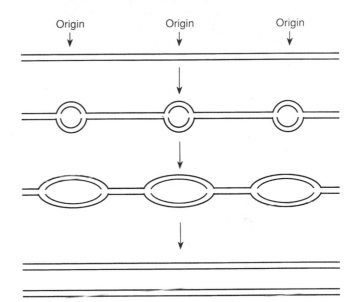

(a)

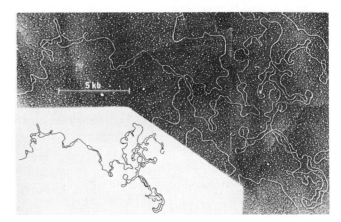

(b)

hypothetical outcomes for either unidirectional or bidirectional replication. By counting silver grains in autoradiographs, Cairns found growth to be bidirectional. This finding has subsequently been verified by both autoradiographic and genetic analysis.

In eukaryotes the DNA molecules (chromosomes) are larger than in prokaryotes and are not circular; there are usually multiple sites of initiation of replication. Thus each eukaryotic chromosome is composed of many replicating units, or **replicons**—stretches of DNA with a single origin of replication. In comparison, the *E. coli* chromosome is composed of only one replicon. In eukaryotes these replicating units form "bubbles" (or "eyes") in the DNA (fig. 9.24) during replication.

THE ENZYMOLOGY
OF DNA REPLICATION

Let us turn now to the details of the processes that take place during DNA replication. Like virtually all metabolic processes, DNA replication is controlled by enzymes. The evidence for the details we describe comes from physical, chemical, and biochemical studies of enzymes and nucleic acids and from the analysis of mutations that influence the replication processes primarily in *E. coli*. More recent techniques of recombinant DNA technology and nucleotide sequencing have allowed the determination of the nucleotide sequences of many of these key regions in DNA and RNA.

There are only three known enzymes that will polymerize nucleotides into a growing strand of DNA in *E. coli*. These enzymes are **DNA polymerase** I, II, and III. DNA polymerase I, discovered by Arthur Kornberg, who subsequently won the Nobel Prize for his work, is primarily utilized in filling in small DNA segments during replication and repair processes. At present, the precise role of DNA polymerase II is not entirely clear, although

Arthur Kornberg (1918–   )
Courtesy of Dr. Arthur Kornberg. Photograph by Karsh.

it is known that it can serve as an alternative repair polymerase if polymerase I is damaged by mutation. DNA polymerase III is the primary active polymerase during normal DNA replication.

In the simplest model for DNA replication, new nucleotides would be added, according to the rules of complementarity, simultaneously on both strands of newly synthesized DNA at the replication fork as the DNA opens up. But a problem exists, created by the antiparallel nature of DNA: The two strands of a DNA double helix run in opposite directions. Going up the duplex, for example, one strand will be a $5' \rightarrow 3'$ strand, whereas the other will be a $3' \rightarrow 5'$ strand. These directions refer to the numbering across the sugar going in one specific direction. In figure 9.25, going from the bottom of the figure to the top, the left-hand strand is a $3' \rightarrow 5'$ strand and the right-hand strand is a $5' \rightarrow 3'$ strand. Since DNA replication involves the formation of two new antiparallel double helices with the old single strands as templates, one new strand would have to be replicated in the $5' \rightarrow 3'$ direction and the other in the $3' \rightarrow 5'$ direction.

However, all the known polymerase enzymes add nucleotides in only the $5' \rightarrow 3'$ direction. That is, the polymerase will catalyze a bond between the first $5'$-PO_4 group of a new nucleotide and the $3'$-OH carbon of the last nucleotide in the newly synthesized strand (fig. 9.25). The polymerases cannot create the same bond with the $5'$ phosphate of a nucleotide already in the DNA and the $3'$ end of a new nucleotide. Thus the simple model needs some revision, unless an undiscovered enzyme exists that can synthesize DNA by adding to the $5'$-PO_4 end. Since most molecular biologists believe that a polymerase with that specificity will not be found, we are left with the need for a better model of DNA replication.

Continuous and Discontinuous DNA Replication

Autoradiographic evidence leads us to believe that replication is occurring simultaneously on both strands. **Continuous** replication is, of course, possible on the $3' \rightarrow 5'$ template strand, which begins with the necessary $3'$-OH **primer.** (Primer is double-stranded DNA—or, as we shall see, a DNA-RNA hybrid—continuing as single-stranded DNA template. The strand being synthesized has a $3'$-OH available; see fig. 9.25.) A **discontinuous** form of replication takes place on the complementary strand, where it occurs in short segments, backward, away from the Y-junction (fig. 9.26). These short segments, called **Okazaki fragments** after R. Okazaki who first saw them, average about 1,500 nucleotides in prokaryotes and 150 in eukaryotes. The strand synthesized continuously is referred to as the **leading strand,** and the strand synthesized discontinuously is referred to as the **lagging strand.**

Once initiated, continuous DNA replication can proceed indefinitely. DNA polymerase III on the leading-strand template has what is called high **processivity:** once it attaches, it doesn't release until the entire strand is replicated. Discontinuous replication, however, requires the repetition of four steps: primer synthesis, elongation, primer removal with gap filling, and ligation.

Primer Synthesis and Elongation

In order for Okazaki fragments to be synthesized, a primer must be created de novo (Latin, from the beginning). None of the DNA polymerases can create that primer. Instead, one of two enzymes, either **RNA polymerase,** the transcribing enzyme (see chapter 10) or, more commonly, **primase,** an RNA polymerase coded for by the dnaG gene, creates the primer. It is from two to sixty nucleotides, depending on the species, at the site of Okazaki fragment initiation (fig. 9.27). The result is a short RNA primer that provides the free $3'$-OH group that DNA polymerase III needs in order to synthesize the Okazaki fragment. DNA polymerase III continues until it reaches the primer RNA of the previously synthesized Okazaki fragment. At that point it stops and releases from the DNA (fig. 9.27).

All three prokaryotic polymerases not only can add new nucleotides to a growing strand in the $5' \rightarrow 3'$ direction but also can remove nucleotides in the opposite $3' \rightarrow 5'$ direction. This property is referred to as *$3' \rightarrow 5'$ exonuclease activity.* Enzymes that degrade nucleic acids are classified as **exonucleases** if they remove nucleotides from the end of a nucleotide strand or as **endonucleases** if they can break the sugar-phosphate backbone in the middle of a nucleotide strand. At first glance, exonuclease activity seems like an extremely curious property for a polymerase to have—curious unless we think about its ability to check complementarity. If the complementarity is improper, which is to say that the wrong nucleotide has been inserted, the polymerase can remove the incorrect nucleotide, put in the proper one, and continue on its way. This is known as the **proofreading** function of the DNA polymerase. In addition, the RNA primers of Okazaki fragments can be removed by exonuclease activity.

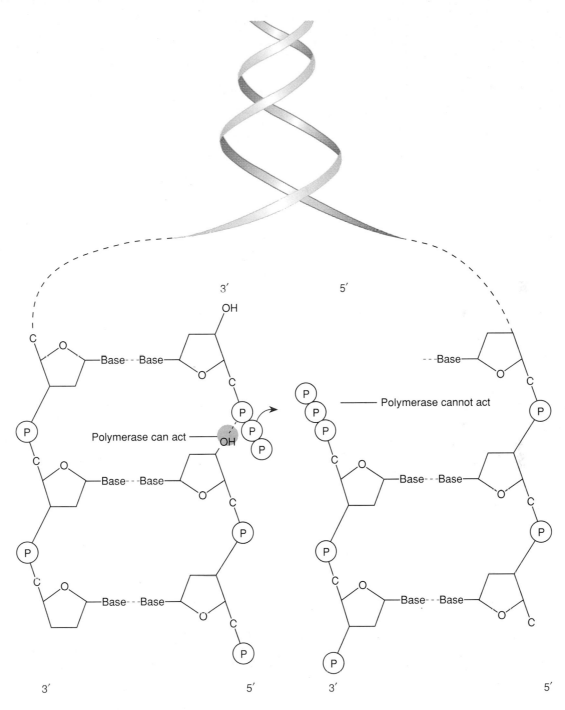

Primer Removal and Gap Filling

DNA polymerase I is a polymerase when it adds nucleotides, one at a time, and an exonuclease when it removes nucleotides one at a time. To complete the Okazaki fragment, DNA polymerase I acts in both capacities. (Mutants of DNA polymerase I cannot properly connect Okazaki fragments.) DNA polymerase I completes the Okazaki fragment by removing the previous RNA primer and replacing it with DNA nucleotides (fig. 9.28). When DNA polymerase I has completed its nuclease and polymerase activity, the two previous Okazaki fragments are almost complete. All that remains is a single phosphodiester bond to be made.

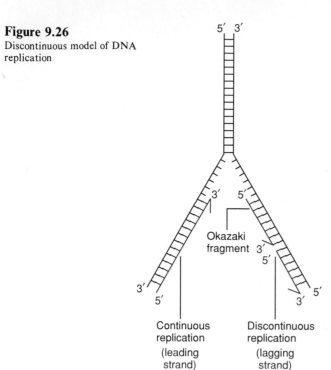

Figure 9.26
Discontinuous model of DNA replication

Ligation

DNA polymerase I cannot make the final bond to join the Okazaki fragment to the previously synthesized DNA. The configuration needing completion is shown in figure 9.29. An enzyme, **DNA ligase,** completes the task by making the final phosphodiester bond in an energy-requiring reaction.

A question of evolutionary interest is why RNA is used for priming of DNA synthesis. Why not use DNA directly and avoid the exonuclease and resynthesis activity seen in figure 9.28? One possible answer is that because priming is inherently more error-prone then regular DNA synthesis, it is best for the cell to have the primer nucleotides removed and replaced by DNA synthesized in a less error-prone fashion before DNA synthesis is completed. If the priming nucleotides are RNA, then DNA polymerase I can recognize and remove them in the final stage of Okazaki fragment synthesis, at which time it replaces them with DNA nucleotides inserted with a low error rate (fig. 9.28).

Figure 9.27
Primer formation and elongation create an Okazaki fragment during discontinuous DNA replication.

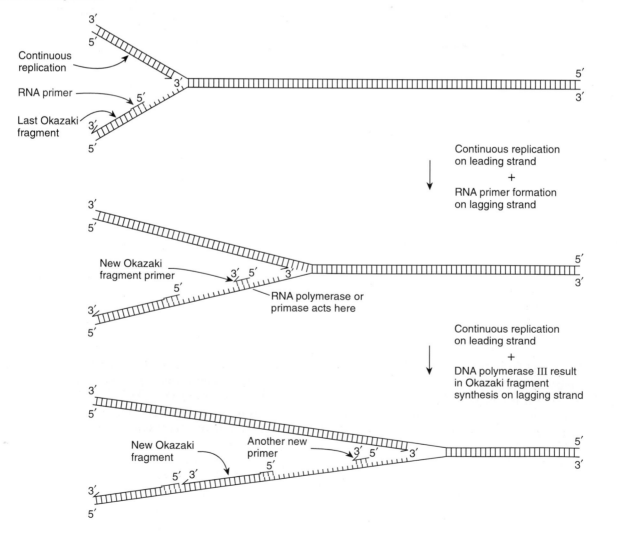

Figure 9.28
The completion of an Okazaki fragment requires removal of RNA primer, base by base, by DNA polymerase I. A final nick in the DNA backbone remains (arrow).

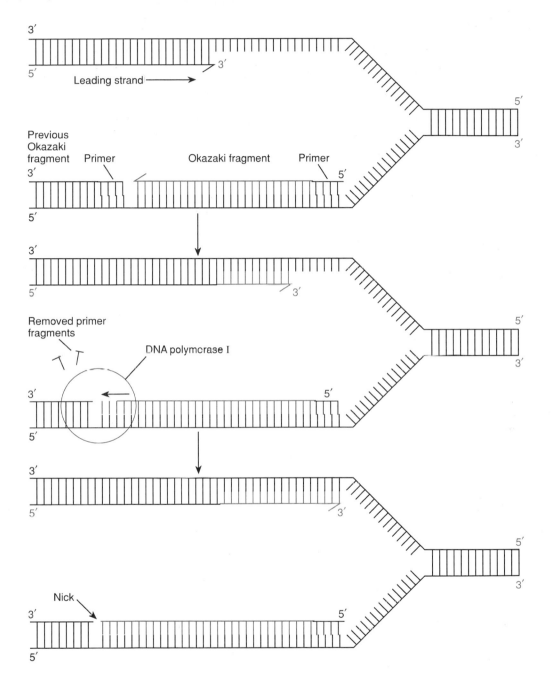

Another question of evolutionary interest is why DNA synthesis cannot take place in the 3′ → 5′ direction. Perhaps the answer has to do with proofreading and the exonuclease removal of incorrect nucleotides. When an incorrect nucleotide is found and removed, the next nucleotide brought in, in the 5′ → 3′ direction, will have a triphosphate end available to provide the energy for its own incorporation (fig. 9.25). Consider what would happen if the polymerase were capable of adding nucleotides in the opposite direction. The energy for the diester bond would be coming from the triphosphate already attached in the growing 3′ → 5′ strand (fig. 9.25). Then, if an error in complementarity were detected and the most recently added nucleotide were removed from the 3′ → 5′ strand by the polymerase, the last nucleotide in the double helix would no longer have a triphosphate available to provide energy for the diester bond with the next nucleotide brought in. Continued polymerization would thus require

Figure 9.29

After DNA polymerase I removes the RNA primer to complete an Okazaki fragment, a final gap remains that is closed by DNA ligase.

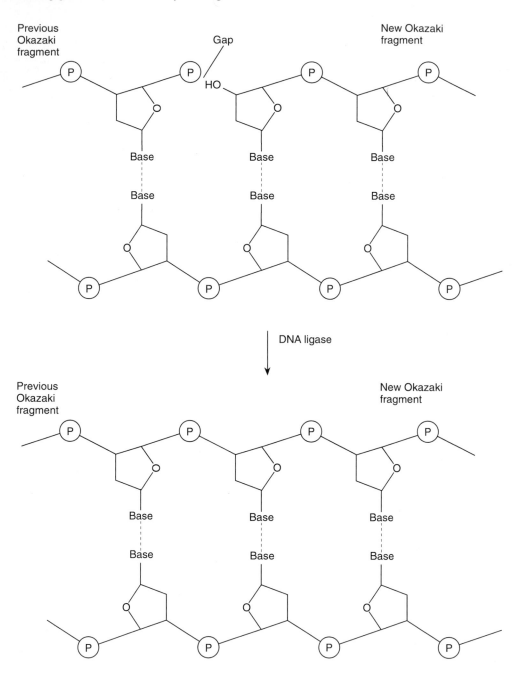

additional enzymatic steps to provide the energy for the process to continue. This could slow the process down considerably. As it is, the process works at a speed of about four hundred nucleotides incorporated per second with an error rate of about one incorrect pairing per one hundred thousand bases, improved to a rate of only one mistake in ten million by exonuclease proofreading. (Other repair systems can improve this error rate another thousandfold, to about one error every 10^{10} times an average base is replicated—see chapter 16.)

The Origin of DNA Replication

Each replicon (e.g., the *E. coli* chromosome or a segment of a eukaryote chromosome) must have a region in which DNA replication is initiated. In *E. coli* this region is referred to as the genetic locus *oriC*. In order for DNA replication to begin, several steps must occur. First, the specific origin site must be recognized by the appropriate proteins. Then the site must be opened and stabilized. And, finally, a replication fork must be initiated in both directions, involving continuous and discontinuous DNA rep-

Figure 9.30
Events at the origin of DNA replication in *E. coli*. The DNA opens up at *oriC* to create two moving Y-junctions. The primosome consists of a helicase and a primase.

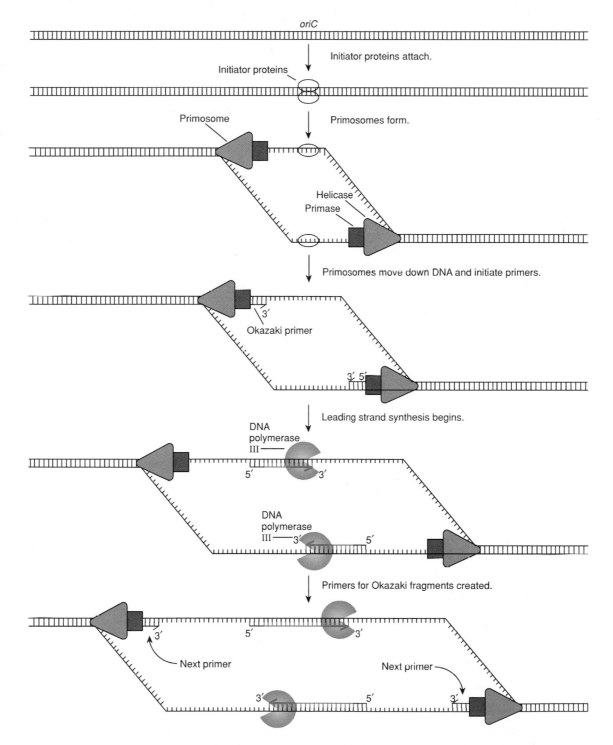

lication. Although many of the proteins involved are known, all the steps at the enzymatic level are not, and hence our understanding is a bit sketchy. Following is a description, most of whose steps are known.

OriC, the origin of replication in *E. coli,* is about 245 base pairs long and is recognized by proteins called **ini-** **tiator proteins** that open up the double helix. The initiator proteins then take part in the attachment of **primosomes,** a complex of two proteins: a primase, which creates RNA primers, and DNA **helicase,** which unwinds DNA at the Y-junction (fig. 9.30). As the primosomes move along, they

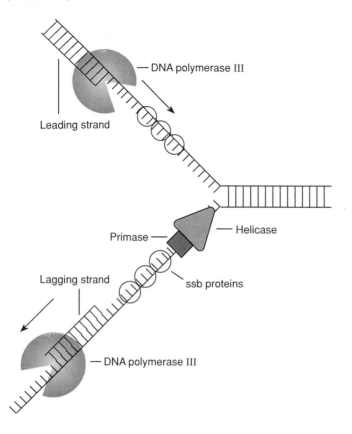

Figure 9.31
Schematic drawing of DNA replication at a Y-junction. Two copies of DNA polymerase III, ssb proteins, and a primosome (helicase + primase) are present.

Leading strand — DNA polymerase III

Primase — Helicase

Lagging strand — ssb proteins

DNA polymerase III

create RNA primers used by DNA polymerase III to initiate Okazaki fragments. At some point, leading-strand synthesis begins and Y-junction activity then proceeds as outlined earlier (figs. 9.27, 9.28, and 9.29).

In some phages and plasmids, the initiation of replication is not with an RNA primer but with a protein. This protein provides the primer configuration with an OH group from an amino acid (see chapter 11). The generality of this type of priming has not been established. Another interesting protein interaction at the origin of replication involves the reverse of initiation of DNA synthesis, the prevention of the initiation of DNA synthesis. This is accomplished by a newly discovered protein, called an "off switch." That is, this protein binds to the DNA at *oriC* and apparently prevents DNA replication from beginning. It does this by its binding activity that prevents the initiator proteins from opening the DNA. Thus this protein may be a very important component in control of the cell cycle, stopping the cell cycle from beginning. Presumably, when the appropriate time comes for the cell cycle to begin, the protein is removed.

Events at the Y-Junction

We now have the image of DNA replication proceeding by a primosome, moving along the lagging-strand template, opening up the DNA (helicase activity), and creating RNA primers (primase activity). One DNA polymerase III moves along the leading-strand template generating the leading strand by continuous DNA replication, whereas a second DNA polymerase III moves backward, away from the Y-junction, creating Okazaki fragments. **Single-strand binding proteins** (ssb proteins) keep single-stranded DNA stabilized (open) during this process, and DNA polymerase I and ligase are connecting Okazaki fragments (fig. 9.31).

This simple picture is slightly complicated by the fact that a single DNA polymerase seems to do the entire lagging strand, rather than dropping off the DNA at the completion of an Okazaki fragment and being replaced by a new one at the newest primer near the Y-junction. In addition, there is evidence that the lagging- and leading-strand synthesis is coordinated. The **replisome** model has arisen in which both copies of DNA polymerase III are attached to each other and work in concert with the primosome at the Y-junction (fig. 9.32). According to this model, a single replisome, consisting of two copies of the DNA polymerase III **holoenzyme** (each actually made of seven subunits), a helicase, and a primase, moves along the DNA. The leading-strand template is immediately fed to a polymerase, whereas the lagging-strand template is not acted on by the polymerase until an RNA primer has been placed on the strand, meaning that a long (fifteen hundred base) single strand has been opened up (fig. 9.32a).

As the replisome moves along, another single-stranded length of the lagging-strand template is formed. At about the time that the Okazaki fragment is completed, a new RNA primer has been created (fig. 9.32b). The Okazaki fragment is released (fig. 9.32c) and a new Okazaki fragment is begun, starting with the latest primer (fig. 9.32d), taking the replisome back to the same configuration as in figure 9.32a, but one Okazaki fragment farther along.

Supercoiling

The simplicity and elegance of the DNA molecule masks an inevitable problem of coiling. Since the DNA molecule is made from two strands that wrap about each other, certain operations, such as DNA replication and its termination, meet topological difficulties. Up to this point, we have seen the circular *E. coli* chromosome in its "relaxed"

Figure 9.32

The replisome consists of two DNA polymerase III holoenzymes and a primosome (helicase + primase). It coordinates replication at the Y-junction.

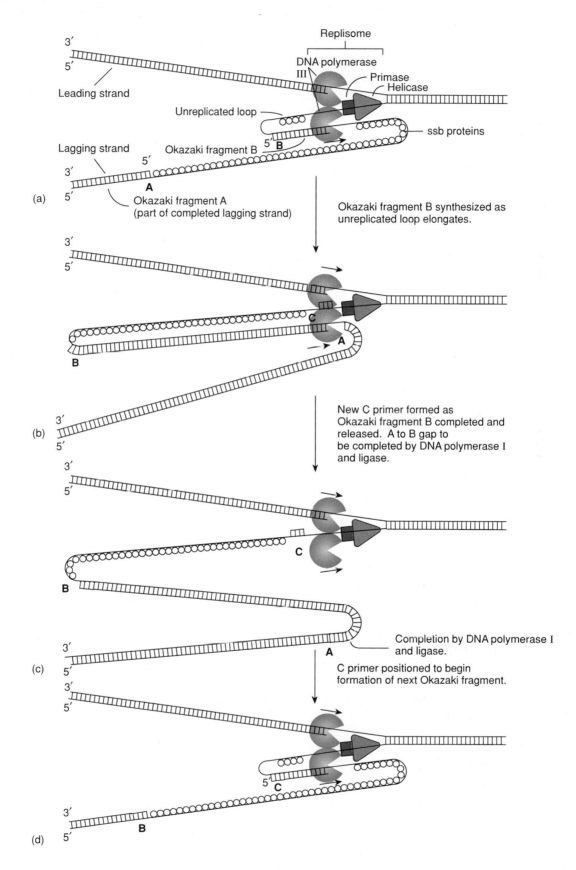

Figure 9.33

Topoisomerases can take relaxed DNA (*center*) and add negative (*left*) or positive (*right*) supercoils. L is the linkage number.

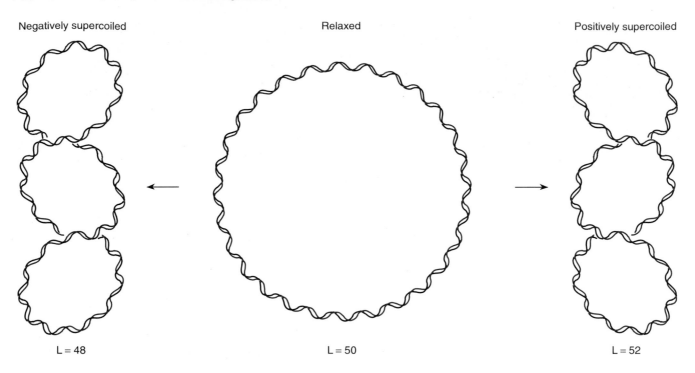

Negatively supercoiled · Relaxed · Positively supercoiled

L = 48 · L = 50 · L = 52

state (e.g., figs. 9.21, 9.22). However, there are enzymes in the cell that cause DNA to become overcoiled (positively **supercoiled**) or undercoiled (negatively supercoiled). Positive supercoiling comes about either from too many turns of the DNA in a given length or from the molecule wrapping around itself (fig. 9.33).

Positive supercoiling comes from having the circular duplex wind about itself in the same direction as the helix twists (right handed), whereas negative supercoiling comes about by having the duplex wind about itself in the opposite direction as the helix twists (left handed). The former state increases the number of turns of one helix around the other (the **linkage number,** L), whereas the latter decreases it. The three forms of DNA in figure 9.33 all have the same sequence yet differ in their linkage number. They are referred to as topological isomers (**topoisomers**). The enzymes that create or alleviate these states are called **topoisomerases.**

Topoisomerases affect supercoiling by either of two methods. Type I topoisomerases break one strand of a double helix and, while binding the broken ends, pass the other strand through the break. The break is then sealed (fig. 9.34). Type II topoisomerases (e.g., **DNA gyrase** in *E. coli*) do the same sort of thing only instead of breaking one strand of a double helix, they break both and pass another double helix through the temporary gap.

As DNA replication proceeds, positive supercoiling builds up ahead of the Y-junction. This is eliminated by the action of topoisomerases that either create negative supercoiling ahead of the Y-junction in preparation for replication or alleviate positive supercoiling after it has been created. The major components of DNA replication in *E. coli* are summarized in table 9.4.

Termination of Replication

The termination of the replication of a circular chromosome presents no major topological problems. The theta-structure replication (see fig. 9.22) finishes with both Y-junctions having proceeded around the molecule. The leading strand on one template closes in on the lagging strand begun in the other direction with the same happening on the other template. The process stops with about twenty-five twists remaining at no particular spot on the chromosome (there is no "termination" locus). A topoisomerase then releases the two circles and DNA polymerase I and ligase close them up (fig. 9.35).

Several different mechanisms have been explored for the termination of the linear chromosomes of some viruses and all eukaryotic genomes. Linear molecules have the problem of completing the last Okazaki fragment. An RNA primer on the very tip of the $3' \rightarrow 5'$ template cannot be replaced by DNA polymerase I (fig. 9.36), assuming even that a final primer can be put on the very tip of the molecule. In eukaryotes, an enzyme, telomerase, attaches repeats of a short sequence at each chromosome tip (telomere; see chapter 14).

Figure 9.34
Topoisomerase I can reduce DNA coiling by passing one strand of the double helix through the other.

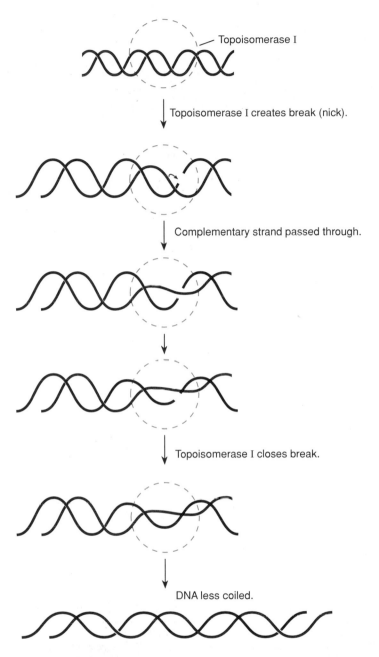

Topoisomerase I

Topoisomerase I creates break (nick).

Complementary strand passed through.

Topoisomerase I closes break.

DNA less coiled.

REPLICATION STRUCTURES

The model of DNA replication that we have presented here comes primarily from evidence gathered in *E. coli,* which replicates by way of the *theta*-structure intermediate (fig. 9.22). However, two other modes of replication occur in circular chromosomes: rolling-circle and D-loop.

Rolling-Circle Model

In the **rolling-circle** mode of replication, a nick (a break in one of the phosphodiester bonds) is made in one of the strands of the circular DNA, resulting in replication of a circle and a tail (fig. 9.37). This form of replication occurs in the Hfr *E. coli* chromosome, or the F plasmid, during conjugation (see chapter 7). The F⁺ or Hfr cell retains the circular daughter while passing the linear tail into the F⁻ cell. This method is also used in several phages, which fill their heads (protein coats) with linear DNA replicated from a circular parent molecule.

In the model for rolling-circle replication, the nick made in one strand creates a free 3′-OH end and a free 5′-PO$_4$ end (fig. 9.37). Synthesis of a new circular strand occurs by addition of nucleotides to the 3′ end using the complementary intact strand as a template. No primer is

TABLE 9.4   Summary of the Enzymes Involved in DNA Replication in E. coli*

Enzyme (Protein)	Genetic Locus	Function
DNA polymerase I	*polA*	Gap filling and primer removal
DNA polymerase II	*polB*	?
DNA polymerase III		
α subunit	*dnaE (polC)*	DNA replication
β subunit	*dnaN*	DNA replication
γ subunit	*dnaX*	DNA replication
δ subunit	?	DNA replication
ε subunit	*dnaQ*	$3' \rightarrow 5'$ exonuclease
θ subunit	?	DNA replication
τ subunit	*dnaX*	DNA replication
Initiator protein	*dnaA*	Binds to origin of replication
RNA polymerase subunits	*rpoA, B, C, D*	RNA primer in some systems
Primase	*dnaG*	RNA primer in some systems
DNA ligase	*lig*	Closes nicked DNA strands
Helicase	*rep*	Unwinds DNA for replication
Ssb proteins	*ssb*	Single-strand stability
DNA topoisomerase I	*topA*	Supercoiling of DNA
DNA topoisomerase II		
α subunit	*gyrA (nalA)*	ATPase
β subunit	*gyrB (cou)*	Cutting, closing of DNA

*Note that some enzymes are composed of subunits, each of which has a specific function.

needed because the original break produces a primer configuration (3'-OH). As nucleotides are added to one end of the broken strand in a continuous fashion, the other end is displaced as a 5'-PO$_4$ tail. As replication of the circular template occurs, the 5'-PO$_4$ tail is replicated in a discontinuous manner, and the resulting double helix can be severed from the double-helical circle by a nuclease. DNA ligase closes the replicated circular strand and can join the ends of the replicated tail into a circle in the F⁻ cell, or can package the linear molecule in a phage head, depending upon which type of circular DNA has been replicated.

D-Loop Model

Chloroplasts and mitochondria have their own circular DNA molecules (see chapter 17) that appear to replicate by a slightly different mechanism than those described. The origin of replication is at a different point on each of the two parental template strands (fig. 9.38). Replication begins on one strand, displacing the other while forming a displacement loop or **D-loop** structure. Replication continues until the process passes the origin of replication on the other strand. Replication is then initiated on the second strand, in the opposite direction. The result is two circles. Some species have chloroplasts and mitochondria with circular DNAs that have multiple D-loops formed.

EUKARYOTIC DNA REPLICATION

As we saw earlier, linear eukaryotic chromosomes usually have multiple origins of replication resulting in figures referred to as "bubbles" or "eyes." Multiple origins allow eukaryotes to replicate their larger quantities of DNA in a relatively short time, even though eukaryotic DNA replication is considerably slowed by the presence of histone proteins associated with the DNA to form chromatin (see chapter 14). For example, the *E. coli* replication fork moves about twenty-five thousand base pairs per minute, whereas the eukaryotic Y-junction moves only about two thousand base pairs per minute. The number of replicons in eukaryotes varies from about five hundred in yeast to as many as sixty thousand in a diploid mammalian cell.

Much less is understood about eukaryotic DNA replication because of the complexity of eukaryotes and the relatively shorter time during which they have been studied effectively. We presume that eukaryotes have solved the same problems faced by prokaryotes in a similar, but not

Figure 9.35

The replication of circular DNA is terminated with topoisomerase activity and gap filling.

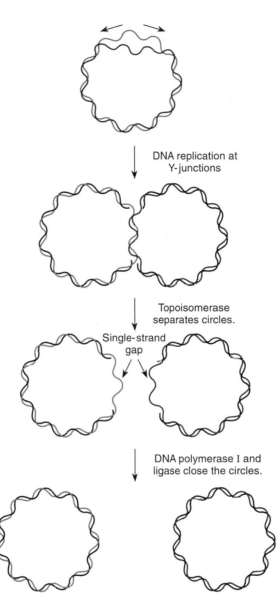

DNA replication at Y-junctions

Topoisomerase separates circles.

Single-strand gap

DNA polymerase I and ligase close the circles.

Figure 9.36

Termination of the replication of a linear DNA molecule results in primer at the tip of each duplex (the primers of the final Okazaki fragments).

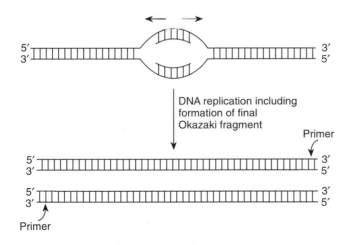

DNA replication including formation of final Okazaki fragment

Primer

Primer

identical, fashion. For example, eukaryotes have five types of DNA polymerases, named DNA polymerase α, β, γ, δ, and ϵ (table 9.5). DNA polymerases γ, δ, and ϵ have exonuclease activity. DNA polymerases α and δ are the major replicating enzymes, with polymerase α replicating the lagging strand and polymerase δ replicating the leading strand. The role of polymerase ϵ is unclear; it seems capable of regular leading- or lagging-strand replication. DNA polymerase β is the major repair polymerase (like polymerase I in prokaryotes). DNA polymerase γ appears to be concerned primarily with mitochondrial DNA replication.

Figure 9.37

Rolling-circle model of DNA replication. The letters A–E provide landmarks on the chromosome.

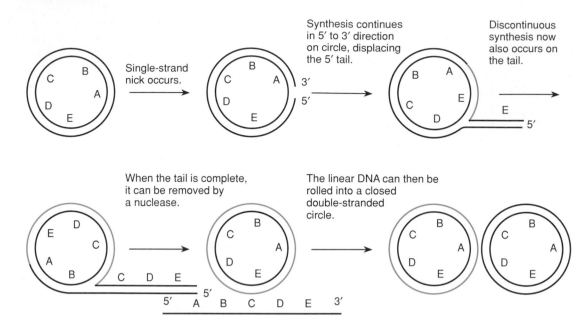

Figure 9.38

D-loops form during mitochondrial and chloroplast DNA replication because the origin of replication is at different places on the two strands of the double helix.

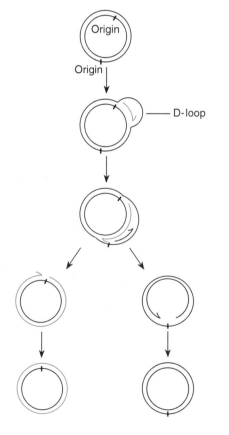

TABLE 9.5 Eukaryotic DNA Polymerases

Enzyme	Function
DNA polymerase α	Replication of nuclear chromosomes (lagging strand)
DNA polymerase β	Repair of nuclear chromosomes
DNA polymerase γ	Replication of mitochondrial chromosomes
DNA polymerase δ	Replication of nuclear chromosomes (leading strand)
DNA polymerase ϵ	Probably replication of nuclear chromosomes

SUMMARY

A genetic material must be able to control the phenotype of a cell or organism (that is, to direct protein synthesis), it must be able to replicate, and it must be located in the chromosomes. Avery and his colleagues demonstrated that DNA was the genetic material when they showed that the transforming agent was DNA. Griffith had originally demonstrated the phenomenon of transformation of *Streptococcus* bacteria in mice. Hershey and Chase demonstrated that it was the DNA of bacteriophage T2 that entered the bacterial cell. Fraenkel-Conrat demonstrated that in viruses without DNA (RNA viruses), such as tobacco mosaic virus, the RNA acted as the genetic material. Thus by 1953 the evidence was strongly supportive of nucleic acids (DNA or, in its absence, RNA) as the genetic material.

Chargaff showed a 1:1 relationship of adenine (A) to thymine (T) and cytosine (C) to guanine (G) in DNA. Wilkins, Franklin, and their colleagues showed, by X-ray crystallography, that DNA was a helix of specific dimensions. Following these lines of evidence, Watson and Crick in 1953 suggested the double-helical model of the structure of DNA. In their model, DNA is made up of two strands, running in opposite directions, with sugar-phosphate backbones and bases facing inward. Bases from the two strands form hydrogen bonds with each other with the restriction that only A and T and only G and C can pair. This explains the quantitative relationships that Chargaff found among the bases. Melting temperatures of DNA also support this structural hypothesis because DNAs with higher G-C contents have higher melting, or denaturation, temperatures; G-C base pairs have three hydrogen bonds versus only two in an A-T base pair. The model of DNA presented is that of the B form. DNA can exist in other forms, including the Z form, a left-handed double helix that may be important in controlling eukaryotic gene expression.

DNA replicates by the unwinding of the double helix, with each strand subsequently acting as a template for a new double helix. This works because of complementarity—only A-T, T-A, G-C, or C-G base pairs form stable hydrogen bonds within the structural constraints of the model. This model of replication is termed *semiconservative*. It was shown to be correct by Meselson and Stahl in an experiment with heavy nitrogen. Autoradiographs of replicating DNA showed that replication proceeds bidirectionally from a point of origin.

Prokaryotic chromosomes are circular with a single initiation point of replication. Eukaryotic DNA is linear with multiple initiation points of replication.

DNA polymerase enzymes add nucleotides only in the $5' \rightarrow 3'$ direction. Replication proceeds in small segments working backward from the Y-junction on the $5' \rightarrow 3'$ template strand. Presumably, the $5' \rightarrow 3'$ restriction has to do with the proofreading ability of DNA polymerases to correct errors in complementarity. Polymerase III is the active replicating enzyme, and polymerase I is involved in DNA repair. Many other enzymes are involved in the creation of Okazaki fragments, the unwinding of DNA, and the release of supercoiling. Less is known about eukaryotic systems.

SOLVED PROBLEMS

1. What evidence led to the idea that DNA was the genetic material?

   ANSWER: Basically, Avery and his colleagues (MacLeod and McCarty), in experiments in which they showed that DNA was the transforming agent, are generally given credit with formalizing the notion that DNA, not protein, is the genetic material. Experimental work, including that of Chargaff, Hershey and Chase, Fraenkel-Conrat, and several others also helped change the general view. At the time that Watson and Crick published their model, the scientific community knew that DNA was the genetic material but didn't know its structure.

2. How does DNA fulfill the requirements of a genetic material?

   ANSWER: DNA is located in chromosomes, has a structure that is easily and accurately replicated, and has the sequence complexity to code for the 50,000 or more genes that a eukaryotic organism has.

3. What enzymes are involved in DNA replication?

   ANSWER: Barring initiation and termination, and looking only at *E. coli,* different processes occur on leading and lagging strands. The process is coordinated at a Y-junction such that a replisome is formed, consisting of a primosome (a primase and a helicase) and two polymerase III holoenzymes. One polymerase acts processively, forming the leading strand, while the other forms Okazaki fragments initiated by primers created by the primase. The Okazaki fragments are completed with DNA polymerase I, which gets rid of the RNA primer of the previous Okazaki fragment by replacing it with DNA. Finally, DNA ligase connects the fragments. Also involved in the process are single-strand binding proteins and topoisomerases that relieve supercoiling of the DNA.

4. What can be concluded about the nucleic acids in the table below?

Molecule	% A	% T	% G	% C	% U
1	28	28	22	22	0
2	31	0	31	17	21
3	15	15	35	35	0

   ANSWER: We must first look to see if U or T is present, for this determines whether the molecule is RNA or DNA, respectively. Molecule 2 is RNA. Now we look at base composition. In double-stranded molecules, A bonds with T (or U) and G bonds with C. This relationship holds for molecules 1 and 3, so they are double stranded, and molecule 2 is single stranded. Finally, the melting temperature increases with the amount of G-C, so the melting temperature of 3 is greater than that of 1.

1. Diagram the results that Meselson and Stahl would have obtained (a) if DNA replication were conservative and (b) if it were dispersive.

2. If the tetranucleotide hypothesis were correct regarding the simplicity of DNA structure, under what circumstances could DNA be the genetic material?

3. Nucleic acids, proteins, carbohydrates, and fatty acids have been mentioned as potential genetic material. What other molecular moieties in the cell could possibly have functioned as the genetic material?

4. In what component parts do DNA and RNA differ?

5. What type of photo would J. Cairns have obtained if DNA replication was conservative? Dispersive?

6. Following is a section of a single strand of DNA. Supply a strand, by the rules of complementarity, that would turn this into a double helix. What RNA bases would primase use if this segment initiated an Okazaki fragment? In which direction would replication proceed?

5'-ATTCTTGGCATTCGC-3'

7. Draw the structure of a short segment of DNA (three base pairs) at the molecular level and indicate the polarity of the strands.

8. Roughly sketch the shape of B and Z DNA remembering that B DNA is a right-handed helix and Z DNA is a left-handed helix.

9. Under what circumstances in DNA would you expect to see a theta structure? D-loop? Rolling-circle? Bubbles? What function does each structure serve?

10. What is a primosome? A replisome? What enzymes make up each? What is the relationship of these structures to each other?

11. What are the differences between continuous and discontinuous DNA replication? Why do both exist?

12. Describe the synthesis of an Okazaki fragment.

13. Describe the enzymology of the origin, continuation, and termination of DNA replication in *E. coli*.

14. Can you think of any other mechanisms besides topoisomerase activity that could release supercoiling in replicating DNA?

15. Draw a diagram showing how topoisomerase II (gyrase) might work.

16. For each of the following nucleic acid molecules, deduce whether it is DNA or RNA and single stranded or double stranded.

Molecule	%A	%G	%T	%C	%U
1.	33	17	33	17	0
2.	33	33	17	17	0
3.	26	24	0	24	26
4.	21	40	21	18	0
5.	15	40	0	30	15

17. A double-stranded DNA molecule is 28% guanosine (G).
    a. What is the complete base composition of this molecule?
    b. Answer the same question, but assume the molecule is double-stranded RNA.

18. The following are melting temperatures for five DNA molecules. Arrange these DNAs in increasing amount of G-C pairs.

    1: 73°C, 2: 69°C, 3: 84°C, 4: 78°C, 5: 82°C.

19. Retroviruses are single-stranded RNA viruses that must insert their genome into the host DNA in order to replicate and produce more viruses. But only double-stranded DNA can be inserted into double-stranded DNA.
    a. Propose a mechanism that could be used by retroviruses.
    b. What novel enzymes might be required by such viruses?

20. Propose a mechanism by which a single-stranded DNA molecule can make copies of itself.

21. We normally think that single-stranded nucleic acids should not melt, but many, in fact, do have a T_m. How can you explain this apparent mystery?

22. Progeria is a human disorder in which affected individuals age prematurely; a nine-year-old often resembles a sixty- to seventy-year-old individual in appearance and physiology. Suppose you extract DNA from a progeric patient and find mostly small DNA fragments rather than the expected long DNA molecules. What enzyme(s) may be defective in patients with progeria?

23. In a DNA molecule, if the amount of G is twice the amount of A, the amount of T is three times the amount of C, and the ratio of pyrimidines to purines is 1.5, what is the base composition of the DNA?

24. A double-strand DNA measures 6.5 μm in length. Approximately how many base pairs does it contain?

25. In developing sea urchins, just after fertilization, the cells divide every 30 to 40 minutes. In the adult, the cells divide once every 10 to 15 hours. The amount of DNA per cell is the same in each case, but the DNA is obviously replicated much faster in developing cells. Propose an explanation to account for the difference in replication time.

Suggested Readings for chapter 9 are on page 645.

Gene Expression: Transcription

The DNA of cyanobacteria have been shown to contain introns. Photo is of the cyanobacterium *Fischerella.*
© Susan Barnes and Norman Pace

All living things synthesize proteins. In fact, the types of proteins that a cell synthesizes determine the kind of cell it is. Hence the genetic material must control the types and quantities of proteins that a cell synthesizes. Proteins (polypeptides) are made up of long strings of amino acids joined together by peptide bonds. Each protein is put together from a choice of only twenty amino acids. The amino acid sequence of a protein is specified by the sequence of nucleotides in DNA or RNA. In all prokaryotes, eukaryotes, and DNA viruses, the gene is a sequence of nucleotides in DNA that codes for the sequence of RNA. That RNA usually determines the amino acids in a polypeptide. RNA usually serves as an intermediary between DNA and proteins; however, the RNA itself can be functional (see below). (In RNA viruses the RNA may serve as a template for the eventual synthesis of DNA, or the RNA may serve as genetic material without DNA ever being formed. We will consider these cases at the end of the chapter.)

Originally, the description of the flow of information involving the genetic material was termed the **central dogma:** DNA transfers information to RNA, which then directly controls protein synthesis (fig. 10.1). DNA also controls its own replication. **Transcription** is the process whereby RNA is synthesized from a DNA template—that is, the DNA information is rewritten, but in basically the same language of nucleotides. The process whereby RNA controls the synthesis of proteins is termed **translation** because information in the language of nucleotides is translated into information in the language of amino acids. In this chapter we concentrate on the events surrounding transcription. At the end of the chapter we will update the central dogma. In the next several chapters we will focus on translation and the various mechanisms of control of both transcription and translation in viruses, prokaryotes, and eukaryotes.

TYPES OF RNA

In the process of protein synthesis, three different kinds of RNA serve in three unique roles. The role of the first type of RNA is as a messenger (**mRNA**) to carry the sequence information of DNA to particles in the cytoplasm known as **ribosomes,** where the messenger will be translated. The role of the second type of RNA is that of a transfer molecule (**tRNA**) to bring the amino acids to the ribosomes, where protein synthesis actually takes place. The role of the third type of RNA is as a structural part of the ribosome. This last type of RNA is called ribosomal RNA (**rRNA**). The general relationship of the roles of these three types of RNA is diagrammed in figure 10.2.

We know that DNA does not directly control protein synthesis because, in eukaryotes, translation occurs in the cytoplasm, whereas DNA remains in the nucleus. We sus-

Figure 10.1

The original central dogma depicting the flow of genetic information

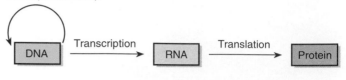

Figure 10.2

Relationship among the three types of RNA during protein synthesis

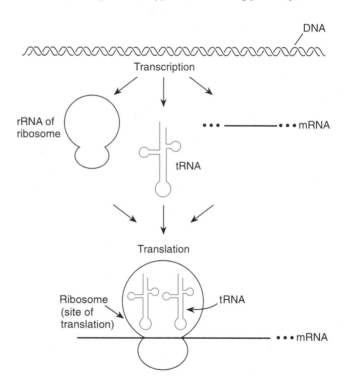

pected for a long time that the genetic intermediate in prokaryotes and eukaryotes is RNA because the cytoplasmic RNA concentration increases with increasing protein synthesis and the cytoplasmic RNAs carry nucleotide sequences complementary to the cell's DNA. Proof came when it was shown that mRNA directs the synthesis of proteins.

TRANSCRIPTION

DNA-RNA Complementarity

What proof do we have that the RNA found in the cytoplasm is complementary to the DNA in the nucleus? Two lines of evidence are important here. First, it has been shown that the RNAs produced by various organisms have

TABLE 10.1	Correspondence of Base Ratios between DNA and RNA of the Same Species	
	RNA % G + C	DNA % G + C
E. coli	52	51
T2 Phage	35	35
Calf Thymus Gland	40	43

base ratios very similar to the base ratios in their DNA (table 10.1). The second line of evidence comes from experiments by Hall, Spiegelman, and others with the procedure of **DNA-RNA hybridization.** In this technique, DNA is denatured by heating, which causes the two strands of the double helix to separate. When the solution is cooled, a certain proportion of the DNA strands will rejoin and rewind—that is, complementary strands "find" each other and re-form double helices. When RNA is added to the solution of denatured DNA and the solution is cooled slowly, some of the RNA will form double helices with the DNA if those fragments of RNA are complementary to a section of the DNA (fig. 10.3). The existence of extensive complementarity between DNA and RNA is a persuasive indication that DNA acts as a template on which complementary RNA is made.

In another experiment, DNA-RNA hybridization was used to show that phage infection led to the production of phage-specific mRNA. Gene-sized pieces of RNA extracted from *Escherichia coli* before and after infection by phage T2 were tested to see if they hybridized with the DNA of the T2 phage or with the DNA of the *E. coli* cell. Interestingly enough, the RNA in the *E. coli* cell was found to hybridize with the *E. coli* DNA before infection but with the T2 DNA after infection. Thus it is apparent that when the phage attacks the *E. coli* cell, it starts to manufacture RNA complementary to its own DNA and stops the *E. coli* DNA from serving as a template.

Having reached the conclusion that protein synthesis is directed by RNA that is transcribed (synthesized) from a DNA template, we look at two questions. Is this RNA single- or double-stranded? Is it synthesized (transcribed) from one or both strands of the parent DNA? For the most part, cellular RNA does not exist as a double helix. It has the ability to form double-helical sections when complementary parts come into apposition (see, e.g., fig. 10.19 or fig. 1 of Box 10.3), but its general form is not a regular double helix. The simplest, and most convincing, evidence for this is that complementary bases of RNA are not found in corresponding proportions (Chargaff's ratios). That is, in RNA, uracil does not usually occur in the same quantity as adenine, nor does cytosine occur in the same quantity as guanine (table 10.2).

Figure 10.3
DNA-RNA hybridization occurs between DNA and complementary RNA.

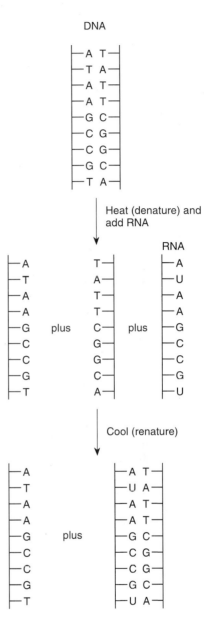

TABLE 10.2	Base Composition in RNA (Percentage)			
	Adenine	Uracil	Guanine	Cytosine
E. coli	24	22	32	22
Euglena	26	19	31	24
Polio Virus	30	25	25	20

The answer to the second question is that RNA is not usually copied from both strands of any given segment of the DNA double helix, although rare exceptions are known. Consider the problem of having a sequence of nucleotides on one strand specify a sequence of amino acids for a protein and the complementary nucleotide sequence also specifying the amino acid sequence of another functional protein. Since most enzymes are three hundred to five hundred amino acids long, the virtual impossibility of this task is obvious. It was, therefore, assumed a priori that, for any particular gene—that is, in any particular segment of DNA—the sequence on only one strand is transcribed and its complementary sequence is not. There is now considerable evidence to support this assumption.

The most impressive evidence that only one DNA strand is transcribed comes from work done with phage SP8, which attacks *Bacillus subtilis*. This phage has the interesting property of having a great disparity in the purine-pyrimidine ratio of the two strands of its DNA. The disparity is significant enough so that the two strands can be separated on the basis of their densities using density-gradient centrifugation. After denaturation and separation of the two strands, DNA-RNA hybridization can be carried out separately on each of the two strands with the RNA that is produced after the virus infects the bacterium. Marmur and his colleagues found that hybridization occurred only between the RNA and the heavier of the two DNA strands. Thus only the heavy strand acted as a template for the production of RNA during the infection process.

That only one strand of DNA serves as a transcription template for RNA has also been verified for several other phages. However, when we get to larger viruses and cells, we find that either of the strands may be transcribed, but only one strand is used as a template in any one region. This was clearly shown in phage T4 of *E. coli,* where certain RNAs hybridize with one strand, and other RNAs hybridize with the other strand. Let us now look at the transcription process itself and then proceed to examine the three types of RNA in detail.

RNA Polymerase

In prokaryotes, RNA transcription is controlled by RNA polymerase. Using DNA as a template, this enzyme polymerizes ribonucleoside triphosphate nucleotides. In eukaryotes there are three RNA polymerases. But since less is known about them, we will focus our attention on prokaryotes, primarily the *E. coli* system; we will relate these findings to eukaryotes later in the chapter. The complete RNA polymerase enzyme—the holoenzyme—is composed of a core enzyme and a **sigma factor.** The core enzyme is composed of four subunits: α (two copies), β, and β'; it is the component of the holoenzyme that actually carries out polymerization. The sigma factor is involved primarily in the recognition of transcription start signals on the DNA. Following initiation of transcription, the sigma factor disassociates from the core enzyme.

Logically, transcription should not be a continuous process like DNA replication is once it is begun. If there were no control of protein synthesis, all the cells of a higher organism would be identical, and a bacterial cell would be producing all of its proteins all of the time. Since some enzymes depend on substrates not present all of the time and since some reactions in a cell occur less frequently than others, the cell—be it a bacterium or a human liver cell—needs to control its protein synthesis to some extent. One of the most efficient ways for a cell to exert the necessary control of protein synthesis is to perform transcription in a selective manner. Transcription of genes coding for unneeded enzymes is a wasteful procedure. Therefore, RNA polymerase should be selective. It should use only those segments (genes or small groups of genes) of the DNA as templates for transcription that are needed by the cell at that particular time.

The mechanisms of transcriptional control need to be examined in two ways. First, we need to understand how the beginnings and ends of transcribable sections (a single gene or a series of adjacent genes) are demarcated. Second, we need to understand how the cell can selectively repress transcription of certain of these transcribable sections. The latter issues are covered in chapters 13 and 15.

RNA polymerase must be able to recognize both the beginnings and the ends of genes (or gene groups) on the DNA double helix in order to initiate and terminate transcription. It must also be able to recognize the correct strand of a gene in order to avoid transcribing the DNA strand that is not informational. This is accomplished by RNA polymerase recognizing certain start and stop signals in DNA, called initiation and termination sequences, respectively.

Initiation and Termination Signals for Transcription

The region on DNA that RNA polymerase associates with immediately before beginning transcription is known as the **promoter.** Without the sigma factor, the core enzyme of RNA polymerase binds randomly along the DNA. Formation of the holoenzyme brings about high affinity of RNA polymerase for DNA sequences in the promoter region. Termination of transcription comes about when the polymerase enzyme recognizes a DNA sequence known as a **terminator sequence.** Let us elaborate on the various stages of transcription.

Promoters

The size of the RNA polymerase molecule is such that it covers a region of about sixty base pairs of DNA. This has been determined by having the polymerase bind to DNA and then digesting the mixture with nucleases. The polymerase "protects" or prevents degradation of the region

BOX 10.1

Observing Transcription in Real Time

The overwhelming evidence that molecular events, such as transcription, take place is from genetic and biochemical analyses, and occasionally an electron micrograph of one type or another. Thus it is refreshing and illuminating to be able to observe some of the processes that we know are taking place in real time; that is, to sit at a microscope and actually see things happen. Such a study on transcription was published in 1991 in *Nature* by four scientists at Washington University in St. Louis.

Normally, we cannot see these events take place; the components are too small. Making them visible in electron microscopes usually requires fixation that destroys the ability of the components to actually continue their tasks. The Washington University group overcame this by attaching a gold particle to DNA, thus rendering the motion of that DNA visible in the light microscope (Box fig. 1a). The scientists immobilized the RNA polymerase to a glass coverslip; thus, as transcription took place, the DNA moved and the length of the tether of the gold particle increased. At first they stopped the process by limiting the concentration of nucleoside triphosphates (NTPs). They could then observe the motion of the gold ball when no transcription was taking place. The scientists predicted that an immobilized gold ball would not move and a tethered gold ball would show a limited amount of brownian motion. That is, it would show a limited amount of blur in light microscope video images averaged over time. However, as soon as NTPs were added, any tethered gold ball would show increased blur as it moved out of the field of vision and eventually was released when transcription was completed. That is exactly what they saw (Box fig. 1b). Thus they succeeded in watching transcription take place in real time.

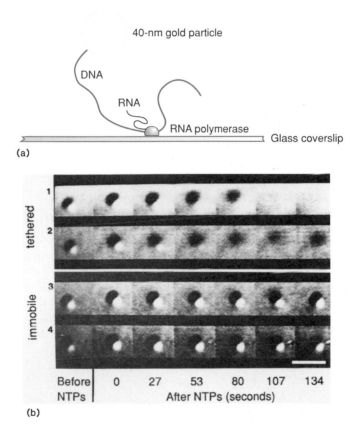

Box figure 1

Visualizing transcription. (*a*) The experimental design in which a RNA polymerase is immobilized on a coverslip waiting for nucleoside triphosphates (NTPs) to be added. The gold particle is tethered by the DNA. (*b*) Enhanced light microscope images of the gold particles. In 3 and 4, presumably immobilized particles showed no change over time. In 1 and 2, brownian motion and hence blur increase through time consistent with lengthening of the tether (transcription). Particle 1 was released after 87 seconds, and 2 after 135 seconds after NTPs were added. Scale bar is 1 μm.

(a) Reprinted from Nature, *Vol. 352:444–448. Copyright © 1991 Macmillan Magazines, Ltd. (b) Transcription by single molecules of RNA polymerase observed by light microscopy. Robert Landick, Department of Biology, Washington University, St. Louis, MO.*

BOX 10.2

Polymerase Collisions: What Can a Cell Do?

Both RNA polymerase and DNA polymerase move along the DNA of a cell during a cell cycle. The DNA polymerase moves along at about ten times the speed of the transcribing enzyme. Since usually many genes are active in a cell, the interaction (collision) of the two enzymes is inevitable. What happens when this inevitable collision takes place? What does the cell do about it? Although we cannot directly observe these interactions, there are various bits of data that suggest that a head-on collision could be fatal to the cell, and patterns of gene placement minimize the chance of a head-on collision.

The problem of the coexistence of these two enzymes was analyzed first by B. Brewer in a paper published in 1988 in the journal *Cell.* In evolutionary terms, the cell could obviate the problems of a head-on collision by either avoiding them or resolving them. Resolution would entail some sort of right-of-way settlement when the two enzymes met; for example, the RNA polymerase could drop off the DNA when a confrontation took place. Avoiding confrontations could be arranged by the orientation of genes such that transcription occurred for the most part in the same direction as DNA replication. That is, DNA replication begins at *oriC* with Y-junctions proceeding to the left and the right until they meet 180° later. Thus, if head-on collisions are to be avoided, genes on the left and right arcs of the bacterial chromosome should be transcribed away from the origin of replication (Box fig. 1).

Brewer presented an analysis of the orientation of genes on the *E. coli* chromosome, and, most recently, D. Zeigler and D. Dean did the same for the chromosome of *Bacillus subtilis.* In *B. subtilis,* 95% (91 of 96) of the genes analyzed were in the proper orientation to avoid a head-on collision of polymerases. Among the exceptions were sporulation genes, genes that would not be transcribed during DNA synthesis and thus whose orientation is not relevant to DNA polymerase activity. In *E. coli,* Brewer found that, overall, 74% (375 of 501) of the genes looked at were oriented in the correct direction to avoid head-on collisions. (Presumably a catch-up collision would mean that the DNA polymerase would slow down until the offending RNA polymerase were finished transcribing, at which point it would drop off the chromosome allowing the DNA polymerase to proceed.) Brewer's data are more impressive when she broke them down according to function and activity of transcription.

For genes that transcribe very actively most of the time, the orientation is about 90% in the "safe" direction. For regulatory genes that are transcribed only very rarely, the orientation is random (50% safe). For other genes the orientation was 72% in the safe direction. Thus there is clearly an organization within the bacterial chromosome to avoid head-on collisions of the two polymerases.

Brewer did provide evidence that a head-on collision of polymerases could be fatal. In studies in which inversions of the *E. coli* chromosome were selected for, it was impossible to isolate inversion mutations that changed the orientation of genes in respect to *oriC.* Thus it appears that a head-on collision of polymerases may not be resolvable by the cell and that evolution has solved the problem by having the transcription of genes generally oriented so that it is in the same direction as DNA replication.

Continued on next page

it is in contact with. The undigested DNA is then isolated and its size determined. This technique is known as **footprinting** (fig. 10.4). Much new information about the nature of recognition regions within promoters has come about with recombinant DNA technology and nucleotide sequencing techniques (see chapter 12). Recent sequencing of numerous promoters has shown that they contain common sequences. If the nucleotide sequences of promoters are aligned with each other, and each has exactly the same series of nucleotides in a given segment, we say that the sequence of that segment is invariant and comprises a **conserved sequence.** If, however, there is some variation in the sequence, but certain nucleotides are present at a high frequency (significantly greater than chance), we refer to those nucleotides as making up a **consensus sequence.** Surrounding a point about ten nucleotides before the first base transcribed is just such a consensus sequence—TATAAT—known as a **Pribnow box** after one of its discoverers (fig. 10.5).

Most of these nucleotides in the Pribnow box are adenines and thymines (fig. 10.5), so the region is primarily held together by only two hydrogen bonds per base pair. Since local DNA denaturation occurs during transcription by RNA polymerase, fewer hydrogen bonds make this process easier energetically. When the polymerase is bound at the promoter region (fig. 10.5), it is in position to begin polymerization six to eight nucleotides down from the Pribnow box.

BOX 10.2—*continued*

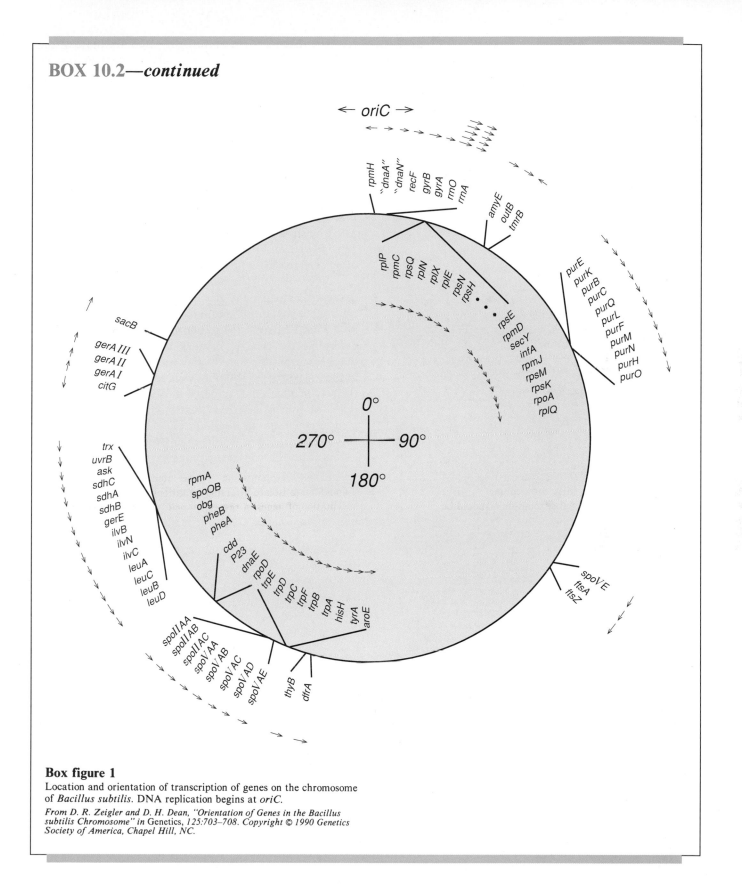

Box figure 1

Location and orientation of transcription of genes on the chromosome of *Bacillus subtilis*. DNA replication begins at *oriC*.

From D. R. Zeigler and D. H. Dean, "Orientation of Genes in the Bacillus subtilis Chromosome" in Genetics, *125:703–708. Copyright © 1990 Genetics Society of America, Chapel Hill, NC.*

Figure 10.4

Footprinting technique. DNA in contact with a protein (e.g., RNA polymerase) is protected from nuclease degradation. The protected DNA is then isolated and characterized.

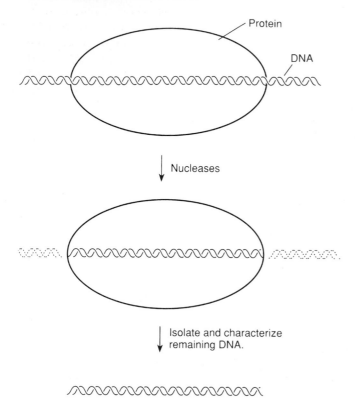

The sequences shown in figure 10.5 are those of the **coding strand** of DNA. It is a general convention to show the coding strand because both that strand and mRNA are complementary to the same **template strand** (also referred to as the **anticoding strand** or **noncoding strand**). The coding strand and the transcribed RNA have the same sequences, substituting U for T in RNA (fig. 10.6). Another convention is to indicate the first base transcribed by number +1 and to use positive numbers to count farther down the DNA in the direction of transcription, the direction referred to as **downstream.** If transcription is proceeding to the right, then the direction to the left is called **upstream,** with bases indicated by negative numbers (fig. 10.7). From this convention, the Pribnow box is often referred to as the −10 sequence.

Figure 10.7 also indicates another region with similar sequences among many promoters centered near −35 and referred to as the −35 sequence. The consensus sequence at −35 is TTGACA. Mutation studies have been done to determine the relative roles of the −10 and −35 sequences in transcription. In other words, mutations of bases in the −10 and −35 regions were looked at to determine the way in which they affected the initiation of transcription. The conclusions from these studies are that both regions contribute to the efficiency of binding by the polymerase. In other words, the more that each sequence differs from the consensus sequence, the less frequently that promoter initiates transcription. The −35 sequence may be the initial point of recognition of the promoter by the holoenzyme, and the −10 sequence is the site of initial melting of the DNA for transcription to begin. Remember that Pribnow boxes are composed mainly of

Figure 10.5

Nucleotide sequences of the promoter region and the first base transcribed on several different DNA sequences. Lambda (λ), T7, and φX174 are phages. *Lac* is an *E. coli* gene and SV40 is an animal virus. Only the SV40 promoter has the consensus sequence of TATAAT. Including other sequenced promoters not shown here, no base is found 100 percent of the time (conserved).

	Promoter region	First base transcribed
λP$_R$	5′ • • • TGGCGGTGATAATGGTTGCATGT • • • 3′
T7A1	• • • CCTATAGGATACTTACAGGCAT • • •
T7A2	• • • CATGCAGTAAGATACAAATCGCTA • • •
φXI74A	• • • TGTATGTTTTCATGCCTCCAAAT • • •
lac	• • • CGGCTCGTATGTTGTGTGGAAT • • •
SV40	• • • TGCAGCTTATAATGGTTACAAATA • • •

Pribnow box

adenine and thymine bases that require the least energy to melt. After transcription begins, the sigma factor is released (fig. 10.8).

Since the holoenzyme recognizes consensus sequences in a promoter, it is not surprising that different promoters are bound more efficiently than others or that there are different sigma factors within a cell. In *E. coli*, the major sigma factor is a protein of 70,000 daltons, referred to as σ^{70}. (One dalton is an atomic mass of 1.0000, approximately equal to the mass of a hydrogen atom.) The existence of other sigma factors provides the cell with a mechanism for the transcription of different genes under different circumstances. For example, in an *E. coli* cell subjected to elevated temperatures, a group of new proteins appears, referred to as **heat-shock proteins,** which probably protect the cell to some extent against the elevated temperatures. These proteins all appear at once because they have promoters recognized by a different sigma factor, one with a molecular weight of 32,000 daltons (σ^{32}). We will discuss heat shock proteins and other systems of transcriptional control in chapters 13 and 15.

Figure 10.6

The template (anticoding) strand of DNA is complementary to both the coding strand and the transcribed RNA. The sequences are from the promoter of the λP$_R$ region (see fig. 10.5).

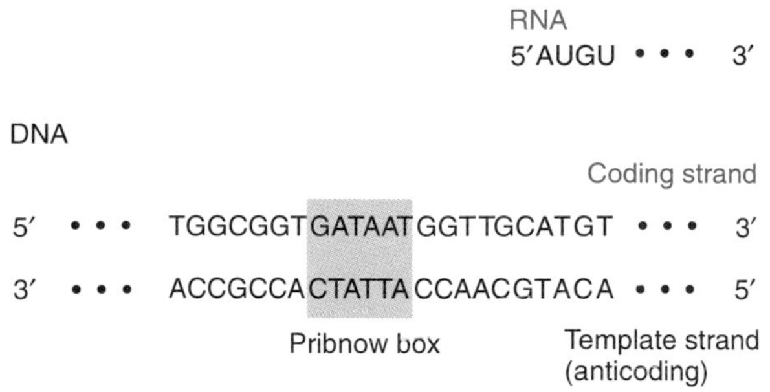

```
                                RNA
                                5′AUGU • • • 3′

DNA
                                         Coding strand

5′ • • •  TGGCGGT GATAAT GGTTGCATGT • • • 3′

3′ • • •  ACCGCCA CTATTA CCAACGTACA • • • 5′

              Pribnow box        Template strand
                                 (anticoding)
```

Figure 10.7

A promoter in the *E. coli oriC* region, with the −35 and −10 sequences shown as well as the first base transcribed. The upstream and downstream directions are indicated.

Transcription, like DNA replication, always proceeds in the 5′ → 3′ direction. That is, new RNA nucleotides are added to 3′-OH free ends as in DNA replication. However, unlike DNA polymerase, RNA polymerase does not seem to proofread as it proceeds—that is, RNA polymerase does not seem to verify the complementarity of the new bases added to the growing RNA strand. This deficiency is not serious since many mRNAs are short-lived and many copies are made from actively transcribed genes. Therefore, an occasional mistake does not produce permanent or overwhelming damage. If a particular RNA is not functional, a new one will be made soon. Evolutionarily speaking, it is probably more important to make RNA quickly than to proofread each RNA made.

As RNA polymerase moves along the DNA, supercoiling in the template DNA molecule is created. The negative supercoiling upstream from the polymerase and the positive supercoiling downstream are presumably alleviated by topoisomerases (fig. 10.9).

Terminators

Transcription continues in a processive manner as nucleotides are added to the growing RNA strand by RNA polymerase according to the rules of complementarity (C, G, A, U of RNA pairing with G, C, T, A of DNA, respectively). The polymerase moves down the DNA until a stop signal, or terminator sequence, is reached by the RNA polymerase. There are two types of terminators, rho-dependent and rho-independent. Their difference lies in their dependency on a protein, the **rho protein** (Greek letter ρ). The functional form of rho is a hexamer, six identical copies of the protein. **Rho-independent terminators** cause termination of transcription even if rho is not present. **Rho-dependent terminators** require the rho protein; without it RNA polymerase continues to transcribe past the terminator, a process known as **readthrough.** Both types of terminators sequenced so far have one thing in common: They

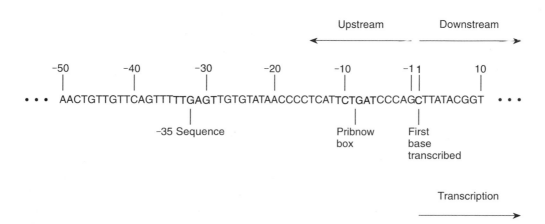

Figure 10.8

Transcription begins after RNA polymerase melts DNA at the −10 sequence. The sigma factor is released as transcription begins.

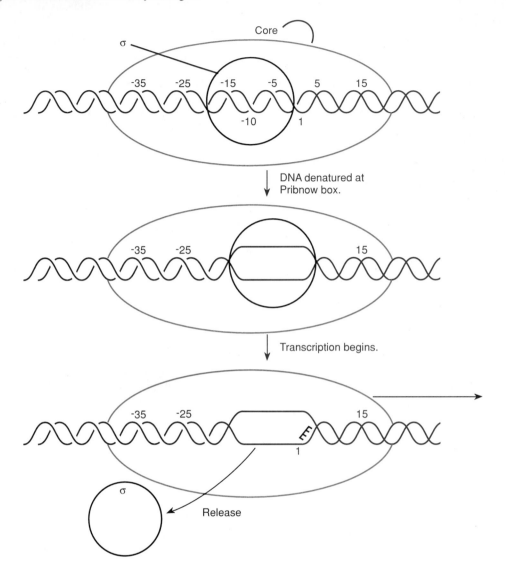

Figure 10.9

RNA polymerase builds up positive supercoils in the downstream direction and leaves negative supercoils in its wake. DNA topoisomerases presumably remove this supercoiling. DNA topoisomerase I tends to alleviate negative supercoils, whereas DNA topoisomerase II tends to alleviate positive supercoils.

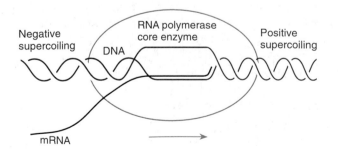

include a sequence and its inverted form separated by a short sequence, a configuration known as an **inverted repeat sequence.** The terminator in figure 10.10 has the sequence AAAGGCTCC, 5′ to 3′, from both the left on the coding strand and from the right on the template strand. There is a four-base-pair sequence separating the repeats. Inverted repeats have the interesting property of being able to form a **stem-loop structure** by the pairing of complementary bases within the transcribed mRNA. In the DNA, a double stem-loop structure, called a *cruciform structure* (cross-shaped), can form (fig. 10.11).

Both rho-dependent and rho-independent terminators have the stem-loop structure in RNA just before the last base transcribed. Rho-independent terminators, as

Figure 10.10

An inverted-repeat base sequence characterizes terminator regions of DNA. Stem-loop structures can occur in DNA or RNA. The double stem-loop in DNA is called a cruciform structure (*cross-shaped*).

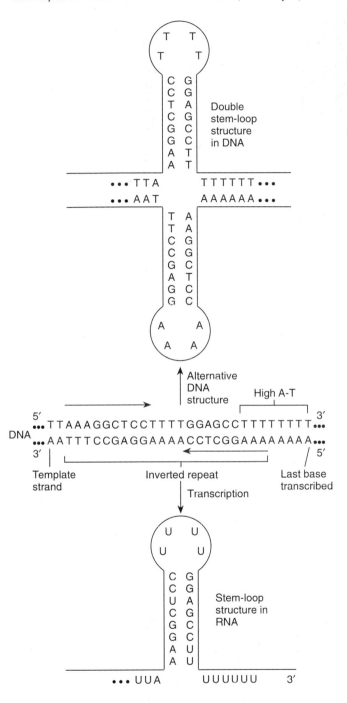

Figure 10.11

Electron micrograph of cruciform DNA

From B. Lewin, 1987. Genes III, fig. 3.7, p. 65. John Wiley & Sons, Inc., New York. Photo by Martin Gellert.

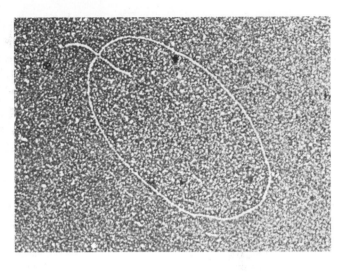

In rho-independent terminators, the pause may occur just after the sequence of uracils is transcribed (fig. 10.12). Uracil-adenine base pairs have two hydrogen bonds and are less stable thermodynamically than guanine-cytosine base pairs. Perhaps during the pause, the uracil-adenine base pairs spontaneously denature, releasing the transcribed RNA. The RNA polymerase then could fall free of the DNA, terminating the process and making the polymerase available for further transcription of other promoters.

Rho-dependent terminators do not have the uracil sequence after the stem-loop structure. Here termination is dependent on the action of rho, which appears to bind to the 5′ end of the newly formed RNA. In an ATP-dependent process, rho travels along the RNA at a speed comparable to the transcription process itself (fig. 10.12). Possibly, when RNA polymerase pauses at the stem-loop structure, rho catches up to the polymerase and unwinds the DNA-RNA hybrid, with the DNA, RNA, and polymerase falling free of each other. It is known that rho can do this because it has DNA-RNA helicase (unwinding) properties.

An overview of transcription is shown in figure 10.13. The information of a gene, coded in the sequence of nucleotides in the DNA, has been transferred in the process of transcription to a complementary sequence of nucleotides in the RNA. This RNA transcript contains a complement of the template strand of the DNA of a gene and thus justifies its being termed a messenger. The transcript contains sequences of nucleotides that will be translated into amino acids as well as segments before and after the coding segment. The translatable segment, or gene, always begins with a three-base sequence, AUG, which is known

shown in figure 10.10, also have a sequence of thymine-containing nucleotides after the inverted repeat, whereas rho-dependent terminators do not. Although the exact sequence of events at the terminator is unknown, it appears that the stem-loop structure of RNA causes the RNA polymerase to pause just after completing it. This pause may then allow termination under two different circumstances.

Figure 10.12

Rho-independent (*top*) and rho-dependent (*bottom*) termination of transcription are preceded by a pause of the RNA polymerase at a terminator sequence.

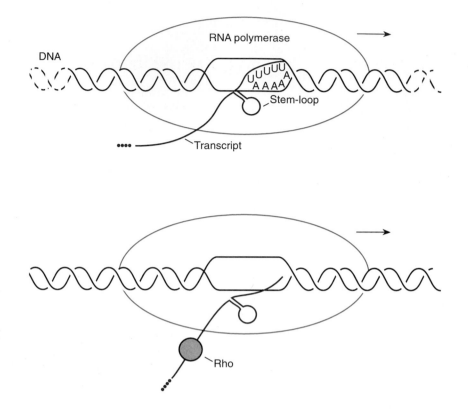

Figure 10.13

Transcription overview and RNA polymerase molecules. RNA polymerase is transcribing near the promoter. The rho factor is shown on the newly formed RNA and the sigma factor is shown nearby, detached from the core polymerase.

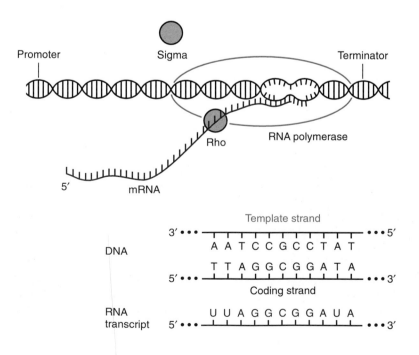

Figure 10.14

Transcribed piece of prokaryotic RNA and its DNA template region. Note the promoter and terminator regions on the DNA and the leader and trailer regions of the RNA. The initiation and nonsense codons are shown.

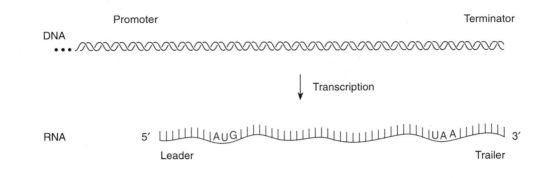

as an initiator codon, and ends with one of the three-base sequences, UAA, UAG, or UGA, known as nonsense codons. (The discussion of these signals will be expanded in chapter 11.)

The portion of the RNA transcript that begins at the start of transcription and goes to the translation initiator codon (AUG) is referred to as a **leader.** The length of RNA from the nonsense codon (UAA, UAG, or UGA) to the last nucleotide transcribed is the **trailer.** These sequences play a role in recognition and structural stability of the mRNA at the ribosome during the process of translation: The leader region can also have regulatory functions (see chapter 13). A complete prokaryotic RNA transcript is diagrammed in figure 10.14. In this simplified drawing the transcript is shown as having only one gene present (AUG → UAA). However, the average prokaryotic transcript has the information for several genes. More will be said about the parts of a transcript later in this chapter and the next. Now we turn our attention to the types of transcripts: ribosomal, transfer, and messenger RNA.

RIBOSOMES AND RIBOSOMAL RNA

Ribosomes are organelles in the cell, composed of proteins and RNA (ribosomal RNA or rRNA), where protein synthesis occurs (the topic of chapter 11). In a rapidly growing *E. coli* cell, ribosomes can make up as much as 25% of the mass of the cell. Ribosomes, as well as other small particles and molecules, have their sizes measured in units that describe their rate of sedimentation during density-gradient centrifugation in sucrose. This technique is used because it gives information on size and shape (due to speed of sedimentation) while simultaneously isolating the molecules being centrifuged. Isolation by centrifugation is a relatively gentle isolation technique; the isolated molecules still retain their biological properties and can then be used for further experimentation. Ultracentrifugation

was developed in the 1920s by the physical chemist Svedberg, and the unit of sedimentation is given his name, the **Svedberg unit,** S.

In sucrose density-gradient centrifugation, the gradient is formed by layering on decreasingly concentrated sucrose solutions. In cesium chloride density-gradient centrifugation, mentioned in the previous chapter, the gradient develops during centrifugation. The sucrose centrifugation is stopped after a fixed time, whereas in the cesium chloride technique the system spins until equilibrium is reached: A substance stays at a particular level in the gradient because of its density rather than its size or shape. The sucrose method tends to be more rapid. Samples can be isolated from a sucrose gradient by punching a hole in the bottom of the tube and collecting the drops in sequentially numbered containers. The first (lowest-numbered) containers will have the heaviest molecules (highest S values).

Ribosomal Subunits

Ribosomes in all organisms are made of two subunits of unequal size. The sedimentation value of the large one in *E. coli* is 50S (50 Svedberg units) and 30S for the smaller one. Together they sediment at about 70S. Eukaryotic ribosomes vary from 55S to 66S in animals and 70S to 80S in fungi and higher plants. Most of our discussion will be confined to the well-studied ribosomes of *E. coli*.

Each ribosomal subunit comprises one or two pieces of rRNA and a fixed number of proteins. The 30S subunit has twenty-one proteins and a 16S molecule of rRNA and the 50S subunit has thirty-four proteins and two pieces of rRNA—one 23S and one 5S section (fig. 10.15). Advances in understanding ribosome structure have come about after protein chemists isolated and purified all the proteins of the ribosome. Completion of this step allowed

Figure 10.15

The *E. coli* ribosome; (*a*) and (*b*) are models of the 70S ribosome showing the relationship of the small and large subunits.

(a) and (b) James A. Lake, Journal of Molecular Biology *105 (1976):131–159. Reproduced by permission of Academic Press.*

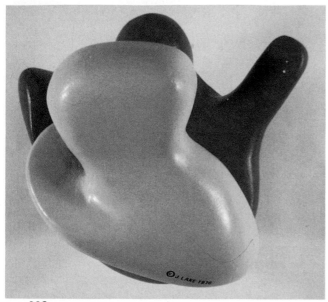

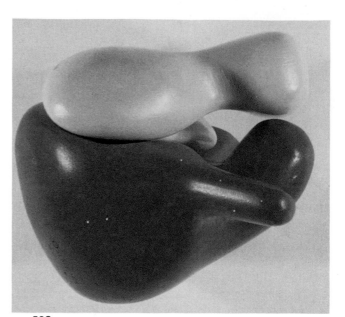

30S
21 proteins
1 16S RNA

50S
34 proteins
1 23S rRNA
1 5S rRNA

Figure 10.16

The *E. coli* transcript that contains the three ribosomal RNA segments also contains four tRNAs and some spacer RNA (*red*) to separate the tRNA and rRNA pieces.

experimentation on the proper sequence needed to assemble the subunits and also allowed immunological techniques to show the positions of many of the proteins in the completed ribosomal subunits.

The Nucleolus in Eukaryotes

In *E. coli,* all three rRNA segments are transcribed as a single long piece of RNA that is then cleaved and modified to form the final three pieces (16S, 23S, and 5S) of rRNA. The region of DNA that contains the three rRNA molecules also contains genes for four tRNAs (fig. 10.16). There appear to be about five to ten copies of this region in each chromosome of *E. coli.* The occurrence of the three rRNA segments on the same piece of RNA assures a final ratio of 1:1:1, the ratio needed in ribosome construction.

In eukaryotes there are four segments of rRNA in the ribosome. The smaller ribosomal subunit has an 18S piece of RNA and the larger subunit has 5S, 5.8S, and 28S seg-

ments. All but the 5S rRNA section are transcribed as part of the same piece of RNA. Eukaryotic cells, however, have many copies of these rRNA genes, depending on the species. For example, the fruit fly, *Drosophila melanogaster,* has about 130 copies of the DNA region from which the larger segments of rRNA are transcribed. These regions occur in tandem on the sex (X and Y) chromosomes and are known collectively as the nucleolar organizer (see chapter 3). The smallest rRNA subunit is also produced from a duplicated gene, but at a different point in the genome. For example, in *D. melanogaster* the 5S subunit is produced on chromosome 2.

Although prokaryotes have two RNA polymerases, only one produces RNA transcripts. The other, primase, functions during DNA synthesis (see chapter 9). Eukaryotes have three RNA polymerases; each is composed of more subunits (8–14) than the prokaryotic form. Eukaryotic RNA polymerase I (or polymerase A) transcribes only the nucleolar organizer DNA. RNA polymerase III (or

TABLE 10.3 Prokaryotic and Eukaryotic
RNA Polymerases

Enzyme	Function
Prokaryotic	
RNA polymerase	Transcription
Primase	Primer synthesis during DNA replication
Eukaryotic	
RNA polymerase I	Transcribes nucleolar organizer
RNA polymerase II	Transcribes most genes
RNA polymerase III	Transcribes 5S rRNA and tRNA genes
Primase	Primer synthesis during DNA replication

Figure 10.17

Transcription in the nucleolus of the newt, *Triturus*. The polarity of
the process (progressing from small to large transcripts) as well as the
spacer DNA (*thin lines* between transcribing areas) is clearly visible.
Magnification 18,000×.

Cover photo Science, *Vol. 164, pages 955–957, 23 May 1969, "Visualization of
Nucleolar Genes." O. L. Miller, Jr. and B. R. Beatty. Copyright 1969 by AAAS.*

polymerase C) transcribes the 5S RNA gene and tRNA
genes. RNA polymerase II (or polymerase B) transcribes
all other genes (table 10.3). In addition, mitochondria,
chloroplasts, and some phages have other RNA polymerases.

At the nucleolar organizer the nucleolus forms, the
familiar dark blob found in the nuclei of eukaryotes. The
nucleolus is the place of assembly of ribosomes. The various ribosomal proteins that have been manufactured in
the cytoplasm migrate to the nucleus and eventually to the
nucleolus, where, with the final forms of the rRNAs, they
are assembled into ribosomes.

In the nucleolar organizer, each repeat of the large
rRNA gene is separated by a region of spacer DNA that
is not transcribed. This is shown in figure 10.17 and diagrammed in figure 10.18. In the electron micrograph of
figure 10.17, the polarity of transcription is evident by the
fact that the RNA is long at one end of the transcribing
segment and short at the other end with a uniform gradation between. Notice that many RNA polymerases are
transcribing each region at the same time. The regions between the transcribed DNA segments are the spacer DNA
regions.

TRANSFER RNA

During protein synthesis (fig. 10.2), amino acids are
brought to the site of protein synthesis, the ribosome, attached to tRNAs. A messenger RNA, with the code of
information from the gene (DNA), will be bound to the
ribosome. The code is read in sequences of three nucleotides, called **codons.** The nucleotides of the codon on
mRNA are complementary to and pair with a sequence
of three bases—the **anticodon**—on a tRNA. Each different tRNA carries a specific amino acid. Thus the specificity of the genetic code is recognized by the tRNA (fig.
10.19).

The correct amino acid is attached to its tRNA by
one of a group of enzymes called **aminoacyl-tRNA synthetases.** There is one specific aminoacyl synthetase for
every amino acid, but the synthetase may recognize more
than one tRNA because there are more tRNAs (and
codons) than there are amino acids. (In chapter 11 we will
discuss the genetic code in more detail.) R. W. Holley, a
Nobel laureate, and his colleagues were the first to discover the nucleotide sequence of a tRNA; in 1964 they
published the structure of the alanine tRNA in yeast (fig.
10.20). The average tRNA is about eighty nucleotides
long.

Similarities of All tRNAs

There are several unusual properties of transfer RNA. For
one, all the different tRNAs of a cell have the same general shape; when purified, the heterogeneous mixture of
all of a cell's tRNAs can form very regular crystals. The
regularity of the shape of tRNAs makes sense. During the
process of protein synthesis, two tRNAs will attach next
to each other on a ribosome and a peptide bond will be
formed between their amino acids. Thus any two tRNAs

Figure 10.18

Details of the transcription of the large rRNA genes shown in figure 10.17. Note the polarity of the process and the spacer DNA.

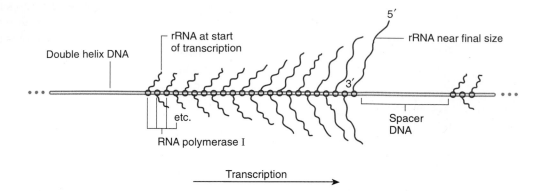

Double helix DNA

rRNA at start of transcription

rRNA near final size

5′

3′

etc.

RNA polymerase I

Spacer DNA

Transcription

Figure 10.19

Specificity of the genetic code manifests itself in the tRNA in which a particular anticodon is associated with a particular amino acid. In this case, glutamic acid is attached to its proper tRNA that has the anticodon CUU.

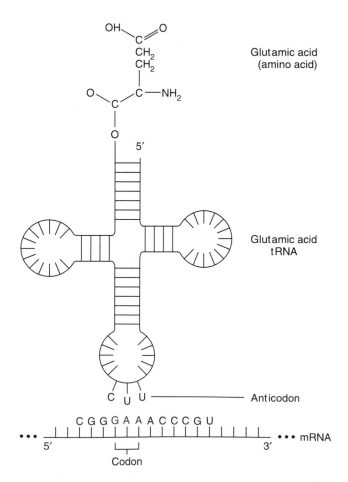

Glutamic acid (amino acid)

Glutamic acid tRNA

Anticodon

C G G G A A A C C C G U

5′   3′   mRNA

Codon

must have the same general dimensions as well as similar structures so that they can be recognized and positioned at the ribosome.

An obvious feature of the tRNA in figure 10.20 is that it has **unusual bases.** This tRNA is originally transcribed from DNA as a molecule about 50% longer than the final eighty nucleotides. In fact, some transcripts contain two copies of the same tRNA, or sometimes several different tRNA genes are part of the same transcript (see fig. 10.16). The transcription of tRNAs is completely regular: It does not involve unusual bases. The transcript is then processed down to the final size of a tRNA by various nucleases that remove trailing and leading pieces of RNA. Then the shortened piece of RNA is further modified, frequently by the addition of methyl groups to the bases already in the RNA (fig. 10.21). Presumably, these unusual bases disrupt normal base pairing and are in part responsible for the loops formed by unpaired bases (fig. 10.20).

Transfer RNA Loops

It is believed that the first loop on the 3′ side (the T- or T-ψ-C-loop) is involved in recognition by the ribosome, which must hold each tRNA in the proper orientation so that complementarity of the anticodon of the tRNA and the codon of the mRNA can be checked. The center loop is the anticodon loop. The aminoacyl-tRNA synthetases seem to recognize many points all over the tRNA molecule (see Box 11.2 on the "second genetic code").

Every tRNA so far examined starts with an adenine-cytosine-cytosine sequence on the 3′ end to which the amino acid is attached. The ribosome-binding loop on all tRNAs has the T-ψ-C-G sequence. The anticodon on all

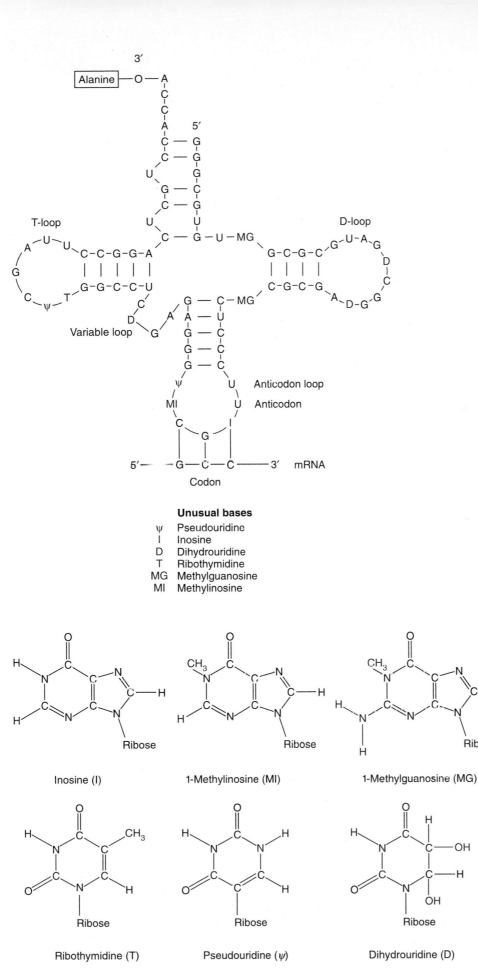

Figure 10.20
Structure and sequences of alanine tRNA in yeast. Note the modified bases in the loops.
Source: Data from Holley, et al., 1965.

Unusual bases

ψ   Pseudouridine
I   Inosine
D   Dihydrouridine
T   Ribothymidine
MG   Methylguanosine
MI   Methylinosine

Figure 10.21
Structures of the modified nucleotides found in alanine tRNA of yeast

Inosine (I)

1-Methylinosine (MI)

1-Methylguanosine (MG)

Ribothymidine (T)

Pseudouridine (ψ)

Dihydrouridine (D)

Figure 10.22
Stereo (*a*) and linear (*b*) views of yeast phenylalanine tRNA. In (*b*)
bases are numbered sequentially from the 5′ end. To make the stereo
view three-dimensional, fuse the images by crossing your eyes.

(*a*) Source: Joel L. Sussman and S. H. Kim, "Three-Dimensional Structure of a
Transfer RNA in Two Crystal Forms," Science vol. 192 (May 28, 1976):853–
858. Figs. 3 and 5 copyright 1976 by the AAAS. Reproduced by permission.
(*b*) From J. L. Sussman and S. H. Kim, "Three-Dimensional Structure of a
Transfer RNA in Two Crystal Forms" in Science, 192, May 28, 1976.
Copyright 1976 by the AAAS.

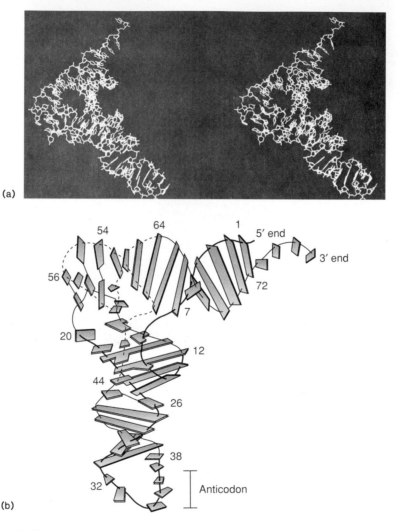

is bounded by uracil on the 5′ side and a purine on the 3′
side. Thus there is a good deal of general similarity among
all the tRNAs, consistent with the fact that they all enter
into protein synthesis in the same way.

The actual shape of the functional tRNA in the cell
is not an open cloverleaf as shown in figure 10.20; rather,
there is helical twisting of the whole molecule. Figure
10.22 shows a three-dimensional structure of a tRNA.

Earlier we considered a rough definition of a gene as
that length of DNA that codes for one protein. But we
have just encountered an inconsistency—both tRNAs and
rRNAs are coded for by genes, yet neither is eventually
translated into a protein. Their transcripts function as final
products without ever being translated. Thus tRNA and
rRNA are the major exceptions to the general rule that a
gene codes for a protein.

EUKARYOTIC TRANSCRIPTION

Although all aspects of the process of transcription differ
to some extent between prokaryotes and eukaryotes, there
are two major differences between the groups that we will
look at here: the coupling of transcription and translation
that is possible in prokaryotes and the extensive **posttranscriptional modifications** that occur in eukaryotic mRNA
(see below). In *E. coli,* translation of the newly transcribed mRNA into a protein can take place before transcription is complete (fig. 10.23). The mRNA is
transcribed in the 5′ → 3′ direction, and it is near the 5′
end that translation begins. As soon as the transcribed 5′
end of the RNA is available, a ribosome attaches to the
mRNA and moves along it in the 5′ → 3′ direction,
lengthening the growing polypeptide as it moves. When

Figure 10.23

(*a*) In prokaryotes, translation of messenger RNA by ribosomes begins before transcription is completed. Ribosomes attach to the growing mRNA strand when the 5′ end becomes accessible. They then move along the RNA as it elongates. When the first ribosome moves from the 5′ end, a second ribosome can attach and so on. (*b*) Electron micrograph of events diagrammed in (*a*). The growing peptides cannot be seen in this preparation. Magnification 44,000×.

(b) Source: Miller, O. L., Jr., B. A. Hamkalo, and C. A. Thomas, Jr. (1970), "Visualization of bacterial genes in action," Science 169: 392–395. Fig. 2. Copyright 1970 by the AAAS.

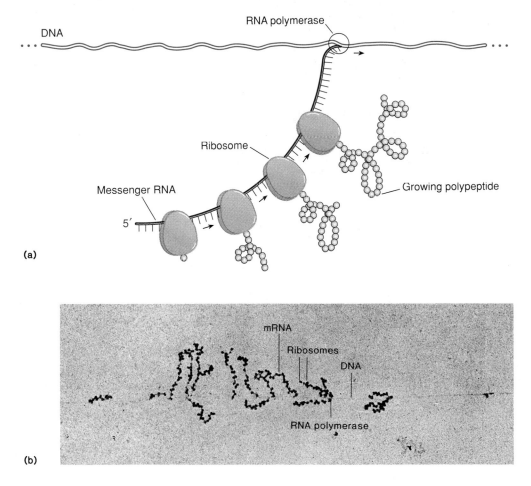

(a)

(b)

the first ribosome moves away from the 5′ end of the transcript, a second ribosome can attach and begin translation. These processes are repetitive. Electron micrographs (fig. 10.23b) show this clearly. In eukaryotes, however, mRNA is synthesized in the nucleus, but protein synthesis takes place in the cytoplasm. We do not have this regional division of labor in *E. coli* because, among other reasons, the bacterium has no nucleus.

Promoters

The promoters in eukaryotes have regions similar to the −10 and −35 sequences in prokaryotic promoters as well as other consensus sequences. In eukaryotes the sequence TATA has been found to be relatively invariant (a con-

sensus sequence) in promoters of RNA polymerase II, the system we will discuss here. This sequence has been called the **TATA box** or **Hogness box,** after its discoverer, D. Hogness. The TATA box is centered at about 25 bases upstream from the start of transcription. Farther upstream, at about −70, is another sequence conserved in many eukaryotic promoters: GG(C or T)CAATCT, called the **CAAT box** (fig. 10.24). These promoter regions have been analyzed by sequencing the DNA, evaluating the effect of mutations, and footprinting. Animal viruses, such as SV40 (simian virus 40) and adenovirus, have proved useful model systems to work with because they are simple, yet since they function within eukaryotic host cells, they are effectively "eukaryotic," just as bacteriophages are effectively "prokaryotic."

Figure 10.24

Promoter sequences recognized by RNA polymerase II in eukaryotes. The CAAT box is at −70 and the Hogness (TATA) box is at about −25.

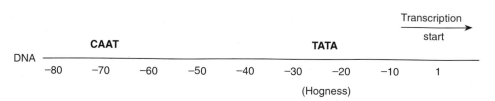

Figure 10.25

The promoter of a eukaryotic gene transcribed by RNA polymerase II. TFIID binds at the TATA box. TFIIA, another transcription factor, seems to bind to TFIID. Two other transcription factors, TFIIB and TFIIE, seem to bind directly to RNA polymerase and perhaps also to each other. The enhancer binding protein might bind to the polymerase, TFIID, or the DNA at the promoter.

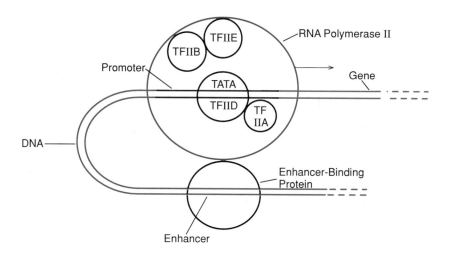

The distance spanned by the CAAT and TATA boxes is larger than the polymerase molecule, leading to an uncertainty as to how the polymerase recognizes both sequences at the start of transcription. In one model, the polymerase attaches to the DNA at the CAAT box and then moves down the DNA to the TATA box where it begins the actual transcription process. Proteins similar to sigma factors, but called **transcription factors,** recognize eukaryotic promoter sequences. There are many eukaryotic transcription factors. For example, in promoters for genes transcribed by polymerase II, a transcription factor, similar to the prokaryotic σ factor, called TFIID, binds at the TATA box. Unlike in prokaryotes, several other transcription factors are also needed by eukaryotic polymerases.

Other upstream DNA sequences (upstream elements) are involved in eukaryotic transcription. Some of these upstream elements, known as **enhancers,** appear to work even though they may be very far removed from the promoter whose activity they affect. For example, a re-peated seventy-two base-pair sequence (sequentially or tandemly repeated) in SV40 enhances transcription even though it is normally two hundred base pairs upstream. This region also promotes transcription even if it is moved still farther upstream to any point on the DNA, or one of the repeats is removed, or, in fact, if the sequence is inverted. Its function and generality are currently being studied. Presumably, proteins that bind to enhancers are interacting with the machinery for transcription by binding either with the TATA region of DNA, TFIID, or the polymerase itself, while the intervening DNA is looped out of the way (fig. 10.25). The control of eukaryotic transcription is a very active area of research that we will return to again in chapter 15.

Little is known about termination of transcription in eukaryotes, although several terminators have been isolated in SV40 and some other organisms. They appear very similar to the rho-independent terminators of prokaryotes, yielding an RNA having stem-loop structure and a sequence of uracil bases.

Figure 10.26

A cap of 7-methyl guanosine is added in the "wrong" direction to the 5′ end of eukaryotic mRNA. In some cases, the 2′-OH groups on the second or second and third riboses (*red*) are methylated.

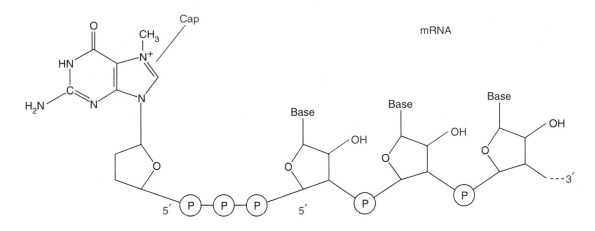

Figure 10.27

Eukaryotic DNA, with intervening sequences, is transcribed and the transcript is then modified in the nucleus before the mRNA is transported into the cytoplasm and translated. Modification consists of splicing, 5′ capping, and 3′ polyadenylation.

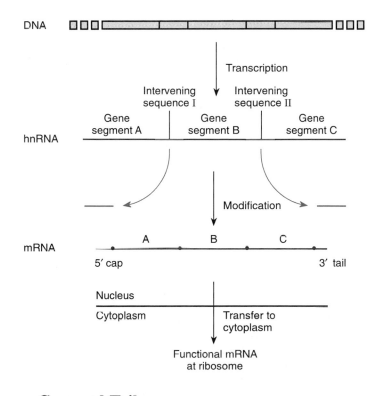

Caps and Tails

Eukaryotic transcription results in a **primary transcript.** In contrast to most prokaryotic transcripts that contain several genes, virtually all eukaryotic transcripts contain the information of just one gene. Three major changes are made to primary transcripts before transport into the cytoplasm: modifications to the 5′ and 3′ ends, and removal of intervening sequences. We refer to these changes as posttranscriptional modifications.

At the 3′ end of most eukaryotic transcripts, a sequence of twenty to two hundred adenine-containing nucleotides, known as a **poly-A tail,** is added by the enzyme poly-A polymerase. Polyadenylation takes place after the 3′ end of the transcript is removed by a nuclease that cuts about twenty nucleotides downstream from the signal 5′-AAUAAA-3′. The tail either adds stability to the molecule or aids somehow in transportation from the nucleus. At the other end, the 5′ end, 7-methyl guanosine is added in the "wrong" direction, 5′ to 5′ (fig. 10.26). This **cap** presumably plays a role in either protecting the mRNA from nuclease degradation, recognition of the mRNA by the ribosome, or transportation of the RNA in or out of the nucleus.

When messenger RNAs were first being studied in eukaryotes, the mRNAs in the nucleus were found to be much larger than those in the cytoplasm and were called **heterogeneous nuclear mRNAs,** or **hnRNAs.** It now turns out that these were primary transcripts, RNAs that had not had any of the major posttranscriptional modifications, including intron removal. In essence, they were pre-mRNAs.

Introns

In eukaryotes, there are segments of DNA within genes that are transcribed into RNA but are never translated into protein sequences. These **intervening sequences,** or **introns,** are removed from the RNA in the nucleus before its transport into the cytoplasm (fig. 10.27). They were first discovered by P. Sharp and his colleagues at MIT and T. Broker and L. Chow and their colleagues at the Cold

Philip A. Sharp (1944–)
Courtesy of Dr. Philip A. Sharp.

Thomas Cech (1947–)
Courtesy of Dr. Thomas Cech. Photo by Ken Abbott.

Thomas Broker (1944–)
Courtesy of Dr. Thomas Broker.

Louise T. Chow (1943–)
Courtesy of Dr. Louise Chow.

Sidney Altman (1939–)
Courtesy of Dr. Sidney Altman. Photo: Michael Marsland, Yale University Office of Public Affairs.

Spring Harbor Laboratory in 1977. An example is given in figure 10.28. The segments of the gene between introns, which are transcribed and translated (expressed), are termed **exons.** The results of intron removal can be seen clearly when an mRNA that has had its introns removed is hybridized with the original gene (fig. 10.29). The DNA forms double-stranded structures with the exons in RNA. The introns in DNA have nothing to pair with in the RNA so they form single-stranded loops. Introns also occur in eukaryotic tRNA and rRNA genes.

For introns to be removed, the ends of the exons must be brought together and connected. This process is called *splicing.* At least two ways of splicing occur, although they may be related: self-splicing and enzyme-mediated splicing.

Self-splicing

Self-splicing by RNA was discovered by Thomas Cech and his colleagues in 1982 building on the work of others, including Sidney Altman, who showed that RNA can have catalytic properties. (Cech and Altman were awarded the 1989 Nobel prize in chemistry.) Working with an intron in the 35S rRNA precursor in the ciliated protozoan, *Tetrahymena,* Cech and his colleagues found that they could get intron removal *in vitro* with no proteins present. A guanine-containing nucleotide (GMP, GDP, or GTP) had to be present. Figure 10.30 is a diagram of how the process of self-splicing occurs. The intron is acting as an enzyme; we call an RNA with enzymatic properties a **ribozyme.**

During self-splicing, the U-A bond at the left (5′) side of the intron is transferred to the GTP. The U that is now unbonded displaces the G at the right (3′) side of the intron, reconnecting the RNA with a U-U connection and releasing the intron (fig. 10.30). Since all bonds are transfers rather than new bonds, no external energy source is required. Self-splicing introns of this type are called **group I introns.** There is an extensive secondary structure (RNA stem-loops) that forms that is important in intron removal (see Box 10.3).

Figure 10.28

Nucleotide sequence of mouse β-globin major gene. The coding DNA strand is shown; cAp (position 79) indicates the start of the capped mRNA; pA indicates the start of the poly-A tail (position 1467); numbers inside the sequence are adjacent amino acid positions; Ter is the termination codon (position 1334). The three-letter abbreviations (e.g., Met, Val, His) refer to amino acids (see chapter 11).

From David A. Konkel, et al., "The Sequence of the Chromosomal Mouse β-Globin Major Gene: Homologies in Capping, Splicing and Poly (A) Sites" in Cell, 15:1125–1132. Copyright © 1978 Cell Press, Cambridge, M.A. Reprinted by permission.

```
                10        20        30        40        50        60        70        80        90
                                                                                      cAp
   0  GGCCAATCTGCTCACACAGGATAGAGAGGGCAGGAGCCAGGCAGAGCATATAAGGTGAGGTAGGATCAGTTGCTCCTCACATTTGCTTCTGACATAGTTG

 100  TGTTGACTCACAACCCCAGAAACAGACATCATGGTGCACCTGACTGATGCTGAGAAGGCTGCTGTCTCTTGCCTGTGGGGAAAGGTGAACTCCATGAAG
                                    MetValHisLeuThrAspAlaGluLysAlaAlaValSerCysLeuTrpGlyLysValAsnSerAspGluV

 200  TTGGTGGTGAGGCCCTGGGCAGGTTGGTATCCAGGTTACAAGGCAGCTCAAGAAGAAGTTGGGTGCTTGGAGACAGAGGTCTGCTTTCCAGCAGACAC
      alGlyGlyGluAlaLeuGlyArg  30

 300  TAACTTTCAGTGTCCCCTGTCTATGTTTCCCTTTTTAGGCTGCTGGTTGTCTACCCTTGGACCCAGCGGTACTTTGATAGCTTTGGAGACCTATCCTCTG
                                 31 LeuLeuValValTyrProTrpThrGlnArgTyrPheAspSerLeuLysGlyTh

 400  CCTCTGCTATCATGGGTAATGCCAAAGTGAAGGCCCATGGCAAGAAGGTGATAACTGCCTTTAACGATGGCCTGAATCACTTGGACAGCCTCAAGGGCAC
      laSerAlaIleMetGlyAsnAlaLysValLysAlaHisGlyLysLysValIleThrAlaPheAsnAspGlyLeuAsnHisLeuAspSerLeuLysGlyTh

 500  CTTTGCCAGCCTCAGTGAGCTCCACTGTGACAAGCTGCATGTGGATCCTGAGAACTTCAGGGTGAGTCTGATGGGCACCTCCTGGGTTTCCTTCCCCTGC
      rPheAlaSerLeuSerGluLeuHisCysAspLysLeuHisValAspProGluAsnPheArg  104

 600  TATTCTGCTCAACCTTCCTATCAGAAAAAAAGGGGAAGCGATTCTAGGGAGCAGTCTCCATGACTGTGTGTGGAGTGTTGACAAGAGTTCGGATATTTTA

 700  TTCTCTACTCAGAATTGCTGCTCCCCCTCACTCTGTTCTGTGTTGTCATTTCCTCTTTCTTTGGTAAGCTTTTTAATTTCCAGTTGCATTTTACTAAATT

 800  AATTAAGCTGGTTATTTACTTCCCATCCTGATATCAGCTTCCCCTCCTCCTTTCCTCCCAGTCCTTCTCTCTCTCCTCTCTCTTTCTCTAATCCTTTCCT

 900  TTCCCTCAGTTCATTCTCTCTTGATCTACGTTTGTTTGTCTTTTTAAATATTGCCTTGTAACTTGCTCAGAGGACAAGGAAGATATGTCCCTGTTTCTTC

1000  TCATAGCTCAAGAATAGTAGCATAATTGGCTTTTATGCAGGGTGACAGGGGAAGAATATATTTTACATATAAATTCTGTTTGACATAGGATTCTTGTGGT

1100  GGTTTGTCCAGTTTAAGGTTGCAAACAAATGTCTTTGTAAATAAGCCTGCAGGTATCTGGTATTTTTGCTCTACAGTTATGTTGATGGTTCTTCCATATT

1200  CCCACAGCTCCTGGGCAATATGATCGTGATTGTGCTGGGCCACCTTGGCAAGGATTTCACCCCCGCTGCACAGGCTGCCTTCCAGAAGGTGGTGGCT
       105 LeuLeuGlyAsnMetIleValIleValLeuGlyHisHisLeuGlyLysAspPheThrProAlaAlaPheGlnLysValValAla

1300  GGAGTGGCCACTGCCTTGGCTCACAAGTACCACTAAACCCCCTTTCCTGCTCTTGCCTGTGAACAATGGTTAATTGTTCCCAAGAGAGCATCTGTCAGTT
      GlyValAlaThrAlaLeuAlaHisLysTyrHisTer                                       pA

1400  GTTGGCAAAATGATAGACATTTGAAAATCTGTCTTCTGACAAATAAAAAGCATTTATGTTCACTGCAATGATGTTTTAAATTATTTGTCTGTGTCATAGA

1500  AGGGTTTATGCTAAGTTTTCAAGATACAAAGAAGTGAGGGTTCAGGTCTCGACCTTGGGGAAATAAA
```

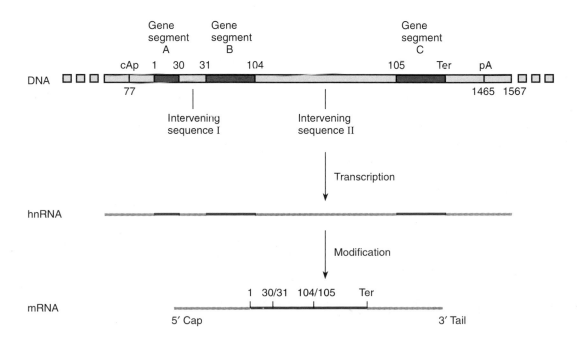

Figure 10.29

The mRNA of adenovirus hybridized with its DNA. Three introns are visible as single-stranded DNA loops. They form single-stranded loops because they have nothing in the RNA molecule to hybridize with. Also visible is the poly-A tail of the mRNA. (*a*) Electron micrograph, (*b*) explanatory diagram.

(a) Courtesy of Louise T. Chow and Thomas Broker.

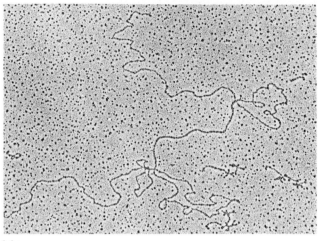

(a)

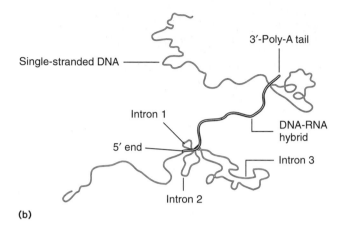

(b)

Figure 10.30

Self-splicing of an rRNA precursor in *Tetrahymena*. An external GTP is required. Two bond transfers result in a shortened RNA and a free intron.

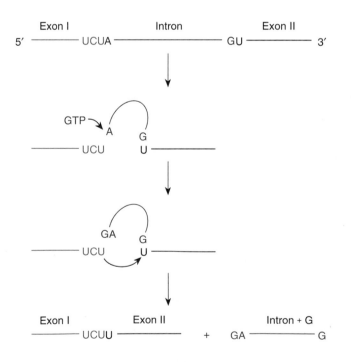

It is noteworthy that the spliced-out intron is capable of continued enzymatic activity using its own sequence as a template and acting on RNAs with specific complementary sequences: It can act like an endonuclease, splitting RNA strands (fig. 10.31), and it can act like a polymerase, transferase, or ligase by binding RNA nucleotides together. The ability of RNA to selectively cut other RNAs has led to the term *molecular scissors*. This phenomenon could lead to clinical treatment. In fact, a hammerhead ribozyme (so named because of its shape—fig. 10.31) has been shown to be capable of disabling the AIDS virus in culture.

The ability of RNA to act as an enzyme has led to another advancement in our thinking. That is, RNA was probably the original genetic material several billion years ago since autocatalytic properties are not present in DNA. We discuss this further in the section on intron function and structure.

Self-splicing has also been found in genes in the mitochondria of yeast. These introns are referred to as **group II introns** because they use a different mechanism of splicing, one that does not require an external nucleotide. Instead, the first bond is transferred to an adenosine within the intron, forming a lariat structure (fig. 10.32). In order for the lariat to form, the ribose of the adenosine must make three phosphodiester bonds (fig. 10.33).

Figure 10.31

So-called "hammerhead" ribozyme (E) and its target substrate (S). The target is cleaved at the arrow. Roman numerals I, II, and III are double-helical segments; boxed areas are conserved sequences.

From W. A. Pieken, et al., "Kinetic Characterization of Ribonuclease-Resistant 2'-Modified Hammerhead Ribozymes" in Science, 253:315. Copyright 1991 by the AAAS.

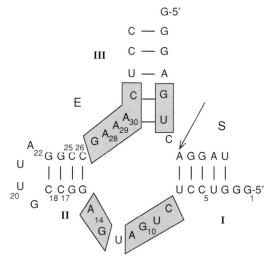

Figure 10.32

Self-splicing of a group II intron results in a lariat configuration of the released intron. No external GTP is required since the first bond transfer takes place with an internal nucleotide, forming the loop of the lariat. A second bond transfer releases the intron.

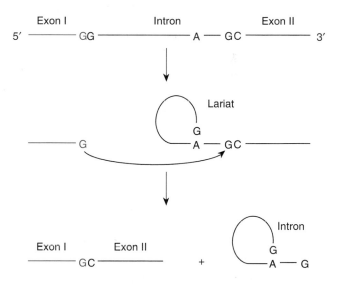

The Spliceosome

Nuclear mRNAs also have their introns removed by way of a lariat structure. Unlike the mitochondrial group II introns, however, nuclear mRNAs have their introns removed with the help of a protein-RNA complex called a **spliceosome,** named by J. Abelson and E. Brody. Let us look at the structure of nuclear introns and then the structure and functioning of the spliceosome.

As mentioned previously, self-splicing seems to be aided by an extensive secondary structure that brings the ends of the introns into close proximity. Spliceosome-dependent introns depend on interactions within the spliceosome itself (spliceosome-mRNA complementarity and protein binding) to ensure accurate splicing. Consensus sequences in nuclear mRNA introns are shown in figure 10.34. At the left (5') side of the intron the GU sequence

BOX 10.3—*continued*

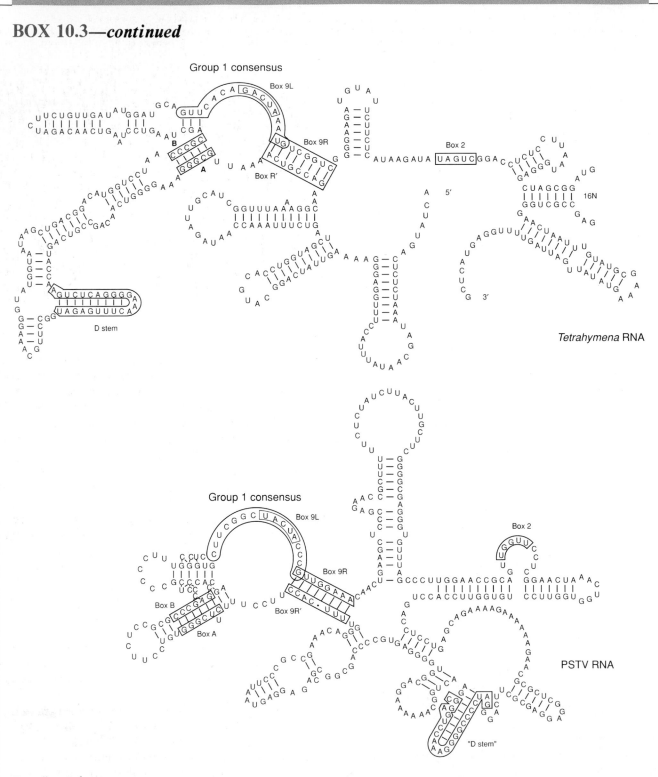

Box figure 1

Self-splicing group I intron of *Tetrahymena* (*top*) and potato spindle tuber viroid (PSTV, *bottom*); 16N in the *upper figure* refers to sixteen nucleotides not shown. Note the similarities around the Group 1 Consensus area.

From G. Dinter-Gottlieb, Proceedings of the National Academy of Sciences, *page 6251, 1986. Reprinted by permission.*

Figure 10.33

The lariat branch point (see fig. 10.32) occurs by three phosphodiester bonds at the same ribose sugar. The lariat loop is formed by the 2' phosphodiester bond.

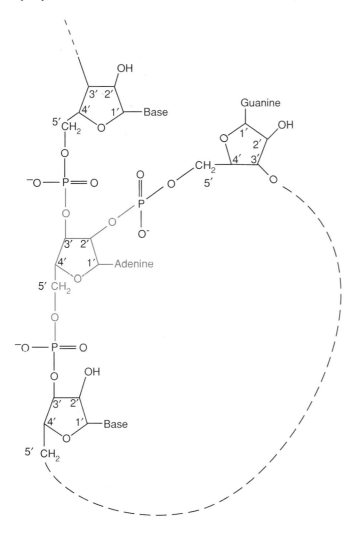

Joan A. Steitz (1941–   )
Courtesy of Dr. Joan A. Steitz.

TABLE 10.4 The Four snRNPs That Make Up the Spliceosome

snRNP RNA	Function
U1	Binds to left end of intron
U2	Binds to branch point
U4 + U6	?
U5	?

Figure 10.34

Consensus sequences of nuclear introns. Letters in *blue* (GU, A, AG) represent invariant bases. The last A of the UACUAAC sequence is the lariat branch point.

Exon	Intron	Exon
──────	AG GUAAGU • • • UACUAAC • • • CAG	──────
5'		3'

is invariant, as is the AG at the right (3') side. The rightmost A of the UACUAAC sequence is the branch point of the lariat and is also invariant. (In DNA nucleotides UACUAAC is TACTAAC; therefore, that region is sometimes referred to as the **TACTAAC box.**)

The splicing apparatus in eukaryotic mRNAs consists of several components called small nuclear ribonucleoproteins, abbreviated as **snRNPs** and pronounced "snurps" (discovered and named by J. Steitz and col-

leagues). Five of these particles take part in splicing, each composed of one or more proteins and a small RNA molecule. The RNA molecules range in size from 100 to 215 bases and have been designated U1, U2, U4, U5, and U6 (table 10.4). Both the U1 and U2 snRNPs use complementarity of their RNAs and that of the mRNA for their specificity. A general model, consistent with the known facts of spliceosome operation, is shown in figure 10.35. At first, two of the snRNPs bind to the 5' end point and the branch point of the intron. Then, U5 and U4/U6 snRNPs join to form the complete spliceosome structure (fig. 10.35). Then the lariat is formed, and the exon's ends are joined to complete the process at which point the snRNPs are free to continue the process on another intron. From recent research we also know that many other proteins are associated with the spliceosome.

It should be noted that in some cases **alternative splicing** has been observed. For example, an intron can be retained occasionally in the mRNA or a splice can be made at an alternative 5' or 3' site, resulting in a single gene producing several different mRNAs. These alternative splicing pathways are often important for the organism, producing proteins that the organism requires. Alternative splicing is seen, for example, in developmentally important processes that we discuss in chapter 15 such as antibody formation and developmental switching in *Drosophila.*

Figure 10.35

Spliceosome removes an intron. Two of the three consensus sequences of figure 10.34 are each recognized by a different snRNP. The spliceosome complex forms, leading to lariat formation and intron removal. The role of the U5 and U4/U6 snRNP, is presently not known.

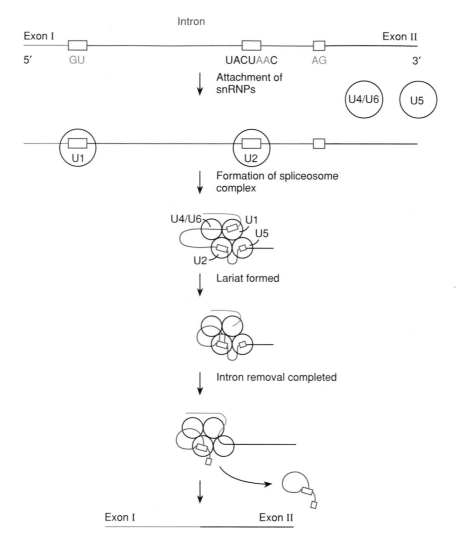

Intron Function and Evolution

Since the discovery of introns, geneticists have been trying to figure out what functions they serve. Several views have arisen. Walter Gilbert suggested that introns separate exons (coding regions) into functional domains of protein—that is, different exons presumably have specific tasks. In a given protein, one exon might code for a membrane binding region, one exon might code for the active site of the enzyme, and one might code for ATPase activity. By recombinational mechanisms, or by excluding an exon during intron removal, **exon shuffling** would allow the rapid evolution of new proteins whose structures would be conglomerates of various functional domains. In fact, in a 1990 article in *Science,* Gilbert, with two colleagues, calculated that all of the proteins in eukaryotes can be accounted for by as few as 1,000 to 7,000 exons; all pro-

teins may be conglomerates of several of this primordial number of exons. However, it should be noted that this view is controversial.

Gilbert's view of exon shuffling has been expanded by J. Darnell and W. F. Doolittle into the *introns-early* view. They suggested that introns arose before the first cells evolved. After eukaryotes evolved from prokaryotes, the prokaryotes lost their introns. This is supported by the notion that, generally, prokaryotes lack introns. This view is also consistent with the opinion that the original genetic material was RNA. In this "RNA world," introns arose as part of the genetic apparatus; they were the first enzymes (ribozymes).

An alternative view is that introns arose late in evolution, after the split of the eukaryotes from the prokaryotes. At first, the justification for this *introns-late* view was that introns evolved late to give the organism the ability to evolve quickly to new environments by an exon-shuffling type of mechanism. However, evolutionary biologists don't accept the rationale of evolution based on future needs. An alternative explanation is that introns are in fact invading "selfish DNA," DNA that can move from place to place in the genome and does not necessarily provide any advantage to the host organism. We call these "jumping genes" transposons and will discuss them at length in chapters 13 and 15. Thus both time frames for the development of introns—late or early—can be supported conceptually.

There is evidence to support all views. Gilbert's exon-shuffling view is supported by the analysis of some genes that do indeed fit the pattern of exons coding for functional domains of a protein. (Analysis consists of DNA sequencing, RNA sequencing, and protein structural analysis.) For example, the second of three exons of the globin gene binds heme. A second example is the human low-density-lipoprotein receptor, which is a mosaic of exon-encoded modules shared with several other proteins. Autocatalytic properties of introns lend credence to the view that RNA was the original genetic material and that introns can move within a genome.

Evidence for the introns-early hypothesis includes the discovery of several introns in phage genes and introns in tRNA and rRNA genes in ancient bacteria (archaebacteria). Until recently, however, there were no introns known in the true bacteria (eubacteria). That changed with recent work from the labs of D. Shub and J. Palmer, who independently discovered an intron in a tRNA gene in seven species of cyanobacteria (blue-green algae of the eubacteria). This intron was suspected because it occurred in the equivalent chloroplast gene; the chloroplast evolved from an invading cyanobacterium. However, this expansion of known places where introns occur has been viewed as supporting both the introns-early and introns-late view. The introns-early supporters say that this evidence confirms that fact that introns arose before the split of the eukaryotes from the prokaryotes. Introns-late supporters say they expect to see some introns in prokaryotes because of the mobility that these bits of genetic material have.

Both the introns-early and the introns-late views may be correct. It is possible that introns arose early, were lost by the prokaryotes in which small genomes and rapid, efficient DNA replication were priorities, and later, in eukaryotes, evolved to produce the exon shuffling suggested by Gilbert.

NEW INFORMATION ABOUT THE FLOW OF GENETIC INFORMATION

The original description of the central dogma was of genetic information flowing from DNA to RNA to protein, with a DNA-DNA loop for self-replication (see fig. 10.1). Until relatively recently, this scheme was believed to be

Figure 10.36
Updated central dogma, showing all the known paths of genetic information transfer. Under laboratory conditions, DNA can itself be translated into protein, but that circumstance apparently does not occur naturally. There is no known information flow beginning with protein, although the exact nature of prions has not been confirmed.

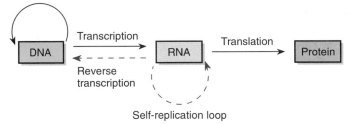

inviolable, but more recent discoveries require that it be modified. In figure 10.36, two new arrows are added to the original central dogma of figure 10.1 to indicate modifications that are necessary to allow for reverse transcription and RNA self-replication.

Reverse Transcription

First, the return arrow from RNA to DNA indicates that RNA can be a template for the synthesis of DNA. All RNA tumor viruses, such as Rous sarcoma virus, as well as the AIDS virus, can make an RNA-dependent DNA polymerase (often referred to as **reverse transcriptase**) that synthesizes a DNA strand complementary to the viral RNA. H. Temin and D. Baltimore received Nobel Prizes for their discovery of this polymerase enzyme. This enzyme is especially important since it is involved in the process of infection of a normal cell by a tumor virus and the transformation of that cell into a cancerous cell. When the viral RNA enters a cell, it brings with it reverse transcriptase. The enzyme synthesizes a DNA-RNA double helix, which then is enzymatically converted into a DNA-DNA double helix that can integrate into the host chromosome. After integration, the DNA is transcribed into copies of the viral RNA, which are both translated and packaged into new viral particles that are released from the cell to repeat the infection process. (We cover this material in more detail in chapters 12 and 15.)

RNA Self-Replication

The second modification to the original central dogma is that RNA can act as a template for its own replication, a process observed in a small class of phages. These **RNA phages,** such as R17, f2, MS2, and Qβ, are the simplest phages known. MS2 contains about thirty-five hundred nucleotides and codes for only three proteins: a coat protein, an attachment protein, and a subunit of the enzyme **RNA replicase.** The coat protein monomer is needed 180 times per new phage. Only one copy of the attachment

BOX 10.4

Guide RNA: Proofreading and Editing the Messenger

Our view of genetic information transfer is that DNA is transcribed into mRNA, which is then translated into protein. Although introns may be removed from the RNA, there is a direct one-to-one relationship among the sequence of nucleotides in the DNA, the sequence of exons, and the sequence of amino acids that appear in the protein. However, in the last several years, examples have arisen in which the DNA sequence does not predict the protein sequence even after introns are removed. In several cases there are many changes in the protein that can only come about by a process of inserting nucleotides into the messenger RNA before it is translated. This insertion is almost exclusively of uridines (U's). The process has been termed **RNA editing** and it is only recently that we have begun to understand it.

RNA editing was particularly evident in the mitochondrial proteins of a group of parasites called trypanosomes (e.g., causing African sleeping sickness); in one case more than 50% of the nucleotides in the messenger were added uridines. These parasites had another mystery attached to them, the existence of minicircles and maxicircles of DNA in specialized mitochondria, called *kinetoplasts*. In the average kinetoplast there are about 50 maxicircles and about 5,000 minicircles, concatenated like chain links (Box fig. 1a). The maxicircles contain genes for mitochondrial function (see chapter 17); as L. Simpson and his colleagues have shown in 1990, both maxicircles and minicircles are templates for **guide RNA (gRNA),** RNA that guides the process of messenger editing.

At first the role of guide RNA (gRNA) could not be ascertained because, after sequencing the nucleotides (see chapter 12), the investigators could not match gRNA with any mRNA using the rules of complementarity. However, when one of the investigators used a rule in which guanine is allowed to pair with uracil (G-U base pair), the complementarity suddenly became visible. (The G-U base pair follows "wobble rules" that we will describe in detail in the next chapter. They occur when the two strands are not in a perfect double helical configuration but rather when one of the strands is curved, allowing different hydrogen bonding than that normally found.)

The model of RNA editing as suggested by Simpson is as follows (Box fig. 1b). The guide RNA (gRNA) forms a double helix with the mRNA. The gRNA has extra bases that do not pair with the mRNA but loop out. At this point, the mRNA is opened with an endonuclease, a complementary base is inserted (U opposite an A), and the mRNA is closed. That constitutes one round of editing and results in the addition of one extra nucleotide in the mRNA. Simpson and others have suggested that the inserted nucleotide may in fact be part of the tail of the gRNA, with the gRNA acting like a ribozyme.

An exciting outcome of this research, aside from a novel mechanism of mRNA processing, is the possibility of clinical rewards. Anytime there is a specialized pathway in a parasite not found in its host, it is possible to use that pathway to selectively attack the parasite. Thus this research might lead us to new ways of combating these trypanosome parasites.

Continued on next page

Howard Temin (1934–   )
Courtesy of Dr. Howard Temin. UW photo media.

David Baltimore (1938–   )
Courtesy of Kucera and Company/Laxenburger Strasse 58.

BOX 10.4—*continued*

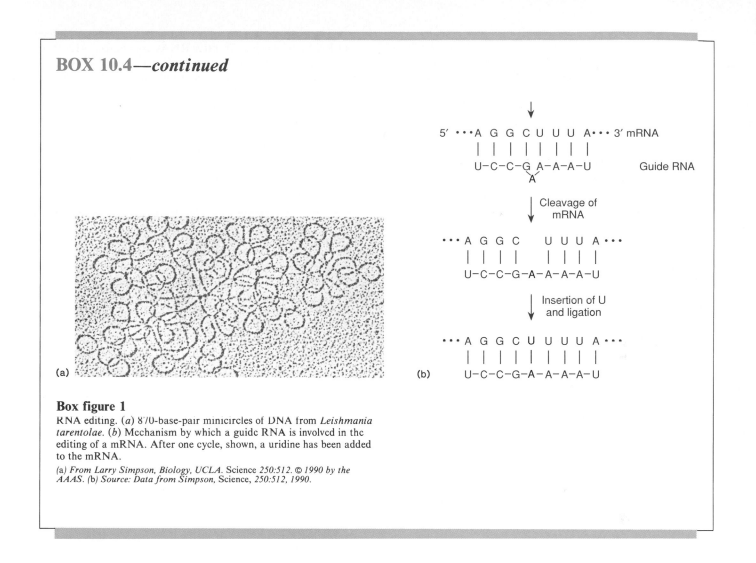

Box figure 1

RNA editing. (*a*) 870-base-pair minicircles of DNA from *Leishmania tarentolae*. (*b*) Mechanism by which a guide RNA is involved in the editing of a mRNA. After one cycle, shown, a uridine has been added to the mRNA.

(a) From Larry Simpson, Biology, UCLA. Science *250:512. © 1990 by the AAAS. (b) Source: Data from Simpson,* Science, *250:512, 1990.*

protein is needed per phage and is responsible for the attachment of the phage and subsequent penetration into the host. The RNA replicase subunit combines with three of the cell's proteins to form the enzyme RNA replicase that allows the single-stranded RNA of the phage to replicate itself.

Since the new protein needed to construct the RNA replicase enzyme must be synthesized before the phage can replicate its own RNA, the phage RNA must first act as a messenger when it infects the cell. Thus we have the situation of protein synthesis without the process of transcription ever taking place. The viral genetic material, its RNA, is first used as a messenger in the process of translation and then used as a template for RNA replication.

Even in our updated central dogma (fig. 10.36), there are no arrows originating at protein. In other words, protein cannot self-replicate, nor can it use amino acid sequence information to reconstruct RNA or DNA. Crick has called these arrows "forbidden transfers." We know of no cellular machinery to control these forbidden pro-

cesses. We await a further analysis of *prions* (infective agents that most likely lack DNA and RNA) before passing a final judgment on forbidden transfers.

The only other possible arrow in figure 10.36 is the one from DNA directly to protein. DNA can act as a messenger under the proper laboratory conditions. It is not, however, known to do so naturally.

In the next chapter we continue this discussion of protein synthesis by describing the process of translation, in which the information in messenger RNA is turned into the sequences of amino acids in proteins.

SUMMARY

The central dogma is a description of the direction of information transfer among DNA, RNA, and protein. In the previous chapter we described the DNA self-replication loop. In this chapter we described the transcriptional process, whereby DNA acts as a template for the production of RNA. Messenger RNA (mRNA) is a complementary copy of the DNA of a gene and migrates to ribosomes, where protein synthesis actually takes place. Transfer RNAs (tRNAs) transport the amino acid building blocks of proteins to the ribosome. Complementarity between the mRNA codon and the tRNA anticodon establishes the amino acid sequence in the synthesized protein specified by the gene. Ribosomal RNA (rRNA) is also involved in this process of gene-directed protein synthesis.

Intracellular RNA is single-stranded, although extensive intracellular stem-loop structures may occur. At any one gene, RNA is transcribed from only one strand of the DNA double helix. The transcribing enzyme is RNA polymerase. In *E. coli* the core enzyme, when associated with a sigma factor, becomes the holoenzyme that recognizes the transcription start signal, the promoter. Several consensus sequences define a promoter. Termination of transcription requires a sequence on the DNA, called the terminator, which causes a stem-loop structure in the RNA. Sometimes the rho protein is required for termination (rho-dependent, as compared with rho-independent, termination). In eukaryotes there are at least three RNA polymerases. Eukaryotic genes have promoters with sequences analogous to those in prokaryotic promoters as well as enhancers that work at a distance. Terminators of eukaryotic genes have not yet been well defined.

The ribosome is made of two subunits, each having protein and RNA components. Transfer RNAs are charged with their particular amino acids by enzymes called aminoacyl-tRNA synthetases. Each tRNA has about eighty nucleotides, including several unusual bases. All tRNAs have similar structures and dimensions. Transfer RNAs and rRNAs are modified from their primary transcripts.

Prokaryotic messenger RNAs are transcribed with a leader before, and a trailer after, the translatable part of the gene. In prokaryotes, translation begins before transcription is completed. In eukaryotes, these processes are completely uncoupled—transcription is nuclear and translation is cytoplasmic. Messenger RNA is modified after transcription: A cap and tail are added and intervening sequences (introns) are removed before transport into the cytoplasm. Introns can be removed by self-splicing or with the aid of the spliceosome, composed of small nuclear ribonucleoproteins (snRNPs). It is not known whether introns arose early or late in evolution or what their functions are.

The study of several RNA viruses has shown that RNA can act as a template to replicate itself and to synthesize DNA. These two processes add new directions of information transfer in the central dogma.

SOLVED PROBLEMS

1. What would be the sequence of segments on a prokaryotic mRNA that had more than one gene present?

   ANSWER: The transcript would have unmodified 3′ (trailer) and 5′ (leader) ends. Reading the sequence of nucleotides on the RNA, you would come across an initiation codon (AUG) and then, after perhaps 900 more nucleotides, a termination codon (UAA, UAG, or UGA). The 900 nucleotides would be those translated into the protein. Then there would be a spacer region of nucleotides followed by another initiation codon, intervening nucleotides that are translated into amino acids, and a termination codon. This sequence of initiation codon, codons that will be translated, a termination codon, and spacer RNA is repeated for as many genes as are present in the mRNA.

2. Can one nucleotide be a conserved sequence?

   ANSWER: Conserved sequences are invariant sequences of DNA or RNA recognizable by either a protein or a complementary sequence of DNA or RNA. However, in group II introns, an adenine near the 3′ end of the intron is needed for lariat formation. Thus this single nucleotide, given its relative position in the intron and possible surrounding bases, is a conserved sequence of one.

3. Why might *E. coli* not have a nucleolus?

   ANSWER: The nucleolus is the site of ribosome construction in eukaryotes, centered at the nucleolus organizer, the tandomly repeated gene for the three larger pieces of rRNA. In *E. coli* there are only five to ten copies of the rRNA gene whereas there is usually an order of magnitude or more copies in eukaryotes. Thus, the simplest reason that a nucleolus is not visible in *E. coli* is because there are too few copies of the gene around which a nucleolus forms.

4. If the following sequence of bases represents the start of a gene, what is the sequence of the transcribed RNA, what is its polarity, and what is the polarity of the DNA?
   G C T A C G G A T T G C T G
   C G A T G C C T A A C G A C

   ANSWER: Begin by writing the complementary strand to each DNA strand: C G A U G C C U A A C G A C for the top and G C U A C G G A U U G C U G for the bottom. Now look for the start codon, AUG. It is present only in the RNA made from the top strand, so the top strand must be transcribed. The polarity of the start codon is 5′-A U G-3′. Since transcription occurs 5′ to 3′, and since nucleic acids are antiparallel, the left end of the top strand is the 3′ end.

1. How could DNA-DNA or DNA-RNA hybridization be used as a tool to construct a phylogenetic (evolutionary) tree of organisms?

2. Given that RNA polymerase does not proofread, do you expect high or low levels of error in transcription as compared with DNA replication? Why is it important for DNA polymerase to proofread, but not RNA polymerase?

3. What are the transcription start and stop signals in eukaryotes and prokaryotes? How are they recognized? Can a transcriptional unit include more than one translational unit (gene)?

4. Diagram the relationships of the three types of RNA at a ribosome. Which relationships make use of complementarity?

5. What is a consensus sequence? A conserved sequence?

6. What would the effect be on transcription if a prokaryotic cell had no sigma factors? No rho protein?

7. Draw a double helical section of prokaryotic DNA containing transcription start and stop information. Give the base sequence of the mRNA transcript.

8. In what ways does the transcriptional process differ in eukaryotes and prokaryotes?

9. Would introns be more or less likely than exons to accumulate mutations through evolutionary time?

10. What would be the effect on the final protein product if an intervening sequence were removed with an extra base? One base too few?

11. The central dogma is frequently diagrammed as a triangle with DNA, RNA, and protein at the apexes. Convert figure 10.36 to this kind of diagram and label and explain all possible arrows.

12. What is heterogeneous nuclear mRNA? Small nuclear ribonucleoproteins?

13. What product would DNA-RNA hybridization produce in a gene with five introns? No introns? Draw these hybrid molecules.

14. What is a stem-loop structure? An inverted repeat? A tandem repeat? Draw a section of a DNA double-helix with an inverted repeat of seven base pairs.

15. What is the function of each of the following sequences: TATAAT, TTGACA, TATA, CAAT, TACTAAC? What is a Pribnow box? A Hogness box?

16. What is footprinting? How did it help define promoter sequences?

17. What are the recognition signals within introns?

18. What are the differences between rho-dependent and rho-independent termination of transcription?

19. What are the differences between group I and group II introns?

20. Diagram ribozyme functioning in a group I intron.

21. How does a spliceosome work? What are its component parts?

22. What are the differences between a σ^{70} and a σ^{32}?

23. What is a transcriptional factor? An enhancer?

24. Draw a typical mature mRNA molecule of a prokaryote and a eukaryote. Label all regions.

25. For the RNA sequence below, determine the sequence of both strands of the DNA from which it was transcribed. Indicate the 5' and 3' ends of the DNA and, with an arrow, which strand was transcribed.
5'-C C A U C A U G A C A G A C C U U G C U A A C G C-3'

26. Below is a DNA fragment isolated from the beginning of a gene. Determine which strand is transcribed, indicate the polarity of the two DNA strands, and then give the sequence of bases in the resultant mRNA and its polarity.
C C G T A C G C C T T T C A G G T T
G G C A T G C G G A A A G T C C A A

27. The following DNA fragment represents the beginning of a gene. Determine which strand is transcribed and indicate polarity of both strands in the DNA.
A T G T A C A T C T A C A T T T A C A T T
T A C A T G T A G A T G T A A A T G T A A

28. In the drawing of a gene below, solid lines (———) represent coding regions and dashed lines (— — —) represent introns. Draw what an RNA-DNA hybrid would look like if cytoplasmic mRNA is hybridized to nuclear DNA.

29. RNA-DNA hybridizations are performed by using mRNA for a given gene that is expressed in the pituitary and the adrenal glands. The DNA used in each case is the full-length gene. Based on the results below, provide an explanation for the different hybrid molecules.
DNA = − − −; RNA = ——————— .

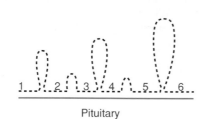

Pituitary

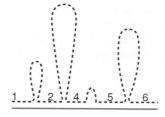

Adrenal

30. Given the following sequence of bases in a DNA molecule that is transcribed into RNA:
C C A G G T A T A A T G C T C C A G T A T G G C
A T G G T A C T T C C G G   ↑
If the T (arrow) is the first base transcribed, determine the sequence and polarity of bases in the RNA, and identify the Pribnow box and the initiator codon.

31. You have isolated a mutant that makes a temperature sensitive rho molecule; rho functions normally at 30°C but not at 40°C. If you grow this strain at both temperatures for a short period of time and isolate total, newly synthesized RNA, what relative size RNA do you expect to find in each case?

32. Suppose you repeat the experiment in problem 31 and find the same size RNA made at both temperatures. Provide two different explanations for this unexpected finding.

33. Enhancers can often exert their effect from a distance; some enhancers are located three hundred or more bases upstream of the promoter. Propose an explanation to account for this observation.

34. Why do you think most promoter regions are A-T rich?

Suggested Readings for chapter 10 are on page 646.

Gene Expression: Translation

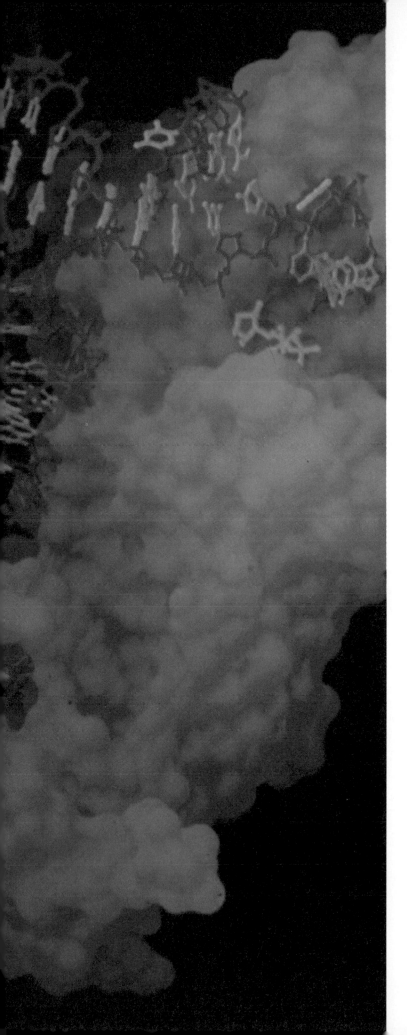

A crystal structure of *E. coli* glutaminyl-tRNA synthetase complexed with tRNA and adenosine triphosphate.

From Dr. M. A. Rould/Cover photo, Science, *vol. 246, Dec. 7, 1989 "Structure of* E. coli *Glutaminyl-tRNA Synthetase: Complexed with tRNA, GLN, and ATP at 2.8 Å Resolution." Copyright 1989 by the AAAS.*

Figure 11.1

The twenty amino acids found in proteins. At physiological pH, the amino acids usually exist as ions. Note the groupings of the various R groups.

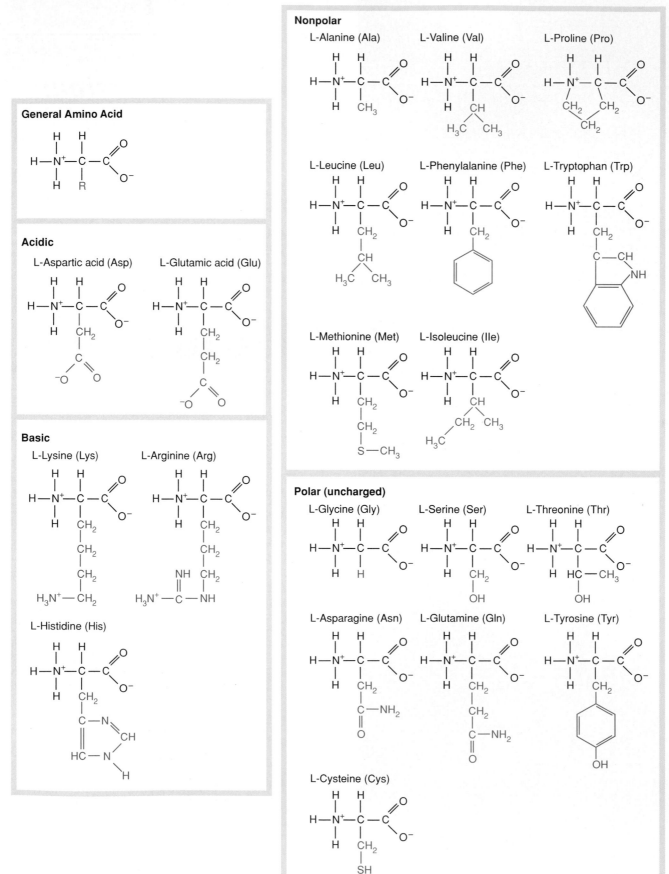

All proteins are synthesized from only twenty naturally occurring amino acids (fig. 11.1). These are called α-amino acids because one carbon, the α carbon, has all groups attached to it: an amino group, a carboxyl (acidic) group, a hydrogen, and one of the twenty different R groups (side chains), imparting the specific properties of that amino acid. (Technically, proline is termed an *imino* acid because of its structure.) Having four groups attached imparts a property known as *chirality* on the amino acid: Like left- and right-handed gloves, the mirror images cannot be superimposed. Because of optical properties, the two forms are referred to as D and L forms, in which D comes from dextrorotatory (right turning) and L comes from levorotatory (left turning). All biologically active amino acids are of the L form and hence we need not refer to this designation. Proteins (polypeptides) are synthesized by having peptide bonds formed between any two amino acids (fig. 11.2). In this manner, long chains of amino acids—called *residues* when incorporated into a protein—can be joined, and all chains will have an amino (N-terminal) end and a carboxyl (C-terminal) end.

The sequence of polymerized amino acids is termed the **primary structure** of a protein. Included in the primary structure is the formation of disulfide bridges between cystcine residues (fig. 11.3). Polypeptides can fold into several structures, the most common of which are α-helices and β-sheets (fig. 11.4). These folding configurations constitute what is referred to as the **secondary structure** of the protein. In some proteins the folding is spontaneous; in some it is guided by other proteins (Box 11.1). Further folding, bringing α-helices and β-sheets into three-dimensional configurations in relation to each other, is referred to as the **tertiary structure** of the protein (fig. 11.5). Many proteins in the active state are composed of several subunits that together are termed the **quaternary structure** of the protein. Translation is the process in which the primary structure of a protein is determined from the nucleotide sequence in a messenger RNA.

INFORMATION TRANSFER

Before proceeding to the details of translation, a sketch of the beginning of the process may be helpful (fig. 11.6). The ribosome with its rRNA and proteins is the site of protein synthesis. The information from the gene is in the form of mRNA, whereas the amino acids are carried to the ribosome attached to tRNAs. A peptide bond will form between the amino acids present at the ribosome, freeing one tRNA (at codon 1 in fig. 11.6) and lengthening the amino acid chain attached to the second tRNA (at codon 2 in fig. 11.6). The mRNA will then move one codon with respect to the ribosome and a new tRNA will attach at codon 3. This cycle is then repeated with the polypeptide lengthening by one amino acid each time. We will begin to look at the details of translation by looking at the tRNAs.

Transfer RNA

Attachment of Amino Acid to tRNA

The function of transfer RNA is to ensure that each amino acid incorporated into a protein corresponds to a particular codon in the messenger RNA. The tRNA serves this function by its structure: it has an anticodon at one end and an amino acid attachment site at the other end. The "correct" amino acid, which is the amino acid corresponding to the anticodon, is attached to the tRNA by enzymes known as aminoacyl-tRNA synthetases (e.g., arginyl-tRNA synthetase, leucyl-tRNA synthetase, etc.). A tRNA with an amino acid attached is said to be charged.

An aminoacyl-tRNA synthetase joins a specific amino acid to its tRNA in a two-stage reaction that takes place on the surface of the enzyme. In the first stage the amino acid is activated with ATP. In the second stage of the reaction, the amino acid is attached with a high-energy bond to the 3′ end of the tRNA. (See fig. 11.7; we denote high-energy bonds as "~." They are high energy in the sense that they liberate a lot of free energy when hydrolyzed.) Thus, during the process of protein synthesis, the energy for the formation of the peptide bond will be present where it is needed. Aminoacyl-tRNA synthetases are a heterogeneous lot, varying from single polypeptides to tetramers made up of two copies each of two different polypeptides.

Figure 11.2

Formation of a peptide bond between two amino acids. The bond is between the carboxyl group of one amino acid and the amino group of the other.

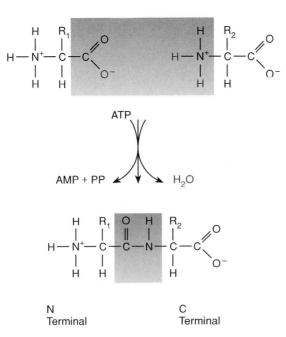

Figure 11.3
Position of disulfide bridges in the protein chymotrypsinogen. The
sequence of amino acids is referred to as the primary structure of the
protein.

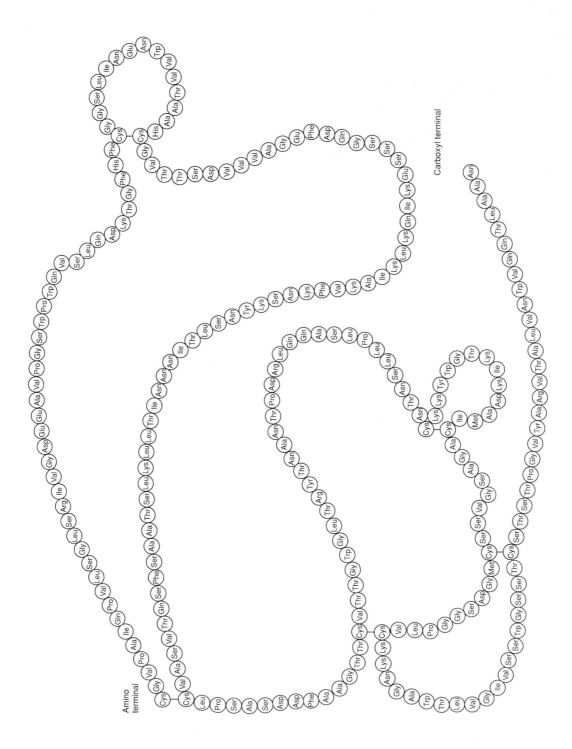

BOX 11.1

MOLECULAR CHAPERONES

Since the biochemist Christian Anfinsen won part of the 1972 Nobel Prize for showing that the enzyme ribonuclease will refold to its original shape after denaturation *in vitro,* it had been believed that the final shape of proteins (secondary and tertiary structure) is formed spontaneously. Recently it has been shown, however, that many proteins do not normally form their final active shape *in vivo* without the help of proteins that have been called **chaperones** or **molecular chaperones.** This term was first used by R. A. Laskey and colleagues to describe the protein nucleoplasmin, which is needed to form nucleosomes from histones (see chapter 14). The term was first applied more generally to the proteins that help in the folding of other proteins by J. Ellis. These molecular chaperones provided a structure on which proper folding of proteins takes place. The chaperones do not appear to provide the three-dimensional structure of the proteins they help, but rather the chaperones seem to bind to a protein in its early stages of folding and prevent unproductive folding. Like human chaperones, they prevent "incorrect interactions," according to Ellis.

Several of the chaperones studied are heat-shock proteins, proteins that are transcribed and translated in cells in response to the elevation of the temperature in cells. Presumably these heat-shock proteins protect the cell against the damage of heat. One way this is accomplished is to prevent heated proteins from denaturing and precipitating. Another way to protect the cell is to provide a framework upon which denatured proteins could be renatured back to their appropriate shapes when the temperature cools.

A well-studied class of chaperones is known as the *chaperonins* or hsp60 proteins because they are heat-shock proteins about 60 kilodaltons (60,000 daltons) in size. They occur in bacteria, chloroplasts, and mitochondria. One of the best studied of these chaperonins is the protein *GroE* of *E. coli.* This protein in its active form is composed of two components, GroEL and GroES. GroEL (hsp60) is composed of two disks, each composed of seven copies of a polypeptide (Box fig. 1). GroES (hsp10) is a smaller component composed of seven copies of a small subunit. Together, GroE is a nonspecific chaperone of proteins within the bacterial cell. Interaction takes place between nascent or denatured proteins and the chaperone, which then releases mature proteins after interaction with ATP takes place (Box fig. 2). About 100 ATPs need to be hydrolyzed per protein that is acted upon by the chaperone.

The study of molecular chaperones might have clinical applications. For example, some viruses of people might need chaperonins in order to function, and some human disease might be caused by mutations of chaperones that lead to an ineffective enzyme.

Continued on next page

Figure 11.4
The general structure of proteins. A α-helix (*left*) and a β-sheet (*right*). These are minimal drawings to indicate shape. Both are stabilized by NH — CO hydrogen bonds. In α-helices they are within the same helix; in β-sheets they are between the adjacent sheets.

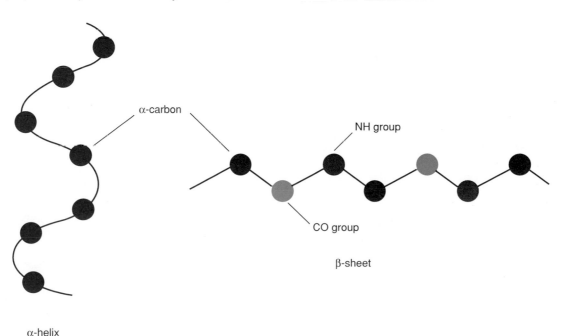

α-carbon

NH group

CO group

β-sheet

α-helix

BOX 11.1—*continued*

Box figure 1
Electron micrograph of a chaperone in protein (*GroEL*) from *E. coli*
Courtesy of Dr. R. W. Hendrix.

Box figure 2
A model by which hsp 60 (similar to *GroEL*) acts as a molecular chaperone within the mitochondrion (after F.–U. Hartl). ATP is needed for the folding and release of the protein.
Source: Celia Hooper, Journal of the National Institutes of Health, *2:64, January/February 1990.*

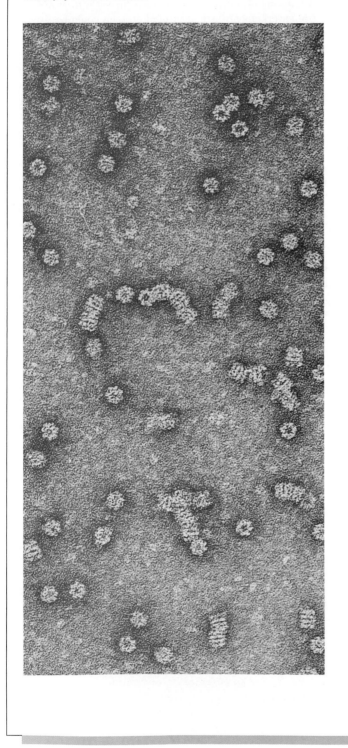

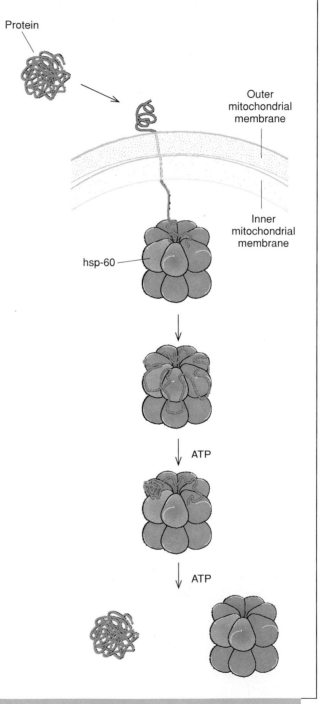

Figure 11.5

Protein structure. A space-filling model (*left*) and a ribbon diagram (*right*) of a subunit of a protein kinase enzyme (adds a phosphate group to its substrate). Two lobes of the protein are in different colors. α-helices are diagrammed as curlicues and β-sheets as flat arrows by a computer program called RIBBON.

(right) *From D. R. Knighton, "Crystal Structure of the Catalytic Subunit of Cyclic Adenosine Monophosphate—Dependent Protein Kinase" in* Science, *253:412, 26 July 1991. Copyright 1991 by the AAAS.*
(left) *From J. M. Sowadski, "Structure of a Peptide Inhibitor Bound to Cayalytic Subunit of Cyclic Adenosine monophosphate Dependent Protein" in* Science, *253, July 26, 1991.*

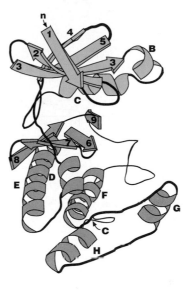

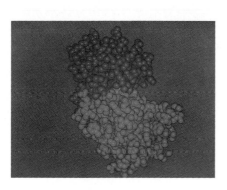

Figure 11.6

View of the initiation of the translation process at the ribosome. Note the two charged tRNAs and the mRNA in position to form the first peptide bond.

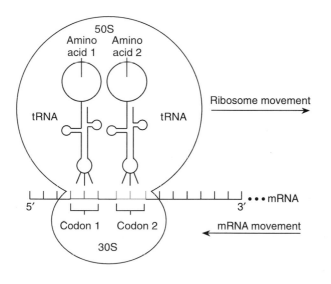

Component Numbers

In bacteria, there are twenty aminoacyl-tRNA synthetases, one for each amino acid. A particular enzyme recognizes a particular amino acid and all the tRNAs that code for that amino acid. In eukaryotes, there is a separate set of twenty cytoplasmic and twenty mitochondrial synthetases, all coded in the nucleus.

Synthetases can initially make errors of attaching the "wrong" amino acid. For example, isoleucyl-tRNA synthetase will attach valine about once in 225 times. However, there is a proofreading step during which only one in 270 to one in 800 of the errors are released intact from the enzyme. The amino acids on the rest of the incorrectly charged tRNAs are hydrolyzed before the tRNAs are released. The overall error rate is the product of the two steps, or only about one incorrectly charged tRNA per 60,000 to 80,000 formed.

There are sixty-four possible codons in the genetic code (four nucleotide bases in groups of three = $4 \times 4 \times 4 = 64$). Three of these codons are used to terminate translation. Thus there is an upper limit of sixty-one tRNAs needed because there are sixty-one different nonterminator codons. About fifty tRNAs are known in bacteria. The number fifty can be explained by the wobble phenomenon, which occurs in the third position of the codon. We will examine this phenomenon in the section on the genetic code. The tRNAs for each amino acid are designated by the convention tRNA^{Leu} (for leucine), tRNA^{His} (for histidine), and so on.

Recognition of the Aminoacyl-tRNA during Protein Synthesis

Although amino acids enter the protein-synthesizing process attached to tRNAs, it was theoretically possible that the amino acid itself was recognized during translation. A simple experiment was done that determined whether the amino acid or the tRNA was recognized.

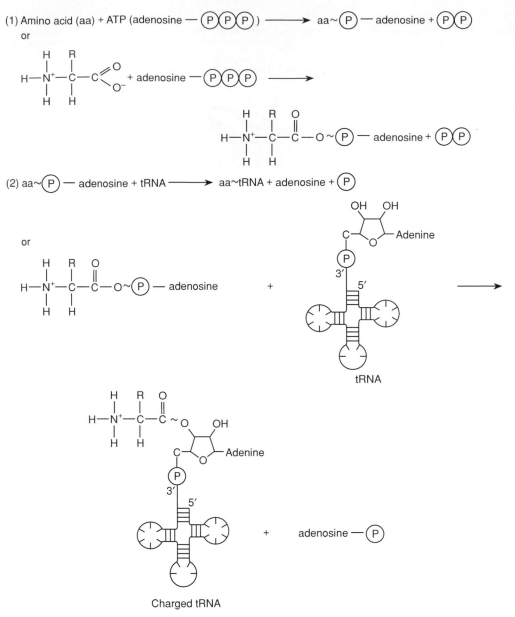

Figure 11.7
It is a two-step process to attach a specific amino acid to its tRNA by an aminoacyl synthetase. High-energy bonds are indicated by ∼. In the first step, an amino acid is attached to AMP with a high-energy bond. In the second step, the high-energy bond is transferred to the tRNA, which is then referred to as charged.

Charged tRNA

In 1962, F. Chapeville and colleagues isolated tRNA that had cysteine attached. They chemically converted the cysteine to alanine by using Raney nickel, a catalytic form of nickel that removes the SH group of cysteine (fig. 11.8). When these tRNAs were used in protein synthesis, alanine was incorporated where cysteine should have been, demonstrating that the tRNA, not the amino acid, was recognized during protein synthesis. The synthetase puts a specific amino acid on a specific tRNA; then, during protein synthesis, information on the tRNA (the anticodon)—not on the amino acid—is checked.

Initiation Complex

Translation can be divided into three stages: initiation, elongation, and termination. Elongation is simply the repetitive process of adding amino acids to a growing peptide chain. There is, however, added complexity in the initiation of protein synthesis, and in its termination. Throughout this chapter we will concentrate on prokaryotic systems, giving relevant information on the less-well-known eukaryotic systems.

It is especially important that the translation process be started precisely. Remember that the genetic code is translated in groups of three nucleotides. If the reading of the messenger RNA begins one base too early or too late, the reading frame is shifted so that an entirely different set of codons is read (fig. 11.9). The protein produced, if any, will probably bear no structural or functional resemblance to the protein originally coded for.

Role of N-Formyl Methionine

The synthesis of every protein in *Escherichia coli* begins with the modified amino acid N-formyl methionine (fig. 11.10). However, none of the completed proteins in *E. coli* contains N-formyl methionine. Many of these proteins do

BOX 11.2

THE SECOND GENETIC CODE

Much research is currently being done by investigators trying to determine the identity signals on tRNAs that are recognized by the aminoacyl-tRNA synthetases. Of course, the integrity of the genetic code requires that a particular tRNA always has the same amino acid attached to it and it is the synthetase enzymes that do the attaching. Scientists in many laboratories are working on the tRNA identity signals. It has been found that recognition of the tRNA may or may not involve the anticodon and, in one case, a change in only one base pair in a tRNA can change the identity of the tRNA (Box fig. 1).

In an article in the scientific journal *Nature,* C. de Duve referred to the identity aspects of the tRNAs as a *second genetic code;* he referred to the individual identity characteristics as *paracodons.* Shortly thereafter, in an article in *Science,* L. Schulman and J. Abelson suggested that the term "second genetic code" was not necessarily a good one because (1) it implied a common set of rules governing recognition and (2) another term is in common usage.

The first reason was suggested because, as can be seen from Box figure 1, virtually any place on the tRNA may be involved in recognition. This differs from the "first genetic code," which has very strict rules in which three bases (one codon) specifies either an amino acid or a stop signal. The second reason was suggested because the term "tRNA identity" is generally recognized and clearly understood. Only time will tell whether the term "second genetic code" becomes established.

Box figure 1

The major identity elements in four tRNAs are shown. Each *circle* represents one nucleotide. *Filled circles* represent nucleotides that are identity elements to the appropriate aminoacyl-tRNA synthetase. Other identity elements on these tRNAs may yet be discovered.

From LaDonne H. Schulman and J. Abelson, "Recent Excitement in Understanding Transfer RNA Identity" in Science, 240: 1592, June 17, 1988. Copyright 1988 by the AAAS.

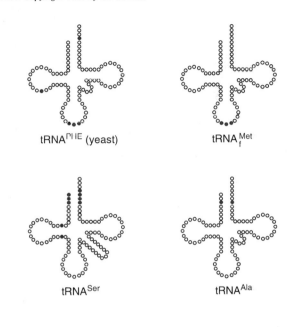

Figure 11.8

Cysteine-tRNA^{Cys} treated with Raney nickel becomes alanine-tRNA^{Cys} by removal of the SH group of cysteine. During protein synthesis, alanine is incorporated in place of cysteine in proteins, indicating that the specificity of amino acid incorporation into proteins resides with the tRNA.

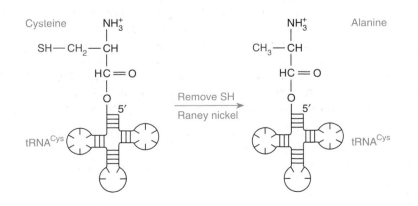

Figure 11.9

(a) In the normal reading of the messenger RNA, codons are read as repeats of CAU, coding for histidine. (b) A shift in the reading frame of the messenger RNA causes the codons to be read as repeats of AUC, coding for isoleucine.

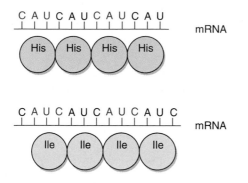

Figure 11.10

The structures of methionine and N-formyl methionine

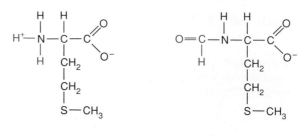

Methionine   N-formyl methionine

Figure 11.11

The two tRNAs for methionine in *E. coli*. (a) The initiator tRNA. (b) The interior tRNA.

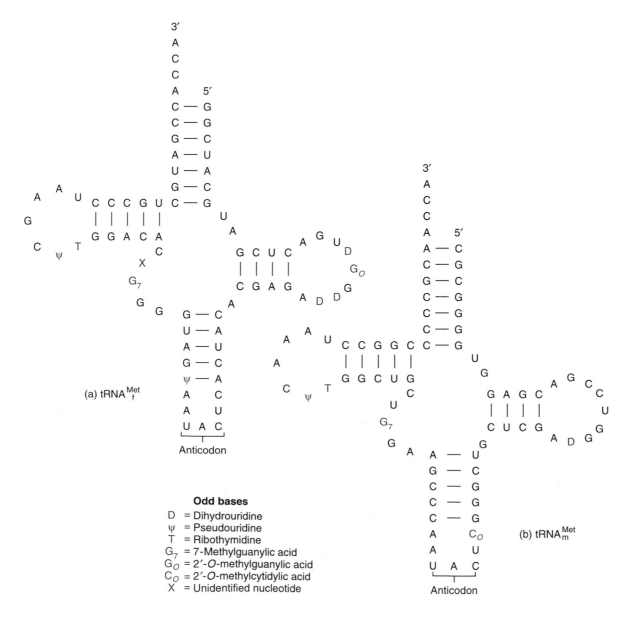

(a) tRNA$_f^{Met}$

(b) tRNA$_m^{Met}$

Odd bases

D = Dihydrouridine
ψ = Pseudouridine
T = Ribothymidine
G$_7$ = 7-Methylguanylic acid
G$_O$ = 2'-*O*-methylguanylic acid
C$_O$ = 2'-*O*-methylcytidylic acid
X = Unidentified nucleotide

not even have methionine as the first amino acid. Obviously, before a protein becomes functional, the initial amino acid is modified or removed. In eukaryotes the initial amino acid, also methionine, does not have an N-formyl group.

Methionine, with a codon of 5'-AUG-3', known as the **initiation codon,** has two tRNAs with the same anticodon (3'-UAC-5') but with different structures (fig. 11.11). One of these tRNAs (tRNA$_f^{Met}$) serves as a part of the initiation complex (see ensuing discussion) and before the initiation of translation will have its methionine chemically modified to N-formyl methionine (fMet). The other tRNA will not have its methionine modified (tRNA$_m^{Met}$). It will be used by the translation machinery to insert methionine into proteins, where called for, in all but the first position. The cell thus has a mechanism to make use of methionine in the normal way as well as to use a modified form of it to initiate protein synthesis. Because of the structure of the prokaryotic initiation tRNA, it can recognize GUG as well as AUG and rarely UUG as initiation codons. In eukaryotes, only AUG serves as an initiation codon. Since the initiation methionine is not formylated in eukaryotes, the eukaryotic tRNA is designated tRNA$_i^{Met}$; there is a separate internal methionine tRNA in eukaryotes, termed tRNA$_m^{Met}$, as in prokaryotes.

30S Ribosomal Subunit

In prokaryotes, the subunits of the ribosome (30S and 50S) are usually dissociated from each other when not involved in translation. To begin translation, an **initiation complex** is formed, consisting of the following components: the 30S subunit of the ribosome, an mRNA, the charged N-formyl methionine tRNA (fMET-tRNA$_f^{Met}$), and three **initiation factors (IF1, IF2, IF3).** These components interact in a series of steps that culminate in the initiation complex. The interaction is not yet understood in its entirety. It is known that IF3 binds to the 30S ribosomal subunit, allowing the 30S subunit to bind mRNA (fig. 11.12, step 1). Meanwhile, a complex has formed with IF2, the charged N-formyl methionine tRNA (fMet-tRNA$_f^{Met}$), and GTP (guanosine triphosphate). IF2 binds to the fMet-tRNA$_f^{Met}$ and GTP (fig. 11.12, step 2). It is IF2 that brings the initiator tRNA to the ribosome. IF2 binds only to the charged initiator tRNA, and, without IF2, the initiator tRNA cannot bind to the ribosome. The final step in initiation-complex formation is bringing together the components (fig. 11.12, step 3).

The hydrolysis of GTP to GDP + P$_i$ (inorganic phosphate, PO$_4^{-3}$—see fig. 9.8) presumably produces conformational changes that allow the initiation complex to join the 50S ribosomal subunit to form the complete ribosome, and then allows the initiation factors and GDP to be released. Frequently, the hydrolysis of a nucleoside triphosphate (e.g., ATP, GTP) in a cell occurs to make available the energy in the phosphate bonds for a metabolic process. In the process of translation, the hydrolysis apparently serves to change the shape of the GTP so that it and the initiation factors can be released from the ribosome after the 70S particle has been formed. Thus hydrolysis of GTP in translation is for conformational change rather than covalent bond formation. The role of IF1 in all of this is not currently known, although it probably either helps the other two initiation factors bind to the 30S ribosomal subunit or stabilizes the 30S initiation complex.

The process in eukaryotes is generally similar, but a bit more complex. There at least eleven initiation factors are involved, including nine initiation factors and two cap-binding proteins that bind to the 5' cap of the eukaryotic mRNA (CBP1 and CBP2). The eukaryotic initiation factors have their abbreviations preceded by an "e" to denote that they are eukaryotic (eIF1, eIF2, etc.).

The prokaryotic messenger RNA is apparently recognized by the ribosome through complementarity of a region at the 3' end of the 16S rRNA and a region slightly upstream from the initiation sequence (AUG) on the mRNA. This idea, the **Shine-Dalgarno hypothesis,** is named after the people who first suggested it (fig. 11.13). The sequence (AGGAGG) of complementarity between the mRNA and the 16S rRNA is referred to as the Shine-Dalgarno sequence. Although there is a good deal of homology between prokaryotic and eukaryotic small ribosomal RNAs, the Shine-Dalgarno region is absent in eukaryotes. The actual mechanism for the recognition of the 5' end of eukaryotic mRNA appears to be based on recognition of the 5' cap of the mRNA by the ribosome, followed by movement down the mRNA by the ribosome. The ribosome scans the mRNA until the initiation codon is recognized, presumably by the presence of the initiation tRNA. This model is referred to as the **scanning hypothesis.** (Recently it has been found that initiation of translation in some eukaryotic mRNAs may be initiated internally, bypassing recognition of the cap and scanning. The generality of this mechanism is not known yet, however.)

Aminoacyl and Peptidyl Sites in the Ribosome

When the 70S ribosome is formed, two sites come into being at the junction of the two subunits. These two sites, or cavities in the ribosome, are referred to as the aminoacyl site (**A site**) and the peptidyl site (**P site;** fig. 11.14). Each site will contain a tRNA just before the formation of a peptide bond: The P site will contain the tRNA with the growing peptide chain (peptidyl-tRNA); the A site will contain a new tRNA with its single amino acid (aminoacyl-tRNA). When the complete 70S ribosome of figure 11.12 is formed, the initiation fMet-tRNA$_f^{Met}$ is placed directly into the P site (fig. 11.14), the only charged tRNA that can be placed directly there.

Figure 11.12

Formation of the prokaryotic initiation complex is a three-step
process. In the first step, the 30S ribosome and the mRNA are
combined. In the second step, the initiator tRNA combines with IF2.
In the final step, the components from steps one and two are
combined, then the 70S ribosome is formed.

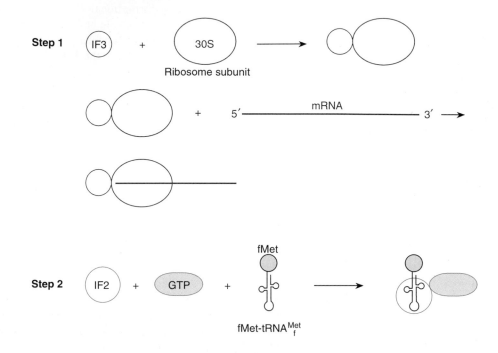

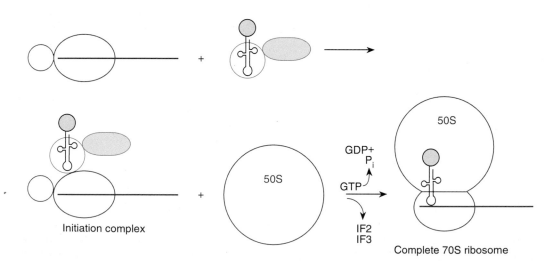

Elongation

Positioning a Second tRNA

The next step in translation is to position the second tRNA,
which is specified by the codon at the A site. The second
tRNA is positioned in the A site of the ribosome so as to
form hydrogen bonds between its anticodon and the second
codon on the mRNA. This step requires the correct tRNA,
another GTP, and (in *E. coli*) two proteins called **elon-**

gation factors (EF-Ts and **EF-Tu). EF-Tu,** bound to GTP,
is required for positioning of a tRNA into the A site of
the ribosome (fig. 11.15). After the tRNA positioning, the
GTP is hydrolyzed to GDP + P$_i$. Upon hydrolysis of the
GTP, the EF-Tu/GDP complex is released from the ribosome. EF-Ts is required to regenerate an EF-Tu/GTP
complex. Apparently EF-Ts displaces the GDP on EF-Tu.
Then a new GTP displaces EF-Ts and now the EF-Tu/
GTP complex can bind another tRNA. Here again the hy-

Figure 11.13
The Shine-Dalgarno sequence (AGGAGG) on mRNA.
Complementarity exists between the 3′ end of the 16S rRNA and a
region of the prokaryotic mRNA, upstream from the AUG initiation
codon (the Shine-Dalgarno hypothesis).

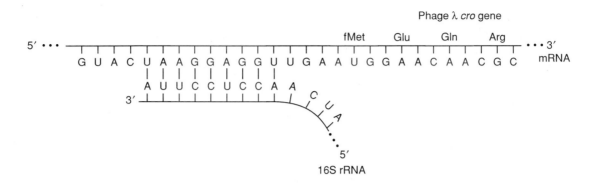

Figure 11.14
The 70S ribosome contains an A site and a P site. These sites are for
tRNAs.

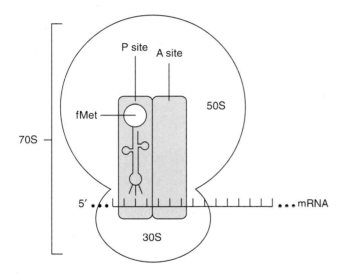

incorporated into protein. The speed of amino acid incorporation is about fifteen amino acids per second in prokaryotes and about two per second in eukaryotes.

Peptide Bond Formation

The two amino acids on the two tRNAs are now in position for formation of a peptide bond between them: Both amino acids should be juxtaposed to an enzymatic center, **peptidyl transferase,** in the 50S subunit. This enzymatic center is an integral part of the 50S subunit, composed of parts of several of the 50S proteins. The enzyme catalyzes a bond transfer from the carboxyl end of N-formyl methionine to the amino end of the second amino acid (phenylalanine in figure 11.16). Every subsequent peptide bond made is identical, regardless of the amino acids involved. The energy used is contained in the high-energy ester bond between the transfer RNA in the P site and its amino acid (fig. 11.17). Immediately after the formation of the peptide bond, the tRNA with the dipeptide is in the A site and a depleted tRNA is in the P site.

Translocation

The next stage in elongation is **translocation** of the ribosome in relation to the mRNA. The translocation process requires another GTP and another elongation factor, **EF-G,** which has also been called **translocase.** The process of translocation results in the movement of the ribosome in relation to the mRNA such that the tRNA with the polypeptide in the A site is moved, with the mRNA, into the P site. When this movement occurs, the depleted tRNA is ejected from the P site and the empty A site becomes available for the next tRNA. The depleted tRNA actually passes through a third site recognized on the ribosome, termed the exit site (**E site).** For simplicity we have not drawn that site. Translocation is completed with the hydrolysis of GTP to GDP + P$_i$ and the release of EF-G.

drolysis of GTP serves the purpose of changing the shape of the GTP so that the EF-Tu/GDP complex can depart from the ribosome after the tRNA in the A site is in place. Figure 11.16 shows the ribosome at the end of this step. EF-Tu does not bind fMet-tRNA$_f^{Met}$, so this blocked (formylated) methionine cannot be inserted into a growing peptide chain.

It takes several milliseconds for the GTP to be hydrolyzed, which allows EF-Tu/GDP to leave the ribosome, and another few milliseconds for the EF-Tu/GDP to actually leave. In those two intervals of time, the codon-anticodon fit of the tRNA is scrutinized. If the correct tRNA is in place, a peptide bond will be formed. If not, the charged tRNA will be released and a new cycle of EF-Tu/GTP-mediated testing of tRNAs will begin. The error rate is only about one mistake in ten thousand amino acids

Figure 11.15

The EF-Ts/EF-Tu cycle. EF-Ts and EF-Tu are required for the attachment of a tRNA to the A site of the ribosome. At *top center*, we have EF-Tu attached to a GDP. The GDP is then displaced by EF-Ts, which in turn is displaced by GTP. A tRNA attaches and is brought to the ribosome. If the codon-anticodon fit is correct, the tRNA attaches at the A site, with the help of the hydrolysis of GTP to GDP + P$_i$. The EF-Tu is now back where we started.

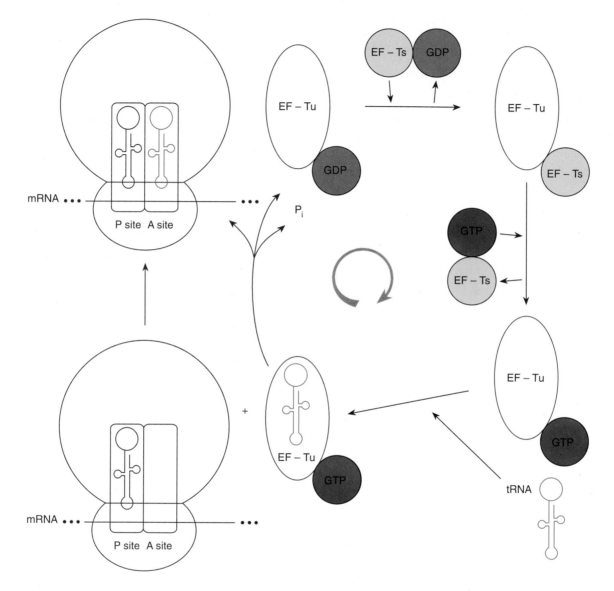

Again, the hydrolysis of GTP is probably for the purpose of a conformational change that will allow the elongation factor and the GDP to be released. In eukaryotes, two elongation factors perform the same tasks that EF-Tu, EF-Ts, and EF-G perform in prokaryotes. The factor, eEF1, replaces EF-Tu and EF-Ts, and eEF2 replaces EF-G.

When translocation is complete, the situation is again as diagrammed in figure 11.14, except that instead of fMet-tRNA$^{Met}_f$, the P site contains the second tRNA (tRNA^{Phe}) with a dipeptide attached to it. The process of elongation is then repeated, with a third tRNA coming

into the A site. The process is repetitive from here to the end (fig. 11.18). By the mechanism of the process, a peptide is synthesized starting from the amino (N-terminal) end and proceeding to the carboxyl (C-terminal) end. During the repetitive aspect of protein synthesis, two GTPs are hydrolyzed per peptide bond: one GTP in the release of EF-Tu from the A site and one GTP in the release of EF-G during translocation of the ribosome after the peptide bond has been formed. In addition, every charged tRNA has had an amino acid attached at the expense of the hydrolysis of an ATP to AMP + PP.

Figure 11.16

A ribosome with two tRNAs attached. In this case, the second codon is for the amino acid phenylalanine. The two amino acids are next to each other.

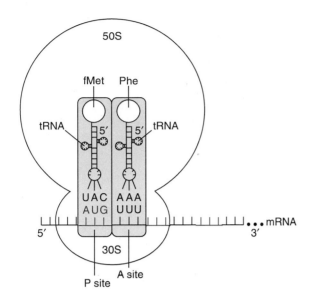

Figure 11.17

Peptide bond formation on the ribosome between N-formyl methionine and phenylalanine. The bond attaching the carboxyl end of the first amino acid to its tRNA is transferred to the amino end of the second amino acid. The first tRNA is now uncharged, whereas the second tRNA has a dipeptide attached.

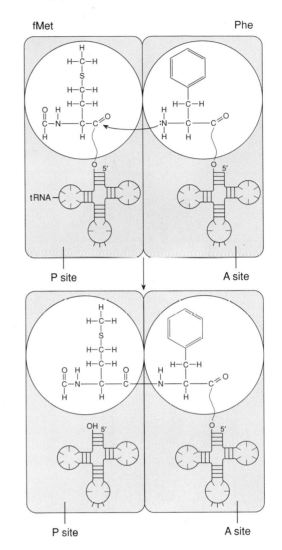

Termination

Nonsense Codons

Termination of protein synthesis occurs when one of three **nonsense codons** appears in the A site of the ribosome. These codons are UAG (sometimes referred to as *amber*), UAA (*ochre*), and UGA (*opal*). ("Amber" is the English translation of the name Bernstein, a graduate student who took part in the discovery of UAG in R. H. Epstein's lab at Caltech. "Ochre" and "opal" are tongue-in-cheek extensions of the first label.) Three proteins called **release factors (RF)** are involved in termination, and a GTP is hydrolyzed to GDP + P_i.

When a nonsense codon enters the A site on the ribosome, it is recognized by a release factor, either RF1 or RF2. RF1 recognizes UAA and UAG; RF2 recognizes UAA and UGA. The result is a blocking of further chain elongation, as shown in figure 11.19. The completed polypeptide and the now-depleted tRNA are released. Then, with the hydrolysis of a GTP, probably mediated by RF3, the mRNA is released, the releasing factors dissociate, and the ribosome dissociates into its component parts. Dissociation is aided by one of the original initiation factors, IF3, which rebinds to the 30S subunit and thus causes dissociation of the 70S ribosome. The ribosomal subunits are now ready to reinitiate protein synthesis with another mRNA. A comparison of prokaryotic and eukaryotic translation is given in table 11.1.

Rate and Cost of Translation

As mentioned before, the average speed of the protein-synthesizing process is about fifteen peptide bonds per second in prokaryotes. Discounting the time for initiation and termination, an average protein of three hundred amino acids is synthesized in about 20 seconds (the released protein will form its final structure spontaneously or will be modified with the aid of other enzymes). An equivalent eukaryotic protein takes about 2.5 minutes to be synthesized. The energy cost is four high-energy phosphate bonds per peptide bond (two from an ATP during tRNA charging and two from GTP hydrolysis during tRNA binding at the A site and translocation) or

Figure 11.18

Cycle of peptide bond formation and translocation on the ribosome. After the peptide bond is transferred (fig. 11.17), the ribosome and mRNA move one codon in relation to each other. Now the tRNA with the peptide is in the P site, and the A site is again open. In this example, the next tRNA that moves into the A site carries glutamic acid.

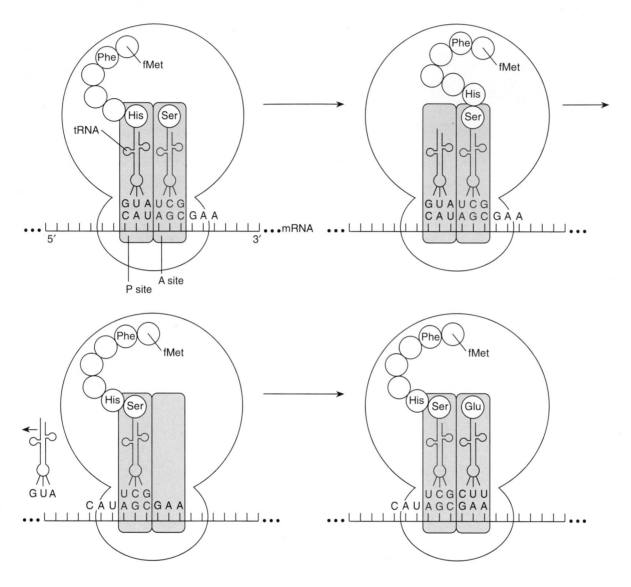

about twelve hundred high-energy bonds per protein. This cost is very high—about 90% of the energy production of an *E. coli* cell goes into protein synthesis. A high energy cost is presumably the price that a living system has to pay for the speed and accuracy of the synthesis of its proteins.

Coupling of Transcription and Translation

In prokaryotes, such as *E. coli,* in which no nuclear envelope exists, translation begins before transcription is completed. Figure 11.20 shows a length of an *E. coli* chromosome. An RNA polymerase can be seen on the DNA transcribing a gene. The mRNA, still being syn-

thesized, can be seen extending away from the DNA. Attached to the mRNA are about a dozen ribosomes. Since translation starts at the same end (5′) of the messenger that is synthesized first, an initiation complex can be formed and translation can begin shortly after transcription begins. As translation proceeds along the messenger, its 5′ end will again become exposed and a new initiation complex can be formed. The occurrence of several ribosomes translating the same messenger is referred to as a **polyribosome,** or simply a **polysome** (fig. 11.21).

Figure 11.19

Chain termination at the ribosome. A nonsense codon in the A site is recognized by one of two release factors. In this case, UAG is recognized by RF1. The complex then falls apart, releasing the peptide.

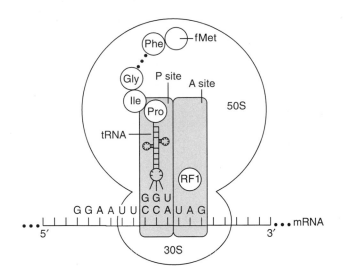

Figure 11.20

A polysome (i.e., multiple ribosomes on the same mRNA). Each ribosome is approximately 250 Å units across. Also visible is the DNA and RNA polymerase.

Reproduced courtesy of Dr. Barbara Hamkalo. International Review of Cytology *(1972) 33: 7, fig. 5. Copyright by Academic Press Inc., Orlando, Florida.*

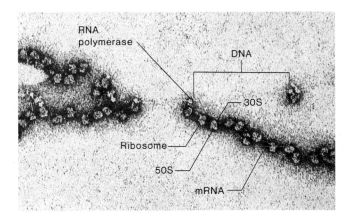

In prokaryotes, most messenger RNAs contain the information for several genes. These RNAs are said to be **polycistronic** (fig. 11.22). (*Cistron,* another term for gene, is defined in chapter 16.) Each gene on the mRNA is translated independently: Each has a Shine-Dalgarno sequence for ribosome recognition (see fig. 11.13) and an initiation codon (AUG) for fMet. The ribosome that completes the translation of the first gene may or may not continue on to the second gene after dissociation. The translation of any of the genes follows all the steps just outlined.

In eukaryotes, however, every messenger RNA so far observed contains only one gene (each is monocistronic). Since ribosomal recognition of a eukaryotic gene depends on the 5′ cap, and since each eukaryotic mRNA has only one cap, only one polypeptide can be translated for any given mRNA. Although certainly not the rule, the translated peptide can be modified or cleaved into smaller functional peptides. For example, in mice a single mRNA codes for a protein that is later cleaved into epidermal growth factor and at least seven other related peptides.

More on the Ribosome

In the last chapter, we briefly discussed the shape and composition of the ribosomal subunits. All of the protein and RNA components are known and have been isolated. Assembly pathways are known. However, there is still much to be learned about the stereochemistry of translational events on the ribosome. At present we can localize some of the functions to certain parts of the ribosome. We know approximately where the mRNA, initiation factors, and EF-Tu are placed on the 30S subunit during translation (fig. 11.23). We also know where peptidyl transferase activity and EF-G reside on the 50S subunit, which has a cleft leading into a tunnel that passes through the structure. At present it seems that the nascent peptide passes through this tunnel, emerging close to a site of membrane binding (fig. 11.23). The tunnel can hold a peptide length of about forty amino acids. Note that, although every ribosome has a membrane-binding site, not all active ribosomes are bound to membranes.

TABLE 11.1   Some Comparisons between Prokaryotic and Eukaryotic Translation

	Prokaryotes	Eukaryotes
Initiation Codon	AUG, GUG, UUG	AUG
Initiation Amino Acid	N-formyl methionine	Methionine
Initiation tRNA	$tRNA_f^{Met}$	$tRNA_i^{Met}$
Interior Methionine tRNA	$tRNA_m^{Met}$	$tRNA_m^{Met}$
Initiation Factors	IF1, IF2, IF3	Nine eIF factors + CBP1, CBP2
Elongation Factors	EF-Tu, EF-Ts	eEF1
Translocation Factor	EF-G	eEF2
Release Factors	RF1, RF2	eRF

Figure 11.21

(a) Protein synthesis at a polysome. Nascent proteins exit from a tunnel in the 50S subunit. Messenger RNA is being translated by the ribosomes while the DNA is being transcribed. (b) An mRNA from the midge, *Chironomus tentans,* showing attached ribosomes and nascent polypeptides emerging from the ribosomes. Note the 5' end of the messenger at the *upper right* (small peptides). Magnification of 165,000×.

(b) Courtesy of S. L. McKnight and O. L. Miller, Jr.

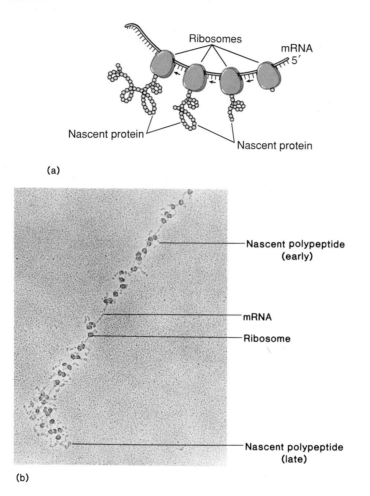

(a)

(b)

Figure 11.22

A prokaryotic polycistronic mRNA. Note the several Shine-Dalgarno sequences for ribosome attachment and the initiation and termination codons marking each gene.

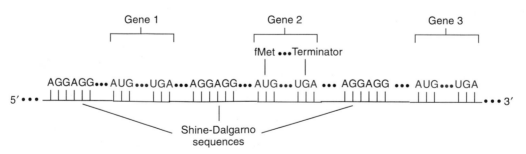

BOX 11.3
AMINO ACID SEQUENCING

Protein-sequencing techniques have been known since 1953 when F. Sanger worked out the complete sequence of the protein hormone insulin. The basic strategy is to purify the protein and then break it up into small peptides several different ways. These peptides are sequenced, and the whole protein sequence can be determined by the overlap pattern of the sequenced subunits.

A protein can be broken into peptide fragments by many different methods, including acid and alkaline hydrolysis. For the most part, proteolytic enzymes (proteases) that hydrolyze the peptides at specific points arc used. *Pepsin,* for example, preferentially hydrolyzes peptide bonds involving aromatic amino acids, methionine, and leucine; *chymotrypsin* hydrolyzes peptide bonds involving carboxyl groups of aromatic amino acids; and *trypsin* hydrolyzes bonds involving the carboxyl groups of arginine and lysine.

The proteolytic digest is usually separated into a *peptide map,* or peptide *fingerprint,* by using a two-dimensional combination of paper chromatography and electrophoresis. (Column chromatography, a technique that we will not consider, is used also.) In two-dimensional chromatography a sample is put onto a piece of paper that is then placed in a solvent system. After an allotted time the paper is dried, turned 90 degrees, and placed in a second solvent system for another allotted time (Box fig. 1). In each solvent, different peptides travel through the paper at different rates. The spots are then developed using ninhydrin, which reacts with the N-terminal amino acid and produces a colored product when heated.

The spots, which represent small peptides, can be cut out of a second, identical chromatogram that has not been sprayed with ninhydrin. These spots can then be sequenced by, for example, the Edman method, whereby the peptide is sequentially degraded from the N-terminal end. Phenylisothiocyanate (PITC) reacts with the amino end of the peptide. When acid is added, the N-terminal amino acid is removed as a PITC derivative and can be identified. The process is then repeated until the whole peptide has been sequenced (Box fig. 2).

Box figure 1
Two-dimensional paper chromatography of a protease digest. Chromatography is done in one solvent system. The plate is then dried, rotated, and placed into a second solvent system. The pattern on the resulting plate is called a fingerprint.

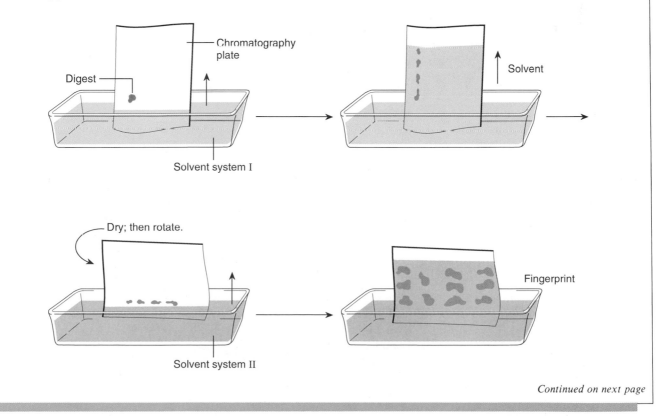

Continued on next page

BOX 11.3—*continued*

Box figure 2

Isolation of amino acids from a peptide for the purposes of sequencing. First, the peptide is bonded to PITC (phenylisothiocyanate) at the amino end. Acid treatment results in a PITC derivative of the amino-terminal amino acid and a peptide that is one amino acid shorter. The PITC derivative can be identified. These steps are then repeated, isolating one amino acid at a time.

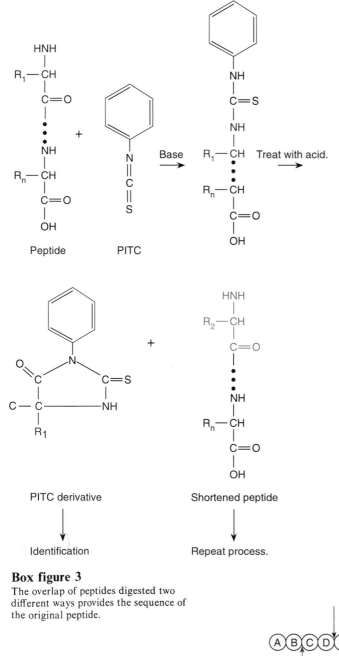

If the fingerprint pattern is worked out for two different digests of the same polypeptide, the unique sequence of the original polypeptide can be determined by overlap. In the illustration (Box fig. 3), the letters A–J represent the amino acids in a polypeptide. A is known to be the first (N-terminal) amino acid since the Edman method sequences peptides from this end. We can thus summarize the methodology as follows:

1. A protein is purified. If it is made up of several subunits, these subunits are separated and purified. (If disulfide bridges exist within a peptide, they must be reduced. They are later determined by digestion, with the bridges intact, followed by resequencing.)
2. Different proteolytic enzymes are used on separate subsamples so that the protein is broken into different sets of peptide fragments.
3. Two-dimensional chromatography, electrophoresis and chromatography, or column chromatography are used to isolate the peptides.
4. The Edman method of sequentially removing amino acids from the N-terminal end is used to sequence each peptide.
5. The amino acid sequence from the N- to C-terminal ends of the protein is deduced from the overlap of sequences in peptide digests generated with different proteolytic enzymes.

Today, protein sequencing can be done automatically by a machine known as an amino acid sequencer (*sequenator*). Taking about two hours per amino acid residue, sequenators can carry out Edman degradation on polypepides up to about fifty amino acids long.

Box figure 3

The overlap of peptides digested two different ways provides the sequence of the original peptide.

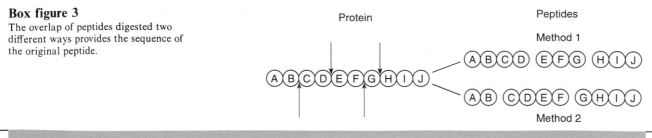

Figure 11.23
Functional sites on the prokaryotic ribosome. The ribosome is
synthesizing a protein involved in membrane passage. Note the
position of the mRNA on the 30S subunit and the cleft, tunnel, and
membrane-binding site on the 50S subunit.

From C. Bernabeu and J. A. Lake, Proceedings of the The National Academy of
Sciences, *79: 3111–3115, 1982. Reprinted by permission.*

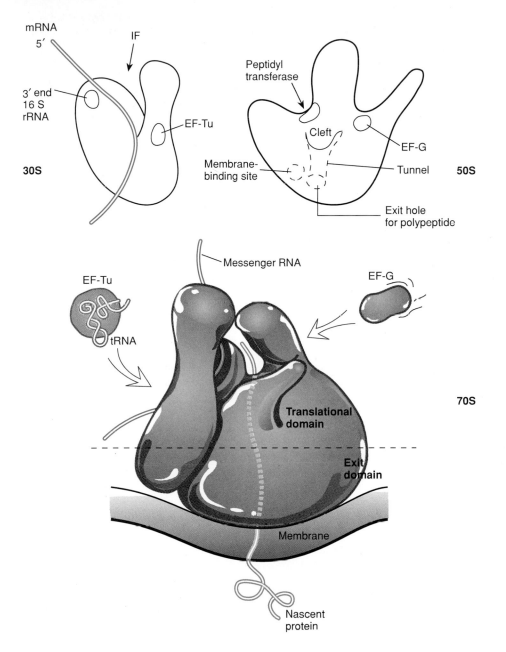

The Signal Hypothesis

Ribosomes are either free in the cytoplasm or associated
with membranes. The choice is determined by the type of
protein being synthesized. Membrane-bound ribosomes,
indistinguishable from free ribosomes, synthesize proteins
that enter membranes. These proteins either become a
part of the membrane or in eukaryotes are passed into
membrane-bound organelles (e.g., the Golgi apparatus,
mitochondria, chloroplasts, vacuoles) or transported outside the cell membrane. The mechanism for membrane
attachment is explained by the **signal hypothesis,** developed by G. Blobel and C. Milstein and their colleagues.
The mechanism is applicable for both prokaryotes and
eukaryotes.

The signal for membrane insertion is coded into the
first one to three dozen amino acids of membrane-bound
proteins. This **signal peptide** takes part in a chain of events
leading to membrane attachment by the ribosome and
membrane insertion of the protein. The first step occurs

Gunter Blobel (1936–   )
Courtesy of Dr. Gunter Blobel, Dept. Cell Biology, Rockefeller University.

Cesar Milstein (1927–   )
Courtesy by the Photographic Department of the MRC Laboratory of Molecular Biology.

Figure 11.24

The signal hypothesis. A signal peptide is recognized by a signal recognition particle (SRP) that draws the ribosome to a docking protein (DP) on the membrane. The peptide synthesized then passes directly into and through the membrane. A signal peptidase (SP) on the other side of the membrane removes the signal peptide, which has completed its function. When translation is completed, the ribosome drops free.

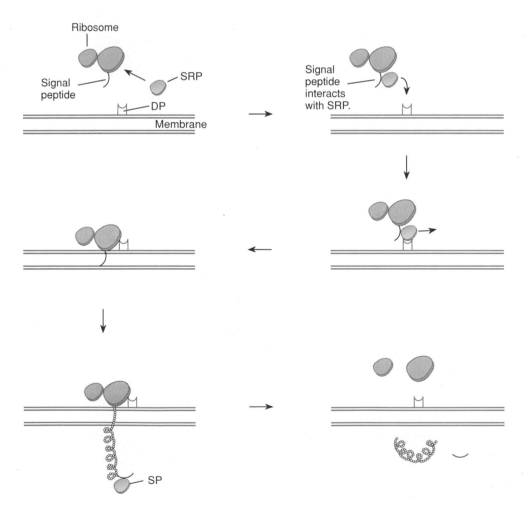

Figure 11.25

Proteins that interact with membranes can eventually have their N-terminal or C-terminal ends on the outside of the membrane, depending on how the protein enters the membrane. An anchor sequence can keep a protein in the membrane.

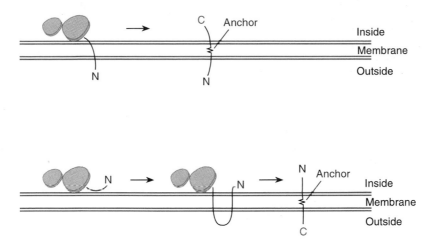

when the signal peptide becomes accessible outside of the ribosome. It is recognized by a ribonucleoprotein particle called the **signal recognition particle** (SRP), which consists of six different proteins and a 7S RNA, which is about three hundred nucleotides long. The complex of SRP, ribosome, and signal peptide then passes, or diffuses, to a membrane in which the SRP binds to a receptor on the membrane, called a **docking protein** (DP; fig. 11.24). During this time, protein synthesis is halted. The ribosome is brought in direct contact with the membrane. Other proteins of the membrane help anchor the ribosome. Protein synthesis then resumes, with the nascent protein passing directly into the membrane, or passed into the membrane shortly after being synthesized. Once through the membrane, the signal peptide is cleaved from the protein by an enzyme called *signal peptidase*. A striking verification of this hypothesis came about through recombinant DNA techniques (chapter 12) in which a signal sequence was placed in front of the α-globin gene whose protein product is normally not transported through a membrane. Translation of this gene resulted in the ribosome becoming membrane bound and the protein passing through the membrane.

Since different proteins enter different membrane-bound compartments (e.g., Golgi apparatus), some mechanism must exist that directs a nascent protein to its proper membrane. This specificity seems to depend on the exact signal sequence and membrane-bound glycoproteins called

signal-sequence receptors. Apparently, after the ribosome binds to the docking protein, the signal peptide interacts with a signal-sequence receptor, which presumably determines whether that protein is specific for that membrane. If it is, the remaining processes continue. If not, the ribosome may be released from the membrane.

Proteins that remain in the membrane can be in a configuration in which the amino-terminal end is either inside or outside of the membrane. Figure 11.25 shows how these two configurations could come about based on whether the N-terminal end (with the signal peptide) is held or passed directly into the membrane. The protein is held within the membrane by a sequence of amino acids called an anchor. In addition, some proteins pass into membranes after their synthesis is completed (e.g., mitochondria-bound proteins). Many proteins, especially those involved in creating channels in cell membranes in which ions pass, traverse the membrane many times (fig. 11.26).

The signal peptide does not seem to have a consensus sequence like the transcription or translation recognition boxes. Rather, similarities (at least for the endoplasmic reticulum and bacterial membrane-bound proteins) include a positively charged (basic) amino acid (commonly lysine or arginine) near the beginning (N-terminal end) followed by about a dozen hydrophobic (nonpolar) amino acids, commonly alanine, isoleucine, leucine, phenylalanine, and valine (table 11.2).

Figure 11.26

Diagram of the α subunit of the adult muscle sodium channel. There are 24 α-helices that reside within the cell membrane. Ion channels form gateways for the passage of ions in and out of cells during polarization and depolarization (electrical impulses) of the membrane.

From R. H. Brown, Jr., Science, *250:1001, 16 November 1990. Copyright 1990 by the AAAS.*

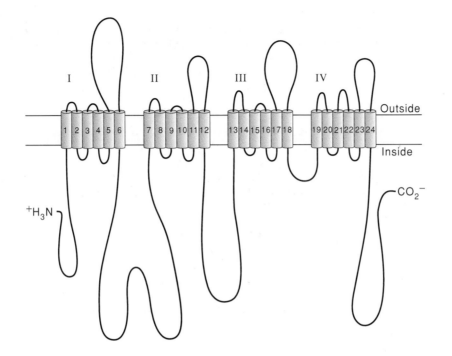

TABLE 11.2   The Signal Peptide of the Bovine Prolactin Protein*

NH_2 – Met Asp Ser Lys Gly Ser Ser Gln Lys Gly Ser Arg Leu Leu Leu Leu Leu Val Val Ser Asn Leu Leu Leu Cys Gln Gly Val Val Ser I Thr Pro Val . . . Asn Asn Cys – COOH

---

*The vertical line separates the signal peptide from the rest of the protein, which consists of 199 residues.

Source: Data from Sasavage, et al., 1982, *Journal of Biological Chemistry* 257:678–681, 1982.

THE GENETIC CODE

Researchers in the mid-1950s assumed that the genetic code would be found to consist of simple sequences of nucleotides specifying particular amino acids. They sought answers to questions such as: Is the code overlapping? Are there nucleotides between code words (punctuation)? How many letters make up a code word (codon)? Logic, along with genetic experiments, supplied some of the answers, but only with the rapidly improving techniques of biochemistry was the genetic language eventually decoded.

Triplet Nature of the Code

From several lines of evidence, it had been thought that the nature of the code was triplet (three bases in mRNA specify one amino acid). If codons contained only one base, they would only be able to specify four amino acids since there are only four different bases in DNA (or mRNA). A couplet code would have $4 \times 4 = 16$ two-base words, or codons, which is still not enough to specify uniquely twenty different amino acids. A triplet code would allow for $4 \times 4 \times 4 = 64$ codons, which are more than enough to specify twenty amino acids.

Evidence for the Triplet Nature of the Code

The triplet-code concept was reinforced by the experimental manipulation of mutant genes primarily by Francis Crick and his colleagues. In these experiments, a chemical mutagen, the acridine dye proflavin, was used to cause inactivation of the rapid lysis (*rIIB*) gene of the bacteriophage T4. Proflavin inactivates the gene by either adding a nucleotide to the DNA or deleting a nucleotide from it (see chapter 16). The *rII* gene controls the plaque morphology of this bacteriophage growing on *E. coli* cells.

Antibiotics, substances that are produced by living organisms that are toxic to other living organisms, are of interest to us for two reasons: They have been extremely important in fighting diseases of human beings and farm animals, and many of them are useful tools for analyzing protein synthesis. Some antibiotics impede the process of protein synthesis in a variety of ways and usually poison bacteria selectively; the effectiveness of antibiotics normally derives from the metabolic differences between prokaryotes and eukaryotes. For example, an antibiotic that blocks a 70S bacterial ribosome without affecting an 80S human ribosome could be an excellent antibiotic.

Puromycin—Puromycin resembles the 3′ end of an aminoacyl-tRNA (Box fig. 1). It is bound to the A site of the bacterial ribosome, where peptidyl transferase creates a bond from the nascent peptide attached to the tRNA in the P site to puromycin. Further elongation can then no longer occur. The peptide chain is then prematurely released and protein synthesis at the ribosome is terminated.

Experiments with puromycin helped demonstrate the existence of the A and P sites of the ribosome. It was found that puromycin could not bind to the ribosome if translocation factor EF-G were absent. With EF-G, translocation took place and puromycin could then bind to the ribosome. Puromycin's ability to bind only after translocation indicated that a second site on the ribosome becomes available after translocation.

Streptomycin, Tetracycline, and Chloramphenicol—Streptomycin, which binds to one of the proteins (protein S12) of the 30S subunit of the prokaryotic ribosome, inhibits initiation of protein synthesis. Streptomycin also causes misreading of codons if chain initiation has already begun, presumably by altering the conformation of the ribosome so that tRNAs are less firmly bound to it. Bacterial mutants that are streptomycin resistant as well as mutants that are streptomycin dependent (they cannot survive without the antibiotic) occur. Both types of mutants have altered 30S subunits, specifically altered protein S12.

Tetracycline blocks protein synthesis by preventing an aminoacyl-tRNA from binding to the A site on the ribosome. Chloramphenicol blocks protein synthesis by binding to the 50S subunit of the prokaryotic ribosome, where it blocks the peptidyl transfer reaction. Chloramphenicol does not affect the eukaryotic ribosome. However, chloramphenicol, as well as several other antibiotics, is used cautiously because the mitochondrial ribosomes within eukaryotic cells are very similar to prokaryotic ribosomes. Some of the antibiotics that affect prokaryotic ribosomes also affect mitochondria. As mentioned before, the similarity of bacteria and mitochondria implies a prokaryotic origin of mitochondria. (Similarities between cyanobacteria and chloroplasts support the idea of a prokaryotic orgin of chloroplasts.)

Continued on next page

Rapid-lysis mutants produce large plaques. The wild-type form of the gene, *rII+*, results in normal plaque morphology.

Figure 11.27 shows the consequences of adding or deleting a nucleotide. From the point of addition or deletion onward, there is a **frameshift** in which new codons are read. If a deletion is combined with an addition to produce a double-mutant gene, the frameshift occurs only in the region between the two mutants. If this region is small enough or does not contain coding for vital amino acids, the function of the gene may be restored. Two deletions or two insertions combined will not restore the reading frame. However, Crick and his colleagues found that the combination of three additions or three deletions also restored gene function. This finding led to the conclusion that the genetic code was a triplet code because a triplet code would be put back into reading frame by three additions or three deletions (fig. 11.28).

Overlap and Punctuation in the Code

The questions still remained as to whether or not the code was overlapping or had punctuation (fig. 11.29). Several logical arguments favored a no-punctuation, nonoverlapping model. An overlapping code would be subject to two restrictions. First, a change in one base (a mutation) could affect more than one codon and thus affect more than one amino acid. But studies of amino acid sequences almost always showed that in various mutants only one amino acid was changed, which argued against codon overlap. Second, there are certain restrictions as to which amino acids occurred next to each other in proteins. For example, the amino acid coded by UUU could never be adjacent to the amino acid coded by AAA because one or both (depending on the number of bases overlapped) of the overlap codons UUA and UAA would always insert other amino acids between them. Overlap, then, appeared not to be the case since, in fact, every amino acid appears next to every other amino acid in one protein or another.

BOX 11.4—*continued*

Box figure 1

Puromycin is bound to the A site of the ribosome. A peptide bond then is formed. Further elongation is prevented, resulting in chain termination.

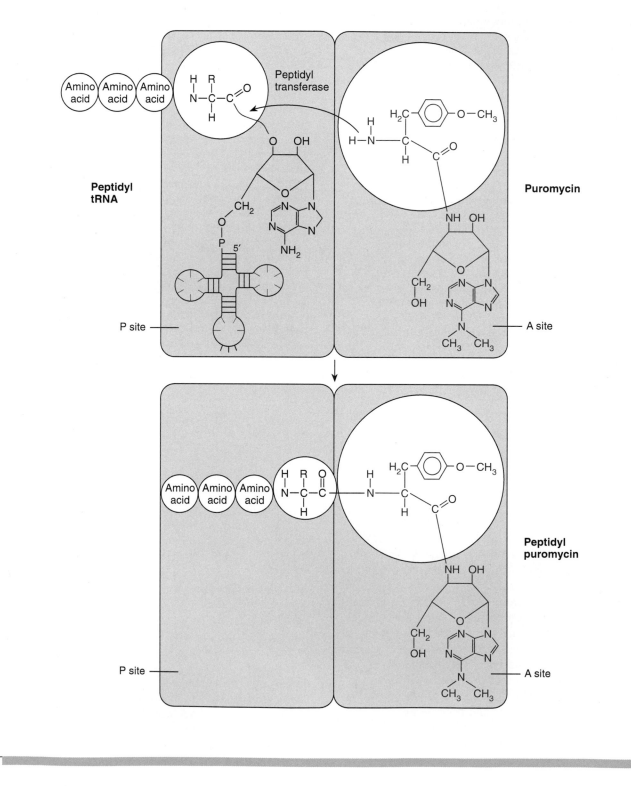

Figure 11.27

Frameshift mutations in a gene result from addition or deletion of one or several nucleotides (not a multiple of three) in the DNA. The mRNA shown normally has a CAG repeat. A single base deletion shifts the three-base reading frame to one of AGC repeats. A later insertion restores the reading frame. *Asterisks* (*) indicate point of deletion or insertion.

Figure 11.29

The genetic code is read as a nonoverlapping code with no punctuation (*top*). Before that was proven, it was suggested that the code could be overlapping by one or two bases (*middle*), or have noncoded bases between code words (punctuation; *bottom*).

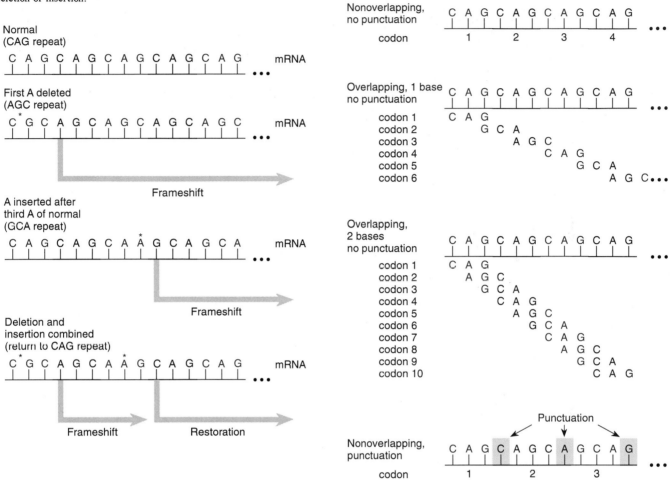

Figure 11.28

The coding frame of CAG repeats is first shifted and then restored by three additions (insertions). *Asterisks* (*) indicate insertion.

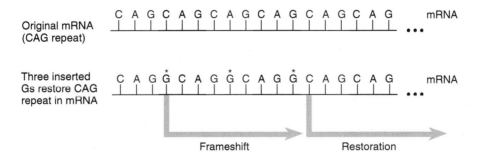

Severo Ochoa (1905–)
Courtesy of Dr. Severo Ochoa.

Marshall W. Nirenberg (1934–)
Courtesy of Dr. Marshall W. Nirenberg.

Punctuation between codons was also tentatively ruled out, in view of the length of the mRNA in the tobacco necrosis satellite virus, which has just about enough codons to specify its coat protein with no room left for a punctuating base or bases between each codon. (But see the information in chapter 12 on phage φX174, in which a given length of genetic material codes for an unusually large number of proteins. That situation appears to be an oddity rather than the rule.)

Breaking the Code

Given that the genetic code is in nonoverlapping triplets, the sixty-four codons still had to be worked out. For example, which amino acid is specified by ACU? The work was done in two stages. In the first stage, M. W. Nirenberg, S. Ochoa, and their colleagues made long artificial mRNAs and determined which amino acids these mRNAs incorporated into protein. In the second stage, specific triplet RNA sequences were synthesized. The amino acid-tRNA complex that was bound by each sequence was then determined.

Synthetic mRNAs

The ability to synthesize long-chain mRNAs resulted from the discovery, by M. Grunberg-Manago and Ochoa in 1955, of the enzyme **polynucleotide phosphorylase,** which

TABLE 11.3 Structure of Artificial mRNA Made by Randomly Assembling Uracil- and Guanine-Containing Ribose Diphosphate Nucleotides with a Ratio of 5U:1G

Codon	Frequency of Occurrence
UUU	$(5/6)^3 = 0.58$
UUG	$(5/6)^2(1/6) = 0.12$
UGU	$(5/6)^2(1/6) = 0.12$
GUU	$(5/6)^2(1/6) = 0.12$
UGG	$(5/6)(1/6)^2 = 0.02$
GUG	$(5/6)(1/6)^2 = 0.02$
GGU	$(5/6)(1/6)^2 = 0.02$
GGG	$(1/6)^3 = 0.005$

will join diphosphate nucleotides into long-chain, single-stranded polynucleotides. Unlike a polymerase, polynucleotide phosphorylase does not need a primer on which to act. This enzyme is found in all bacteria. (Its main function in the cell is probably the reverse of the use made of it here. It most likely serves as an exonuclease, degrading mRNA.) In 1961, Nirenberg and J. H. Matthei added artificially formed RNA polynucleotides of known composition to an *E. coli* ribosome system and looked for the incorporation of amino acids into proteins.

The system just described is called a **cell-free system,** a mixture primarily of the cytoplasmic components of cells, such as *E. coli,* but missing nucleic acids and membrane components. These systems are relatively easy to create by disruption and then fractionation of whole cells. They hold the advantage of containing virtually all the components needed for protein synthesis except the messenger RNAs. Their disadvantages are that they are relatively short-lived (several hours) and they are relatively inefficient in translation. However, an added benefit to the *E. coli* cell-free system is that it will translate, albeit inefficiently, RNAs that normally are not translated *in vivo* because they lack the signals for the initiation of translation. This feature of the system is what allowed these scientists to use artificial mRNAs that contained no Shine-Dalgarno sequence for ribosome binding.

Nirenberg and Matthei found that when uridine diphosphates were made into a messenger (poly-U) by the enzyme polynucleotide phosphorylase in the *E. coli* cell-free system, phenylalanine residues were incorporated into a polypeptide. Thus the first code word established was UUU for phenylalanine. The work was continued by Nirenberg and Ochoa and their associates. They found that AAA was the code word for lysine, CCC was the code word for proline, and GGG was the code word for glycine.

They then made synthetic mRNAs by using mixtures of the various diphosphate nucleotides in known proportions. An example is given in table 11.3. From their ex-

Phillip Leder (1934–)
Courtesy of Dr. Phillip Leder.

periments it was possible to determine the bases used in many of the code words, but not their specific order. For example, cysteine, leucine, and valine are all coded by two U's and a G, but the experiment could not sort out the order of these bases (5'-UUG-3', 5'-UGU-3', or 5'-GUU-3') for any one of them. Determining the order required a step in sophistication—that is, being able to synthesize known trinucleotides.

Synthetic Codons

When trinucleotides of known composition could be manufactured, Nirenberg and P. Leder in 1964 developed a "binding assay." They found that isolated *E. coli* ribosomes, in the presence of high-molarity magnesium chloride, could bind trinucleotides as if they were messengers. Also bound would be the tRNA that carried the anticodon complementary to the trinucleotide. It was thus possible, using radioactive amino acids, to determine which mRNA trinucleotide coded for a particular amino acid. A given synthetic trinucleotide was mixed with ribosomes and aminoacyl-tRNAs, including one amino acid that was radioactively labeled. The reaction mixture was passed over a filter that would allow everything except the large trinucleotide + ribosome + aminoacyl-tRNA complex to pass through. If the radioactivity passed through the filter, it meant that the radioactive amino acid was not associated with the ribosome. The experiment was then repeated with another labeled amino acid. When the radioactivity appeared on the filter, the investigators knew that the amino acid was affiliated with the ribosome. Thus that amino acid was coded by the trinucleotide codon. In other words, the radioactive amino acid was attached to a tRNA whose anticodon was complementary to the trinucleotide codon and thus bound at the ribosome.

For example, in figure 11.30, the trinucleotide is 5'-CUG-3'. The tRNA with the anticodon 3'-GAC-5' is charged with leucine. The mixture is passed through a filter. If threonine, or any other amino acid except leucine, is radioactive, the radioactivity will pass through the filter. When the experiment is repeated with radioactive leu-

Figure 11.30
The binding assay to determine the amino acid associated with a given trinucleotide codon. Transfer RNAs with noncomplementary codons will pass through the membrane. Transfer RNAs with anticodons complementary to the trinucleotide will bind to the ribosome and not pass through the filter. When the tRNA is charged with a radioactive amino acid, the radioactivity is trapped in the filter.

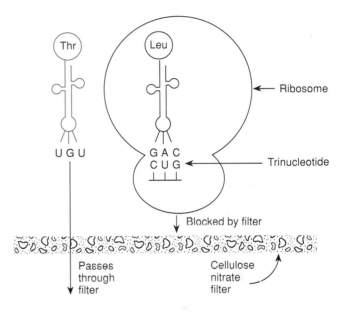

cine, the leucine, and hence the radioactivity, is trapped by the filter. In a short period of time, all of the codons were deciphered (table 11.4).

Wobble Hypothesis

The genetic code is a **degenerate code** in that a given amino acid may have more than one codon. As can be seen from table 11.4, eight of the sixteen boxes contain just one amino acid per box. (A box is determined by the first and second positions, e.g., the UUX box, in which X is any of the four bases.) Therefore, for these eight amino acids the codon need only be read in the first two positions because the same amino acid will be represented regardless of the third base of the codon. These eight groups of codons are termed **unmixed families** of codons. An unmixed family is the four codons beginning with the same two bases that specify a single amino acid. For example, the codon family GUX codes for valine. **Mixed families** code for two amino acids, or stop signals and one or two amino acids.

Six of the mixed-family boxes are split in half so that the codons are differentiated by the presence of a purine or a pyrimidine in the third base. For example, CAU and CAC both code for histidine; in both, the third base, U (uracil) or C (cytosine), is a pyrimidine. Only two of the families of codons are split differently.

TABLE 11.4   The Genetic Code

First Position (5′ End)	Second Position				Third Position (3′ End)
	U	C	A	G	
U	Phe	Ser	Tyr	Cys	U
	Phe	Ser	Tyr	Cys	C
	Leu	Ser	stop	stop	A
	Leu	Ser	stop	Trp	G
C	Leu	Pro	His	Arg	U
	Leu	Pro	His	Arg	C
	Leu	Pro	Gln	Arg	A
	Leu	Pro	Gln	Arg	G
A	Ile	Thr	Asn	Ser	U
	Ile	Thr	Asn	Ser	C
	Ile	Thr	Lys	Arg	A
	Met (start)	Thr	Lys	Arg	G
G	Val	Ala	Asp	Gly	U
	Val	Ala	Asp	Gly	C
	Val	Ala	Glu	Gly	A
	Val	Ala	Glu	Gly	G

The lesser importance of the third position in the genetic code ties in with two facts about transfer RNAs. First, although there would seem to be a need for sixty-two tRNAs—since there are sixty-one codons specifying amino acids and an additional codon for initiation—there are actually only about fifty different tRNAs in an *E. coli* cell. Second, a rare base such as inosine can appear in the anticodon, usually in the position that is complementary to the third position of the codon. These two facts lead to the concept that some kind of conservation of tRNAs is occurring and that rare bases may be involved.

It should be mentioned, to avoid confusion, that both mRNA and tRNA bases are usually numbered from the 5′ side. Thus the number one base of the codon is complementary to the number three base of the anticodon (fig. 11.31). Thus the codon base of lesser importance is the number three base, whereas its complement in the anticodon is the number one base.

Since the first position of the anticodon is not as constrained as the other two positions, a given base at that position may be able to pair with any of several bases in the third position of the codon (fig. 11.32). Crick characterized this ability as **wobble.** Table 11.5 shows the possible pairings that would produce a tRNA system compatible with the known code. For example, if an isoleucine tRNA has the anticodon 3′-UAI-5′, it is compatible with the three codons for that amino acid (table 11.4): 5′-AUU-3′, 5′-AUC-3′, and 5′-AUA-3′. That is, inosine

Figure 11.31

Codon and anticodon base positions are numbered from the 5′ end. The 3′ position in the codon (5′ in the anticodon) is the wobble base.

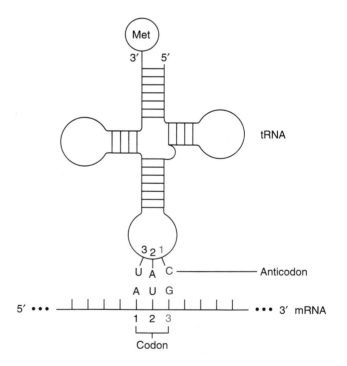

Figure 11.32

Base pairing possibilities of guanine and inosine in the third (3′) position of a codon. In the wobble position, guanine can form base pairs with both cytosine and uracil. Inosine, in the wobble position, can pair with cytosine, adenine, and uracil.

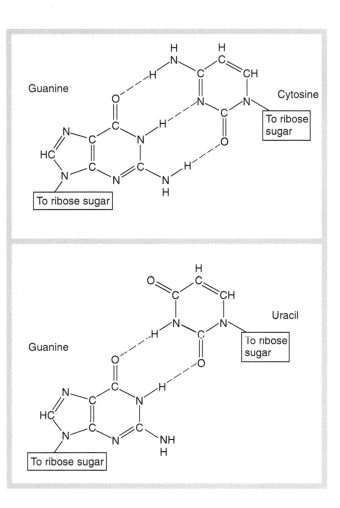

in the first (5′) position of the anticodon can recognize U, C, or A in the third (3′) position of the codon, and thus one tRNA will complement all three codons for isoleucine.

Universality of the Genetic Code

Until 1979, scientists had concluded from all experiments that the genetic code was universal. That is, the codon dictionary (table 11.4) was the same for *E. coli,* human beings, and oak trees, as well as all other species studied

TABLE 11.5 Pairing Combinations at the Third Codon Position

No. 1 Base in tRNA (5′ End)	No. 3 Base in mRNA (3′ End)
G	U *or* C
C	G
A	U
U	A *or* G
I	A, U, *or* C

TABLE 11.6   The Genetic Code Dictionary of Yeast Mitochondria*

First Position (5′ End)	Second Position				Third Position (3′ End)
	U	C	A	G	
U	Phe AAG	Ser AGU	Tyr AUG	Cys ACG	U / C
	Leu AAU		stop	Trp ACU	A / G
C	Thr GAU	Pro GGU	His GUG	Arg GCA	U / C
			Gln GUU		A / G
A	Ile UAG	Thr UGU	Asn UUG	Ser UCG	U / C
	Met UAC		Lys UUU	Arg UCU	A / G
G	Val CAU	Ala CGU	Asp CUG	Gly CCU	U / C
			Glu CUU		A / G

Source: Data from Bonitz, et al., 1980.
*Anticodons (3′ → 5′) are given within boxes. (The ACU Trp anticodon is predicted.)

to that time. The universality of the code had been demonstrated, for example, by taking the ribosomes and mRNA from rabbit reticulocytes and mixing them with the aminoacyl-tRNAs and other translational components from *E. coli*. Rabbit hemoglobin was synthesized.

In 1979 and 1980, however, discrepancies were noted when mitochondrial genes for structural proteins were sequenced (see chapters 12 and 17). It was discovered that there were two kinds of deviations from universality in the reading of the code by mitochondrial tRNAs. First, fewer tRNAs were needed to read the code. Second, there were several instances in which a codon was interpreted differently by the mitochondrial and cellular systems.

According to Crick's wobble rules (table 11.5), thirty-two tRNAs (including one for initiation) can complement all sixty-one nonterminating codons. Unmixed families require two tRNAs and mixed families require one, two, or three tRNAs, depending on the family. In the yeast mitochondrial coding system, apparently only twenty-four tRNAs are needed. The reduction in numbers is accomplished primarily by having each unmixed family recognized by only one tRNA (table 11.6: cf. table 11.4). Because mitochondrial tRNAs for unmixed families of codons have a U in the first (wobble) position of the anticodon, apparently, given the structure of the mitochondrial tRNAs, the U can pair with U, C, A, or G. Presumably, there has been evolutionary pressure to minimize the number of tRNA genes in the DNA of the mitochondrion, perhaps because of its small size. Reduction

from thirty-two to twenty-four is a 25% savings. (Recent evidence suggests that mammalian mitochondria may need only twenty-two tRNAs.)

It has also been found that yeast mitochondria read the CUX family as threonine rather than as leucine (tables 11.4 and 11.6) and the terminator UGA (opal) as tryptophan rather than as termination. However, in the reading of the CUX family, there appear to be differences among different groups of organisms. Human and *Neurospora* mitochondria appear to read the CUX codons as leucine, just as cellular systems do. Of the groups so far analyzed, only yeast reads the CUX family as threonine. Human and *Drosophila* mitochondria read AGA and AGG as stop signals rather than as arginine (table 11.7).

In 1985, it was discovered that *Paramecium* species read the UAA and UAG stop codons as glutamine within the cell. In addition, a prokaryote (*Mycoplasma capricolum*) reads UGA as tryptophan. We do not yet know how general this latest finding is: The genetic code of very few species has been scrutinized. We can thus conclude that the genetic code seems to have universal tendencies among prokaryotes, eukaryotes, and viruses. Mitochondria, however, read the code slightly differently. Different wobble rules apply, and at least one terminator and one unmixed family of codons are read differently by mitochondria and cells. Also, the mitochondrial discrepancies are not universal among all types of mitochondria. Further work, involving the sequencing of more mitochondrial DNAs, should elucidate the pattern of discrepancies

TABLE 11.7 Common and Alternative Meanings of Codons

Codon	General Meaning	Alternative Meaning
CUX	Leu	Thr in yeast mitochondria
AUA	Ile	Met in mitochondria of yeast, *Drosophila,* and vertebrates
UGA	*stop*	Trp in mycoplasmas and mitochondria other than higher plants
AGA/AGG	Arg	*Stop* in mitochondria of yeast and vertebrates
		Ser in mitochondria of *Drosophila*
CGG	Arg	Trp in mitochondria of higher plants
UAA/UAG	*stop*	Gln in ciliated protozoa

among the mitochondria of diverse species. We also now know that not every organism reads all codons in the same way. Ciliated protozoa and a mycoplasma read some stop signals as coding for amino acids.

One other class of variation of codon reading occurs, referred to as *site-specific variation,* in which the interpretation of a codon depends on its specific location. We are already familiar with the fact that GUG and, rarely, UUG can be prokaryotic initiation codons. This means that they will be recognized by tRNA$_f^{Met}$. However, they are not recognized by tRNA$_m^{Met}$ (i.e., GUG and UUG are not misread internally in messenger RNAs). In some cases, two of the termination codons (UGA and UAG, but not UAA) are misinterpreted as codons for amino acids. That is, termination will not occur at the normal place, resulting in a longer than usual protein. In some cases these "readthrough" proteins are vital—the organism depends on their existence. For example, in the phage Qβ, the coat-protein gene is read through about 2% of the time. Without this small number of readthrough proteins, the phage coat cannot properly be constructed.

One last example of site-specific variation involves the amino acid selenocysteine (cysteine with a selenium atom replacing the sulfur: fig. 11.1). Although many proteins have unusual amino acids, almost all are due to posttranslational modifications of normal amino acids. However, the amino acid selenocysteine seems to be inserted directly into some proteins, such as formate dehydrogenase in *E. coli,* which has selenium in its active site. Selenocysteine is inserted into the protein by a novel tRNA that recognizes the termination codon, UGA, if that codon is involved in a particular stem-loop secondary structure in the mRNA. In other words, the termination codon, UGA, is normally read as a stop codon; however, if involved in a particular stem-loop structure, a selenocysteine is inserted instead of termination. The selenocysteine tRNA is originally charged with a serine that is then modified to a selenocysteine.

Evolution of the Genetic Code

It has been theorized that the genetic code has wobble in it because it originally arose as a couplet code for the small number of amino acids that were in use several billion years ago. As new amino acids with useful properties became available, they were incorporated into proteins by a code modified by adding a third letter with less specificity. This theory is especially easy to envision if we see the code as originally punctuated with a base between each codon.

The fact that all sixty-one possible codons are used means that biological systems are protected to some extent against some mutations. For example, in the unmixed codon family 5'-CUX-3', any mutation in the third position will produce another codon for the same amino acid. Wobble in the third position and codon arrangement ensures that less than half of the mutations in the third position of codons result in a different amino acid being specified.

There are also patterns in the genetic code in which the mutation of one codon to another will result in an amino acid of similar properties. A high probability exists that a functional protein will be produced by such a mutation. All the codons with U as the middle base, for example, are for amino acids that are hydrophobic (phenylalanine, leucine, isoleucine, methionine, and valine). Mutation in the first or third positions for any of these codons will still code a hydrophobic amino acid. Both of the two negatively charged amino acids, aspartic acid and glutamic acid, have codons that start with GA. All of the aromatic amino acids—phenylalanine, tyrosine, and tryptophan (see fig. 11.1)—have codons that begin with uracil. Such patterns minimize the negative effects of mutations.

This chapter completes the discussion of the mechanics of gene expression. The next chapter deals with recombinant DNA technology, followed by several chapters concerned with the control of gene expression in both prokaryotes and eukaryotes.

SUMMARY

A charged tRNA has an anticodon at one end and a specific amino acid at the other end. The tRNAs are charged with the proper amino acid by aminoacyl-tRNA synthetase enzymes that incorporate the energy of ATP into amino acid-tRNA bonds. Hence no additional source of energy is needed during peptide bond formation. During protein synthesis the translation apparatus at the ribosome recognizes the transfer RNA. Through complementarity, the anticodon pairs with an mRNA codon.

An initiation complex forms at the start of translation. In prokaryotes it consists of the mRNA, the 30S subunit of the ribosome, the initiator tRNA with N-formyl methionine (fMet-tRNA$_f^{Met}$), and the initiation factors IF1, IF2, and IF3. The 50S ribosomal subunit is then added and A and P sites form in the resulting 70S ribosome. The charged N-formyl methionine tRNA is in the P site. A GTP is hydrolyzed and the initiation factors are released.

A tRNA enters the A site, which requires the involvement of elongation factors EF-Ts and EF-Tu (in *E. coli*). Another GTP hydrolysis releases the elongation factor, EF-Tu, which had originally brought the charged tRNA to the ribosome. Peptidyl transferase, a component of the 50S ribosomal subunit, transfers the amino acid from the tRNA in the P site to the amino end of the amino acid on the tRNA in the A site.

With the help of elongation factor G (EF-G), the ribosome translocates in relation to the mRNA. The depleted tRNA is released from the P site, and the tRNA with the growing peptide is moved into the P site. A GTP is hydrolyzed and EF-G is released. Elongation and translocation continue until a nonsense codon enters the A site. With the aid of the release factors RF1 and RF2, the protein is released, and the mRNA-ribosome complex dissociates. Eukaryotes have slightly more complex processes involving several more proteins.

Many antibiotics interfere with translation in prokaryotes. Puromycin, streptomycin, tetracycline, and chloramphenicol all act at the ribosome. Studying the mode of action of these antibiotics has provided insights into the mechanism of the translation process.

The genetic code was first assumed to be triplet on the basis of logical arguments regarding the minimum size of codons. With his work on deletion and insertion mutants, Crick provided evidence that the code was triplet. Part of the code was worked out initially with the synthesis of long, artificial mRNAs and then with the synthesis of specific trinucleotide codons. Crick's wobble hypothesis accounts for the fact that fewer than sixty-one tRNAs can read the entire genetic code. The reduction of tRNAs can happen because additional complementary base pairings occur in the third position (3′) of the codon.

The rule of universality of the genetic code has to be modified in light of findings regarding mitochondrial tRNAs, in which only twenty-four are needed to read the code. In addition, some sense codons are interpreted differently in mitochondrial systems; some nonmitochondrial systems read stop codons differently (a mycoplasma and ciliated protozoa); and some site-specific variation in codon reading also occurs. The structure of the code in both cells and mitochondria seems to protect the cell against a good deal of potential mutation.

SOLVED PROBLEMS

1. What is the energy requirement of protein biosynthesis?

   ANSWER: The cost of adding one amino acid to a growing polypeptide is four high-energy bonds: two from an ATP during the charging of the tRNA, and two from the hydrolysis of GTPs during tRNA binding to the A site of the ribosome and during translocation. Thus, for an average protein of 300 amino acids, there is a cost of 1,200 high-energy bonds.

2. What are the start and stop signals of translation?

   ANSWER: Once a messenger RNA is attached at the ribosome, the start signal is the methionine initiation codon (AUG), whereas the stop signal is one of the three nonsense codons (UAA, UAG, and UGA). Binding to the ribosome in order to position the mRNA in relation to the A and P sites differs in prokaryotes and eukaryotes. In prokaryotes, the Shine-Dalgarno sequence allows the 16S rRNA and the mRNA to form hydrogen bonds, thus locating the beginning of the mRNA at the ribosome. In eukaryotes, the 5′ cap is recognized by the ribosome and the ribosome then proceeds to scan the mRNA for the initiation codon.

3. What amino acids could replace methionine by a one-base mutation?

   ANSWER: The codon for methionine (internal as well as initiation) is AUG. If the A is replaced, we would get UUG (Leu), CUG (Leu), and GUG (Val); if the U is replaced we would get AAG (Lys), ACG (Thr), and AGG (Arg); and if the G is replaced, we would get AUA (Ile), AUU (Ile), and AUC (Ile). Hence a one-base change in the codon for methionine could result in six different amino acids.

1. Given the following piece of DNA that will be transcribed and then translated into a pentapeptide, provide the base sequence for its mRNA. Give the anticodons on the tRNAs by making use of wobble rules. What amino acids are incorporated? Draw the actual structure of the pentapeptide.
   3'-TACAATGGCCCTTTTATC-5'
   5'-ATGTTACCGGGAAAATAG-3'

2. Give an alternative mechanism of translation that would need only one site on the ribosome.

3. If DNA contained only the bases cytosine and guanine, how long would a code word have to be? How could we tell if this DNA were double-stranded?

4. If an artificial mRNA contains two parts uracil to one of cytosine, name the amino acids and the proportions in which they should be incorporated into protein.

5. Draw the details of a moment in time during the translation at the ribosome of the mRNA produced in problem 1. Include in the diagram the ribosomal sites, the tRNAs, and the various nonribosomal proteins involved.

6. How do prokaryotic and eukaryotic ribosomes recognize the 5' end of mRNAs? Could eukaryotic mRNAs be polycistronic?

7. How many aminoacyl-tRNA synthetases are there? What do they use for recognition signals? What is the "second genetic code"?

8. What are the similarities and differences among the three nonsense codons? Draw their theoretical anticodons by using the wobble rules.

9. Describe an experiment that demonstrates that the tRNA and not its amino acid is recognized at the ribosome during translation.

10. Other than the antibiotics named in the chapter, suggest five "theoretical" antibiotics that could interfere with the prokaryotic translation process.

11. How many single-base deletions are required to restore the reading frame of an mRNA? Give an example.

12. A "nonsense mutation" is one in which a codon for an amino acid changes to one for chain termination. Give an example. What are its consequences?

13. The reverse situation to question 12 is a mutation from a nonsense codon to a codon for an amino acid. Give an example. What are its consequences?

14. What would be proved or disproved if an organism were discovered that did not follow any of the rules of the codon dictionary? Do we expect organisms from another galaxy (if they exist) to use our codon dictionary?

15. What are the consequences of having a prokaryotic initiation tRNA recognized by an internal methionine codon?

16. What is the role of EF-Ts in elongation? EF-Tu? What are their eukaryotic equivalents?

17. What are the roles of RF1 and RF2 in chain termination? What are their eukaryotic equivalents?

18. What is a signal peptide? What role does it play in eukaryotes? What is its fate?

19. Why doesn't puromycin disrupt eukaryotic translation?

20. What would the genetic code dictionary (table 11.4) look like if wobble occurred in the second position rather than the third (i.e., if an unmixed family of codons were of the form GXU)?

21. A peptide, fifteen amino acids long, is digested by two methods, and each segment is sequenced according to the Edman degradation technique (see Box 11.3). The fifteen amino acids are denoted by the letters A through O, with F as the N-terminal amino acid. If the segments are as given below, what is the sequence of the original peptide?

    Method 1 CABHLN; FGKI; OEDJM

    Method 2 KICAB; JM; FG; HLNOED

22. In experiments using repeating polymers, (GCGC)$_n$ incorporates alanine and arginine into polypeptides, and (CGGCGG)$_n$ incorporates arginine, glycine, and alanine. What codon can definitely be assigned to glycine?

23. If poly-G is used as an mRNA in an incorporation experiment, glycine is incorporated into a polypeptide. If poly-C is used, proline is incorporated. If both poly-G and poly-C are used, no amino acids are incorporated into protein. Why?

24. A protein has leucine at a particular position. If the codon for leucine is CUC, how many different amino acids will appear as a result of single base substitutions?

25. In human hemoglobin, the β chain is 146 amino acids long. What is the minimum length of RNA needed to make this protein?

26. A transcribed DNA strand has the following sequence: 3'-TACTAACTTACGCTCGCCTCA-5'
    a. What is the sequence of RNA made from this strand?
    b. What is the sequence of amino acids made by the RNA?

27. Polymers of (GUA)$_n$ result in the incorporation of only two different amino acids rather than three seen for most other three-base polymers. Why?

28. The sixth amino acid in the β chain of normal human hemoglobin is glu. Two different mutants have val and lys respectively. What is the likely codon for glu?

29. A normal protein has the following C-terminal amino acid sequence: *ser-thr-lys-leu*-COOH. A mutant is isolated with the following sequence: *ser-thr-lys-leu-leu-phe-arg*-COOH. What has probably happened to produce the mutant protein?

30. A segment of a normal protein and three different mutants appear below

normal	_ _ _ gly-ala-ser-his-cys-leu-phe _ _ _
mutant 1	_ _ _ gly-ala-ser-his
mutant 2	_ _ _ gly-ala-ser-leu-cys-leu-phe _ _ _
mutant 3	_ _ _ gly-val-ala-ile-ala-ser _ _ _

What is the probable sequence of bases in the normal RNA?

31. A normal protein has his in a given position. Four mutants are isolated and determined to have the following amino acids instead of his: tyr, gln, pro, or leu. What are the possible codon assignments, and what codon is probably used for his?

Suggested Readings for chapter 11 are on page 648.

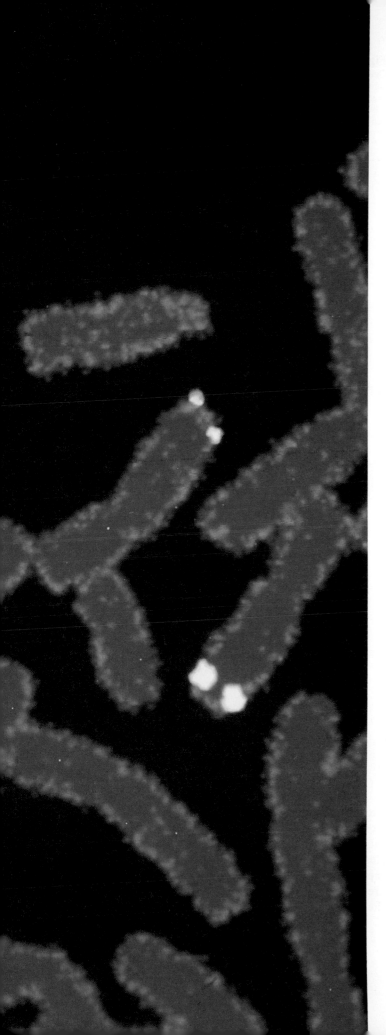

12

DNA Cloning and Sequencing

OBJECTIVES

- ♦ To envision the concepts behind gene cloning
- ♦ To look at the techniques of gene cloning
- ♦ To study the methods of DNA sequencing
- ♦ To look at the practical benefits and human issues of genetic engineering

Fluorescent probes of almost any DNA (*bright spots*) can be used to locate DNA sequences in human chromosomes (*red*) with new, nonradioactive techniques.

© *Peter Menzel/Photographed at Yale University Medical School*

Since the mid-1970s, the field of molecular genetics has undergone an explosive growth, noticeable not only to geneticists but also to medical practitioners and researchers, agronomists, animal scientists, venture capitalists, and the public in general. From the perspective of medical practitioners and researchers, new treatments for diseases have become available. Agronomists see the possibility of greatly improved crop yields, and animal scientists see the possibility of greatly improved production of food from domesticated animals. Geneticists and molecular biologists are gaining major new insights into understanding gene expression and its control.

The new techniques of DNA manipulation, centered around the isolation, amplification, sequencing, and expression of genes, are based on the insertion of a particular piece of DNA into a vehicle or vector—a plasmid or phage. A plasmid is placed into a host cell, either prokaryotic or eukaryotic, which then replicates (producing a clone), and the vector with its foreign piece of DNA also replicates. A phage simply multiplies in host cells. In both cases the foreign gene becomes amplified in number; it can be expressed (transcribed and translated into a protein) when in a plasmid in a host cell. A commonly used host cell is *E. coli* (fig. 12.1). Following its amplification, the foreign DNA can be purified and its nucleotide sequence can be determined. When it is expressed, large quantities of the gene product can be obtained. The new technology is variously referred to as **gene cloning, recombinant DNA technology,** or **genetic engineering.** In this chapter we look in detail at the methods and procedures of recombinant DNA technology, including DNA sequencing.

DNA CLONING

Restriction Endonucleases

The 1978 Nobel Prizes in physiology and medicine were awarded to W. Arber, H. Smith, and D. Nathans for their pioneering work in the study of **restriction endonucleases.** These are enzymes that bacteria use to destroy foreign DNA, usually of viruses. The enzymes recognize certain nucleotide sequences found on the foreign DNA, usually from four to ten base pairs, then cleave the DNA at all sites containing that specific sequence. (Restriction endonucleases originally were so named because they "restricted" phage infection among strains of bacteria. Phages that could survive in one strain could not survive in other strains with different restriction enzymes.)

Three types of restriction endonucleases are known. Their grouping is based on the types of sequences recognized, the nature of the cut made in the DNA, and the enzyme structure. Types I and III restriction endo-

Figure 12.1
Overview of recombinant DNA techniques. A hybrid vector is created, containing an insert of foreign DNA. The vector is inserted into a host organism. Replication of the host results in many copies of the foreign DNA and, if the gene is expressed, production of quantities of the gene product.

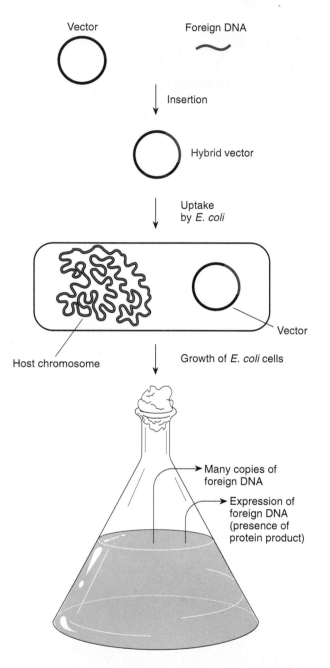

nucleases are not useful for gene cloning because they cleave DNA at sites other than the recognition sites and thus cause random cleavage patterns. Type II endonucleases, however, recognize specific sites and cleave at just these sites. The sites recognized by type II endonucleases are inverted repeats; they have twofold symmetry. To see the symmetry, you must read outward from a central axis

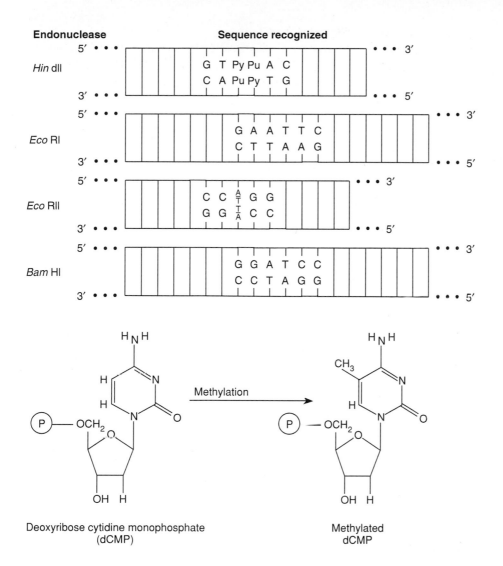

Endonuclease

Sequence recognized

Hin dII

5′ • • •   • • • 3′

        G T Py Pu A C
        C A Pu Py T G

3′ • • •   • • • 5′

Eco RI

5′ • • •   • • • 3′

            G A A T T C
            C T T A A G

3′ • • •   • • • 5′

Eco RII

5′ • • •   • • • 3′

        C C ^A_T G G
        G G ^T_A C C

3′ • • •   • • • 5′

Bam HI

5′ • • •   • • • 3′

            G G A T C C
            C C T A G G

3′ • • •   • • • 5′

Figure 12.2
Sequences cleaved by various type II restriction endonucleases. Py is pyrimidine and Pu is purine. *Color* denotes places where endonucleases cleave the DNA. Not all restriction endonucleases make staggered cuts.

Deoxyribose cytidine monophosphate
(dCMP)

Methylation →

Methylated
dCMP

Figure 12.3
The addition of a methyl group to cytosine by a methylase enzyme converts it to 5-methylcytosine.

on opposite strands of the DNA. For example, the type II restriction endonuclease *Bam* HI recognizes

$$5'\text{-GGA} \mid \text{TCC-}3'$$
$$3'\text{-CCT} \mid \text{AGG-}5'$$

Reading out from the center (vertical line) is AGG on the top strand and AGG on the bottom strand. The sequence is, in a sense, a **palindrome,** a sequence read the same from either direction. (Palindrome is from the Greek *palindromos,* which means "to run back." The name Hannah and the numerical sequence 1238321 are palindromes.) In figure 12.2 are some palindromic sequences recognized by type II restriction endonucleases; well over one hundred type II enzymes are known now.

The host cell protects its own DNA, not by being free of these sequences, but usually by methylating its own DNA in these regions (fig. 12.3). The same sequences that are attacked by the endonucleases in the unmethylated condition are protected when methylated. After host DNA replication, new double helices are hemimethylated; that is, the old strand is methylated but the new one is not. In this configuration the new strand is quickly methylated (fig. 12.4).

Restriction endonucleases are named after the bacterium from which they were isolated. *Bam* HI came from *Bacillus amyloliquefaciens,* strain H. *Eco* RI came from *E. coli,* strain RY13. *Hin* dII came from *Haemophilus influenzae,* strain Rd. *Bgl* I came from *Bacillus globigii.* From here on we will refer to type II restriction endonucleases simply as restriction enzymes.

Restriction enzymes cut the DNA in two different ways. For example, *Hin* dII cuts the recognition sequence down the middle, leaving "blunt" ends on the DNA (see fig. 12.2). We will discuss how pieces of DNA with blunt ends can be used in cloning. The staggered cuts made, for

Figure 12.4

Host DNA is methylated in the *Hpa* II restriction site. *Asterisks* indicate methyl groups on cytosines. After DNA replication, the DNA is hemimethylated; the new strands have no methyl groups. Hemimethylated DNA is then fully methylated.

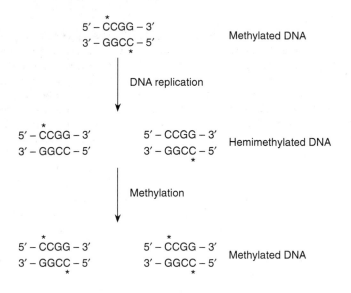

example, by *Bam* HI leave "sticky" ends that can spontaneously reanneal by hydrogen bonding of the complementary bases (see fig. 12.2). The ability to reanneal these sticky ends opened up the field of gene cloning.

Hybrid Vectors

With current technology it is feasible to join together, *in vitro*, DNAs from widely different sources. In figure 12.5 we see how a circular DNA molecule cleaved by a specific restriction enzyme can recircularize if it is cleaved in only one place, or how different molecules with the same free ends can anneal to form hybrid molecules. Only the action of a DNA ligase is needed to make the molecules complete (see chapter 9).

One of the pieces of DNA involved in the annealing can be a plasmid, a piece of DNA that can replicate in a cell independently of the cellular chromosome. The **hybrid plasmid** (fig. 12.6) can be transferred into a cell. A hybrid plasmid is also known as a **hybrid vehicle, hybrid vector,** or **chimeric plasmid.** The latter is named after the *chimera,* a mythological monster with a lion's head, a goat's body, and a serpent's tail. A variety of procedures exist whereby this chimeric plasmid can be introduced into a host cell. For example, a bacterial cell can be made permeable to this, or any, plasmid by the addition of a dilute solution of calcium chloride. The foreign DNA will be replicated each time the plasmid DNA replicates.

Note that in the process of inserting a piece of foreign DNA, the restriction site is duplicated, with one copy at either end of the insert. This property makes it easy to remove the cloned insert at some future time, if needed, since it is enclosed by restriction sites.

Restriction Cloning

A few conditions have to be met in order to succeed in cloning DNAs from different sources. The cleavage site of the plasmid must occur in a nonessential region of the plasmid or the plasmid will be rendered ineffective by the process. Also, the **passenger** DNA (foreign DNA) that is being fused into the vehicle plasmid should be a functional unit—that is, it should be at least a complete gene or intact region of interest. Other conditions also prevail.

A plasmid vehicle must be cleaved at only one point by the endonuclease. If it is cleaved at more than one point, it will be fragmented during the experiment. Some phage vehicles must be cleaved at two points so that the foreign DNA can be substituted for a length of the phage DNA rather than simply inserted. Common vehicles, derivatives of phage λ, have been named **Charon phages** (pronounced karon) after the mythical boatman of the River Styx. (See chapter 13 for a detailed discussion of phage λ.)

During normal phage infection (see chapter 7), only DNA within a certain size range is packaged into λ heads. Thus for λ to be a useful vector, part of its DNA must be replaced by the foreign DNA. We note that λ can function quite well as a hybrid vehicle with a 15,000-base-pair (15 kilobases or 15-kb) section replaced by foreign DNA because that section of phage DNA is used for integration into the *E. coli* chromosome, a nonessential phage function. Genetic engineers have created a λ DNA molecule with the nonessential region missing and containing an *Eco* RI cleavage site. Only hybrid DNA can thus be incorporated into phage heads because the diminished phage DNA without an insert is too small to be packaged properly.

One disadvantage of cloning with normal *E. coli* plasmids is that they are unstable if the foreign DNA is very large, greater than 15 kb. That is, if a large chromosomal segment is cloned, the plasmid will tend to selectively lose parts of the clone as the plasmid replicates. Primarily for this reason, geneticists began using phage λ as a vector (see fig. 7.24) because these phages could successfully maintain foreign DNA as large as 24 kb. The phage chromosome is about 50 kb of DNA; within the phage head it is linear, within the cell it is circular. The DNA to fill the phage head is recognized because it has a small segment of single-stranded DNA at either end called a *cos* site (twelve bases). Reannealing of *cos* sites allows λ chromosomes to circularize when they enter a host cell. Cutting the DNA at the *cos* site opens the circle into a linear molecule (fig. 12.7). (*Cos* is derived from the term "cohesive ends.") Geneticists have taken advantage of these *cos* sites to clone even larger segments of foreign DNA because it turns out that even 24 kb is not adequate to study some eukaryotic genes or gene groups. Many eukaryotic genes are very large because of their introns and transcriptional control segments. DNA up to 50 kb can be cloned if *cos* segments are attached to either end with a plasmid origin of DNA replication. These *cos*-

Figure 12.5

Circular plasmid DNA with a palindrome recognized by *Eco* RI. After the DNA is cleaved by the endonuclease, it has two exposed ends that can join to recircularize the molecule or to unite two or more linear molecules of DNA cleaved by the same restriction endonuclease. The final nicks are closed with DNA ligase.

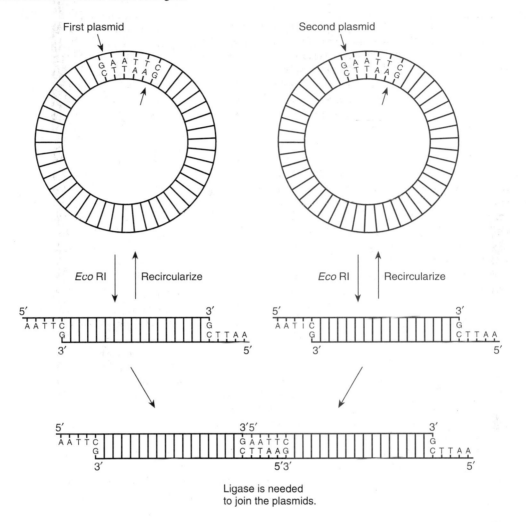

Ligase is needed
to join the plasmids.

site-containing plasmids are called **cosmids** (fig. 12.7). Not only do these cosmids allow the cloning of very large pieces of DNA, they actually select for large segments of foreign DNA because small cosmids are not incorporated into phage heads. Thus a range in size of foreign DNA, from 2.5 to 50 kb, can be cloned using plasmids, Charon phages, or cosmids.

Selecting for Hybrid Vectors

In the methods we have been describing, restriction enzymes separately cut both vector and foreign DNA. The two are then mixed in the presence of ligase. Many products are created, which can be generally divided into three categories: vectors with appropriate foreign DNA; vectors with inappropriate foreign DNA; and junk, including uncut vectors and vector fragments. In a later sec-

tion we will discuss methods of finding a particular piece of foreign DNA in a vector. Here we will point out ways that vectors with inserts of any kind are selected.

Charon phages are selected simply by their ability to infect *E. coli* cells. As we mentioned above, only λ DNA with a foreign insert will be packaged. Plasmids (including cosmids) that contain foreign DNA can be selected through screening for antibiotic resistance. A widely used cloning plasmid is pBR322. It contains genes for tetracycline and ampicillin resistance and various restriction sites. There is, for example, a *Bam* HI site in the tetracycline-resistance gene (fig. 12.8). After cloning, there will be plasmids present with the foreign DNA and those without an insert. *E. coli* cells are then exposed to these plasmids and plated on a medium without antibiotics. Replica plating is then done onto plates with one or both antibiotics. Colonies resistant to both antibiotics have cells

Figure 12.6

Formation of a hybrid plasmid. The same restriction endonuclease, in this case *Eco* RII, is used to cleave both host and foreign DNA. Some of the time, cleaved ends will come together to form a plasmid with an insert of the foreign DNA. Ligase seals the nicks.

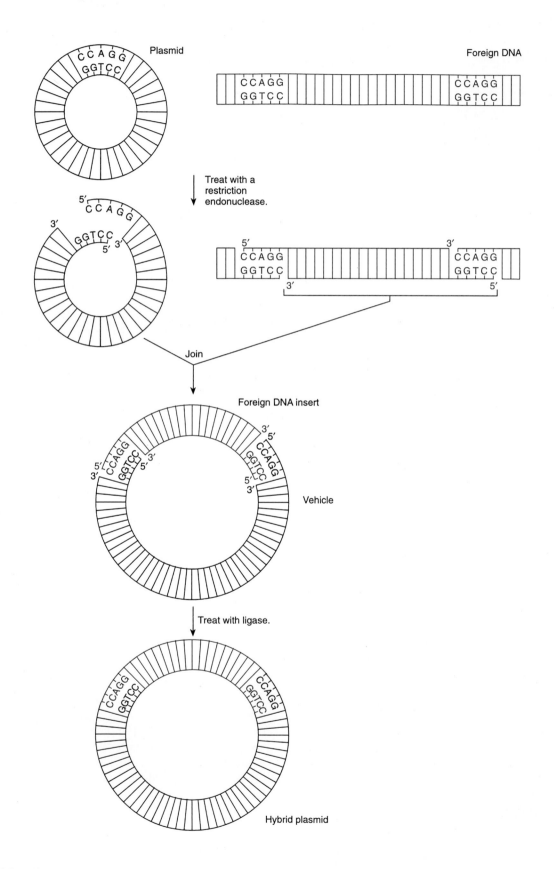

Figure 12.7

A cosmid. It is a plasmid with *cos* sites and is transferred between bacteria within phage lambda heads. The *cos* sites are single-stranded; they reanneal to a *circle* when inside the host.

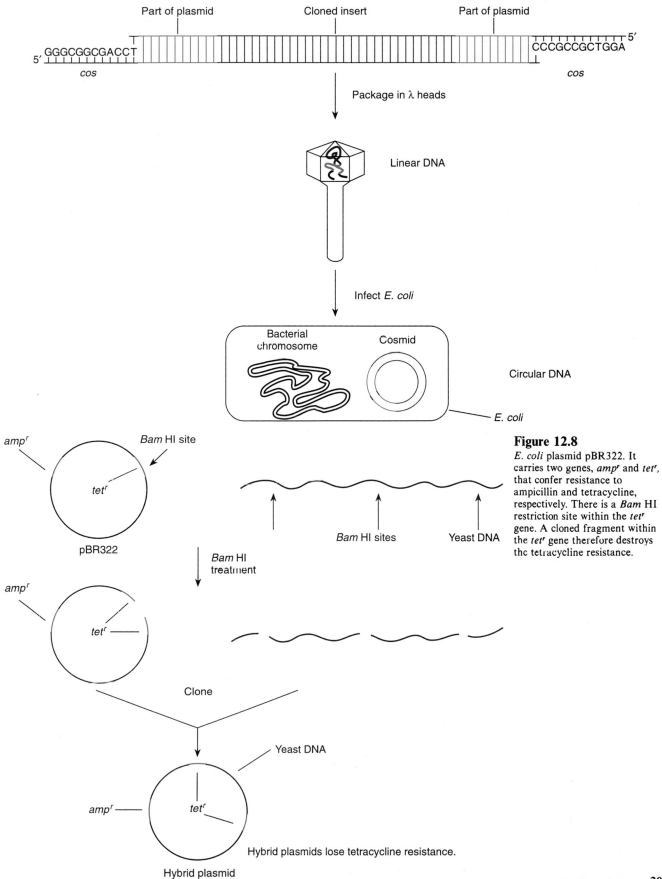

Part of plasmid   Cloned insert   Part of plasmid

5′ GGGCGGCGACCT   CCCGCCGCTGGA 5′

cos   *cos*

Package in λ heads

Linear DNA

Infect *E. coli*

Bacterial chromosome   Cosmid

Circular DNA

E. coli

amp^r   *Bam* HI site

tet^r

pBR322

Bam HI sites   Yeast DNA

Figure 12.8

E. coli plasmid pBR322. It carries two genes, *amp^r* and *tet^r*, that confer resistance to ampicillin and tetracycline, respectively. There is a *Bam* HI restriction site within the *tet^r* gene. A cloned fragment within the *tet^r* gene therefore destroys the tetracycline resistance.

amp^r

tet^r

Bam HI treatment

Clone

Yeast DNA

amp^r   tet^r

Hybrid plasmids lose tetracycline resistance.

Hybrid plasmid

Figure 12.9

Poly-dA/poly-dT technique for producing hybrid plasmids. One of the nucleotides is added to the 3' ends of the plasmid, and the complementary nucleotides are added to the 3' ends of the foreign DNA. The open plasmid and the foreign DNA then have complementary ends and can form a circle. Repair enzymes, culminating in DNA ligase, fill in the gaps.

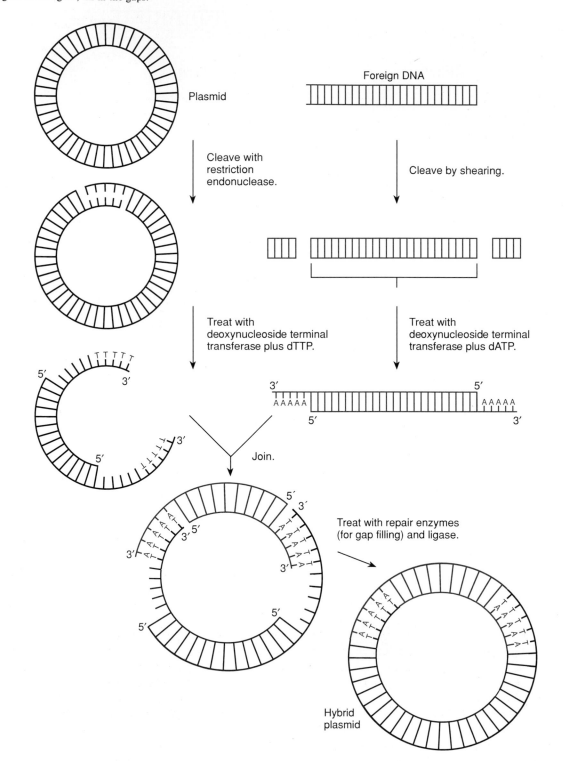

Figure 12.10

Linkers. They are small segments of DNA with an internal restriction site. They can be added to blunt-ended DNA by T4 ligase. Treatment with the restriction enzymes creates DNA with ends compatible with any DNA cut by the same restriction enzyme.

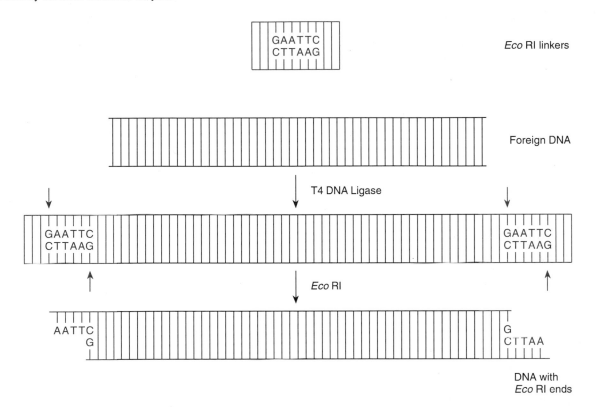

Eco RI linkers

Foreign DNA

T4 DNA Ligase

Eco RI

DNA with *Eco* RI ends

with plasmids having no inserts; those resistant only to ampicillin have a plasmid with an insert. Colonies resistant to neither antibiotic have cells with no plasmids.

Poly-dA/Poly-dT and Blunt-End Ligation Methods of Cloning

Restriction endonuclease treatment may not suffice for cloning: An endonuclease cut may fall in the wrong place, say in the middle of a desired gene, or the foreign DNA may have been isolated by other methods, such as physical shearing. In these cases, several other methods of cloning can be used. In the **poly-dA/poly-dT technique,** an addition of poly-dA (a sequence of deoxyadenylate nucleotides) is made to the 3′ ends of the passenger DNA, and poly-dT (deoxythymidylate nucleotides) is added to the 3′ ends of the vehicle, creating complementary "sticky ends." To accomplish this, the foreign DNA is treated with the enzyme deoxynucleoside terminal transferase and dATP, whereas the cleaved plasmids are treated with deoxynucleoside terminal transferase and dTTP. (This method also works using dCTP/dGTP [deoxycytidylate and deoxyguanylate nucleotides].) Gaps in the overlap regions are filled in by gap-filling repair enzymes (chapter 16). The process is illustrated in figure 12.9.

A more popular method of joining passenger and vehicle molecules that do not have sticky ends is called **blunt-end ligation.** Without adding poly-dA or poly-dT tails, blunt-ended DNA can be joined by the phage enzyme, T4 DNA ligase. Blunt ends can be generated when segments of DNA to be cloned are created by physically breaking the DNA or by certain restriction endonucleases, such as *Hin* dII (see fig. 12.2), which form blunt ends. Since the ligase is nonspecific about which blunt ends it joins, many different, unwanted products result from its action. Cloning using restriction enzymes that produce sticky ends is the preferred method.

A variation of blunt-end ligation uses linkers—short, artificially synthesized pieces of DNA containing a restriction endonuclease recognition site. When these linkers are attached to blunt pieces of DNA and then treated with the restriction endonuclease, sticky ends are created. In figure 12.10, the linkers are 12 bp (base pair) segments of DNA with an *Eco* RI site in the middle. They are attached to the DNA to be cloned with T4 DNA ligase. Subsequent treatment with *Eco* RI will result in DNA with *Eco* RI sticky ends.

DNA for cloning can generally be obtained in two ways: (1) a desired gene or DNA segment can be synthesized or isolated or (2) the genome of an organism can be

Figure 12.11

(a) An mRNA converted into double-stranded DNA for cloning using reverse transcriptase. (b) A poly-T segment is added, which is complementary to the poly-A tail of a eukaryotic mRNA. (c) Reverse transcriptase then uses this primed configuration to synthesize DNA from the RNA template. At the end of the RNA, the enzyme forms a hairpin loop, which, after the degradation of the original RNA (d), forms a primer for DNA polymerase I to form double-stranded DNA (e). The enzyme S1 nuclease removes the hairpin. (f) The result is cDNA.

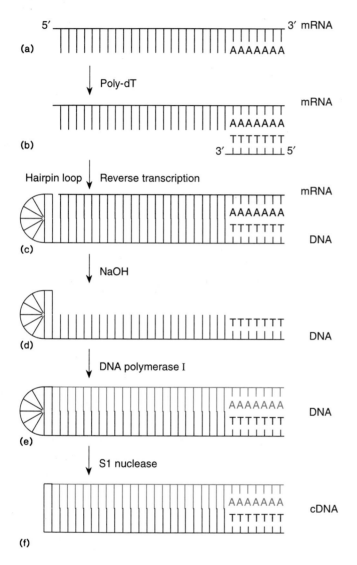

Creating a Clonable Product

Creating DNA to Clone

In order to clone a particular gene (or DNA segment), and only that gene, a scientist must have a purified double-stranded piece of DNA containing that gene. There are numerous ways of obtaining that DNA, but most entail creating or isolating a single-stranded mRNA that is then enzymatically converted into double-stranded DNA. (One of these methods is described in the following paragraphs.) The problem is then reduced to obtaining the desired mRNA.

The mRNA for a particular gene can be obtained several different ways, depending on the particular gene. If there are large quantities of the RNA available from a particular cell, it can be isolated directly. For example, mammalian erythrocytes have abundant quantities of α- and β-globin mRNAs. Also, ribosomal RNA and many transfer RNAs are relatively easy to isolate in quantities adequate for cloning.

Double-stranded DNA for cloning is made from the purified RNA with the aid of the enzyme reverse transcriptase, isolated from RNA tumor viruses (see chapter 10). We describe here the conversion of RNA to DNA using a eukaryotic mRNA with a 3' poly-A tail (fig. 12.11). In the first step, a poly-T primer is added, which base-pairs with the poly-A tail of the mRNA. This short double-stranded region is now a primer for polymerase activity: a free 3'-OH exists. The primed RNA is then treated with the enzyme reverse transcriptase, which will polymerize DNA nucleotides using the RNA as a template. The result is a DNA-RNA hybrid molecule (fig. 12.11c).

Reverse transcriptase has the interesting property *in vitro* of creating a hairpin turn at the 5' end of the template. It begins to double back on itself, displacing the RNA and using the newly synthesized DNA as a template. The hairpin loop is useful because it provides a 3'-OH structure, a primer, for continued DNA synthesis after the mRNA is removed with NaOH, a treatment that does not affect DNA (fig. 12.11d). The resulting single-stranded DNA is referred to as **complementary DNA (cDNA)**.

Since a primer structure exists, DNA polymerase I can use the cDNA as a template to create a double-stranded DNA (fig. 12.11e). The hairpin loop can be removed by the enzyme S1 nuclease, which degrades single-stranded DNA (fig. 12.11f). Hence, starting with a piece of single-stranded mRNA, we have generated a piece of double-stranded DNA, which itself is often referred to as cDNA. This double-stranded DNA can then be cloned using the blunt-end methods described earlier.

broken into small pieces and the small pieces can be randomly cloned (originally referred to as a "shotgun" experiment). Then the desired DNA segment must then be "fished" for from among the various clones created. We look first at synthesizing or isolating a desirable gene before cloning it, and then we look at the process of locating a desired gene after it has been cloned.

Figure 12.12

The first cycle in automated, solid-phase DNA synthesis. Nucleotide 1 is bound to a silica substrate. Nucleotide 2, a DMT derivative, forms a phosphodiester bond with the first nucleotide. No further nucleotides can be added until the DMT is removed. After the DMT is removed, the dinucleotide is ready for a new cycle. The R_1 protecting groups (CH_3) are removed after the process is complete. The oligonucleotide is then separated from the silica substrate.

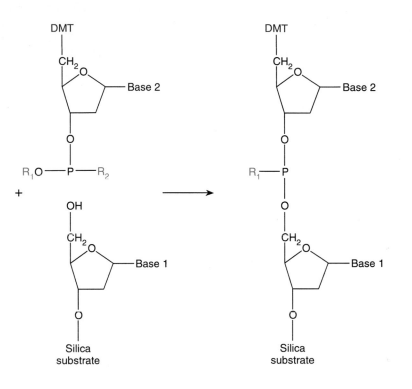

If the RNA is not available in large enough quantities, it is possible to directly synthesize DNA *in vitro* if the amino acid sequence of its expressed protein is known. The nucleotide sequence can be obtained from the genetic code dictionary (see table 11.4) if the sequence of amino acids is known from the protein product of the gene. This method will probably not re-create the original DNA because there is redundancy in the genetic code. In other words, when a leucine is encountered in a protein, it could have been coded by any one of six different codons. Despite an element of guesswork, it is possible to synthesize a DNA that will code for a particular protein.

Currently, DNA sequences of over one hundred bases can be synthesized by automated machines adding one base at a time in 10-minute cycles. A nucleotide is anchored to a silica substrate by its 3′ end. In the first cycle, a second nucleotide is added. To avoid multiple (additional) attachments, the second nucleotide has its 5′ end blocked by a chemical group, DMT (dimethoxytrityl). Thus only one nucleotide can be added per cycle (fig. 12.12). Other groups on the nucleotides, such as the phosphates, are protected during the various chemical reactions with various R-groups (usually CH_3). The cycle is completed by washing out excess nucleotide and removing

the DMT group. The dinucleotide is now ready for the next cycle. After the desired oligonucleotide is synthesized, the R-groups are removed and the oligonucleotide is released from the substrate.

You will notice that any method that uses the genetic code dictionary to produce DNA to clone will be missing the promoter of the gene and its transcriptional control sequences as well as other areas of the DNA not translated. Although techniques exist to attach a promoter to a synthetic gene, we will describe techniques to overcome this problem in which random pieces of the genome are cloned.

Direct Formation of Clonable DNA

The previous methods cannot be used when an RNA is not available; when desirable sections of the DNA, such as promoter sequences, are not transcribed or translated; or when the protein product itself has not been sequenced. In these cases another method is used to clone a gene. The DNA of an organism is broken into small pieces from which the desired gene, or DNA fragment, is isolated. The desired DNA can be isolated either before or after cloning. The DNA is referred to as genomic DNA to differentiate it from cDNA.

Figure 12.13

Creating a genomic library using the shotgun approach in creating inserts. First, the genome is fragmented. The fragments are then randomly cloned in vectors. The collection of these vectors is referred to as a genomic library.

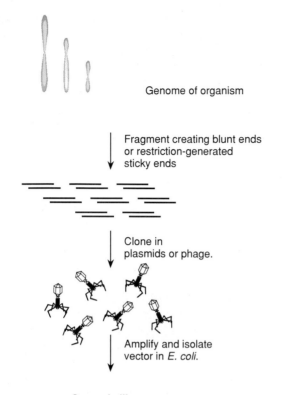

Genome of organism

Fragment creating blunt ends or restriction-generated sticky ends

Clone in plasmids or phage.

Amplify and isolate vector in *E. coli.*

Genomic library

Figure 12.14

Locating the rabbit β-globin gene within a DNA digest using the Southern blotting technique. The rabbit DNA (*a*) is segmented (*b*) and then electrophoresed on agarose gels (*c*). Southern blotting follows (*d*) in which the DNA is transferred to nitrocellulose filters. A radioactive probe (β-globin mRNA) locates the DNA fragment with the gene after autoradiography (*e*). (The researchers refer to the appearance of the radioactive band as the "lighting up" of the band.)

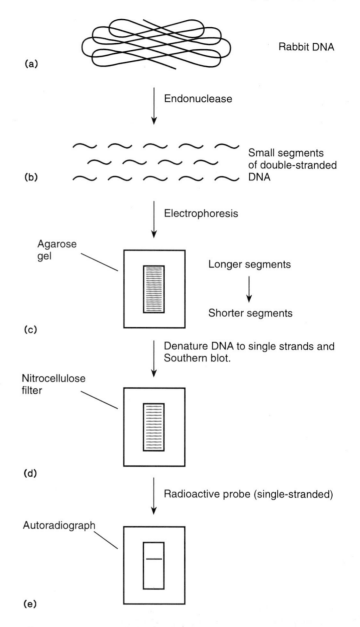

(a)   Rabbit DNA

Endonuclease

(b)   Small segments of double-stranded DNA

Electrophoresis

Agarose gel

Longer segments

Shorter segments

(c)

Denature DNA to single strands and Southern blot.

Nitrocellulose filter

(d)

Radioactive probe (single-stranded)

Autoradiograph

(e)

If the original DNA is isolated before cloning, then only that DNA need be cloned. Alternatively, a "shotgun" approach can be used in which the entire genome of an organism can be cloned (in small pieces, of course), creating a **genomic library,** a complete set of cloned fragments of the original genome of a species (fig. 12.13). In the case of a genomic library, a desired gene can be located after it is already cloned.

PROBING FOR A SPECIFIC GENE

When DNA segments are generated randomly by endonuclease digests or by physical shearing, a desired gene must be located. As mentioned above, we can look for the gene either before or after it is cloned. We look first at locating a specific gene in a DNA digest, before the DNA has been cloned.

Southern Blotting

To locate a specific gene in the midst of a DNA digest, a specific **probe** must be constructed. Probes are usually radioactively labeled nucleic acids whose complementarity precisely locates a particular DNA sequence. Thus if we wished to locate the gene for β-globin, we could use ra-

dioactively labeled β-globin mRNA or radioactively labeled cDNA. RNA-DNA or DNA-DNA hybrids would form between the specific gene and the radioactive probe. Autoradiography would then locate the radioactive probe.

Let us assume that we wanted to clone the rabbit β-globin gene. First, we would create a restriction digest or a DNA digest (fig. 12.14). We would then subject this digest to electrophoresis on agarose to separate the various fragments according to their sizes. Agarose is a good medium for separating DNA fragments of a wide variety of sizes.

Figure 12.15

Arrangement of gel and filters in the Southern blotting technique. The salt buffer is drawn upwards by the dry filter paper, transferring the DNA from the agarose gel to the nitrocellulose filter.

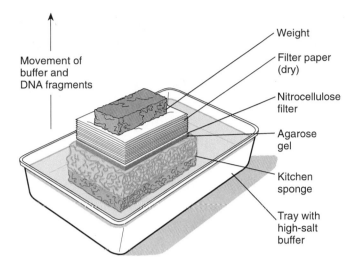

Movement of buffer and DNA fragments

Weight

Filter paper (dry)

Nitrocellulose filter

Agarose gel

Kitchen sponge

Tray with high-salt buffer

In a digest of this kind, however, there are usually so many fragments that the result is simply a smear of oligonucleotides, from very small to very large. To proceed further, we have to transfer the electrophoresed fragments to another medium for probing or the DNA fragments would diffuse off the agarose gel. Nitrocellulose filters are excellent for hybridization because the DNA fragments bind to the nitrocellulose under high salt concentrations and will not diffuse off. The transfer procedure, first devised by E. M. Southern, is called **Southern blotting.** In this technique the double-stranded DNA on the agarose gel is first denatured to single-stranded DNA, usually with NaOH. Then the agarose gel is placed directly against a piece of nitrocellulose filter, which is then placed agarose-side down on a wet sponge. Dry filter paper is placed against the nitrocellulose side. The high-salt buffer in the sponge is then drawn toward the dry filter paper, carrying the DNA segments from the agarose to the nitrocellulose (fig. 12.15). The DNA digest fragments are then permanently bound to the nitrocellulose filter by heating. DNA-DNA hybridization takes place on the filter. (A similar technique can be performed on RNA in which it is called, tongue-in-cheek, **northern blotting.** Immunological techniques, not involving nucleotide complementarity, can be used to probe for proteins in an analogous technique called **western blotting.**)

A radioactive probe can be obtained several different ways. In this example the easiest way to obtain a radioactive probe would be to isolate β-globin mRNA from rabbit reticulocytes and construct cDNA using the reverse-transcriptase method described previously. The deoxyribonucleotides used during reverse transcription would contain radioactive phosphorus, ^{32}P. As figure 12.14 shows, a single radioactive band locates a DNA segment with the β-globin gene. It should be noted that the probe,

originating from mRNA, will lack the introns present in the gene. However, probing is successful as long as there is a reasonable partial match.

To clone the β-globin gene, a second agarose gel would be run with a sample of the digest used in figure 12.14. That gel, not subject to DNA-DNA hybridization, would have the β-globin segment in the same place. The band, whose location is known from the autoradiograph, could be cut out of the agarose gel to isolate the DNA, which could then be cloned by methods discussed earlier in the chapter.

Probing for a Cloned Gene

Dot Blotting

The methods previously described are also useful in locating genes already cloned within plasmids, for example, after a genomic library has been constructed. To isolate yeast tRNA genes, for example, a library of the yeast genome was created using the *E. coli* plasmid pBR313. In this case the electrophoresis step is skipped since whole clones will be probed for a particular sequence of DNA rather than DNA segments within a digest.

In one particular experiment, yeast cells were grown on a medium that contained ^{32}P in order to produce radioactive tRNA—the probes—that were then isolated. Meanwhile 1,140 *E. coli* colonies, each resistant to ampicillin but sensitive to tetracycline—indicating the presence of a hybrid plasmid—were grown and transferred directly to twelve nitrocellulose filters, each holding 95 clones. In preparation for probing, the cells were lysed and their DNA denatured. The plasmid DNA within the cells of each clone was then hybridized with the radioactive tRNA probes. Figure 12.16 is an autoradiograph of the 1,140 clones. Dark spots (radioactivity) indicate clones carrying yeast tRNA genes (those clones "light up" autoradiographically). This technique, hybridization of cloned DNA without an electrophoretic-separation step, is referred to as **dot blotting.**

Western Blotting

An entirely different method used to locate particular cloned genes utilizes the actual expression of the cloned genes in the plasmid-containing cells. If a eukaryotic gene is cloned in an *E. coli* plasmid downstream from an active promoter, that gene may be expressed (transcribed and translated into protein). Plasmids that allow the expression of their foreign DNA are termed **expression vectors.** A particular protein product can be located by a method completely analogous to either Southern blotting or dot blotting, called western blotting. In this technique, probing is done with antibodies specific for a particular protein, rather than a radioactive oligonucleotide probe. These antibodies are located by a second antibody specific to the first, which has either a radioactive label or a color marker. For example, assume we are looking for the expression of

Figure 12.16

Dot blot autoradiograph of 1,140 clones carrying yeast DNA fragments obtained by restriction endonuclease activity (*Hin* dIII). Clones were hybridized with radioactive yeast tRNA. *Dark spots* indicate clones carrying yeast tRNA genes.

Source: J. S. Beckmann, "Cloning of Yeast Transfer RNA Genes in Escherichia coli," *Science vol. 196 (8 April 1977): pages 205–208, Fig. 1. Copyright 1977 by the AAAS.*

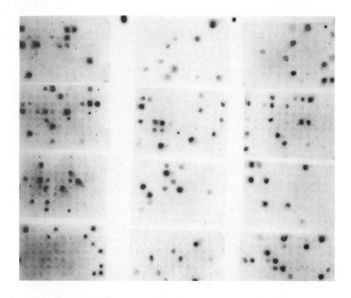

Figure 12.17

Western blot technique to locate an expressed protein from among many cloned genes using radioactive antibodies. Clones that may carry the expressed protein are lysed. Antibodies are applied that locate the protein either through autoradiography (*shown*) or with a color marker.

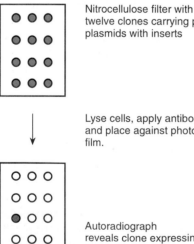

Nitrocellulose filter with twelve clones carrying pBR313 plasmids with inserts

Lyse cells, apply antibodies, and place against photographic film.

Autoradiograph reveals clone expressing the probed-for gene.

a particular protein in the clones of figure 12.16. The clones would be transferred to a nitrocellulose membrane where they would be lysed (e.g., with chloroform vapor). Then an antibody, specific for the particular protein, would be applied. A second antibody, specific for the first antibody and radioactively labeled, would be applied to the filters.

Figure 12.18

Chromosome walking technique to locate overlapping cloned inserts. A whole chromosome region is made available for study. Insert A is fragmented and made into subclones that are used to probe for regions that overlap A and the next region down (B). A second clone detected with part of A and part of the next region (B) is itself fragmented and subcloned to repeat the process.

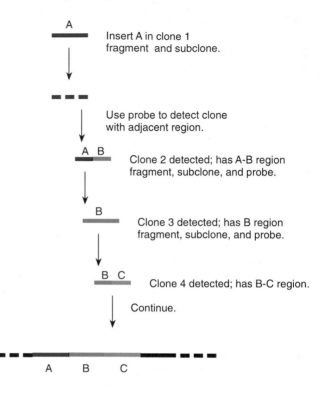

A — Insert A in clone 1 fragment and subclone.

Use probe to detect clone with adjacent region.

A B — Clone 2 detected; has A-B region fragment, subclone, and probe.

B — Clone 3 detected; has B region fragment, subclone, and probe.

B C — Clone 4 detected; has B-C region.

Continue.

A B C

Autoradiography would then locate the presence of the antibodies and thus indicate which of the clones contains the particular gene and is expressing it (fig. 12.17).

Chromosome Walking

Despite the limited size of any one inserted piece of foreign DNA, it is possible to learn about longer stretches of DNA by using a technique of overlapping clones called **chromosome walking.** For example, let's say that a particular gene (in region A) is located in clone 1, as discovered through probing. The cloned insert can be removed, using the same restriction enzyme used to insert it in the vector initially, and broken into small pieces that are subcloned and used as probes themselves. The idea is to locate another clone with an inserted region that overlaps the first one (fig. 12.18). The second clone is now treated the same way—subcloned with segments used to probe for yet another overlap farther down the chromosome. In this way, relatively long segments of a chromosome can be available for study in overlapping clones.

One obvious use of chromosome walking is to discover what genes lie next to each other on eukaryotic chromosomes, assuming that a genomic library exists. To date, regions as long as 250 kb have been walked. The technique is very tedious and is halted at certain areas not amenable to walking such as repeated sequences that are found in the DNA of eukaryotes (see chapter 14). Once an overlapping probe contains a commonly repeated sequence, it hybridizes to many clones that do not contain adjacent segments. This "cross-referencing" lessens the value of the technique. Currently, newer techniques (termed **chromosome jumping**), designed to bypass regions not amenable to walking, are being developed. These techniques depend on the ability to locate the two ends of a segment without having to walk through the middle. Ends of a segment can be located if the region has been inverted or if a large region is cloned and the middle part later removed, leaving just the ends. A probe of the ends allows the investigator to locate clones with first one end and then the other, effectively jumping over the intervening region.

Heteroduplex Analysis

Another useful tool in the molecular geneticist's bag of techniques is **heteroduplex analysis** or **heteroduplex mapping.** This procedure allows the direct electron microscopic visualization of cloned products by observing the heteroduplex DNA formed when the clone is hybridized with another piece of nucleic acid, usually its probe or the plasmid without any clone.

Heteroduplex mapping was used to study strains of the bacterium *Clostridium tetani,* which causes tetanus by producing a protein neurotoxin. Based on the known amino acid sequence of the amino end of the toxin, an eighteen-nucleotide DNA probe was synthesized. This probe hybridized with DNA in a plasmid from a toxic strain but did not hybridize with DNA in plasmids from nontoxic strains; the plasmid containing the toxin gene was designated pCL1. In the course of the work, a second plasmid, designated pCL2, was derived from pCL1. The pCL2 hybridized with the probe but did not produce the tetanus toxin. Researchers hypothesized that pCL2 arose as a deletion of part of the tetanus toxin gene in the plasmid. This conjecture was verified using heteroduplex analysis.

First, both pCL1 and pCL2 were treated with *Bam* HI endonuclease, which cleaves each plasmid only once. Before hybridization studies are done, circular plasmids are usually made linear so they can unwind fully and hybridize. The result of the hybridization is shown in figure 12.19. Clearly, pCL2 is a plasmid with a region of about 22 kb missing. The location of the toxin gene is thus localized on the plasmid. (Figure 12.19 shows recognition sites for four restriction endonucleases and the DNA lengths between them in kb [inner numbers]. These values were determined using methods we outline under restriction mapping.)

EUKARYOTIC VECTORS

The work we have described so far involves introducing chimeric plasmids into bacteria, primarily *E. coli.* However, there are several reasons why we want to extend these techniques to eukaryotic cells. First, a prokaryote like *E. coli* is not capable of fully expressing some eukaryotic genes since it lacks enzyme systems necessary for some posttranscriptional and posttranslational modifications such as intron removal and some protein modification. Second, we also wish to study the organization and expression of the eukaryotic genome *in vivo* (in the living system), something that can only be accomplished by working directly with eukaryotic cells. Finally, we wish to learn how to manipulate the genomes of eukaryotes for medical as well as economic reasons. To these ends we discuss eukaryotic plasmids and the direct manipulation of eukaryotic genomes *in vivo.*

Yeast Vectors

Yeasts, small eukaryotes that can be manipulated in the lab like prokaryotes, have been extensively studied. Baker's yeast, *Saccharomyces cerevisiae,* has a naturally occurring plasmid about 2 μm in length. In addition, bacterial plasmids have been introduced into yeast. Unfortunately, these plasmids eventually tend to be lost from the cells. This tendency has been overcome, however, by constructing bacterial plasmids that contain a piece of yeast centromere (CEN) and the origin of yeast DNA replication (ARS for autonomously replicating sequence; fig. 12.20). The plasmids are then carried from one generation to the next by the yeast. The plasmids can have telomeric sequences inserted and can then be linearized by cutting with an endonuclease at the telomeric sequences, or the plasmids can be linearized first and then have telomeric sequences added to their ends. These plasmids are then called **yeast artificial chromosomes (YACs).** The particular advantage of the YACs is that they are capable of having very large pieces of DNA inserted. Remember that a cosmid can hold about 50 kb; a YAC can hold as much as 800 kb. The ability to clone this much DNA is valuable when working with the large genes found in eukaryotes.

Recombinant DNA studies in yeast have increased our knowledge of gene regulation in eukaryotes, of the way in which the centromere works, and of the way in which the tips of eukaryotic linear chromosomes are replicated. In addition, YACs have allowed us to analyze and sequence very large segments of eukaryotic DNA. (In chapter 15 we look at the control of gene expression in yeast.)

Figure 12.19

Heteroduplex mapping of plasmids pCL1 and pCL2. The latter is missing a region found in the former. The DNA of both plasmids is cut once to form linear molecules that are then hybridized, forming heteroduplexes. (*a*) Electron micrograph. (*b*) Explanatory diagram. (*c*) Diagram of the plasmid showing restriction sites.

(b,c) From C. Finn, et al., "The Structural Gene for Tetanus Neurotoxin Is On a Plasmid" in Science, *224:881–884. Copyright 1984 by the AAAS.*

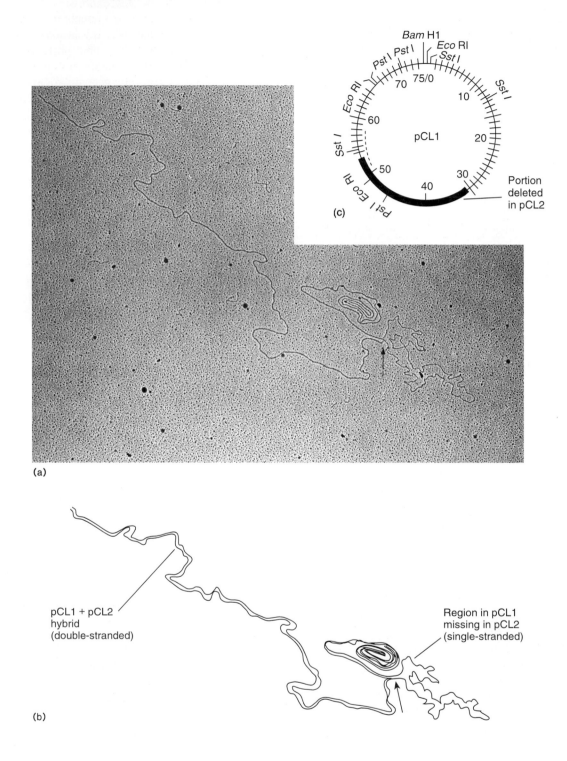

(a)

(b)

pCL1 + pCL2 hybrid (double-stranded)

Region in pCL1 missing in pCL2 (single-stranded)

(c)

Bam H1
Eco RI
Pst I *Pst* I
Sst I
Eco RI
70 75/0
60
Sst I
50
Eco RI
Pst I
40
pCL1
10
Sst I
20
30
Portion deleted in pCL2

Figure 12.20

E. coli plasmid pBR322 modified for use in yeast. This plasmid will survive and replicate in both yeast and E. coli because it has the origin of replication for both, as well as a yeast centromeric region. When linearized and having telomeres added, the YAC is now suitable for the cloning of large pieces of DNA.

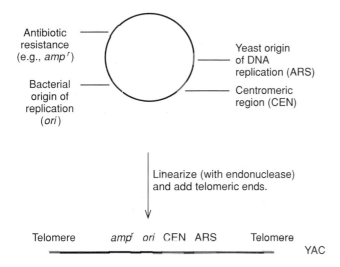

Animal Vectors

The vehicle most commonly used in higher animals is the DNA tumor virus SV40. (SV, or simian vacuolating virus, was first isolated in monkeys; however, it can transform normal mouse, rabbit, and hamster cells. Unlike the use of the word transformation in bacteria [chapter 7], transformation in eukaryotes refers to the changing of a normal cell into a rapidly growing, cancerous one.) SV40 is an icosahedral particle with a small (5,224-bp) chromosome, which is a circular, double-stranded DNA molecule.

Like λ vectors, SV40 virions can have part of their DNA replaced by foreign DNA. The viruses can then be used in recombinant DNA studies in one of two ways (fig. 12.21). They can replicate and complete their life cycle with the help of nonrecombinant viruses, or they can replicate in the host without making active virus particles by existing as circular plasmids in the cytoplasm or by being integrated into the host's chromosomes. SV40 has become a valuable tool in mammalian genome studies. For example, the rabbit β-globin gene was cloned in SV40, and enhancer sequences (see chapter 10) were discovered in SV40. Much new understanding of transformation (oncogenesis) has come from using DNA tumor viruses as vehicles. (See further discussion on SV40 in chapter 15, under viral oncogenesis.)

Plant Vectors

The best studied system of introducing foreign genes into plants is the naturally occurring crown gall tumor system.

The soil bacterium *Agrobacterium tumefaciens* causes tumors, known as crown galls, in many dicotyledonous plants (fig. 12.22). In essence, the crown gall is made of transformed plant cells. These cells have been transformed by a plasmid within the bacterium called the *tumor-inducing,* or *Ti,* plasmid. Transformation occurs when a piece of the plasmid called T-DNA (for transferred DNA) is integrated into the host (plant cell) chromosome. Crown gall cells produce odd amino acid derivatives, termed opines, which are used by the A. tumefaciens cells. By manipulating this system, geneticists have begun to understand the transformation process in plants as well as to develop a manipulatable system for introducing foreign genes into plants (see below).

The study of genetics in plants has been helped a great deal by the availability of model organisms similar to E. coli, yeast, and fruit flies. Recently, much attention has been turned to the meadow weed, *Arabidopsis thaliana.* This small plant (fig. 12.23) is ideal for studying the genetics of plants because the genome is small, approximately 100 million base pairs located in only five chromosomes ($2n = 10$). This is about five times the genome of yeast or twenty times the genome of E. coli. Thus, in terms of size of genome, it is quite manageable and joins the ranks of organisms whose genomes are slated to be completely sequenced in the near future (see below). The plants are also easy to grow in very large numbers, and each plant produces as many as 10,000 seeds. Hence this organism compares very favorably with fruit flies and yeast as an organism to study questions of gene control in a eukaryote, in this case a plant.

Expression of Foreign DNA in Eukaryotic Cells

Foreign DNA can be introduced into eukaryotic cells in methods similar to bacterial transformation, but is called **transfection** because, as we just described, the term transformation in eukaryotes has come to mean cancerous growth. Eukaryotic organisms that take up this foreign DNA are referred to as **transgenic.** Most of the techniques described here transcend taxonomic lines.

Animal cells, or plant cells with their walls removed (protoplasts), can take up foreign chromosomes or DNA directly from the environment with a very low efficiency (in the presence of calcium phosphate). The efficiency can be greatly improved by directly injecting the DNA. For example, transgenic mice are now routinely prepared by injecting either oocytes or one- or two-celled embryos. They are obtained from female mice after appropriate hormone treatment. After injection of about two picoliters (2×10^{-12} liters) of cloned DNA, the cells are reimplanted into receptive female hosts.

Figure 12.21

SV40 virus can be used as a gene-cloning vehicle. Although part of
the virus is replaced by inserted DNA during cloning, it can still
replicate with the aid of normal helper viruses (nonrecombinant
SV40). Without the aid of helper viruses, it can either replicate as a
plasmid or integrate into the host chromosome.

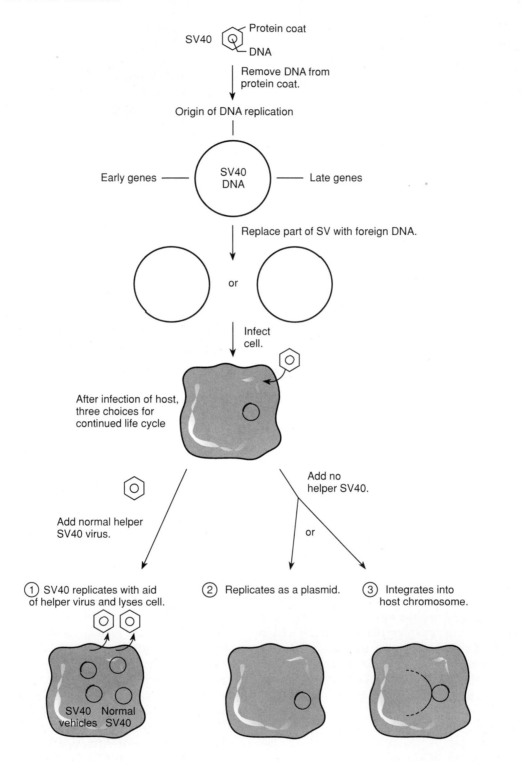

Figure 12.22

Crown gall on tobacco plant (*Nicotiana tabacum*) produced by *Agrobacterium tumefaciens* containing Ti plasmids

Courtesy of Robert Turgeon and B. Gillian Turgeon, Cornell University.

Figure 12.23

A dwarf form of the plant *Arabidopsis thaliana*

Source: Science, *Vol. 243, March 10, 1989, cover. © 1989 AAAS, Washington D.C. Photo by DeVere Patton. Courtesy of E. I. DuPont de Nemours and Company.*

Figure 12.24

Injection of DNA into the nucleus (germinal vesicle) of a mouse oocyte. The oocyte is held by suction from a pipette.

© John Gardon/Phototake

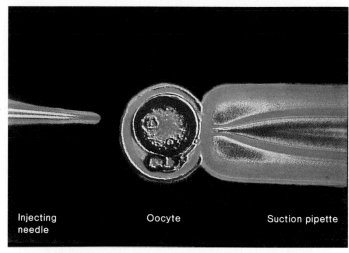

Using modifications of nuclear transplant techniques developed in the 1950s (see chapter 15), DNA is injected directly into the mouse oocyte or embryo (fig. 12.24). In about 15% of these injections, the foreign DNA is incorporated into the embryo. Transgenic animals are useful both to study the expression and control of foreign eukaryotic genes and also in their own right as recipients of successful therapy and manipulation. For example, a mutant mouse lacking the gonadotropin-releasing hormone, and thus sterile, was restored to normalcy by being transfected with an intact gonadotropin-releasing hormone gene that was incorporated into its genome and its germ line.

In 1988, a transgenic mouse that is prone to cancer was the first genetically engineered animal to be patented. This mouse provides an excellent model for studying cancer (see chapter 15). A controversy has arisen as to whether engineered higher organisms should be patentable (see Box 12.1). Currently, work is under way to transfect agriculturally important animals (cattle, sheep, pigs) with growth hormone genes, producing "super" organisms. Mice have already been successfully transfected with a rat growth hormone gene (fig. 12.25), and transgenic sheep have been produced that express the gene for a human clotting factor.

Transfection can also be mediated by retroviruses (RNA viruses containing the gene for reverse transcriptase). For example, human white blood cells lacking the enzyme adenosine deaminase were repaired by infection with a retroviral vector. A retrovirus responsible for a form of leukemia in rodents, the Moloney murine leukemia virus, was engineered such that all genes of the virus were

BOX 12.1

THE RECOMBINANT DNA DISPUTE

Paul Berg shared the 1980 Nobel Prize in chemistry for creating the first recombinant DNA molecule, a hybrid λ phage that contained the genome of the simian tumor virus, SV40. The fact that he could do this work was worrisome to many people, himself included. The recombinant DNA dispute was underway. Berg voluntarily stopped inserting tumor virus genes into phages that attack the common intestinal virus, *E. coli.*

People continue to worry about the dangers of recombinant DNA work. There is the immediate and obvious concern that cancer or toxin genes will "escape" from the laboratory. In other words, recombinant DNA technology could create a bacterium or plasmid that contained toxin or tumor genes. The modified bacterium or plasmid could then accidentally infect people. A 1974 report by the National Academy of Sciences led to a February 1975 meeting, which took place at the Asilomar Conference Center, south of San Francisco. Berg convened this meeting, which was attended by over one hundred molecular biologists. The recommendations of the Asilomar Committee later formed the basis of official guidelines by the National Institutes of Health (NIH). In essence, NIH established guidelines of containment.

Containment means physical and biological barriers to the escape of dangerous organisms. Four levels of risk were recognized by the NIH guidelines, from minimal to high, for which four levels of containment were outlined (called P1 through P4). The most hazardous experiments, dealing with the manipulation of tumor viruses and toxin genes, required extreme care, which included negative-pressure air locks to the laboratory and experiments done in laminar-flow hoods, with filtered or incinerated exhaust air.

Biological containment means the development of host cells and manipulated vectors that are incapable of successful reproduction outside the lab even if they should escape. High-risk work was done with host cells or vectors that were modified. For example, a bacterium of the *E. coli* strain EK2 cannot survive in the human gut because it has mutations that do not permit it to synthesize thymine or diaminopimelate. The lack of thymine-synthesizing ability is lethal because the cell cannot replicate its DNA. The diaminopimelate is a cell-wall constituent without which the cells burst. These bacteria also carry mutations that make them extremely sensitive to bile salts. Thus, if by accident the cells were to escape, they would pose virtually

Paul Berg (1926–   )
Courtesy of Dr. Paul Berg.

no threat. Fewer than one in 10^8 of these cells can survive in the human gut for 24 hours. The plasmids used for recombinant research were modified so that they could not be transferred from one cell to the next. Again, if containment failed, neither the host cells nor their plasmids would survive.

In 1979, the guidelines were relaxed. Although it was wise to be cautious, it appears that initial fears were somewhat unwarranted. Recombinant DNA work now seems to pose little danger: Containment works very well and engineered bacteria do very poorly under natural conditions. *E. coli* has been living in mammalian guts for millions of years, in which time it has had numerous opportunities to incorporate mammalian DNA into its

Continued on next page

BOX 12.1—*continued*

Box figure 1

A field test of virus-resistant plants. Tomato plants were made resistant to tobacco mosaic virus (TMV) infection by introducing the TMV coat protein gene into the plant using the Ti-plasmid containing *Agrobacterium tumefaciens* system. Plants that express the coat protein gene are resistant to attack by the virus, presumably by a mechanism that involves preventing the native viral RNA from being released from its coat protein. Greenhouse grown plants were transplanted to the field in 1988. Control (*right*) and engineered plants (*left*) were infected with the virus two weeks after planting. The tomato yield from the engineered plants was more than three times that of the controls.

Courtesy of Monsanto Company.

genome (intestinal cells are dying and sloughing off into the gut all the time). No "Andromeda strain" has arisen, nor do we foresee one in the future.

Currently, concern is focused primarily in the area of application of recombinant DNA technology. The medical community is showing extreme caution in using recombinant DNA technology to treat people. And others are very wary of the general application of recombinant DNA methods. In August of 1984, an attempt to test the ability of engineered bacteria (*Pseudomonas syringae*) to protect crop plants against frost was blocked by a federal judge in a suit initiated by concerned social activists (Box fig. 1). In April of 1987, the tests went ahead. The tests have proven to be quite safe and others are in progress.

More recently, however, the recombinant DNA dispute has taken a whole new twist. It now has surfaced as a conflict between academic freedom and industrial secrecy. It seems that recombinant DNA technology is very lucrative. Numerous academic scientists have either begun genetic engineering companies or become affiliated with pharmaceutical companies. However, the philosophies of private enterprise and academia are in conflict. Academic

Stanley N. Cohen (1935–   )
Courtesy of Stanford University.

endeavors are presumably done openly with free exchange of information among colleagues, whereas private enterprise entails some degree of secrecy at least until patents are obtained to protect the investments of the companies. Thus a basic conflict exists for scientists trained in gene cloning. The conflict has been prevalent, especially since late 1980, when the first patent for recombinant DNA techniques was awarded to Stanford University and the University of California for techniques developed by Stanley Cohen and Herbert Boyer. The humorist Art Buchwald entitled a newspaper column: "Splicing the genes and slicing the profits of Free Enterprise U." It is not clear what the outcome of these conflicts will be for American and international research institutions.

Figure 12.25

Mouse littermates. The *larger one* is a transgenic mouse containing the rat growth hormone gene.

Source: Photograph by Dr. Ralph L. Brinster, "Dramatic growth of mice that develop from eggs microinjected with metallothionein growth hormone fusion genes" by Richard D. Palmiter, Ralph L. Brinster, et al. Nature 300 (16 December) Cover, 1982. Copyright © 1982 Macmillan Magazines Limited.

Figure 12.26

The Moloney murine leukemia virus (*top*) contains only three genes: *gag, pol,* and *env.* The *gag* gene will yield capsid proteins, *pol* will yield reverse transcriptase and an integration enzyme, and *env* will yield the envelope glycoprotein. These genes are removed and replaced with a neomycin-resistance gene (*neo^R*) for selection purposes, and the human adenosine deaminase gene (h*ADA*) with an SV40 protein (*bottom*).

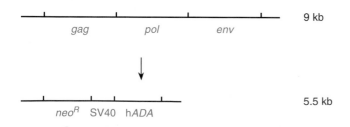

removed and replaced with an antibiotic marker (neomycin resistance) and the human adenosine deaminase gene (fig. 12.26). The virus binds to the cell surface and is taken into the cell, its RNA is converted to DNA by reverse transcription, and the DNA is incorporated into one of the cell's chromosomes. It is not possible for this highly modified virus to attack and damage the cells unless a helper virus is added. Unlike the SV40 viruses in figure 12.21, the modified Moloney viruses cannot initiate a successful infection without the helpers because vital genes have been removed.

Three other recent techniques for the delivery of recombinant DNA to eukaryotic cells are worth mentioning: electroporation, liposome-mediated transfer, and "biolistic" transfer. In **electroporation,** exogenous DNA is taken up by cells subjected to a brief exposure of high-voltage electricity. Presumably, this electric field creates transient micropores in the cell membrane, allowing exogenous DNA to enter.

Liposome-mediated transfection is a technique in which foreign DNA is encapsulated in artificial membrane-bound vesicles, called liposomes. The liposomes are then used to deliver their DNA to target cells. In one experiment, 50% of mice injected with these DNA-containing liposomes were successfully transfected—they expressed the proteins encoded by the transfecting DNA.

Last is the development of techniques to deliver foreign DNA into mitochondria and chloroplasts, which have proven difficult targets for genetic engineering because, among other reasons, they have double-membrane walls that have not proven amenable to delivery of recombinant DNA. Recently, transfection has been successful in both mitochondria and chloroplasts using a **biolistic** (biological ballistic) process in which recombinant DNA was delivered coated on tungsten microprojectiles that were literally shot into these organelles.

We conclude this section by discussing how plants are transfected with the Ti plasmids of *Agrobacterium tumefaciens,* as alluded to earlier. When a plant is infected with *A. tumefaciens* containing the Ti plasmid, a crown gall tumor is induced when the Ti plasmid transfects the host plant, transferring the T-DNA region. Those cells transfected with the T-DNA are induced to grow as well as produce opines that the bacteria feed on. Much recent research has been concentrated on engineering Ti plasmids to contain other genes that are also transferred to the host plants during infection, creating transgenic plants. One series of experiments has been especially fascinating.

Tobacco plants have been transfected by Ti plasmids containing the luciferase gene from fireflies. The product of this gene catalyzes the ATP-dependent oxidation of luciferin, which emits light. When a transfected plant is watered with luciferin, it glows like a firefly (fig. 12.27). The value of these experiments is not the production of glowing plants, but rather the use of the glow to "report" the action of specific genes. In further experiments, the promoters and enhancers of certain genes were attached to the luciferase gene. Luciferase thus would only be produced when these promoters were activated, thus the glowing areas of the plant show where the transfected gene is active. The luciferase gene is thus a "reporter" system used to monitor the activity of the gene of interest, and thus this system has much promise.

Figure 12.27
Luminescent transgenic tobacco plant containing the firefly luciferase gene. The plant was watered with luciferin resulting in a firefly glow.
D. W. Ow, Keith V. Wood, Marlene DeLuca, Jeffrey R. Dewet, Donald R. Helsinki, Stephen H. Howell, "Transient and Stable Expression of the Firefly Luciferase Gene in Plant Cells and Transgenic Plants" Science 234 (Nov. 14, 1986):856–859, Figure 1. Copyright 1986 by the AAAS.

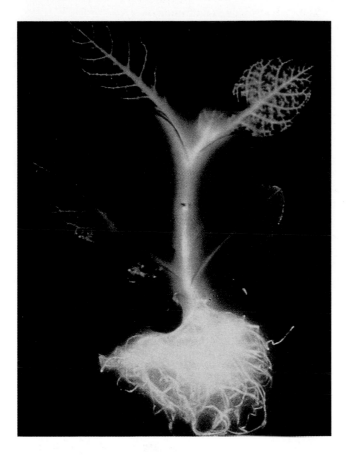

Agrobacterium tumefaciens does not successfully attack all plants, being notably ineffective in attacking cereals. However, transfection has been done successfully in cereals (including transfecting the luciferase system) by using electroporation. Hence the same experiments can be done in almost any plant, using Ti-plasmid transfection in some and electroporation in others. Soybeans have also been transfected by the use of particle acceleration (biologistics) using DNA-coated gold particles.

RESTRICTION MAPPING

The number of cuts that a restriction enzyme makes in a segment of double-stranded DNA depends on the size of that DNA, its sequence, and also the number of base pairs in the recognition sequence of the particular enzyme. That is, one with only three base pairs in the recognition sequence will cut more times than one with six base pairs in the sequence since the probability of a sequence occurring by chance is a function of the length of that sequence.

A sequence of three bases occurs more often by chance ($1/4^3 = 1/64$ bp) than a sequence of six bases ($1/4^6 = 1/4,096$ bp). *Hin* dII, for example, cuts the circular DNA of the tumor virus SV40 into eleven pieces; some restriction enzymes can cut *E. coli* DNA into hundreds of pieces. The product of the action of a restriction enzyme is called a **restriction digest.**

Using electrophoresis, we can separate the fragments of a restriction digest by size. With techniques described below, we can locate the restriction sites on the original gene or piece of DNA. That is, we can construct a map of the restriction recognition sites that will give us the physical distance between sites, in base pairs (fig. 12.28). This **restriction map** is extremely valuable for several reasons.

Since it is a representation of the physical structure of the DNA, various sites can be localized to their physical positions on the DNA. For example, when tritiated thymidine, a radioactive nucleotide, was added for a very short period of time during the beginning of DNA replication in SV40 viruses, the radioactivity always appeared in only one restriction fragment. This demonstrated that SV40 replication started from a single, unique point.

In addition, a restriction map often allows researchers to make a correlation between the genetic map and the physical map of a chromosome. Certain large physical changes in the DNA, such as deletions, insertions, or nucleotide changes at restriction sites, can be localized on the genetic map. These changes can be seen as changes in size, or the absence, of certain restriction fragments when compared with wild-type DNA that has not suffered these alterations. This information not only allows us to see changes in the DNA but also has been found useful in looking at the evolutionary distances between species (see chapter 21). The differences in fragment sizes are called **restriction fragment length polymorphisms (RFLPs)**—usually pronounced "ruflups" or "riflips"—and have proven valuable in pinpointing the exact location of genes and determining identity or relatedness of individuals. A restriction digest is also useful for isolating short segments of DNA that can be sequenced easily.

Constructing a Restriction Map

How do we construct a restriction map? At the top of figure 12.28 is a hypothetical piece of DNA with three cuts made by restriction enzyme "A." Below this map is a diagram of the electrophoresed digest on agarose gels, which are usually used because their porosity allows the movement of DNA fragments of relatively large size. The restriction enzyme makes three cuts in the DNA, generating fragments that are 200, 50, 400, and 100 base pairs (bp) long. The banding pattern on the gel at the left of figure 12.28 is the result of the electrophoresing of that digest. (Note that smaller segments move faster than larger segments.)

Figure 12.28

Restriction map from electrophoresis of a restriction endonuclease digest. (a) Original piece of DNA showing restriction sites marked by "A." (b) Agarose gels showing bands of total and partial restriction digests. *Asterisks* refer to radioactive bands produced by end-labeled segments. At the *left* are molecular weight markers in base pairs (e.g., 800 bp, 700 bp). The total digest produces fragments that are 400, 200, 100, and 50 bp—of which the 200- and 100-bp fragments are end labeled. The partial digest yields six additional bands.

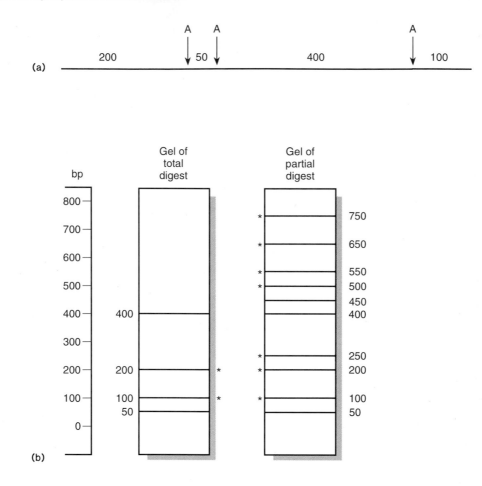

The sizes of the segments are determined by comparison with standards of known size (not shown, although the scale is indicated on the left). The order of these segments on the chromosome is not obvious from the gel. Several methods can be used to determine the exact order of the restriction segments on the original piece of DNA.

Before restriction enzyme digestion, the 5' ends of the DNA can be radioactively labeled with ^{32}P using the enzyme polynucleotide kinase. Since the enzyme is acting on double-stranded DNA, both ends will be labeled. Upon electrophoresis after digestion of the DNA in figure 12.28, the 200-bp and 100-bp bands will be radioactively labeled, indicating that these segments are the termini of that piece of DNA.

The order of the other segments can be determined by slowing down the digestion process to produce a **partial digest.** If the reaction is cooled or allowed to proceed for only a short time, not all the restriction sites will be cut.

Some pieces of DNA will not be cut at all, some will be cut once, some twice, and some cut at all three restriction sites. The result of electrophoresis of this partial digest is seen at the right of figure 12.28b. From this gel we can reconstruct the order of segments. This gel contains the four original segments plus six new segments each containing at least one uncut restriction site.

From the total digest gel, we know that the 200- and 100-bp segments are on the outside because they were radioactively labeled and the 50- and 400-bp segments are on the inside. In the partial digest, we find a 250-bp segment but not a 150-bp segment, which tells us that the 50-bp segment lies just inside and next to the 200-bp terminus (fig. 12.29b). There is a 500-bp segment but not a 600-bp segment, which tells us that the 400-bp segment lies adjacent to the 100-bp terminus (fig. 12.29c). An unlabeled 450-bp segment confirms that the 400- and 50-bp segments are adjacent and internal in the DNA. We thus

Figure 12.29

Steps in the reconstruction of the DNA of figure 12.28 from the gels of the total and partial restriction endonuclease digests. *Asterisks* refer to ^{32}P end labels. From the total digest, the 100- and 200-bp segments are established as the end segments (*a*). Since there are also 50- and 400-bp fragments within the DNA (established from the total digest), only certain fragments are possible from the partial digest that establish that the 50-bp fragment is adjacent to the 200-bp fragment and the 400-bp fragment is adjacent to the 100-bp end segment (steps *b* and *c*). The occurrence of an unlabeled 450-bp fragment in the partial digest verifies the relationship of the 50- and 400-bp fragments (*d*) yielding the final structure (*e*). All the fragments in the partial digest are consistent with this arrangement.

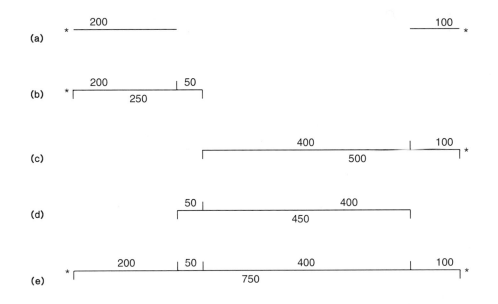

unequivocally reconstruct the original DNA (compare fig. 12.29*e* with fig. 12.28*a*), giving a map of sites of restriction enzyme recognition regions separated by known lengths of DNA.

Double Digests

In practice, restriction mapping is usually done with several different restriction enzymes. Figure 12.30 is of the DNA of figure 12.28, with the recognition sites of a second endonuclease, "B," included. Using the same methodology just outlined, we can show that the order of the B segments is 350, 250, and 150 bp arising from two cuts by endonuclease B. What we do not know is how to overlay the two maps. Do the B segments run left to right or right to left with respect to the A segments (fig. 12.30*a* and *b*)? We can determine the unequivocal order by digesting a sample of the original DNA with both enzymes simultaneously, thus producing a **double digest.**

The two orders shown in figure 12.30*a* and *b* are used to make different predictions about the double digest. From the first order (*a*), we predict a 200-bp end segment, radioactively labeled. From the second order (*b*), we predict that the labeled 200-bp segment will be cut back to 150

bp: There should not be a labeled 200-bp segment. The double digest shows a labeled 200-bp segment, verifying order (*a*).

Restriction mapping thus provides us with a physical map of a piece of DNA, showing restriction endonuclease sites separated by known lengths of DNA. This technique gives us short DNA segments of known position that can be sequenced as well as a physical map of the DNA that can be compared with the genetic map and upon which mutations and other particular markers can be located.

Restriction Fragment Length Polymorphisms

DNA Fingerprinting

Restriction fragment length polymorphisms (RFLPs), obtainable from restriction digests, are proving to be very valuable genetic markers in two areas of study: human gene mapping and forensics. In a restriction digest of the whole human genome, there might be thousands of fragments from a single restriction enzyme. Probes have been developed from virtually any small segment of DNA for Southern blotting of these digests. Most probes recognize

Figure 12.30
Overlay of the sites of recognition by two different restriction endonucleases on the same piece of DNA (A and B). (a) Actual arrangement. (b) Hypothetical alternative arrangement. (c) Electrophoresis of the total restriction digests by A alone, B alone, and both together. *Asterisks* indicate radioactive end-labeled bands. Order (a) is seen to be consistent with all the bands found in all the digests, whereas order (b) is not. For example, in order (b) an internal (unlabeled) 150-bp fragment is predicted, which is not found in the total digest.

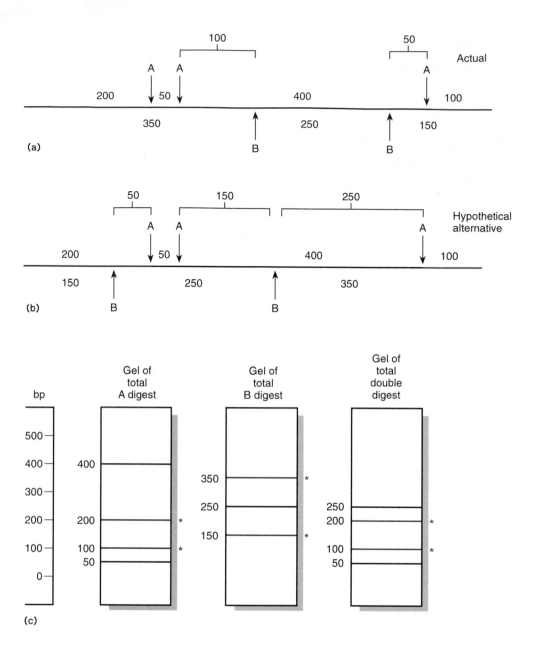

a single band. Genetic variation usually comes in the form of a second allele that, due to a mutation, lacks a restriction site and is therefore part of a larger piece of DNA (fig. 12.31). RFLPs with only two alleles do not have the genetic variation needed for the studies described in the following paragraphs. However, some probes have uncovered **hypervariable loci** with many alleles (any one person has, of course, only two of the many possible alleles). The variation within a population is generated because these hypervariable loci contain many tandem repeats of short (10- to 60-bp) segments. Due presumably to unequal crossing over (see chapter 8), much variation is generated by one of these loci, termed **variable-number-of-tandem-repeats (VNTR) loci.**

As a result of this variation, probing for one of these VNTR loci in a population will reveal many alleles. In cases in which a number of different loci are recognized

Alec Jeffreys (1950–   )
Courtesy of Dr. Alec Jeffreys.

by the same probe, each individual will have many bands on a Southern blot, with most people having unique patterns. In one system, developed by A. Jeffreys, a single probe locates fifty or more variable bands per person. The Southern blots create a genetic "fingerprint" of extreme

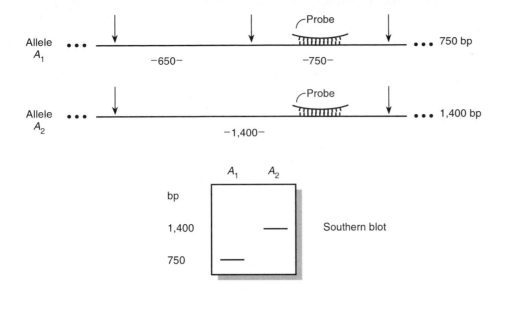

Figure 12.31
RFLP analysis. Allele A_1 is a gene segment 750 bp long identified by probe binding. In allele A_2 the restriction site to the left of allele A_1 has been changed. The probe thus recognizes a 1,400-bp fragment instead of the 750-bp fragment. During Southern blotting, two different bands are seen from the two alleles. *Arrows* indicate restriction sites.

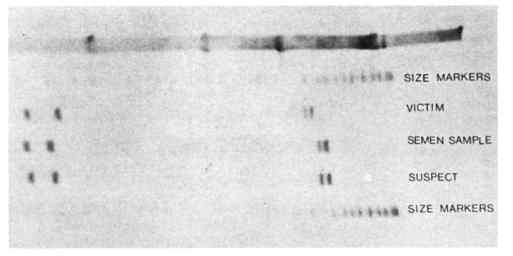

Figure 12.32
Southern blot of DNA from victim and suspect and DNA from semen sample left by a rapist-murderer, showing the match. For courtroom purposes, several additional probes would be used to increase the specificity of the match because the probe shown distinguishes (lights up) only a few bands. (Note slight band distortion, tilting bands.)
Photo by Lifecodes Corporation.

value in forensics. DNA extracted from blood or semen samples left by the offender can be compared with DNA patterns of suspects. If Jeffreys' probes are used to compare the patterns, the likelihood that the two patterns would randomly match is infinitesimally small. This technique thus has a greater power to positively identify individuals than using their fingerprints (fig. 12.32).

Recently, DNA fingerprinting has been accepted in some cases in courts of law and used to positively identify rapists and murderers. Although the theory of using this technique is not in doubt, courts have been somewhat reluctant to accept it because of its practical application. Notably, one of the companies doing the tests has been criticized for some of its methods in declaring matches between suspects and samples when the Southern blots were flawed. On an electrophoretic gel, or the Southern blot from that gel, differences in quantity of samples in different lanes or variation in the gels can cause differences in migration or distortion of bands, and thus cor-

rective analyses frequently are done. These latter techniques have caused DNA fingerprinting evidence to be dismissed in some cases. With refinements of techniques and greater sophistication by courts, these techniques should become used routinely in legal situations.

Linkage Analysis

RFLPs are now proving useful in locating genes in the human genome through linkage analysis. RFLPs are especially valuable when a genetic disease is known but its protein product and gene location are unknown. By locating the gene (to within about a million base pairs), techniques such as chromosome walking will allow the eventual localization and isolation of the gene. With the gene in hand, its sequence and protein product can be determined, a first step in medical treatment.

To locate a gene this way, pedigrees are needed of families with a genetic disease, as well as RFLPs of known location. By linkage analysis of the disease gene and the

Figure 12.33

(a) Southern blot of DNA from six persons cut with *Msp* I and probed with LDR152, a segment of DNA from chromosome 19. Two alleles are pointed out: allele 1 at 1.3 kb and allele 2 at 2.3 kb.

(b) Pedigree of family showing occurrence of myotonic muscular dystrophy and alleles of the LDR152 probe (genotypes 12, 22, and 11). Note that *squares* represent males and *circles* females. The dystrophy is associated with allele 1 of the LDR152 probe (genotypes 11 and 12).

(b) From R. J. Bartlett, et al., "A New Probe for the Diagnosis of Myotonic Muscular Dystrophy" in Science, 235:1648–1650, March 27, 1987. Copyright 1987 by the AAAS.

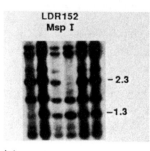

(a)

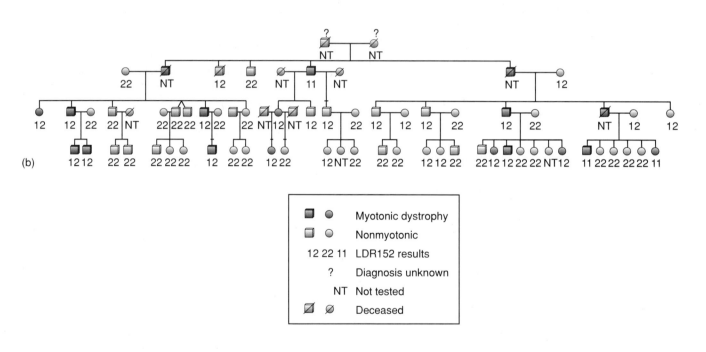

(b)

RFLP variants, the disease gene can be located. Currently the genes for cystic fibrosis, Alzheimer's disease, Marfan syndrome, neurofibromatosis (NF), fragile-X syndrome, and myotonic muscular dystrophy (among other genes) are being localized with these methods. For example, in figure 12.33 we see a Southern blot of DNA cut with restriction enzyme *Msp* I, and probed with LDR152, a segment of DNA from chromosome 19. The blot reveals two alleles, labeled "1" (of a 1.3-kb fragment) and "2" (of a 2.3-kb fragment). In the pedigree in figure 12.33, we see that in every case in which genotypes have been determined, a person with myotonic muscular dystrophy also has the "1" allele. Thus, in this pedigree, the "1" allele is very tightly linked to the LDR152 probe and therefore the gene for myotonic muscular dystrophy must be very close to this probe on chromosome 19. Localizing genes this way is of great medical value and also a forerunner of the DNA sequencing of the whole genome (see Box 12.2).

BOX 12.2

MAPPING AND SEQUENCING THE HUMAN GENOME

Rationale

During the past several years, a debate has arisen in the genetics community as to whether it is worth mapping and sequencing the entire human genome of three billion base pairs. This is no small project. Estimated to consume about thirty thousand person-years of effort over the next fifteen years and cost $3 billion, the project has its proponents and detractors.

The value of having the complete map of the human genome is not really in doubt. A complete sequence does not have the same level of support. A complete map would facilitate the identification of disease-causing genes. Not only would this be a first step in developing treatments, but the study of mutant genes could lead to much insight into normal gene structure and function. Molecular probes could identify parents at high risk for children with particular genetic diseases. Further knowledge could be gained in cancer study, epidemiology, and evolution. These are but a few of the many benefits that knowing the map and sequence of the human genome can bring.

A committee of the National Academy of Sciences published a 1988 report unanimously recommending the support of both mapping and sequencing projects, with support by the federal government of $200 million per year. Those against this project believe that the normal method of biomedical research in the United States—the single investigator following his or her own interests—is preferable to creating a large bureaucracy. They also feel that the mobilization and cost required to obtain the complete sequence are excessive. The money may come at the expense of other ongoing research efforts rather than in addition to it, justifying to some extent the concerns of those who do not favor this "big-science" endeavor.

It has been pointed out, however, that the discovery of the cystic fibrosis gene, a recent accomplishment, cost approximately $170 million. If we multiply this by the 50,000 to 100,000 genes that are estimated to be in our genome, the cost is close to a trillion dollars. Thus, it is argued, it is much cheaper to simply sequence the whole genome, noting the various genes in it as the work progresses, rather than have individual research groups isolate single genes of interest.

Methods

There are many levels of maps and sequences. In chapter 6 we developed the early stages of the human chromosome map. Generally, a locus was located on a particular chromosome by tissue-culture techniques (somatic-cell hybridization). Loci could be pinpointed further using aberrant chromosomes, for example, those with deletions. If a locus was present when the intact human chromosome was present but absent if the human deletion chromosome was present, the gene could be localized to that deleted region.

Further sophistication required the creation of genomic libraries of particular chromosomes. The ideal situation is to have the entire chromosome contained in overlapping clones, with overlap established through methods such as chromosome walking. Then the goal is to be able to identify each cloned segment with a particular "tag," a RFLP that is specific to that clone. Some parameters to consider are that each crossover map unit represents approximately one million base pairs. Thus it would be hoped to have, minimally, ten or so overlapping clones per centimorgan. These clones then need to be subcloned for sequencing.

At that point, known genes can be localized to specific clones by probes created from their RNA or DNA. To complete the process, sequencing can be carried out from either direction of each known gene, with the new sequences searched for genes (Box fig. 1).

Several other approaches, however, are also advocated. We mention two. Scientists who are currently working on a particular gene of interest or importance can search for that gene directly, as we described for the muscular dystrophy gene. There, the investigator needs to associate the disease phenotype with a marker, an allele of a RFLP. With that localization, the library clone that contains that gene can then be sequenced and studied.

Other scientists suggest more efficient ways of proceeding. For example, J. C. Venter and colleagues suggest that instead of sequencing the whole genome, we really are interested only in expressed genes. He and his colleagues therefore are beginning by isolating mRNAs and using them to create cDNAs (complementary DNAs), and then using the cDNAs as probes to locate the expressed genes in a genomic library. Thus they will be working only with genes that are active and of the most interest. So far they have isolated cDNAs of more than 600 genes active in the human brain.

Along with the human genome initiative are coordinated initiatives on other organisms, with the hope that technological advances coming from the human genome project will prove useful. Organisms scheduled to have their genomes sequenced are those of agricultural importance (pig, cow, sheep, chicken) and those of theoretical interest (*E. coli*, yeast, mouse).

Ethics

In addition to the expected scientific and medical information that we will gain from the human genome project are ethical problems. When the human genome is sequenced, we will shortly thereafter have the ability to test people for various genes that we cannot test for now, such as genes for latent diseases (heart conditions, cancer). Can insurance companies then demand to test individuals for a whole battery of genes and decide afterward whether that

Continued on next page

BOX 12.2—*continued*

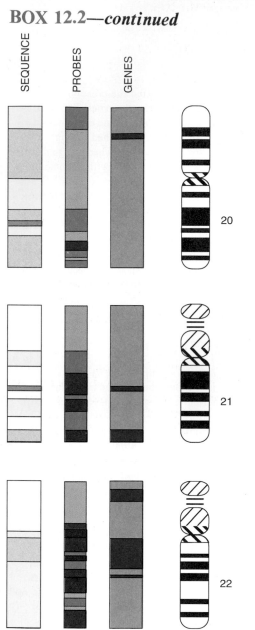

Box figure 1

Status of the human genome map as of July 31, 1990. Shown are chromosomes 20–22. The various bars show progress in sequences, probes, and genes in each band, using a color code shown in the legend.

From B. R. Jasny, supplement to Science, *12 October 1990. Copyright 1990 by the AAAS.*

LEGEND

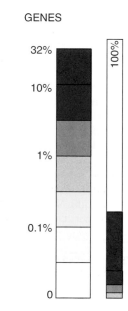

SEQUENCE   PROBES   GENES

Jonathan Beckwith (1935–   )
Courtesy of Dr. Jonathan R. Beckwith.

person is insurable, or what that person's insurance rates should be? Will many persons find themselves uninsurable because they have genes that might predispose them to cancer? Will individuals find themselves unemployable because of similar problems? What should doctors do about diagnosing a genetic disease (such as Huntington's disease) that has no cure? Should the patient be told? Another ethical issue is to what extent should genetic intervention be used to change the course of a person's life. With the knowledge of the sequence and location of our genes and the technology to transfer genes into people, will transgenic people be the norm? Should we not only cure diseases this way but tailor a person to some whim? Will genetic intervention into our basic blueprint be routine?

These are not trivial questions; they are of concern to both the scientists involved in the genome project and the agencies that are funding it. An ethics panel has been set up as part of the human genome project and has been promised 3% of the budget to study the ethical issues involved in mapping our genome. On that panel are people like J. Beckwith, a molecular geneticist who has devoted much time to reasonable criticism of genetic technology.

In the fall of 1988, the Nobel laureate James Watson was appointed to head the new National Institutes of Health (NIH) Office of Human Genome Research (he has since stepped down); $28 million was set aside for this effort, and further cooperation and funding has been provided by the Department of Energy. An international effort has begun. The program is under way.

Figure 12.34

Polymerase chain reaction. DNA is denatured, a primer oligonucleotide is added, and replication proceeds. After replication, a new cycle proceeds.

From J. L. Marx, "Multiplying Genes by Leaps and Bounds" in Science, *240:1408–1410, June 10, 1988. Copyright 1988 by the AAAS.*

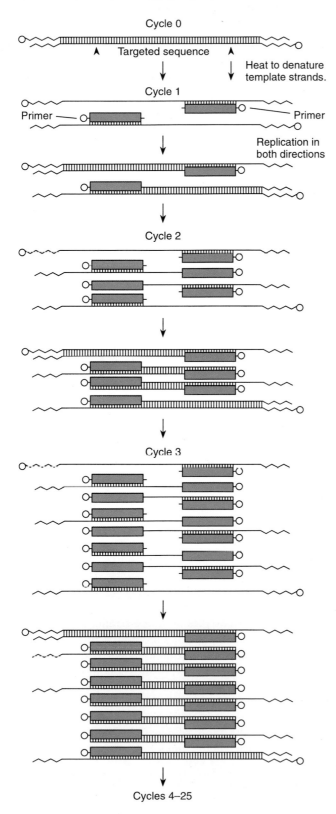

Cycle 0

Targeted sequence

Heat to denature template strands.

Cycle 1

Primer — Primer

Replication in both directions

Cycle 2

Cycle 3

Cycles 4–25

Polymerase Chain Reaction

In many instances, a DNA sample is available, but it is in such small quantity or is so old (museum specimens, dried specimens) as to be considered useless for further study. That situation was changed in 1983 when Kary Mullis, a biochemist working for the Cetus Corporation, devised the technique we now refer to as the **polymerase chain reaction** (PCR). It can be used to amplify whatever DNA is present, however small in quantity or poor in quality. The only requirement is that the sequence of nucleotides on either side of the sequence of interest is known. That information is needed to construct primers on either side of the sequence of interest. Once that is done, the sequence between the primers can be amplified.

In the technique, the primers and the ingredients for DNA replication are added to the sample. Then, the mixture is heated. This separates the DNA strands leading to their replication, starting from either side of the sequence of interest. After a few minutes of DNA replication, the process is stopped by heating; the mixture is denatured to separate the strands and a new cycle of replication is initiated (fig. 12.34). The various stages in the cycle are controlled by changes in temperature since the temperatures for denaturation, primer annealing, and DNA replication are different. In about twenty cycles, a million copies of the DNA are made; in thirty cycles, a billion copies are made. The technique is aided by using DNA polymerase from a hot-springs bacterium, *Thermus aquaticus,* that can withstand the denaturing temperatures. Thus after each cycle of replication new components do not have to be added to the reaction mixture. Rather, the cycling can be continued without interruption in PCR machines that are simply programmable water baths that can accurately and rapidly change the water temperature that surrounds the reaction mixture.

Depending on the availability of primers, any DNA can be amplified in this technique. For example, mitochondrial DNA from the remains of a 40,000-year-old woolly mammoth was amplified using PCR. The mammoth's mitochondrial DNA was very similar in sequence to that of modern elephants. Currently a project is underway to analyze Abraham Lincoln's genetic makeup. There still exist several dried blood and tissue samples from clothing taken at the time of Lincoln's assassination, as well as skeletal remains. Those samples have enough DNA to determine, for example, whether Lincoln had Marfan's syndrome, a connective-tissue disease whose symptoms include some physical attributes that Lincoln had. Although not necessarily lethal, the disease can lead to death from heart and aortic problems. In a more mundane fashion, PCR (voted "the molecule of the year" by *Science* in 1989) is now a routine tool in the laboratories of molecular geneticists who use it to rapidly amplify DNA regions of interest.

Frederick Sanger (1918–)
Courtesy of Dr. Frederick Sanger.

Walter Gilbert (1932–)
Photo: Rick Stafford.

We now turn our attention to a major result of recombinant DNA technology, DNA sequencing. Recombinant DNA technology, with its ability to isolate and amplify small, well-defined regions of chromosomes, has allowed the development of these techniques.

DNA SEQUENCING

Paul Berg of Stanford, Walter Gilbert of Harvard, and Fred Sanger of the Medical Research Council in Cambridge, England, shared the 1980 Nobel Prize in chemistry. Berg won for creating the first recombinant DNA molecules when he spliced the SV40 genome into phage λ. Gilbert and Sanger were awarded the prize for independently developing methods of sequencing DNA. Gilbert, along with Allan Maxam, developed a method of DNA sequencing called the chemical method. It involves chemically breaking down the DNA at specific bases. Sanger, who won a Nobel Prize in 1959 for sequencing the insulin protein, later took part in developing methods for sequencing RNA. His method of sequencing DNA, developed with Alan Coulson, involved DNA synthesis and was called the plus-and-minus method. Recent advances, developed by Sanger, Coulson, and S. Nicklen, using specific chain-terminating nucleotides, have led to a modification of the plus-and-minus method known as the dideoxy method, which we will examine in detail.

Figure 12.35
Dideoxy nucleotides cause chain termination during DNA replication. The dideoxy primer configuration lacks the 3'-OH group needed for chain lengthening.

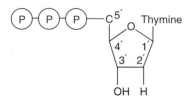

Deoxythymidine triphosphate (dTTP)

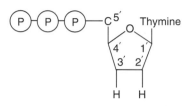

Dideoxythymidine triphosphate (ddTTP)

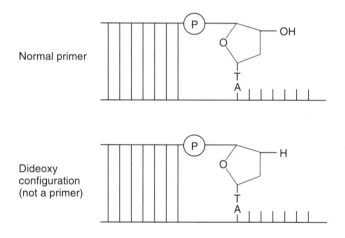

The Dideoxy Method

In the **dideoxy method,** newly synthesized DNA is sequenced. Remember from chapter 9 that DNA synthesis occurs at a primer configuration, one in which double-stranded DNA ends with a 3'-OH group on one strand. The other strand continues as single-stranded DNA (fig. 12.35, *middle*). In the dideoxy method, a primer configuration of the DNA to be sequenced is created and replication proceeds. A trick, using chain-terminating nucleotides, is used to stop DNA synthesis at known positions. Chain termination is achieved by using nucleotides whose sugars lack OH groups at both the 2' and 3' carbons (hence the term "dideoxy"). Without a 3'-OH group a dideoxynucleotide cannot be used for further DNA polymerization (fig. 12.35).

Because of chain-terminating nucleotides, synthesis can be stopped at a known base. The sample is elongated separately in four different reaction mixtures, each having

Figure 12.36

Initial steps in the dideoxy method of DNA sequencing. The *asterisks* indicate the dideoxynucleotides. The DNA to be sequenced is placed into a primer configuration (*a,b*). Four reaction mixtures are created, each with all four normal nucleotides plus one of the dideoxynucleotides. Thus DNA synthesis in each reaction mixture is stopped a percentage of the time when the complement to the dideoxynucleotide appears in the template (*c*).

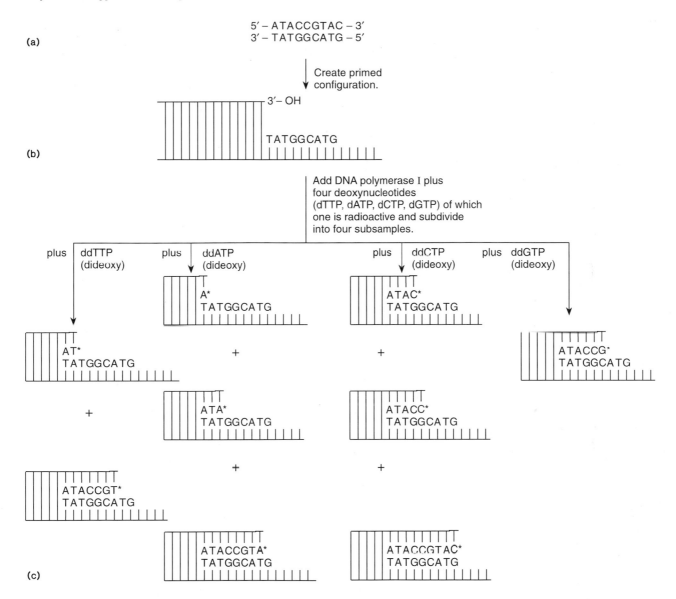

all four normal nucleotides but also having a proportion of one of the chain-terminating dideoxy nucleotides. For example, if the pool of thymine precursors contains a portion of the dideoxythymidine triphosphate molecules, then synthesis of the growing strand will be terminated some of the time when adenine appears on the template, thus creating fragments that end in thymine. Similar reactions are carried out for each of the other nucleotides, producing fragments that terminate when the respective complementary nucleotide is present. The resulting fragments from each reaction are electrophoresed, generating a pattern from which the sequence of the newly synthesized DNA can be read directly off the gel. Let's go through an example.

In figure 12.36a we show the DNA to be sequenced, a small segment of nine base pairs. In order to sequence this segment, it is necessary to get one strand of this double-stranded segment into the configuration shown in figure 12.36b. The DNA to be sequenced must be the template for new DNA synthesis. (We will discuss how we obtain the required configuration—see "Creating a General-Purpose Primer.") Having created the necessary primer

Figure 12.37

Electrophoresis of segments produced by the dideoxy method of DNA sequencing that allows direct reading of the sequence. The *asterisks* indicate the dideoxynucleotides. The newly synthesized reaction products seen in figure 12.36 are isolated by removal of the primer and template. Each reaction mixture (e.g., ddTTP is the mixture containing dideoxythymines) produces specific products of specific lengths that can be determined by electrophoresis. In the case of the ddTTP mixture, two fragments ending in thymine are possible; one is two bases long, the other seven bases long. Thus thymine appears in positions two and seven of the original piece of DNA.

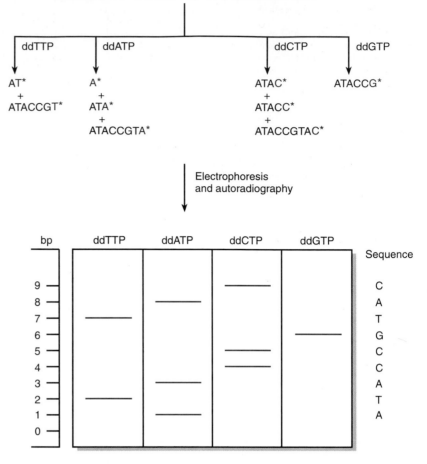

configuration, we take four subsamples of it, each of which includes all four nucleoside triphosphates plus DNA polymerase I. At least one of the nucleoside triphosphates is radioactively labeled, usually with ^{32}P. This label will allow us to identify newly synthesized DNA by autoradiography.

To each of the four subsamples one of the dideoxynucleotides is added—one subsample gets ddTTP, one gets ddATP, one gets ddCTP, and one gets ddGTP. These dideoxynucleotides are added in addition to the regular deoxynucleotides so that there will be some probability that chain termination will occur at every appropriate position. If the dideoxynucleotide were added in place of the deoxynucleotide, then the chain would be terminated the first time—and only the first time—that the complement of that base appeared in the template strand. By mixing the dideoxy- and the deoxynucleotides, we are assured that termination will occur in every appropriate position.

For example, if in reaction subsample 1, 50% of the thymine-containing nucleotides were dTTP and 50% were ddTTP, then there would be a 0.5 probability of chain termination every time a thymine was required during DNA synthesis (see fig. 12.36c). If the ratio were shifted to more dTTP, then the average synthesized segment would be larger. If the ratio were shifted toward ddTTP, then the

average synthesized segment would be shorter. Note that the template has two adenines. Therefore, in the ddTTP reaction mixture, adenine's complement (thymine) is needed twice. There are thus two possible points for ddTTP to be incorporated, two possible chain terminations, and therefore two fragments possible ending in dideoxythymidine, of two and seven bases, respectively. Similarly, there are three possible fragments ending in adenine, of one, three, and eight bases; three ending in cytosine, of four, five, and nine bases; and one ending in guanine, of six bases (fig. 12.36c and fig. 12.37, *top*).

After DNA synthesis is completed, the old primer is removed by methods mentioned below, leaving only newly synthesized DNA fragments (fig. 12.37). Newly replicated segments of varying length from each reaction mixture are placed in separate slots and then electrophoresed on polyacrylamide gels to determine the lengths of the segments present. Since only newly synthesized DNA segments are radioactive, radioautography lets us keep track of newly synthesized DNA. As you can see from the autoradiograph of the gel in figure 12.37, each subsample will produce segments that begin at the primer configuration (beginning of synthesis) and end with the chain-terminating dideoxy base. By starting at the bottom and reading up back and forth across the gel, the exact se-

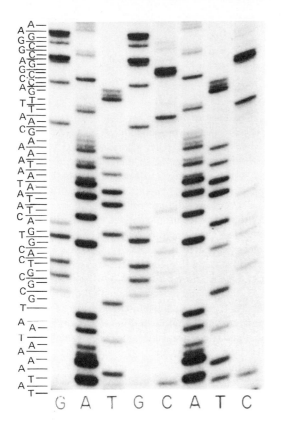

Figure 12.38
Autoradiograph of a dideoxy sequencing gel. The letters G, A, T, and C at the bottom refer to the ddGTP, ddATP, ddTTP, and ddCTP reaction mixtures, respectively. Lanes are repeated for easier identification of the bands. Verification is also done by sequencing the complementary strand and checking for agreement.
From Dr. T. Gingerast, Science 209 *(September 19, 1980):1322–28. "Steps toward Computer Analysis of Nucleotide Sequences." Copyright 1980 by the AAAS.*

quence of the DNA segment can be determined directly. Because they have the appearance of stepladders in each lane (fig. 12.38), the gels are usually referred to as **stepladder gels** or **ladder gels.**

This technique (in the form of the original plus-and-minus method) was first used to sequence the genome of the DNA phage φX174 (see Box 12.3). That phage was used because it lent itself to the sequencing method. It has single-stranded DNA within the phage coat, yet its DNA becomes double-stranded once it enters the bacterium. Creating a primer configuration was thus relatively easy. The double-stranded circle from within the host could be treated with a restriction endonuclease to produce double-stranded fragments (fig. 12.39). These fragments could then be denatured. From this mixture, a particular fragment could be isolated by electrophoresis. The isolated strand would reanneal to the single-stranded DNA taken from phage heads, forming a primer for new growth. The same restriction endonuclease would free the new growth after it had taken place. Thus the dideoxy method previously described was relatively easy to apply to the 5,387-base chromosome of φX174.

Creating a General-Purpose Primer

In order for the dideoxy method to be generally applicable, a routine method to create a primer was needed. A general primer was created for routine sequencing work by recombinant DNA engineering of an *E. coli* vector, the single-stranded DNA phage M13. This phage is similar to φX174 in that both are packaged as single-stranded DNA and both are replicated to double helices within the host. Therefore, the double-stranded form within the host, called the replicating form (RF), can be engineered by standard methods and the single-stranded form can be used for sequencing. The system works as follows.

By very clever engineering, J. Messing and his colleagues created cloning sites for a variety of restriction enzymes in a bacterial gene (*lacZ*) that had been inserted into M13 (fig. 12.40). The gene is for the β-galactosidase enzyme that normally breaks down lactose. It also breaks down an artificial substrate of the enzyme, X-gal, which is normally colorless. When cleaved by β-galactosidase, X-gal becomes blue. Thus, in the presence of the functional *lacZ* gene, M13 plaques are blue. If the gene is disrupted by a cloned insert, X-gal is not broken down, and hence plaques are colorless. (M13 doesn't form true plaques because it doesn't lyse the *E. coli* cells. It does form turbid sites due to reduced growth of the bacteria.)

After cloning, phages released from the host contain single-stranded DNA. An oligonucleotide primer has been synthesized that is complementary to a region of the phage DNA upstream from the cloning sites. Single-stranded phage DNA containing a cloned insert is isolated and hybridized with the synthetic oligonucleotide. This operation creates the primer configuration for dideoxy

BOX 12.3

GENES WITHIN GENES

Complete sequencing of a DNA genome using Sanger and Coulson's plus-and-minus method (the forerunner to the dideoxy method) was first accomplished with φX174, a virus that contains a single-stranded DNA circle of 5,387 bases within its protein capsule. Once injected into the host, the DNA is replicated to form a double helix that then proceeds in a normal viral fashion to replicate itself, manufacture its own coat proteins, lyse the cell, and escape. This virus has nine genes. The virion is a small, twenty-faced polyhedron with a small spike at each of its twelve vertices. It is this spike that attaches φX174 to *E. coli*. The coat accounts for one protein and the spike accounts for two. Thus three of the virus's nine genes manufacture coat proteins. Box figure 1 illustrates the location of the genes in φX174, obtained through standard mapping methods.

From the information obtained from the sequencing of MS2, an RNA virus, it was believed that there should be a nontranslated sequence between each gene, presumably for the purpose of controlling expression of each gene. However, careful perusal of the nucleotide sequence of φX174 provided several surprises. First, the ends of three genes overlapped the beginnings of the next genes (A-C, C-D, and D-J); in the first two cases, the initiation codon is entirely within the end of the previous gene *but* read in a different frame of reference. In the sequence ATGA, the ATG is the initiation of the next gene, whereas the TGA is the termination of the previous gene. In the D-J interface, one A is shared: TA*A*TG (Box fig. 2). It is the number 3 base of the termination codon and the number 1 base of the initiation codon. The surprises did not end there.

At first, with the sequence of nucleotides spread out in front of them, the researchers could not find the B and the E genes—these genes appeared to be missing. Upon careful analysis, however, they found that the B gene was entirely within the A gene and the E gene was entirely within the D gene (Box fig. 3). Their finding went against all theory. We are led to believe, from logical arguments, that genes cannot substantially overlap. There would be too much of a constraint on function: The functional sequence of one gene would have to also be a functional sequence in the other. Similarly, there would be an evolutionary constraint involved. The genes would have to evolve together. But here we have two cases in which genes do overlap. How could overlapping genes come about?

Box figure 1

Presumed location of the nine genes of φX174 on its circular chromosome. Transcription begins at three different places, each marked *p,* for promoter. The function of each gene is given within the circle.

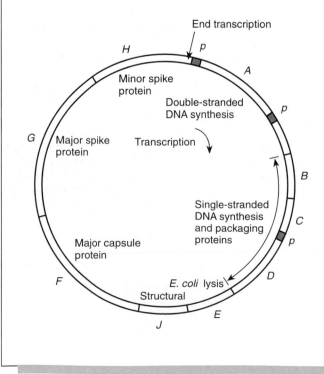

Box figure 2

Sequence, shown as ribose nucleotides, where genes E and D end and gene J begins. Each is out of register with the other two. The A of AUG for gene J is the second A of the UAA terminator of gene D.

```
J                                      Met  Ser
5' • • •  AAGGAGUGAUGUAAUGUCU  • • •  3'
E            Lys  Glu  Stop
D            Glu  Gly  Val  Met  Stop
```

Box figure 3

The actual map of the nine genes of φX174. Note that B is entirely within A and E is entirely within D.

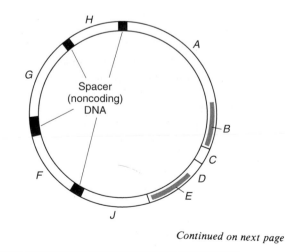

Continued on next page

BOX 12.3—_continued_

There are a large number of thymine bases in the ϕX174 genome. In the D gene particularly, many of the codons end with thymine. The imbedded E gene is read on a shifted frame within D so that the terminal base of D's codons is the middle base of E's. A look at the genetic code (see table 11.4) will show that the codons with U in the middle (E's codons) are mainly for hydrophobic amino acids. Thus E is a protein with detergent properties. In fact, it is the protein responsible for the dissolution of the outer cell wall of the host bacterium, a process that can be accomplished _in vitro_ by a detergent. The properties of the E gene, then, are more the properties of its individual amino acids rather than their exact sequence.

In the A-B case, there is an indication that the two genes were once autonomous. This indication is based on the patterns of the codons in which A's codons tend to end in thymine before the overlap, but thereafter, in the region of overlap, B's codons end in thymine whereas A's codons do not. Presumably, a mutational event tagged the B material onto the end of the earlier, shorter A gene and improved its enzymatic ability. We can, however, only speculate.

The amazing arrangement of this viral DNA is one of extreme economy. The protein package is small, yet a minimum of nine genes had to be packed into it. We have seen this kind of economy before in the codon usage of mitochondrial DNA (see chapter 11). In ϕX174, not only are beginnings and ends of some genes overlapping, but genes also occur within genes.

As more sequencing has taken place, other novel overlap situations have been seen. For example, a case was discovered in the rat of two genes transcribed from opposite strands of the same region of DNA. On one strand the gonadotropin-releasing hormone (GnRH) is located. On the other is a gene (RH) that produces a protein of unknown function that is expressed in the heart.

Overlap of genes is known to occur in bacteria as well. In _E. coli,_ the promoter for the _ampC_ gene (coding for the enzyme β-lactamase) begins within the last ten codons for the _frdC_ gene, which codes for a subunit of the enzyme fumarate reductase. There is evidence that in this arrangement the _frdC_ terminator can have some regulatory control of _ampC_ transcription. (See chapter 13 for a discussion of regulatory processes in eukaryotes.)

With DNA sequence data, including the complete maps of other small genomes such as SV40 and mitochondrial chromosomes, we have accumulated much information about gene arrangements. Overlap to one degree or another has been found in small viruses (ϕX174, SV40), large viruses (λ), mitochondrial chromosomes, bacterial DNA, and even eukaryotes, where several cases are now known in which genes are located within introns of other genes. In one of the few examples known, three genes are located in an intron of the neurofibromatosis gene, a gene that causes a disfiguring neurological disease. Although serving economic ends, overlap and embedding of genes may have some regulatory role in transcription.

---

sequencing of the cloned DNA. Virtually any clonable segment of DNA can be sequenced using this very general method. Theoretically, that segment could be any size.

Stepladder gels, however, are effective only up to about 400 bp. To sequence regions larger than that requires sequencing overlapping segments and reconstituting the sequence by the overlap pattern, similar to the methods we described for amino acid sequencing (chapter 11, Box 11.3). Overlapping segments of DNA to sequence are obtained either by using two or more restriction enzymes or by sequencing segments created by shearing the DNA into small, blunt-ended pieces and then using linkers or blunt-end ligation to clone them.

The most recent innovation in DNA sequencing involves using fluorescent dyes. Four dyes are used, each fluorescing at a different wavelength (505, 512, 519, and 526 nm). The methodology is essentially the same as described previously except that each of the four dideoxy nucleotides has a different one of the four dyes attached. After the newly synthesized fragments are isolated, the products from all four reactions are run together in the same lane of a polyacrylamide gel. The gel is then scanned with an argon laser that excites the dye molecules. An instrument records the color of the peaks, reading the sequence directly and automatically (fig. 12.41). This method greatly simplifies sequencing since it is automated. It also alleviates the necessity for radioactive tags.

Even when automated DNA sequencers are not used, radioactive tags can now be replaced with chemiluminescent tags of the dideoxy nucleotides, again alleviating the need for radioactive isotopes. In addition, it is possible to combine PCR and dideoxy sequencing to get simultaneous amplification and sequencing of a particular DNA sequence. Innovative techniques make it possible to sequence DNA more quickly, cheaply, and efficiently.

Figure 12.39

The genome of φX174 lent itself to the dideoxy method (originally the plus-and-minus method) of DNA sequencing. Because the phage occurs in both the single- and double-stranded forms, it can be manipulated for sequencing. The double-stranded form is fragmented with an endonuclease. One fragment is isolated by electrophoresis and hybridized to the single-stranded form creating a primer for new DNA synthesis and thus dideoxy sequencing. Newly synthesized DNA can be isolated by treating with the same restriction enzyme, which will create the same cut it made originally.

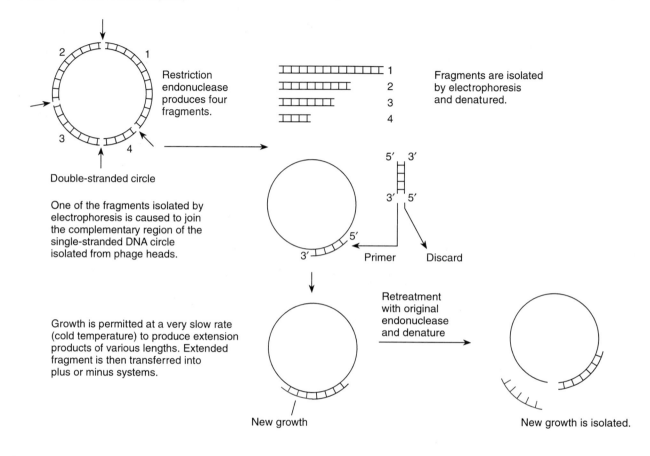

Restriction endonuclease produces four fragments.

Double-stranded circle

One of the fragments isolated by electrophoresis is caused to join the complementary region of the single-stranded DNA circle isolated from phage heads.

Growth is permitted at a very slow rate (cold temperature) to produce extension products of various lengths. Extended fragment is then transferred into plus or minus systems.

Fragments are isolated by electrophoresis and denatured.

Primer   Discard

Retreatment with original endonuclease and denature

New growth

New growth is isolated.

Either the dideoxy method, or the other method of sequencing, the chemical method, which we have not discussed, allows us to read the sequence of hundreds of nucleotides on a single gel. Whole viral genomes, prokaryotic and eukaryotic genes, and numerous regions of interest in prokaryotes, eukaryotes, and viruses have been sequenced. Sequence information on about one million base pairs per year is accumulating. As W. Gilbert said in his Nobel Prize acceptance speech in 1981, "When we work out the structure of DNA molecules, we examine the fundamental level that underlies all processes in living cells."

PRACTICAL BENEFITS FROM GENE CLONING

Throughout the chapter we have mentioned applications of genetic engineering. Here we summarize some of the accomplishments and future directions in the medical, agricultural, and industrial arenas.

Medicine

In biomedicine, genetic engineering has had remarkable successes. On the one hand, basic knowledge about how genes work (and don't work) has been advanced tremendously. On the other hand, recombinant DNA methodology has made available large quantities of substances previously in short supply. The latter include insulin, interferon (an antiviral agent), growth hormone, growth factors, blood-clotting factors, and vaccines for diseases such as hepatitis B, herpes, and rabies. Advances in AIDS and cancer research will be discussed in chapter 15. Genetic engineering is making it possible to manufacture antibodies to order and to diagnose and treat diseases. The sequencing of the human genome (Box 12.2) will further aid medicine by identifying the genes for various diseases, a first step in discovering cures. We also pointed out the use of restriction fragment length polymorphisms and polymerase chain reaction as techniques of tremendous power in the identification of individual people.

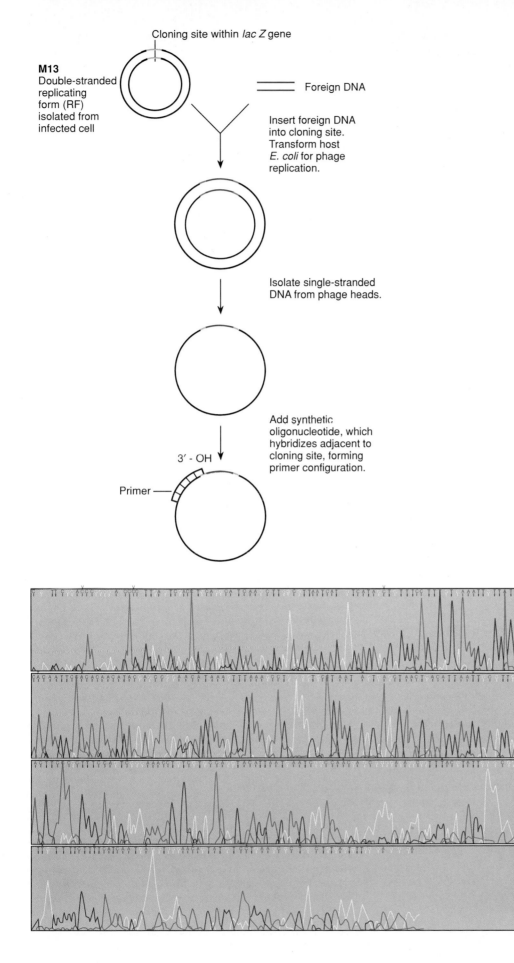

Cloning site within *lac Z* gene

M13
Double-stranded
replicating
form (RF)
isolated from
infected cell

Foreign DNA

Insert foreign DNA
into cloning site.
Transform host
E. coli for phage
replication.

Isolate single-stranded
DNA from phage heads.

Add synthetic
oligonucleotide, which
hybridizes adjacent to
cloning site, forming
primer configuration.

3' - OH

Primer

Figure 12.40
Phage M13, a useful vector for
sequencing a piece of cloned
DNA by the dideoxy method
since it also exists in both single-
and double-stranded forms. In
addition, it contains restriction
sites within a copy of the *lacZ*
gene present (*blue*). This property
allows for the selection of clones
with inserted pieces of foreign
DNA. An artificial
oligonucleotide, which hybridizes
adjacent to the cloning site,
provides the primer configuration
needed for new synthesis.

Figure 12.41
Processed data from automated
analysis using fluorescent dyes.
The sequence (97.5 percent
accurate) is read left to right, top
to bottom. Each dideoxy base
fluoresces a different color in laser
light, making it possible for a
machine to read the products of
the dideoxy sequencing process
directly, with all the products in
one lane rather than four.
*Reprinted from L. Johnston-Dow, et
al., 1987, BioTechniques, 5:754–765,
with permission of the authors and
Eaton Publishing.*

On another front, transgenic mice have shown that genetic engineering can be applied to higher organisms. The use of this as a treatment for human diseases, however, is only just beginning. In July of 1990 two committees of the National Institutes of Health approved gene therapy treatments on people. In the first, a tumor necrosis factor gene was to be injected into people with cancer. The necrosis factor often shrinks tumors. The gene is in lymphocytes that have been engineered. The first group to be tested will include those with malignant melanoma, an incurable form of skin cancer. In another study, a child was infused with cells to replace a gene for the enzyme adenosine deaminase, an enzyme whose absence results in a dysfunctional immune system. AIDS, hemophilia, and diabetes are other disease conditions that should be amenable to gene therapy in the near future.

Agriculture

In agriculture, several areas of research are being actively pursued. Bacteria have been modified to impart frost protection in plants (see Box 12.1). *Pseudomonas syringae* has been modified by deleting a gene for a protein that nucleates ice-crystal formation. If these bacteria successfully replace the normal ones, ice will not form as readily when the temperature drops below freezing. Other bacterial projects include amplification of nitrogen fixation in *Rhizobium meliloti* and the introduction of insect toxins into *Pseudomonas fluorescens* to protect plant roots. Other techniques used to protect plants include the development of bacterial and viral strains to kill various pests such as the cotton bollworm and the Douglas fir tussock moth.

As we mentioned earlier, plants and livestock are being engineered directly. Plants are being engineered for drought resistance, virus resistance, and better qualities desired by consumers. For example, tomatoes with reduced activity of the enzyme polygalacturonase, which breaks down cell walls upon ripening, have a longer shelf life. Livestock are being engineered for growth and disease-resistance traits.

Industry

Industrial applications of biotechnology include the engineering of bacteria to break down toxic wastes, the utilization of cellulose by yeast to produce glucose and alcohol for fuel, the use of algae in mariculture (the cultivation of marine organisms in their natural environments) to produce both food and other useful substances, and the use of genetic engineering in the food industry for better processing methods and waste conversion. As an example, baker's yeast (*Saccharomyces cerevisiae*) has been modified with a plasmid that contains two cellulase genes, an endoglucanase and an exoglucanase, that convert cellulose to glucose. The glucose is then converted to ethyl alcohol by the yeast. These yeasts are now capable of digesting wood (cellulose) and converting it directly to alcohol. The potential exists to harvest the alcohol produced by the yeast as a fuel to replace fossil fuels that are in dwindling supplies and are polluting the planet.

As you can see, there is no one direction that biotechnology is going. Many advances are being made that can affect every person's life in a beneficial way. Cautious optimism is certainly in order.

SUMMARY

Recombinant DNA techniques revolve around the cloning of foreign DNA in a vector (a plasmid or phage). Cloned DNA can be amplified, expressed, and sequenced. Gene cloning techniques came about with the discovery of restriction endonucleases. Type II restriction endonucleases cleave DNA at palindromic regions, those with twofold symmetry.

Hybrid vectors can be constructed several different ways. Foreign and vector DNA can be made compatible by treating each with the same restriction endonuclease—each will then have the same sticky ends. If that does not work, incompatible ends can be made compatible with the poly-dA/poly-dT cloning method. Either the foreign DNA or the vector is enzymatically given a 5′ poly-dA tail; the other is given a 5′ poly-dT tail. The now-complementary ends will anneal. Alternatively, blunt ends can be joined by T4 DNA ligase. In a variation of this method, linkers containing restriction sites are added to vehicle and foreign DNA. These linkers are then treated with a restriction endonuclease that gives the DNA sticky ends.

DNA to be cloned can be synthesized from an RNA template (cDNA) or isolated by various techniques. If mRNA is available, it can be converted into a clonable complementary DNA with the use of the enzyme reverse transcriptase. If DNA is to be isolated directly, either by endonuclease treatment or physical shearing, it must be identified among all the other DNA fragments created. Locating a desirable piece of DNA is done with probes, radioactive complementary nucleic acids. Southern blotting, a diffusion technique, is used first, followed by DNA-DNA or DNA-RNA hybridization and autoradiography. If the DNA is cloned first, as in the creation of a genomic library, probes can be created as before or expression of the cloned gene can be determined. Chromosome walking allows the analysis of long stretches of DNA. Heteroduplex analysis, the electron microscope observation of hybrid nucleic acid molecules, is a useful adjunct technique in the analysis of hybrid vectors.

Eukaryotic vectors have been developed, including yeast plasmids, tumor virus vehicles in animals, and crown gall tumor plasmids in plants. Eukaryotes can be transfected by foreign DNA and express it as transgenic organisms. DNA can be injected, shot in on projectiles, electroporated, or introduced by the use of viruses, plasmids, or liposomes.

Restriction digests can be separated by electrophoresis from which a restriction map can be constructed. The latter is a physical map of the DNA showing the location of restriction enzyme recognition sites. The genetic maps, generated by mating analysis, can then be superimposed on the restriction maps, locating regions of interest on the physical map. Restriction fragment length polymorphisms (RFLPs) provide a tool for locating genes through linkage analysis and are also valuable in forensic science. Polymerase chain reaction (PCR) is a technique to rapidly amplify particular segments of DNA.

DNA is usually sequenced by one of two methods. The dideoxy method developed by Sanger and his colleagues requires the synthesis of DNA in the presence of chain-terminating (dideoxy) nucleotides. Electrophoresis followed by autoradiography allows the direct determination of the sequence of nucleotides synthesized. Fluorescent labeling allows computerized sequence determinations. The phage ϕX174 was sequenced in its entirety by the forerunner of this technique, the plus-and-minus method. Gilbert and Maxam's chemical method also is used widely.

We closed by summarizing the applied fronts of genetic engineering. Medical, agricultural, and industrial applications are becoming accepted.

SOLVED PROBLEMS

1. A piece of eukaryotic DNA is obtained by shearing. How could we get this piece of DNA into a *Bam* HI site in plasmid pBR322, and how would we know when the foreign DNA has been cloned?

   ANSWER: Since the two pieces of DNA (the eukaryotic piece and the plasmid) have different ends, they must be made compatible before cloning. The simplest way would be to attach blunt-ended linkers to the foreign DNA with phage T4 DNA ligase (see fig. 12.10). The linkers, of course, would have a *Bam* HI site within. After the linkers are attached to the foreign DNA, it would be treated with the *Bam* HI restriction enzyme, giving the foreign DNA *Bam* HI ends. The plasmid is then also treated with the restriction enzyme and the two (the foreign DNA and the cut plasmid) are now mixed together in the presence of *E. coli* DNA ligase, which seals up the plasmids, with or without cloned inserts (see fig. 12.6). Since they have compatible ends, some of the time a piece of foreign DNA will be inserted into a plasmid. The plasmids are then taken up by *E. coli* cells that are grown overnight in an incubator. The bacterial colonies are then replica plated on media with the antibiotics ampicillin or tetracycline. Colonies that are resistant to ampicillin but sensitive to tetracycline are assumed to be of bacteria containing plasmids with cloned inserts (see fig. 12.8).

2. How does a reporter system work?

   ANSWER: A reporter system is a genetically manipulated system that displays a particular phenotype or reaction when a desired event has taken place. In this chapter we discussed the firefly luciferase reporter system in which the desired result (transcription of a particular promoter) causes a transgenic tobacco plant to glow. Let us say that we are studying the control of transcription of a particular eukaryotic gene. We could attach the promoter of that gene to the firefly luciferase gene in a Ti plasmid by various cloning techniques. The plasmid could then transfect tobacco plants. We could then continue our experiment to determine whether the promoter under study is active under various conditions. We would know whether it was active by watering the plants with luciferin. If the plant glows, then the luciferase gene

product is present, which means that the promoter under question was active. In other words, the glowing of the plant "reports" the action of the promoter under question; the promoter was active because it allowed the transcription of the luciferase gene.

3. A piece of DNA has the sequence 3'-CCGCATAAG-5'. It is sequenced using the dideoxy method. How many bands are found on the ladder gel? How many bands and of what size are found for each reaction mixture?

   ANSWER: Since the piece of DNA is nine bases long, the total number of bands in all four lanes of a sequencing gel will add up to nine (see fig. 12.37). By each reaction mixture, we mean the four reaction mixtures each with one of the dideoxynucleotides. In the reaction mixture with ddATP, chain termination occurs at the thymine in the piece of DNA; that is, a DNA segment was synthesized that is six bases long. In the reaction mixture with ddTTP, chain termination occurs opposite each of the adenines, producing DNA segments of five, seven, and eight nucleotides. In the reaction mixture with ddCTP, chain termination occurs opposite the guanines in positions three and nine. And, in the reaction mixture with ddGTP, chain termination occurs after synthesis of segments one, two, and four bases long (see figure).

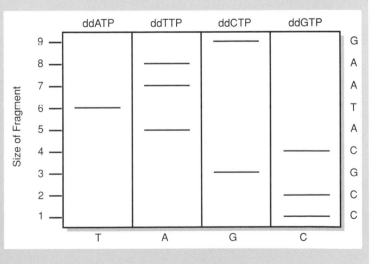

4. A linear DNA molecule 1,000 bp long is digested with the following restriction enzymes with the following results:

Eco RI	400 bp, 600 bp
Bgl II	250 bp, 750 bp
Eco RI + *Bgl* II	250, 350, 400 bp

Determine the restriction map.

ANSWER: Each enzyme alone produces two fragments, so the molecule has one site for each enzyme. Since we get different size fragments with each enzyme, the sites must be located asymmetrically along the DNA. Draw these sites:

Eco RI
400 | 600 and *Bgl* II
250 | 750

The *Eco* RI fragment that lacks a *Bgl* II site should appear in the double digest. If *Bgl* II cuts within the 400 bp fragment, we expect to see 150, 250, and 600 bp fragments. We don't see this, so the *Bgl* II site is not within the 400 bp *Eco* RI fragment. Thus the map looks like

Eco RI   *Bgl* II
400 ↓   350   ↓250

EXERCISES AND PROBLEMS

1. What specific properties of type II endonucleases make them useful in gene cloning?

2. The following is a single strand of DNA. What, if any, are potential restriction enzyme recognition sequences in the double-helical form of the DNA?

5'-TAGAATTCGACGGATCCGGGGCATGCAGATCA-3'

3. Assuming a random arrangement of nucleotides on a piece of DNA, what is the probability that a restriction endonuclease whose recognition site consists of four bases (a four-cutter) will cut the DNA? What is the probability for a six-cutter? An eight-cutter?

4. Under what circumstances are restriction endonucleases unsuitable for cloning a piece of foreign DNA?

5. What methods exist to create sticky ends or create ends for joining two pieces of DNA? When is each method favored?

6. Diagram a possible heteroduplex between two phage λ vectors, one with and one without a cloned insert.

7. What are the differences among plasmid, cosmid, and expression vectors? Under what circumstances are each useful?

8. What are the steps by which mRNA can be converted into cDNA? How would we obtain radioactive cDNA? Radioactive mRNA?

9. What is chromosome walking? When is it used?

10. How would we isolate a human alanine tRNA gene for cloning? How would we locate a clone with a human alanine tRNA gene in a genomic library?

11. What are the differences among Southern, western, northern, and dot blotting?

12. How would you develop a probe for a gene whose mRNA could not be isolated? How could an expression vector be used to isolate a cloned gene?

13. How are *E. coli* plasmids manipulated to survive in yeast? How can virus genomes, such as SV40 and phage λ, survive as functioning vectors when parts of their genomes are replaced by cloned DNA?

14. What methods are used to get foreign DNA into eukaryotic cells? What is transfection? What is a transgenic mouse?

15. What is hypervariable DNA? A RFLP? A VNTR locus?

16. How are DNA fingerprints useful in forensic cases? Could they be used in paternity exclusion? How?

17. What is PCR? When is it used?

18. The segment of DNA shown below is cut four times by the restriction endonuclease *Eco* RI at the places shown. Diagram the gel banding that would result from electrophoresis of the total and partial digests. Note the end-labeled segments and regions where several segments form bands at the same place on the gel.

100   ↓   300   ↓50↓   250   ↓   150

19. On the following page are gels of a total and partial digest of a DNA segment treated with *Hin* dII. End-labeled segments are noted by asterisks. Draw the restriction map of the original segment.

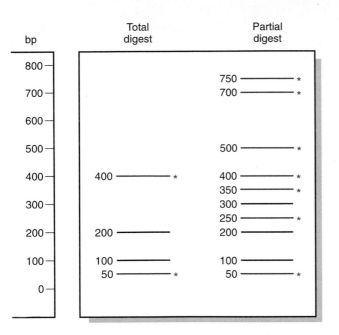

20. Several mutants of the DNA segment shown in problem 18 were isolated. They gave the following gel patterns when the total digests were electrophoresed. Asterisks denote the end-labeled segments. Can you determine the nature of the mutations?

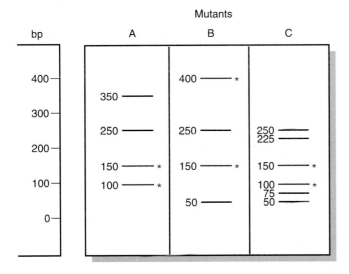

21. Restriction maps of a segment of DNA were worked out separately for *Bam* HI and *Taq* I. Two overlays of the maps are possible. The double-digest gel is shown (asterisks denote end labels). Which overlay is correct?

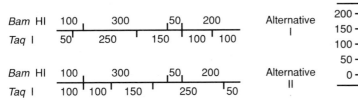

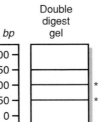

22. What are the steps in the dideoxy method of DNA sequencing? How has the technique been improved with fluorescent dyes?

23. The following diagram is of a dideoxy sequencing gel. What is the sequence of the DNA under study?

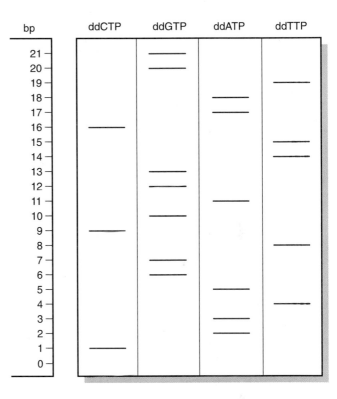

24. How can a particular piece of DNA be manipulated so as to be in the appropriate configuration for dideoxy sequencing?

25. Provide, if possible, DNA sequences that can mark the termination of one gene and the initiation of another, given that the genes overlap in one, two, three, four, five, six, or seven bases.

26. Describe some areas of practical benefit of genetic engineering. Why might some people be concerned about its widespread use?

27. A linear DNA molecule 1,000 bp long gives the following size fragments when treated with the following restriction enzymes. Derive a restriction map.

Eco RI:	300 bp, 700 bp
Bam HI:	150 bp, 200 bp, 250 bp, 400 bp
Eco RI + *Bam* HI:	50 bp, 100 bp, 200 bp, 250 bp, 400 bp

28. A linear DNA molecule is cut with *Eco* RI and yields fragments of 3 kb, 4.2 kb, and 5 kb. What are the possible restriction maps?

29. You have double-stranded DNA that you radioactively label at the 5' ends. Digestion of this molecule with either *Eco* RI or *Bam* HI yields the following fragments. The numbers are in kb, and an asterisk indicates the fragments that are labeled.

Eco RI: 2.8, 4.6, 6.2*, 7.4, 8.0*

Bam HI: 6.0*, 10.0*, 13.0

If unlabeled DNA is digested with both enzymes simultaneously, the following fragments appear: 1.1, 2.0, 2.8, 3.5, 6.0, 6.2, 7.4. What are the restriction maps for the two enzymes?

30. A 12 kb DNA molecule cut with *Eco* RI yields one 12 kb fragment. When the original molecule is cut with *Bam* HI, three fragments of 2 kb, 4.5 kb, and 5.5 kb are produced. When the fragment from *Eco* RI is treated with *Bam* HI, four fragments of 2 kb, 2.5 kb, 3.0 kb, and 4.5 kb are produced. Draw a restriction map.

31. Exonuclease III is an enzyme that sequentially removes bases from the 5' end of double-stranded DNA. It requires a free 5' phosphate to work. The following two molecules, each 100 bp long, are digested with exonuclease III. Molecule 1 is completely digested; molecule 2 is only partially digested. Explain these results.

Molecule 1:	CGTTCAG . . .
	GCAAGTC . . .
Molecule 2:	A A A A A A A A A A . . .
	T T T T T T T T T T . . .

32. Draw the expected gel pattern derived from the dideoxy sequencing method for a template strand that has the following sequence:

5'—CAGCGAATGCGGAA—3'

33. A plasmid that contains an *Eco* RI site within an ampicillin resistance gene is digested with *Eco* RI, and then re-ligated. This plasmid is used to transform *E. coli,* and the plasmid reisolated from the ampicillin resistant colonies. The reisolated plasmids from two different colonies are electrophoresed, and the results appear in the next column.

undigested plasmid	colony 1 digested, re-ligated	colony 2 digested, re-ligated
		———
———	———	———

How do you account for the two bands in colony 2?

34. Most human genes contain one or more introns. Since bacteria cannot excise introns from mRNA, how can bacteria be used to make large quantities of a human protein?

35. A plasmid that is 3 kb in length contains a gene for ampicillin resistance and a gene for tetracycline resistance. The plasmid has a single site for each of the following enzymes: *Eco* RI, *Bgl* II, *Hin* dIII, *Pst* I, and *Sal* I. If DNA is cloned into the *Eco* RI site, resistance to either antibiotic is not affected. DNA cloned into *Bgl* II, *Hin* dIII, or *Sal* I sites abolishes tetracycline resistance, and DNA inserted into the *Pst* I site eliminates ampicillin resistance. If the plasmid is digested completely with enzyme mixes, the following size fragments result:

Mixture	Fragment size (kb)
Eco RI + *Pst* I	0.7, 2.3
Eco RI + *Bgl* II	0.3, 2.7
Eco RI + *Hin* dIII	0.08, 2.92
Eco RI + *Sal* I	0.85, 2.15
Eco RI + *Bgl* II + *Pst* I	0.3, 0.7, 2.00

Draw a restriction map of the plasmid and indicate the locations of the resistance genes and the sites of enzymatic cleavage.

36. A gene has the following *Eco* RI restriction map:

1.0 kb	0.7 kb	2.0 kb

Draw the gel pattern expected from
a. a mutant that has lost the site that generated the 1.0 and 0.7 kb fragments.
b. a mutant that has a new site within the 2.0 kb fragment.

37. A DNA fragment 8 kb in size is labeled with ^{32}P at the 5' ends. It is then digested with *Eco* RI, *Bgl* II, or a mixture of both enzymes. The size of the fragments and the labeled fragments (*) appear below; sizes are in kb.

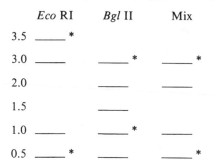

Which of the two maps on page 347 is consistent with the above results?

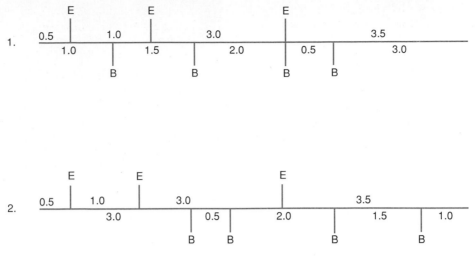

1.

2.

38. You now take an unlabeled molecule from problem 37, digest it with *Hin* dIII, and get two fragments, 5.5 and 2.5 kb in size. If the *Hin* dIII does not cut within the 3.5 kb *Eco* RI fragment, what size fragments do you expect in a double digest of *Hin* dIII and *Eco* RI?

39. A DNA strand with the sequence: 3'-GACTATTCCGAAAC-5' is sequenced by the dideoxy method. If the reaction mixture contains all four radioactive deoxynucleotide triphosphates and 10% dideoxythymidine, what size labeled bands do you expect to see on the gel?

40. Two normal individuals have a child with Down syndrome. RFLP analysis with a chromosome 21 probe is performed on all three individuals and the results of the gels appear at right. Based on these results, what can you conclude about the origins of the 21st chromosomes?

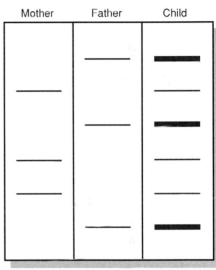

Suggested Readings for chapter 12 are on page 649.

13

Gene Expression: Control in Prokaryotes and Phages

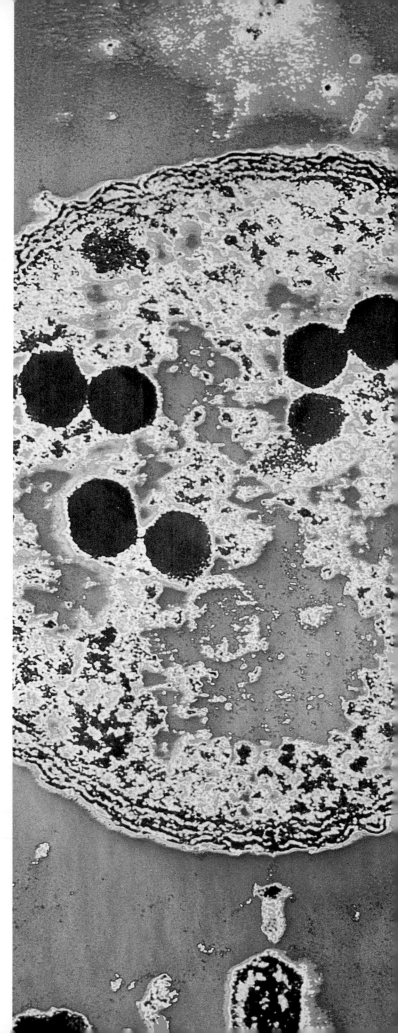

T2 bacteriophages (with polyhedral heads) being synthesized within an *Escherichia coli* cell.

© *Lee Simon/Stammers/Science Photo Library/Photo Researchers, Inc.*

enes are transcribed into RNA, which, for the most part, is then translated into protein. Control mechanisms are exercised along the way. Without some control of gene expression, an *Escherichia coli* cell, for example, would produce all its proteins in large quantities all the time, and all the cells in a eukaryotic organism would be identical. Although most control mechanisms are negative (preventing something from happening), controls can also be positive (causing some action to occur or enhancing some action). This chapter is devoted to analyzing control processes in prokaryotes and phages; in chapter 15 we will examine control processes in eukaryotes.

In the process leading from a sequence of nucleotides in DNA to a protein, there are many places in which control can be exerted. In general, control of gene expression can take place at the levels of transcription, translation, or protein functioning.

One of the best-understood mechanisms exerts control of transcription, in which the production of mRNA is regulated according to need. *E. coli* mRNAs are short-lived *in vivo:* They are degraded enzymatically within about two minutes. A complete turnover (degradation and resynthesis) in the cell's mRNA occurs rapidly and continually, and this rapid turnover is a prerequisite for transcriptional control, a central feature of regulation of prokaryotic gene expression.

THE OPERON MODEL

Not all of the proteins prokaryotes can produce are needed in all circumstances in the same quantity. For example, the enzymes in many synthetic pathways are in low concentration or absent when an adequate quantity of the end product of the pathway is already available to the cell. That is, if the cell encounters an abundance of the amino acid tryptophan in the environment or if it is overproducing tryptophan, the cell stops the manufacture of tryptophan until a need arises again. A **repressible system** is a system of enzymes wherein synthesis of the enzymes is repressed, and production of the end product stops when it is no longer needed. Repressible systems are repressed by the appearance in the cell of an excess of the end product of their synthetic (anabolic) pathway.

On the other hand, some metabolites, such as lactose, that the cell breaks down for energy and as a carbon source, may not always be present in the cell's environment. If a given metabolite is not present, enzymes for its breakdown are not useful, and synthesizing these enzymes is wasteful. If the cell produces enzymes for the degradation of a particular carbon source only when this carbon source is present in the environment, the enzyme system is known as an **inducible system.** Inducible enzymes are synthesized when the environment includes a substrate for those enzymes that will then be catabolized (broken down).

The best-studied inducible system is the ***lac* operon** in *E. coli.* Since the term *operon* refers to the control mechanism, we will defer a definition until we describe the mechanism.

LAC OPERON (INDUCIBLE SYSTEM)

Lactose Metabolism

Lactose (milk sugar—a disaccharide) is a β-galactoside that can be used by *E. coli* for energy and as a carbon source after it is broken down into glucose and galactose. The enzyme that performs the breakdown is **β-galactosidase** (fig. 13.1). (The enzyme can additionally convert lactose to allolactose, which, as we will see, is also important.) There are very few molecules of β-galactosidase in a wild-type *E. coli* cell grown in the absence of lactose. Within minutes after adding lactose to the medium, however, this enzyme appears in quantity within the bacterial cell. When the synthesis of β-galactosidase (encoded by the *lacZ,* or *z* gene) is induced, the production of two additional enzymes is also induced: **β-galactoside permease** (encoded by the *lacY,* or *y* gene) and **β-galactoside acetyltransferase** (encoded by the *lacA,* or *a* gene). The permease is involved in concentrating lactose in the cell. The transferase is believed to protect the cell from the buildup of toxic products created by β-galactosidase acting on other galactosides. By acetylating galactosides other than lactose, the transferase prevents β-galactosidase from cleaving them.

Regulator Gene

Not only are the three *lac* genes (*z, y, a*) induced together, but they are adjacent to one another in the *E. coli* chromosome; they are, in fact, transcribed on a single, polycistronic mRNA (fig. 13.2). Induction involves the protein product of another gene, called the **regulator gene,** or *i* gene (*lacI*). Although the regulator gene is located adjacent to the three other *lac* genes, it is a totally independent transcriptional entity. The regulator specifies a protein called a **repressor,** which interferes with the transcription of the genes involved in lactose metabolism.

Operator

In order for the repressor protein to exert its influence over transcription, there must be a control element (receptor site) located near the beginning of the β-galactosidase (*lacZ*) gene. This control element is a region referred to as the **operator,** or operator site (fig. 13.2). The operator site is a sequence of DNA that is recognized by the product of the regulator gene, the repressor. When the repressor is bound to the operator, it interferes with binding of RNA

Figure 13.1

The enzyme β-galactosidase hydrolytically cleaves lactose into glucose and galactose (a). The enzyme can also convert lactose to allolactose (b).

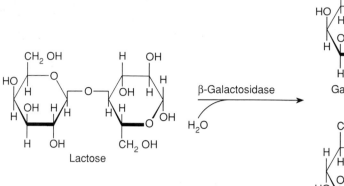

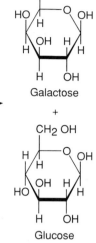

Galactose

+

Glucose

(a)

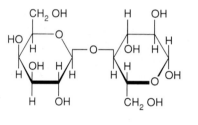

(b) Lactose

Allolactose

Figure 13.2

The *lac* operon is transcribed as a multigenic (polycistronic) mRNA. The *z*, *y*, and *a* indicate the *lacZ*, *lacY*, and *lacA* loci. The mRNA transcript is then translated as individual proteins. The *lac* operon regulator gene is denoted as *i*; the *o* stands for operator and the *p* for promoter. Both the operon and the regulator gene have their own promoters.

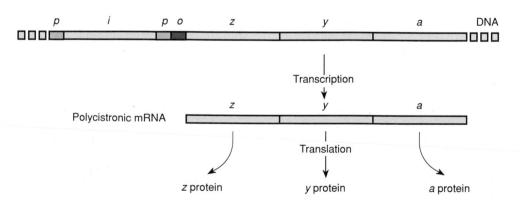

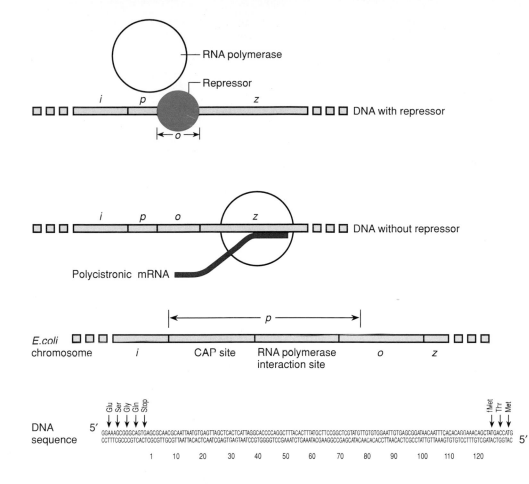

Figure 13.3
The repressor. By binding to the operator, the repressor prevents RNA polymerase from binding to the promoter and transcribing the *lac* operon. When repressor is not present, transcription takes place. The functional repressor is a tetramer.

Figure 13.4
The *lac* operon promoter and operator regions. The CAP site is described later. The base sequence corresponds to the diagram above it. The terminal amino acids of the *i* gene are shown as well as the initial amino acids of the *lacZ* gene.

From R. C. Dickson, et al., "Genetic Regulation: The Lac Control Region" in Science, *187:27–35, January 10, 1975. Copyright 1975 by the AAAS.*

polymerase (see chapter 10) and thus transcription of the operon is prevented (fig. 13.3). The repressor is released when it combines with an *inducer,* a derivative of lactose called allolactose (see fig. 13.1).

The nucleotide sequence of the *lac* operator is shown in figure 13.4. Note that here the concept of the promoter is expanded to be the region of DNA that not only is recognized by RNA polymerase but also has other controlling functions in the immediate vicinity of the initiation site of transcription. We can now define an **operon** as a sequence of adjacent genes all under the transcriptional control of the same promoter and operator.

Induction of the *lac* Operon

Under conditions of repression, before the operon can be "turned on" to produce lactose-utilizing enzymes, the repressor will have to be removed from the operator. The repressor is an **allosteric protein,** which has the general property that when it binds with one particular molecule, the shape of the protein is changed, and thus changes its ability to react with a second particular molecule. Here the first molecule is the inducer allolactose and the second molecule is the operator DNA. The operator region has twofold symmetry, allowing two of the four subunits of

the tetrameric repressor to bind to the operator, one on each side of the axis of symmetry. The shape of the repressor is such that it fits into the grooves of the DNA to locate the exact base sequence of the operator; it then binds at that point. The other two subunits of the repressor tetramer provide the binding site for allolactose, and when it is bound, it causes the repressor to lose its affinity for the operator (fig. 13.5).

With allolactose bound to the repressor, the ability of the repressor to bind to the operator is greatly reduced, by a factor of 10^3. Since no covalent bonds are involved, the repressor simply dissociates from the operator. After the repressor releases from the operator, RNA polymerase can now bind and begin transcription. The three *lac* operon genes are then transcribed and subsequently translated into their respective proteins.

This system of control is very efficient. The presence of the lactose molecule permits transcription of the genes of the *lac* operon, which act to break down the lactose. After all the lactose is metabolized, the repressor then returns to its original shape and can again bind to the operator. The system is once again "turned off." Using very elegant genetic analysis, details of this system were worked out by François Jacob and Jacques Monod, who subsequently won the Nobel Prize for their efforts.

Figure 13.5
The *lac* operator DNA is a palindrome. Two of the four subunits of the repressor bind to it. Base pairs that do not follow rules of palindrome symmetry are in *color*. Numbers refer to bases transcribed.

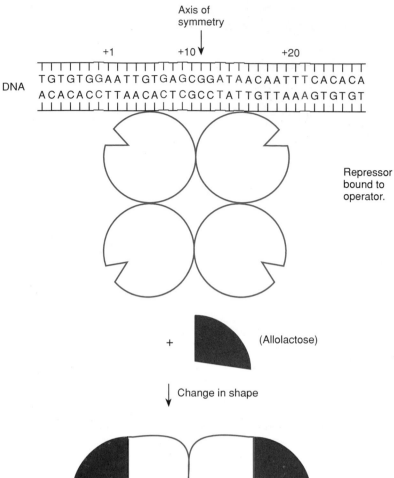

Axis of symmetry

+1   +10   +20

DNA

TGTGTGGAATTGTGAGCGGATAACAATTTCACACA
ACACACCTTAACACTCGCCTATTGTTAAAGTGTGT

Repressor bound to operator.

+   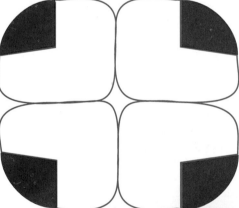   (Allolactose)

Change in shape

Repressor falls free of operator.

François Jacob (1920–   )
Courtesy of Dr. Francois Jacob.

Jacques Monod (1910–1976)
Archives Photographiques, Musée Pasteur

Figure 13.6

A *lac* operon merozygote of *E. coli*. The chromosome of the F′ factor has a regulator constitutive mutation. However, the cell has a wild-type phenotype; both operons are repressed because enough repressor is produced by the chromosomal allele to bind to both operators.

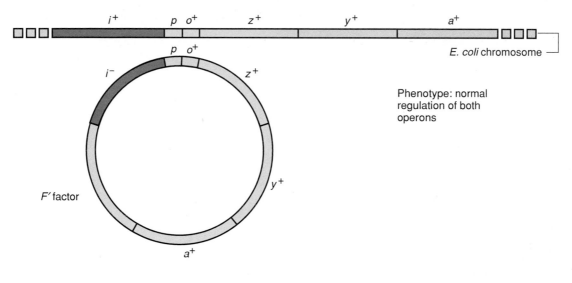

Phenotype: normal regulation of both operons

Lac Operon Mutants

Merozygote Formation

Discovery and verification of this system came about through the use of mutants and partial diploids of the *lac* operon. The structural (enzyme-specifying) genes of the *lac* operon, *z, y,* and *a,* all have known mutant forms in which the particular enzyme does not perform its function. These mutant forms are designated z^-, y^-, and a^-. The alleles responsible for normal forms of the enzymes are z^+, y^+, and a^+.

Partial diploids in *E. coli* can be created through sexduction (chapter 7) because some strains of *E. coli* have the *lac* operon incorporated into an F′ factor. Since F^+ strains can pass the F′ particle into F^- strains, *lac* operon diploids (also called merozygotes, or partial diploids) can be formed. By careful manipulation, various combinations of mutations can be looked at in the diploid state.

Constitutive Mutants

Constitutive mutants are mutants in which the three *lac* operon genes are transcribed at all times—that is, they are not turned off even in the absence of lactose. Inspection of figure 13.3 will show that constitutive production of the enzymes can come about in several ways. A defective repressor, produced by a mutant regulator gene, will not turn the system off, nor will a mutant operator that will no longer bind the normal repressor. The regulator constitutive mutants are designated i^-; the operator constitutive mutants are designated o^c. Both types of mutants produce the same phenotype: constitutive expression of the three *lac* operon genes.

When a new mutant is isolated, it is possible to determine whether it is caused by a regulator or operator mutation. For example, the exact location of a mutation on the bacterial chromosome can be determined by standard mapping techniques (see chapter 7) or, more recently, by DNA sequencing (see chapter 12). Alternatively, the Jacob and Monod model predicts different modes of action for the two types of mutations. In merozygotes, a constitutive operator mutation will affect only the operon of which it is physically a part. Operator mutations are therefore called **cis-dominant.** However, a constitutive *i*-gene mutation, since it works through an altered protein, will be recessive to a wild-type regulator gene in the same cell, regardless of which operon (chromosomal or F′ factor) the mutation is on. Constitutive regulator mutations are, therefore, **trans-acting.** (If two mutations are on the same piece of DNA, they are in the *cis* configuration. If they are on different pieces of DNA, they are in the *trans* configuration.) *Trans-acting* mutations work through a product, usually a protein, that diffuses through the cytoplasm. *Cis-acting* mutants are changes in recognition sequences on the DNA.

In figure 13.6, the F′ factor has a regulator constitutive mutation (i^-). The cell, however, will have the normal (inducible) phenotype because the chromosomal i^+ allele is dominant to the mutation—the i^+ regulates both the chromosomal and F′ operons. Hence both operons are inducible. In figure 13.7, however, there are both regulator and operator constitutive mutations in the merozygote cell. The cell, however, has the constitutive phenotype because the *lac* operon on the bacterial chromosome will be continually transcribed. Regardless of the alleles of the regulator genes, the chromosomal operon has a *cis-dominant* constitutive operator mutation.

Figure 13.7

A *lac* operon merozygote of *E. coli*. The chromosomal operon has an operator constitutive mutation; it is transcribed constitutively. The F′ factor has a regulator constitutive mutation; however, the F′ factor operon is repressed because there is enough wild-type repressor produced by the chromosomal regulator gene.

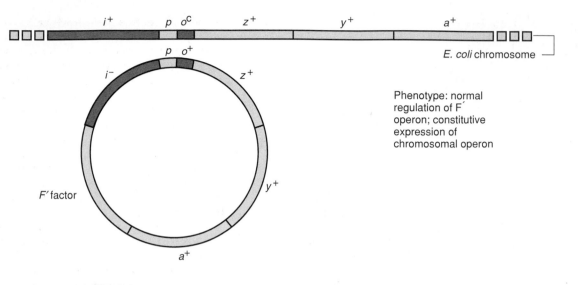

Phenotype: normal regulation of F′ operon; constitutive expression of chromosomal operon

Mark Ptashne (1940–   )
Courtesy of Dr. Mark Ptashne.

Other lac *Operon Control Mutations*

Other mutations have also been discovered that support the Jacob and Monod operon model. A super-repressed mutation, i^s, has been isolated. This mutation represses the operon even in the presence of large quantities of the inducer. Thus the repressor seems to have lost the ability to recognize the inducer. Basically, the *i*-gene product is acting as a constant repressor rather than as an allosteric protein. In an i^s/i^+ merozygote, both operons are repressed because the i^s repressor binds to both operators. Another mutation, i^Q, produces much more of the repressor than is normal and is presumably a mutation of the promoter region of the *i* gene.

In 1966, W. Gilbert and B. Müller-Hill isolated the *lac* repressor and thereby provided the final proof of the validity of the model. At about the same time, M. Ptashne and his colleagues isolated the repressor for phage λ operons. Control of gene expression in phage λ is discussed later in this chapter.

CATABOLITE REPRESSION

An interesting property of the *lac* operon and other operons that code for enzymes that catabolize certain sugars (e.g., arabinose, galactose) is that they are all repressed by the presence of glucose. That is, glucose is catabolized in preference to other sugars; the mechanism (**catabolite repression**) involves **cyclic AMP** (cAMP; fig. 13.8). In eukaryotes, cAMP acts as a *second messenger,* an intracellular messenger regulated by certain extracellular hormones. Geneticists were surprised to discover cAMP in *E. coli,* where it works in conjunction with another regulatory protein, the **catabolite activator protein (CAP),** to control the transcription of certain operons.

In the absence of glucose, cAMP combines with CAP and the CAP-cAMP complex binds to a distal part of the promoter of operons with CAP sites (e.g., the *lac* operon; see fig. 13.4). This binding apparently enhances the affinity of RNA polymerase for the promoter, because without the binding of the CAP-cAMP complex to the promoter, the transcription rate is very low. The uptake of glucose by *E. coli* cells causes the loss of cAMP from the cell, probably by the inhibition of adenylcyclase (fig. 13.8), and thus lowers the CAP-cAMP level. The transcription rate of operons with CAP sites will, therefore, be reduced (fig. 13.9). The same reduction of transcription rates is noticed in mutant strains of *E. coli* when this part of the distal end of the promoter is deleted. The binding of CAP-cAMP to the CAP site causes the DNA to bend by more than 90° (fig. 13.10). This bending, by itself, may be the cause for enhanced transcription, making the DNA more available to RNA polymerase.

Figure 13.8
Structure of cyclic AMP
(cAMP). Uptake of glucose
lowers the quantity of cyclic AMP
in the cell, by inhibiting the
enzyme adenylcyclase that
converts ATP to cAMP.

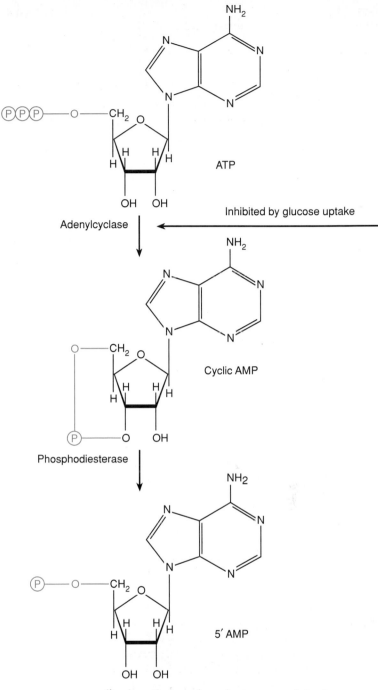

Catabolite repression is an example of positive regulation: Binding of the CAP-cAMP complex at the CAP site enhances the transcription rate of that transcription unit. Thus the *lac* operon is both positively and negatively regulated: The repressor exerts negative control and the CAP-cAMP complex exerts positive control of transcription.

TRP OPERON (REPRESSIBLE SYSTEM)

The inducible operons are activated when the substrate that is to be catabolized enters the cell. Anabolic operons function in a reverse manner. They are turned off (re-

pressed) when their end product accumulates in excess of the needs of the cell. Transcription of repressible operons appears to be controlled by two entirely different, although not mutually exclusive, mechanisms. The first mechanism follows the basic scheme of inducible operons and involves the end product of the pathway. The second mechanism involves secondary structure in mRNA transcribed from an attenuator region of the operon.

Tryptophan Synthesis

One of the best-studied repressible systems is the tryptophan, or *trp*, operon in *E. coli*. The *trp* operon contains the five genes that code for the synthesis of the enzymes

Figure 13.9

Catabolite repression. When cAMP is present in the cell (no glucose is present), it binds with CAP protein and together they bind to the CAP site in various sugar-metabolizing operons, such as the *lac* operon shown here. The CAP-cAMP complex enhances the transcription of the operon. When glucose is present, it inhibits the formation of cAMP. Thus no CAP-cAMP complex is formed, and therefore transcription of the same operons is not enhanced.

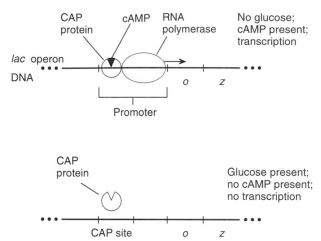

Figure 13.10

CAP–DNA interaction. Model of cap protein and DNA. The cAMP-binding domain is *blue*, DNA-binding domain is *purple*, and the cAMP are *red*.

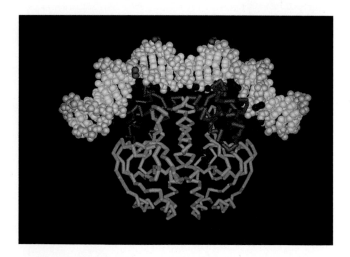

that build tryptophan starting with chorismic acid (fig. 13.11). It has a promoter-operator sequence (*p, o*) as well as its own regulator gene (*trpR*).

Operator Control

In this repressible system the product of the *trpR* gene, the repressor, is inactive by itself; it does not recognize the operator sequence of the *trp* operon. The repressor only becomes active when it combines with tryptophan. Thus,

Figure 13.11

Genes of the tryptophan operon in *E. coli*. The enzymes they produce control the conversion of chorismic acid to tryptophan. The symbol *o* on the chromosome refers to the *trp* operator, which has its own repressor, the product of the *trpR* gene.

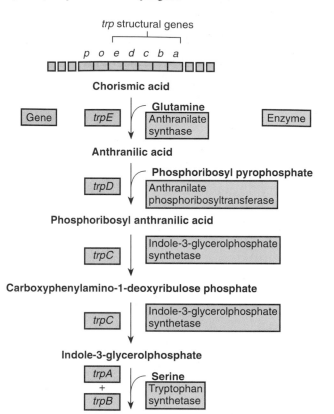

when there is an excess of tryptophan, enough will be available to bind with and activate the repressor. Tryptophan is thus referred to as the **corepressor.** The corepressor-repressor complex then recognizes the operator, binds to it, and prevents transcription by RNA polymerase.

After the available tryptophan in the cell is used up, eventually the last one will diffuse from the repressor, which will then detach from the *trp* operator. The transcription process will be blocked no longer and can proceed normally (the operon is now **derepressed**). Transcription will continue until enough of the various enzymes have been synthesized to produce a sufficient quantity of tryptophan. Then there will again be an excess of tryptophan. Some will be available to bind to the repressor and make a functional complex. Thus the operon will again be shut off and the process will repeat itself, assuring that tryptophan is being synthesized when it is needed (fig. 13.12). This regulation is modified, however, by the existence of the second mechanism for regulating repressible operons, that of attenuation.

Figure 13.12
The repressor-corepressor complex binds at the operator and prevents the transcription of the *trp* operon in *E. coli*. Without the corepressor, the repressor cannot bind and therefore transcription is not prevented from occurring.

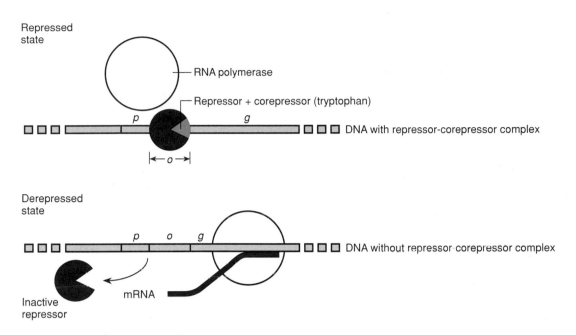

Charles Yanofsky (1925–   )
Courtesy of Dr. Charles Yanofsky.

TRP OPERON (ATTENUATOR-CONTROLLED SYSTEM)

Details of the second control mechanism of repressible operons have been elucidated primarily by C. Yanofsky and his colleagues, who worked with the tryptophan operon in *E. coli*. This type of operon control, control by an **attenuator region,** has been demonstrated for at least five other amino acid-synthesizing operons, including the leucine and histidine operons. This regulatory mechanism may be the same for most operons involved in the synthesis of an amino acid.

Leader Transcript

In the *trp* operon there is an attenuator region between the operator and the first structural gene (fig. 13.13). The mRNA transcribed from the attenuator region, termed the **leader transcript,** has been sequenced, with two surprising and interesting facts emerging. First, four subregions of the mRNA are defined by the fact that they have base sequences that are complementary to each other such that three different stem-loop structures can form in the mRNA (fig. 13.14). Depending on circumstances, regions 1–2 and 3–4 can form two stem-loop structures, or region 2–3 can form a single stem-loop. When one stem-loop structure is formed, the others are preempted. As we will see, the particular combination of stem-loop structures determines whether transcription will continue.

Leader Peptide Gene

The second fact obtained by sequencing the leader transcript is that there is a small gene coding information for a peptide from bases 27 to 68 (fig. 13.15). The gene for this peptide is referred to as the **leader peptide gene.** It codes for fourteen amino acids, of which two adjacent ones are tryptophan. These adjacent tryptophan codons are critically important in attenuator regulation. The proposed mechanism for this regulation is as follows.

Figure 13.13

Attenuator region of the *trp* operon, which contains the leader peptide gene. This region is transcribed into the leader transcript.

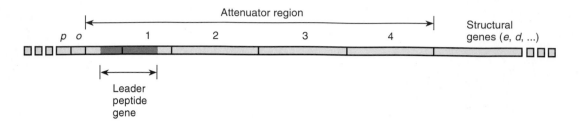

Figure 13.14

Nucleotide sequence of part of the leader transcript of the *trp* attenuator region (bases 50 to 140). Stem-loops 1–2 and 3–4, or stem-loop 2–3 can form because of complementarity of the nucleotides. All possible base-pairings are shown in the middle of the figure.

From D. L. Oxender, et al., "Attenuation in the Escherichia coli *Tryptophan Operon: Role of RNA Secondary Structure Involving the Tryptophan Codon Region" in* Proceedings of the National Academy of Sciences, *76:5524–5528, 1979. Reprinted by permission.*

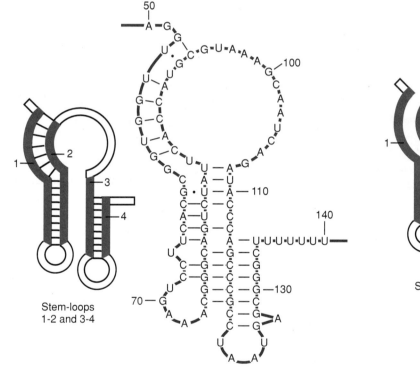

Figure 13.15

Base sequence of the *trp* leader peptide gene and the amino acids coded for by these nucleotides. Note the presence of adjacent tryptophan codons.

From D. L. Oxender, et al., "Attenuation in the Escherichia coli *Tryptophan Operon: Role of RNA Secondary Structure Involving the Tryptophan Codon Region" in* Proceedings of the National Academy of Sciences, *76:5524–5528, 1979. Reprinted by permission.*

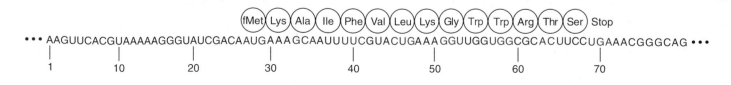

Figure 13.16

Model for attenuation in the *E. coli trp* operon. The *circle* represents
the ribosome; the *strand* is the leader transcript of figure 13.14. Under
conditions of excess tryptophan, the 3–4 stem-loop forms, terminating
transcription. Under conditions of tryptophan starvation, the ribosome
is stalled, and the 2–3 stem-loop forms, allowing continued
transcription. Under general starvation there is no translation resulting
in the formation of the 1–2 and 3–4 stem-loops, which again results in
the termination of transcription.

From D. L. Oxender, et al., "Attenuation in the Escherichia coli *Tryptophan
Operon: Role of RNA Secondary Structure Involving the Tryptophan Codon
Region" in* Proceedings of the National Academy of Sciences, 76:5524–5528,
1979. Reprinted by permission.*

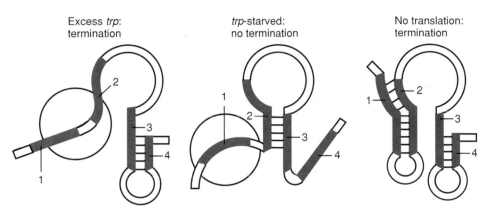

Excess *trp*:
termination

trp-starved:
no termination

No translation:
termination

Excess Tryptophan

Assuming that the operator site is available to RNA polymerase, transcription of the attenuator region will begin.
As soon as the 5′ end of the mRNA for the leader peptide
gene has been transcribed, a ribosome will attach and begin
the process of translation of this mRNA. Depending on
the levels of amino acids in the cell, three different outcomes of this translation process can take place. If the
concentration of tryptophan in the cell is such that abundant tryptophanyl-tRNAs exist, translation will proceed
down the leader peptide gene. The moving ribosome will
overlap regions 1 and 2 of the transcript and allow the
3–4 stem-loop to form as shown in the configuration at the
far left of figure 13.16. This stem-loop structure is referred to as the **terminator** or **attenuator stem,** and will
cause transcription to be terminated. Hence when existing
quantities of tryptophan, in the form of tryptophanyl-
tRNA, are adequate for translation of the leader peptide
gene, transcription is terminated.

Tryptophan Starvation

If there is a lowered quantity of tryptophanyl-tRNA, the
ribosome will have to wait at the first tryptophan codon
until it acquires a Trp-tRNA^Trp. This is shown in the configuration in the middle part of figure 13.16. The stalled
ribosome will permit the 2–3 stem-loop to form, which
precludes the formation of the terminator (3–4) stem-loop.
In this configuration, transcription is not terminated, so
that, eventually, the whole operon is transcribed and
translated, which will raise the level of tryptophan in the
cell. The 2–3 stem-loop structure is referred to as the
preemptor stem.

General Starvation

A final configuration is possible, as shown on the far right
in figure 13.16. Here no ribosome interferes with stem formation and, presumably, the 1–2 and 3–4 (terminator)
stem-loops will form. This configuration will also terminate transcription because of the existence of the terminator stem. It is believed that this configuration will occur
if the ribosome is stalled on the 5′ side of the *trp* codons,
which will happen when the cell is starved for other amino
acids. Presumably, it makes no sense to manufacture tryptophan when other amino acids are in short supply. Hence
the cell can carefully bring up the levels of the various
amino acids in the most efficient manner.

Redundant Controls

It is not completely obvious why there are apparently re-
dundant controls (repression and attenuation) of tryptophan biosynthesis. Some amino acid operons are controlled
only by attenuation. One example is the *his* operon in *E.
coli,* in which the leader peptide gene contains seven histidine codons in a row. This *E. coli* system has a more
stringent control than the *trp* operon in which the cell is
testing both the tryptophan levels (tryptophan is the corepressor) and the tryptophanyl-tRNA levels (in the attenuator control system). It is not precisely clear, however,
why the cell needs to "know" both these levels. The attenuator system also allows the cell to regulate tryptophan
synthesis on the basis of the shortage of other amino acids.
For example, when there is a shortage of both tryptophan
and arginine, operator control will allow transcription to
begin, but attenuator control will terminate transcription

Figure 13.17

Genetic map of phage λ. There are four operons present: the repressor, left, right, and late operons. The prophage, a linear form integrated into the bacterial chromosome, begins and ends at *att*. The mature phage, another linear form found packed into the phage heads, begins and ends at *Nu1* (*cos* sites).

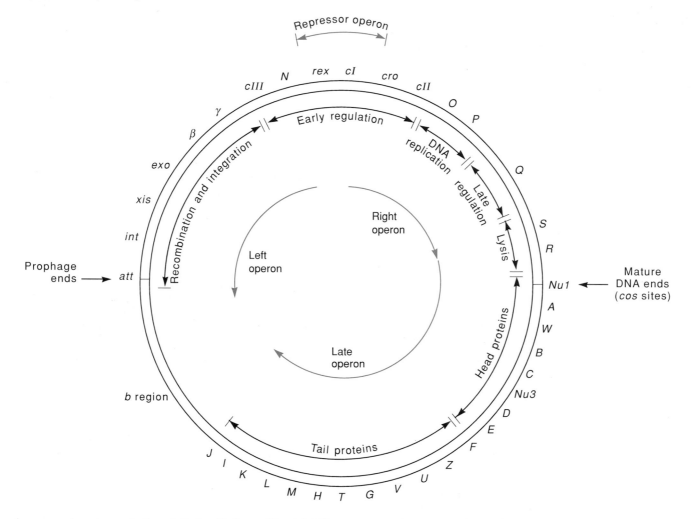

because stem-loops 1–2 and 3–4 will form (fig. 13.16). More research will clarify our understanding of why the cell has these interesting and efficient mechanisms of operon control.

LYTIC AND LYSOGENIC CYCLES IN PHAGE λ

When a phage infects a cell, it must express its genes in an orderly fashion; there are genes whose products are needed early in infection as well as genes whose products are not needed until late in infection. Early genes usually control phage DNA replication; late genes usually determine phage coat proteins and the lysis of the bacterial cell. A phage is most efficient if it expresses the early genes first and the late genes last in the infection process. Also, temperate phages have the option either of entering into lysogeny with the cell or of lysing the cell; and, here too,

control processes determine which path will be taken. One generalization that holds for most phages is that their genes are clustered into early and late operons, with separate transcriptional control mechanisms for each.

Phage λ is perhaps the best-studied bacteriophage. Since it is a temperate phage, it can exist either vegetatively or as a prophage, integrated into the host chromosome. This phage warrants our attention because of the interesting and complex way that the life-cycle choice is determined. It is a model system of operon controls. The complexity is a result of having two conflicting life-cycle choices.

Briefly, the expression of one of the two life-cycle alternatives, lysogenic or lytic cycles, depends on access for operator sites by two repressors, called *cI* and *cro*. The *cI* repressor acts to favor lysogeny and it represses the lytic cycle. The *cro* repressor favors the lytic cycle and it represses lysogeny. The operator sites, when bound by either

Figure 13.18

The two linear forms of λ phage. The prophage is flanked by *E. coli* DNA (*bio* and *gal* loci). The mature linear DNA (found within phage protein coats) is flanked by *cos* sites.

Mature DNA   *cos*   A   W   B   • • •   Q   S   R   *cos*

Prophage   *gal*   *int*   *xis*   *exo*   • • •   K   I   J   b   *bio*
      • • •   • • •
    E. coli   *E. coli*

cI or *cro*, can either enhance or repress transcription. Other control mechanisms are also involved in determining aspects of the λ life cycle, including antitermination and multiple promoters for the same genes.

Phage λ Operons

Phage λ (see fig. 7.24) exhibits a complex system of controls of both early and late operons as well as controls for the decision of lytic infection versus lysogenic integration. The genes of λ are grouped into four operons: left, right, late, and repressor (fig. 13.17). The left and right operons contain the genes for DNA replication and recombination and phage integration. The late operon contains the genes that determine phage head and tail proteins and lysis of the host cell. The sequence of events following phage infection is relatively well known.

The map of λ (fig. 13.17) is a circle, but the λ chromosome has two linear stages in its life cycle. It is packed within the phage head in one linear form and it integrates into the host chromosome to form a prophage in another linear form (fig. 13.18). Those two linear forms do not have the same ends (figs. 13.17 and 13.18). The mature DNA, which is packed within the phage heads before lysis of the cells, is flanked by *cos* sites (chapter 12) and results from a break in the circular map between the *A* and *R* loci. The prophage is integrated at the *att* site and hence the circular map is broken there at integration.

The homologous integration sites on both λ and the *E. coli* chromosome consist of a 15-bp core sequence (called "O" in both), flanked by different sequences on both sides in both the bacterium and the phage (fig. 13.19). In the phage, the region is referred to as POP', where P and P' (P for phage) are two different flanking regions of the O core on the phage DNA. In the bacterium the region is called BOB', where B and B' (B for bacterium) are two different flanking regions of the O core on the *E. coli* chromosome. Integration, which is a part of the lysogenic life cycle, requires the product of the λ *int* gene, a protein known as *integrase,* and is referred to as **site-specific recombination.** Later excision of the prophage, during induction when the phage leaves the host chromosome to enter the lytic cycle, requires both the integrase and the protein product of the neighboring *xis* gene, *excisionase.*

After infection of the *E. coli* cell by a λ phage, the phage DNA circularizes using the complementarity of the *cos* sites. Transcription begins, and within a very short time the phage is guided toward either entering the lytic cycle and producing virus progeny, or entering the lysogenic cycle and integrating into the host chromosome. What events lead up to this "decision" as to which path to take?

Early and Late Transcription

When the phage first infects an *E. coli* cell, transcription of the left and right operons begins at the left (p_L) and right (p_R) promoters, respectively: The *N* (left) and *cro* (right) genes are transcribed. Transcription then stops on both operons at rho-dependent terminators (t_{RI}, t_{LI}). Transcription cannot continue until the protein product of the *N* gene is produced. This protein is called an **antiterminator protein.** When it binds at sites upstream from the terminators, called *nutL* and *nutR* (*nut* stands for *N* utilization; *L* and *R* stand for left and right), the polymerase reads through the terminators and continues on to transcribe the left and right operons (fig. 13.20).

Transcription then continues along the left and right operons through the *cII* and *cIII* genes (see fig. 13.17). Later, if the lytic response is followed, the *Q* gene, which codes for a second antiterminator protein, in the right operon, will have the same effect on the late operon as the *N* gene did on the two early operons: Without the *Q*-gene product, transcription of the late operon proceeds about two hundred nucleotides and then terminates. With the *Q*-gene product, the late operon is transcribed. Hence, in phage λ, general control of transcription is mediated by proteins that allow RNA polymerase to proceed past termination signals. If only the previously described events were to transpire, the lytic cycle would always be followed. However, a complex series of events can also take place in the repressor region that may lead to a "decision" to follow the lysogenic cycle instead.

Repressor Transcription

The *cIII*-protein product inhibits a host-cell protease, called *HflA,* that would break down the *cII*-gene product. The *cII*-gene product binds at two promoters, enhancing their availability to RNA polymerase, just as the CAP-cAMP product enhances the transcription of the *lac* operon. The *cII* protein binds at the promoters for *cI* transcription and for *int* transcription (fig. 13.21). At this point, the phage can still "choose" between either the lytic or the lysogenic cycles. Integrase (the product of the *int* gene) and *cI* (repressor) proteins are now produced, favoring lysogeny, as well as the *cro*-gene product, the *antirepressor,* which is a repressor of *cI* and therefore favors the lytic pathway. (*Cro* stands for *control of repression and other*

Figure 13.19

Integration of λ phage into the *E. coli* chromosome requires a crossover between the two attach sites, called POP′ (phage) and BOB′ (bacteria). (*a*) General pattern of this site-specific attachment. (*b*) Nucleotide sequences of the various components.

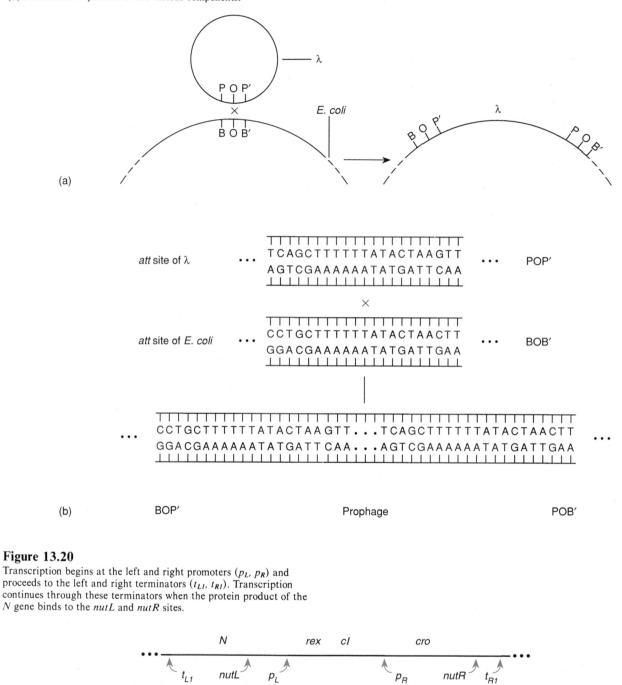

(a)

att site of λ   ···   TCAGCTTTTTTATACTAAGTT   ···   POP′
                      AGTCGAAAAAATATGATTCAA

×

att site of *E. coli*   ···   CCTGCTTTTTTATACTAACTT   ···   BOB′
                              GGACGAAAAAATATGATTGAA

···   CCTGCTTTTTTATACTAAGTT...TCAGCTTTTTTATACTAACTT   ···
      GGACGAAAAAATATGATTCAA...AGTCGAAAAAATATGATTGAA

(b)   BOP′   Prophage   POB′

Figure 13.20

Transcription begins at the left and right promoters (p_L, p_R) and proceeds to the left and right terminators (t_{L1}, t_{R1}). Transcription continues through these terminators when the protein product of the *N* gene binds to the *nutL* and *nutR* sites.

N   rex   cI   cro

t_{L1} nutL p_L p_R nutR t_{R1}

◄──── Transcription   Transcription ────►

things; the "c" of *cI*, the repressor, stands for "clear," which is the appearance of λ plaques that have *cI* mutations. These mutants can only undergo lysis without the possibility of lysogeny. Normal λ infections produce turbid plaques, accounted for by lysogenic bacterial growth within the plaques.) We now focus further on the repressor region with its operators and promoters.

Maintenance of Repression

The *cI* gene, with the aid of the *cII*-gene product, is transcribed from a promoter known as p_{RE}, the *RE* standing for *repression establishment* (fig. 13.22). Once *cI* is transcribed, it is translated into a protein called the λ *repressor*, which interacts at the left and right operators, o_L

Figure 13.21
The *cII* gene product of phage λ binds to the *cI* promoter (*pRE*) and the *int* promoter (*pI*), enhancing transcription of those genes. The *cIII* protein breaks down the *HflA* protease that normally would break down the *cII* protein.

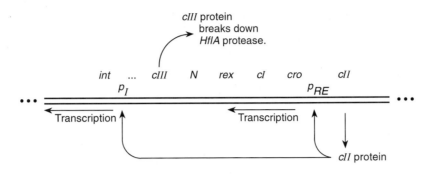

Figure 13.22
Early regulation region of phage λ. The *cI* (and *rex*) genes are transcribed by two promoters, *pRE* and *pRM*. The left operator overlaps the left promoter and the right operator overlaps both the right promoter and the maintenance-of-repression promoter.

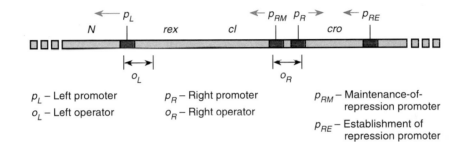

p_L – Left promoter	p_R – Right promoter
o_L – Left operator	o_R – Right operator

p_{RM} – Maintenance-of-repression promoter

p_{RE} – Establishment of repression promoter

Figure 13.23
The right operator on the phage λ chromosome overlaps the *pRM* and *pR* promoters. There are three repressor recognition sites within the operator: o_{R1}, o_{R2}, and o_{R3}. Preferential binding by the *cro* repressor to o_{R3} and the *cI* repressor to o_{R1} determines whether transcription will occur to the left or the right.

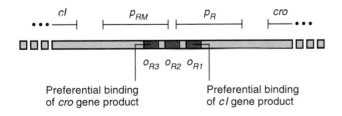

Preferential binding of *cro* gene product

Preferential binding of *cI* gene product

and o_R, of the left and right operons. When these operators are bound by *cI* protein, transcription of the left and right operons (and therefore also the late operon) ceases. There are several ramifications of the repression. First, lysogeny can be initiated because the *int* gene has been transcribed at the early stage of infection. Second, since *cII* and *cIII* are no longer being synthesized, *cI* transcription from the *pRE* promoter is stopped. However, *cI* can

still be transcribed because there is a second promoter, p_{RM} (*RM* stands for *repression maintenance*), that will allow low levels of transcription of the *cI* gene.

The *cI* gene can further control its own concentration in the cell. When the right and left operators were sequenced, each was discovered to have three sites of repressor recognition (fig. 13.23). On the right operator, for example, the rightmost site (o_{R1}) was found to be most efficient at binding repressor. When repressor was bound at this site, the right operon was repressed and transcription of *cI* was enhanced (again, in a way similar to enhancement of transcription by binding of CAP-cAMP at the CAP site in the *lac* operon). Excess repressor, when present, however, was also bound by the other two sites within o_R. The foregoing process results in the repression of the *cI* gene itself. Hence maintenance levels of *cI* can be kept within very narrow limits.

A third ramification of repression is the prevention of superinfection. That is, bacteria lysogenic for λ phage are protected from further infection by other λ phages because repressor is already present in the cell. These bacteria are also protected from infection by T4 phage with *rII* mutants. This protection is controlled by the *rex*-gene product, the other gene in the repressor operon.

Figure 13.24

The λ repressor. (*a*) The λ repressor is a dimer with each subunit having helical amino- and carboxyl-terminal ends. The helical structure of the amino-terminal ends binds in the major groove of DNA. (*b*) A computer-generated view of repressor binding to DNA. (*c*) Space-filling model of repressor binding to DNA.

(*b*) From S. R. Jordan and C. O. Pabo, "Structure of Llambda Complex at 2.5 Å Resolution: Details of the Repressor Operator Interactions" in Science, 242: 893–899, November 11, 1988. Copyright 1988 by the AAAS. (c) From J. F. Reidnaar-Olson and R. T. Sauer, "Combinatorial Cassette Mutagenesis as a Probe of the Informational Content of Protein Sequences," Science, Vol. 241, July 1, 1988. © 1988 by AAAS, Washington, D.C.

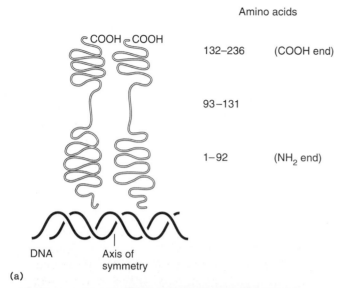

Amino acids	
132–236	(COOH end)
93–131	
1–92	(NH$_2$ end)

(a)

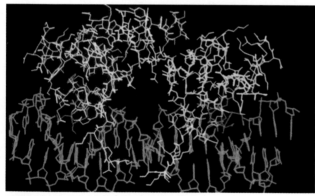

(b)

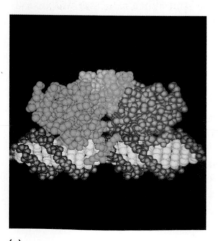

(c)

The promoters for maintenance and establishment of repression differ markedly in their control of repressor gene expression. When p_{RE} is active, a very high level of repressor is present, whereas p_{RM} produces only a low level of repressor. The level of repressor is due to the length of the leader RNA transcribed on the 5′ side of the *cI* gene. The p_{RE} promoter transcribes a very long leader RNA and is very efficient at translation of the *cI* region. In contrast, the p_{RM} promoter begins transcription at the initiation codon of the protein. This leaderless mRNA is translated very inefficiently into *cI*.

The λ repressor is a dimer of two identical subunits (fig. 13.24). Each subunit is composed of two domains, or "ends." The carboxyl- and amino-terminal ends are separated by a relatively open region, susceptible to attack by proteases. The alpha-helical regions of the amino-terminal ends interdigitate into the major groove of the DNA to locate the specific sequences making up the left and right operator subsections. As described earlier for the *lac* operator, o_{R1}, o_{R2}, and o_{R3} each have twofold symmetry.

The binding of the λ repressor in o_{R1} enhances the binding of another molecule of repressor into o_{R2}. Together, they enhance p_{RM} transcription, presumably through contact with RNA polymerase. The repressors also block p_R transcription (fig. 13.23).

Lysogenic versus Lytic Response

We have described the mechanism by which λ establishes lysogeny. How then does λ turn toward the lytic cycle? Here control is exerted by the *cro*-gene product, another repressor molecule that works at the left and right operators in a manner antagonistic to the way that the *cI* repressor works. In other words, using the right operator as an example, *cro*-gene product binds preferentially to the leftmost of the three sites within o_R and represses *cI* but enhances the production of *cro* (fig. 13.23).

The *cro*-gene product can direct the cell toward a lytic response if it occupies the o_R and o_L sites before the λ repressor, or if the λ repressor is removed. From the point of view of phage λ, when would be a good time for the *cI* repressor to be removed? Thinking in evolutionary terms, we would expect that a prophage might be at an advantage if it left a host's chromosome and began the lytic cycle when it "sensed" damage to the host. In fact, one of the best ways to induce a prophage to enter the lytic cycle is to direct ultraviolet (UV) light at the host bacterium. UV light causes damage to DNA and induces several repair systems. One, called SOS repair (see chapter 16), makes use of the protein product of the *recA* gene. Among the activities of this enzyme is to cleave the λ repressor in the susceptible region between domains. The cleaved repressor falls free of the DNA, thereby making the operator sites available for the *cro*-gene product. The lytic cycle then follows.

Initially, however, when the phage first infects an *E. coli* cell, the "decision" for lytic versus lysogenic growth is probably a function of the *cII* gene product. This protein is, as we mentioned, susceptible to bacterial proteases, which in turn are indicators of cell growth. When *E. coli* growth is limited, its proteases tend to be limited, a circumstance that would favor lysogeny for the phage. It is the *cII* protein that when active favors lysogeny and when inactive favors the lytic cycle. Thus, under active bacterial growth, the *cII* protein is more readily destroyed, it thus fails to enhance *cI* transcription and lysis follows. When bacteria are not growing actively, the *cII* protein is not readily destroyed, it enhances *cI* transcription, and lysogeny results. Thus, under initial infection, lysogeny or the lytic cycle will depend primarily on the *cII* protein, which gauges the health and activity of the host. After lysogeny is established, it can be reversed by processes that inactivate the *cI* protein, indicating genetic damage to the bacterium (the SOS response) or an abundance of other hosts in the environment (zygotic induction, see chapter 7).

All the details regarding the *cI-cro* competition are not known, but an understanding of the relationship of lytic and lysogenic life cycles and the nature of DNA-protein recognition has emerged (fig. 13.25 and table 13.1).

TRANSPOSABLE GENETIC ELEMENTS

Up until this point, we have thought of the genome in fairly conservative terms. If we map a gene today, we expect to see it in the same place tomorrow. However, our discovery of mobile genetic elements has modified that view to some extent. We now know that some segments of the genome can move quite readily from one place to another. The moving of those elements has effects on the phenotype of the organism, primarily at the transcriptional level. We thus begin our discussion of mobile genetic elements here, and we conclude it in chapter 15, because mobile elements also affect phenotypes of eukaryotes.

IS Elements

Transposable genetic elements, transposons, or even *jumping genes,* are regions of the genome that can move from one place to another. In some cases, transposable elements move, whereas in others a copy is inserted at a new place with the original still existing in its original place of insertion. Barbara McClintock first discovered transposable

elements in corn in the 1940s (see chapter 15); they were discovered in prokaryotes in 1967. Transposons first showed up as **polar mutants** in the *lac* operon of *E. coli* in which no genes of an operon were expressed past the point of the polar mutant. This effect was explained by assuming that the transposon brought with it a transcription stop signal. The presence of an inserted piece of DNA in these polar mutants was verified by heteroduplex analysis (fig. 13.26).

The first transposable elements discovered in bacteria were called **insertion sequences** or **IS elements.** It turns out that these are the simplest transposons. The IS elements consist of a central region of about 700 to 1,500 bp surrounded by an inverted repeat of about 10 to 30 bp. (These numbers are constants for any given IS element.) Presumably, the inverted repeats signal the transposing enzyme that it is at the ends of the IS element. The central region of the IS element contains a gene or genes for the transposing event: No bacterial genes are carried by the relatively small IS elements (fig. 13.27).

The target site, to which the transposable element will be moved, is not a specific sequence, as with the *att* site of λ. It becomes a direct repeat flanking the IS element only after insertion, giving rise to a possible model of insertion (fig. 13.28). The target site is cut in a staggered fashion, leaving single-stranded ends. The IS element is then inserted between the single-stranded ends. Repair processes convert the two single-stranded tails to double-stranded segments and hence to direct flanking repeats. When DNA is sequenced, the pattern of a direct flanking repeat surrounding an inverted repeat, with a segment in the middle, signals the existence of a transposable element.

Composite Transposons

After the discovery of IS elements, a more complex type of transposable element, a **composite transposon,** was discovered. A composite transposon consists of a central region surrounded by two IS elements. The central region usually contains bacterial genes, frequently antibiotic resistance loci. For example, the composite transposon, Tn3, contains the genes for transposase and resolvase, as well as the bacterial gene for β-lactamase, which confers resistance to ampicillin (fig. 13.29). Arrangements of composite transposons can vary quite a bit. The IS elements at the two ends can be identical or different; they can be in the same or different orientations; they can be similar to known IS elements or IS elements that have been modified such that they are different from any freely existing IS elements. In the latter case they are called IS-like elements.

Figure 13.25

Summary of regulation of phage λ life cycles. (1) In the initial infection, transcription begins in *cro* and *N,* but terminates shortly thereafter at left and right terminators. (2) The product of the *N* gene allows transcription through the initial terminators; in essence, all genes can be transcribed now. (3) Lysogeny will occur if the *cI* protein gains access to the right and left operators; the lytic cycle will prevail if the *cro*-gene product gains access to those two operators.

(1) Initial infection. Transcription from p_R and p_L through *cro* and *N*. Termination at t_{R1} and t_{L1}.

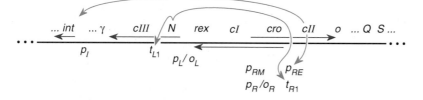

(2) *N* protein allows antitermination at t_{L1} and t_{R1}. Transcription continues through *cII* and *cIII*. Protein product of *cII* allows transcription at p_I and p_{RE}.

(3) Repressor and antirepressor (*cI* and *cro*-gene products) compete for o_R and o_L sites.

Lytic growth

Antirepressor (*cro* protein) gains access to o_{R3}, o_{R2}, o_{L3}, and o_{L2}. Right, left, and late operons transcribed. Repressor region (*cI, rex*) repressed.

Lysogeny

Repressor (*cI* protein) gains access to o_{R1}, o_{R2}, o_{L1}, and o_{L2}. Right, left, and late operons repressed. Transcription at p_{RM} enhanced.

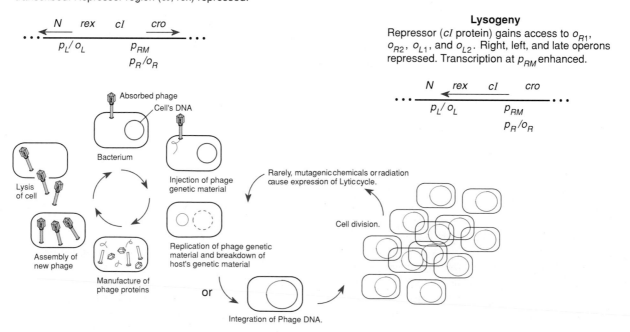

TABLE 13.1 Elements in Phage λ Infection

Gene Products

cI	Repressor protein whose function favors lysogeny
cII	Enhances transcription at the p_I and p_{RE} promoters
cIII	Inhibits the *HflA* protease
cro	Antirepressor protein that favors lytic cycle
N	Antiterminator acting at *nutR* and *nutL*
rex	Protects bacterium from infection by T4 *rII* mutants
int	Integrase for prophage integration
Q	Antiterminator of late operon
HflA	Bacterial protease that degrades *cII* protein

Promoters of

P_R	Right operon
p_L	Left operon
p_{RE}	Establishment of repression at repressor region
p_{RM}	Maintenance of repression at repressor region
$p_{R'}$	Late operon
p_I	*int* gene

Terminators

t_{RI}	Terminates after *cro* gene
t_{LI}	Terminates after *N* gene

Antiterminators

nutR	In *cro* gene
nutL	In *N* gene

Two IS elements can transpose virtually any region between them. In fact, composite transposons most likely came into being when two IS elements became located near each other. We can see this very clearly in a simple experiment. In figure 13.30, there is a small plasmid constructed with transposon Tn10 in it. The "reverse" transposon, consisting of the two IS elements and the plasmid genes, or the normal transposon, could each transpose.

Mechanism of Transposition

Although we do not know the exact mechanism of transposition, we do know that transposition does not use the normal recombination machinery of the cell (see chapter 16). A model by J. Shapiro, which is widely recognized, explains the fact that many transposons in the process of transposition go through a **cointegrate** state (fig. 13.31), a state in which there is a fusion of two elements. During the process of transposition (in this case from one plasmid to another), an intermediate cointegrate stage is formed, made up of both plasmids and two copies of the transposon. Then, through a process called *resolution*, the

Figure 13.26

Heteroduplex analysis revealing a transposon. (*a*) Two plasmids were hybridized, one with and one without a transposon. (*b*) The transposon is seen as a single-stranded loop; it has nothing to pair with in the heteroduplex.

(*a*) *Courtesy of M. M. Schwesinger. From "Plasmids," by R. Novick. Copyright © 1980 by Scientific American, Inc.*

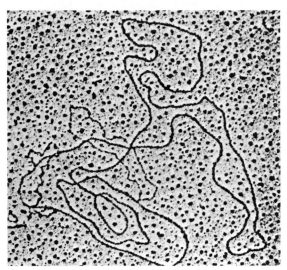

(a)

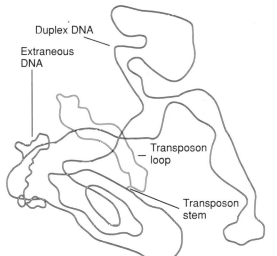

Duplex DNA

Extraneous DNA

Transposon loop

Transposon stem

(b)

J. A. Shapiro (1943–   )
Courtesy of Dr. J. A. Shapiro.

Figure 13.27

An IS element inserted into a target site in a bacterial chromosome creates a direct repeat on either side of the IS element. The explanation is shown in figure 13.28.

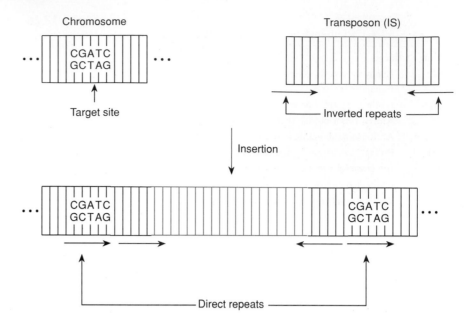

Figure 13.28

Insertion of an IS element results in a direct flanking repeat surrounding the transposon in the host chromosome. This occurs because the insertion takes place at a point in which a staggered cut in the host DNA is made, leaving complementary regions on either side of the transposon. Repair replication results in two copies of the flanking sequence.

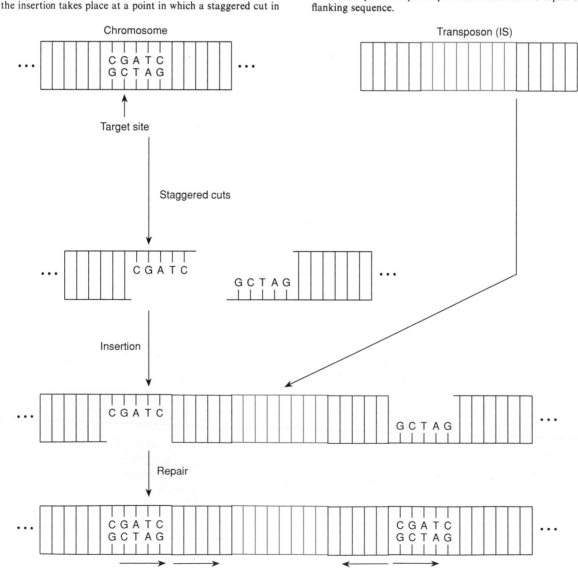

Figure 13.29

A composite transposon consists of a central region flanked by two IS elements. Transposon Tn3 contains the transposase and resolvase enzyme genes as well as the bacterial gene, β-lactamase, which protects the cell from the antibiotic ampicillin.

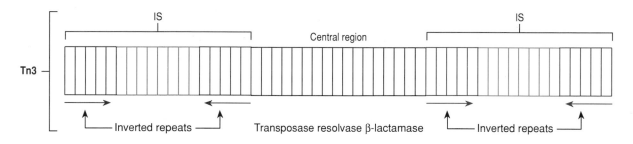

Figure 13.30

Two IS elements in a plasmid can transpose virtually any region between them. In the case shown, either the Tn10 transposon or the "reverse" transposon is transposed.

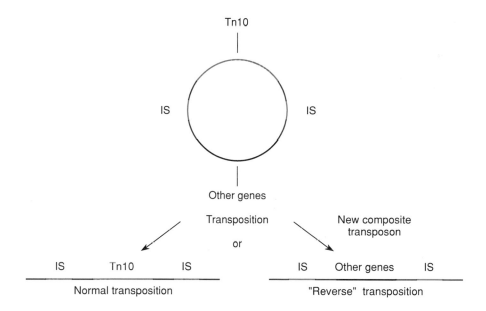

cointegrate is reduced back to the two original plasmids, each now containing a copy of the transposon. A diagram of Shapiro's mechanism is given in figure 13.32.

At first the donor and recipient DNA molecules are given staggered cuts (fig. 13.32*a* and *b*). Then nonhomologous ends are joined in such a way that they are connected by only one strand of the transposon (fig. 13.32*c*). This process is presumably controlled by the transposon-coded *transposase* enzyme. Repair-DNA replication now takes place to fill in the single-stranded segments. The result is a cointegrate of the two plasmids with two copies of the transposon. The last step is a recombination event at a homologous site within the two transposons. This is catalyzed by a *resolvase* enzyme, which resolves the cointegrate into the original two plasmids, each with a copy of the transposon (fig. 13.32*e*).

Phenotypic and Genotypic Effects of Transposition

Transposition can have several effects on the phenotype and genotype of an organism. If transposition takes place into a gene or its promoter, it can disrupt the expression of that gene. Depending on the orientation of a transposon, it can prevent the expression of genes. A transposon can also cause deletions and inversions.

Direct repeats on a chromosome can come about, for example, by the sequential transposition of the same IS or transposon, in the same orientation. Pairing followed by recombination will result in a deletion of the section between the repeats (fig. 13.33). In the case of inverted repeats, pairing followed by recombination will result in an inversion of the section between the repeats.

Figure 13.31
Transposition frequently goes through an intermediate cointegrate stage. In this case, the transposon is copied from one plasmid to another, with an intermediate stage consisting of a single large plasmid.

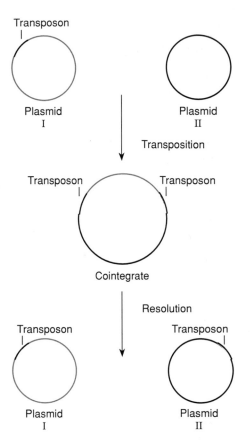

Figure 13.32
The Shapiro mechanism of transposition. Staggered cuts are made at the site of transposon insertion and at either side of the transposon itself (*a, b*). Nonhomologous single strands are joined, resulting in two single-stranded copies of the transposon in the cointegrate (*c*). Repair replication of these single strands results in two copies of the transposon being present (*d*). A crossover at the transposon resolves the cointegrate into two plasmids, each with a copy of the transposon (*e*).

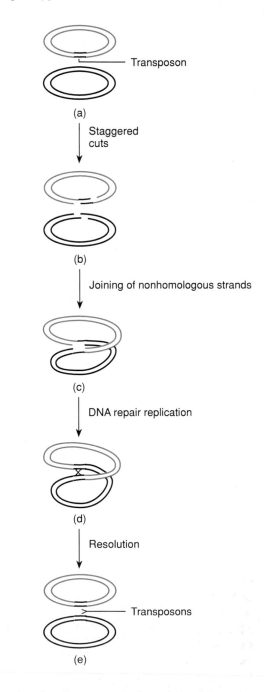

A well-known case of transposon orientation controlling a phenotype in bacteria occurs in *Salmonella typhimurium.* The flagella of this bacterium occur in two types. Any particular bacterium has either type 1 or type 2 flagella (called phase 1 or phase 2 flagella). The difference is in the flagellin protein of which the flagella are composed. Phase 1 flagella are determined by the *H1* gene and phase 2 flagella are determined by the *H2* gene. The change from one phase to another occurs at a rate of about 10^{-4} per cell division. After extensive genetic analysis, the following scheme was suggested, and later verified using recombinant DNA techniques.

The *H1* and *H2* genes are at separate locations on the bacterial chromosome (fig. 13.34). *H2* is part of an operon that also contains the gene *rH1*, the repressor of *H1*. The promoter of this operon lies within a transposon upstream of the operon. When the promoter is in the proper orientation, the *H2* operon is expressed, resulting in phase 2 flagella. The *rH1* gene product represses the *H1* gene (fig. 13.34*a*). If the inverted repeat ends of the transposon undergo recombination, the transposon will be inverted (see fig. 13.33), causing the promoter to be in an incorrect ori-

entation for the transcription of the *H2* operon. No *H1* repressor will be made, with the result that the *H1* gene will be expressed (fig. 13.34*b*).

As recently summarized by N. Kleckner, transposons can have marked effects on the phenotype by their actions in transposition and by the fact that they may carry genes

Figure 13.33

Pairing and recombination in repeats in DNA. (*a*) Direct repeats can result in deletion (in the form of a *circle*). (*b*) Inverted repeats can result in an inversion of the region between the repeats.

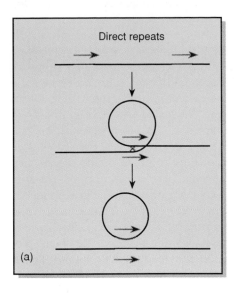

Direct repeats

(a)

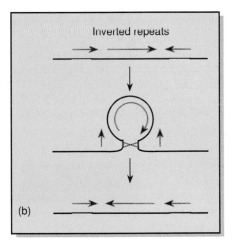

Inverted repeats

(b)

Nancy Kleckner (1947–   )
Courtesy of Nancy Kleckner. Photo by Stu Rosner.

Figure 13.34

Arrangement of flagellin genes on the *Salmonella* chromosome. The promoter (*p*) is within a transposon. In one orientation (*a*), the *H2* operon is transcribed, which results in *H2* flagellin and *rH1* protein, the repressor of the *H1* gene. In the second orientation (*b*), the *H2* operon is not transcribed, resulting in uninhibited transcription of the *H1* gene.

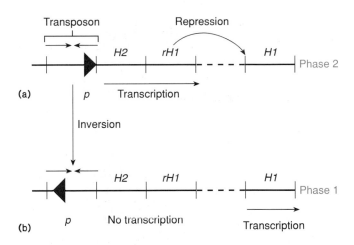

valuable to the cell. However, they can also exist without any noticeable consequences. This fact has led some evolutionary geneticists to suggest that transposons are an evolutionary accident that, once created, are self-maintaining. Since they may exist without a noticeable benefit to the host's phenotype, transposons have been referred to as **selfish DNA.** In recent theoretical and experimental studies, however, some scientists have suggested that transposons improve the evolutionary fitness of the bacteria that have them (see chapter 20).

OTHER TRANSCRIPTIONAL CONTROL SYSTEMS

Transcription Factors

Phage T4

Phage T4, with seventy-three genes, has its transcription controlled by the nature of particular RNA polymerase specificity factors. Early T4 genes have promoters whose specificity of recognition depends on the sigma factor of the host. However, middle and late operons of T4 have promoters whose specificity is determined by other proteins that are synthesized during the early infection process. For example, late promoters require the bacterial RNA polymerase plus the products of genes 33 and 55 of the T4 chromosome. Some proteins function both early and late and are specified by genes that have several promoters, with each promoter being recognized by a different specificity factor.

Heat-Shock Proteins

A response to elevated temperature, found in both prokaryotes and eukaryotes, is the production of heat-shock proteins (see chapter 10). In *E. coli,* elevated temperatures result in the general shutdown of protein synthesis concomitant with the appearance of seventeen heat-shock proteins. These proteins help protect the cell against the consequences of elevated temperatures; some appear to be protein chaperones (see chapter 11). The production of these proteins is the direct result of the gene product of the *HtpR* gene, which codes for a sigma factor, σ^{32}, a 32,000-dalton protein. The normal sigma factor, σ^{70}, the product of the *rpoD* gene, is a 70,000-dalton protein; the heat-shock genes have promoters recognized by σ^{32} rather than σ^{70}. Heat shock somehow causes the *HtpR* gene to become active, activating the heat-shock genes, and somehow reduces the activity of the *rpoD* gene. From DNA sequence data, the difference in promoters between normal genes and heat-shock genes seems to lie in the -10 consensus sequence (Pribnow box). In normal genes it is TATAAT; in heat-shock protein genes it is CCCCATXT, in which X is any base.

Promoter Efficiency

In addition to the mechanisms previously described, there are other ways to regulate the transcription of mRNA. One way is to control the efficiency with which various processes take place. For example, we know that the promoter sequence of different genes in *E. coli* is different. Since the affinity for RNA polymerase is different for the different sequences, the rate of transcription of the genes will also vary. The more efficient promoters will be transcribed at a greater rate than the less efficient promoters. An example is the promoter of the *i* gene of the *lac* operon. This promoter is for a constitutive gene that usually produces only about one mRNA per cell cycle. However, mutants of the promoter sequence are known that produce up to fifty mRNAs per cell cycle. Here, then, the transcriptional rate is controlled by the efficiency of the promoter in binding RNA polymerase. Efficiency can be controlled by the direct sequence of nucleotides (i.e., differences from the consensus sequence) or distance between consensus regions. For example, there is variation among promoters in the number of bases between the -35 sequence and the -10 sequence. Seventeen seems to be the optimal number of bases separating the two. Presumably, more or less than seventeen will reduce the efficiency of transcription.

TRANSLATIONAL CONTROL

When considering the topic of control of gene expression, it is important to remember that all control mechanisms are aimed at exerting an influence on either the amount, or the activity, of the gene product. Therefore, in addition to transcriptional controls, which influence the amount of mRNA produced, there are also translational controls affecting how efficiently the mRNA is translated. In prokaryotes, translational control is of lesser importance than transcriptional control for two reasons. First, mRNAs are extremely unstable; with a lifetime of only about two minutes, there is little room for controlling the rates of translation of existing mRNAs because they simply do not last very long. Second, although there are some indications of translational control in prokaryotes, such control is inefficient—energy is wasted synthesizing mRNAs that may never be used.

Translational control can be exerted on a gene if the gene occurs distally from the promoter in a polycistronic operon. The genes that are transcribed last appear to be translated at a lower rate than the genes transcribed first. The three *lac* operon genes, for instance, are translated roughly in a ratio of 10:5:2. This ratio is due to the polarity of the translation process. That is, in prokaryotes, translation is directly tied to transcription—an mRNA can have ribosomes attached to it well before transcription is finished. Thus genes at the beginning of the operon will be available for translation before genes at the end. In addition, exonucleases seem to degrade mRNA more efficiently from the 3′ end. Presumably, in evolutionary time, natural selection has ordered the genes within operons such that those determining enzymes needed in greater quantities will be at the beginning of an operon.

Translation can also be regulated by RNA-RNA hybridization. RNA complementary to the 5′ end of a messenger RNA can prevent the translation of that mRNA. Several examples of this type of regulation are known. The regulating RNA is called **antisense RNA.** In figure 13.35, the mRNA from the *ompF* gene in *E. coli* is prevented from being translated by complementary base-pair binding with an antisense RNA, called *mic*F RNA (*mic* stands for *m*RNA-*i*nterfering *c*omplementary RNA). The *ompF* gene codes for a membrane component called a *porin,* which, as the name suggests, provides pores in the cell membrane for transport of materials. Surprisingly, a second porin gene, *ompC,* seems to be the source of the *mic*F RNA. Transcription of the opposite DNA strand (the one not normally transcribed), near the promoter of the *ompC* gene yields the antisense RNA. One porin gene thus seems to be regulating the expression of another porin gene, for reasons that are not completely understood. Antisense RNA has also been implicated in such phenomena as the control of plasmid number and the control of transposon Tn10 transposition. Control by antisense RNA is a fertile field for gene therapy because antisense RNA can be artificially synthesized and then injected into eukaryotic cells.

A third translational control mechanism consists of the efficiency with which the messenger RNA is bound to the ribosome. This efficiency is related to some extent to the sequence of nucleotides at the 5′ end of the mRNA

Figure 13.35

(a) Complementarity between the RNA of the *ompF* gene and antisense RNA, *micF*. The region of overlap includes the Shine-Dalgarno sequence and the initiation codon, effectively preventing ribosome binding and translation of the *ompF* RNA. Notice the stem-loops on each side of the overlap. (b) The introduction of antisense RNA into a petunia plant (*Petunia hybrida*). At the top is a normal red flower. At the bottom is a flower with very little red because of the presence of antisense RNA for the chalcone synthase gene, responsible for an enzyme controlling the basic pigment production in the plant. An antisense gene produces an antisense RNA that forms a duplex with the normal RNA, preventing translation. The result is a much reduced production of the enzyme and of pigment. The lower plant was made transgenic using the *Agrobacterium tumefaciens* system.

(a) *Reproduced, with permission, from the* Annual Review of Biochemistry, *Volume 55,* © *1986 by Annual Reviews, Inc.* (b) *From Alexander R. van der Krol, J. N. M. Mol, and A. R. Stuitje, "Modulation of eukaryotic gene expression by complementary RNA or DNA sequences,"* Bio Techniques *6 (1988):958–976, figure 3, parts 1 and 4.*

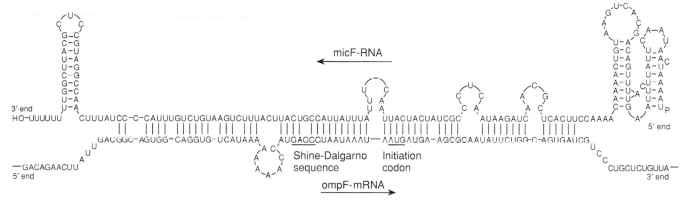

(a)

(b)

BLE 13.2 Codon Distribution in MS2, an RNA Virus

First Position	U		C		A		G		Third Position
U	Phe	10	Ser	13	Tyr	8	Cys	7	U
	Phe	13	Ser	10	Tyr	13	Cys	4	C
	Leu	11	Ser	10	stop	1	stop	0	A
	Leu	4	Ser	13	stop	1	Trp	14	G
C	Leu	10	Pro	7	His	4	Arg	13	U
	Leu	14	Pro	3	His	4	Arg	11	C
	Leu	13	Pro	6	Gln	10	Arg	6	A
	Leu	6	Pro	5	Gln	16	Arg	4	G
A	Ile	8	Thr	14	Asn	11	Ser	4	U
	Ile	16	Thr	10	Asn	23	Ser	8	C
	Ile	7	Thr	8	Lys	12	Arg	8	A
	Met	15	Thr	5	Lys	17	Arg	6	G
G	Val	13	Ala	19	Asp	18	Gly	17	U
	Val	12	Ala	12	Asp	11	Gly	11	C
	Val	11	Ala	14	Glu	9	Gly	4	A
	Val	10	Ala	8	Glu	14	Gly	4	G

that is complementary to the 3′ end of the 16S rRNA segment in the ribosome (the Shine-Dalgarno sequence). Variations from the consensus sequences will have different efficiencies of binding and, therefore, will be translated at different rates.

The redundancy in the genetic code can also play a part in translational control of some proteins since different tRNAs occur in the cell in different quantities. Genes with abundant protein products may have codons that specify the more common tRNAs. This concept is called **codon preference.** In other words, certain codons are preferred; they specify tRNAs that are abundant. Genes that code for proteins not needed in high abundance could have several codons specifying the rarer tRNAs, which would slow down the process of translation of these genes. The codon distribution of the phage MS2 in table 13.2 shows that, with the exception of the UGA stop codon, every codon is used. However, the distribution is not random for all amino acids. For example, the amino acid glycine has two common codons and two rarer codons. The same holds for arginine but not, for example, valine.

A final translational control mechanism, called the **stringent response,** occurs when prokaryotic cells are starved of amino acids. When an uncharged tRNA finds its way into the A site of the ribosome, an event likely under general amino acid starvation, it causes an **idling reaction** by the ribosome, which entails the production of the odd nucleotide, guanosine tetraphosphate (3′-ppGpp-5′; fig. 13.36). This nucleotide, originally called "magic spot" because of its sudden appearance on chromatograms, is produced by a protein, called the **stringent factor,** which is the product of the relA gene. (The gene is called rel from the **relaxed mutant** that does not have the stringent response.) The stringent factor is associated with the ribosome, although it is not one of the structural proteins of either ribosomal subunit.

Precisely what the odd nucleotide does is unknown. The stringent response, however, includes several major changes in bacterial physiology, including an almost complete cessation of the transcription of rRNA and tRNA and a major cessation in transcription of mRNAs. This is a radical attempt by bacteria to wait out bad times until the environment of the cell once again contains nutrients. It perhaps could be called bacterial "hibernation."

POSTTRANSLATIONAL CONTROL

Feedback Inhibition

Even after a gene has been transcribed and the mRNA translated, a cell can still exert some control over the functioning of the enzymes produced if the enzymes are allosteric proteins. We have discussed the activation and deactivation of operon repressors (e.g., lac, trp) owing to

Figure 13.36

The idling reaction. The stringent factor catalyzes the conversion of GDP to ppGpp. The added pyrophosphate groups come from ATP.

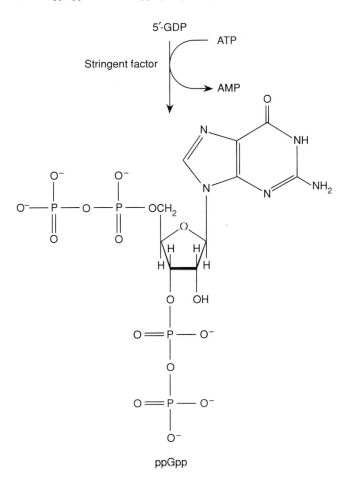

ppGpp

Figure 13.37

Aspartate transcarbamylase catalyzes the first step in pyrimidine biosynthesis. An end product, CTP, inhibits the enzyme.

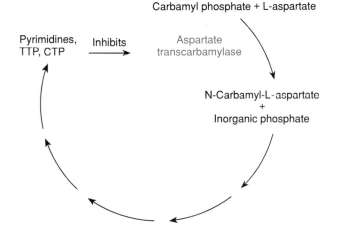

their allosteric properties. Similar effects occur with other proteins. The need for posttranslational control is apparent because of the relative longevity of proteins as compared with RNA. When an operon is repressed, it no longer transcribes mRNA; however, the mRNA that was previously transcribed has been translated into protein, and this protein is still functioning. Thus, during the process of operon repression, it would also be efficient for the cell to control the activity of existing proteins.

An example of posttranslational control occurs with the enzyme aspartate transcarbamylase, which catalyzes the first step in the pathway of pyrimidine biosynthesis in *E. coli* (fig. 13.37). An excess of one of the end products of the pathway, cytidine triphosphate (CTP), inhibits the functioning of aspartate transcarbamylase. This method of control is called **feedback inhibition** because a product of the pathway is the agent that turns off the pathway.

Aspartate transcarbamylase is an allosteric enzyme. Its active site is responsible for the condensation of carbamyl phosphate and L-aspartate (fig. 13.37). However, it also has regulatory sites that have an affinity for CTP. When CTP is bound in a regulatory site, the conformation of the enzyme changes and the enzyme has a lowered affinity for its normal substrates; recognition of CTP inhibits the condensation reaction normally carried out by the enzyme (fig. 13.38). Thus allosteric enzymes provide a mechanism for control of protein function after the protein has been synthesized. This mechanism applies not only to regulatory proteins, such as the *lac* and *trp* repressors, but also to regular enzymatic proteins.

Protein Degradation

A final area of control affecting the amount of gene product in a cell is control of the rate at which proteins are degraded. There is a great deal of variation in the normal life spans of proteins. For example, some proteins last longer than a cell cycle, whereas others may be broken down in seconds. Two research groups have suggested models of control of protein degradation, the **N-end rule,** and the **PEST hypothesis.**

According to the N-end rule, the amino acid at the amino-, or N-terminal, end of a protein is a signal to proteases that control the average length of life of a protein. In recent experiments, almost complete predictability was achieved in determining the life span of the β-galactosidase protein based on its N-terminal amino acid. Protein life spans range from 2 minutes for those with N-terminal arginine to greater than 20 hours for those with N-terminal methionine or five other amino acids (table 13.3).

Figure 13.38

Stereo views of aspartate transcarbamylase. *Top:* the enzyme bound with CTP. *Bottom:* the enzyme not bound with CTP. Notice the large difference in shape with and without CTP. With CTP the enzyme literally closes up.

From Kantrowitz and Lipscomb. Science *24 (August 5, 1988):669–74, figure 5. "*Escherichia coli* Aspartate Transcarbamylase: The Relation between Structure and Function. Copyright 1988 by the AAAS.*

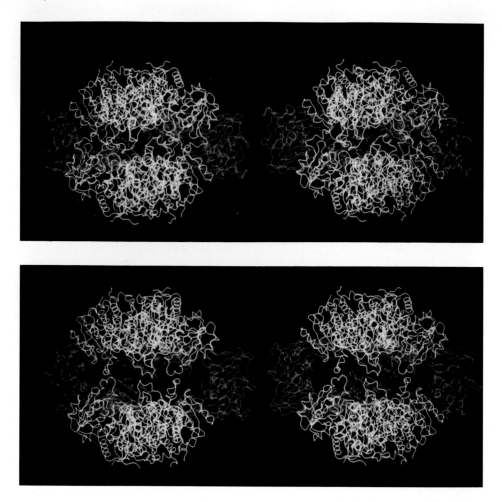

TABLE 13.3 Relationship Between N-Terminal Amino Acid and Half-Life of *E. coli* β-Galactosidase Proteins with Modified N-Terminal Amino Acids

N-Terminal Amino Acid	Half-Life
Met, Ser, Ala, Thr, Val, Gly	>20 hours
Ile, Glu	30 minutes
Tyr, Gln	10 minutes
Pro	7 minutes
Phe, Leu, Asp, Lys	3 minutes
Arg	2 minutes

Source: Data from Bachmair, Finley, and Varshavsky, *Science*, 234:179–186, 1986.

According to the PEST hypothesis, protein degradation is determined by regions rich in one of four amino acids: proline, glutamic acid, serine, and threonine. (The one-letter abbreviations of these four amino acids are P, E, S, and T, respectively.) Proteins that have these regions tend to be degraded in less than 2 hours (table 13.4). In one study of thirty-five proteins with half-lives of between 20 and 220 hours, only three contained a PEST region. We see that not only are different proteins programmed to survive different times in the cell, but that programming seems to be based on the N-terminal amino acid as well as various regions rich in the PEST amino acids within the proteins.

TABLE 13.4 Amino Acid Sequences Showing Extremes of PEST Amino Acid Concentrations*

Protein	Segment	Half-Life (Hours)	Sequence
E1A	177–202	0.5	RTCGMFVYSPVSEPEPEPEPEPEPAR
c-myc	241–269	0.5	HEETPPTTSSDSEEEQEDEEEIDVVSVEK
c-fos	128–139	0.5	KVEQLSPEEEEK
α-casein	58–79	2–5	KEMEAESISSSEEIVPNSVQEK
β-casein	1–25	2–5	RELEELNVPGEIVESLSSSEESITR

*The one-letter amino acid code is A, alanine; C, cytosine; D, aspartic acid; E, glutamic acid; F, phenylalanine; G, glycine; H, histidine; I, isoleucine; K, lysine; L, leucine; M, methionine; N, asparagine; P, proline; Q, glutamine; R, arginine; S, serine; T, threonine; V, valine; W, tryptophan; and Y, tyrosine. E1A is adenovirus early protein; c-myc and c-fos are oncogenes (cancer-causing genes).
Source: Data from S. Rogers, et al., *Science,* 234:364–368, 1986.

SUMMARY

Most bacterial genes are organized into operons, which either can be repressed or induced. Transcription begins in inducible operons, such as *lac,* when the metabolite upon which the operon enzymes act appears in the environment. The metabolite (or a derivative), the inducer, combines with the repressor (the product of the independent regulator gene) and renders the repressor nonfunctional. In the absence of the inducer, the repressor binds to the operator, a segment between the promoter and the first gene of the operon. When in place, the repressor blocks transcription. After combining with the inducer, the repressor diffuses from the operator and transcription proceeds.

All operons responsible for the breakdown of sugars in *E. coli* are inducible. In the presence of glucose, other inducible sugar operons (such as the arabinose and galactose operons) are repressed, even if their sugar appears in the environment. This process is called *catabolite repression.* Cyclic AMP and a catabolite activator protein (CAP) enhance the transcription of the nonglucose sugar operons. Glucose lowers the level of cyclic AMP in the cell and thus prevents the enhancement of transcription of these other operons.

Repressible operons, such as the *trp* operon in *E. coli,* have the same basic components as an inducible operon—polycistronic transcription controlled by an operator site between the promoter and the first structural gene. However, the repressor protein, controlled by an independent regulator gene, is functional in blocking transcription only after it has combined with the corepressor, which is the end product of the operon's pathway or some form of the end product (tryptophan in the *trp* operon). In addition, amino acid-synthesizing operons usually have an attenuator region. The ability of a ribosome to translate a leader peptide gene determines the secondary structure of the mRNA transcript. If the ribosome can translate the leader peptide gene, then there must be adequate quantities of the amino acid present, and a terminator stem and loop will form in the mRNA causing termination of transcription.

Control of gene expression in λ phage is well known. The "decision" for lytic versus lysogenic response is determined by competition between two repressors, *cI* and *cro.*

Transposons are mobile genetic elements, copies of which can be inserted at other places in the genome. Their ends are inverted repeats. Upon insertion, they are flanked by short direct repeats. They can be simple (IS elements) or complex. Their presence can cause inversions or deletions. The flagellar phase in *Salmonella* is controlled by the orientation of a transposon.

Different promoters for early and late operons in phages are another form of transcriptional control, as seen in phage T4 transcription and the transcription of heat-shock proteins. Translational control can be exercised through a gene's position in an operon (genes at the beginning are transcribed most frequently), through redundancy of the genetic code, or through a stringent response that shuts down most transcription during starvation. Posttranslational control is primarily by feedback inhibition. The rate of protein degradation is programmed by the N-terminal amino acid or particular regions within the proteins.

SOLVED PROBLEMS

1. How could you determine whether the genes for the breakdown of the sugar arabinose are under inducible control in *E. coli*?

ANSWER: Inducible means that the genes to break down the substrate—arabinose, a five-carbon sugar in this case—are not active in the absence of the inducer (again, arabinose). Therefore, in the absence of the arabinose in the environment of the cells, there should be no activity of the arabinose utilization enzymes within the bacterial cells, but after arabinose is added

to the medium, then the enzymes should be present. We thus need to assay the contents of the cells before and after arabinose is added to the medium. The cells should be assayed after they are broken open and the DNA destroyed so as not to confound the experiment. Using a standard biochemical analysis for arabinose, we should find that the bacterial cell is incapable of metabolizing arabinose before induction but capable of metabolizing it afterward. If the cells were capable of metabolizing arabinose in both cases we would say that arabinose utilization is constitutive. If the cells were incapable of utilizing arabinose in both cases we would conclude that the bacterium is incapable of using the sugar arabinose as an energy source. (In fact, arabinose utilization is inducible.)

2. Why would the RecA protein of *E. coli* cleave the λ repressor?

ANSWER: Since the cleaving of the λ repressor is a signal to begin the lytic phase of the life cycle of the phage, it seems odd that the bacterial cell would be an accomplice to its own destruction. However, the phenomenon makes much more sense if we realize that the RecA protein has several other functions critically important to the bacterial cell (see chapter 16). The λ phage has, however, evolved the ability to take advantage of the existence of the RecA protein by evolving a repressor that is sensitive to it. Evolutionary biologists view this as "coevolution," two competing organisms trying to evolve to take advantage of or minimize properties of the other. The bacterium, however, might be at a disadvantage. Since RecA has many other functions involving interactions with other proteins, it may be highly limited in how it can change. This is one plausible explanation as to why RecA liberates phage λ.

3. What are the differences in action of the λ promoters P_{RE} and P_{RM}?

ANSWER: The promoters P_{RE} and P_{RM} are both promoters of the repressor operon of phage λ. Transcription from these promoters allows production of the *cI* repressor protein, the repressor that favors lysogeny. Initially, the promoter P_{RE} is activated. For it to be a site of transcription, it must be activated by the product of the *cII* gene, which lies in the right operon.

This promoter produces a mRNA that has a long leader and is translated very efficiently. Once the repressor binds at the operators of the left and right operons, the *cII* gene is no longer transcribed, and therefore P_{RE} is no longer a site for transcription. However, the repressor gene can still be transcribed from the P_{RM} promoter, which does not need the product of the *cII* gene. But, this promoter produces a transcript with virtually no leader, and thus is translated very inefficiently. At that point, however, with lysogeny in progress, only a very small quantity of repressor is needed to maintain lysogeny. Thus the two promoters are the sites for the initiation of the repressor operon under different circumstances: one early in the infection stage and one after lysogeny is under way.

4. What are the phenotypes of the following partial diploids for the *lac* operon in *E. coli* in the presence and absence of lactose?
   a. (F′) $i^+ o^+ p^+ z^-/i^- o^+ p^+ z^+$ (chromosome)
   b. (F′) $i^+ o^c p^+ z^+/i^+ o^+ p^- z^+$ (chromosome)

ANSWER: Consider one DNA molecule at a time. If one DNA molecule can never make the enzyme, it can be ignored. In (a) the top DNA (F′) will never make enzyme (it is z^-), and the bottom one (chromosome) will never make repressor (it is i^-). The functional repressor for the top molecule (i^+) will bind to both DNAs and hence the bottom operon will not be transcribed in the absence of lactose and will be induced to transcribe in the presence of lactose. In (b) the top DNA (F′) will always be transcribing (constitutive) because the repressor can never bind the operator (o^c); hence the operon will be transcribed all the time. The bottom DNA (chromosome) can never make RNA (it is p^-).

EXERCISES AND PROBLEMS

1. Are the following *E. coli* cells constitutive or inducible for the *z* gene?
   a. $i^+ o^+ z^+$
   b. $i^- o^+ z^+$
   c. $i^- o^c z^+$
   d. $i^+ o^c z^+$
   e. $i^s o^+ z^+$
   f. $i^Q o^+ z^+$

2. Given the following *lac* operon merozygotes, determine which form or forms of the *z* gene (z^+, z^-) are transcribed either (a) with or (b) without inducer present.
   a. $i^+ o^+ z^+/F′ i^+ o^+ z^+$
   b. $i^- o^+ z^+/F′ i^\pm o^+ z^-$
   c. $i^+ o^+ z^+/F′ i^- o^+ z^-$
   d. $i^- o^c z^-/F′ i^+ o^+ z^+$
   e. $i^- o^+ z^-/F′ i^+ o^c z^+$

3. Construct a merozygote of the *trp* operon in *E. coli* with two forms of the first gene (*e* gene) in the operon. Describe the types of *cis* and *trans* effects that are possible, given mutants of any component of the operon. Can this repressible system work for any type of operon other than those controlling amino acid synthesis?

4. Describe the interaction of the attenuator and the operator control mechanisms in the *trp* operon of *E. coli* under varying concentrations of tryptophan in the cell. How does attenuator control react to shortages of other amino acids?

5. What is the fate of a λ phage entering an *E. coli* cell that contains quantities of λ repressor? What is the fate of the same phage entering an *E. coli* cell that contains quantities of the *cro*-gene product?

6. Describe the fate of λ phages during the infection process with mutants in the following genes: *cI, cII, cIII, N, cro, att, Q.*

7. What is the fate of λ phages during the infection process with mutants in the following areas: o_{RI}, o_{R3}, p_L, p_{RE}, p_{RM}, p_R, t_{LI}, t_{RI}, *nutL, nutR*?

8. What are the three different physical forms that the phage λ chromosome can take?

9. How does ultraviolet (UV) damage induce the lytic life cycle in phage λ?

10. The λ prophage is sometimes induced into the lytic life cycle when an Hfr lysogen (lysogenic cell) conjugates with a nonlysogenic F⁻ cell. How might induction come about in this instance?

11. List the steps from transcription through translation to enzyme function and note all the points at which control could be exerted.

12. Why are IS elements sometimes referred to as selfish DNA?

13. What are the differences among an IS element, a transposon, an intron, a plasmid, and a cointegrate?

14. Describe the Shapiro model of transposition. What are the roles of transposase, DNA polymerase I, ligase, and resolvase?

15. Why are transposons flanked by direct repeats?

16. How do transposons induce deletions? Inversions?

17. Describe how a transposon controls the expression of the flagellar phase in *Salmonella*.

18. What is a polar mutation? What can cause it?

19. What are the advantages of transcriptional control over translational control?

20. Describe the role of cyclic AMP in transcriptional control in *E. coli*.

21. What is feedback inhibition? What other roles do allosteric proteins play in regulating gene expression?

22. Operon systems exert negative control in the sense that they act through inhibition. The CAP system exerts positive control because it acts through enhancement of transcription. Describe how an operon could work if it were dependent only upon positive control.

23. How are heat-shock proteins induced?

24. What controls the rate of degradation of proteins?

25. What is the stringent response? How does it work? What is an idling reaction?

26. What is antisense RNA? How does it work? What is the obvious source of this regulatory RNA? How could this RNA be used to treat a disease clinically?

27. The tryptophan operon is under negative control; it is on in the absence of tryptophan and off in the presence of tryptophan. The symbols *a, b,* and *c* represent the gene for tryptophan synthetase, the operator region, and the repressor—but not necessarily in that order. From the following data, determine which letter is the gene, the repressor, and the operator (+ = tryptophan synthetase activity; − = no activity).

Strain	Genotype	Tryptophan absent	Tryptophan present
1	$a^- b^+ c^+$	+	+
2	$a^+ b^+ c^-$	+	+
3	$a^+ b^- c^+$	−	−
4	$a^+ b^- c^+/a^- b^+ c^-$	+	+
5	$a^+ b^+ c^+/a^- b^- c^-$	+	−
6	$a^+ b^+ c^-/a^- b^- c^+$	+	−
7	$a^- b^+ c^+/a^+ b^- c^-$	+	+

28. J. Beckwith isolated point mutations that were simultaneously uninducible for the *lac, ara, mal,* and *gal* operons, even in the absence of glucose. Provide two different functions that could be missing in these mutants.

29. You have isolated a repressor for an inducible operon and have determined that it has two different binding sites, one for the inducer and one for the operator. Mutants of the repressor result in three different phenotypes as far as binding is concerned. What are these phenotypes?

30. The tryptophan operon is a repressible operon. The corepressor is charged $tRNA^{His}$, the gene for which is not part of the operon. For the following mutants, tell whether the enzymes of the operon will be made; then tell whether each mutant would be *cis* dominant in a partial diploid.
   a. RNA polymerase cannot bind the promoter.
   b. The repressor cannot bind operator DNA.
   c. The repressor cannot bind charged $tRNA^{His}$.

31. An *E. coli* strain is isolated that produces β-galactosidase (*lac z*) and permease (*lac y*) constitutively. Provide two possible mutations that could cause the above phenotype, then describe how each mutation would behave in a partial diploid in which one DNA is wild-type for the entire *lac* system.

32. A temperature-sensitive mutant of the λ *cI* gene has been isolated. At 30°C it binds λ DNA, but it cannot bind DNA at 42°C. What will be the consequence of incubating *E. coli* that are lysogenic for this λ mutant at 42°C?

33. The mutant in problem 32 is heated to 42°C for 5 minutes, cooled to 30°C, and grown for one hour so that the cells will divide once or twice. The temperature is then raised to 42°C and you wait for lysis. Many of the cells are not lysed, and are in fact able to form colonies. Explain these results.

34. You have isolated two *E. coli* mutants that synthesize β-galactosidase constitutively.
   a. If these mutants affect different functions, in what two functions could they be defective?
   b. You can make a partial diploid of the mutants, but you can use only one strain. What strain do you use, and what result do you expect for each mutant?

35. In T4 phage, the genes *rIIA* and *rIIB* lie adjacent to each other in the T4 chromosome. During the early phase of infection, *rIIA* and *rIIB* products are present in equimolar amounts. In the late phase of infection, the amount of *rIIB* is 10 to 15 times higher than *rIIA*. Nonsense mutations in *rIIA* eliminate early but not late *rIIB* production. In the mutants that contain small deletions near the end of *rIIA*, the amount of *rIIA* is always equal to the amount of *rIIB*, regardless of the time of infection. Based on the above information, devise a map of the *rII* region. Include the location(s) of the promoter(s).

36. A hypothetical operon has a sequence of sites Q R S T U in the promoter region, but the exact location of the operator and promoter consensus sequences have not been identified. Various deletions of this operator region are isolated and mapped. Their locations appear below, in which a "/" represents a deleted region.

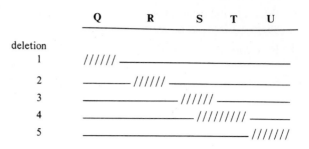

Deletions 3 and 4 are found to produce constitutive levels of RNA of the operon, and deletion 1 is found to never make RNA. Where are the operator and promoter consensus sequences probably located?

Suggested Readings for chapter 13 are on page 651.

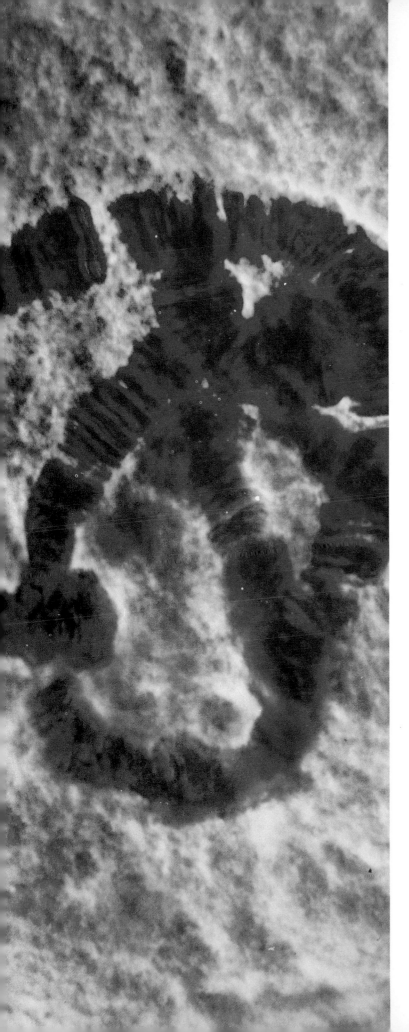

The Eukaryotic Chromosome

A giant salivary gland chromosome from the midge, *Chironomus* sp. (150×).

© *F. & A. Michler/Peter Arnold, Inc.*

In chapter 13 we looked at the control of gene expression in prokaryotes and bacteriophages. Compared with eukaryotes, bacteriophages and prokaryotes are relatively simple. Of fundamental importance is that, in these lower forms, the operon model of induction and repression of transcription is a unifying theme for control of gene expression. Despite nuances such as catabolite repression and attenuator control, the operon model provides a relatively clear picture of how genes are turned on and off in phages and prokaryotes. No such model currently exists for eukaryotes. In attempting to elucidate models for control of gene expression in eukaryotes, one very important factor that must be taken into account is the complexity of the structure of the eukaryotic chromosome. In this chapter, we will cover the current understanding of how these very large structures are organized.

THE EUKARYOTIC CELL

Eukaryotes and prokaryotes are the two superkingdoms of organisms. The following comparisons, using *E. coli* as a general model for prokaryotes, show generally how much more complex eukaryotes are than prokaryotes:

1. An *E. coli* chromosome contains approximately 4.2×10^6 bp of DNA. The haploid human genome contains nearly one thousand times as much DNA.

2. Eukaryotic DNA is in the form of **nucleoprotein,** a DNA-histone protein complex. Although a few histone-like proteins have been found in *E. coli,* its chromosomal DNA is not complexed with protein to anywhere near the same extent as eukaryotic DNA.

3. An *E. coli* cell has very little internal structure. Eukaryotes have a number of internal organelles, and an extensive lipid membrane system, including the nuclear envelope itself.

4. An *E. coli* cell is small (0.5–5.0 μm in length for bacteria). Eukaryotic cells are generally larger than prokaryotes (10–50 μm in length for animal tissue cells).

5. The mRNA of *E. coli* is translated while it is being transcribed. Eukaryotic mRNA is modified within the nucleus before being transported out for translation in the cytoplasm.

6. No mRNA isolated from eukaryotic cells, including the mRNA of animal viruses, has been found to be polycistronic (containing many genes). Most prokaryotic mRNAs are polycistronic.

7. Most *E. coli* genes are parts of inducible or repressible operons. With minor exceptions, there is no evidence for operons in eukaryotes.

8. *E. coli* exists as a simple, single cell. Although some prokaryotes do aggregate, sporulate, and show a few other limited forms of differentiation, they are primarily one-celled organisms. And, although some eukaryotes are single-celled (e.g., yeast), the essence of eukaryotes is differentiation. In human beings, a zygote gives rise to every other cell type in the body in a relatively predictable manner.

To appreciate fully the complexity of eukaryotes, we will begin by looking at the eukaryotic chromosome. In the next chapter, we will look at the patterns of development in eukaryotes and some possible mechanisms of control of gene expression.

THE EUKARYOTIC CHROMOSOME

DNA Arrangement

Evidence that the eukaryotic chromosome is **uninemic**—that is, contains one double helix of DNA—comes from several sources. The best data are provided by radioactive-labeling studies, first done by J. Taylor and his colleagues in 1957. If a eukaryote is allowed to undergo one DNA replication in the presence of tritiated (^{3}H) thymidine, each of the daughter chromatids would be expected to contain a double helix with one unlabeled DNA template strand and one labeled strand of newly synthesized bases (fig. 14.1). The configuration is expected on the basis of semiconservative replication, with each chromatid containing one double helix. A second round of DNA replication, in the absence of ^{3}H-thymidine, should produce chromosomes in which, before cell division, one chromatid would have unlabeled DNA and one would have labeled DNA (fig. 14.1). Figure 14.2 shows the chromosomes after a division in nonlabeled media. As expected, one chromatid of every figure is labeled and one is not.

In another kind of experiment, R. Kavenoff, L. Klotz, and B. Zimm demonstrated that *Drosophila* nuclei contained pieces of DNA of the size predicted from their DNA content, based on the premises that each chromosome contains one DNA molecule. They isolated the DNA and measured the size of the largest DNA molecules using the *viscoelastic* property of DNA, the rate at which stretched molecules relax. From other sources, primarily UV absorbance studies, it was estimated that the largest *Drosophila* chromosome had about 43×10^9 daltons of DNA. Results from the viscoelastic measurements indicated the presence of DNA molecules of between 38 and 44×10^9 daltons. Viscoelastic measurements of inversions, which changed the ratio of the arms but not the overall size of the chromosome, yielded similar results. However, a translocation that radically changed the size of the chromosome to 59×10^9 daltons resulted in an equivalent change in the viscoelastic estimates to between 52 and 64×10^9 daltons.

Figure 14.1

Radioactive labeling of a uninemic eukaryotic chromosome following semiconservative replication. Replication occurs first in the presence of ³H-thymidine and then in its absence. *Color* represents labeling. After the second round of replication, one chromatid of each chromosome is labeled whereas the other is not, confirming that there is only one DNA molecule per chromatid (uninemic).

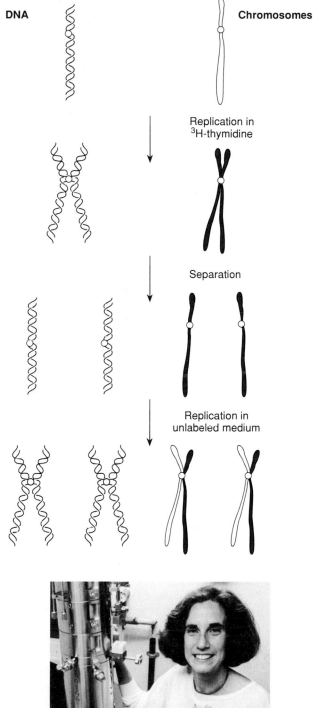

DNA   **Chromosomes**

Replication in
³H-thymidine

Separation

Replication in
unlabeled medium

Ruth Kavenoff (1944–   )
Courtesy of Dr. Ruth Kavenoff.

Figure 14.2

Second metaphase in hamster cells in culture after one replication in the presence of ³H-thymidine followed by one in cold medium. Cases in which the label apparently switches from one chromatid to the other are caused by sister chromatid exchanges (at arrows).

Source: G. Marin and D. M. Prescott, "The Frequency of Sister Chromatid Exchanges Following Exposure to Varying Doses of H³-Thymidine or X-ray," Reproduced from the Journal of Cell Biology *21 (1964): 159–167 by copyright permission of the Rockefeller University Press.*

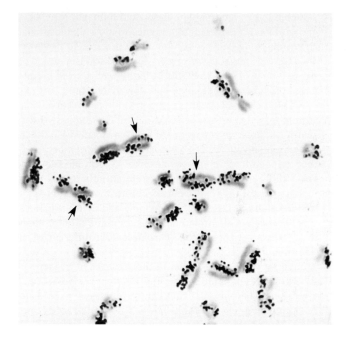

The conclusion from these studies is that the largest chromosome of *Drosophila,* and by extension every eukaryotic chromosome, contains a single DNA molecule running from end to end, encompassing both arms. The viscoelastic values were corroborated by carefully isolating and measuring the lengths of long DNA molecules, an especially difficult task given the propensity of DNA to break. The longest molecule that the investigators found was 1.2 cm long, equivalent to between 24 and 32×10^9 daltons (fig. 14.3), close to the predicted size. Thus the evidence is in complete concordance with the simple uninemic model of eukaryotic chromosome structure.

Nucleoprotein Composition

Nucleosome Structure

Since each eukaryotic chromosome consists of a single, relatively long piece of duplex DNA, the average diploid cell contains many of these long pieces of DNA. In order for chromosomes to be properly distributed to each daughter cell during mitosis and meiosis, they must be condensed into structures that are more easily managed. The mechanism used by eukaryotes is to wrap the DNA around "spools" of protein. Wrapping the DNA around these spools constitutes the first step in a series of coiling and folding processes that eventually result in the fully compacted chromosome that we see at metaphase.

BOX 14.1

HIGH-SPEED CHROMOSOMAL SORTING

In order to facilitate creating recombinant genomic libraries, for mapping purposes and for other reasons, it is valuable to be able to isolate individual human chromosomes. Purified chromosomes would make localizing genes, mapping, and sequencing more precise. To these ends, several methods have been developed to isolate chromosomes. Here we discuss a high-speed sorting method based on fluorescent staining and flow cytometry.

DNA can be treated with several fluorescent dyes. Chromosomes can then be recognized individually by their relative fluorescent intensities. The dyes Hoechst 33258 and chromomycin A3 are a valuable combination because they respond to different wavelengths of light and they bind DNA differently. Hoechst binds preferentially to DNA rich in adenine and thymine, whereas chromomycin binds preferentially to DNA rich in guanine and cytosine. Thus, since every human chromosome has a unique ratio of bases, the relative intensity of each chromosome is different when fluorescing.

Chromomycin fluoresces in the presence of a laser tuned to 458 nm, and Hoechst fluoresces in the presence of a UV laser. The chromosomes can be identified when their relative fluorescence in the two lasers is plotted, producing a *flow karyotype* (Box fig. 1). Modern flow cytometry techniques then allow these identified chromosomes to be isolated.

In practice, chromosomes are isolated in large numbers from cells that have been arrested in metaphase by treatment with colcemid, which inhibits spindle formation. These chromosomes are then isolated in buffer and treated with the two dyes. The chromosomes are then separated at high speed (200 per second) in a flow cytometry device (Box fig. 2). As the chromosome-containing buffer passes through the laser beams, identification is made. The liquid is then forced to form minute droplets (215,000 per second) by passing through a vibrator. Based on chromosome identification, specific droplets carrying the identified chromosomes are then charged, either positively or negatively, and passed between deflection plates. Positively charged droplets pass one way and negatively charged droplets pass the other way, thus allowing the simultaneous isolation of two different chromosomes. At a rate of 200 chromosomes per second, it

Box figure 1

Flow karyotype of human chromosomes at very high resolution, measured under low-speed sorting (15–35 chromosomes per second). The ordinate is Hoechst 33258 fluorescence intensity and the abscissa is chromomycin A3 fluorescence intensity. All chromosomes are resolved except numbers 9–12.

From J. W. Gray, et al., "High-Speed Chromosome Sorting" in Science, *238:323–329. Copyright 1987 by the AAAS.*

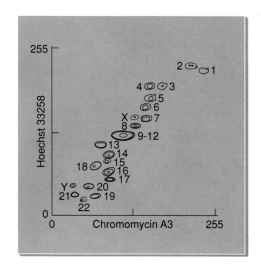

is possible to isolate 0.1 μg of DNA in less than an hour; 0.1 μg of DNA is adequate for library construction and represents about 5×10^5 average chromosomes.

The technique is not perfect. During isolation, debris and clumps of chromosomes are produced that cause contamination problems. Then, some chromosomes are so similar in their fluorescence as to be hard to separate. This is true, for example, for chromosomes 9 to 12. Also, chromosome 21 is hard to separate because its fluorescence tends to fall into the debris area.

Some of these problems, however, can be overcome by using hybrid cell lines of hamsters, for example, containing only one human chromosome. It is much easier to isolate the human chromosome from the hybrid line. Purity values of 90% are not unreasonable, with some in excess of 95%.

Interphase nuclei can be disrupted by being placed in a hypotonic solution such as water. When this happens, chromatin material is released. When this material is observed under the electron microscope, small particles, called **nucleosomes,** can be seen (fig. 14.4). These are the spools upon which the DNA is wrapped. They are made of **histone** proteins and associated DNA (table 14.1). The histones, a group of arginine- and lysine-rich basic proteins, have been relatively well characterized. They are especially well suited to bind to the negatively charged DNA. At first, geneticists thought that the selective binding of these proteins to DNA might be the mechanism of transcriptional control. We now know that histones are too homogeneous to act as selective control proteins (table 14.2).

Box figure 2

The flow cytometry device used to separate chromosomes at high speed. A buffer with chromosomes enters the device. Lasers cause fluorescence that is analyzed with the aid of the photomultiplier tubes. Droplet formation is induced, and, based on flow rate of 50 m/sec, appropriate drops are charged. Charged drops are then separated by charged deflection plates and collected. Uncharged droplets pass through.

From J. W. Gray, et al., "High-Speed Chromosome Sorting" in Science, *238:323–329. Copyright © 1987 by the AAAS.*

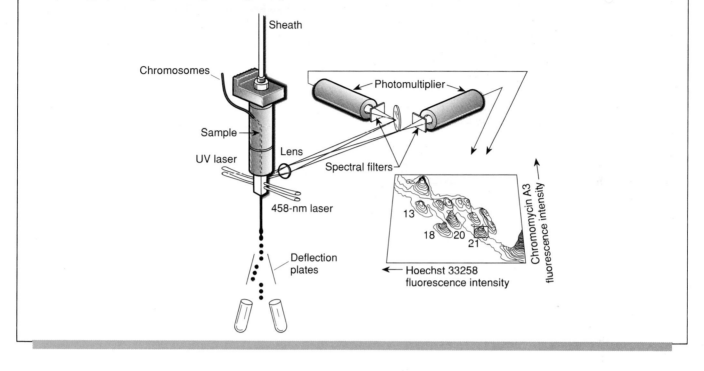

Figure 14.3

Autoradiograph of a 1.2-cm radioactive DNA molecule carefully isolated from *Drosophila melanogaster*. Drops of DNA solution were placed on microscope slides that were tilted to allow the DNA to spread slowly down the slide. A photographic emulsion was applied and later developed after a five-month exposure period.

From Ruth Kavenoff, Lynn C. Klotz, and Bruno H. Zimm, Symposia on Quantitative Biology *(Cold Spring Harbor) 38 (1973):4.*

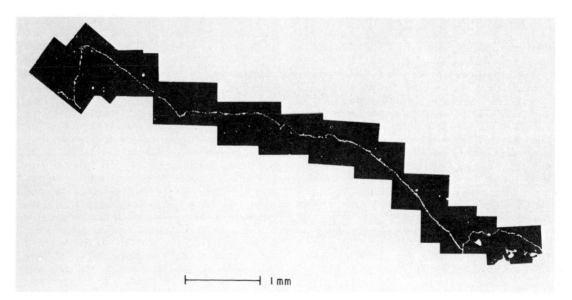

Figure 14.4

Electron micrograph of chromatin fibers. Photo shows nucleosome structures (*spheres*) and connecting strands of DNA called linkers. The bar is 100 nm.

Source: D. E. Olins and A. L. Olins, "Nucleosomes: The Structural Quantum in Chromosomes," American Scientist 66 (November 1978): 704–711. Reproduced by permission.

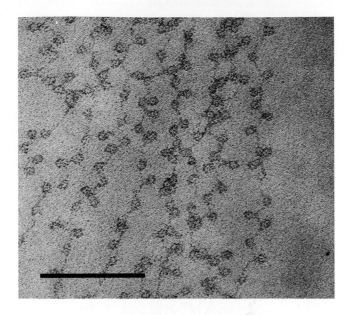

TABLE 14.1 The Constituency of Calf Thymus Chromatin

Constituent	Relative Weight*
DNA	100
Histone proteins	114
Nonhistone proteins	33
RNA	1

*Weight relative to 100 units of DNA.

When chromatin is treated with micrococcal nuclease, individual nucleosomes can be isolated, indicating that the DNA between nucleosomes is accessible to digestion. If nuclease digestion continues, all unprotected DNA is digested, leaving only DNA protected from digestion by its interaction with the histones. The results of these studies indicate that a length of 146 base pairs (bp) of DNA, the core DNA, is intimately associated with the nucleosome, and another 50 to 75 bp, depending on species, connects the nucleosomes (linker DNA; fig. 14.5). When the quantities of the various histones were measured, there were two each of histones H2A, H2B, H3, and H4 per nucleosome and only one molecule of histone H1. Reconstitution and degradation studies have indicated that histone H1 is not a necessary component in the formation of nucleosomes. A current model has histone H1 associated with the linker DNA as it enters and emerges from the nucleosome (fig. 14.6). Since the length of DNA involved in a nucleosome is short, and since the arrangement of histones is so regular, nucleosomes are clearly nonspecific arrangements of the DNA. Nucleosomes, then, are a first-order packaging of DNA; they reduce its length and undoubtedly make the coiling and contraction required during mitosis and meiosis more efficient.

Although the bulk of eukaryotic DNA seems to be arranged in nonspecific nucleosome binding, there are regions of the DNA, known as **nuclease hypersensitive sites,** that appear to be nucleosome free. These sites, usually multiples of a nucleosomal region of about 200 bp, are particularly sensitive to digestion by different nucleases. When these regions are isolated, they usually have sequences indicating functions in the replication, transcription, or other activities of DNA. For example, numerous promoter regions in *Drosophila,* mouse, and human DNA are in nuclease hypersensitive sites. Hence some specific DNA sequences are kept free of nucleosomes, and these sequences appear to be those that are recognized by various enzymes such as RNA polymerase. How these regions are recognized and kept free of nucleosomes is

TABLE 14.2 Composition of Histones

Fraction	Class	Number of Amino Acids	Percentage of Basic Amino Acids
H1	Very lysine rich	213	30
H2A	Lysine, arginine rich	129	23
H2B	Moderately lysine rich	125	24
H3	Arginine rich	135	24
H4	Arginine, glycine rich	102	27

Figure 14.5

The eukaryotic chromosome is associated with histone proteins to form nucleosomes. The protein core is wrapped with two loops of DNA and connected with a length of DNA called a linker.

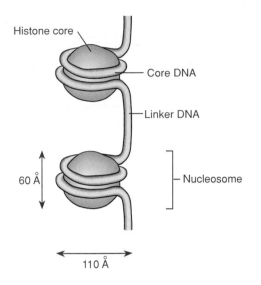

Figure 14.6

Nucleosome structure. (*a*) Schematic comparison of the eight histones comprising the nucleosome in salt solution. A dimer consists of an H2A and an H2B histone molecule; a tetramer consists of two each of H3 and H4 histones. DNA fits in surface grooves on the more compacted structure found in physiological conditions (*b*). The diagram in (*c*) shows the presumed position of the H1 histone.

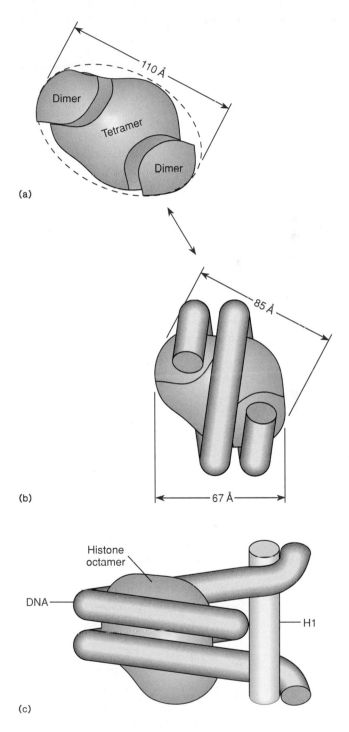

presently unknown. It is, however, known from current research that DNA replication proceeds through nucleosomed DNA without displacing the histones from the DNA.

Higher-Order Structure of Chromatin

Since the nucleosome has a width of only 110 Å, and metaphase chromosomes appear to be constructed of a fiber having a diameter of about 2,000 Å (fig. 14.7), there are several additional levels of chromatin compaction leading to the metaphase chromosome. Various experiments, involving changing the ionic strength that the chromatin is subjected to, indicate that the 110-Å DNA spontaneously forms a 300-Å, solenoid-like fiber with increased ionic strength. It seems that this fiber results from the coiling of the nucleosomal DNA (fig. 14.8). This 300-Å fiber is not, however, the final form of the DNA. We can account for the contraction of the 300-Å fiber to the 2,000-Å fiber found in metaphase chromosomes by the formation of a second solenoid-like structure from the winding of the 300-Å fiber (fig. 14.9).

If the histones are removed from a chromosome, the DNA billows out, leaving a proteinaceous structure termed a **scaffold** (fig. 14.10). This scaffold structure is formed from **nonhistone proteins,** which are composed of twelve to twenty or more types of proteins with very limited heterogeneity. Thus the major nonhistone proteins are probably also involved in chromosome structure rather than in genetic control. However, the analytical methods used are incapable, for the most part, of separating proteins that make up less than 1% of the chromatin protein. Thus gene expression may be controlled by proteins that exist in very small quantities in the chromatin.

Figure 14.7

Chinese hamster chromosome. Note the fibers making up the chromosome; they are approximately 2,000 Å in diameter. Magnification 11,800×.

Source: Courtesy of Dr. Hans Ris.

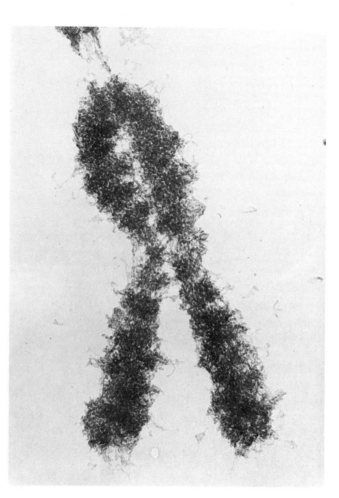

Figure 14.8

Solenoid model for the formation of the 300-Å chromatin fiber. Nucleosomal DNA wraps in a helical fashion, forming a hollow core.

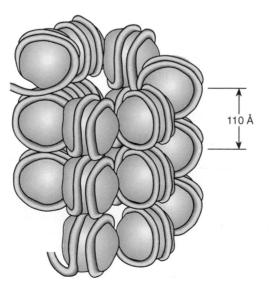

110 Å

Figure 14.9
The 2,000-Å fiber of the eukaryotic chromosome is a solenoidlike structure formed by the coiling of the 300-Å fiber, which itself is a solenoid.

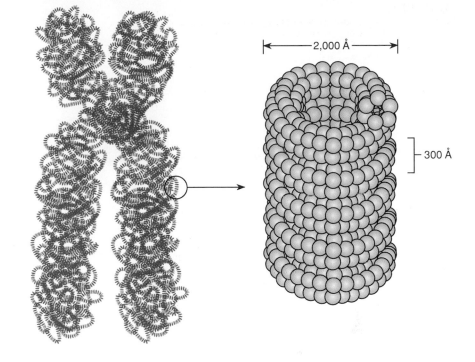

Figure 14.10
Scaffold protein. When the histones are removed from a eukaryotic chromosome, a fibrous scaffold remains. The DNA loops out from this scaffold. The bar is 2 μm.

J. Paulson and U. Laemmli, "The Structure of Histone Depleted Metaphase Chromosomes," Cell 12 (1977):817–828. Cell is copyrighted by the MIT Press. Micrograph courtesy of James R. Paulson.

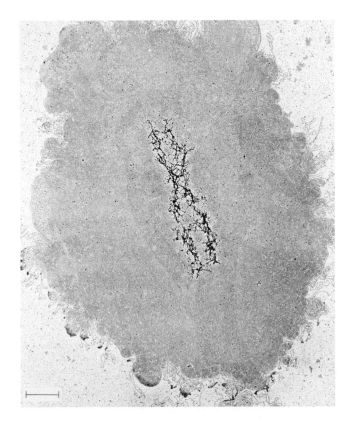

Polyteny, Puffs, and Balbiani Rings

The salivary glands, as well as some other tissues of *Drosophila* and other diptera, contain giant, banded chromosomes (fig. 6.12) that are the result of the synapsis of homologues followed by replication of the chromosomes without cell division (endomitosis). These chromosomes consist of more than one thousand copies of the same chromatid and appear as alternating dark bands and lighter interband regions. The dark bands are referred to as *chromomeres*. Also seen are diffuse areas referred to as **chromosome puffs** (fig. 14.11). Chromosome puffs are also referred to as **Balbiani rings,** which were originally defined as puffs specifically in the midge, *Chironomus,* whose polytene chromosomes were discovered by E. G. Balbiani in 1881. Currently, the term is used synonymously with all puffs, or at least the larger puffs, in all species with polytene chromosomes.

The structure of the polytene chromosome can be explained by the diagram in figure 14.12. Dark bands (chromomeres) are due to tight coiling of the 300-Å fiber; light interband regions are due to looser coiling. The figure also shows how chromosome puffs would come about by the unfolding of fibers in regions of active transcription.

Staining with reagents specific for RNA, such as toluidine blue, or autoradiography with tritiated (^{3}H) uridine, has been used to demonstrate that there is active transcription going on in the puffs but not in neighboring regions of the polytene chromosomes. The mRNA isolated from cells with puffs has also been shown to hybridize only to the puffed regions of the chromosomes. Thus these regions of the DNA are complementary to the

Figure 14.11

A chromosome puff on the left arm of chromosome 3 of the midge,
Chironomus pallidivittatus
Jan-Erik Edström, et al. 1982. Developmental Biology *91:131–137, figure 1B.*

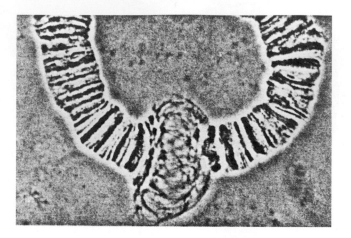

Figure 14.12

Polytene chromosome with bands and a puff. Three of the
approximately one thousand synapsed chromatids are shown
diagrammatically on the *right*. Each chromatid is in the form
of a 300-Å fiber.

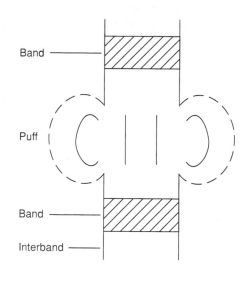 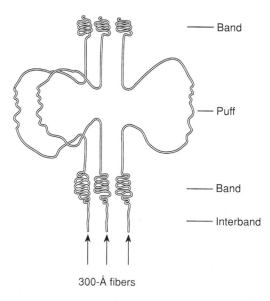

mRNA (fig. 14.13) and represent areas of active transcription. With modern recombinant DNA techniques, it has also been shown that each puff probably represents the transcription of only one gene.

Puffs generally fall into four categories. *Stage-specific puffs* appear during a certain stage of development, such as molting. *Tissue-specific puffs* are active in one tissue but not another. (In dipteran larvae, tissues other than the salivary glands, such as the midgut and Malpighian tubules, have polytene chromosomes.) *Constitutive puffs* are active almost all the time in a specific tissue.

And *environmentally induced puffs* appear after some environmental change, such as heat shock (fig. 14.14). In *Drosophila,* about 80% of the puffs are stage specific; in *Chironomus,* only about 20% are. For example, at the time of molt in insects, the hormone ecdysone is secreted by the prothoracic gland. At the same time, many puff patterns change (fig. 14.15). Similar changes in puff patterns can be induced by the injection of ecdysone. Hence molting, a stage-specific developmental sequence, is related to a sequential transcription sequence in the chromosomes.

Figure 14.13
Hybridization at a *Chironomus tentans* salivary gland chromosome
puff. The chromosomal DNA is hybridized with labeled RNA
transcribed from the locus whose activity is forming the puff.

*Reprinted by permission from B. Lambert, "Repeated DNA sequences in a
Balbiani Ring," Journal of Molecular Biology 72 (1972):65–75. Copyright by
Academic Press Inc. (London) Ltd.*

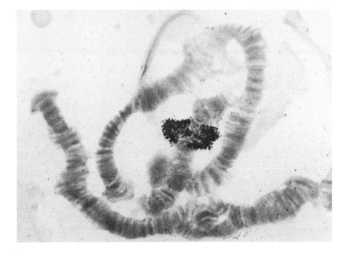

Figure 14.14
Puff 4–81B of the salivary gland in *Drosophila hydei* is induced by
heat shock (37° C for one half hour). (*a*) Normal activity.
(*b*) Temperature shock *in vitro*, resulting in the puff.

*Source: H. D. Berendes et al. 1965. "Experimental Puffs in Salivary Gland
Chromosomes of* Drosophila Hydei*," Chromosoma (Berl.) 16:35–46,
Figure 4a-b, page 40.*

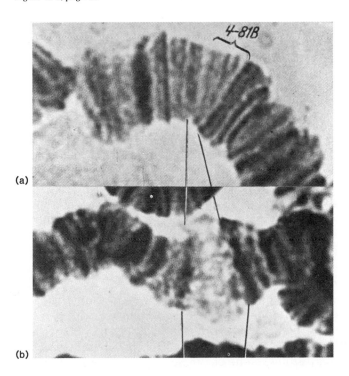

Figure 14.15
Puff patterns on a segment of a salivary gland chromosome of
Chironomus tentans during molt. As time proceeds, puffs appear and
disappear and change in size.

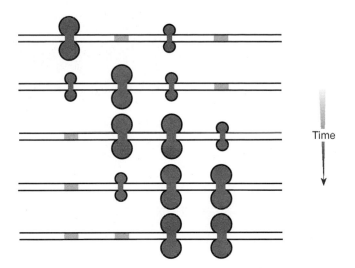

Lampbrush Chromosomes

Lampbrush chromosomes occur in amphibian oocytes and
are so named because their looped-out configuration has
the appearance of a brush for cleaning lamps, now a relatively uncommon household item (fig. 14.16). The loops
of the lampbrush chromosomes are covered by an RNA
matrix and are undoubtedly the sites of active transcription. Presumably, the loops are unwindings of the single
chromosome, similar to the unwindings in the polytene
chromosome shown in figure 14.12. Thus under certain
circumstances, such as in polytene chromosomal puffs and
in lampbrush chromosomes, active transcription can be
seen. Since only certain bands puff at any one moment in
polytene chromosomes and the loops of lampbrush chromosomes are of various sizes (with some regions not looped
at all), we have evidence of specific transcription with no
indication, so far, of the nature of the control of that transcription.

Chromosomal Banding

There are several chromosome-staining techniques that
reveal consistent banding patterns. By means of these patterns, all of the human chromosomes can be differentiated
(see fig. 5.1). Of possibly greater importance is the fact
that these staining techniques have provided some insight
into the structure of the chromosome. The techniques for
staining the C, G, and R chromosome bands will serve as
an illustration.

G-bands are obtained with **Giemsa stain,** which is a
complex of stains specific for the phosphate groups of
DNA. Treatment of fixed chromatin with hot salts brings

Figure 14.16

Lampbrush chromosome of the newt, *Triturus viridescens.*
Centromere is at the *arrow;* the two long homologues are held
together by three chiasmata. Magnification 238×.

*Source: Joseph G. Gall, figure 2 in D. M. Prescott, ed., Methods in Cell
Physiology, vol. 2 (New York: Academic Press, 1966), p. 39. Reproduced by
permission.*

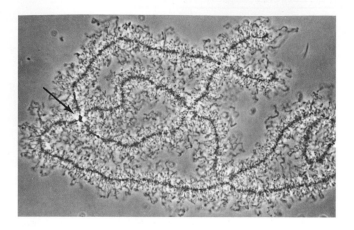

Figure 14.17

Model of eukaryotic (mammalian) chromosomal banding. G-bands are
chromomere clusters, which result from the contraction of smaller
chromomeres, which, in turn, result from looping of the 300-Å fiber.

Reproduced with permission, from the Annual Review of Genetics, *Volume 12,*
© *1978 by Annual Reviews Inc.*

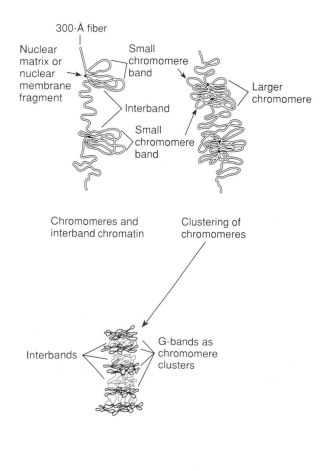

David E. Comings (1935–   )
Courtesy Dr. David E. Comings.

out the G-bands. Giemsa stain enhances banding that is
already visible in mitotic chromosomes. The banding pattern is caused by an arrangement of chromomeres. Under
careful observation, the major G-bands can be seen to
consist of many smaller chromomeres. This banding appearance has led D. Comings to suggest the mechanism
of chromosome folding shown in figure 14.17.

C-bands are Giemsa-stained bands after treatment of
the chromosomes with NaOH. The "C" is for "centromere," because these bands represent constitutive heterochromatin surrounding the centromeres (fig. 14.18). The
DNA is also usually satellite rich. **Satellite DNA** differs
in buoyant density from the major portion of cellular DNA.
When eukaryotic DNA is isolated and centrifuged in CsCl,
forming a density gradient, the majority of the DNA forms
one band in the gradient at a single buoyant density. The
buoyancy is determined by the G-C content of the DNA.
However, smaller secondary bands are also usually present
indicating regions of DNA having sequences different from
the majority of the cell's DNA (fig. 14.19). DNA isolated
this way is referred to as satellite DNA because of the
secondary, or satellite, bands they form in the density gradient. As we will see, this DNA is found primarily around
centromeres and consists of numerous repetitions of a short
sequence.

R-bands are visible with a technique that stains the
regions between G-bands. Since the dark-light pattern is
the opposite of the G-band pattern, these bands are called
reverse bands.

From the information supplied by these staining techniques, D. Comings distinguished between three basic
chromatin types: euchromatin, constitutive heterochromatin, and intercalary heterochromatin (table 14.3). Presumably, the only chromatin involved in transcription is
euchromatin. Constitutive heterochromatin surrounds the
centromere and is rich in satellite DNA. **Intercalary
heterochromatin** is dispersed. Thus it becomes apparent
that the eukaryotic chromosome is a relatively complex
structure.

Figure 14.18

(a) C banding of chromosomes from a cell in the bone marrow of the house mouse, *Mus musculus*. The *arrow* indicates that the Y chromatids have already separated into two chromosomes. (b) Yellow fluorescence indicates probe of satellite DNA in human chromosomes (centromeres).

(a) B. Vig, "Sequence of Centromere Separation: Role of Centromeric Heterochromatin," Genetics 102 (1982): 795–806. (b) Photograph Courtesy of Oncor, Inc. Gaithersburg, Maryland.

(a)

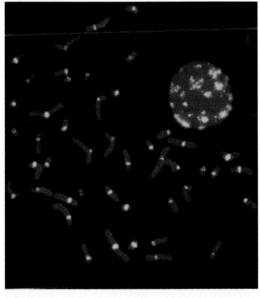

(b)

Centromeres and Telomeres

Two regions of the eukaryotic chromosome have specific functions—the centromere and the telomeres. The centromere is involved in chromosome movement during mitosis and meiosis, whereas the telomeres terminate the chromosomes. As we pointed out in chapter 3, the

Figure 14.19

Satellite DNA in *Drosophila virilis*. The quantity of DNA is graphed against the buoyant density, resulting in four peaks of DNA. The large peak (at *left*) is the major DNA component of the cell; the other three bands are satellite DNA.

From J. Gall, et al., Cold Spring Harbor Laboratory Symposia on Quantitative Biology, 38:417–421. Copyright © 1974 Cold Spring Harbor Laboratory Press, Cold Spring Harbor, NY. Reprinted by permission.

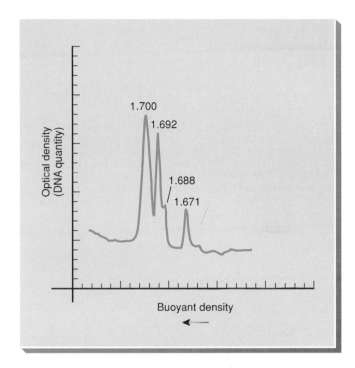

terms centromere and kinetochore are frequently used interchangeably. The kinetochore is technically the interface between the constriction in the chromosome (the centromere) and the microtubules of the spindle. The kinetochore of higher organisms (e.g., mammals) contains proteins and some RNA. Microscopically, it is a trilaminar structure, attached to chromatin at the inner layer and to microtubules at the outer layer (fig. 14.20).

Most of our knowledge of the genetics of centromeres has come from work in yeast (*Saccharomyces cerevisiae*). Most artificially created yeast plasmids are not maintained by cells because they are lost during mitosis. However, plasmids can be isolated that do replicate normally during cell division. Presumably, they contain centromeres (*CEN* regions), allowing them to replicate and move in synchrony with the host's chromosomes. Through further genetic engineering it has been possible to isolate smaller and smaller regions that can serve as centromeres. After sequencing the centromeres of ten yeast chromosomes, it was possible to conclude that the centromere from yeast is about 220 bp long with three consensus regions (fig. 14.21).

The 220-bp length of *CEN* regions of yeast chromosomes is about 150 to 200 Å, the same as the diameter of a microtubule (200 Å), indicating that only one microtubule attaches to each centromere during mitosis or

TABLE 14.3   The Three Major Types of Chromatin in Eukaryotic Chromosomes

	Centromeric Constitutive Heterochromatin	Intercalary Heterochromatin	Euchromatin
Relation to Bands	In C-bands	In G-bands	In R-bands
Location	Usually centromeric	Chromosome arms	Chromosome arms
Condition during Interphase	Condensed	Condensed	Usually dispersed
Genetic Activity	Inactive	Probably inactive	Usually active
Relation to Chromomeres	Centromeric chromomere	Intercalary chromomeres	Interchromomeric

Figure 14.20

The kinetochore of a metaphase chromosome of the rat kangaroo. *IL, ML,* and *OL* refer to *inner, middle,* and *outer* layers, respectively, of the kinetochore. Note the microtubules attached to the kinetochore and the large mass of dark-staining chromatin, making up most of the figure. Magnification 30,800×.

From B. R. Brinkley and J. Cartwright, Jr. 1971. J. Cell Biology *50:416–431.*

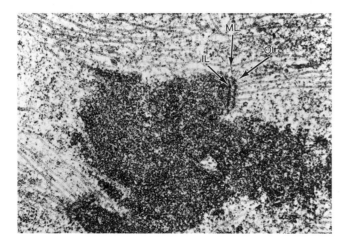

meiosis in a yeast cell. Higher eukaryotes have more microtubules attached during these division processes (fig. 14.20; see also fig. 3.9). Yeast centromeres differ from those in higher eukaryotes in other ways also. Specifically, no centromeric constriction is visible in yeast chromosomes nor does yeast DNA appear to have centromeric heterochromatin (satellite DNA). Hence the picture we have of centromere organization based on yeast is simple compared with higher eukaryotes. The yeast centromeres also appear to be nucleosome free (nuclease hypersensitive sites), and bound by proteins, as yet unisolated, that bind microtubules, and thus would be defined as kinetochores (fig. 14.22).

Since eukaryotic chromosomes are linear, each has two ends, referred to as **telomeres,** that not only mark the termination of the linear chromosome, but also have several specific functions (fig. 14.23). Telomeres must prevent the chromosome ends from acting in a "sticky" fashion, the

way that broken chromosome ends act (see chapter 8). Telomeres must also prevent the ends of chromosomes from being degraded by exonucleases and must allow chromosome ends to be properly replicated.

All telomeres so far isolated are repetitions of sequences of five to eight bases. In human beings, the telomeric sequence is TTAGGG, repeated 250 to 1,000 times at the end of each chromosome. The human telomere was discovered by R. Moyzis and his colleagues when they probed the highly repetitive segment of human DNA. (Highly repetitive DNA, as its name implies, consists of numerous copies of a single sequence, and usually comprises the satellite components of the cell's DNA; see next section.) When a probe for this sequence was applied to human chromosomes, it was found at the tip of each chromosome in roughly the same quantity (fig. 14.24). This is a highly conserved sequence, found in all vertebrates studied as well as unicellular trypanosomes. Similar sequences are found in various other eukaryotes (table 14.4), the first sequence being isolated by Blackburn and Gall in 1978.

When a linear DNA molecule is being replicated, the $3' \rightarrow 5'$ strand can be replicated to the end (see chapter 9). The $5' \rightarrow 3'$ strand, however, is replicated with RNA primers that are then degraded, leaving a short gap on the progeny strand (fig. 14.25). It is always the G-rich strand of telomeric DNA that ends up single stranded, forming a $3'$ overhang of twelve to sixteen nucleotides. Thus the normal replication process of a linear DNA molecule leaves an incomplete terminus. Hence we suspect a unique mechanism for the replication of telomeres.

Telomeric sequences appear to be added de novo without DNA template assistance, by an enzyme called **telomerase,** discovered by E. Blackburn and her colleagues. This was seen when telomeres from another species were engineered into yeast cells. After a cell cycle, the yeast telomeric sequence had been added on at the chromosome ends. It seems that the number of repeats of the telomeric sequence is variable, capable of change with each cell cycle. Some unknown mechanism, however, keeps the repeat number within limits.

Figure 14.21

Consensus sequence for the three regions (I–III) of ten yeast centromeres. Pu represents any purine, Py represents any pyrimidine, and X represents any base. The *arrows* are over inverted repeat sequences.

Source: Data from L. Clarke and J. Carbon, "The Structure and Function of Yeast Centromeres" in Annual Review of Genetics, *19:29–56, 1985.*

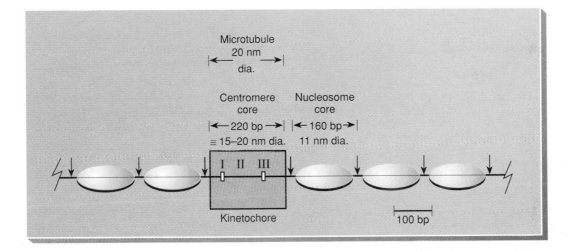

        I   II   III

PuTCACPuTG —— 78–86 bp —— TGTTTPyTGXTTTCCGAAAXXXXAAA
                         91–95% AT

Figure 14.22

Schematic view of a yeast centromeric region. The *arrows* are the nuclease hypersensitive sites. A microtubule is about the same width as the centromeric region.

Reproduced, with permission, from the Annual Review of Genetics, *Volume 19. © 1985 by Annual Reviews Inc.*

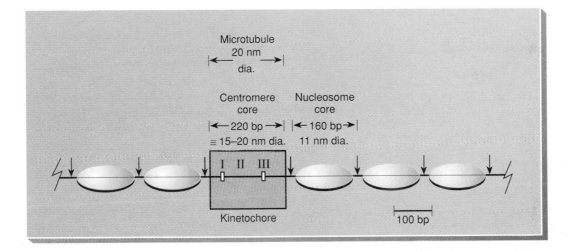

Figure 14.23

Telomere of polytene chromosome from the salivary gland of *Chironomus thummi,* using the technique of laser-scanning differential contrast imaging. Chromatin structure, possibly looping, is visible.

From Donna J. Arndt-Jovin, Michel Robert Nicoud, Stephen J. Kaufman, Thomas M. Jovin, "Fluorescence Digital Imaging Microscopy in Cell Biology, Science *(1985) 230:247–256 fig. 7C, pg. 253. Copyright 1985 by the AAAS.*

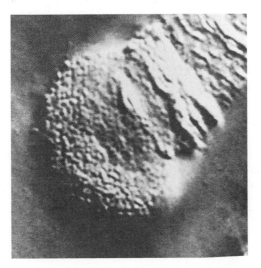

Figure 14.24

The human genome probed for the telomeric sequence, TTAGGG, using fluorescent staining techniques. The *yellow dots* at the tips of the chromosomes are the probes.

From Robert K. Moyzis et al., 1988. Proceedings of the National Academy of Science. USA *85:6622–6626, Figure 4, left.*

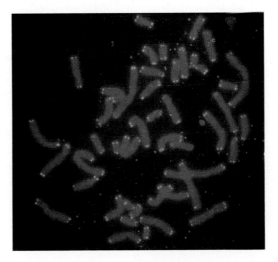

TABLE 14.4 Telomeric Sequences in Eukaryotes. The G-Rich Strand of the Double Helix Is Shown

Organism	Telomeric Repeat
Human beings, other mammals, birds, reptiles	TTAGGG
Trypanosomes	TTAGGG
Holotrichous ciliates (*Tetrahymena*)	GGGGTT
Hypotrichous ciliates (*Stylonychia*)	GGGGTTTT
Yeast	GT, GGT, and GGGT

Figure 14.25

Removal of final primers after the replication of linear DNA creates single-stranded ends.

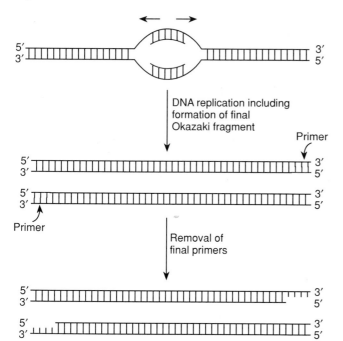

DNA replication including formation of final Okazaki fragment

Primer

Primer

Removal of final primers

Elizabeth H. Blackburn (1948–   )
Courtesy of Dr. Elizabeth H. Blackburn.

Another aspect of telomeric function seems to be in their secondary structure. Hairpin configurations seem to be a regular part of telomere structures as seen with several methods of DNA analysis, including nondenaturing polyacrylamide-gel electrophoresis. The hairpins are especially novel since they seem to involve guanine to guanine base pairing, a non-Watson and Crick configuration. Constitutive single-strand breaks also seem to be a part of telomere structure. Their purpose is unknown, although their formation at the ends of telomeres could protect the telomere.

When telomerase was isolated by Blackburn and her colleagues, they discovered that a segment of RNA, about 160 bp, is an integral part of the enzyme. Under normal circumstances, that RNA has a region that is complementary to the G-rich repeat of the telomeric DNA sequence of the species. After careful experimentation, including modifying the gene for the telomerase RNA, Blackburn and her colleagues concluded that telomerase uses the RNA as a template from which to add telomere repeats to the ends of chromosomes. Telomerase is thus a reverse transcriptase, using RNA nucleotides as a template to polymerize DNA nucleotides.

In a preliminary model, Blackburn and her colleagues proposed that the first step in telomere extension is hybridization of the 3′ end of the telomere with the RNA component of telomerase (fig. 14.26a). Then, with the telomerase RNA as a template, the 3′ end of the telomerase is extended (fig. 14.26b). Finally, a translocation step takes place in which the telomere is displaced in reference to the RNA, returning to the configuration at the beginning of the process (fig. 14.26c). The single-stranded C-rich strand is then synthesized with DNA polymerase I and DNA ligase. In this way, the human telomere is extended to about 10 kb (10,000 bases).

Telomere length may simply be a compromise between synthesis processes that elongate the telomere and exonuclease processes that shorten the telomere. Currently, investigators are looking into the role of telomerase length in senescence (aging) of cells and tumor formation. Perhaps the length of the telomere is an indicator of cell age and may be a clue to the general aging process, although that notion is quite speculative at the moment.

DNA Repetition in Eukaryotic Chromosomes

DNA-DNA Hybridization

We can investigate the repetitiveness of the DNA within the eukaryotic genome, a concept developed by R. Britten and his colleagues. The nature of the DNA arrangement in eukaryotic chromosomes can be studied by using the technique of **DNA-DNA hybridization.** When DNA is heated, it denatures or unwinds into single strands. When it is cooled, it renatures. The rate of renaturation depends

Figure 14.26
Telomerase extends telomeres using telomerase RNA as a template.
Gap filling by DNA polymerase I and ligase complete the double
helix.
Source: Data from Shippen-Lentz and Blackburn, Science, *247:550, 1990.*

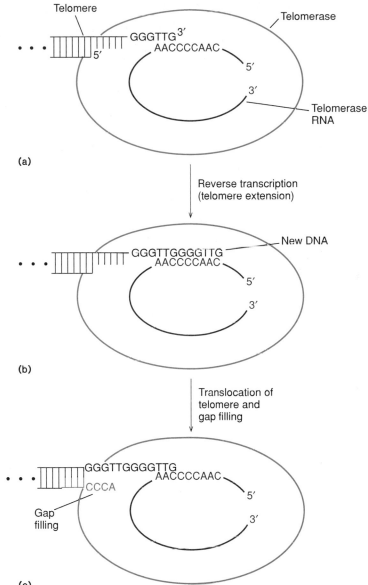

Roy J. Britten (1919–   )
Courtesy of Dr. Roy J. Britten.

on the concentration of nucleotide strands and their sequences, given that the temperature of renaturation is kept
constant and the sample is broken into small, uniform
pieces.

Cot Curves

If C_o is the original concentration of single-stranded
(denatured) DNA in moles per liter and t is elapsed time
in seconds, then their product provides a scale of renaturation called $C_o t$ (or cot). When **cot values** are plotted
against the quantity of remaining single-stranded DNA,

Figure 14.27

Cot curves for renaturation of DNA. At the *top,* "nucleotide pairs" refers to unique sequences, the number of base pairs in the unique sequence of the DNA (or RNA). The *arrows* are over the $cot_{1/2}$ values.

From R. Britten and D. Kohne, "Repeated Sequences in DNA" in Science, *161:529–540. Copyright 1968 by the AAAS.*

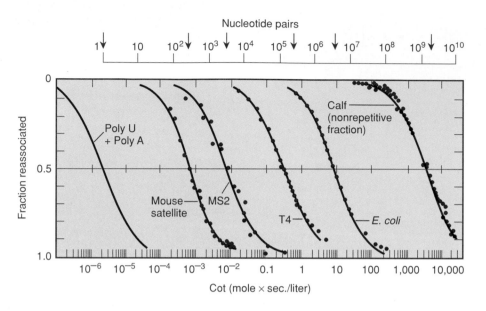

Figure 14.28

Cot curve for mouse DNA. *Arrows* indicate approximate $cot_{1/2}$ values of the three apparent segments that make up this single cot curve: highly repetitive (*a*), moderately repetitive (*b*), and nonrepetitive (unique) DNA (*c*).

From Betty L. McConaughy and Brian J. McCarthy, "Related Base Sequences in the DNA of Simple and Complex Organisms. VI. The Extent of Base Sequence Divergence among the DNAs of Various Rodents" in Biochemical Genetics, *4:425–446. Copyright © 1970 Plenum Publishing Corporation, New York, NY.*

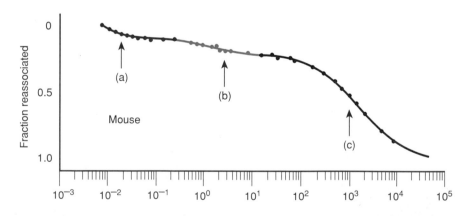

the curve (cot curve or cot plot) is informative. The midpoint, referred to as the **$cot_{1/2}$** value, estimates the amount of homology within the DNA or, more precisely, the length of unique DNA in the sample. By **unique DNA,** we mean the length of DNA in a sample that has no repeated sequences. We assume, for example, that the *E. coli* chromosome is virtually unique. In figure 14.27, fractions of reassociated (renatured) DNA are plotted against cot values.

As indicated by this figure, the samples of DNA shown produce cot curves of approximately the same shape, although DNAs of different complexities (different unique lengths) are located at different points along the abscissa. The farther to the right, the more slowly the single-stranded nucleic acids reassociate. Furthermore, we can relate the $cot_{1/2}$ value to length of unique sequence. For example, *E. coli,* with a chromosome of 4.2×10^6 base pairs, has a $cot_{1/2}$ of about 10. Poly U + poly A strands

Figure 14.29

Frequency of repetition of mouse DNA segments. *Dotted regions* are less certain. Peaks for satellite, intermediately repetitive, and unique DNA are seen.

From R. Britten and D. Kohne, "Repeated Sequences in DNA" in Science, *161:529–540. Copyright 1968 by the AAAS.*

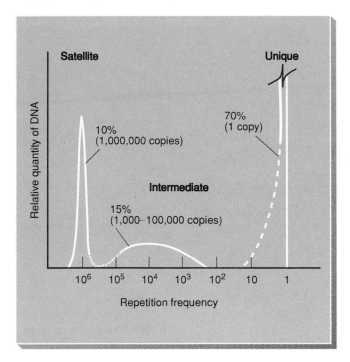

have a unique length of only one base pair, and this nucleic acid has a $cot_{1/2}$ value of about 2×10^{-6}. From these values and values for the nucleic acids of phages MS2, T4, and others, the "nucleotide pairs" axis above the curves of figure 14.27 was obtained.

Unique and Repetitive DNA

Eukaryotic satellite DNA has a low $cot_{1/2}$ value. Since the basic length of unique satellite DNA is about two hundred base pairs (fig. 14.27), and since the quantity of this DNA per cell is much more than this, satellite DNA must be a highly **repetitive DNA,** containing many copies of the same nucleotide sequence. Given the quantity of satellite DNA per cell, there must be more than one million repetitions of the sequence of two hundred nucleotide pairs.

The cot curve for the whole genome of a eukaryote is different from the curves of figure 14.27. The cot curve of mouse DNA is shown in figure 14.28. Note that this curve is actually made up of three separate cot curves whose $cot_{1/2}$ values are indicated by arrows. On the basis of the $cot_{1/2}$ values and the proportion of the genome that each segment comprises (ordinate), the degree of repetitiveness of each segment can be determined. There is a highly repetitive segment (satellite), a segment that is of intermediate repetitiveness, and a segment of unique DNA. These segments make up about 10%, 15%, and 70%, respectively, of the total mouse DNA (fig. 14.29).

The highly repetitive satellite DNA is found primarily around centromeres and telomeres; it is not known to be transcribed. The unique DNA makes up the structural genes; much of it is transcribed.

Intermediately Repetitive DNA

Intermediately, or moderately, repetitive DNA in eukaryotes occurs in at least three categories in the genome: (1) dispersed, probably nontranscribed DNA such as the Alu family (see below); (2) transcribed genes in many copies that are virtually identical, such as rRNA and histone genes; and (3) transcribed genes in many copies that have diverged from each other, such as antibody, collagen, and globin genes. We use the term **gene family** to refer to genes that have arisen by duplication, with or without divergence, from an ancestral gene.

In many mammals, a large portion of the intermediately repetitive DNA is composed of many copies of a short sequence dispersed throughout the genome. For example, in human beings, there are about three hundred thousand copies of a 300–bp sequence. Because this sequence is cleaved by the restriction endonuclease *Alu* I, it is called the **Alu family.** The exact role of this DNA is not clear at the moment. The possibility exists that some of the members of this family are transcribed into small, 7S, RNAs. In some cases, larger RNAs contain Alu sequences. Alu may be involved in the numerous origins of DNA replication along eukaryotic chromosomes (see fig. 9.24) or it may be involved in creating secondary structure in mRNAs.

Several types of genes create a product that is needed in such large quantity that one copy of the gene could not fulfill the cell's needs. We are familiar with the nucleolus, the site of the ribosomal RNA genes (see fig. 10.17). Human beings have about two hundred copies of the major rRNA gene and about two thousand copies of the 5S rRNA gene. Fruit flies have about two hundred and one hundred copies, respectively, of the two genes.

In some cases, the normal number of multiple copies of a gene are not enough. The cell must then resort to **gene amplification,** a process whereby the cell increases the number of copies of the gene. For example, during oogenesis, rRNA genes (rDNA) are often amplified. In *Xenopus,* rDNA is amplified about one thousand times, which allows an oocyte to accumulate about 10^{12} ribosomes. The amplified DNA is in the form of small, circular, extrachromosomal molecules of DNA. Several models have been proposed as to how cells actually amplify their DNA. One model relies on unequal crossing over (as in *Bar* eye in *Drosophila*), whereas another model is based on unscheduled extra DNA replication in a region followed by recombinational events that generate linear and circular forms of the excess DNA. It is not clear at the present moment which model is correct.

Figure 14.30

The arrangement of histone genes (*color*) within the five-gene cluster in sea urchins and fruit flies. *Arrows* indicate the direction of transcription. Spacer DNA (*black*) separates the genes.

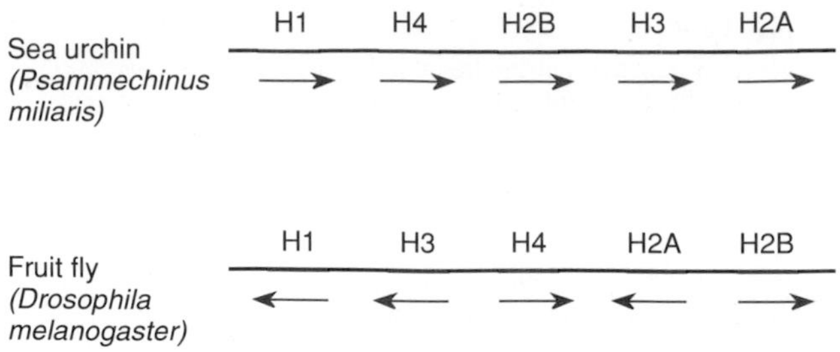

Sea urchin
(*Psammechinus
miliaris*)

H1	H4	H2B	H3	H2A
→	→	→	→	→

Fruit fly
(*Drosophila
melanogaster*)

H1	H3	H4	H2A	H2B
←	←	→	←	→

TABLE 14.5 Types of Human Hemoglobin

Type	Generally When Present	Composition
Embryonic	Up until 8 weeks of gestation and beyond	$\zeta_2\epsilon_2$
Fetal (Hb F)	8 weeks to birth	$\alpha_2\gamma_2$
Adult (Hb A)	Just before birth and beyond	$\alpha_2\beta_2$
Adult (Hb A$_2$)	In immature cells	$\alpha_2\delta_2$

Note: Subscripts refer to the numbers of subunits present.

In addition to rRNA genes, other genes are repeated, assuring adequate gene products. The number and location of repeated genes is usually discovered by hybridization studies using probes, similar to the way that telomeric DNA was shown to be at the tips of the chromosomes (fig. 14.24). Repeated genes include the genes for tRNAs and histones. The average tRNA is repeated about a dozen times in *Drosophila*. Human beings have over thirteen hundred copies of tRNA genes in the haploid genome. In many species, the five histone genes form a repeated cluster, although each gene is transcribed independently (fig. 14.30), as compared with prokaryotic operons that are transcribed as a unit. The arrangement of histone genes may be more complex in higher forms. There are indications in mammals that histone genes may lie in small groups or even as individual genes.

Several types of genes occur in similar but not identical forms—that is, an original gene was duplicated but, unlike histone or rRNA genes, the copies diverged in function. These gene families include globin genes, immunoglobulin genes (see chapter 15), chorion protein (insect eggshell) genes, and *Drosophila* heat-shock genes.

The Globin Gene Family

Globins are oxygen-transporting and storage molecules found in animals, some plants, and microorganisms. In higher vertebrates there are two types of globins: myoglobin, which stores oxygen in muscles, and hemoglobin, found in red blood cells. Myoglobins function as single molecules, whereas hemoglobins occur as tetramers, two each of two protein chains. Evolution in the globin gene family can be traced by comparative studies of globins among species as well as molecular studies of globins

within a species (see chapter 21). Studying hemoglobins has provided a great deal of information on gene expression and evolution. We turn our attention to the globin gene family in human beings.

During human development, four major hemoglobins appear: embryonic hemoglobin, Hb F, Hb A, and Hb A$_2$ (table 14.5). Structurally, the ζ (Greek, zeta) subunit (a component of embryonic hemoglobin) is α-like, whereas the rest are β-like (fig. 14.31; see also fig. 10.28). Fetal hemoglobin has a higher affinity for oxygen than does adult hemoglobin, thus allowing fetuses to draw oxygen from their mother's blood. From a comparative study of the DNA sequences, the evolution of the various hemoglobin genes has been inferred (fig. 14.32).

The α genes are located in a cluster on chromosome 16; the β genes are located in a cluster on chromosome 11 (fig. 14.33). These two clusters provide a clear case history of gene duplication, presumably by unequal crossing over, followed by divergence. Having a second or third copy of a gene allows one of the duplicates to diverge (and perhaps to become nonfunctional in the process), whereas the original still performs the required function.

Many diseases of genetic interest involve the hemoglobins. In fact, hemoglobinopathies, including sickle-cell anemia and the thalassemias, are the most common genetic disorders in the world population. The best-known mutation of a hemoglobin gene itself is the one that causes sickle-cell anemia, a mutation of the sixth amino acid of the β chain. In the homozygous state, the disease is usually fatal. However, heterozygotes show an increased resistance to malaria. One of the ramifications of that fact is that the sickle-cell allele is maintained at relatively high frequencies in malarial regions (see chapter 21).

Figure 14.31

The structure of adult α- and β-globin genes. The *numbers* refer to amino acids (or translated codons).

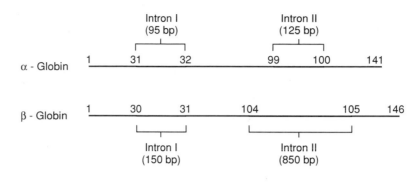

Figure 14.32

The evolution of the various human globin genes from an ancestral primitive gene. The diagram represents a branching tree that begins on the *left* and progresses to the *right*. Each branch point is an evolutionary step in which the genes presumably became duplicated and then either diverged or simply remained as duplicates, as in present-day genes (on the *right*).

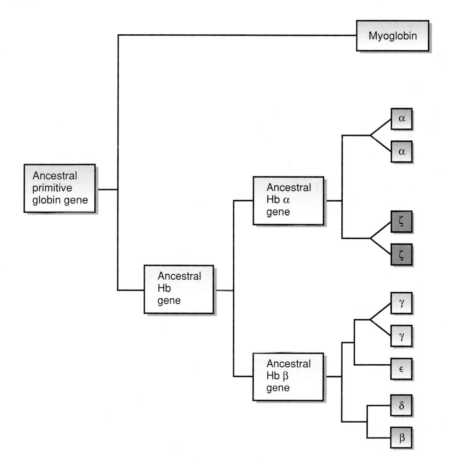

The *thalassemias* are a group of diseases that affect the regulation of the α and β hemoglobin genes. (Thalassemia comes from the Greek for "sea blood," because the disease is best known in individuals living around the Mediterranean Sea.) In α and β thalassemias the α or β subunit, respectively, is present in very low quantities, or absent entirely. Many of the genetic defects are deletions, possibly due to unequal crossing over within the globin gene complexes. T. Maniatis showed that β thalassemia is caused by a mutation in the β globin gene that disrupts RNA splicing. The body compensates by forming γ_4 or β_4

Figure 14.33

The α- and β-globin clusters in human beings. The ψβ1 and 2 and the ψα1 refer to nontranscribing genes (pseudogenes). Mutation has rendered the pseudogenes inactive. Within each gene box, *solid regions* refer to exons and *open regions* refer to introns.

Reproduced, with permission, from the Annual Review of Genetics, *Volume 14,* © *1980 by Annual Reviews Inc.*

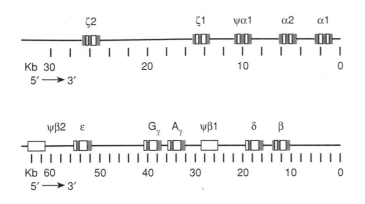

Tom Maniatis (1943–   )
Courtesy of Dr. Tom Maniatis.

hemoglobin in α thalassemias, or $\alpha_2\gamma_2$ or $\alpha_2\delta_2$ in β thalassemias. These are relatively unsuitable or inefficient responses; the diseases range from very mild to very severe, frequently fatal. More information is needed regarding the control of hemoglobin production in the thalassemias.

SUMMARY

To study developmental control in eukaryotes, we must understand the eukaryotic chromosome, which is apparently uninemic: It consists of one DNA double helix per chromosome. Nucleoprotein is composed of DNA, histones, and nonhistone proteins. The nucleosome, a uniform packaging of the DNA, is made of histones. The majority of the nonhistone proteins create the scaffold structure of the chromosome and are not involved in regulation. Presumably, undetectably small quantities of the nonhistone proteins take part in regulation of transcription.

Core DNA, wrapped around nucleosomes, is separated by linker DNA between nucleosomes. There are regions of DNA, vulnerable to nucleases, that do not contain nucleosomes. They are referred to as nuclease hypersensitive sites. The 110-Å nucleosomed DNA forms a 300-Å fiber by coiling into a solenoid-like structure. Coiling of this fiber presumably forms the thick, 2,000-Å fiber seen in metaphase chromosomes.

The centromere and telomeres are specific functional regions of a chromosome. Centromeres isolated from yeast chromosomes have three consensus areas. Telomeres are tandem repeats of a short (5- to 8-bp) segment. Telomeric sequences are added to the ends of chromosomes without the use of a DNA template. The number of repeats varies, and the mechanism controlling the number is unknown.

Substructuring in the eukaryotic chromosome is demonstrated by G-, C-, and R-banding techniques. C-bands (constitutive heterochromatin) appear to be around the centromeres. These bands consist primarily of satellite DNA, which seems to have a structural role in the chromosome. G-bands (Giemsa bands) presumably represent intercalary heterochromatin and, also presumably, do not have an active transcriptional role. R-bands (reverse bands) appear between the G-bands and represent intercalary euchromatin, the site of transcribed, structural genes.

The cot measure of DNA-DNA hybridization reveals that about 10% of the eukaryotic chromosome consists of highly repetitive nucleotide sequences (satellite DNA), about 15% consists of DNA that is intermediately repetitive, and about 70% consists of unique DNA in which there is no repetition of nucleotide sequences. Intermediately repetitive DNA is made up of dispersed DNA, such as the Alu family in human beings; genes needed in many copies, such as histone, rRNA, and tRNA genes; and genes that have duplicated and diverged, such as genes of the globin family.

SOLVED PROBLEMS

1. Why is higher-order chromosomal structure expected in eukaryotes but not prokaryotes?

   ANSWER: The simplest explanation is the difference in amount of the genetic material in prokaryotes and eukaryotes. Since the average human chromosome has several centimeters of DNA, that DNA must be contracted down to a size in which it can be moved during the processes of mitosis and meiosis without tangling and breaking it. Nucleosomes provide the first order of coiling, and then several levels of coiling of the nucleosomed DNA bring it down to a manageable size for nuclear divisional processes.

2. Why might we expect to see chromosomal puffs that are tissue- and stage-specific, constitutive, and environmentally induced?

   ANSWER: The various patterns of chromosomal puffing are expected because chromosomal puffing indicates transcription, the activity of specific genes. Thus, since various tissues are different because they have different proteins, each tissue is expected to have a unique suite of active genes and thus a unique suite of puffs. Similarly, different stages in an insect's development would require different genes to be active and therefore different puffs should appear at different stages of development. Correspondingly, some genes will be active all the time because they specify proteins, such as ribosomal protein genes, that are needed all the time. Finally, environmental insults such as heat shock are known to induce a group of genes that are needed to react to the specific insult, resulting in a suite of puffs that respond consistently to an environmental insult.

3. A DNA sample of unknown origin has a $cot_{1/2}$ of 500. What is the length of unique DNA specific to that sample?

ANSWER: To relate $cot_{1/2}$ to size of unique DNA, simply use the graph of figure 14.27. From that you can see that a $cot_{1/2}$ of 500 falls somewhere between the $cot_{1/2}$ of *E. coli* and that of calf. If you then use the upper scale of that figure, you can see that a $cot_{1/2}$ of 500 corresponds to a unique sequence of something less than 2×10^8 base pairs.

EXERCISES AND PROBLEMS

1. Summarize the major differences between eukaryotes and prokaryotes, including the structures of their DNAs.

2. Summarize the evidence that the eukaryotic chromosome is uninemic.

3. What results would you get in the experiment shown in figure 14.1 if the eukaryotic chromosome were not uninemic but instead had some other number of complete DNA molecules (e.g., binemic)?

4. What are the major protein components of the eukaryotic chromosome? What are their functions?

5. What is the evidence used to determine the length of DNA associated with a nucleosome? What is a nucleosome hypersensitive site? What functions are associated with these sites?

6. What is the protein composition of a nucleosome? What function does histone H1 have?

7. What are the relationships among the 110-Å, 300-Å, and 2,000-Å fibers of the eukaryotic chromosome?

8. Draw a mitotic chromosome during metaphase. Diagram the various kinds of bands that can be brought out by various staining techniques. What information about the DNA content of these bands is known?

9. Give a 300-Å-fiber model of the chromosome to account for G-bands.

10. Give a 300-Å-fiber model of the chromosome to account for polytene chromosomal puffs.

11. What are the differences among polytene chromosomes, lampbrush chromosomes, puffs, and Balbiani rings? Draw an example of each.

12. Under what circumstances does a chromosomal puff occur? What does it signify?

13. What is satellite DNA? What does it signify?

14. What is a centromere? A kinetochore? What do we know about the sequences within a yeast centromere?

15. What is a telomere? What are its functions? What is its structure?

16. What is a cot curve? Draw some cot curves of various types of DNA. What do the $cot_{1/2}$ values mean? How can you determine the degree of repetitiveness of mouse DNA from cot curves?

17. What functions exist in unique, repetitive, and highly repetitive DNAs?

18. How would you use Southern blot techniques to locate the number and position of Alu members in the human chromosomes?

19. How could you use modern technology to determine the direction of transcription of the histone genes in figure 14.30?

20. How many functional globin genes are there in mammals?

21. How could you determine, using modern techniques, that the α- and β-globin pseudogenes exist?

22. Kavenoff *et al.* determined the size of DNA in *Drosophila* chromosomes in two ways: 1) Spectrophotometric measurements were made on the largest intact chromosome. These measurements were then used to calculate the amount of DNA in each chromosome. 2) Nuclei were gently lysed. The lengths of the longest molecules were measured, and those lengths were used to determine the amount of DNA in each molecule. What results for each method would you expect if
   a. the chromosomes contain one DNA molecule?
   b. the chromosomes contain more than one DNA molecule?

23. What can be said about the base composition of the satellite with a density of 1.671 in figure 14.19?

24. When chromatin is partially digested with an endonuclease, the proteins removed, and the DNA separated in a sizing gel, DNA fragments in multiples of 200 bp are found. Provide an explanation for these observations.

25. If chromatin is digested with an endonuclease to produce 200-bp fragments, and these fragments are then used for transcription experiments, very little RNA is made. Provide an explanation for this observation.

26. Can nucleosomes contain the DNA for one gene? Explain.

27. If radioactive probes are made from highly repetitive DNA, these probes hybridize *in situ* mainly to centromeric and telomeric regions. What does this result suggest about the organization of chromosomes?

28. When DNA renaturation experiments are done using 400-bp fragments, 48% of the DNA behaves as repetitive DNA. If 4,000-bp fragments are used instead, 80% of the DNA behaves as repetitive DNA. Propose an explanation to reconcile these two results.

29. Total DNA is isolated from two plants of the same species, one of which was kept in the dark until the leaves turned white, and the other of which was kept in the light. When cot curves are prepared for these two DNAs, the curves are identical. What conclusions can be drawn from this experiment?

30. Draw a cot curve for an organism that has a biphasic curve with the following parameters: One part of the curve has $cot_{1/2} = 10^{-2}$ and represents 40% of the DNA. The other part of the curve has $cot_{1/2} = 10^3$ and represents 60% of the DNA. The fraction reassociated is 100% when $cot = 10^4$.

31. How many base pairs are there in a molecule with $cot_{1/2} = 10^3$? (Refer to figure 14.27.)

The following two questions refer to this equation:

$$c/c_o = \frac{1}{1 + k cot}$$

where c = concentration of single strand DNA at time t

c_o = total or original DNA concentration

t = incubation time in seconds

k = rate constant

32. Calculate c/c_o for cot = 0.01, 0.05, 0.1, 0.5, 1.0, 5.0, 10.0, 50, 100. Assume $k = 1$.

33. A particular bacterial species has $k = 0.15$. What is the $cot_{1/2}$ for its DNA?

Suggested Readings for chapter 14 are on page 652.

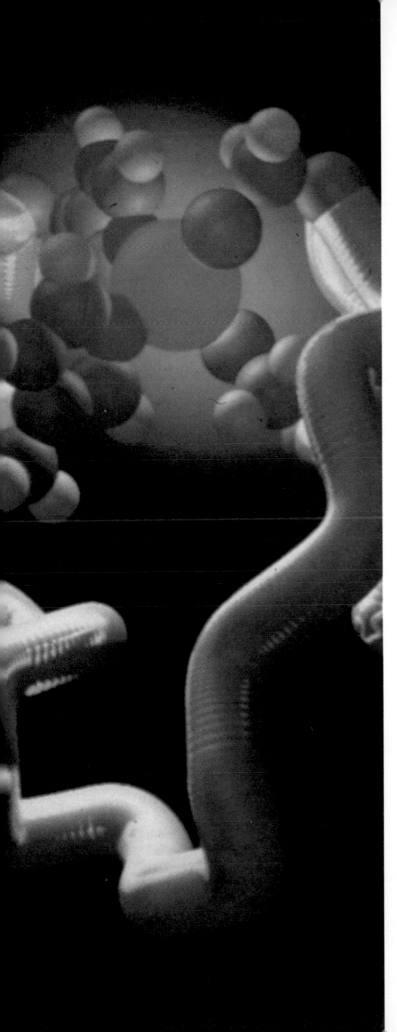

15

Gene Expression: Control in Eukaryotes

Three-dimensional computer drawing of a zinc finger. The amino acid backbone is purple; parts of the histidine and cysteine amino acid residues are shown around the zinc ion (*blue*).

Photograph provided by Michael Pique and Peter E. Wright, Department of Molecular Biology, Research Institute of Scripps Clinic. 1 cover photo, Science, *vol. 245, 11 August 1989. Copyright 1989 by the AAAS.*

In this chapter we turn our attention to the control of gene expression in eukaryotes. We look at patterns in development, and the control of gene expression that is presumably determining those patterns. We then look at two genetic systems in development, immunogenetics—specifically the way in which antibody diversity is generated in higher forms—and cancer, growth and development gone awry.

PATTERNS IN DEVELOPMENT

Development is the orderly sequence of change that produces increased complexity during the growth of an organism. Presumably, it is controlled by the orderly, sequential expression of genes. A favored approach to understanding the genetic control of this process in higher organisms requires first learning the details of the normal developmental process in an organism and then studying the disruption of this normal process by mutation or experimental manipulation.

Differentiated Nuclei Can Be Totipotent

At one point it was believed that development might take place through permanent changes in chromosomes. The idea that perhaps there are subtle changes in chromosomes during development that are not observable by karyotyping a cell has been explored by the method of nuclear transplantation. In this technique, nuclei of differentiated cells are put into zygotes that are about to begin development. The support of normal development by the transplanted nuclei is a demonstration that development does not proceed by permanent chromosomal differentiation. In other words, the nuclei are **totipotent:** Any nucleus can give rise to any, and therefore all, cell types.

A technique in frogs to explore totipotency was developed by R. Briggs and T. King. A frog's egg can be activated to begin development (e.g., by pin prick) and then all of its original genetic material can be removed or destroyed. The method was developed further by J. B. Gurdon who worked with African clawed toads (*Xenopus*). As shown in figure 15.1a, an ultraviolet light is used to destroy the nucleus. Nuclei from tissues of more advanced embryos can then be inserted into the enucleated egg. Figure 15.1b illustrates how a nucleus from a differentiated tissue cell is drawn into a pipette that has an inner diameter smaller than the cell. The cell is destroyed, but the nucleus remains intact. The nucleus is then injected into the enucleated egg. The process produces an egg developing under the control of a foreign nucleus. In the African clawed toad, nuclei from the differentiated intestine will support normal growth. F. Steward and his colleagues have shown that single cells from the petiole or the root of a carrot, when grown in cell culture, will produce embryoids, embryo-like structures that will produce perfectly normal carrots when transplanted to soil.

Figure 15.1
Technique of nuclear transplantation. (*a*) Destruction of the host nucleus with ultraviolet light. (*b*) Obtaining and inserting a foreign nucleus. A differentiated cell is drawn into a pipette whose diameter is narrower than the cell. The nucleus survives but the cell is destroyed. The pipette is then blown out into the enucleated egg, depositing the nucleus.

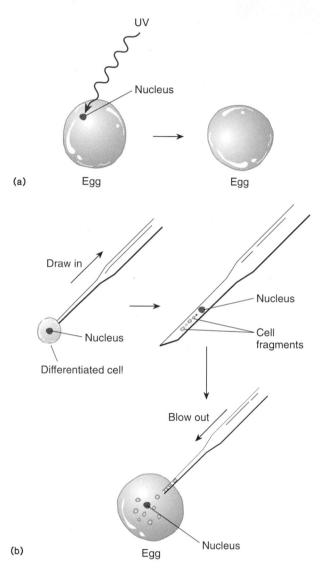

Several conclusions can be drawn from these and similar studies. They demonstrate that cell differentiation can and does take place without any permanent change in the genetic material. In the case of the frog, however, nuclei from cells later in development have a declining success rate of supporting development. Thus permanent changes to the nucleus can occur during development. Although these changes may only be age-related phenomena, whereby the nuclei from more differentiated tissues do not survive the experimental technique or do not undergo mitosis at a fast enough rate, we must be careful about oversimplifying our conclusions regarding development. Nevertheless, it does seem safe to suggest that simple loss or permanent change of genetic material fails to account generally for organismal development.

Figure 15.2

Fate map of the right side of a *Drosophila* blastoderm (a very early stage in development) before gastrulation, the process in which cells that will later form the internal organs move to the inside of the developing embryo. The *rectangles* labeled dorsal and abdomen refer to sternites and tergites, subsections of the abdomen.

From Y. Hotta and S. Benzer, "Mapping Behavior in Drosophila *Mosaics" in F. H. Ruddle, editor,* Genetic Mechanisms of Development. *Copyright © 1973 by Academic Press, Orlando, FL. Reprinted by permission.*

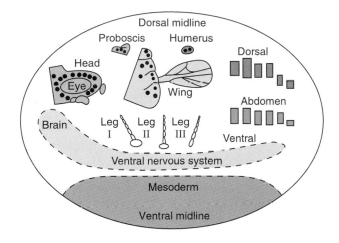

Sydney Brenner (1927–   )
Courtesy of Dr. Sydney Brenner.

Development in *Caenorhabditis elegans*

The orderly sequence of development has been worked out in many organisms by observing the movement pattern of cells in developing embryos. Often, simple observation is augmented by placing dyes that do not damage cells (vital dyes) on parts of the developing embryo and watching where these dyed regions or cells go. From these observations, it has been possible to construct **fate maps,** which indicate the ultimate fate of various regions of a zygote or early embryo (fig. 15.2). The construction of fate maps is one part of the description of the gross aspects of development at the cellular level.

Historically, amphibians were the focus of developmental research because they have large eggs that are easily observed and manipulated. Then attention focused on *Drosophila* because, in addition to ease of manipulation, there is a tremendous amount of genetic information available on the fruit fly. We will turn our attention to it in the following pages. However, recently, the nematode, *Caenorhabditis elegans,* has emerged as another model organism for developmental studies because of its simplicity. The species consists of males and self-fertilizing hermaphrodites (fig. 15.3). Hermaphrodites have two sex chromosomes (XX), whereas males have only one (X0). Each individual consists of only about one thousand cells; its life cycle lasts only 3.5 days, and with only 8×10^7 bp of DNA, it has the smallest genome of any multicellular organism. In 1963, S. Brenner proposed learning the lineage of every cell in the adult. With the efforts of nu-

merous colleagues, that work has now been completed. From the fertilized egg to the adult, the division and fate of every cell of this nematode worm is known.

Cells were followed throughout development; as cell division took place cell movements were noted (fig. 15.4). Starting with a few original cells, denoted by uppercase letters (fig. 15.5*a*), observers kept track of daughter cells by appending a letter indicating relative position after each division. The designations were: a, anterior; p, posterior; l, left; r, right; d, dorsal; and v, ventral. Thus, in a newly hatched larva, one particular pharyngeal motor neuron cell is AB.araapapaav (fig. 15.5*b*). The process was found to be relatively invariant from one individual to the next.

Developmental Steps

The first stage in our understanding of the genetics of development is now known for this species. How much has this description of development helped us understand the mechanism of developmental processes? Several interesting concepts have emerged.

First has been the discovery that the fate of early embryonic layers is not absolute. In higher eukaryotes, early development gives rise to three embryonic tissue layers: the mesoderm, endoderm, and ectoderm. These embryonic layers then give rise to all the adult tissues in a pattern that was believed to be invariant. All of the nervous system, for example, was believed to arise only from the ectoderm. In *C. elegans,* however, some neurons can arise from mesodermal cells. A progenitor cell can divide, producing one daughter that will become a neuron (ectoderm) and one that will become a muscle cell (mesoderm). This indicates that there is more flexibility in development than was previously recognized.

Second, cell death is often programmed into the developmental process. During development, one in six of all cells produced dies (fig. 15.5*b*). These deaths are sex-specific and consistent. For example, there are cases in which a particular division leads to the death of a cell in one sex but not in the other.

Figure 15.3

The roundworm *Caenorhabditis elegans*. (*a*) Self-fertilizing hermaphrodite. (*b*) Male. The worms are about 0.3 mm long.

(line art) From J. Sulston and H. Horvitz, "Post-Embryonic Cell Lineages of the Nematode, Caenorhabditis elegans" *in* Developmental Biology, 56:110–156. *Copyright © 1977 Academic Press, Orlando, FL. Reprinted by permission. (photos) J. E. Sulston and H. R. Horvitz. "Post-Embryonic Cell Lineages of the Nematode* Caenorhabditis Elegans," *Developmental Biology 100 (1977): 110–156.*

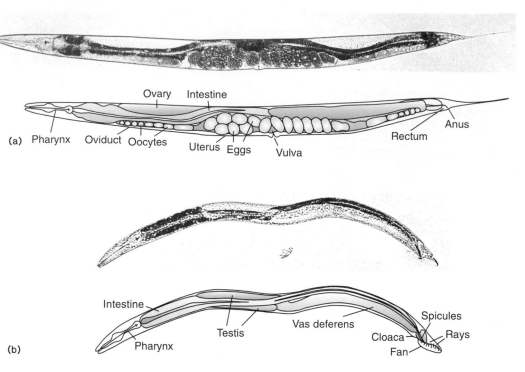

Figure 15.4

Early stages of *Caenorhabditis elegans*. (*a*) Maternal and paternal pronuclei are about to fuse. The bar is 10 μm long. (*b*) Beginning of gastrulation. Ea and Ep are intestinal precursor cells; D is a mesodermal precursor cell; P$_4$ gives rise to germ cells. (*c*) After five hundred minutes pharynx (ph) and intestine (int) are visible. The *black arrows* point to germ cells, the *white arrow* points to the mouth.

J. Sulston, et al., *"The Embryonic Cell Lineage of the Nematode,* Caenorhabditis elegans." *Developmental Biology 100 (1983):64–119.*

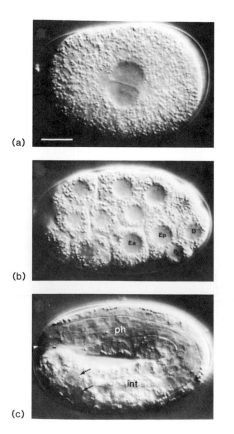

Third, symmetry is not an accidental process. We might think that bilateral symmetry of an organism results from a mirror-image process of sister cells. In other words, we might predict that symmetry would be a natural outcome of cell divisions by daughter cells on opposite sides of the developing organism. This is not always the case. Sometimes, symmetry is the result of movement and division of cells that are not symmetrically related (fig. 15.6). Thus bilateral symmetry is sometimes actively created.

Aberrant Development

The next stage in understanding the developmental process is to observe aberrant development—development that is induced either by physical manipulation of the developing organism or by developmental mutants. In some of the experiments on developing *C. elegans*, specific cells were destroyed (ablated) by a laser microbeam. The conclusion from these studies is that cells are relatively fixed in their developmental pathways: If a precursor cell was ablated, its derivative cell types were missing in the adult. For example, when muscle cell precursors were ablated, the muscle cells that would have come from the destroyed cells never appeared; other cells did not adjust their developmental pathways to "fill in" for the missing tissue. The neighboring cells seemed to be programmed for a specific developmental fate, rather than for responsiveness to the cellular milieu.

There were, however, several exceptions to this rule, such that an ablated cell was replaced by a cell that would normally have had a different fate. When AB.plapa was ablated, AB.plapaapa could not form. However, another cell, AB.prapaapa, took its place. Our conclusions from

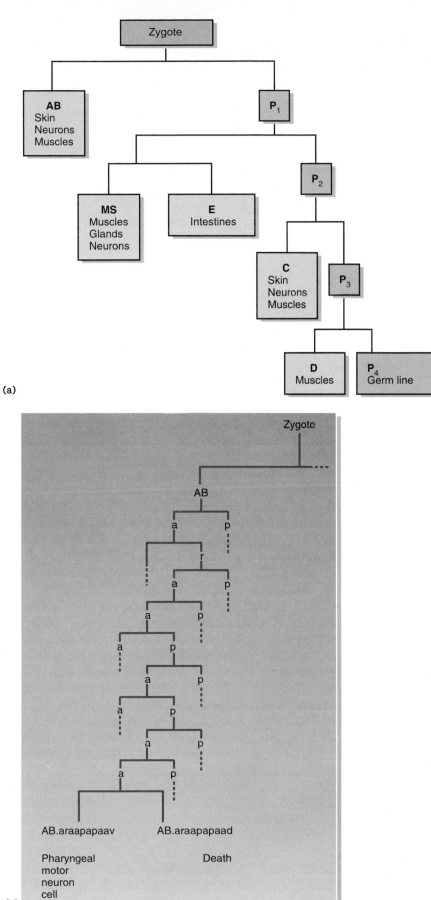

Figure 15.5

Cell lineage of *Caenorhabditis elegans*. (*a*) The zygote gives rise to six founding cells (AB, MS, E, C, D, and P₄), which then give rise to all the cell types of the adult. (*b*) The lineage of cell AB.araapapaav, a pharyngeal motor neuron cell, and its sister, AB.araapapaad, which dies. *Dashed lines* indicate continued divisions. All the descendants of AB.araa are either pharyngeal motor neurons or deaths. All the descendants of its sister, AB.arap, are either pharyngeal neurons, other neurons, or deaths. (a, anterior; p, posterior; r, right; d, dorsal; v, ventral.)

From J. Sulston, et al., "The Embryonic Cell Lineage of the Nematode, Caenorhabditis elegans*" in* Developmental Biology, *100:64–119. Copyright © 1983 Academic Press, Orlando, FL. Reprinted by permission.*

Figure 15.6

In the AB group of cells of *Caenorhabditis elegans*, bilateral symmetry is generated in an orderly fashion posteriorly (*p*) but not anteriorly (*a*). Cells that generate analogous (symmetrically located) groups of cells on the left (*l*) and right (*r*) sides of the body are connected by *dotted lines*. Posteriorly, simple, orderly cell divisions produce symmetry, whereas anteriorly, complex division patterns and cell movements produce symmetry.

From B. Lewin, "The Continuing Tale of a Small Worm" in Science 225:153–156. Copyright 1984 by the AAAS.

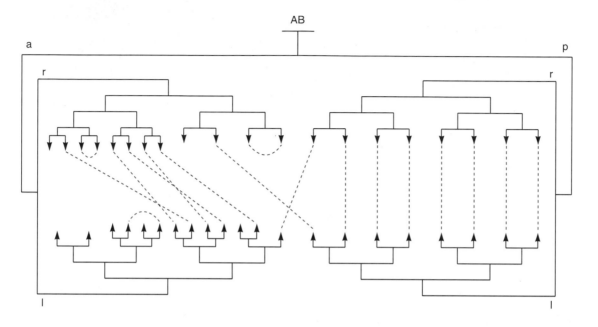

these experiments are that, for the most part, cells follow a relatively fixed path in development (the pattern is *hardwired*), although some cells can respond to a changing environment by altering their developmental fates. The cells of higher eukaryotes may have more flexibility in determining their fates.

Homeotic Mutants

The study of mutants has also been valuable in understanding the control of developmental pathways. Over five hundred genes have been identified in *C. elegans* from studies of thousands of mutations. One group of mutants that is extremely interesting is that of **homeotic mutants,** in which one cell type follows the developmental pathway normally followed by other cell types. These mutants are important because they seem to indicate the presence of *master-switch* genes, genes whose function it is to regulate the development of whole groups of other genes. These genes presumably act as binary switches: They define whether development goes in one direction or another.

Although no operons are known in eukaryotes, the idea that one or a very few genes may control the expression of many genes in development is an attractive idea to geneticists. Mutations in homeotic genes seem to indicate that they may be these master genes. Two major homeotic gene complexes are known in *Drosophila melanogaster:*

Walter J. Gehring (1939–   )
Courtesy Dr. Walter J. Gehring.

Edward B. Lewis (1918–   )
Courtesy of Dr. Edward B. Lewis.

the *Antennapedia* complex, worked on extensively by T. Kaufman, W. Gehring, and their colleagues, and the *bithorax* complex, analyzed extensively by E. Lewis, D. Hogness, and their colleagues. Genes in the *Antennapedia* complex control the fate of the anterior development of the fruit fly (head and anterior thorax), whereas genes in the *bithorax* complex control the fate of posterior development (posterior thorax and abdomen). Mutations in genes of these complexes can change the fate of development of whole sections of the fly. For example, *Nasobemia,* an *Antennapedia*-complex mutant, results in legs

Figure 15.7

Nasobemia, a mutation that causes legs to grow in the place of antennae on the head of a *Drosophila*
Courtesy of Dr. Walter J. Gehring.

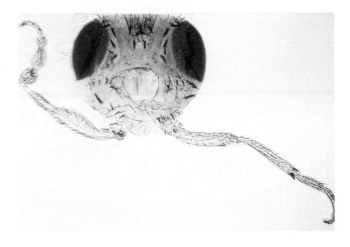

Figure 15.8

A normal fruit fly (*a*) and a bithorax mutant (*b*). The bithorax mutant is actually a combination of three mutations that produce a fly with an almost perfect second thorax with its own set of wings.
J. L. Marx, "Genes That Control Development," Science 213 (Sept. 25, 1981): 1486. Courtesy of E. B. Lewis, Caltech.

(a)

(b)

growing where antennae would normally be located (fig. 15.7); and *bithorax,* a *bithorax*-complex mutant, produces flies with two thoraxes (four-winged diptera; fig. 15.8).

Homeotic mutants have also been studied in *Caenorhabditis.* For example, mutations of the *lin-12* locus affect the normal development of the vulva in the *Caenorhabditis* hermaphrodite (fig. 15.9). Two classes of mutations have been described. (1) Zero-class mutations (*lin-12*[0]), which are recessive mutations causing loss of gene activity, result in a nonfunctional cellular protrusion in place of the normal vulva (fig. 15.9*c*). These mutants are sterile. (2) The d-class mutations (*lin-12*[d]), which are semidominants, seem to elevate gene activity. Certain of these mutations result in worms with many vulvae (fig. 15.9*b*), in which a series of nonfunctional pseudovulvae are produced. These mutants are fertile because larvae hatch internally and eventually escape through the adult's body wall.

These mutations support the notion of a master-switch gene that determines normal vulval development. Inactivity of this gene results in a sterile organism, with merely a vulval protrusion; normal activity results in a normal vulva; excess activity can cause a series of pseudovulvae. Hence the level of *lin-12* activity determines which cell fates will be realized.

The Homeo Box

Drosophila development takes place through segmentation. As the embryo develops, three anterior segments appear that will form the head, three middle segments appear that will form the thorax, and eight posterior segments appear that will form the abdomen (fig. 15.10). In addition to the homeotic mutations, several other classes of mutations radically affect development. *Fushi tarazu,*

Japanese for "not enough segments," is an example of a mutation in a gene affecting segmentation. Whereas homeotic mutations tend to change the developmental pathway of one segment into another, segmentation genes affect the actual number of segments (fig. 15.11). Using recombinant DNA techniques, W. Gehring and his colleagues found a common 180-bp sequence of DNA in genes of the *Antennapedia* complex, the *bithorax* complex, and the *fushi tarazu* gene. Further probing localized this same segment of 180 bp to about a dozen genes in *Drosophila,* all with homeotic or segmentation properties. They thus called it the **homeo box.**

Figure 15.9

Examples of *lin-12* mutants of *Caenorhabditis elegans*. (*a*) Wild-type.
Arrowhead indicates vulva; the worms are about 0.3 mm long. (*b*) A
d-class mutant with five pseudovulvae noted by *lines*. (*c*) A zero-class
mutant with a nonfunctional protrusion in place of the vulva.

I. S. Greenwald, P. W. Sternberg, and H. R. Horvitz, "The lin-12 *Locus
Specifies Cell Dates in* Caenorhabditis elegans," *Cell 34 (1983):435–444. Cell is
copyrighted by the MIT Press.*

(a)

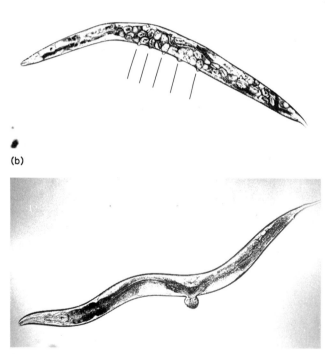

(b)

(c)

Figure 15.10

Scanning electron micrograph of a *Drosophila* embryo of about
fourteen hours, about to hatch. The embryo is about 0.4 mm long; the
eggshell has been removed to reveal the structures. T1–T3 and A1–A8
represent the three thoracic and eight abdominal segments,
respectively. The head segments have migrated internally.

Photograph courtesy of Drs. F. R. Turner and A. P. Mahowald.

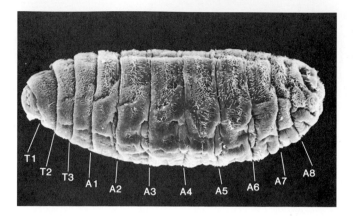

Figure 15.11

Expression of *fushi tarazu* mutations lead to fewer segments in
Drosophila embryos. The *white stripes* on the embryos are of small
bumps called denticles. They mark the anterior edge of each segment.
The thoracic and abdominal segments are visible in the normal
embryo (*top*). The lower embryo, a *fushi tarazu* mutant, has half the
number of segments.

Courtesy of Walter J. Gehring, October, 1985, Scientific American, *pg. 158.
Copyright © 1985 Scientific American, Inc. All Rights Reserved.*

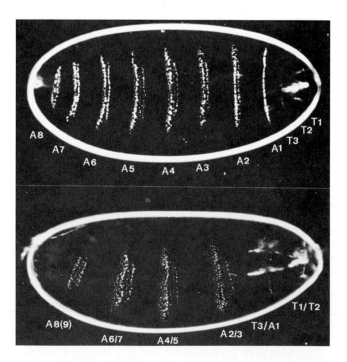

Using a recombinant probe for the homeo box, or a
computer search for the sequence, researchers have found
it in genes in plants, yeast, sea urchins, frogs, and human
beings. This high degree of sequence conservation across
widely divergent groups of organisms suggests that the sequence is crucial to the functioning of homeotic genes and
that the mechanism arose early in evolutionary time.

The nucleotides of the homeo box are translated into
a sixty-amino acid peptide region called the **homeo domain**

Figure 15.12
The homeo domain of five genes: the *MO*-10 gene from the mouse (*Mus*), the *MM*3 gene from the frog (*Rana*), and three homeo and segment genes from *Drosophila*. The *Antennapedia* gene is considered the consensus sequence; red amino acids differ from this sequence.

From Walter J. Gehring, Scientific American, *October 1985. Copyright © 1985 by Scientific American, Inc. Reprinted by permission of the author.*

	1																			20
Antennapedia	Arg	Lys	Arg	Gly	Arg	Gln	Thr	Tyr	Thr	Arg	Tyr	Gln	Thr	Leu	Glu	Leu	Glu	Lys	Glu	Phe
Mouse MO-10	Ser	Lys	Arg	Gly	Arg	Thr	Ala	Tyr	Thr	Arg	Pro	Gln	Leu	Val	Glu	Leu	Glu	Lys	Glu	Phe
Frog MM3	Arg	Lys	Arg	Gly	Arg	Gln	Thr	Tyr	Thr	Arg	Tyr	Gln	Thr	Leu	Glu	Leu	Glu	Lys	Glu	Phe
Fushi tarazu	Ser	Lys	Arg	Thr	Arg	Gln	Thr	Tyr	Thr	Arg	Tyr	Gln	Thr	Leu	Glu	Leu	Glu	Lys	Glu	Phe
Ultrabithorax	Arg	Arg	Arg	Gly	Arg	Gln	Thr	Tyr	Thr	Arg	Tyr	Gln	Thr	Leu	Glu	Leu	Glu	Lys	Glu	Phe

	21																			40
Antennapedia	His	Phe	Asn	Arg	Tyr	Leu	Thr	Arg	Arg	Arg	Arg	Ile	Glu	Ile	Ala	His	Ala	Leu	Cys	Leu
Mouse MO-10	His	Phe	Asn	Arg	Tyr	Leu	Met	Arg	Pro	Arg	Arg	Val	Glu	Met	Ala	Asn	Leu	Leu	Asn	Leu
Frog MM3	His	Phe	Asn	Arg	Tyr	Leu	Thr	Arg	Arg	Arg	Arg	Ile	Glu	Ile	Ala	His	Val	Leu	Cys	Leu
Fushi tarazu	His	Phe	Asn	Arg	Tyr	Ile	Thr	Arg	Arg	Arg	Arg	Ile	Asp	Ile	Ala	Asn	Ala	Leu	Ser	Leu
Ultrabithorax	His	Thr	Asn	His	Tyr	Leu	Thr	Arg	Arg	Arg	Arg	Ile	Glu	Met	Ala	Tyr	Ala	Leu	Cys	Leu

	41																			60
Antennapedia	Thr	Glu	Arg	Gln	Ile	Lys	Ile	Trp	Phe	Gln	Asn	Arg	Arg	Met	Lys	Trp	Lys	Lys	Glu	Asn
Mouse MO-10	Thr	Glu	Arg	Gln	Ile	Lys	Ile	Trp	Phe	Gln	Asn	Arg	Arg	Met	Lys	Tyr	Lys	Lys	Asp	Gln
Frog MM3	Thr	Glu	Arg	Gln	Ile	Lys	Ile	Trp	Phe	Gln	Asn	Arg	Arg	Met	Lys	Trp	Lys	Lys	Glu	Asn
Fushi tarazu	Ser	Glu	Arg	Gln	Ile	Lys	Ile	Trp	Phe	Gln	Asn	Arg	Arg	Met	Lys	Ser	Lys	Lys	Asp	Arg
Ultrabithorax	Thr	Glu	Arg	Gln	Ile	Lys	Ile	Trp	Phe	Gln	Asn	Arg	Arg	Met	Lys	Leu	Lys	Lys	Glu	Ile

Figure 15.13
The helix-turn-helix motif of a DNA-binding protein. The two helices are pictured as cylinders. The α-helix 1 recognizes the DNA sequence in the major groove; the α-helix 2 stabilizes the configuration.

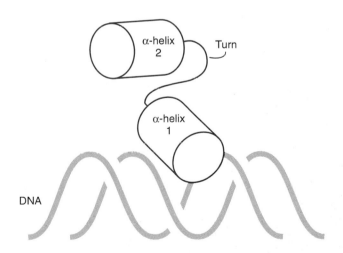

(fig. 15.12). Analysis of the amino acid sequence of this homeo domain suggests that the protein functions by binding to DNA. That conclusion was reached by noticing a particular conformation of amino acids in the homeo domain. Amino acids 31 to 38 and 41 to 50 form α helices. The configuration of two α helices in a protein separated by a short segment (called a "turn") has been found in many proteins that bind to DNA (e.g., cro protein, λ re-pressor, and CAP protein). It is called the **helix-turn-helix motif.** One α helix recognizes a DNA sequence by fitting into the major groove, and the other helix stabilizes the configuration (fig. 15.13). This speculation has been confirmed by localizing homeo domains to proteins that definitely bind DNA, including human transcription factors. The ability of the homeo domain protein to bind to DNA is consistent with its role as part of the protein product of a master gene—by binding to various DNA regions it could control the expression of many genes.

The homeo box within homeotic and segmentation genes in *Drosophila,* and throughout the animal kingdom, is most likely a controlling element within major-switch genes. When mutant homeotic genes are introduced into normal organisms through recombinant techniques, the major effects of these genes on development are realized, further supporting their roles as major switches in development. This is a new and exciting trail in discovering the genetic control of development.

CONTROL OF TRANSCRIPTION IN EUKARYOTES

The homeo domain, by binding to DNA, probably affects eukaryotic development by controlling transcription. In chapter 10, we described the process of transcription and identified various recognition regions in the eukaryotic

BOX 15.1

ZINC FINGERS, LEUCINE ZIPPERS, AND COPPER FISTS

The helix-turn-helix motif, of two α helices separated by a short turn, is found in some proteins that bind to DNA. However, different motifs have also been found in other proteins that also bind to DNA. Three recently discovered motifs are the **zinc finger,** the **leucine zipper,** and the **copper fist.** The zinc finger, a fingerlike projection of amino acids, whose base consists of cysteine and histidine residues binding a zinc ion, was first discovered in 1985 by A. Klug and his colleagues in the transcription factor TFIIIA in *Xenopus* (Box fig. 1). These fingers are referred to as C_2H_2 proteins because two cysteines (C) and two histidines (H) are involved. There are also C_x proteins in which x is either 4, 5, or 6, referring to the number of cysteines involved in the chelation of the zinc ion.

In analyzing a DNA-binding protein from rat liver nuclei, scientists noticed that in α-helical regions of the protein, there was a repetition of leucines every seven residues for sequences as long as forty-two residues. In a helical configuration, these leucines would line up on one side of the protein (Box fig. 2). When a computer search for sequences of this type was done, several other proteins, believed to bind to DNA, showed up with this configuration, including three cancer-causing genes, c-*myc,* *fos,* and *jun,* and a transcription-regulating protein in yeast. Using the computer, the scientists developed the leucine-zipper model in which two helices with leucine repeats would interdigitate the leucines, in zipper fashion, to form a stable molecule (Box fig. 3), one that could

Box figure 1

The zinc-finger configuration of the TFIIIA protein. Zinc chelates with cysteines and histidines to form the base of the finger structure.

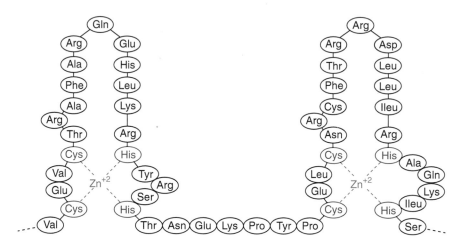

Box figure 2

Three-dimensional model of the leucine-zipper region of the enhancer-binding protein (EBP) isolated from rat liver cells. The leucine residues, which line up, are in *blue.*

From Dr. W. H. Landschulz, et al., "The Leucine Zipper: A Hypothetical Structure Common to a New Class of DNA Binding Proteins. Science 240 (24 June 1988):1759–64, fig. 4. Copyright 1988 by the AAAS.

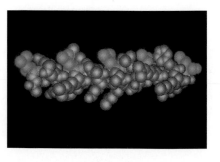

Continued on next page

BOX 15.1—*continued*

Box figure 3

Diagram showing how two α-helical strands can interdigitate to form a closed leucine zipper. *Tubes* represent the α-helical protein; "*lollipops*" represent leucine residues.

From W. H. Landschulz, et al., in Science, *240:1763. Copyright 1988 by the AAAS.*

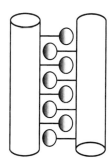

Box figure 4

(*a*) The copper-fist structure of the yeast ACE1 protein. *Solid circles* are copper ions, "S" indicates cysteine residues, (+) indicates basic amino acids involved in DNA binding, (−) indicates acidic amino acids involved in transcription, and "C" and "N" are the carboxyl and amino ends of the protein, respectively. (*b*) Three copper fists are bound to the promoter of the metallothionein gene (*cup1*) at the UASd, UAScL, and UAScR sites.

Source: Deborah M. Barnes, "News: Meeting Highlights: Regulating Transcription with a Copper Fist" in The Journal of the National Institutes of Health, *Vol. 1, 1989.*

provide a scaffolding for other amino acids that could then recognize specific DNA sequences in order to perform their functions.

The most recently discovered DNA-binding motif has been named the copper fist. This structure is formed by the ACE1 regulatory protein in yeast. Presumably, when copper, a potential heavy-metal poison, enters the cell in excess quantities, it causes ACE1 to form the "fist" structure around eight copper ions, like a fist formed around a copper penny. Cysteine residues in the ACE1 protein actually bind the copper ions (Box fig. 4). The "knuckles" of the fist are positively charged and interact with DNA. In the fist configuration, the ACE1 protein binds at the promoter and acts to enhance the transcription of the yeast metallothionein gene, *cup1,* whose protein product binds excess copper ions, protecting the cell from them. Hence excess copper ions are neutralized by a series of events triggered by the ACE1 protein forming a copper fist. We await the discovery of additional motifs of DNA-binding proteins and the interesting names to be given to them.

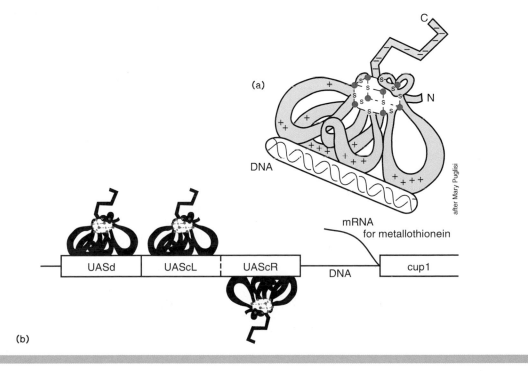

promoter. In chapter 12, we mentioned enhancer sequences, which also influence the transcription of eukaryotic genes. We believe that gene expression in eukaryotes is most likely controlled at the transcriptional stage as in prokaryotes and bacteriophages. We have numerous examples of the interaction of proteins and promoters that result in the induction of transcription. For example, as we mentioned in earlier chapters, heat-shock genes are transcribed when a novel transcription factor (a sigma factor) appears in the cell. This sigma factor binds to a consensus sequence in the promoter and allows transcription of the two dozen or so genes that then appear in response to the heat shock. However, the signal for the production of the transcription factor itself is unknown, resulting in an incomplete story. We are thus still seeking general patterns of control of eukaryotic transcription; methylation has been implicated as one mechanism affecting the control of eukaryotic gene transcription.

Methylation and Z DNA

Recently, there has been a great deal of research into the role of methylation in controlling transcription in eukaryotes. The importance of methylation in DNA-protein interactions is well known. In chapter 12, we showed that a particular DNA sequence could be protected from restriction endonucleases if it was methylated. A few percent of cytosine residues are methylated in many eukaryotic organisms, mainly in CG sequences (see fig. 12.3).

The level of methylation of DNA can be determined by using **isoschizomers,** restriction endonucleases, such as *Hpa* II and *Msp* I, that recognize the same sequence, in this case CCGG. *Hpa* II will cleave CCGG but not C^{m}CGG (where ^{m}C refers to 5-methyl cytosine). *Msp* I, however, will cleave both CCGG and C^{m}CGG. Thus the differential action of these endonucleases in a restriction digest indicates the level of methylation of certain sequences; restriction digests with *Msp* I give all CCGG sequences, whereas digests with *Hpa* II uncover only unmethylated sites. The difference is the methylated sites.

The degree of methylation of DNA is related to its transcriptional state. Genes that are dormant in one cell type but active in another, or genes that are dormant at one stage of development but active in another, are usually undermethylated when active and more fully methylated when inactive. For example, adenovirus, a transforming virus, has been observed in many eukaryotic cell lines. In most lines in which the adenovirus DNA has integrated into the host chromosome, late viral genes are turned off. These genes are highly methylated at their CCGG or GCGC sites.

In addition, chemicals that prevent methylation frequently activate previously dormant genes. For example, 5-azacytidine inhibits methylation; X-chromosome genes, which are normally deactivated, can be reactivated by treatment with 5-azacytidine. There are numerous other examples of the activation of genes after treatment with this chemical. The activated genes can be shown to lack methylated cytosines that had previously been methylated.

Further interest has recently been generated in the role of methylation in controlling gene expression by the discovery of Z DNA, and the fact that Z DNA can be stabilized by methylation (see chapter 9). This observation has led to a model of transcriptional regulation based on alternative structures of DNA. Sequences (such as CG repetitions) that could exist as Z DNA exist as B DNA when being transcribed. If the gene is to be repressed (turned off), the CG sequences are converted to stable Z DNA by methylation, which then blocks transcription. There have also been suggestions that methylation could control nucleosome formation. Nucleosomes could be prevented from forming at regions to be transcribed. Perhaps this control is exerted through methylation.

Several lines of evidence, however, lead us to believe that methylation may not be the sole basis of control of gene expression in eukaryotes. Lower vertebrates have only about a fourth of the methylation that higher vertebrates have, and *Drosophila,* for example, does not have methylated DNA at all. In addition, in about 20% of the cases examined, methylation was not related to gene action. Research has been hampered to some extent by the failure to isolate enzymes responsible for de novo methylation that would be involved in control of gene function. The enzymes that have been isolated are the ones responsible for maintenance methylation of hemimethylated DNA (see chapter 12). We await further research in this exciting area of study.

Transposons and Transcriptional Control

We have already shown (chapter 13) that transposons can affect gene expression in prokaryotes, as for example in controlling the flagellar phase in *Salmonella.* Here we present several examples of transposons controlling eukaryotic gene expression.

Barbara McClintock discovered transposons in the 1940s, without the aid of tools of molecular genetics. She won the Nobel Prize for her work in 1983. She observed corn kernels that were streaked or spotted, indicating a high mutation rate. After careful genetic analysis, she showed that the mutability that she observed was due to transposons, which she called *controlling elements.* She discovered several families of transposons, and other researchers have discovered still more.

The Ac-Ds *System*
The *Ac-Ds* system consists of two transposons. McClintock referred to the *Ac* (*activator*) transposon as an autonomous element and referred to the *Ds* (*dissociation*) transposon as a nonautonomous element. *Ds* cannot transpose until *Ac* enters the genome. At that time, *Ds* can

Figure 15.14

The *Ac-Ds* mutability system in corn. (*a*) A corn kernel that is purple because it has the dominant *C* allele. (*b*) A corn kernel in which the dominant *C* allele has been inactivated by a previous transposition of the *Ds* element into the *C* allele, disrupting it. In this case, the *Ac* gene is not present; thus, the *Ds* element cannot leave its position within the *C* gene. (*c*) When *Ac* is again present (e.g., brought in by a genetic cross), *Ds* can leave its position and the *C* allele in those cells is then restored. The result is the spotted corn kernel.

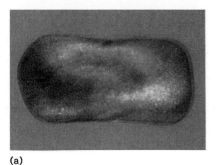
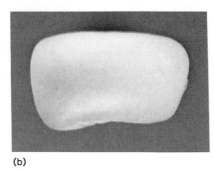

(a) (b) (c)

transpose, be excised, or cause the chromosome on which it occurs to break. *Ds* affects the phenotype by blocking expression of the genes it transposes into as well as by causing the loss of alleles in acentric chromosome fragments lost when *Ds* breaks its chromosome.

Figure 15.14*a* is a purple corn kernel; a single dominant allele at the *C* locus is required for purple pigment production. In figure 15.14*b*, the *Ds* element has transposed into the *C* allele, inactivating it and rendering the kernel white. This corn kernel is from a plant that did not have the *Ac* element. Thus *Ds* could not move out of *C*. In figure 15.14*c*, the *Ac* element has again entered the genome by way of another genetic cross. In the presence of *Ac*, *Ds* leaves its site in some of the cells, restoring *C*-allele activity and therefore producing colored spots where those cells and their progeny are located in the kernel. Recently, *Ds* and *Ac* elements have been cloned and sequenced. They are typical transposons and very similar to each other. As might be expected, however, *Ds* has a deletion that prevents it from producing transposase. In order for *Ds* to transpose, *Ac* must provide the transposase. Several *Ds* elements that have been sequenced apparently arose from *Ac* elements by deletion.

It is interesting to note that one of Mendel's original seven characteristics of pea plants, wrinkled peas (*rr:* fig. 2.3), is caused by a transposon that inserts in the gene for starch-branching enzyme I. With this gene functional, branch-chained amylopectins are produced in addition to straight-chained amylose. With a failure in starch production, more sugar is present in these seeds, leading to greater osmotic pressure and therefore greater water content. Upon maturation, more water is lost from these seeds, resulting in greater shrinkage and wrinkling as compared to the wild-type (*RR* and *Rr*). The transposon that disrupts this gene is about 800 bp long and is very similar to the *Ds* transposon in maize.

In a sense, the *Ac-Ds* system represents "accidental" control of transcription by an invasive element that seems harmful (or at best neutral) to the organism. The next example, however, represents a highly evolved system whose alternative expressions are advantageous to the organism.

Control of Mating Type in Yeast

The mating type in yeast is determined by transposons. Haploid yeast cells exist in one of two mating types, *a* and *α*, determined by the *MATa* and *MATα* alleles. Homothallic strains of yeast switch mating types, as often as every generation. (The term **homothallic,** a misnomer, means that every cell is alike—each can mate with any other. The term was applied before it was realized that the cells were changing mating types.) Homothallism is determined by a dominant *HO* allele that codes for an endonuclease that initiates transposition. Strains that do not change mating type are **heterothallic,** determined by the recessive *ho* allele; that is, no active endonuclease is present to allow transposition and thus there is no change in mating type.

The ability to switch mating types in a single cell implies that both forms of the mating-type gene are present in each cell. In 1971, Y. Oshima and I. Takano proposed that mating type was controlled by a transpositional event, similar to the *Ac-Ds* system in corn or the flagellar phase in *Salmonella*. Later genetic and recombinant DNA studies confirmed the exact mechanism. Transposition is involved, but not the way Oshima and Takano originally suggested.

The third chromosome in yeast contains the mating-type locus (*MAT*). Silent (unexpressed) copies of the mating-type alleles are found on the left and right arms of the same chromosome (fig. 15.15). *HML* contains the silent *α* allele and *HMR* contains the silent *a* allele. In

Gene Expression: Control in Eukaryotes **417**

Figure 15.15

Role of transposition in control of mating type in yeast. (*a*) Mating-type loci on the third chromosome. *MAT* is the active mating-type locus. *HML* and *HMR* are silent loci, carrying the two mating-type alleles, α and a, respectively. (*b*) Transposition of *HML* to *MAT* results in the *MAT*α allele at the *MAT* site and the α mating type. (*c*) Transposition of *HMR* to *MAT* results in the *MAT*a allele being active, giving the a mating type.

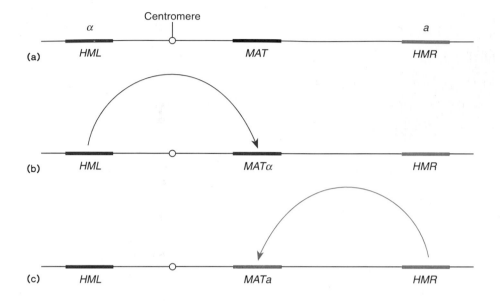

transposition, one or the other (*HMR* or *HML*) moves to the *MAT* site, replacing whatever allele was there to begin with. The *HMR* or *HML* alleles do not change, though. This mechanism of mating-type control in yeast has been called a **cassette mechanism.** The *MAT* site is analogous to a cassette player with *HMR* and *HML* as cassette tapes. The act of transposition brings a new tape to the cassette player.

*MAT*a and *MAT*α each begin a genetic cascade wherein certain genes are activated and others are repressed. For example, *MAT*α transcribes the *MAT*α1 gene that activates the transcription of an α-factor (a pheromone) gene and an a-factor (pheromone) receptor gene. (**Pheromones** are chemical signals, analogous to hormones, that convey information between individuals.) Conjugation requires the emitting of one type of pheromone and the reception of the other type: An α cell emits α factor and is receptive to a factor; an a cell emits a factor and is receptive to α factor. This switching of types, from an α to an a cell or vice versa is similar to homeotic control. It should therefore not be too surprising to know that the *MAT* genes have homeo boxes.

The effects of insertion by transposable elements can be demonstrated in other eukaryotes. *Drosophila* is proving to be useful in this area of research; not only does it have several classes of transposons, but *Drosophila* provides a good organism for cytological studies. Radioactive transposons used as probes can hybridize with the banded polytene chromosomes, showing exactly where these elements are located. In summary, then, transposons can

affect eukaryotic gene expression. However, with the exception of a few systems like mating-type determination in yeast, transposons appear to have a random, disruptive effect on developmental processes.

We now turn our attention to two specific genetic topics in development that are at the cutting edge of modern research, with great medical importance: immunogenetics and cancer. In immunogenetics, we are concerned with how a relatively small number of genes protects us from a myriad of different foreign agents; in cancer we are concerned with how the normal mechanisms controlling cell growth are lost.

IMMUNOGENETICS

Vertebrates have evolved the ability to protect themselves against invading bacteria and viruses and their own cancer cells. **Immunity** is this ability of an animal to resist infection. (We will limit our discussion to mammals, primarily mice and people.) The foreign substance from the bacteria, virus, or cancer cell that evokes an immune response is called an **antigen.** The immune response itself is a complex interaction of various cell types and other components. The immune system of an organism can selectively destroy thousands of kinds of antigens without harming its own cells, quite an amazing accomplishment.

The two major components of this system are the B and T lymphocytes, white blood cells originating in bone marrow and maturing in either the bone marrow (B cells) or the thymus gland (T cells). The B cells are responsible for producing very specific proteins called **antibodies,** or **immunoglobulins (Igs),** that protect the organism from antigens. Immunoglobulins can coat antigens so that they are more readily engulfed by phagocytes (white blood cells that engulf foreign material); immunoglobulins can combine with the antigens—for example, by covering the membrane-recognition sites of a virus—and thereby directly prevent their ability to function; or, in combination with complement, a blood component, immunoglobulins can lead to death if the antigen is an intact cell.

Whereas the B cells are producing immunoglobulins, one type of T cell is concerned with locating and destroying infected cells. The **cytotoxic T lymphocyte** attacks host cells that have been infected by a virus or bacterium. Thus infected cells are destroyed before new viruses or bacteria can be produced, helping to terminate the infection. Cytotoxic T lymphocytes recognize infected host cells by receptors on the surface of the T cells called **T-cell receptors.** These receptors recognize an infected host cell by two aspects of the infected cell's surface: **major histocompatibility complex** (MHC) gene products and antigens. The MHC proteins tell the T-cell receptor that it is dealing with a cell of its own body, and the antigen tells the T cell that the host cell is infected. All host cells have MHC components on their surfaces; an infected cell has the ability to cause part of the antigen to appear on its surface with the MHC protein, as if the MHC protein were "presenting" the antigen to the T-cell receptor (fig. 15.16).

The dual attack by B and T cells has three main components of genetic interest: antibodies (immunoglobulins), T-cell receptors, and products of the MHC. These three protein families are evolutionarily related to each other and each provides a tremendous diversity of protein products. We know the most about immunoglobulins and therefore we turn our attention to them.

Immunoglobulins

Immunoglobulins, produced by the B cells, are large protein molecules composed of two identical light (L) polypeptide chains (about 214 amino acids) and two identical heavy (H) chains (about 440 amino acids), held together by sulfhydryl bonds (fig. 15.17). Each polypeptide chain has a variable (V) and a constant (C) region of amino acid sequences. The variable regions recognize the antigens and thereby give specificity to the immunoglobulin (fig. 15.18). There are five major types of heavy chains (γ, α, μ, δ, and ϵ), giving rise to five types of immunoglobulins: IgG, IgA, IgM, IgD, and IgE. Each has slightly different properties; for example, only IgG can cross the placenta, giving immunity to the fetus. In addition, every immunoglobulin has one of two types of light chains, κ or λ (kappa or lambda).

Mutations of the constant region of the chains are called **allotypes** and follow the rules of Mendelian inher-

Figure 15.16
When a mammal (e.g., mouse) is infected by a virus, part of the viral coat is recognized as an antigen, triggering an immune response. B cells produce antibodies that specifically attach to the viral antigen. Infected cells "present" the antigenic part of the viral coat to the outside at the major histocompatibility complex protein on the cell surface. This MHC-antigen complex is recognized by the T-cell receptors, which then trigger the destruction of the infected cell.

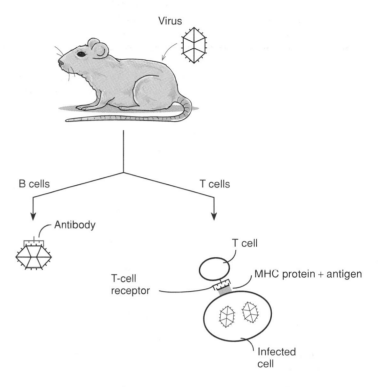

itance. In the variable region, however, diversity is much greater than two alleles per individual. The variation in this region is referred to as **idiotypic variation.** The average individual has the potential to express between 10^6 and 10^8 different immunoglobulins, each with a different amino acid sequence. The lower limit, 10^6, is arrived at through the study of persons with multiple myeloma, a malignancy in which one lymphocyte divides over and over until it makes up a substantial portion of that person's lymphocytes. From these persons we can isolate a relatively purified immunoglobulin that is the product of a single clone of cells and is referred to as a **monoclonal antibody.** A very low proportion of a normal person's lymphocytes produces any one specific immunoglobulin.

Multiple myeloma cells can be fused to spleen cells. The resulting cells, called **hybridomas,** producing monoclonal antibodies, can be perpetuated in tissue culture indefinitely, thus providing a ready supply of specific monoclonal antibodies. Recent work with hybridomas has allowed us to locate, isolate, and sequence immunoglobulin genes. How can one genome produce 10^8 different antibody molecules?

Figure 15.17

Schematic view of an immunoglobulin protein (IgG). V = variable region; C = constant region; L = light chain; H = heavy chain. The S-S bonds are sulfhydryl bridges across two cysteines. The NH_2 ends of the molecule form the antigen recognition parts. The internal sulfhydryl bonds roughly mark areas called domains, two each on the light chains and four each on the heavy chains. Similar domains are found in the T-cell receptors and the MHC proteins. These domains indicate the evolutionary relatedness of these three types of molecules.

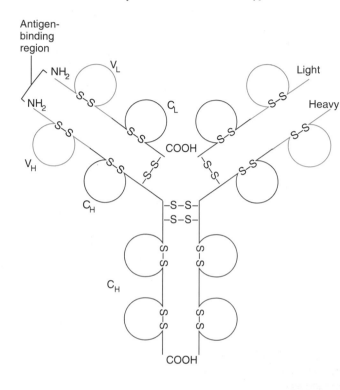

Antibody Diversity

Since the mammalian genome does not have 10^8 genes (even 10^6 genes is too high an estimate), different models were suggested to explain antibody diversity. In 1965, W. J. Dreyer and J. C. Bennett suggested that a given chain of an immunoglobulin was not the result of one gene but of a combination of genes, one for the constant region and one for the variable region. In addition, they suggested that a particular organism, in its haploid genome, had only one constant gene but several hundred or thousand variable genes. The final product would be the result of the combination of one of the variable genes and the constant gene. Modern recombinant DNA technology has verified the essence of that model.

In reality, several genes go into forming the variable regions of the heavy and light chains, given that we are using the word gene for DNA segments that will code for a part of the final heavy or light chain of the immunoglobulin. Genes for the κ, λ, and heavy chains are located on chromosomes 2, 22, and 14, respectively, in human beings and chromosomes 6, 16, and 12, respectively, in mice. Each is a multigene complex. Let us examine the κ light-chain gene complex in mice as an example of how the DNA must be modified in order to produce the final protein product (fig. 15.19).

The first step in DNA rearrangement is the joining of a V (variable) and a J (joining) gene in a B cell (fig. 15.19), a process called **V-J joining.** Since any one of three hundred V genes can combine with any one of five J genes,

Figure 15.18

The interaction of an antigen-binding region of an immunoglobulin with an antigen (the protein lysozyme, in this case). Note how the antigen (green) "fits" into the variable end of the immunoglobulin (a). In (b) the antigen has been pulled away. In (c) the connecting faces of the molecules have been rotated to show areas of fit or interaction. The part of the immunoglobulin molecule shown consists of one light chain (yellow) and the equivalent length of heavy chain (blue), as shown in (d).

(a–c) Courtesy of Dr. A. G. Amit.
Science 233 (August 1986):749, fig. 3.
"Three-Dimensional Structure of an Antigen-Antibody Complex at 2.8 Revolutions." Copyright 1986 by the AAAS.

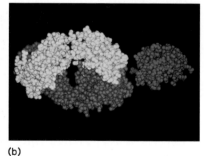

(a)

(b)

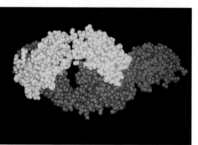

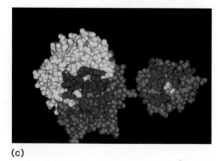

(c)

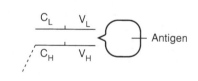

(d)

Figure 15.19

The complex for the *k* light chain in mice is composed of about three hundred variable genes ($V_{\kappa 1} - V_{\kappa 300}$), five joining genes ($J_{\kappa 1} - J_{\kappa 5}$), and one constant gene, C_κ, in the undifferentiated cell (germ line). The final *k* light chain will be composed of the products of one variable gene, one joining gene, and the constant gene.

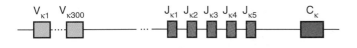

fifteen hundred different combinations are possible. Since we expect this to be another example of site-specific recombination as we saw with phage λ integration in chapter 13, there must be recombinational signal sequences present. Through DNA sequencing, these signals have been determined to be a heptamer (7 bp) and a nonamer (9 bp), separated by 12 bp on one side and 23 bp on the other (fig. 15.20). The two signals are oriented in opposite directions so that a stem-loop structure can form in either of the DNA strands. A single crossover then frees a circle of DNA containing intervening V and J genes between the two that are joined (fig. 15.20).

The point of crossover at the V-J junction is itself variable, generating **junctional diversity.** Not only are any two V and J genes capable of being brought together, but the sequence at the junction of the two genes can vary. For example, we see in figure 15.21 that the junction in the protein at amino acids 95 and 96 can be Pro-Trp, Pro-Arg, or Pro-Pro, depending on exactly where the crossover occurred.

In figure 15.22, we see the DNA after joining of $V_{\kappa 50}$ and $J_{\kappa 4}$. This gene is now transcribed. The region between $J_{\kappa 4}$ and C_κ (the constant gene) is then removed by RNA splicing, leaving the final mRNA product, which is then translated into a *k* light chain. (Note that the mRNA splicing can be done several different ways. Here an exon—J_5—is removed, an example of alternative splicing, mentioned in chapter 10.) In this cell the homologous *k* region is repressed as well as both λ regions, a phenomenon known as **allelic exclusion.** Thus this cell produces only one light chain, the $V_{\kappa 50}$-$J_{\kappa 4}$-C_κ protein.

Similar types of events take place in the heavy-chain gene, or in the λ light-chain gene if it had been active. There are some differences, however (fig. 15.23). The λ complex in the mouse has two variable genes with four each of J and C genes. The heavy-chain complex has about two hundred V genes, four J genes, and eight C genes of five major types ($\alpha, \gamma, \delta, \epsilon, \mu$), accounting for the five types of immunoglobulins (IgA, IgG, IgD, IgE, IgM). In addition, heavy-chain regions have another set of genes, called diversity (D) genes. There are at least a dozen such genes in the mouse heavy-chain complex, and they add still another variable region to the final protein. In the heavy chain, first D-J joining takes place, then V-DJ joining, and lastly splicing creates the final heavy-chain product (fig. 15.24).

In addition, further junctional diversity seems to be added in heavy-chain recombination by the addition of nucleotides in a template-free fashion. In other words, some mechanism, possibly involving the enzyme terminal transferase, adds nucleotides, called **N segments,** at the junctions, which are not specified in the DNA. For example, in one case, the sequence GTGGGGGCC (three codons long) was found at a J-D junction but not seen in the undifferentiated (germline) genome.

There is a final place in which variability is generated. Recent sequencing studies indicate that mutation occurs in variable regions after recombination has taken place. The mechanism of this specific mutagenesis, called **somatic hypermutation,** is not known. Given the number of variable, constant, joining, and diversity genes, as well as the diversity generated at the various joining junctions, it is easy to see how at least 10^8 different immunoglobulin combinations could be generated (table 15.1).

T-Cell Receptors and MHC Proteins

As we mentioned earlier, genetic diversity also exists in the T-cell receptor and the MHC (major histocompatibility complex) system. From its function (recognizing both the antigen and the MHC "self" gene product), it seems evident that the T-cell receptor must show the type of diversity that immunoglobulins have. In fact, the T-cell receptor genes are very similar to the immunoglobulin genes. T-cell receptors are composed of α and β subunits; there are V, J, and C components of the α subunit and V, J, D, and C components of the β subunits (fig. 15.25). (Other proteins, γ and δ, seem to be important developmentally but rare in adults.)

The MHC genes (called the H-2 locus in mice and the HLA locus in human beings) comprise a region of from 2,000 to 4,000 kb made up of many genes. The genes are generally referred to as class I, II, and III genes. Class I genes code for the MHC proteins that we have discussed, involved in cytotoxic T-cell recognition of host cells that have been infected, and also, parenthetically, involved in transplant rejection. Class II genes produce proteins involved in intercellular communication among B and T cells and other immune functions. Class III genes code for proteins in the complement system, which is involved in the destruction of foreign cells. A class I MHC protein (α protein) associates with a separate gene product, β_2 macroglobulin, to form the membrane-bound MHC protein

Figure 15.20

In order for V-J joining to take place (site-specific recombination), there must be a signal at the V side and another at the J side. One signal is a heptamer (7 bp) and a nonamer (9 bp) separated by 23 bp, and the other signal is the same heptamer and nonamer separated by 12 bp, in the reverse orientation. $V_{\kappa n}$ represents any of the three hundred V genes and $J_{\kappa m}$ represents any of the five J genes. The signals allow single strands of the DNA to form stem-loop structures that can be processed to cut out the intervening V and J genes.

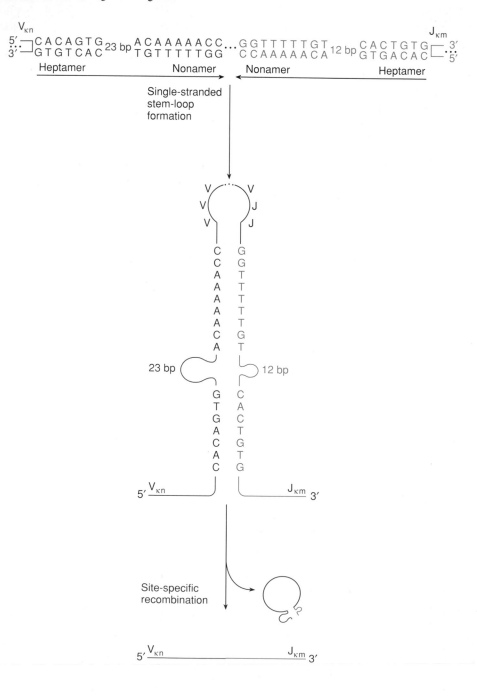

Figure 15.21

Variability in crossing over during the process of V-J joining generates junctional diversity. In this case $V_{\kappa 41}$ and J_5 of the mouse are shown. Amino acid codons 94 and 97 are always the same, TCT and ACG, respectively, as are the first two bases of codon 95, CC. Depending on the exact point of crossover, five different codon pairs can be generated. Codons for Pro are the first of the pair (95) and codons for Trp, Arg, or Pro are the second (96). Matching numbered arrows indicate crossover points for the five possibilities.

Source: Data from E. E. Max, et al., "Sequences of Five Potential Recombination Sites Encoded Close to an Immunoglobulin κ Constant Region Gene" in Proceedings of the National Academy of Sciences, 76:3450–3454, 1979.

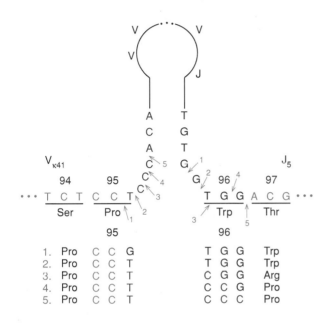

Figure 15.22

V-J joining in the κ region of the mouse. In this example $V_{\kappa 50}$ is joined to J_4 and then to C_κ. First, V-J joining takes place using the heptamer-nonamer signals shown in figure 15.20. Then transcription of the region from the $V_{\kappa 50}$ to the C_κ genes takes place. Splicing the RNA removes the region containing the extra J gene, J_5. The final RNA, containing $V_{\kappa 50}$-J_4-C_κ, is then translated into the κ light chain.

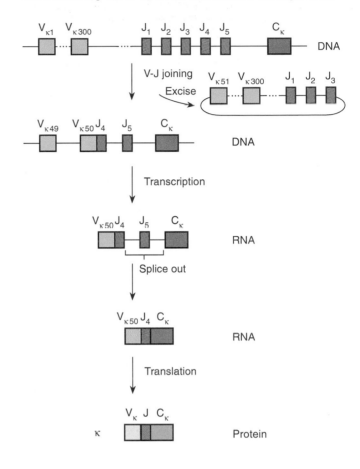

Figure 15.23

Arrangement of the genes in the light and heavy chains of the immunoglobulin complexes of the mouse. Note that the λ complex has the genes in a more integrated, rather than grouped, arrangement. Also the constant region of the heavy chain is made up of eight genes, including variants of the γ gene.

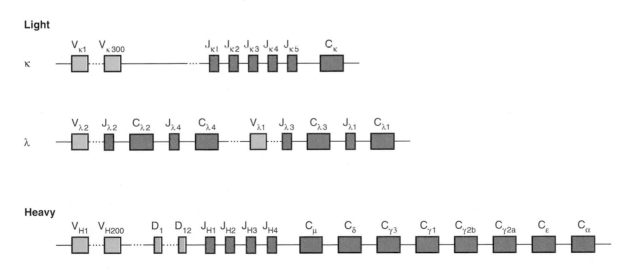

Figure 15.24

Formation of an immunoglobulin heavy chain (IgM). First, D-J joining takes place followed in a similar manner by V-DJ joining. In each case intervening DNA is spliced out by the site-specific recombination forming the join. Then, as in light-chain formation (fig. 15.22), the modified region is transcribed; RNA processing (splicing) then brings the final regions together, which, when translated, form the V-D-J-C heavy chain.

Source: Data from F. W. Alt, et al., "Development of the Primary Antibody Repertoire" in Science, 238:1079–1087, 1987.

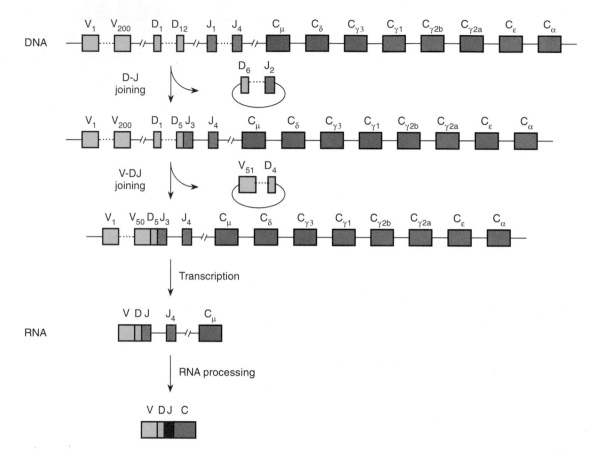

TABLE 15.1   Three Hundred Immunoglobulin Genes Can Generate Eighteen Billion Different Antibodies

Source	Factor
Light Chains	
V genes	150×
J genes	5×
V-J recombination	10× = 7,500×
Heavy Chains	
V genes	80×
D genes	50×
J genes	6×
V-D, D-J recombination	100× = 2,400,000 ×
Total	7,500 × 2,400,000 = 18 billion

Source: After P. Leder, 1982.

(fig. 15.26). In figure 15.16, we saw how the T-cell receptor on a cytotoxic T cell might recognize an antigen presented by the MHC class I protein on an infected cell.

We still have much to learn about control mechanisms involved in the exact formation of antibodies and general control mechanisms at the cellular level by which the entire immune system responds to antigen invasion. The immune system, however, sometimes cannot protect us from our own cells when control of their growth is lost.

CANCER

Cancer is an informal term for a diverse class of diseases marked by abnormal cell proliferation. Control of normal cell development is lost, and cells proliferate at an inappropriate rate to form growths known as **tumors (neoplasms).** Benign tumors grow only in one place and do not invade other tissues. The cells of malignant tumors not

Figure 15.25
The T-cell receptor is made up of two protein chains, α and β, anchored in the cell membrane. Each protein is similar to an immunoglobulin chain and is created the same way, with V-J types of joining taking place. (*a*) The α gene complex is composed of numerous V and J genes and one constant gene. The β gene complex has V, D, J, and C genes. (*b*) Each protein chain has two domains, similar to the domains of the immunoglobulins, indicating a common evolutionary ancestry.

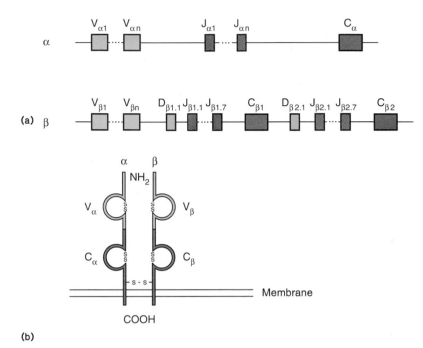

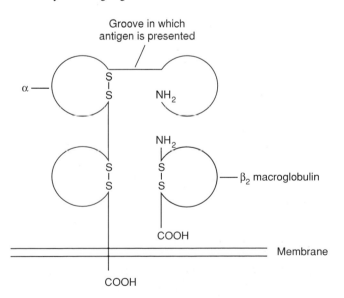

(b)

Figure 15.26
The MHC (major histocompatibility complex) class I protein is composed of two protein chains. The α chain is composed of three domains similar to the immunoglobulins and T-cell receptors. The second chain is β_2 macroglobulin. Antigens are presented by the MHC class I proteins to the T-cell receptors to signal that the cell has been invaded by the foreign agent.

only continue to proliferate, but also invade nearby tissues or, by a process called **metastasis,** spread to distant parts of the body through blood or lymph vessels and start new centers of uncontrolled cell growth wherever they go.

Cancers are generally divided into four groups, dependent on the type of cells originally involved. Two types of cancer cause overproduction of white blood cells. **Leukemias** are diseases of the bone marrow that cause excessive production of leukocytes, which originate in the bone marrow. **Lymphomas** are diseases of the lymph nodes and spleen that cause excessive production of lymphocytes, which originate in the lymph nodes and spleen. **Sarcomas** are tumors of tissue that arise from the embryological mesoderm such as muscle, bone, and cartilage. About 85% of cancers are **carcinomas,** tumors arising from epithelial tissue such as glands, breast, skin, and linings of the urogenital and respiratory systems.

Most evidence indicates that cancers are *clonal*—that is, they arise from a single aberrant cell that then proliferates. Analyzing the cause, or causes, of cancer comes down to trying to understand how one cell is changed, or *transformed,* from a normal cell to a cancerous one.

Currently, there are two theories regarding the causes of cancer: the mutation theory and the virus theory. The mutation theory states that most cancers are caused by

TABLE 15.2 Neoplasms with a Known Consistent Chromosomal Defect*

Disease	Chromosomal Defect	Breakpoints or Deletion
Leukemias		
Chronic myelogenous leukemia	t(9; 22)	9q34.1 and 22q11.21
Acute nonlymphocytic leukemia	t(9; 22)	9q34.1 and 22q11.21
Chronic lymphocytic leukemia	+12	
Acute lymphocytic leukemia	t(9; 22)	9q34.1 and 22q11.21
Lymphomas		
Burkitt's	t(8; 14)	8q24.13 and 14q32.33
Small cell lymphocytic	+12	
Carcinomas		
Neuroblastoma, disseminated	del 1p	1p31p36
Small cell lung carcinoma	del 3p	3p14p23
Papillary cystadenocarcinoma of ovary	t(6; 14)	6q21 and 14q24
Constitutional retinoblastoma	del 13q	13q14.13
Wilms' tumor	del 11p	11p13

From J. Yunis, "The Chromosomal Basis of Human Neoplasia" in *Science*, 221: 227–236. Copyright © 1983 by the AAAS.
*Subbands, denoted by a decimal-digit system, were defined by using high-resolution banding techniques; t refers to translocation, del to deletion, and "+" to an additional chromosome.

genetic mutation. The viral theory states that most cancers are viral in origin. These theories are becoming one unified theory. In essence, cancers are genetic diseases; they result from the inappropriate activity of certain genes whether those genes changed from mutation or were imported or activated by viruses.

Mutation Theory of Cancer

Mutations, both point and chromosomal, have been implicated in carcinogenesis. For example, the disease **xeroderma pigmentosum** in human beings is caused by mutations that inactivate the UV-mutation repair system (chapter 16) so that exposure to the sun results in skin lesions that often become malignant. A related disease, **ataxia-telangiectasia,** is a defect in X-ray-induced repair mechanisms. Persons with this defect are at risk for acute and chronic leukemia, lymphomas, and, in women, ovarian cancers.

People with certain other diseases have a higher than normal risk of developing cancer. **Fanconi's anemia** is a syndrome of malformations of the heart, kidney, and extremities, pigmentary changes of the skin, and changes in the bone marrow. It is associated with acute leukemia and with cancers of the skin, liver, and esophagus. Relatives of people with Fanconi's anemia are also prone to malignancies; the risk of cancer mortality before age forty-five for heterozygotes is increased three- to sixfold. On the basis of risk and gene frequency, as many as 1% of persons dying before age forty-five from any malignancy may be heterozygous for Fanconi's anemia, and as many as 5% may be heterozygous for ataxia-telangiectasia.

Most cancers are associated with chromosomal defects; improved chromosomal banding techniques have demonstrated that a specific chromosomal defect is often associated with a specific cancer (table 15.2; fig. 15.27). The implication is that because of the new location of a gene (by translocation or deletion of intervening material) that gene may fall under the control of more powerful promoters or hybrid promoters outside the range of that gene's normal control. As we shall see, very often genes that are known to be able to transform cells (**oncogenes**) are the ones that are relocated into regions of new control. These oncogenes then become more active and transformation follows.

Cancer-Family Syndromes

Except for malignancies that are known to be controlled by a single gene, such as xeroderma pigmentosum, clinicians disagree as to whether or not a predisposition for malignancies is inherited. One view holds that the human population is heterogeneous, with a low-risk group and a high-risk group. When four thousand clinic registrants were interviewed, almost half reported virtually no family history of cancer, whereas about 7% reported that many family members had cancer. This 7% was considered cancer prone on the basis of the fact that three or more close relatives of the interviewed person had cancer. The interpretation of the study is that some families are predisposed toward cancer but most are not, rather than that everyone in the population has a uniform and low probability of developing cancer. Lending support to this interpretation are the **cancer-family syndromes,** in which family members seem to inherit a nonspecific predisposition

Figure 15.27

High resolution G-banded chromosomes from patients with (*a*) non-Burkitt small-cell lymphoma showing the translocation of material from chromosome 8 to chromosome 14 [t(8; 14)]; (*b*) follicular small-cleaved-cell lymphoma showing the translocation of material from chromosome 18 to chromosome 14 [t(14; 18)]; (*c*) chronic myelogenous leukemia showing the translocation of material from chromosome 22 to chromosome 9 [t(9; 22)]; (*d*) constitutional retinoblastoma showing the deletion of band 13q14; and (*e*) constitutional Wilm's tumor showing the deletion of band 11p13. *Arrows* show break and translocation points and *brackets* indicate normal regions (*wide brackets*) and regions of band deletions (*narrow brackets*).

J. Yunis, "The Chromosomal Basis of Human Neoplasia." Science 221 (July 15, 1983):227–236. Copyright 1983 by the AAAS.

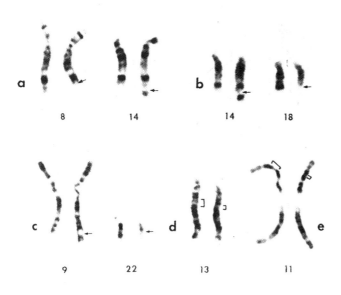

Figure 15.28

Pedigree of a type I cancer-family syndrome. This is interpreted as the inheritance of the propensity toward cancer rather than the inheritance of any specific type of cancer.

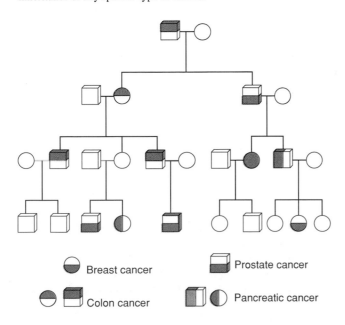

◐ Breast cancer

◑▤ Colon cancer

▥ Prostate cancer

▯◑ Pancreatic cancer

toward tumors of various types. At least two cancer-family syndromes are known. In figure 15.28, we see a pedigree of a cancer-family syndrome in which the predisposition for several different types of cancers, rather than a particular type of cancer, seems to be inherited. Women get breast, colon, and pancreatic cancers, whereas men get colon, prostate, and pancreatic cancers. There also seem to be genes that suppress cancer.

Tumor-Suppressor Genes

Recently, a new class of cancer-related genes has been discovered, the so-called **tumor-suppressor genes** (also called anti-cancer or **anti-oncogenes**). These genes seem to act by suppressing malignant growth; in the homozygous recessive state, however, cancer ensues. The first tumor-suppressor gene to be isolated was the gene for **retinoblastoma,** a tumor of retinoblast cells, which are precursors to cone cells in the retina of the eye. This is a disease of young children because after the retinoblast cells differentiate, they no longer divide and apparently can no longer form tumors. The disease occurs both in a hereditary and a sporadic form. Both forms are presumably due to the recessive homozygous (or, in some cases, hemizygous) state of the locus. In the hereditary form, individuals inherit one mutant allele; a second mutation generates the disease. In the sporadic form (with identical symptoms), apparently both alleles have mutated spontaneously in the somatic tissue of the retina. Alfred Knudsen made this suggestion in 1971, that two mutations would be required for retinoblastoma. The retinoblastoma gene has also been implicated in other cancers including sarcomas and carcinomas of the lung, bladder, and breast.

How do we know that retinoblastoma results from the loss of suppression rather than simply the activity of an oncogene? This question was answered by J. Yunis, who examined cells from several retinoblastoma patients. He found that there was frequently a deletion of part of chromosome 13, specifically band q14 (see fig. 15.27*d*). Yunis noticed that the exact points of deletion varied from individual to individual, indicating that the phenomenon was due to loss of gene action rather than enhancement of gene activity due to the new placement of genes previously separated by the deleted material.

Further support of tumor-suppressor genes came from work by E. Stanbridge and his colleagues with another childhood cancer, **Wilm's tumor,** a kidney cancer that is also believed to be caused by the loss of activity of a tumor-suppressor gene. It is associated with the loss of band p13 on chromosome 11 (see fig. 15.27*e*). Researchers introduced a normal chromosome 11 into Wilm's tumor cells growing in culture. The result was normal cell growth, exactly what would be predicted if the normal chromosome introduced a suppressor—that is, if the normal gene was a tumor-suppressor gene.

Figure 15.29
Malignancy can result from lack of the retinoblastoma repressor protein either through homozygous deletion (or other mutagenesis) or through inactivation of the repressor by a viral oncogene protein. If the retinoblastoma gene itself is inactivated, retinoblastoma will result during early development of the individual. If the repressor is inactivated later in development, the inactivating agent (e.g., adenovirus) can cause other types of malignancies. For example, the E1A protein of adenovirus will inactivate the retinoblastoma repressor, leading to malignant growth in tissue culture.

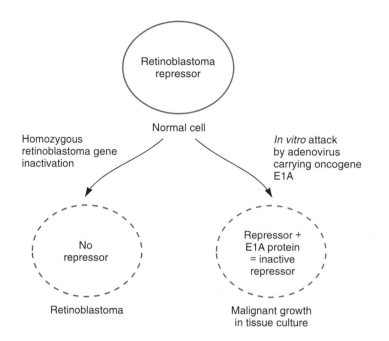

We should mention here that other loci, not located on the short arm of chromosome 11, have been implicated in causing Wilm's tumor. Also, children with the condition frequently have had a loss of the maternally inherited chromosome rather than either maternal or paternal chromosome randomly. This fact is consistent with a phenomenon, currently being investigated, called **imprinting** (or **molecular** or **parental imprinting**) in which there is differential expression of a gene depending on whether that gene was maternally or paternally inherited. It has been seen in a few examples in *Drosophila,* mice, and, as above, in human beings. We discuss it in greater detail in chapter 17.

The retinoblastoma gene has been isolated and cloned. The gene specifies a 105-kilodalton protein (p105) found in the nucleus, as would be expected if it were a suppressor of DNA transcription. It binds with at least three known oncogene proteins: the E1A protein of adenovirus, the SV40 large T antigen, and the 16E7 protein of human papilloma virus, a virus associated with 50% of cervical carcinomas. The implications of these findings are that these three viruses may use a similar mechanism in transformation, and this mechanism may involve inactivation of the retinoblastoma p105 protein (fig. 15.29).

A third tumor-suppressor gene is the *p53* gene, named for its 53 kilodalton protein product. This gene has been implicated in a wide range of cancers including leukemia, carcinomas of the lung and breast, brain tumors, bone cancers, and soft-tissue sarcomas. In fact, it seems at the moment to be the most common genetic change found in human cancers. The gene is located in band p13 of chromosome 17. Recently, mutation of the *p53* gene was shown to be the cause of a cancer-family syndrome, resulting in the range of tumors mentioned above in different individuals in the pedigree.

Viral Theory of Cancer

Retroviruses
Animal viruses come in many different varieties, with DNA or RNA as their genetic material (fig. 15.30). Several classes of viruses, both DNA and RNA, can transform cells. Transformation may or may not be caused by an oncogene carried by the virus. Some DNA viruses carry oncogenes, such as the adenovirus mentioned previously that carries the gene for the E1A protein, which may act by binding to the retinoblastoma repressor protein. Oncogenes, however, were originally discovered in retroviruses, a group of very simple RNA viruses that contain the enzyme reverse transcriptase that, after the virus enters the host cell, converts the viral RNA into DNA. In 1910,

Figure 15.30

Representatives of families of animal viruses. The abbreviations ss and ds refer to single-stranded and double-stranded, respectively. The viruses are drawn to size. The plus (+) and minus (−) signs define the strand of an RNA virus. Plus strands can serve as messengers; minus strands are complementary to the messenger strand. The retrovirus is about 100 nm in diameter.

From R. E. F. Matthews, Intervirology, *12:3–5. Copyright © 1979 S. Karger AG, Basel, Switzerland as appeared in Watson, et al.,* Molecular Biology of the Gene, *Fourth Edition., Volume II. Copyright © 1987 Benjamin/ Cummings Publishing Company, Menlo Park, CA. Reprinted by permission.*

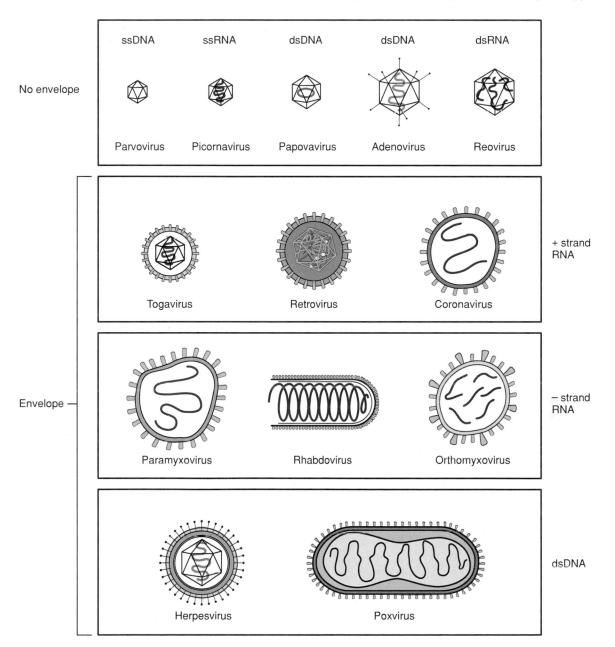

Peyton Rous, who later won the Nobel Prize for his work, discovered that a sarcoma in chickens could be induced by a cell-free extract from a tumor in another chicken. The transmitted agent was found to be a retrovirus, later named Rous sarcoma virus. This was the first retrovirus to be discovered.

The retrovirus, which usually carries only three genes, integrates into the host genome in a series of steps (fig. 15.31). When the virus enters the host, it is in the form of a plus (+) RNA strand (capable of acting as an mRNA; the minus [−] strand is the complement to the + strand). At either end is a repeated sequence (R) located outside two unique sequences (U3 and U5). Through the process of reverse transcription, using the reverse transcriptase brought in by the virus, the viral RNA is converted to a double-stranded DNA. During that process, the ends of the DNA take on the configuration of long terminal repeats (LTRs), repetitions of U3-R-U5. The linear DNA

Figure 15.31

A retrovirus RNA genome. R is a repeated sequence; U3 and U5 are unique sequences; PBS is the primer-binding site, LTR is the long terminal repeat; *gag, pol,* and *env* are viral genes. During the process of reverse transcription, the LTRs are created at the ends of the DNA. Direct host repeats are created when the viral DNA integrates into the host chromosome.

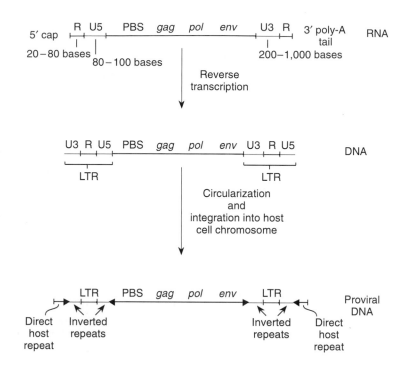

Peyton Rous (1879–1970)
Courtesy of Rockefeller University Archives.

then circularizes and integrates into the host genome in the same manner that a transposon does, generating short direct repeats at either end. Since some transposons have been found to involve an intermediate RNA stage, like retroviruses, together these transposons and retroviruses are referred to as **retrotransposons (retroposons).**

In figure 15.32 we outline the relatively complex steps by which the viral RNA is converted to double-stranded DNA by reverse transcription. In the process, the long terminal repeat is created at each end. In a novel twist, a tRNA acts as the primer to DNA replication, being complementary to the primer-binding site (PBS) of the viral

genome. This tRNA is usually brought in with the virion from the previous host, already base paired at the PBS. In frame 1 of figure 15.32, we show the viral RNA in circular form, although the ends are not connected. The tRNA primer is base paired at the PBS, and the first region of DNA has been synthesized (minus strand). The R and U5 regions of the viral RNA are then degraded by RNase activity (frame 2). The R region of the newly synthesized DNA then hybridizes to the R region of the 3′ end of the viral RNA in what is called the first "jump" (frame 3). The DNA strand is then elongated all around the viral RNA, ending in the PBS, thereby displacing the tRNA primer (frame 4). You can see that the DNA strand now has a U3-R-U5 sequence at one end, the sequence referred to as the long terminal repeat (LTR).

The original RNA is now further degraded from both ends (frame 5). The remaining RNA acts as a primer for the beginning of synthesis of the plus DNA strand. The LTR is first replicated, followed by the PBS using the PBS-complementary portion of the tRNA as a template (frame 6). The tRNA and the rest of the original viral RNA is now degraded (frame 7). The two DNA strands are now formed into a double helix at their PBSs (the second "jump," frame 8). Held together only at the PBS region (frame 9), both strands of DNA undergo replication, each using the other as a template. In frame 10 we see the process completed, and in frame 11 the double-stranded DNA

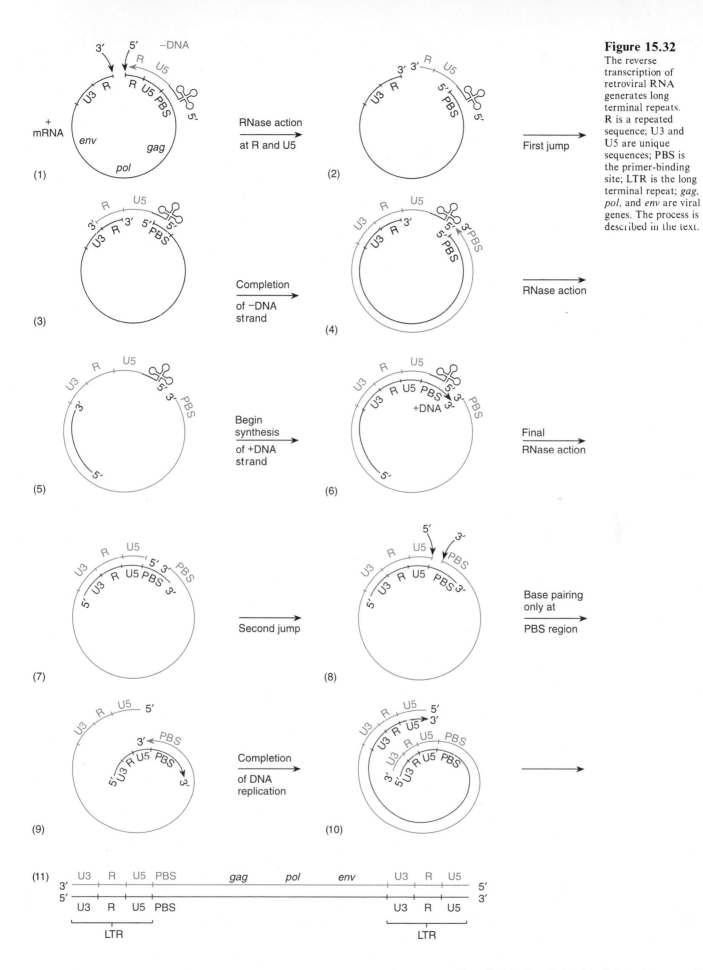

Figure 15.32
The reverse transcription of retroviral RNA generates long terminal repeats. R is a repeated sequence; U3 and U5 are unique sequences; PBS is the primer-binding site; LTR is the long terminal repeat; *gag, pol,* and *env* are viral genes. The process is described in the text.

Figure 15.33

Expression of a retroviral mRNA. Translation occurs of the *gag* gene and occasionally, due to read-through, of the *gag-pol* genes. The result is core virion proteins and the enzymes reverse transcriptase, protease, and integrase. Splicing must take place before *env* can be translated. Cleavage of the primary transcript and some modification results in envelope glycoproteins.

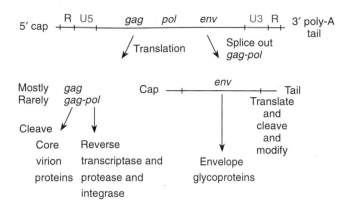

is again depicted linearly. Circularization followed by integration (see fig. 15.31) then converts the double-stranded form of the retrovirus to an integrated provirus. As a provirus, the retrovirus is replicated from generation to generation with the host chromosome. By transcription, the provirus produces RNA that either acts as an mRNA for viral function or is incorporated into new viral particles that are then released from the infected cell.

As mentioned in chapter 12, a retrovirus minimally contains only the *gag, pol,* and *env* genes. Depending on the particular virus, different mechanisms of gene expression are possible. One mode is depicted in figure 15.33. The viral mRNA is translated starting with *gag.* There is a translation termination signal at the end of the *gag* gene that is occasionally read through, resulting in a *gag-pol* protein. The *env* gene is translated only after the viral RNA is spliced to remove the *gag-pol* region. The protein products of all three genes are further modified by cleavage and other changes (phosphorylation and glycosylation), resulting in core virion proteins from *gag*, reverse transcriptase, protease, and integrase from *pol,* and envelope glycoproteins from *env.*

As we mentioned before, retroviruses can cause cellular transformation directly from their integration or from the oncogenes they carry. Transformation from integra-

TABLE 15.3   Oncogenes, Their Origin, and Their Protein Products

Oncogene	Virus and Species of Origin	Function of Products
abl	Abelson murine leukemia virus (mouse)	Tyrosine kinase
fes*	ST feline sarcoma virus (cat)	
fps*	Fujinami sarcoma virus (chicken)	
fgr	Gardner-Rasheed feline sarcoma virus (cat)	
ros	UR II avian sarcoma virus (chicken)	
src	Rous sarcoma virus (chicken)	Tyrosine kinase
yes	Y73 sarcoma virus (chicken)	
erbB	Avian erythroblastosis virus (chicken)	Tyrosine kinase
fms	McDonough feline sarcoma virus (cat)	Growth factor
raf*	3611 Murine sarcoma virus (mouse)	
mil(mht)*	MH2 virus (chicken)	
mos	Avian myeloblastosis virus (chicken)	Protein kinase
sis	Simian sarcoma virus (woolly monkey)	Growth factor
Ha-*ras*	Harvey murine sarcoma virus (rat)	Bind guanosine triphosphate
Ki-*ras*	Kirsten murine sarcoma virus (rat)	Bind guanosine triphosphate
fos	FBJ osteosarcoma virus (mouse)	Bind DNA, controls transcription
myb	Avian myeloblastosis virus (chicken)	DNA binding, transcription regulation
myc	MC29 myelocytomatosis virus (chicken)	
*erb*A	Avian erythroblastosis virus (chicken)	DNA binding
ets	E26 virus (chicken)	
rel	Reticuloendotheliosis virus (turkey)	Bind, DNA—transcription factor
ski	Avian SKV770 virus (chicken)	
jun	Avian sarcoma virus 17 (chicken)	Transcription factor

*fes and *fps* are feline and avian versions of the same oncogene; *raf* and *mil(mht)* are murine and avian oncogene counterparts.
From J. Marx, "What Do Oncogenes Do?" in *Science,* 223:673–676, February 17, 1984. Copyright 1984 by the AAAS.

tion comes about presumably because the provirus somehow either inactivates a tumor-suppressor gene or activates an oncogene in a process called **insertion mutagenesis.** The U3 region of the retrovirus contains both an enhancer and a promoter. Since there is an LTR at either end of the provirus, cellular genes can be turned on when the virus integrates. Currently, however, most research effort in this area is centered on the oncogenes themselves.

Oncogenes

By genetic analysis and recombinant DNA studies, Rous sarcoma virus was found to transform cells through the action of a single gene. This gene, called *src* for sarcoma, was the first viral oncogene discovered. Since then at least fifty have been discovered, and each has been given a three-letter designation (table 15.3). Unlike tumor-suppressors, which lead to cancer when in the homozygous mutant condition, oncogenes act in a dominant fashion: Only one copy of the activated gene need be present for transformation to occur.

With the viral oncogene in hand, researchers could create a probe for the gene and look within the DNA of the host organism. To the surprise of virtually everyone, these oncogenes were found in untransformed cells. Since cellular oncogenes have introns and viral oncogenes do not, and since transforming viruses can function quite well as viruses without their oncogenes, geneticists generally accept the theory that these oncogenes originated in the host and were picked up, presumably as mRNAs, by the retroviruses. In other words, genes that function normally within a host organism were picked up by retroviruses, and upon infecting another host, transformed that host cell into a cancerous cell.

To distinguish oncogenes within viruses and hosts, we prefix the name of a viral oncogene, such as *src,* with a v (v-*src*) and a cellular oncogene with a c (c-*src*). Cellular oncogenes within a nontransformed cell are also called **proto-oncogenes.** How are proto-oncogenes induced to become oncogenes, and what do proto-oncogenes normally do in the cell?

Oncogene Induction

Proto-oncogenes can be induced, we believe, three different ways. First, a mutation can cause a proto-oncogene to transform its host cell. For example, a *ras* proto-oncogene (table 15.3) was converted to an oncogene when one codon, GGC (glycine), was converted to GTC (valine). Second, a proto-oncogene can be activated if it is moved to a region with a strong promoter or enhancer. Burkitt's lymphoma, for example, is associated with a translocation involving the proto-oncogene c-*myc,* which is normally located on chromosome 8 (fig. 15.34). When translocated to chromosome 14, c-*myc* is placed contiguous with the immunoglobulin IgM constant gene (*Ig-C$_\mu$*). This gene is very active in lymphocytes. Hence c-*myc* is now transcribed at a much higher rate than normal, re-

sulting in cellular transformation. The c-*myc* gene normally occurs near a **fragile site,** a region of a chromosome that has a tendency to break. Many proto-oncogenes occur near fragile sites on chromosomes. The simple capture of a gene by a retrovirus might be enough for transformation since the gene is brought under the influence of viral transcriptional control. However, not all genes captured this way are oncogenes. Third, a proto-oncogene can be activated if it is amplified. Several cases are known in which amplified genes (e.g., c-*ras* and c-*abl*) or genes on trisomic chromosomes are related to transformation.

Viral oncogenes can cause transformation by the same mechanisms. Either a mutation of the oncogene itself or the placement of the gene next to an active viral promoter can cause high levels of transcription of the oncogene and hence transformation of the cell. What are the gene products of proto-oncogenes?

Oncogene Function

We know that proto-oncogenes are important to the cell because they have been conserved evolutionarily. For instance, c-*src* is found in fruit flies (*Drosophila*) as well as in vertebrates; c-*ras* is found in yeasts and in human beings. We believe that they are all genes that can promote growth of cells. At the present moment, the known protein products of oncogenes can be classified into at least four categories: tyrosine kinases, growth factors, GTP-binding proteins, and DNA-binding proteins (see table 15.3). As proto-oncogenes, they seem to function normally at low levels. In transformed cells, oncogenes function at high levels. Can these proteins explain cancerous growth?

Tyrosine kinases are enzymes that add a phosphate group to tyrosine residues in proteins. Most kinases phosphorylate serine and threonine. (These three amino acids have OH groups available for phosphorylation.) Proteins that are phosphorylated at their tyrosine residues appear to be involved in cytoskeleton shape (transformed cells are shaped differently than normal cells) and in glycolysis (cancer cells tend toward the anaerobic glycolytic pathway). Overactivation of the cellular oncogene could result in inappropriate kinase activity, thereby changing many of the cellular activities leading to the condition we call cancer.

The c-*sis* gene seems to encode platelet-derived growth factor, which stimulates cells to grow. Its potential in transformation is obvious. GTP-binding proteins, the product of v-*ras,* for example, can play a role in transmitting endocrine signals across membranes. Perhaps increased quantities of GTP-binding proteins send continuous or amplified signals to certain cells and thus enhance growth. The v-*myc* gene product is a protein that binds to DNA. Its role is not known, but if it were to signal continuous DNA replication, it too could induce transformation.

BOX 15.2

AIDS

"It is a modern plague: the first great pandemic of the second half of the 20th century. The flat clinical-sounding name given to the disease by epidemiologists—acquired immune deficiency syndrome—has been shortened to the chilling acronym AIDS." So began a 1987 article by Robert C. Gallo of the National Cancer Institute, codiscoverer, with Luc Montagnier of the Pasteur Institute of Paris, of the causative agent of AIDS. Gallo isolated the first human retrovirus, HTLV-I (human T-lymphotropic virus type I), in 1980. HTLV-I causes adult T-cell leukemia. Together with Montagnier, Gallo discovered HIV, the human immunodeficiency virus, in 1983 (Box fig. 1). It is a new retrovirus, never before seen, causing a disease first diagnosed in 1981 among young male homosexuals in the United States. Although it is not a cancerous disease, it is caused by a retrovirus, and thus merits our attention here.

The AIDS virus attacks a subset of T cells, T4 cells (lymphocytes that mature in the thymus gland), vital to the immune system. A particular protein on the surface of these T cells, called CD4, is a receptor for the HIV virus coat protein, gp120 (Box fig. 2). Also attacked are macrophages, another group of cells involved in the immune response. With destruction of the T cells, a person's immune system loses the ability to fight off common diseases. Persons who develop the disease frequently fall victim to other, opportunistic diseases such as pneumonia caused by the protozoan *Pneumocystis carinii;* Kaposi's sarcoma, a rare cancer found in people taking immunosuppressive drugs; and several other conditions, normally rare but found in people with suppressed immune systems. These conditions collectively became known as the acquired immune deficiency syndrome.

AIDS has spread throughout most of the world. It is believed to have originated in Africa and may have been localized to small isolated populations for twenty to one hundred years. Presumably, modern migration patterns brought some affected people into cities and from there the disease spread. There seem to be two worldwide patterns of spread of AIDS, which does not appear to be spread by casual contact. In the New World, Australia, and Western Europe, the disease is spread primarily by homosexual men and intravenous drug users. They are the groups at highest risk. In Africa and the Caribbean, the disease is primarily spread during heterosexual sex. At the moment, Eastern Europe, Asia, and North Africa have relatively low infection rates. In the United States, there were over

Box figure 1
An AIDS virus buds from an infected cell, resulting in a free viral particle (*bottom*).
J. Marx, "The AIDS Virus—Well-Known But a Mystery," Science 236 (April 24, 1987):391. Copyright 1987 by the AAAS.

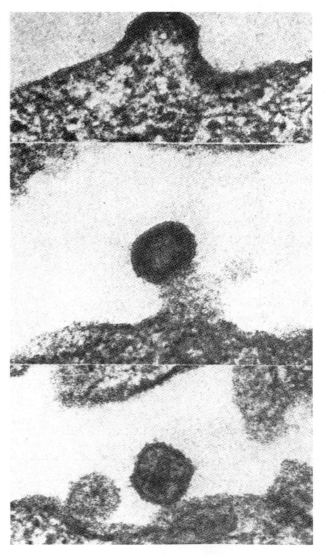

Robert C. Gallo (1937–   )
Courtesy of Dr. Robert Gallo.

Luc Montagnier (1932–   )
Courtesy of Dr. Luc Montagnier.

Continued on next page

BOX 15.2—*continued*

170,000 individuals with AIDS reported as of mid-1991, with estimates that there are about one million persons infected with the disease. Most of those who got the disease before 1986 have died. However, the infection rate seems to have peaked in the United States in 1985; unfortunately, the only mode of increasing infections is heterosexual sex. There is currently a second form of the disease caused by the HIV-2 virus, and similar viruses have been isolated from other primates.

The HIV retrovirus is complicated. Not only does it have the *gag, pol,* and *env* genes, but it has at least eight other genes of which at least three are regulators (Box fig. 3). There is much overlap in the genome, with different genes at the same place translated in different reading frames. The regulatory genes *tat* and *rev* are each made of two exons with an intron removed to form the final protein. The protein *tev/tnv* is composed of sequences in the *tat,*

Box figure 2
AIDS virion structure. Numbers associated with proteins are kilodalton masses (e.g., gp120 is a 120-kilodalton protein).

From "What Science Knows about AIDS," by George Kelvin. Copyright © 1988 by Scientific American, Inc. All rights reserved.

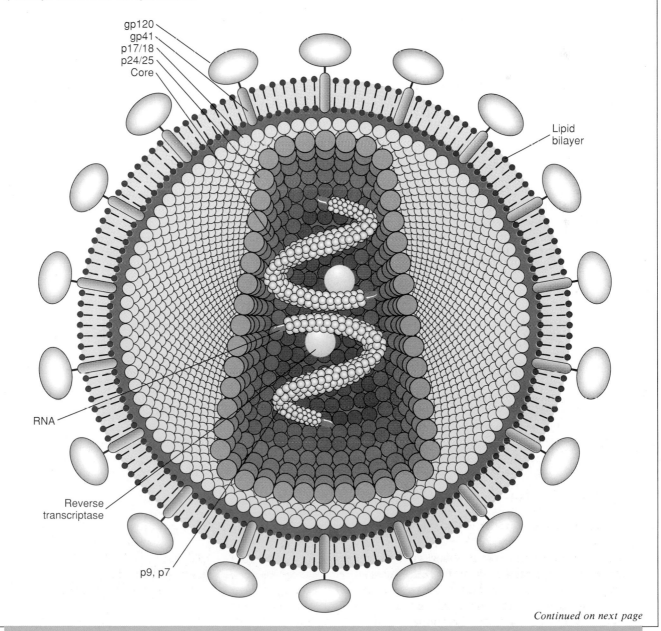

Continued on next page

Gene Expression: Control in Eukaryotes   **435**

BOX 15.2—continued

Box figure 3

The genome of HIV-1. Boxes represent different genes. *tev/tnv, tat,* and *rev* are separated into parts by introns. TAR is in the long terminal repeat (LTR) and RRE is in *env*. Data from Vaishnav and Wong-Staal, 1991.

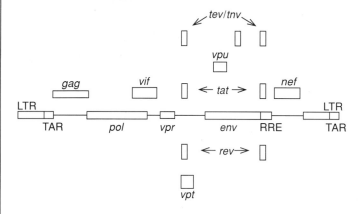

Box figure 4

AZT; it differs from deoxythymidine monophosphate at the 3′ position of the sugar.

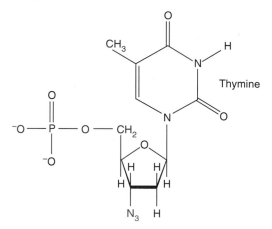

env, and *rev* genes. The gene *vpt* has no initiation codon but is presumably translated by a frame shift after initiation at the *tat* gene. Given all this complexity, the actual role of none of these gene products has been determined with certainty.

The main regulatory protein appears to be *tat* (for *trans-*activating *transcription* factor), which binds at a sequence in the long terminal repeat named TAR, for *trans-*activating *response* element. Presumably *tat* enhances transcription of the viral DNA and the efficiency of the translation of the RNA produced. The *rev* gene product binds at a region in the *env* gene called RRE for *rev response element* and possibly enhances the transport of viral mRNAs into the cytoplasm. Together, *tat* and *rev* are responsible for the major expression of viral structural genes (*gag, pol,* and *env*). The *nef* protein, the third regulatory protein, appears to be a GTP-binding protein; its function is uncertain as is the function of most of the remaining genes, termed accessory genes (*vif, vpr, vpu, vpt,* and *tev/tnv*).

The various repressor activities explain to some extent the rather unpredictable nature of the disease, frequently remaining dormant for long periods of time, and also perhaps explaining the lethality of the disease, a behavior counter to most other retroviruses that seem to encourage cell growth rather than destroy the cell.

Testing for AIDS is done by various techniques (such as western blots) looking for antibodies to the AIDS proteins, usually gp120, gp41, and reverse transcriptase. Several fronts are being opened to combat the disease. Currently at least one drug, 3′-azido-2′,3′-Dideoxythymidine (AZT, Box fig. 4), has some promise. It

is a thymidine analogue without a 3′-OH group, meaning that it will result in chain termination during DNA replication. It seems that during the reverse transcription process, reverse transcriptase preferentially chooses AZT over normal thymidine-containing nucleotides, whereas mammalian DNA polymerases have the opposite preference. Thus AZT preferentially prevents the reverse transcription of the HIV RNA at levels that are not toxic to the cell. Dideoxyinosine has the same effect and also has been licensed as a treatment of AIDS. Other strategies to combat the disease are under consideration, including the possibility of administering the free receptor protein, CD4, which should tie up the virus without damaging cells.

Currently, the prognosis is not bright because heterosexual transmission is climbing and there is no cure for the disease. Educational efforts are being expanded to dissuade people from high-risk behavior. Although intravenous drug users and prostitutes are difficult groups to reach through educational programs, the current program does seem to have raised the awareness level of the general public. Currently, the United States government is spending $1.3 billion on AIDS research, more than is being spent on heart research. Despite the gloomy predictions, it is worth noting that evolutionary biologists believe that even the most toxic diseases evolve toward lower toxicity. Not only does the susceptible population develop some immunity, but evolutionary forces tend to favor less toxic disease organisms that have a chance to infect more hosts if they do less damage to their current host. Thus the long-term prognosis is favorable, although that is little solace to those who have the disease or who are at high risk for getting it now.

Figure 15.34
Translocation associated with Burkitt's lymphoma. The proto-oncogene c-*myc*, on the end of chromosome 8q, is translocated to the end of chromosome 14q, adjacent to the immunoglobulin constant gene, Ig-C_μ. Presumably, location of c-*myc* next to the active immunoglobulin promoter (in B cells) results in transformation of the cell.

From J. Yunis, "The Chromosomal Basis of Human Neoplasia" in Science, *221:227–236. Copyright © 1983 by the AAAS.*

Burkitt's lymphoma
Translocation 8;14 = c–myc;Ig rearrangement

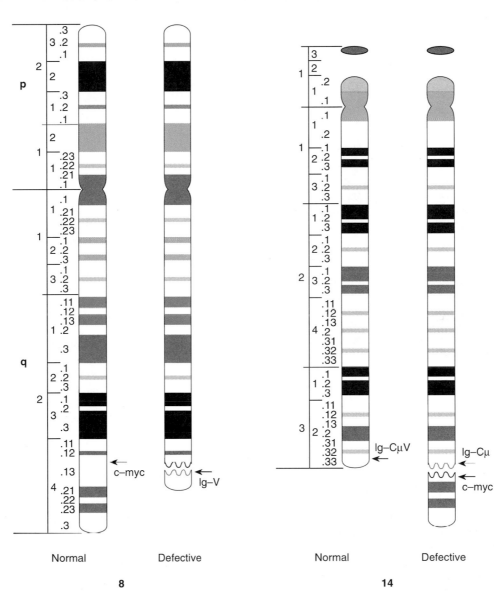

We seem to be at an exciting threshold in our understanding of cellular transformation (cancer). Although clinicians recognize over a hundred types of cancers, almost all may be the result of inappropriate activity of oncogenes or proto-oncogenes whose mechanism of action may fall into as few as four different categories. Chromosomal translocations, deletions, or simple mutations may be the major factor in activating these oncogenes or proto-oncogenes, thus tying together the mutation and viral theories of transformation.

Cancer-family syndromes may be the result of the inheritance of the propensity for chromosome breakage, or they may be the result of the inheritance of one mutation in a two- or multistep process. However, knowing the action

Figure 15.35

The various steps in converting a normal colon cell into a cancerous one. Four genes are involved, two oncogenes and two tumor-suppressor genes (*DCC* and *p53*).

From J. Marx, "Possible New Colon Gene Found" in Science, *251:1317, March 1991. Copyright 1991 by the AAAS.*

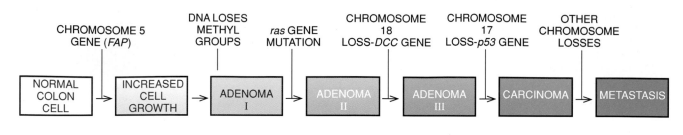

of transforming genes may not make for easy cures of cancer since they are "normal" genes that function properly in the untransformed cell.

On the other hand, we should not oversimplify the problem. Many cancers are the results of several different steps that lead eventually to full-blown cancerous growth. This is especially true in many of the cancers that produce solid tumors. The colon cancer named familial adenomatous polyposis (FAP) requires at least four genetic steps. Two are the mutation of oncogenes, the *FAP* gene on chromosome 5 and the *ras* gene, and the other two are inactivation of tumor suppressors, the *DCC* gene on chromosome 18 and *p53* on chromosome 17 (fig. 15.35). In each stage, the disease progresses to a new level, although it is not clear that the steps have to occur in order. Thus, some cancers, although still of genetic origin, require many steps for the culmination of the disease.

Environmental Causes of Cancer

Environment plays a major role in carcinogenesis, and many environmental carcinogenic agents are known (table 15.4). Many of the agents in table 15.4 are also mutagens (see chapter 16). Avoidable substances in the environment and the diet are estimated to cause 80 to 90% of all cancers, although the exact mechanisms by which these agents induce transformation are generally unknown. Perhaps the most effective cancer prevention program would be simply to remove as many carcinogens from the environment as possible.

SUMMARY

Development is the orderly sequence of changes that give rise to a complex organism. The ultimate goal of the developmental geneticist is to understand the role of genes in controlling the processes of development. Development does not have to proceed by permanently changing chromosomes. Nuclear transplantation has shown that differentiated nuclei can be totipotent. The nematode *Caenorhabditis elegans* provides a model for developmental

TABLE 15.4 Carcinogenic Substances in the Environment

Carcinogen	Cancer Site(s)
Aromatic amines	Bladder
Arsenic	Liver, lung, skin
Asbestos	Lung
Benzine	Bone marrow
Chromium	Lung, nose, nasopharynx sinuses
Cigarettes	Lung
Coal products	Bladder, lung
Dusts	Lung
Ionizing radiation	Bone, bone marrow, lung
Iron oxide	Lung
Isopropyl oil	Nasopharynx sinuses, nose
Mustard gas	Lung
Nickel	Lung, nasopharynx sinuses, nose
Petroleum	Lung
Ultraviolet irradiation	Skin
Vinyl chloride	Liver
Wood and leather dust	Nasopharynx sinuses, nose

studies because the lineage of every cell is known. Homeotic genes are believed to be master switches in development. They are under active study in fruit flies and nematodes.

Methylation may play a role in control of gene expression in eukaryotes since the methylation level is high in nontranscribed genes in these cells. The stability of Z DNA is also dependent on methylation. Decreases in methylation of certain DNA sequences may be related to the conformational change of the DNA from the Z to the B form. The presence of Z DNA may prevent transcription of a gene. Transposons can control gene expression. Mutable loci in corn and mating type in yeast are both determined by transposition.

Immunoglobulins are not made from individual genes, but from recombined pieces of different genes. About 10^8 different antibodies can be generated by the human genome. T-cell receptors and products of the major histocompatibility complex are other genetically variable components of the immune system.

Cancer is associated with chromosomal breaks and oncogenes. Oncogenes, which were originally discovered in retroviruses and later found in untransformed cells, determine enzymes that are tyrosine kinases, growth factors, or GTP- or DNA-binding proteins. The genes can be activated by mutation, amplification, or transposition to an active promoter or enhancer. Tumor-suppressor genes, such as *p53* and the gene for retinoblastoma, are recessive genes that normally repress malignant growth. Most cancers may be caused by oncogenes that are activated by various mutational events.

SOLVED PROBLEMS

1. Relate the homeo box, homeo domain, and master-switch concepts.

   ANSWER: A master-switch gene is a gene in a eukaryote that functions like a repressor in prokaryotes in the sense that it controls many genes. In a prokaryote, this control is achieved with operon organization. That is, many genes controlling the same function are transcribed as a unit. Thus a gene that represses transcription of an operon represses all of the genes in that operon. A master-switch gene is viewed in a similar manner given that polygenic transcripts are virtually absent from eukaryotes. A master-switch gene might be one that translates to a repressor-like protein, one that acts to control transcription of many genes. For this to happen, the master-switch gene would need to interact with DNA. Thus the finding of a homeo box that transcribes a homeo domain in genes that control large phenotypic changes is consistent with this view, the homeo domain being the part of the repressor that binds to the DNA.

2. What are the various stages in the formation of an immunoglobulin molecule in which diversity is generated?

   ANSWER: Variability is generated through four general processes: choice of which subunit genes to combine, choice of how to combine these subunit genes, de novo generation of diversity at junctions, and unusually high mutation rates. Thus, in our description of the formation of a kappa chain, diversity is added by: (1) the choice of which variable and joining genes to combine; (2) recombinational variability at the point of recombination; (3) the creation of N segments at the junctions; and (4) somatic hypermutation.

3. How can you reconcile the viral and mutation theories of cancer?

   ANSWER: The two theories are reconciled given that both define cancer as being caused by the inappropriate action of genes. In the mutation theory, inappropriate activity is generated by a mutation of a gene. In the viral theory, the inappropriate activity is generated by a gene brought into the cell by a virus.

EXERCISES AND PROBLEMS

1. How would you construct a fate map in *Xenopus*?

2. What is the lineage of the intestinal cell E.plppa in *Caenorhabditis elegans?*

3. What is a homeotic mutant? Why are genes that give rise to homeotic mutants thought of as binary switches?

4. Describe the phenotype of several homeotic and segmentation gene mutants.

5. What is a homeo box? How were they discovered? What is their significance? What is a homeo domain?

6. How might symmetry of the adult organism come about during development?

7. What is the value to an investigator of killing selected cells during development or of observing the effects of mutations on development?

8. What is the helix-turn-helix motif of DNA binding? What other motifs are known for DNA-binding proteins?

9. An investigator removes the nucleus from a frog's egg and replaces it with an early blastula nucleus from *Xenopus*. What alternative predictions can you make about the future course of development? Will a frog or *Xenopus* result, if anything? Will it have frog or *Xenopus* germ cells?

10. Why do you suppose so much research on developmental genetics has been done with amphibians?

11. Diagram the sequence on the yeast third chromosome as the mating type changes from *a* to *α* and back again.

12. What is the general mechanism whereby an antibody "recognizes" an antigen?

13. What components go into making an Ig light chain? A heavy chain?

14. How many different antibodies does a B lymphocyte produce? How many can it potentially produce before it differentiates?

15. What are the nucleotide recognition signals in V-J joining?

16. What enzymes would be required for the steps of figure 15.22 to take place?

17. What are B and T lymphocytes? What roles do they have in the immune response?

18. What is a T-cell receptor?

19. What is the major histocompatibility complex?

20. What gross chromosomal abnormalities are associated with cancers?

21. From the pedigree of figure 15.28, what modes of inheritance would be consistent with each type of cancer assuming that each was controlled by a single gene?

22. What chromosomal abnormality is associated with retinoblastoma? With Wilm's tumor?

23. What is the proposed mechanism of action of the retinoblastoma gene? What evidence supports this mechanism? Why is it called an anti-oncogene?

24. Retinoblastoma has been called a recessive oncogene. Explain.

25. What are the general forms of animal viruses? What types of genetic material do they have?

26. What is the minimal genetic complement of a retrovirus? What does each of the genes code for?

27. What translation mechanisms exist for the expression of the genes of a retrovirus?

28. How are long terminal repeats generated at the ends of retroviral DNA? What signals and sequences do they encode?

29. What are the differences among v-*src*, c-*src*, and proto-*src* genes?

30. How can the proto-*src* gene be activated?

31. What is the evidence that the c-*src* gene came before the v-*src* gene?

32. How does translocation activate the c-*myc* gene in Burkitt's lymphoma?

33. A given DNA molecule, when digested with either *Hpa* II or *Msp* I, gives three and five fragments, respectively. Given that these two enzymes recognize the same sequence, how can you account for the different numbers of fragments produced?

34. Tissue culture cells are exposed for five minutes to radioactive dUTP in the presence or absence of 5-azacytidine. Radioactivity in RNA is determined to be 1,500 counts per minute without azacytidine and 27,300 in the presence of azacytidine. Propose an explanation to account for these results.

35. A retrovirus, lacking a cellular oncogene, is shown to be integrated 3 kb from a cellular oncogene. When the RNA for this oncogene is quantified, infected cells are found to have ten times more oncogene-specific mRNA than uninfected cells. How can you account for this increase in RNA synthesis?

36. A disorder of the immune system results in the complete lack of antibody production. Provide two possible molecular defects that could result in such a condition.

37. If drugs that inhibit transcription are injected into fertilized frog eggs, early cell division and protein synthesis still occur. Why?

38. Many polycistronic mRNAs have been found in prokaryotes, and all have an initiation codon for each gene of the operon. Polycistronic mRNA, with multiple initiation codons, does not seem to exist in eukaryotes, but many eukaryotic mRNAs contain more than one gene. How can you account for this apparent enigma?

39. Many alleles for the genes for the constant region of antibodies have been found. Suppose that two such alleles for the λ light chain are called c_1 and c_2. In a heterozygote, c_1/c_2, some cells are found to make only c_1, and others only c_2. Propose an explanation to account for this observation.

40. A cDNA probe for a cellular oncogene is made. Cellular DNA from normal cells and from two different clones of cells infected with a retrovirus that lacks the oncogene is digested with a particular restriction enzyme. The DNA is separated in a gel and hybridized with the radioactive probe. The results appear below.

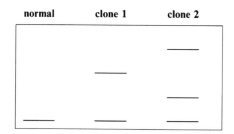

Interpret these results by describing where the retrovirus has inserted.

41. Complementary DNA is made from mRNA for the light chain of an antibody molecule. DNA from embryonic cells and from mature B lymphocytes is isolated and digested with a restriction enzyme, and the fragments are separated in a gel. Radioactive cDNA is used to probe this gel, and the results appear below. Provide an explanation for these results.

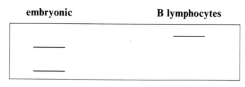

42. Assume that a particular oncogene produces a growth factor.
    a. How could a retrovirus affect the oncogene so that the cell is now cancerous?
    b. How could you test your hypothesis?

43. Suppose a cellular oncogene makes a protein that controls the progression of normal cells through the cell cycle. What would be the consequence of a viral insertion within the gene?

Suggested Readings for chapter 15 are on page 653.

DNA: Its Mutation, Repair, and Recombination

An ear of corn showing a single kernel (*red*) with a color mutation.
© *William E. Ferguson*

The mutation, repair, and recombination of DNA are treated together in this chapter because the three processes have much in common. The physical alteration of DNA is involved in each; repair and recombination share some of the same enzymes. We progress from mutation, the change in DNA, to repair of damaged DNA, and, finally, to recombination, the new arrangement of pieces of DNA.

MUTATION

The concept of mutation (a term coined by de Vries, a rediscoverer of Mendel) is pervasive in genetics. **Mutation** is both the process by which a gene (or chromosome) changes structurally and the end result of that process. Without alternative forms of genes, the biological diversity that exists today would not have evolved. Without alternative forms of genes, it would have been virtually impossible for geneticists to determine which of an organism's characteristics are genetically controlled. The background for our current knowledge in genetics was provided by studies of mutation.

Fluctuation Test

In 1943, Salvador Luria and Max Delbrück published a paper entitled "Mutations of Bacteria from Virus Sensitivity to Virus Resistance." This paper ushered in the era of bacterial genetics by demonstrating that the phenotypic variants found in bacteria are actually due to mutations rather than to induced physiological changes. Very little work had previously been done in bacterial genetics because of the feeling that bacteria did not have "normal" genetic systems like the systems of fruit flies and corn. Rather, bacteria were believed to respond to environmental change by physiological adaptation, a non-Darwinian view. As Luria said, bacteriology remained "the last stronghold of Lamarckism" (the belief that acquired characteristics are inherited).

What Causes Genetic Variation?

Luria and Delbrück studied the Ton^r (phage T1-resistant) mutants of a normal Ton^s (phage T1-sensitive) *Escherichia coli* strain. They used an enrichment experiment, as described in chapter 7, wherein a petri plate is spread with *E. coli* bacteria and T1 phages. Normally no bacterial colonies will grow on the plate: All the bacteria will be lysed. However, if one of the bacterial cells is resistant to T1 phages, it will produce a bacterial colony, and all descendants of the cells from this colony will be T1 resistant. There are two possible explanations for the appearance of T1-resistant colonies:

1. Every *E. coli* cell is capable of being induced to be resistant to phage T1, but only a very small number actually are induced. That is, all cells are

Salvador E. Luria (1912–1991)
Courtesy of Dr. S. E. Luria.

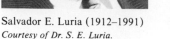

Max Delbrück (1906–1981)
Courtesy of Dr. Max Delbrück.

genetically identical, each with a very low probability of acting in a resistant manner in the presence of T1 phages. When resistance is induced, the cell and its progeny remain resistant.

2. In the culture, a small number of *E. coli* cells exist that are already resistant to phage T1; in the presence of phage T1, only these cells survive.

If the presumed rates of induction and mutation are the same, determining which of the two mechanisms is operating is difficult. Luria and Delbrück, however, developed a means of distinguishing between these mechanisms. They reasoned as follows: If T1 resistance was physiologically induced, the relative frequency of resistant *E. coli* cells in a culture of the normal (Ton^s) strain should be a constant, independent of the number of cells in the culture or the length of time that the culture has been growing. If resistance was due to random mutation, the frequency of mutant (Ton^r) cells will depend on when the mutations occurred. In other words, the appearance of a mutant cell would be a random event. If a mutation occurs early in the growth of the culture, then many cells will descend from the mutant cell and therefore there will be many resistant colonies. If the mutation does not occur until late in the growth of the culture, then the subsequent number of mutant cells will be few. Thus, if the mutation hypothesis is correct, there will be considerable fluctuation from culture to culture in the number of resistant cells present (fig. 16.1).

Results of the Fluctuation Test

To distinguish between these hypotheses, Luria and Delbrück developed what is known as the **fluctuation test.** They counted the mutants both in small ("individual") cultures and in subsamples from a single large ("bulk") culture. All subsamples from a bulk culture should have the same number of resistant cells, differing only because of random sampling error. If, however, mutation occurs, the number of resistant cells among the individual cultures should vary considerably from culture to culture; it

Figure 16.1

Occurrence of *E. coli* *Ton*ʳ colonies in *Ton*ˢ cultures. Ten cultures of *E. coli* cells were grown from a standard inoculum in each test tube in the absence of phage T1 and then spread on petri plates in the presence of phage T1. The resistant cells appear as colonies on the plates. We expect a uniform distribution of resistant cells if the physiological induction hypothesis is correct (*a*) or a great fluctuation in the number of resistant cells if the random mutation hypothesis is correct (*b*).

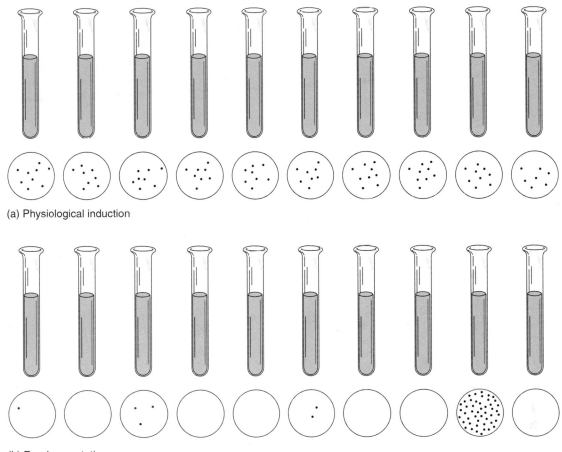

(a) Physiological induction

(b) Random mutation

would be related to the time that the mutation occurred during the growth of each culture. If mutation arose early, there would be many resistant cells. If it arose late, there would be relatively few resistant cells. Under physiological induction, the distribution of resistant colonies should not differ between the individual and bulk cultures.

Luria and Delbrück inoculated twenty individual cultures and one bulk culture with *E. coli* cells and incubated them in the absence of phage T1. Each individual culture was then spread out on a petri plate containing a very high concentration of T1 phages; ten subsamples from the bulk culture were plated in the same way. We can see from the results (table 16.1) that there was minimal variation in the number of resistant cells among the bulk-culture subsamples but a very large amount of variation, as predicted for random mutation, among the individual cultures.

Given that bacteria have "normal" genetic systems that undergo mutations, bacteria could then be used, along with higher organisms, to answer genetic questions. As we have pointed out, the era of molecular genetics began with the use of prokaryotic and viral systems in genetic re-

search. In the next section, we turn our attention to several basic questions about the gene, questions whose answers were made possible in several instances only because prokaryotic systems were available.

Genetic Fine Structure

How do we determine the relationship among several mutations that cause the same phenotypic change? What are the smallest units of DNA capable of mutation and recombination? Are the gene and its protein product colinear? The answers to the latter two questions are important from a historical perspective. The answer to the first question is relevant to our current understanding of genetics.

Complementation

If two recessive mutations arise independently and both have the same phenotype, how do we know whether they are both mutations of the same gene? That is, are they

TABLE 16.1 Results from the Luria and Delbrück Fluctuation Test

Individual Cultures		Samples from Bulk Culture	
Culture Number	Ton^r Colonies Found	Sample Number	Ton^r Colonies Found
1	1	1	14
2	0	2	15
3	3	3	13
4	0	4	21
5	0	5	15
6	5	6	14
7	0	7	26
8	5	8	16
9	0	9	20
10	6	10	13
11	107		
12	0		
13	0		
14	0		
15	1		
16	0		
17	0		
18	64		
19	0		
20	35		
Mean ($\bar{n}$)	11.4		16.7
Standard deviation	27.4		4.3

Source: Data from S. E. Luria and M. Delbrück, *Genetics*, 28: 491, 1943.
*Each culture and sample was 0.2 ml containing about 2×10^7 *E. coli* cells.

Seymour Benzer (1921–   )
Dr. Seymour Benzer, 1970.

alleles? To answer this question, we must construct a heterozygote and determine if there is **complementation** between the two mutations. A heterozygote with two mutations of the same gene will produce only mutant mRNAs that will result in mutant enzymes (fig. 16.2*a*). If, however, the mutations are not allelic, the gamete from the a_1 parent will also contain an a_2^+ allele, whereas the gamete from the a_2 parent will also contain the a_1^+ allele (fig. 16.2*b*). If the two mutant genes are truly alleles, then the phenotype of the heterozygote should be mutant. If, however, the two mutant genes are nonallelic, then the a_1 mutant will have contributed the wild-type allele at the A_2 locus and the a_2 mutant will have contributed the wild-type allele at the A_1 locus to the heterozygote. Thus the two mutations will complement each other and the wild-type results. Mutations that fail to complement each other are termed **functional alleles.** The test for defining alleles strictly on this basis of functionality is termed the *cis-trans* **complementation test.**

There are two different configurations in which a heterozygous double mutant can be formed (fig. 16.3). In the *cis-trans* complementation test, only the *trans* configuration was used to determine whether the two mutations were allelic. In reality, the *cis* configuration is never tested; it is the conceptual control, wherein wild-type activity (with recessive mutations) is always expected. The test is thus sometimes simply called a *trans* test. Functional alleles produce a wild-type phenotype in the *cis* configuration but a mutant phenotype in the *trans* configuration. This difference in phenotypes is called a *cis-trans* position effect.

From the terms *cis* and *trans*, Seymour Benzer coined the term **cistron** for the smallest genetic unit (length of genetic material) that exhibits a *cis-trans* position effect. We thus have a new word for the gene, one in which function is more explicit. We have, in essence, refined Beadle and Tatum's one-gene-one-enzyme hypothesis to a more accurate one-cistron-one-polypeptide concept. The cistron is thus the smallest unit that codes for a messenger RNA that is then translated into a single polypeptide.

From functional alleles we can go one step further in recombinational analysis by determining whether two allelic mutations occur at exactly the same place in the cistron. In other words, when two mutations are found to be functional alleles, are they also **structural alleles?** The methods used to analyze complementation at the beginning of this section can be used here. Crosses are carried out to form a mutant heterozygote whose offspring are then tested for recombination between the two mutational sites. If no recombination occurs, then the two alleles probably contain the same structural change (involving the same base pair) and are thus structural alleles. If a small amount of recombination occurs that generates wild-type offspring, then the two alleles are not mutations at the same point (fig. 16.4). Alleles that were functional but not

Figure 16.2

The complementation test defines allelism. Are two mutations (a_1, a_2) affecting the same trait allelic? Mutant homozygotes are crossed to form a heterozygote. (*a*) If the mutations are allelic, then (in the heterozygote) both copies of the gene are mutant, resulting in the mutant phenotype. (*b*) If the mutations are nonallelic, then there is a wild-type allele of each gene present in the heterozygote, resulting in the wild-type. (The two loci need not be on the same chromosome.)

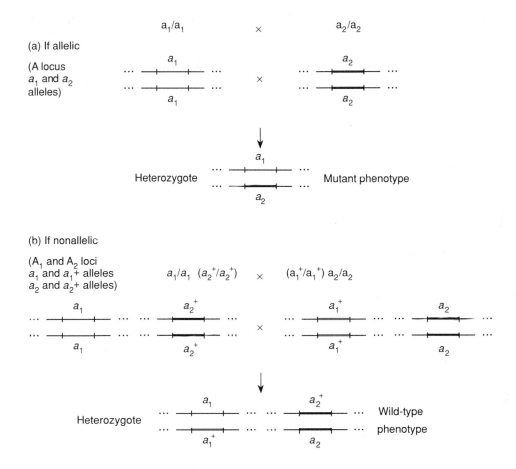

Figure 16.3

A heterozygote of two recessive mutations can have the *trans* or *cis* arrangement. In the *trans* position, functional alleles produce a mutant phenotype. (*Colored marks* represent mutant lesions.) In the *cis* position, the wild-type phenotype is produced because a wild-type copy of the locus is present. The *cis-trans* position effect defines functional alleles.

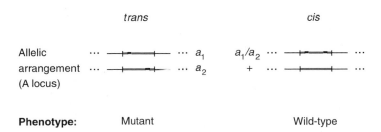

structural were first termed **pseudoalleles** because it was believed that loci were made of subloci. Fine-structure analysis led to the understanding that a locus is a length of genetic material divisible by recombination rather than a bead on a string.

Eye-color mutants of *Drosophila* can be studied by complementation analysis. The white-eye locus has a series of alleles producing varying shades of red. This locus is sex linked, at about map position 3.0 on the X chromosome. (There are several other eye-color loci on the X chromosome not relevant to this cross—e.g., prune and ruby.) If an apricot-eyed female is mated with a white-eyed male, the female offspring are all heterozygous and have mutant light-colored eyes (fig. 16.5). Thus apricot and white are functional alleles: They do not complement (table 16.2). To determine whether apricot and white are structural alleles, light-eyed females are crossed with white-eyed males, and the offspring are observed for the presence of wild-type or light-eyed males. Their rate of appearance is less than 0.001%, which is, however, significantly above the mutation rate. The conclusion is that apricot and white are functional, but not structural, alleles.

Figure 16.4

Functional alleles may or may not be structurally allelic. (*Colored marks* represent mutant sites.) Functional alleles that are not also structural alleles can have recombination between the mutant sites resulting in occasional wild-type (and double mutant) offspring. Structural alleles (which are also always functional alleles) are defective at the same base pairs.

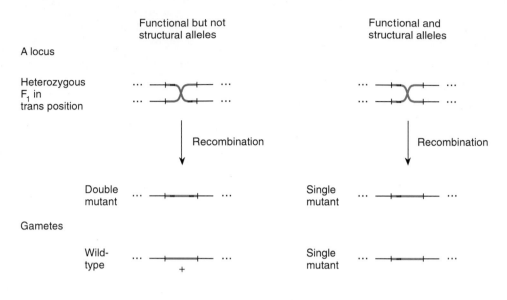

TABLE 16.2   Complementation Matrix of X-Linked *Drosophila* Eye-Color Mutants*

	White	Prune	Apricot	Buff	Cherry	Eosin	Ruby
White (w)	−	+	−	−	−	−	+
Prune (pn)		−	+	+	+	+	+
Apricot (wᵃ)			−	−	−	−	+
Buff (wᵇᶠ)				−	−	−	+
Cherry (wᶜʰ)					−	−	+
Eosin (wᵉ)						−	+
Ruby (rb)							−

*Plus sign indicates that female offspring are wild-type; minus sign indicates that they are mutant.

Fine-Structure Mapping

After Beadle and Tatum established in 1941 that the gene controls the production of an enzyme that then controls a step in a biochemical pathway, Benzer used analytical techniques to dissect the fine structure of the gene. Fine-structure mapping means an examination of the size and number of sites within a gene that are capable of mutation and recombination. In the late 1950s, when biochemical techniques were not available for DNA sequencing, Benzer used classical recombination and mutation techniques with bacterial viruses to provide reasonable estimates to the details of fine structure and to give insight into the nature of the gene. He coined the terms **muton** for the smallest mutable site and **recon** for the smallest unit of recombination. It is now known that a single base pair is both the muton and the recon.

Before Benzer's work, genes were thought of as beads on a string. Analysis of mutation sites within a gene by means of recombination was hampered by the very low rate of recombination between sites within a gene. If two mutant genes are functional alleles (involving different sites on the same gene), a distinct probability exists of getting both mutant sites (and both wild-type sites) on the same chromosome by recombination (see fig. 16.4); but, in view of the very short distances within a gene, this probability is very low. Although it certainly seemed desirable to map sites within the gene, the problem of finding an organism that would allow fine-structure analysis remained until Benzer decided to use phage T4.

***r*II Screening Techniques.** —Benzer used the T4 bacteriophage because of the growth potential of phages, in which a generation takes about an hour and the increase

Figure 16.5

Crosses demonstrating that apricot and white eyes are functional, but not structural, alleles in *Drosophila*. Light-eyed females are heterozygous for both alleles. When testcrossed, they produce occasional offspring that are wild-type or light eyed, indicating a crossover between the two mutant sites (white and apricot) in the heterozygous females.

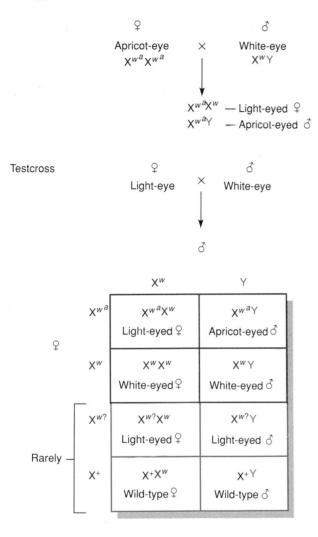

Figure 16.6

Using *E. coli* K12 and B strains to screen for recombination at the *r*II locus of phage T4. Two *r*II mutants are crossed by having both phage infect the same B-strain bacteria. The offspring are plated on a lawn of K12 bacteria in which only wild-type phage can grow. The technique thus selects only wild-type recombinants.

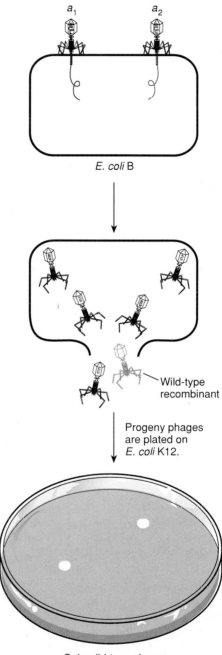

in numbers per generation is about a hundredfold. Actually, any prokaryote or virus should suffice, but Benzer made use of other unique screening properties of the phage that made it possible to recognize one particular mutant in about a billion phages. Benzer used *r*II mutants of T4. These mutants produce large, smooth-edged plaques on *E. coli,* whereas the wild-type produces smaller, less-smooth-edged plaques (see fig. 7.9).

The screening system employed by Benzer made use of the fact that *r*II mutants do not grow on *E. coli* strain K12, whereas the wild-type can grow on K12. The normal host strain, *E. coli* B, allows growth of both the wild-type and *r*II mutants. Thus various mutants can be crossed by mixed infection of *E. coli* B cells with two mutants, and Benzer could screen for wild-type recombinants by plating the resultant phages on *E. coli* K12 (fig. 16.6) in which

only a wild-type recombinant will produce a plaque. Thus it is possible to detect about one recombinant in a billion phages, all in an afternoon's work. This ability to detect recombinants at such a low level allowed Benzer to see recombinational events between positions very close together on the DNA, events that would normally occur at a frequency too low to detect.

TABLE 16.3 Complementation Matrix of 10 rII Mutants*

	1	2	3	4	5	6	7	8	9	10
1	−	−	+	−	−	−	+	−	−	−
2		−	+	−	−	−	+	−	−	−
3			−	−	+	+	−	+	−	+
4				−	−	−	−	−	−	−
5					−	−	+	−	−	−
6						−	+	−	−	−
7							−	+	−	+
8								−	−	−
9									−	−
10										−

*Plus sign indicates complementation; minus sign indicates no complementation. The two cistrons are arbitrarily designated *A* and *B*. Mutants 4 and 9 must be deletions that cover parts of both cistrons. Alleles: *A* cistron: 1, 2, 4, 5, 6, 8, 9, 10; *B* cistron: 3, 4, 7, 9.

Benzer sought to map the number of sites capable of recombination and mutation within the *r*II region of T4. He began by isolating independently derived *r*II mutants and crossing them among themselves. The first thing he found was that the *r*II region was composed of two cistrons: Almost all of the mutations belonged to one of two **complementation groups.** The *A*-cistron mutations would not complement each other but would complement the mutations of the *B* cistron. The exceptions were mutations that seemed to belong to both cistrons and were soon found to be deletions in which part of each cistron was missing (table 16.3).

Deletion Mapping. —As the number of independently isolated mutations of the *A* and *B* cistrons increased, it became obvious that to make every possible pairwise cross would entail millions of crosses. To overcome this problem, Benzer isolated mutants that had partial or complete deletions of each cistron. Deletion mutations were easily discovered because they acted like structural alleles to alleles that were not themselves structurally allelic. In other words, if mutations *a, b,* and *c* are functional, but not structural, alleles of each other, and mutation *d* is a structural allele to *a, b,* and *c, d* must have a deletion of the bases mutated in *a, b,* and *c.* Once a sequence of deletion mutations covering the *A* and *B* cistrons was isolated, a minimum of crosses was required to localize a new mutation to a portion of one of the cistrons. A second series of smaller deletions within each region was then isolated, and further localization was accomplished (fig. 16.7).

Then each new mutant was crossed with each of the other mutants isolated in its subregion to localize the relative position of the new mutation. If the mutation was structurally allelic to a previously isolated mutation, it was scored as an independent isolation of the same mutation. If it was not a structural allele of any of the known mu-

tations of the subregion, it was added to this region as a new mutation point. The exact position of each new mutation within the region was determined by the relative frequency of recombination between it and the known mutations of this region (see chapter 7). Benzer eventually isolated about 350 mutations from eighty different subregions defined by deletion mutations. An abbreviated map is shown in figure 16.8.

What conclusions did Benzer draw from his work? First, he concluded that since all of the mutations in both *r*II cistrons can be ordered in a linear fashion, the original Watson-Crick model of DNA as a linear molecule was correct. Second, he concluded that reasonable inroads had been made toward saturation of the map, which would occur when at least one mutation had been located at every mutable site. Benzer reasoned that since many sites were represented by only one mutation, sites must occur represented by zero mutations (i.e., not yet represented by a mutation). Given that he had mapped about 350 sites, he calculated that there were at least another 100 sites still undetected by mutation. We now know that 450 sites is an underestimate. However, since the protein products of these cistrons were not isolated, there were no independent estimates of the number of nucleotides in these cistrons (number of amino acids times three nucleotides per codon). Thus, although Benzer had not saturated the map with mutations, he certainly had made respectable progress in dissecting the gene and demonstrating that it was not an indivisible unit, a "bead on a string."

Hot Spots. —Benzer also looked into the lack of uniformity in the occurrence of mutations (note two major "hot spots" at B4 and A6c of fig. 16.8). Presuming that all base pairs are either AT or GC, this lack of uniformity was unexpected. Benzer suggested that spontaneous mutation is not just a function of the base pair itself, but is affected by the surrounding bases as well. This concept still holds.

To recapitulate, Benzer's work supports the model of the gene as a linear arrangement of DNA whose nucleotides are the smallest units of mutation. The link between any adjacent nucleotides can be broken in the recombinational process. The smallest functional unit, determined by a complementation test, is the cistron. Mutagenesis is not uniform throughout the cistron but may depend on the particular arrangement of bases in a given region.

Intra-allelic Complementation

Benzer warned that certainty is elusive in the complementation test because sometimes two mutations of the same functional unit (cistron) can result in partial activity. The problem can be traced to the interactions of subunits at the polypeptide level. That is, some proteins are made up of subunits, and it is possible that certain mutant combinations produce subunits that interact to restore the enzymatic function of the protein (fig. 16.9). This

Figure 16.7

Localization of an rII mutation by deletion mapping. Newly isolated mutants are crossed with mutants with selected deletions to quickly localize the new mutation to a small region of the cistron. If the new mutant (e.g., r960) is located in the A5c2a2 region, it would not produce the wild-type by recombination with r1272, r1241, rJ3, rPT1, or rPB242 (the *solid red part of the bar* indicates deleted segments). It would produce the wild-type by recombination with rA105 and r638 and thus the mutation would be localized to the A5 region. When crossed with r1605, r1589, and rPB230, it would produce only the rare recombinant with rPB230, indicating the mutation is in the A5c region. When crossed with r1993, r1695, and r1168, the mutant would produce the wild-type by recombination with r1993 and r1168 and the mutation would be localized to the A5c2a2 region. Finally the mutant would be crossed pairwise with all the known mutants of this region to determine relative arrangement and distance.

From "The Fine Structure of the Gene" by Seymour Benzer. Copyright © 1962 by Scientific American, Inc. All rights reserved.

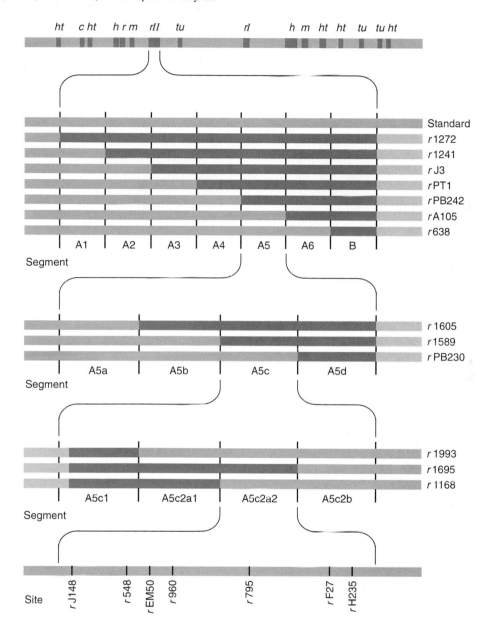

phenomenon is known as **intra-allelic complementation.** With this problem in mind, geneticists use the complementation test to determine functional relationships among mutations.

Colinearity

Next we look at the colinearity of the gene and the polypeptide. Benzer's work established that the gene was a linear entity as had been predicted by Watson and Crick.

However, Benzer could not demonstrate the colinearity of the gene and its protein product. To do this, it is necessary to show that for every mutational change of the DNA, a corresponding change takes place in the protein product of the gene. Colinearity would be established by showing that nucleotide and amino acid changes occurred linearly and in the same order in the protein and in the cistron.

Ideally, Benzer himself might have solved the colinearity issue. He was halfway there, with his 350 or so isolated mutations of phage T4. However, Benzer did not

Figure 16.8

Abbreviated map of spontaneous mutations of the *A* and *B* cistrons of the *r*II region of T4. Each square represents one independently isolated mutation. Note the "hot spots" at A6c and B4.

From Seymour Benzer, "On the Topography of the Genetic Fine Structure" in Proceedings of the National Academy of Sciences, *47:403–415, 1961. Reprinted by permission.*

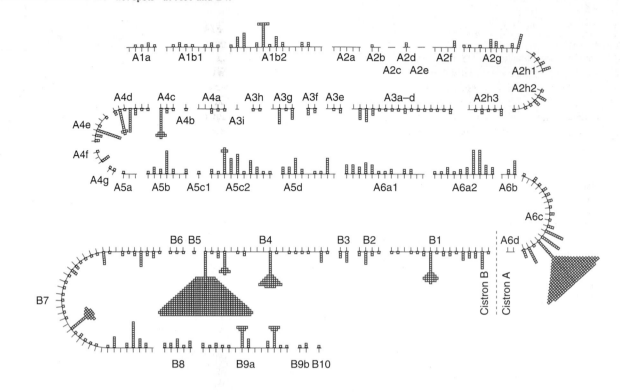

Figure 16.9

Intra-allelic complementation. With certain mutations, it is possible to get enzymatic activity in a heterozygote for two nonfunctional alleles, if the two polypeptides form a functional enzyme. (Active site is shown in *color*.)

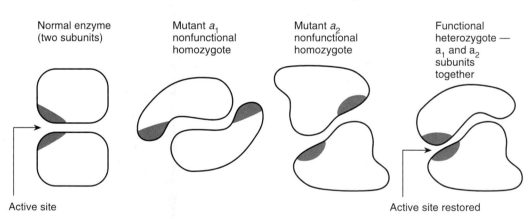

have a protein product to analyze; no particular mutant protein had been isolated from *r*II mutants. In the midst of competition to find just the right system, Charles Yanofsky of Stanford University and his colleagues emerged in the mid-1960s with the required proof that the ordering of a polypeptide's amino acids corresponded to the nucleotide sequence in the gene that specified it. Yanofsky's success rested with his choice of an amenable system, one using the enzymes from a biochemical pathway.

Yanofsky did his research on the tryptophan biosynthetic pathway in *E. coli*. The last enzyme in the pathway, tryptophan synthetase, catalyzes the reaction of indole-3-glycerol-phosphate and serine to tryptophan and 3-phosphoglyceraldehyde. The enzyme itself is made of four subunits specified by two separate cistrons with each polypeptide present twice.

Yanofsky and his colleagues concentrated on the *A* subunit. They mapped *A*-cistron mutations with transduction (chapter 7) using the transducing phage P1. Each new mutant was first tested against a series of deletion mutants to establish the region in which the mutation was located. Then mutants for a particular region were crossed among themselves to establish relative positions and distances.

The protein products of the bacterial genes were isolated using electrophoresis and chromatography to establish the fingerprint patterns of the proteins (see chapter 11). Assuming a single mutation, a comparison of the mutant and the wild-type fingerprints would show a difference of just one amino acid spot (fig. 16.10), a process that avoided the necessity of having to sequence the entire protein. The mutant amino acid was identified by analysis of just this one spot. The detail of nucleotide and amino acid changes are shown for nine of the mutations in this 267-amino acid protein in figure 16.11.

We can see from this figure that nine mutations in the linear A cistron of tryptophan synthetase are colinear with nine amino acid changes in the protein itself. In two cases, two mutations mapped so close as to be almost indistinguishable. In both cases, the two mutations proved to be

Figure 16.10
Difference in "fingerprints" between mutant and wild-type polypeptide digests. The single spot that differs in the mutant can be isolated and sequenced, avoiding the need to sequence the whole protein.

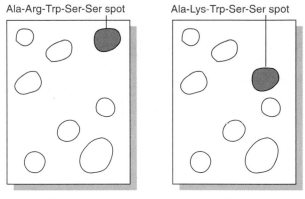

Ala-Arg-Trp-Ser-Ser spot   Ala-Lys-Trp-Ser-Ser spot

Wild-type   Mutant

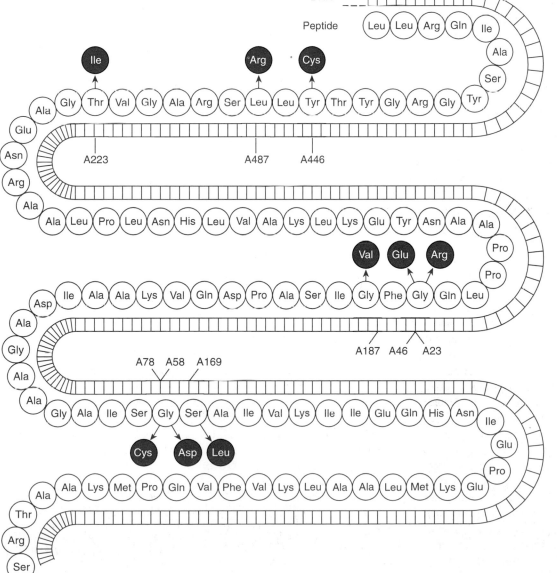

Figure 16.11
Amino acid sequence of the carboxyl terminal end of the tryptophan synthetase A protein and its DNA. Mutations are shown on the DNA (e.g., A446) as are the changed amino acids of those mutations (in *color*). DNA and protein changes are colinear.

Hermann J. Muller (1890–1967)
Courtesy of National Academy of Sciences.

Lewis J. Stadler (1896–1954)
Genetics 41 (1956): frontispiece.

Figure 16.12
Types of point mutations of DNA. Single-step changes are replacements, additions, or deletions. A second point mutation in the same gene can result either in a double mutation, reversion to the original, or intragenic suppression. In this case, intragenic suppression is illustrated by an addition of one base followed by a nearby deletion of a different base.

Original sequence

```
AAACCCGGG
TTTGGGCCC
```

Single-step

Replacement

```
AAACTCGGG
TTTGAGCCC
```

Addition

```
AAACCCTGGG
TTTGGGACCC
```

Deletion

```
AAACCGGG
TTTGGCCC
```

Double-step

Second independent change (single-step replacement, then second single-step replacement)

```
AAACTCGAG
TTTGAGCTC
```

Back mutation (of single-step replacement)

```
AAACCCGGG
TTTGGGCCC
```

Intragenic suppressor (single-step addition, then single-step deletion)

```
AACCCTGGG
TTGGGACCC
```

in the same codon: The same amino acid position was altered in each (A23-A46, A58-A78). Thus, exactly as predicted and expected, there is colinearity between the gene and protein. This work was independently confirmed at the same time by Brenner and his colleagues, who used head-protein mutants of phage T4.

Spontaneous versus Induced Mutation

H. J. Muller won the Nobel Prize for demonstrating that X rays can cause mutations. This work was published in 1927 in a paper entitled "Artificial Transmutation of the Gene." At about the same time, L. J. Stadler induced mutations in barley with X rays. The basic impetus for their work was the fact that mutations occur so infrequently that genetic research was hampered by the inability to obtain mutants. Muller exposed flies to varying doses of X rays and then observed their progeny. He came to several conclusions. First, X rays greatly increased the occurrence of mutations. Second, the inheritance patterns of X-ray-induced mutations and the phenotypes of organisms with them were similar to those that resulted from natural, or "spontaneous," mutations.

Mutation Rates

The **mutation rate** is the number of mutations that arise per cell division in bacteria and single-celled organisms, or the number of mutations that arise per gamete in higher organisms. Mutation rates vary tremendously depending upon the length of genetic material, the kind of mutation, and other factors to be discussed. Luria and Delbrück, for example, found that in *E. coli* the mutation rate per cell division of Ton^s to Ton^r was 3×10^{-8}, whereas the mutation rate of the wild-type to the histidine-requiring phenotype (His$^+$ to His$^-$) was 2×10^{-6}. The rate of **reversion** (return of the mutant to the wild-type) was 7.5×10^{-9}. The mutation and reversion rates differ because the His$^-$ phenotype can be caused by many different mutations, whereas reversion requires specific, and hence less

probable, changes to correct the His$^-$ phenotype back to the wild-type. The lethal mutation rate in *Drosophila* is about 1×10^{-2} per gamete for the total genome. This number is relatively large because, as with His$^-$, many different mutations produce the same phenotype (lethality in this case).

Point Mutations

The mutations of primary concern in this chapter are **point mutations,** which consist of single changes in the nucleotide sequence. (Chromosomal mutations, changes in the number and visible structure of chromosomes, were treated in chapter 8.) If the change is a replacement of some kind, then a new codon will be created. In many cases this new

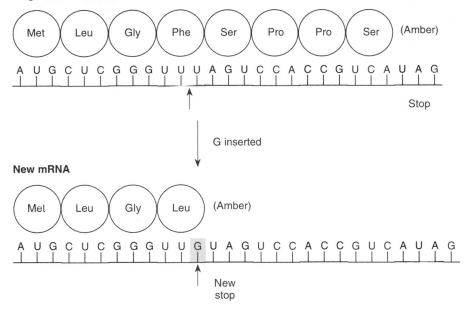

Original short mRNA

Met — Leu — Gly — Phe — Ser — Pro — Pro — Ser (Amber)

A U G C U C G G G U U U A G U C C A C C G U C A U A G

Stop

↑ G inserted

New mRNA

Mel — Leu — Gly — Leu (Amber)

A U G C U C G G G U U G U A G U C C A C C G U C A U A G

New stop

Figure 16.13
Possible effects of a frameshift mutation. The insertion of a single base results in the creation of a new stop sequence (*amber*). The result will be premature termination of translation.

codon, upon translation, will result in a new amino acid. As discussed in chapter 11, one of the outcomes of redundancy in the genetic code is partial protection of the cell from the effects of mutation: Common amino acids have the most codons, similar amino acids have similar codons, and the wobble position of the codon is the least important position in translation. However, when base changes result in new amino acids, new proteins appear that can alter the morphology or physiology of the organism and result in phenotypic novelty or lethality.

Frameshift Mutation

A point mutation may consist of replacement, addition, or deletion of a base (fig. 16.12). Point mutations that add or subtract a base are, potentially, the most devastating in their effects on the cell or organism because they change the reading frame of a gene from the site of mutation onward (fig. 16.13). A frameshift mutation causes two problems. First, all the codons from the frameshift on will be different and thus yield (most probably) a useless protein. Second, stop-signal information will be misread. One of the new codons may be a nonsense codon, which will cause a premature stoppage of translation. Or, if the translation apparatus reaches the original nonsense codon, it will no longer be recognized as such because it is in a different reading frame, and therefore, the translation process will continue beyond the end of the gene.

Back Mutation and Suppression

A second point mutation in the same gene can have one of three possible effects (see fig. 16.12). First, the mutation can result in either another mutant codon or one codon that has experienced two changes. Second, if the change is at the same site, the original sequence can be returned, an effect known as **back mutation:** The gene then becomes

a revertant with the original function restored. Third, **intragenic suppression** can take place. Intragenic suppression occurs when a second mutation in the same gene masks the occurrence of the original mutation without actually restoring the original sequence. The new sequence is a double mutation that appears to have the original (unmutated) phenotype. In the example of figure 16.12, a T addition is followed by an A deletion that substitutes the AACCCT sequence for the original AAACCC. These sequences, when transcribed (UUGGGA, UUUGGG) are codons for leucine-glycine and phenylalanine-glycine, respectively. Intragenic suppression occurs if the new codons were for either different amino acids or the same amino acids, as long as the phenotype of the organism is reverted approximately to the original. Suppressed mutations can be distinguished from true back mutations either by subtle differences in phenotype, by genetic crosses, by changes in the amino acid sequence of a protein, or by DNA sequencing.

Conditional Lethality

A class of mutants that has been very useful to geneticists is the conditional-lethal mutant, a mutant that is lethal under one set of circumstances but not under another set of circumstances. **Nutritional-requirement mutants** are good examples (see chapter 7). **Temperature-sensitive mutants** are conditional lethals that have made it possible for geneticists to work with genes that control vital functions of the cell, such as DNA synthesis. Many temperature-sensitive mutants are completely normal at 25° C but cannot synthesize DNA at 42° C. Presumably, temperature-sensitive mutations result in enzymes with amino acid substitutions that lead to protein denaturation at temperatures lower than the normal denaturation temperature.

Figure 16.14
Normal and tautomeric forms of
DNA bases. Adenine and cytosine
can exist in the amino, or the rare
imino, forms; guanine and
thymine can exist in the keto, or
rare enol, forms.

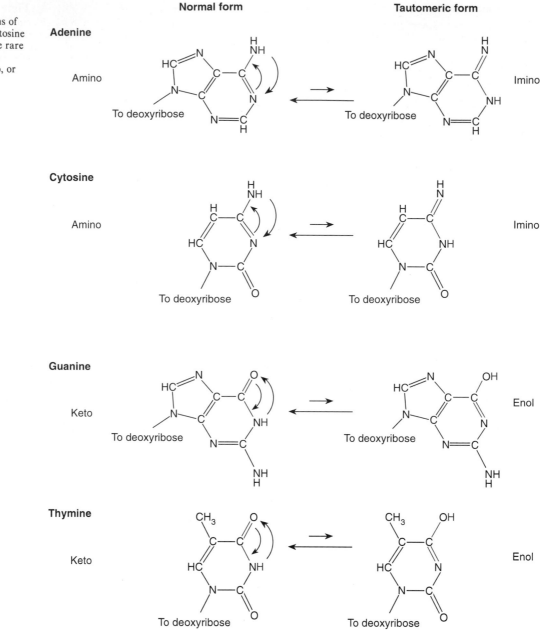

Thus the enzyme has normal function at 25° C, the **permissive temperature,** but is nonfunctional at 42° C, the **restrictive temperature.**

The interesting thing about most conditional-lethal mutants of *E. coli* that cannot synthesize DNA at the restrictive temperature is that they have a completely normal DNA polymerase I. From this information we inferred that polymerase I is not normally the enzyme used by the *E. coli* cell for DNA replication. When an organism with a conditional mutation of polymerase I was isolated, it was found to replicate its DNA normally, but it was unable to repair damage to the DNA. From this it was concluded that polymerase I is primarily involved in repair rather than replication of DNA. It is thus apparent that conditional-lethal mutations make it possible to do genetic analysis on genes that could not otherwise be studied.

Spontaneous Mutagenesis

Watson and Crick originally suggested that mutation could occur spontaneously during the process of DNA replication if pairing errors occurred. If a base of the DNA underwent a proton shift into one of its rare tautomeric forms (**tautomeric shift**) during the replication process, an inappropriate pairing of bases would occur. Normally, adenine and cytosine are in the amino (NH_2) form. Their tautomeric shifts are to the imino (NH) form. Similarly, guanine and thymine go from a keto (C = O) form to an enol (COH) form (fig. 16.14). Table 16.4 shows the new base pairings that would occur following tautomeric shifts of the DNA bases. Figure 16.15 provides an example of the molecular structure of one of these tautomeric pairings.

Figure 16.15

Tautomeric forms of adenine. In the common amino form, adenine does not base pair with cytosine; in the tautomeric imino form, it can.

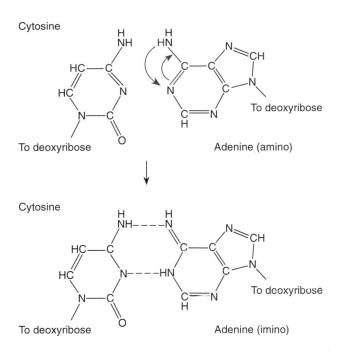

Figure 16.16

Tautomeric shifts result in transition mutations. The tautomerization can occur in the template base or in the substrate base. Tautomeric shifts are shown in *red;* the resulting transition in *blue.* The transition shows up after a second generation of DNA replication.

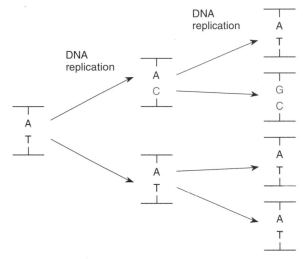

Template transition—tautomerization of adenine in the template

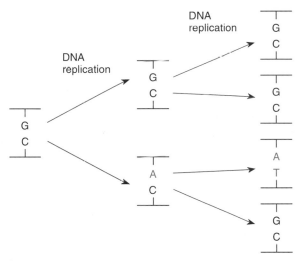

Substrate transition—tautomerization of incoming adenine

TABLE 16.4	Pairing Relationships of DNA Bases in the Normal and Tautomeric Forms	
Base	**In Normal State Pairs with**	**In Tautomeric State Pairs with**
A	T	C
T	A	G
G	C	T
C	G	A

During DNA replication, a tautomeric shift in either the incoming base (*substrate transition*) or the base already in the strand (*template transition*) will result in mispairing. The mispairing will be permanent and will result in a new base pair after an additional round of DNA replication. The original strand is unchanged (fig. 16.16).

In the example in figure 16.16, the replacement of one base pair by another maintains the same purine-pyrimidine relationship: AT is replaced by GC and GC by AT. In both examples, a purine-pyrimidine combination is replaced by a purine-pyrimidine combination. (Or, more specifically, a purine replaces another purine: Adenine is replaced by guanine in the first example and guanine by adenine in the second.) The mutation is referred to as a **transition mutation:** A purine (or pyrimidine) is replaced by another purine (or pyrimidine) through a transitional state involving a tautomeric shift. The form of replacement in which a purine replaces a pyrimidine or vice versa is referred to as a **transversion mutation.** How transversion mutations occur is less clear.

Transversions may arise by a combination of two events, a tautomerization and a base rotation. (We saw base rotations in the formation of Z DNA in chapter 9.) For example, an AT base pair can be converted to a TA base pair (a transversion) by an intermediate AA pairing (fig. 16.17). Adenine can pair with adenine if one of the bases undergoes a tautomeric shift while the other rotates about its base-sugar (glycosidic) bond (fig. 16.18). The normal configuration of the base is referred to as the *anti* configuration, whereas the rotated form is referred to as the *syn* configuration. Since it is now believed that as many as 10% of bases may be in the *syn* configuration at any moment, the transversion mutagenesis rate should be about 10% of the transition mutagenesis rate, a value not inconsistent with current information.

THE AMES TEST FOR CARCINOGENS

Which chemicals cause cancer in human beings? It is difficult to determine whether any substance is a carcinogen because tests for carcinogenicity usually involve administering the substance in question to laboratory rats or mice to determine whether the substance actually causes cancer in these animals. Even these tests, however, are not absolute predictors of cancer in people. Tests of this nature are very expensive ($1 million to $2 million each) and very time-consuming (three to four years). Since more than fifty thousand different chemical compounds are used in industry, with thousands more being added continually, the task of making the working environment as well as the environment in general safe from cancer seems overwhelming. We can, however, make a preliminary determination about the cancer-causing properties of any substance very quickly because there is a relationship between mutagenicity and carcinogenicity. Many substances in the environment that can cause cancer also can cause mutations. Both kinds of effects are related to DNA damage.

Bruce Ames, at the University of California at Berkeley, developed a routine screening test for mutagenicity of a substance. Substances that prove positive in this test are suspected of being carcinogens and would have to be tested further to determine their abilities to cause cancer in mammals.

Ames worked with a strain of *Salmonella typhimurium* that requires histidine to grow. This strain will not grow on minimal medium. However, the strain will grow if a mutagen is added to the medium causing the defective gene in the histidine pathway to revert to the wild-type. (Mutagens inducing gross chromosomal damage, such as deletions or inversions, will not be detected.) Under normal circumstances, there is a background mutation rate: A certain number of *Salmonella* cells will revert spontaneously, and therefore a certain number of colonies will grow on the minimal medium. A mutagen, however, will increase the number of colonies growing on minimal medium. This procedure is, therefore, a rapid, inexpensive, and easy test for a substance's mutagenicity.

Bruce Ames (1928–   )
Courtesy Dr. Bruce Ames.

To improve this test's ability to detect carcinogens, Ames added a supplement of rat liver extract to the medium. It is known that, although many substances are themselves not carcinogens, the breakdown of these substances in the liver creates substances that are carcinogenic. Rat liver enzymes will act on a substance the same way human livers will, converting a noncarcinogenic primary substance into a possible carcinogen. The liver enzymes can also make a mutagen nonmutagenic.

Other short-term tests are in use that effectively duplicate the Ames test. These include tests for mutagenicity in mouse lymphoma cells and two tests in Chinese hamster ovary cells: a test for chromosomal aberrations and a test for sister-chromatid exchanges. None of these tests is better than the Ames test, which has scored better than 90% correct when tested with hundreds of known carcinogens. Thousands of other substances have been subjected to this test; many have been found to be mutagenic. These substances are usually withdrawn from the workplace or home environment. From time to time we read that a certain substance is believed to be carcinogenic and is being taken off grocery store shelves. Examples have included hair dyes, food preservatives, food-coloring agents, and artificial sweeteners. Many of these first were suspected after they failed the Ames test.

Chemical Mutagenesis

Muller demonstrated that X rays can cause mutation. Mutation can also be induced by certain chemical and temperature treatments. Determining the mode of action of various chemical mutagens has provided insight into the mutation process as well as the process of transformation (see Box 16.2). In addition, knowing the mode of action of chemical mutagens has allowed geneticists to produce large numbers of certain types of mutations at will.

Transitions

Transitions are routinely produced by base analogues. Two of the most widely used base analogues are the pyrimidine analogue 5-bromouracil (5BU) and the purine analogue 2-aminopurine (2AP; fig. 16.19). The mutagenic mechanisms of the two are similar. The 5BU is incorporated into DNA in place of thymine; it acts just like thymine in DNA replication and, since hydrogen bonding is not changed, it should induce no mutation. However, it seems that the bromine atom causes 5BU to tautomerize more

Figure 16.17

A model for transversion mutagenesis. An AT base pair can be converted to a TA base pair (a transversion) by way of an intermediate AA base pair. One of the *red* bases is in the rare tautomeric form, while the other is in the *syn* configuration. After a second round of DNA replication, one DNA duplex will have a transversion at that point (*blue*).

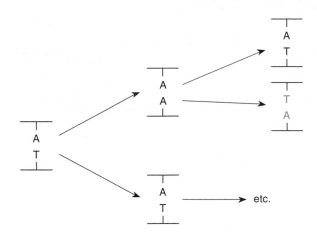

Figure 16.18

Transversion mutagenesis. An AA base pair can form if one base undergoes a tautomeric shift while the other rotates about its glycosidic (sugar) bond. In (*a*), both bases are in their normal configurations; there is no hydrogen bonding. In (*b*), hydrogen bonds are possible.

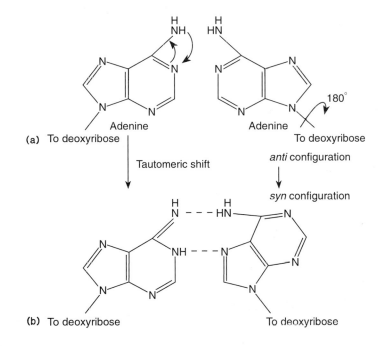

Figure 16.19

Structure of the base analogues 5-bromouracil (5BU) and 2-aminopurine (2AP)

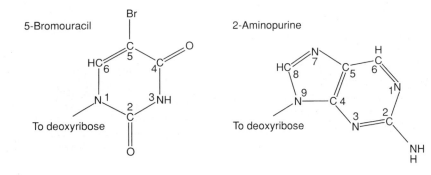

BOX 16.2

IN VITRO SITE-DIRECTED MUTAGENESIS

One way of studying the way that proteins work is to change the sequence of the amino acids of the protein. For example, if a scientist were working on the active site of a particular enzyme, he or she could learn how the enzyme modifies its substrates by changing one or a few amino acids. Changes could be made in order to study the role of shape or charge on the functioning of the enzyme. Aside from chemically induced mutagenesis, advances in recombinant DNA techniques have made it possible for a research scientist to create exactly the changes he or she wants in a protein.

To begin with, the gene for the protein or enzyme must be cloned so that it can be manipulated. Once cloned, deletions are easy to create with restriction endonucleases. If a particular endonuclease cuts the gene in two places, the intervening segment can be spliced out (Box fig. 1a; see chapter 12). If the endonuclease cuts only once, the ends of the cut can be digested away by exonucleases, extending the deletion away from the cut in both directions (Box fig. 1b). Insertions can be created by either cutting the gene and repairing the single-stranded ends (Box fig. 1c), or by creating a linker with the desired sequence and inserting the linker at the site of an endonuclease cut (Box fig. 1d).

Far more impressive, however, is the ability to change a single specific codon in order to replace any amino acid in the protein with any other amino acid. The process involves directed mutagenesis using artificially created oligonucleotides.

Basically, a short sequence of DNA (an oligonucleotide) is synthesized complementary to a region of the cloned gene, but having a change in one or more bases of a codon to specify a different amino acid. That oligonucleotide is then hybridized with the single-stranded form of the clone (Box fig. 2). Although one or more bases will not match, hybridization can usually be made to take place by adjusting pH or ionic strength of the solution. The hybridized oligonucleotide is then used as a primer for DNA replication; the whole plasmid is replicated, resulting in heteroduplex DNA. In subsequent DNA replications of the heteroduplex, both the original clone and the mutated DNA will be produced. The latter can be isolated by appropriate selection methods; it is a plasmid with a cloned gene that has the exact mutation that the researcher wanted. Using techniques of this type, advancements have been made in understanding exactly the way in which various components of an enzyme contribute to its functioning.

Box figure 1

A cloned gene can be mutated several ways. (a) If a restriction endonuclease has two sites in the gene, the intermediate piece can be spliced out. (b) If the endonuclease has only one site, the gene can be opened at that site and limited digestion by exonucleases will delete part of the gene.

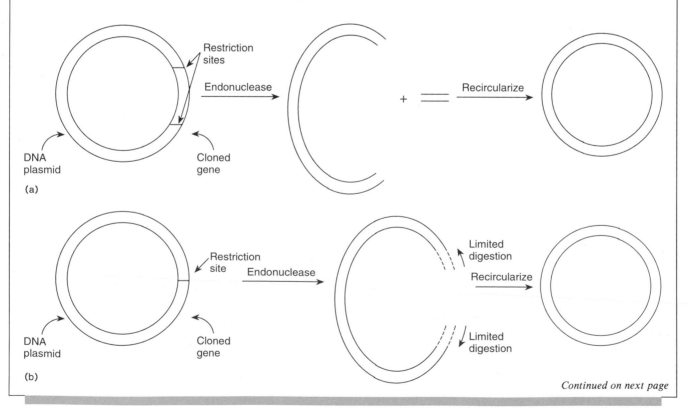

Continued on next page

BOX 16.2—*continued*

(*c*) If an endonuclease has an offset region between its splice points of three or six nucleotides (one or two codons), that length (TAT in this case) can be inserted by repair of single-stranded ends after cutting by the endonuclease. The resulting blunt ends can be spliced together. (Note that actually an ATT region has been converted to an ATTATT region. If reading codons along the DNA, the actual insertion is of a TAT codon.) (*d*) A linker of any length (usually the length of a specific number of codons) can be inserted at a restriction site.

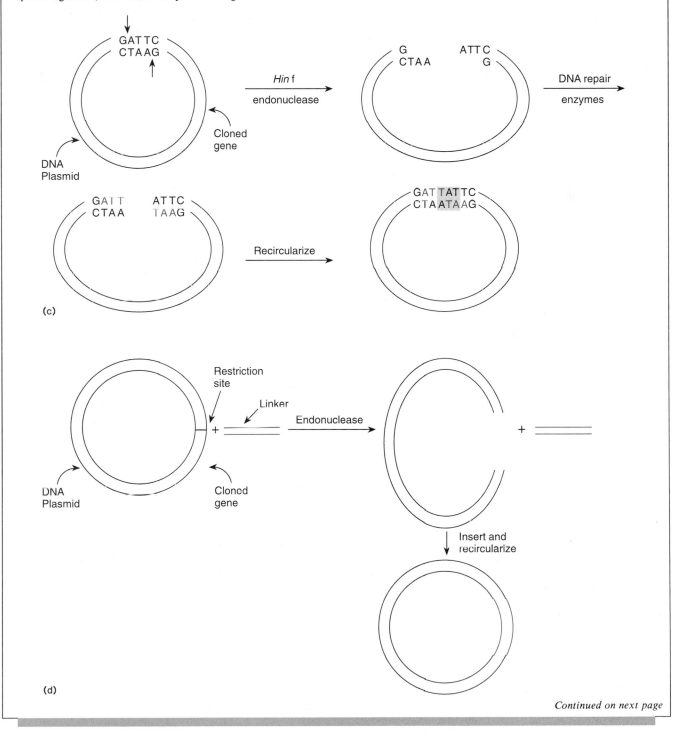

Continued on next page

BOX 16.2—*continued*

Box figure 2

Site-directed mutagenesis can be of any nucleotide(s). In this case an inserted gene with an Ile-Gly sequence is converted, at the direction of the investigator, to an Ile-Ala sequence. A single-stranded form of the plasmid is isolated. A synthetically prepared oligonucleotide (twenty-three bases in this example) is added. It will hybridize at the complementary site despite differing by three bases. Then DNA replication to form a double helix is carried out using the oligonucleotide configuration as a primer. After the strands of the duplex are separated, the original plasmid as well as the mutated plasmid can be isolated. (Note that the investigators changed two codons, although they changed only one amino acid because they also wanted to introduce an Alu site at that point for future studies.)

From J. E. Villafranca, et al., "Directed Mutagenesis of Dihydrofolate Reductase," in Science 222:782–788. Copyright 1983 by the AAAS.

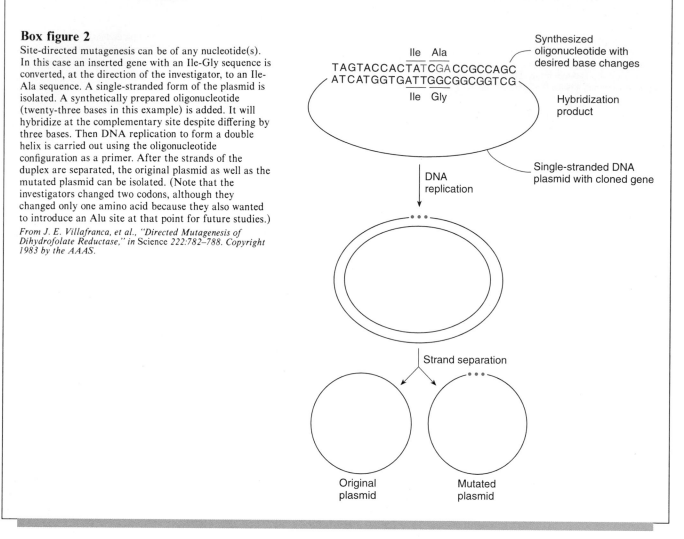

readily than thymine does. Thus 5BU goes from the keto form (fig. 16.19) to the enol form (as does thymine in fig. 16.14) more readily than thymine. Frequent transitions result when the enol form of 5BU pairs with guanine.

The 2AP is mutagenic by virtue of the fact that it can, like adenine, form two hydrogen bonds with thymine. When in the rare state it can pair with cytosine (fig. 16.20). Thus at times it replaces adenine and at times guanine. It promotes transition mutations.

Nitrous acid (HNO_2) also readily produces transitions by replacing amino groups on nucleotides with keto groups ($—NH_2$ to $=O$), with the result that cytosine is converted to uracil, adenine to hypoxanthine, and guanine to xanthine. As can be seen from figure 16.21, transition mutation results from two of the changes. Uracil base pairs with adenine instead of guanine, thus leading to a UA base pair replacing a CG base pair; hypoxanthine (H) pairs with cytosine instead of thymine, with which the original adenine paired. Thus an HC base pair replaces an AT base pair. Both of these base pairs (UA and HC) are transition mutations. Xanthine, however, pairs with cytosine just as guanine does. Thus the change of guanine to xanthine does not result in changes in base pairing.

Like nitrous acid, heat can also deaminate cytosine to form uracil and thus bring about transitions (CG to TA). Heat apparently can also bring about transversions by an unknown mechanism.

Transversions

Ethyl methane sulfonate ($CH_3SO_3CH_2CH_3$) and ethyl ethane sulfonate ($CH_3CH_2SO_3CH_2CH_3$) are agents that cause the removal of purine rings from DNA by a multistep process that begins with the ethylation of a purine ring and ends with the hydrolysis of the glycosidic (purine-

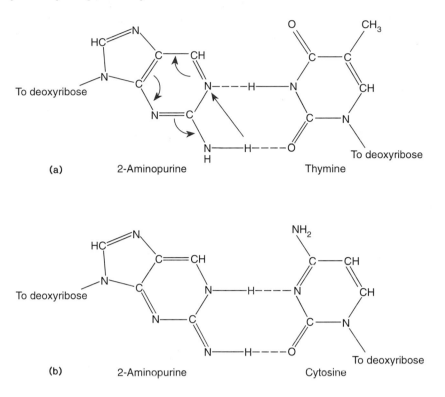

(a)   2-Aminopurine   Thymine

(b)   2-Aminopurine   Cytosine

deoxyribose) bond, causing the loss of the base. These sites are referred to as AP (apurinic-apyrimidinic) sites. When DNA replication takes place at the AP sites, DNA polymerase III is assumed to be free to insert any of the four possible bases into the new strand as a complement to the gap created when these alkylating agents removed the purine (fig. 16.22). If thymine is placed in the newly formed strand, then the original base pair will be restored; insertion of cytosine will result in a transition mutation; insertion of either adenine or guanine will result in a transversion mutation. Of course, the gap is still there and will continue to generate new mutations each generation until it is repaired. During DNA replication in *E. coli*, there is a tendency for the polymerase to place adenine opposite the gap more frequently than it places other bases.

Insertions and Deletions

The molecules of the acridine dyes, such as proflavin and acridine orange (fig. 16.23), are flat. Presumably, they initiate mutation by inserting into the DNA double helix, causing a buckling of the helix in the region of insertion, which could lead to additions and deletions of bases during DNA replication. Crick and Brenner used acridine-induced mutations to demonstrate both that the genetic code was read from a fixed point and that it was triplet (chapter 11).

Misalignment Mutagenesis

Additions and deletions in DNA can also come about by misalignment of a template strand and the newly formed (progeny) strand in a region in which there is a repeated sequence. For example, in figure 16.24 we expect the progeny strand to contain six adjacent adenines because the template strand contains six adjacent thymines. Misalignment of the progeny strand results in seven consecutive adenines: six thymines replicated plus one already replicated but misaligned. Misalignment of the template strand results in five consecutive adenines because one thymine is not available in the template. Regions like this of long runs of a particular base may be very mutation prone. Their existence may explain the "hot spots" observed by Benzer (see fig. 16.8) and others.

Intergenic Suppression

When a critical mutation occurs in a codon, several routes can lead to survival of the individual; simple reversion and intragenic suppression have already been considered. A third route is that of **intergenic suppression**—restoration of the function of a mutated gene by changes in a different gene, called a **suppressor gene.** Suppressor genes are usually tRNA genes. When mutated, intergenic suppressors change the way in which a codon is read.

Figure 16.21
Nitrous acid converts cytosine to uracil, adenine to hypoxanthine, and guanine to xanthine. Uracil pairs with adenine whereas its progenitor, cytosine, normally pairs with thymine; hypoxanthine pairs with cytosine whereas its progenitor, adenine, normally pairs with thymine; and xanthine pairs with cytosine, the same base that guanine, its progenitor, pairs with. Thus, only the first two result in transition mutations.

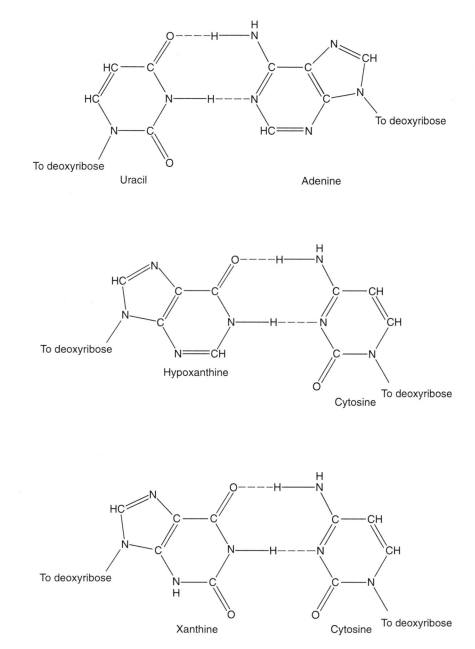

Suppressor genes can restore proper reading to nonsense, missense, and frameshift mutations. **Nonsense mutations** convert a codon that originally specified an amino acid into one of the three nonsense codons. **Missense mutations** change a codon so that it specifies a different amino acid. Frameshift mutations, by additions or deletions of nucleotides, cause an alteration in the reading frame of codons. A frameshift mutation, caused by the insertion of a single base, can be suppressed by a tRNA that has an added base in its anticodon (fig. 16.25a). It reads four bases as a codon and thus restores the original reading frame.

The tRNA produced by a nonsense suppressor gene reads the nonsense codon as if it were a codon for an amino acid; an amino acid is placed into the protein, and reading of the mRNA continues. At least three suppressors of the

Figure 16.22

Four possible outcomes after treatment of DNA with an alkylating agent, which removes the purine, adenine, in this example. The bases shown in *red* are the four bases that DNA polymerase may insert opposite the gap. After another round of DNA replication, the gap remains to generate further mutations, whereas the inserted base forms a base pair (*blue*), which can be a restoration or a transition or transversion mutation.

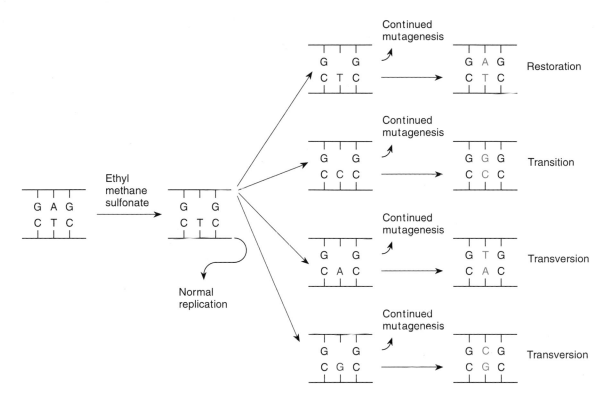

Figure 16.23

Structure of two acridine dyes, proflavin and acridine orange

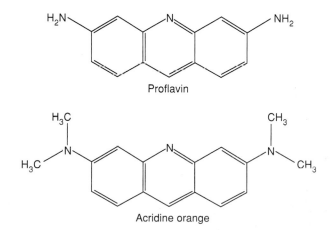

mutant amber codon (UAG) are known in *E. coli.* One suppressor puts tyrosine, one puts glutamine, and one puts serine into the protein chain at the point of an amber codon. Normally tyrosine tRNA has the anticodon 3'-AUG-5'. The suppressor tRNA that reads amber as a tyrosine codon has the anticodon 3'-AUC-5', which is complementary to amber. Hence a mutated tyrosine tRNA reads amber as a tyrosine codon (fig. 16.25b).

The following question arises: If the amber nonsense codon is no longer read as a stop signal, then won't all the genes terminating in the amber codon continue to be translated beyond their ends, thus resulting in the death

Figure 16.24

Misalignment of a template or progeny strand during DNA synthesis. If the progeny strand is misaligned after DNA replication has begun, the resulting progeny strand will have an additional base. If the template strand is misaligned during DNA replication, the resulting progeny strand will have a deletion of a base. These changes will show up after another round of DNA replication.

Reprinted by permission of American Scientist, journal of Sigma Xi, The Scientific Research Society.

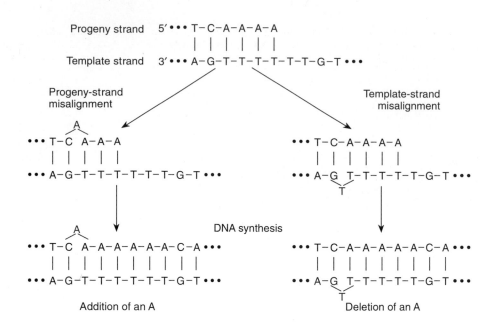

Figure 16.25

Frameshift and nonsense suppression by mutant tRNAs. In (*a*) a thymine has been inserted into DNA resulting in a frameshift. However, a tRNA with four bases in the codon region reads the inserted base as part of the previous codon in the mRNA. The frameshift thus does not occur. In (*b*) an *amber* mutation (UAG), which normally results in chain termination, is read as tyrosine by a mutant tyrosine tRNA that has the anticodon sequence complementary to the *amber* codon.

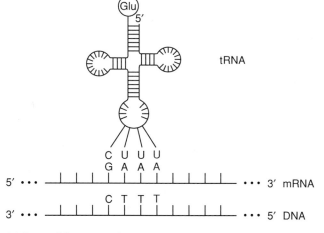

(a) Frameshift suppression

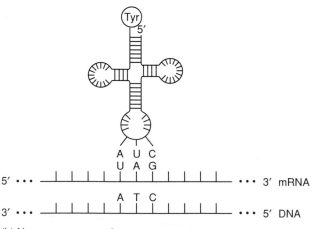

(b) Nonsense suppression

of the cell? In the tyrosine case, two genes for tyrosine tRNA were found: One contributes the major fraction of the tRNAs and the other, the minor fraction. It is the minor-fraction gene that mutates to act as the suppressor. Thus most mRNAs are translated normally and most amber mutations result in premature termination, although a sufficient number are translated (suppressed) to ensure the viability of the mutant cell. In general, intergenic suppressor mutants would quickly be eliminated in nature because they are inefficient—the cells are not healthy. In the laboratory, special conditions can be provided that allow them to be grown and studied.

Mutator and Antimutator Mutations

Whereas intergenic suppressors are mutations that "restore" the normal phenotype, mostly through mutation of tRNA loci, mutations known as **mutator** and **antimutator mutations** cause an increase or decrease in the overall mutation rate of the cell. They are frequently mutations of DNA polymerase, which, as you remember, not only polymerizes DNA nucleotides 5' to 3' complementary to the template strand but also check to be sure that the correct base was put in (proofread). If, in the proofreading process, the polymerase discovers an error, it can correct this error with its 3' to 5' exonuclease activity. Mutator and antimutator mutations can involve changes in the proofreading ability (exonuclease activity) of the polymerase.

Phage T4 has its own DNA polymerase with known mutator and antimutator mutants. Mutator mutants are found to be very poor proofreaders (they have low exonuclease to polymerase ratios) and thus introduce mutations throughout the phage genome. Antimutator mutants, however, have exceptionally efficient proofreading ability (high exonuclease to polymerase ratios) in their DNA polymerase and, therefore, result in a very low mutation rate for the whole genome.

DNA REPAIR

Radiation, chemical mutagens, heat, enzymatic errors, and spontaneous decay are constantly damaging DNA. For example, it is estimated that several thousand bases in DNA are spontaneously lost each day in every mammalian cell due to spontaneous decay. In the long evolutionary challenge to prevent mutation, cells have evolved numerous mechanisms to repair damaged or incorrectly replicated DNA. Many enzymes, acting alone or in concert with other enzymes, repair DNA. Repair systems are generally placed in three broad categories: damage reversal, excision repair, and postreplicative repair.

Figure 16.26

UV-induced dimerization of adjacent thymines in DNA. The *red lines* represent the dimer bonds in the adjacent thymines.

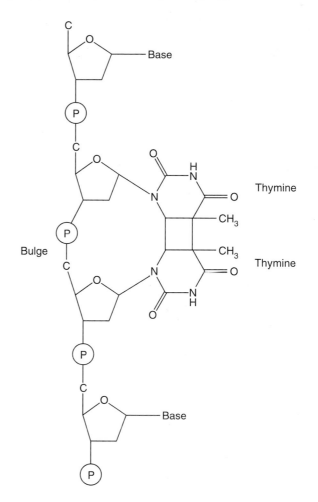

Damage Reversal

Ultraviolet (UV) light causes linkage, or **dimerization,** of adjacent pyrimidines in DNA (fig. 16.26). Although cytosine-cytosine and cytosine-thymine dimers are produced occasionally, the principal product of UV irradiation is thymine-thymine dimers. These can be repaired several different ways, of which the simplest is the reversal of the dimerization process to restore the original unlinked thymines.

In *E. coli* there is an enzyme, DNA photolyase, the product of the *phr* gene (for **photoreactivation**), that will bind in the dark to dimerized thymines. When light shines on the cell, the enzyme breaks the dimer bonds with light energy. The enzyme then falls free of the DNA. This enzyme thus reverses the UV-induced dimerization. Another example of direct DNA repair is O^6-mGua DNA

Figure 16.27

The structure of O⁶-methylguanine. The *red color* shows the modification of guanine, in which the normal configuration is a double-bonded oxygen.

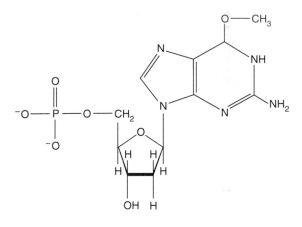

Figure 16.28

Mechanism of excision repair of thymine dimers. A 12-bp segment of DNA with the dimer is removed by an excinuclease and a helicase. Repair by DNA polymerase I and DNA ligase patch the duplex.

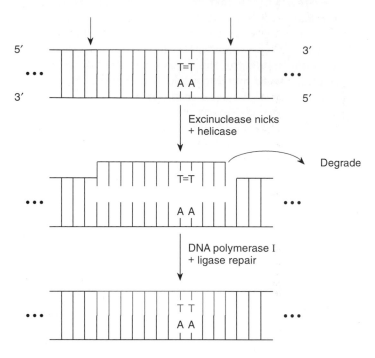

methyltransferase, an enzyme that removes the methyl groups from O⁶-methylguanine, the major product of DNA-methylating agents (fig. 16.27).

Excision Repair

Excision repair refers to the general mechanism of DNA repair that works by removing the damaged portion of the DNA. Various endonuclease enzymes can sense the distortion in the DNA double helix caused by damage of one sort or another. During the process of excision repair, bases are removed from the damaged strand. The gap is then patched using complementarity with the remaining strand. We can broadly categorize these systems as UV damage repair, AP repair, and mismatch repair.

UV Damage Repair

If DNA is exposed to UV light, thymine dimerization will occur. In the absence of photoreactivation (in *phr⁻* cells), thymine dimers are found in the cytoplasm after about half an hour. This experiment indicates that excision of the dimers rather than an undoing of the dimerization occurs. This excision is accomplished in the following manner. An endonuclease that can detect thymine dimers makes a nick in the DNA strand on both sides of the dimer (fig. 16.28). This endonuclease is composed of subunits coded by the *uvrA, uvrB,* and *uvrC* genes (for *UV r*epair) and called the ABC excinuclease. The enzyme nicks (hydrolyzes) the damaged strand at the eighth phosphodiester bond on the 5′ side and the fourth phosphodiester bond on the 3′ side of the dimer. DNA helicase II, the product of the *uvrD* gene, separates the strands to release the 12-bp segment, which is eventually degraded. DNA polymerase I and ligase then fill in the short gap, repairing the DNA. The ABC excinuclease also removes other types of damaged DNA in small patches, all of about a dozen

bases long. Thus a relatively simple system exists for locating, excising, and repairing certain kinds of damaged DNA.

Excision repair systems are found in virtually all organisms, from the largest viruses to eukaryotes. In human beings, the autosomal recessive trait, xeroderma pigmentosum, is due to an inability to repair thymine dimerization induced by UV light. Persons with this trait freckle heavily when exposed to the UV rays of the sun, and they have a high incidence of skin cancer. There appear to be nine complementation groups (loci) involved in UV damage repair in human beings.

AP Repair

AP repair refers to the repair of apurinic and apyrimidinic sites on DNA. These are sites in which the base has been removed (see fig. 16.22). A base can be removed from a nucleotide within DNA either by direct action of an agent such as radiation, or by **DNA glycosylases,** enzymes that sense damaged bases and remove them. For example, uracil-DNA glycosylase, the product of the *ung* gene, recognizes uracil within DNA and cleaves it out at the base-sugar (glycosidic) bond. **AP endonucleases** then initiate excision repair at the AP site. Class I AP endonucleases nick the DNA at the 3′ side of the AP site, whereas class II AP endonucleases nick DNA at the 5′ side of the site. An exonuclease then removes a short region of DNA including the AP site (two to four bases). DNA polymerase I and ligase then repair the gap with a short patch (fig. 16.29).

Figure 16.29

A uracil in DNA is repaired. First, the uracil is removed by a glycosylase. The newly formed AP site (apurinic/apyrimidinic) is repaired first by the removal of a small segment with the AP site, followed by repair of the site by DNA polymerase I and DNA ligase.

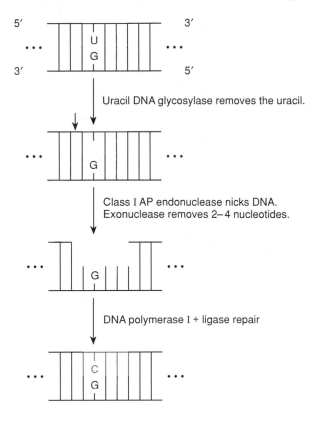

Mismatch Repair

A final form of excision repair is referred to as **mismatch repair.** It is responsible for about 99% of all repairs to DNA. As DNA polymerase replicates DNA, some errors are made that are not corrected by the proofreading capability of the polymerase. For example, a template G can be paired with a T rather than a C in the progeny strand. The GT base pair does not fit correctly in the DNA duplex. It is recognized by the mismatch repair system, which follows behind the replicating fork. This system, whose members are specified by the *mutH, mutL, mutS,* and *mutU* genes, is responsible for the removal of the incorrect base by an excision repair process. (The genes are called *mut* for *mut*ator because mutations of these genes cause high levels of spontaneous mutation in the cells.) The mismatch repair enzymes initiate the removal of the incorrect base by nicking the DNA strand on both sides of the mismatch.

We might wonder how the mismatch repair system recognizes the progeny, rather than the template, base as the wrong one. After all, in a mismatch there are no defective bases and theoretically either partner could be the "wrong" base. In *E. coli,* the answer lies in the methylation state of the DNA. DNA methylase, the product of the *dam* locus, methylates 5′-GATC-3′ sequences at the adenine residue, which are relatively common in the DNA of *E. coli.* Since the mismatch repair enzymes follow the replication fork of the DNA, they usually reach the site of mismatch before the methylase. Template strands will be methylated, whereas progeny strands, being newly synthesized, will not be methylated. Thus the methylation state of the DNA cues the mismatch repair enzymes as to which base to attack. After the methylase passes by, both strands of the DNA will be methylated.

The mismatch repair system may be able to operate with and without methylation cues. There is also some doubt as to whether the GATC sites are the actual places in which the DNA is nicked. In figure 16.30 we present one model that assumes that the DNA is nicked at the GATC sequences. The mismatch is recognized by the products of the *mutL* and *mutS* genes. The product of the *mutH* gene nicks the DNA; the product of the *mutU* gene (also called *uvrD*), DNA helicase II, unwinds the nicked region so it falls free. DNA polymerase I and ligase then fill in the patch, which can be over one thousand bases long. Single-stranded binding proteins usually attach to the single-stranded parent and progeny strands during the unwinding process to stabilize the single strands.

Postreplicative Repair

When DNA polymerase III encounters certain damage in *E. coli,* such as thymine dimers, it cannot proceed. Instead the polymerase stops DNA synthesis and, leaving a gap, skips down the DNA to resume replication as far as eight hundred or more bases away. If allowed to remain, this gap will result in DNA that is deficient and broken. Since part of one strand is absent and the other has damage, there appears to be no viable template upon which new DNA can be replicated. However, there is an undamaged copy of this region on the other newly replicated daughter duplex. A group of enzymes, with one specified by the *recA* locus having central importance, repairs the gap. Since the repair takes place at a gap created by the failure of DNA replication, the process is called **postreplicative repair.** The *recA* locus was originally discovered in another process, recombination. In fact, postreplicative repair is sometimes called recombinational repair and shares many enzymes with recombination.

The RecA Protein

The RecA protein has two major properties. First, it coats single-stranded DNA (fig. 16.31) and causes that single-stranded DNA to invade double-stranded DNA (fig. 16.32). By invasion we mean that the single-stranded DNA will attempt to form complementary base pairs with one of the strands of the double-stranded DNA while displacing the other strand of that double helix. One possible mechanism for this activity, assuming two sites on the

Figure 16.30

A model of mismatch repair. *Asterisks* show the methylated adenines in the GATC sequences. The mismatched progeny section is removed by *mut*H and helicase after *mut*S and *mut*L recognize the mismatch. The patch is repaired and the progeny strand is methylated.

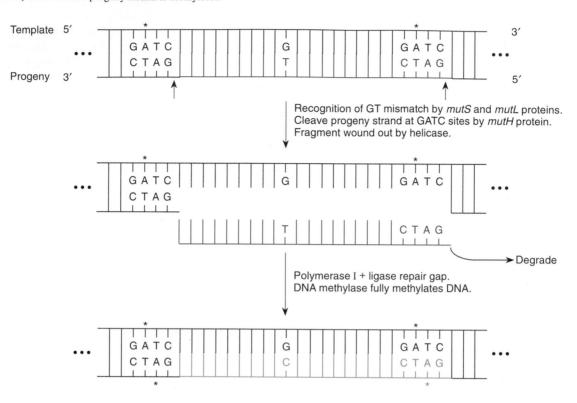

Recognition of GT mismatch by *mut*S and *mut*L proteins. Cleave progeny strand at GATC sites by *mut*H protein. Fragment wound out by helicase.

Polymerase I + ligase repair gap. DNA methylase fully methylates DNA.

Figure 16.31

Scanning tunneling microscope picture of single-stranded DNA coated with RecA protein (*large arrow*). The *small arrow* indicates uncoated double-stranded DNA. Bars denote 20 nm in each direction. (In fig. 16.33 we show how the very large coated DNA can invade the very small uncoated DNA.)

From Amrein, et al., 1988. Science 240:22 April 1988, 515, Figure 2. *"Scanning Tunneling Microscopy of RecA-DNA Complexes Coated with a Conducting Film." Copyright 1988 by the AAAS.*

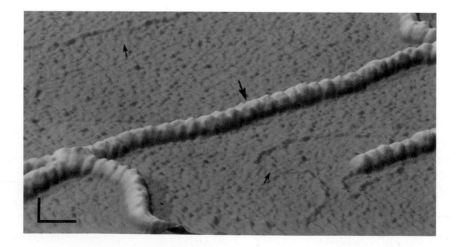

Figure 16.32

One property of the RecA protein. It causes single-stranded DNA to invade double-stranded DNA and to move along the double-stranded DNA until a region of complementarity is found.

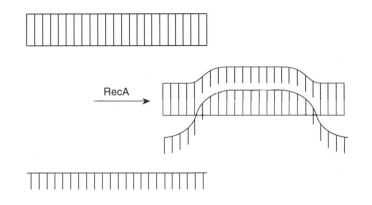

Figure 16.33

A model of how the RecA protein can cause insinuation of single-stranded DNA into a double-stranded molecule. (*I*) Axial view of one nucleotide (with two phosphate groups) of single-stranded DNA is shown attached at site I in this cross-sectional diagram of the RecA protein, which is about 60 percent larger than actually shown. (*II*) Duplex DNA is bound at site II of RecA. (*III*) RecA protein rotates the bases such that the single-stranded DNA forms a complementary base pair with one strand of the duplex, leaving the other strand of the duplex unpaired (see fig. 16.32).

Reprinted by permission from Nature, *Vol. 309, p. 217. Copyright © 1984 Macmillan Magazines Ltd.*

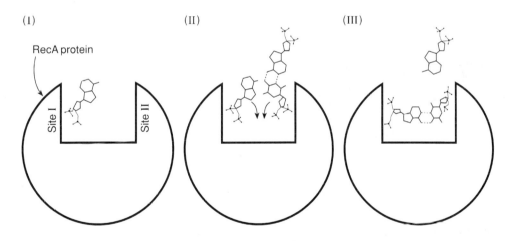

enzyme, is shown in figure 16.33. RecA will continue to move the single-stranded DNA along the double-stranded DNA until a region of homology is found.

The second major property of the RecA protein is that, when stimulated by the presence of single-stranded DNA, it acts either as a protease wherein it cleaves the λ cI repressor protein (see chapter 13) or acts to cause autocatalysis of another repressor, called LexA, and thus initiates several sequences of reactions. We begin by looking at the direct role of the RecA protein in postreplicative repair.

The RecA protein is responsible for filling a postreplicative gap in newly replicated DNA with a strand from the undamaged sister duplex. Gap-filling processes then complete both strands. In figure 16.34a we see a replication fork in which DNA polymerase III has left a gap in the progeny strand in the region of a thymine dimer. The RecA protein is responsible for the damaged single strand to invade the sister duplex (fig. 16.34b). Endonuclease activity then frees the double helix containing the thymine dimer (fig. 16.34c). DNA polymerase I and DNA

Figure 16.34

RecA-dependent postreplicative repair of DNA. DNA polymerase III skips past a thymine dimer during DNA replication (*a*). With the help of RecA, the single strand with the thymine dimer invades the normal sister duplex (*b*). An endonuclease nicks the new duplex at either side of the thymine dimer site, freeing the new duplex with the thymine dimer and leaving the sister duplex single-stranded (*c*). Repair enzymes then create two intact daughter duplexes (*d*).

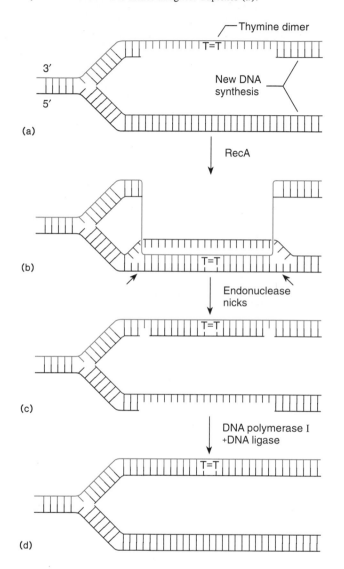

ligase return both daughter helices to the intact state (fig. 16.34*d*). The thymine dimer still exists, but now its duplex is intact and another cell cycle is available for photoreactivation or excision repair to remove the dimer.

The SOS Response

Postreplicative repair is part of a cell reaction called the **SOS response.** When an *E. coli* cell is exposed to excessive quantities of UV light, other mutagens, or agents that damage DNA (such as alkylating or cross-linking agents), or when DNA replication is inhibited, gaps are created in the DNA. In the presence of this single-stranded DNA, the RecA protein interacts with the LexA protein, the product of the *lexA* gene. The LexA protein is a repressor that normally represses about eighteen genes, including itself. The other genes include *recA, uvrA, uvrB,* and *uvrD;* two genes that inhibit cell division, *sulA* and *sulB;* and several others. Each of these genes has a consensus sequence in its promoter called the **SOS box:** 5'-$CTGX_{10}CAG$-3' (where X_{10} refers to any ten bases). The LexA protein normally binds at the SOS box, limiting the transcription of these genes. When *recA* is activated by single-stranded DNA, it interacts with the LexA protein in such a way as to trigger the autocatalytic properties of LexA. Although this fact has not been completely established—RecA might in fact cleave LexA—there is evidence that the LexA protein catalyzes its own breakdown after interacting with RecA that has been activated by single-stranded DNA (fig. 16.35). Transcription then follows from all the genes having an SOS box. The two inhibitors of cell division, the product of the *sulA* and *sulB* genes, presumably increase the amount of time that the cell has to repair the damage before the next round of DNA replication.

Eventually, the DNA damage is repaired. There is no single-stranded DNA to activate RecA and therefore LexA is no longer destroyed. LexA again represses the suite of proteins involved in the SOS response, and the SOS response is over. Table 16.5 summarizes some of the enzymes and proteins involved in DNA repair.

As we mentioned in chapter 13, λ prophage can be induced into vegetative growth by UV light. This is another effect of the SOS response. RecA not only causes the LexA protein to be inactivated, but it also directly inactivates the λ repressor, the product of the λ *cI* gene. From an evolutionary point of view, it makes sense for phage λ to have evolved a repressor protein that is inactivated by the RecA protein. As a prophage, λ is dependent on the survival of the host cell. In cases in which that survival might be in jeopardy, the prophage would be at an advantage if it could sense these situations and make copies of itself that could leave the host. One of these times might be when the host has suffered a lot of DNA damage. The SOS response is a signal to a prophage that the cell has received just that damage. Hence the prophage is induced when RecA acts as a protease; the λ repressor is destroyed, the cro protein becomes dominant, and vegetative growth follows. From an evolutionary perspective, the *E. coli* cell has not created an enzyme (RecA) that seeks out the λ repressor for the benefit of λ. Rather, the λ repressor has evolved for its own advantages to be sensitive to RecA.

The process of SOS repair seems to result in many mutations. Whether these mutations are due to inherent errors in RecA postreplicative repair or whether other last-ditch repair processes might come into play to handle postreplicative gaps is not completely clear.

Figure 16.35

The LexA protein represses itself, *recA*, and several other loci (*uvrA*, *uvrB*, *uvrD*, *sulA*, and *sulB*) by binding at the SOS box in each of the loci. Activated RecA protein causes autocatalysis of LexA, eliminating repression of all the loci that are then transcribed and translated.

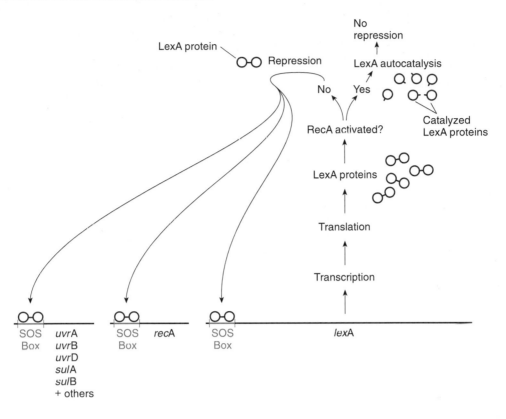

TABLE 16.5   A Summary of Some of the Enzymes and Proteins Involved in DNA Repair in *E. coli*, Not Including DNA Polymerase I, DNA Ligase, and Single-Strand Binding Proteins

Enzyme	Gene	Action
Damage Reversal		
DNA photolyase	*phr*	Undimerizes thymine dimers
DNA methyltransferase	*ada*	Demethylates guanines in DNA
Excision Repair		
ABC excinuclease	*uvrA, uvrB, uvrC*	Nicks both sides of thymine dimers
DNA helicase II	*uvrD (mutU)*	Unwinds DNA
Uracil-DNA glycosylase	*ung*	Removes uracils from DNA
Endonuclease IV	*nfo*	Nicks AP sites on the 5′ side
DNA methylase	*dam*	Methylates 5′-GATC-3′ DNA sequences
MutL, MutS proteins	*mutL, mutS*	Mismatch recognition in DNA
MutH endonuclease	*mutH*	Cleaves mismatched progeny DNA
Postreplicative Repair		
RecA protein	*recA*	Invasion of single-stranded DNA; cleavage of LexA; protease
LexA protein	*lexA*	Repressor of SOS proteins
SulA, SulB proteins	*sulA, sulB*	Inhibit cell division

RECOMBINATION

Since Rec⁻ *E. coli* cells lack the ability for both recombination and postreplicative repair of mutation damage, they provide some insight into the type of mechanism involved in the recombination process. Although recombination, the nonparental arrangement of alleles in progeny, can come about both by independent assortment and crossing over, we will be concerned here with recombination due to crossing over between homologous pieces of DNA (**homologous recombination**). We briefly discussed transpositional recombination in chapter 13 and site-specific recombination (e.g., λ integration) in chapters 7, 13, and 15.

Recombination is a **breakage-and-reunion** process. Homologous parts of chromosomes come into apposition, at which point both strands are broken and then reconnected in a crosswise fashion (see fig. 6.4). This general model fits what we know about the concordance of recombination and repair: Both involve breakage of the DNA and a small amount of repair synthesis, and both involve some of the same enzymes.

R. Holliday (1932–   )
Courtesy of James L. German, III, M.D.

Holliday Mechanism of Breakage and Reunion

The currently accepted model of recombination was put forth in 1964 by R. Holliday: It includes the formation of hybrid DNA molecules between the two strands about to undergo recombination. The occurrence of complementary base pairings is consistent with the fact that recombination is an exact process—rarely does it result in additions or deletions of nucleotides.

The first step in recombination is the breakage of two homologous double helices, each at the same place and each on only one strand of each duplex (fig. 16.36*b*). The broken strands then pair with their complement on the other duplex and covalent bonds are formed (fig. 16.36*c–e*). The crossover point can slide down the duplexes, a process known as **branch migration** (fig. 16.36*f*). In order to release the cross-linked duplexes, a second cut in each double helix is required.

In order to visualize fully the fact that two alternative sets of second cuts can be made, the model of figure 16.36*g* is rotated about its branch point (fig. 16.36*h*) to give an open view of the structure (fig. 16.36*i*). If the second cuts occur on the same strands that were originally cut, there will be no recombination of loci to the sides of the hybrid piece (patch formation; fig. 16.36*j–l, left*). If, however, the cuts are in the previously uncut strands, there will be

a reciprocal recombination of loci at the ends (splice formation; fig. 16.36*j–l, right*). The structure of figure 16.36*i* is called a Holliday intermediate structure and can be seen in the electron microscope (fig. 16.37).

The initial steps of recombination involve the RecA protein: A single strand of DNA invades a double helix, displacing the other strand in the double helix (fig. 16.36*c–f*). How then are single strands of DNA formed and how are the nicks in the DNA made? These steps in the recombination process are carried out by a second protein, the RecBCD protein (also known as exonuclease V) whose subunits are the products of the *recB*, *recC*, and *recD* loci. This protein has helicase properties—it unwinds DNA—and nuclease properties.

The RecBCD protein enters a DNA double helix from one end and travels along it in an ATP-dependent process. As it travels along the DNA it unwinds it. Rewinding takes place in its wake. However, the unwinding process at 300 bp per second is faster than the rewinding process at about 200 bp per second, resulting in the emergence of single-stranded loops at a rate of about 100 bp per second (fig. 16.38). Hence one activity of the RecBCD protein generates single-stranded loops (fig. 16.39). Nicking of one strand of the DNA double helix occurs at specific sequences, called **chi sites** (fig. 16.38). They have the sequence 5′-GCTGGTGG-3′. Hence the RecBCD protein

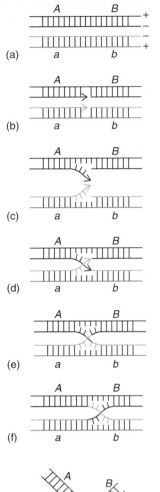

(a)

(b)

(c)

(d)

(e)

(f)

(g)

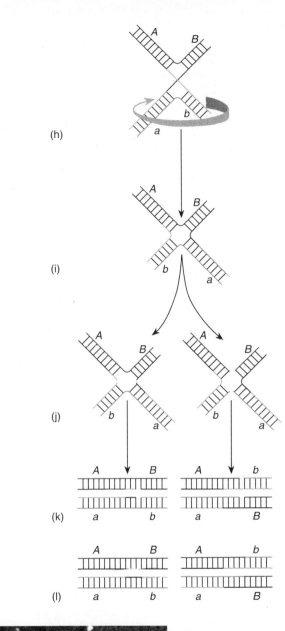

(h)

(i)

(j)

(k)

(l)

Figure 16.36

The Holliday model of genetic recombination. Two homologous duplexes of the four present in a meiotic tetrad are shown. The pluses and minuses indicate the antiparallel strands. (*a*) *A, a, B,* and *b* are alleles of the A and B loci. In (*b*), one strand of each duplex is broken and each broken strand invades the other duplex (*c*). Covalent bonds are formed (*d* and *e*), crosslinking the duplexes. Branch migration then occurs (*f*). In (*g*), (*h*), and (*i*), a rotation of one segment of the crosslinked duplexes is shown to clarify how a second set of cuts will either form a patch (*j–l, left*) or a splice (*j–l, right*).

From H. Potter and D. Dressler, "On the Mechanism of Genetic Recombination: Electron Microscopic Observation of Recombination Intermediates" in Proceedings of the National Academy of Sciences, *73:3000–3004, 1976. Reprinted by permission.*

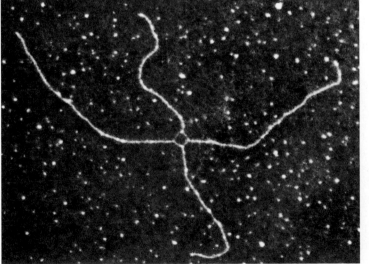

Figure 16.37

A Holliday intermediate structure, equivalent to the structure seen in figure 16.36*i.* Each arm is about 1 micron long.

H. Potter and D. Dressler, "DNA Recombination: In Vivo and In Vitro Studies" in Cold Spring Harbor Symposium on Quantitative Biology, *Volume XLIII (1979), pp. 969–985.*

Figure 16.38

RecBCD enters a DNA double helix at one end and travels along it at a rate of 300 bp per second, creating single-stranded loops. At chi sites the protein cuts one of the strands of the DNA. The RecBCD protein can thus create single-stranded DNA that can be coated with RecA protein for invasion of homologous duplexes; its nuclease activity can free the newly formed duplexes (e.g., fig. 16.40).

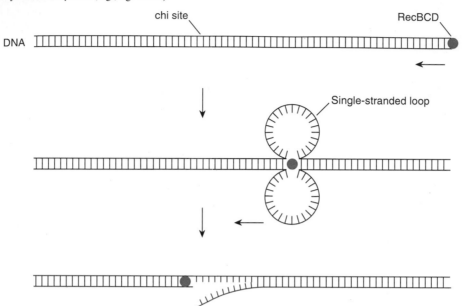

Figure 16.39

(*a*) Electron micrograph of single-stranded loops of DNA created by RecBCD protein and stabilized by single-stranded binding proteins (ssbs). The bar is 0.5 microns long. (*b*) Diagrammatic explanation of the electron micrograph.

(a) *From Dr. A. Taylor and Dr. G. R. Smith, 1980. Unwinding and rewinding of DNA by the RECBCD-enzyme.* Cell 22 (Nov.):447–457. Cell *is copyrighted by the MIT Press.*

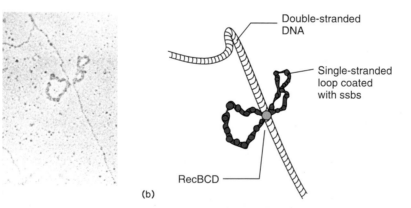

(a)   (b)

produces both single-stranded DNA that can invade its homologue after being coated by RecA, as well as nicks in the DNA to free that strand for invasion. There are about one thousand chi sites on the *E. coli* chromosome.

The branch migration step of recombination (fig. 16.36*e–f*) is mediated by the RecA-coated DNA filament. This, too, is an ATP-requiring process. Hydrolysis of ATP by the RecA protein provides the energy for the rotational movement of the DNA strands in respect to each other and the continued progress of the branch point.

Bacterial Recombination

In bacteria, recombination is a process of integrating foreign DNA into the host chromosome, a nonreciprocal process (see fig. 7.11). The Holliday model is easily adapted to this nonreciprocal incorporation of genetic material (fig. 16.40). Here, a strand of the exogenote invades the double helix of the *E. coli* chromosome, a process that the RecA and RecBCD proteins can mediate. After this pairing, the

Figure 16.40

Holliday model adapted to nonreciprocal incorporation of DNA into the *E. coli* chromosome. With the help of the RecBCD and RecA proteins, a strand of the exogenote invades the bacterial chromosome. Nucleases then free the newly formed heteroduplex; extraneous pieces of DNA are degraded. DNA ligase then completes the process.

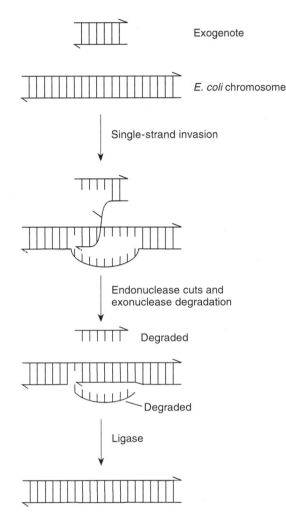

Exogenote

E. coli chromosome

Single-strand invasion

Endonuclease cuts and exonuclease degradation

Degraded

Degraded

Ligase

unpaired segments of the double helix of the bacteria and the exogenote are both degraded. Finally, the remaining double helix is sealed by DNA ligase.

Hybrid DNA

The result of bacterial recombination or meiotic recombination with branch migration is a length of **hybrid DNA.** This hybrid DNA, also called **heterozygous DNA** or **heteroduplex DNA,** has one of two fates, if we assume a difference in base sequences in the two strands. Either the heteroduplex can separate unchanged at the next cell division, or the cell's mismatch repair system can repair it (fig. 16.41). Without appropriate methylation cues, the mismatch repair system can convert the CA base pair to either a CG or a TA base pair. If TA were the original bacterial base pair, conversion to CG would be a successful recombination, whereas return of the CA to TA would be restoration rather than recombination.

Recombination in yeast, or any other eukaryote, will generate two heteroduplexes. The repair process can cause **gene conversion** (fig. 16.42), the alteration of progeny ratios indicating that one allele was converted to another. The mismatched AC (fig. 16.42) will be changed to an AT or a GC base pair; the mismatched TG base pair will be changed to TA or CG. The result of the repair, as shown in the bottom of figure 16.42, can be gene conversion in which an expected ratio of 2:2 (*a a* + +) is converted to a 3:1 ratio (*a a a* +) or a 1:3 ratio (*a* + + +). If the heteroduplexes are not repaired, then a single cell will generate both kinds of offspring after one round of DNA replication. Thus the colony from the cell will be half wild-type (+) and half mutant (*a*). It is only in an Ascomycete, such as yeast, in which all the products of a single meiosis remain together, that we can see this phenomenon.

SUMMARY

In 1943, Luria and Delbrück demonstrated that bacterial changes are true mutations similar to mutations in higher organisms. They showed that a high variability occurs in the number of mutants in small cultures as compared with the number of mutants in repeated subsamples of a large culture. Mutations occur spontaneously and are caused by mutagens, which include chemicals and radiation. This chapter is primarily concerned with point mutations rather than changes in whole chromosomes or chromosomal parts.

Allelism is defined by the *cis-trans* complementation test. Complementation implies independent loci, or nonalleles. The lack of complementation implies allelism. Functional alleles that differ from each other at the same nucleotides are also called structural alleles. Fine-structure studies using complementation testing and deletion mapping were done by Benzer using T4 phages. Colinearity of the gene and protein was demonstrated by Yanofsky, who had the advantage of working with a gene whose protein product was known.

After a mutation, the normal phenotype, or an approximation of it, can be restored either by back mutation or, alternatively, by suppression. Intragenic suppression occurs when a second mutation within the same gene causes a return of normal or nearly normal function. Intergenic suppression occurs when a second mutation happens, usually in a tRNA gene. Nonsense, missense, and frameshift mutations can all be suppressed.

Spontaneous mutation occurs primarily because of tautomerization of the bases of DNA. If a base is in the rare form during DNA replication, it can form unusual base pairings that result in mutation. The mechanisms by which the most commonly used mutagens work have been outlined.

DNA repair processes can be divided into three categories: damage reversal, excision repair, and postreplicative repair. Photoreactivation is an example of damage reversal. Thymine dimers

Figure 16.41

Fate of a heteroduplex DNA. Recombination results in heteroduplex DNA with mismatched bases. Replication without repair produces two different daughter molecules. Repair converts the mismatched base pair to one or the other normal base pair.

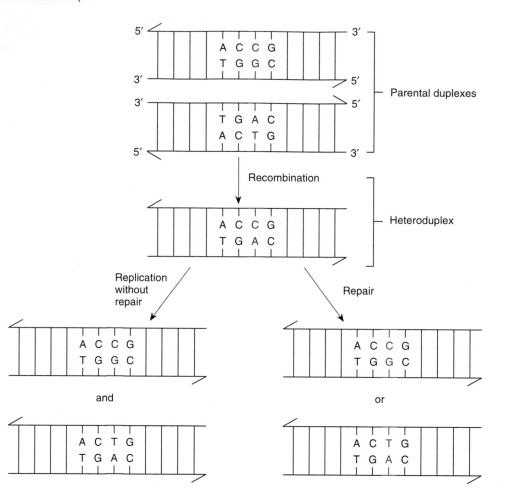

are undimerized by a photolyase enzyme in the presence of light energy. Excision repair is a process in which a damaged section of a strand of DNA is removed. The gap is filled in by DNA polymerase I. Excision repair is initiated by many endonucleases that recognize damaged DNA. Thymine dimers, improper bases, apurinic and apyrimidinic sites, and mismatches all signal particular endonucleases to initiate excision repair. The methylation state of *E. coli* DNA signals to the mismatch repair system which base to remove.

Postreplicative repair fills in gaps left by DNA polymerase III. The RecA protein is central to this process. A single strand from the undamaged duplex is used to fill the gap in the dam-

aged duplex. Single-stranded DNA induces the SOS response, in which LexA-mediated repression is temporarily eliminated.

According to the Holliday model, the process of recombination uses many of the steps in postreplicative repair. In addition to the RecA protein, the RecBCD protein has several roles in recombination. Other models and modifications of the Holliday model contribute to an understanding of nonreciprocity during recombination. Most recombination models agree on a heterozygous intermediate stage. Repair of this can lead to gene conversion. Thus, briefly, there is a battery of enzymes within the cell that can modify DNA. These enzymes serve in DNA replication, repair, and recombination.

Figure 16.42

Gene conversion can be caused by recombination and repair. During recombination, heteroduplex DNA is formed, containing mismatched base pairs. Without methylation cues, repair enzymes convert the mismatch to a complementary base pair, in a random fashion—that is, an AC base pair can be converted to either an AT or a GC base pair. Two of the four possible repair choices create 3:1 ratios of alleles rather than the expected 2:2 ratios in the offspring. The 3:1 ratio represents gene conversion.

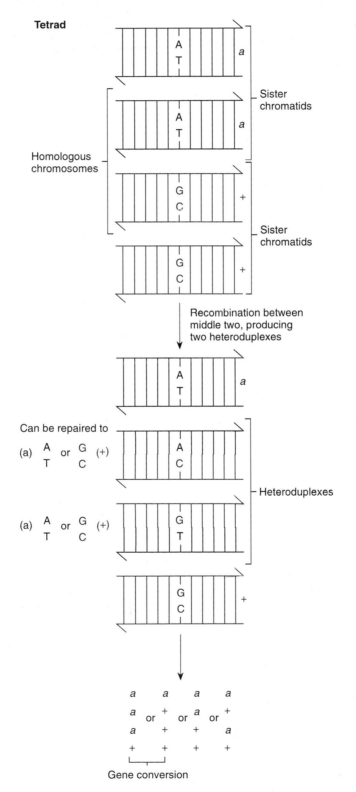

SOLVED PROBLEMS

1. An investigator isolates two wing mutants of *Drosophila melanogaster*. The flies differ in the patterns of veins on the wing. Are the mutations that cause these variants allelic?

   ANSWER: To verify allelism, the investigator must create a heterozygote of the two mutations by either mating the flies if of opposite sexes or breeding each mutant into a separate stock from which matings can be made. If the heterozygotes are of the wild-type, then the mutations are not allelic. If the heterozygote has a mutant phenotype, then we presume that the mutations are functional alleles. (Allelism should be verified in females to be sure that the locus is not on the X chromosome of which males have only one.) If the mutations are functional alleles, then it is possible to determine whether they are also structural alleles by looking for wild-type offspring of the heterozygote. If they occur at a rate higher than the background mutation rate, then the alleles are not structural alleles. If the occurrence of wild-type offspring is at the mutation rate only, then the alleles are presumably structural.

2. What is the difference between mismatch repair and AP repair?

   ANSWER: Both processes are similar in that they entail removal of an incorrect base in a DNA double helix by an excision process followed by a repair process. The processes differ in the event that triggers them. Mismatch repair is triggered by a base pair that does not occupy the correct space in the double helix— that is, by a non-Watson and Crick pairing (not A-T or G-C). AP repair is triggered by enzymes that recognize a missing base.

3. What role does the RecBCD protein play in recombination?

   ANSWER: In order for recombination to take place, a single strand of DNA from one sister chromatid must insinuate itself into the double helix of the other with the help of the RecA protein. It is the RecBCD protein that creates the single-stranded DNA. It does so by traveling down the double helix creating single strands in its wake. At chi sites it cuts the DNA, freeing one of the strands. This single-stranded loop or single-stranded end can then be acted upon by RecA to initiate recombination.

EXERCISES AND PROBLEMS

1. Construct a data set that Luria and Delbrück might have obtained that would prove the mutation theory wrong.

2. What types of enzymatic functions are best studied using temperature-sensitive mutations?

3. Seven arginine-requiring mutants of *E. coli* were independently isolated. All pairwise matings were done (by transduction) to determine the number of loci (complementation groups) involved. If a (+) indicates growth and a (−) no growth on minimal medium, how many complementation groups are involved here? Why is only "half" a table given? Must the upper left to lower right diagonal be all (−)?

	1	2	3	4	5	6	7
1	−	+	+	+	+	−	−
2		−	+	+	−	+	+
3			−	−	+	+	+
4				−	+	+	+
5					−	+	+
6						−	−
7							−

4. Several *r*II mutations (M to S) have been localized to the *A* cistron because of their failure to complement with a known deletion of the *A* cistron. The phages carrying these mutations are then mated pairwise with the following series of subregion deletions. The mating is done on *E. coli* B and plated out on *E. coli* K12. A (+) shows the presence of plaques on K12, whereas a (−) shows an absence of growth. Localize each of the *r*II *A* mutations.

Mutant	Deletion			
	1	2	3	4
M	+	+	−	+
N	+	−	−	+
O	+	+	+	−
P	+	+	−	−
Q	+	−	+	+
R	−	−	+	+
S	−	−	−	+

Relative positions of deletions

5. A *Drosophila* worker isolates four eye-color forms of the fly: wild-type, white, carmine, and ruby. (The worker does not know that white, carmine, and ruby are three separate loci on the X chromosome.) What crosses should be made to determine allelic relations of the genes? What results would be expected? A new mutant, eosin, is isolated. What crosses should be carried out to determine that eosin is an allele of white?

6. Define structural and functional alleles. What is the *cis* part of a *cis-trans* complementation test?

7. Did Benzer and Yanofsky work with genes that had intervening sequences? What is the relevance to their work of the occurrence of introns?

8. How can intra-allelic complementation result in incorrect conclusions about allelism?

9. *E. coli* bacteria of strain K12 are lysogenic for phage λ. Why do *r*II mutants of phage T4 not grow on these bacteria?

10. Diagram the tautomeric base pairings in DNA. What base pair replacements occur because of the shifts?

11. What is the difference between a substrate and a template transition mutation?

12. Describe two mechanisms for transversion mutagenesis.

13. 5BU, 2AP, proflavin, ethyl ethane sulfonate, and nitrous acid are chemical mutagens. What does each do?

14. A point mutation occurs in a particular gene. Describe the types of mutational events that can restore a functional protein, including intergenic events. Consider missense, nonsense, and frameshift mutations.

15. By what mechanism does misalignment result in addition or deletion of bases?

16. What are the differences and similarities between intergenic and intragenic suppression?

17. UV light causes thymine dimerization. Describe the mechanisms, in order of efficiency, that can repair the damage. Name the enzymes involved.

18. What types of damage are recognized by excision repair endonucleases?

19. What are the functions of the RecA protein? How is it involved in phage λ induction?

20. Diagram, in careful detail, a recombination by way of the Holliday model. What enzymes are required at each step?

21. What are the different enzymes that are involved in reciprocal and nonreciprocal recombination?

22. Eight independent mutants of *E. coli,* requiring tryptophan (*trp*⁻), are isolated. Complementation tests are performed in all pairwise combinations. Based on the results at the top of the next column, determine how many genes you have identified and which mutants are in which genes.
(+ = complementation, − = no complementation)

	1	2	3	4	5	6	7	8
1	−	+	−	−	+	+	+	−
2		−	+	+	−	+	+	+
3			−	−	+	−	−	−
4				−	+	+	+	−
5					−	+	+	+
6						−	−	+
7							−	+
8								−

23. Complementation tests are usually done with recessive mutations, for if the mutations were dominant, all progeny, regardless of whether genes are allelic or not, will be mutant. Suppose you have isolated in a diploid species two independent dominant mutations that each confer resistance to the drug cycloheximide. Call these mutations *Chx-1* and *Chx-2*. What crosses can you perform to determine whether the mutations are allelic? Your crosses should allow you to determine whether the mutations are allelic, nonallelic and unlinked, or nonallelic and linked.

24. A series of overlapping deletions in phage T4 are isolated. All pairwise crosses are performed, and the progeny scored for wild-type recombinants. In the table below, + = wild-type progeny recovered; − = no wild-type progeny recovered.

	1	2	3	4	5
1	−	+	−	−	−
2		−	+	+	−
3			−	+	+
4				−	+
5					−

 a. Draw a deletion map of these mutations.
 b. A point mutation, 6, is isolated and crossed with all of the deletion strains. Wild-type recombinants are recovered only with strains 2 and 3. What is the location of the point mutation?

25. Hydroxylamine is a chemical that causes exclusively C → T transition mutations. Can nonsense mutations be reverted with hydroxylamine? Explain.

26. A nonsense suppressor is isolated and is shown to involve a tyrosine tRNA. When this mutant tRNA is sequenced, the anticodon is found to be normal, but a mutation is found in the dihydrouridine loop. What does this finding suggest about how a tRNA interacts with the mRNA?

27. Devise selection-enrichment procedures for isolating the following kinds of mutants:
 a. extra-large bacterial cells
 b. nonmotile ciliated protozoans

28. Two chemically induced mutants, *x* and *y*, are treated with the following mutagens to see if revertants can be produced: 2-amino purine (AP), 5-bromouracil (BU), acridine (AC), hydroxylamine (HA), and ethylmethanesulfonate (EMS). In the table below, + = revertants and − = no revertants. For each mutation, determine the probable base change that occurred to change the wild-type to the mutant.

Mutant	AP	BU	Chemical AC	HA	EMS
x	−	+	−	+	+
y	+	−	−	−	−

29. What situation will lead to a false positive in a complementation test or, in other words, indicate two genes when in fact the mutations are in the same gene?

30. What situation will lead to false negatives in a complementation test or, in other words, indicate mutations are in the same gene when in fact they are in different genes?

31. Suppose you repeat the Luria-Delbrück fluctuation test, but this time you look for *lac⁻* colonies. Your "individual cultures" give the following numbers of *lac⁻* colonies: 20, 25, 22, 18, 24, 19, 17, 25, 26, and 18. Subsamples from the bulk culture give identical results to those above. What can you conclude from these results?

32. You have isolated a new histidine auxotroph, and, despite all efforts, you cannot produce any revertants. What probably happened to produce the original mutant?

Suggested Readings for chapter 16 are on page 656.

Non-Mendelian
Inheritance

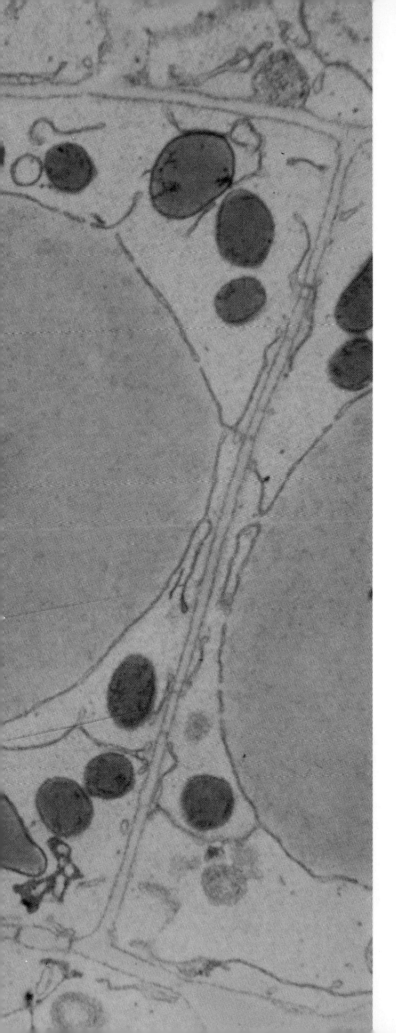

A plant cell of an African violet (*Saintpaulia ionantha;* sporogenous tissue). Note the cell wall, nucleus, and chloroplasts (2,250×).
© *Biophoto Associates/Science Source/Photo Researchers, Inc.*

We have already observed that the phenotype can be controlled by chromosomal genes and the environment. In this chapter we deal with another mode of inheritance, namely, non-Mendelian inheritance (also called extrachromosomal, cytoplasmic, and nonchromosomal inheritance, or maternal effects). **Maternal effects** are the influences of a mother's genotype on the phenotype of her offspring (examples of which are snail coiling and moth pigmentation); **cytoplasmic inheritance** is controlled by nonnuclear genomes, found in chloroplasts and mitochondria, infective agents, and plasmids. Therefore, non-Mendelian inheritance does not follow simple ratios of phenotypes controlled by loci located on nuclear chromosomes segregating and assorting in the manner described by Mendel.

Maternal effects result from the asymmetric contribution of the female parent to the development of zygotes. Although both male and female parents contribute equally to the zygote in terms of chromosomal genes (with the exception of sex chromosomes), the sperm rarely contributes anything to development other than chromosomes. The female parent usually contributes the initial cytoplasm and organelles of the zygote. Zygotic development, therefore, usually begins within a maternal milieu, so that the maternal cytoplasm directly affects zygotic development.

Cytoplasmic inheritance refers to the inheritance pattern of organelles and parasitic or symbiotic particles that have their own genetic material. Chloroplasts, mitochondria, bacteria, viruses, and of course plasmids all have their own genetic material. These genomes are open to mutation. As we shall see, their inheritance pattern will not follow Mendel's rules for chromosomal genes.

DETERMINING NON-MENDELIAN INHERITANCE

How does one determine that a trait is inherited? The question does not have as obvious an answer as we might expect. Environmentally induced traits can mimic inherited phenotypes such as with phenocopies discussed in chapter 5. For example, the inheritance of vitamin D-resistant rickets is mimicked by lack of vitamin D in the diet. Determining that the dietary rickets is not inherited is possible by simply administering adequate quantities of vitamin D. Inherited rickets does not respond to vitamin D until about 150 times the normally adequate amount is administered.

Some environmentally induced traits persist for several generations. For example, a particular *Drosophila* strain that normally grows at 21° C was exposed to 36° C for 22 hours. Dwarf progeny were produced. When they were mated among themselves, fewer and fewer dwarfs appeared in each generation, but smaller-than-normal flies

were seen in the fifth generation. The appearance of an environmentally induced trait that persists for several generations has been termed **dauermodification.**

Extrachromosomal inheritance is usually identified by the odd results of reciprocal crosses. If the progeny of reciprocal crosses are not followed for several generations, the results can be misleading when extrachromosomal inheritance is involved. Where feasible, the technique of nuclear transplantation has proved useful in identifying extrachromosomal inheritance. In this technique, the nucleus of a cell, such as an amoeba or frog egg, is removed by microsurgery or destroyed by radiation, and another nucleus is substituted. Thus not only can a nucleus be isolated from its cytoplasm, but various nuclei can be implanted in the same cytoplasm.

A similar experiment, called a *heterokaryon test,* can be done with various fungi such as *Neurospora* and *Aspergillus:* mycelia can fuse and form a heterokaryon, which is a cell containing nuclei from different strains. Thus nuclei of both strains exist in the mixed cytoplasm. Subsequently, spores (conidia) that have one or the other nucleus in the mixed cytoplasm can be isolated. The phenotype of the colonies produced from these isolated conidia will show whether the trait under observation is controlled by the nucleus or the cytoplasm.

Chromosomal genes in a particular cytoplasm can also be isolated by repeated backcrossing of offspring with the male-parent type. In each cross, the content of the female chromosomal genes will be halved, but, presumably, the cytoplasm will remain similar to the female line. Thus, after several generations, male genes can be isolated in female cytoplasm. The phenotypic results of the final cross will indicate whether inheritance was chromosomal or extrachromosomal.

MATERNAL EFFECTS

Snail Coiling

Snails are coiled either to the right (dextrally) or to the left (sinistrally) as determined by holding the snail with the apex up and looking at the opening. The snail is dextrally coiled if the opening comes from the right-hand side and sinistrally coiled if it comes from the left-hand side (fig. 17.1). The inheritance pattern of the coiling is at first perplexing.

In the left half of figure 17.1, a dextral snail provides the eggs and a sinistral snail provides the sperm. The offspring are all dextral; presumably, therefore, dextral coiling is dominant. When the F_1 are self-fertilized (snails are hermaphroditic), all the offspring are dextrally coiled. The result is unexpected. Nevertheless, when the F_2 are self-fertilized, one fourth will produce only sinistral offspring and three fourths will produce only dextral off-

Figure 17.1

Inheritance of coiling in the pond snail, *Limnaea peregra*. Reciprocal crosses (*D* = dominant dextral and *d* = recessive sinistral) are shown (*DD* mated with *dd* in each case). The F₁ in both crosses have the *Dd* genotype, but reflect the mother's genotype in respect to coiling; *DD* mothers produce dextrally coiled offspring, whereas *dd* mothers produce sinistrally coiled offspring. The F₂ in both cases are identical because the genotypes of the F₁ mothers are identical (*Dd*). The coiling of a snail's shell is determined by its mother's genotype.

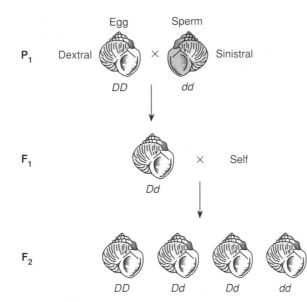
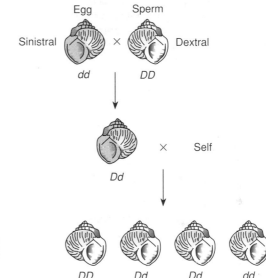

spring. If self-fertilization is continued through ensuing generations, this 3:1 phenotypic ratio will be revealed as a Mendelian 1:2:1 genotypic ratio, thereby reaffirming the notion of a single locus with two alleles of which dextral is dominant. However, something interfered with the expected inheritance pattern.

When the reciprocal cross is made (fig. 17.1, *right*), the F₁ have the same genotype as just described but are coiled sinistrally, as is the female parent. From here on, the results are exactly the same for both crosses. In both cases, the F₁ are phenotypically similar to the female parent even though they have the same genotype. The explanation is that the genotype of the maternal parent determines the phenotype of the offspring, with dextral dominant. Thus the *DD* mother in figure 17.1 produces F₁ progeny that are dextral with a *Dd* genotype, and the *dd* mother produces progeny with the same *Dd* genotype but sinistral because the mother was *dd*. Why does this pattern occur?

A process of **spiral cleavage** takes place in the zygote of mollusks and some other invertebrates. The spindle at mitosis is tipped in relation to the axis of the egg. If the spindle is tipped one way, a snail will be coiled sinistrally; if it is tipped the other way, the snail will be coiled dextrally. The direction of tipping is determined by the maternal cytoplasm, which is under the control of the maternal genotype. Obviously, maternal control affects only one generation—in each generation the coiling is dependent on the maternal genotype.

Moth Pigmentation

There are other examples of maternal effects in which the cytoplasm of the mother, under the control of chromosomal genes, controls the phenotype of her offspring. In the flour moth, *Ephestia kühniella,* kynurenin, which is a precursor for pigment, is accumulated in the eggs. The recessive allele, *a,* when homozygous, results in a lack of kynurenin. Reciprocal crosses give different results for larvae and adults. When a nonpigmented female is crossed with a pigmented male, the results are strictly Mendelian; but when the mother is pigmented (a^+/a), all the larvae are pigmented regardless of their genotype (fig. 17.2). The initial larval pigmentation comes from residual kynurenin in the eggs, which is then diluted out so that an adult's pigmentation conforms to its own genotype.

Imprinting

Although sex-linkage occurs, we do not expect different inheritance patterns from genes located on autosomal chromosomes dependent on which parent the gene came from. That is, the genotype of an offspring should be predicted by the alleles present regardless of which parent a particular allele came from. That understanding has now been shown to be incorrect for a group of genes whose phenotypic effects depend on which parent a particular gene was inherited from. In chapter 8 we discussed the fragile-X syndrome in which the chromosome usually had

Figure 17.2

Inheritance pattern of larval and adult pigmentation in the flour moth, *Ephestia kühniella.* The presence (a^+) or absence (a) of kynurenin is controlled by a single locus. In the cross on the *left,* the mother is aa (nonpigmented). Her aa offspring, in both the larval and adult stages, are nonpigmented. In the reciprocal cross (*right*), the mother has the a^+a genotype and is pigmented. Her aa offspring are nonpigmented as adults but are pigmented as larvae because of residual kynurenin from the egg that is eventually diluted out.

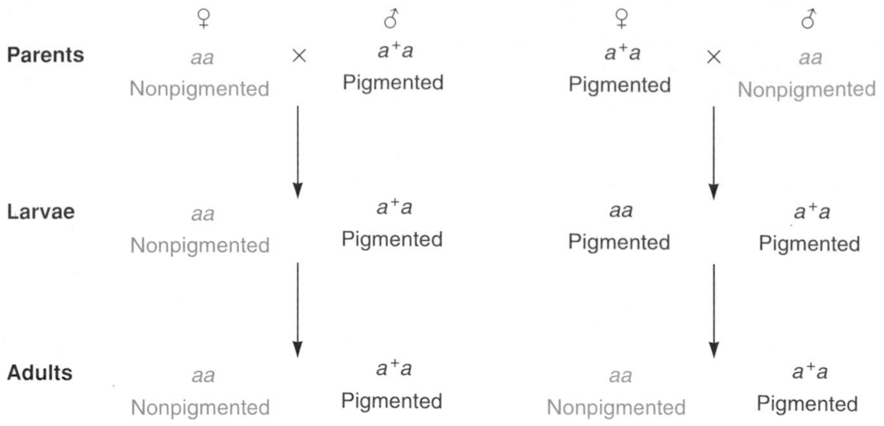

	♀	♂	♀	♂
Parents	aa Nonpigmented	× a^+a Pigmented	a^+a Pigmented	× aa Nonpigmented
Larvae	aa Nonpigmented	a^+a Pigmented	aa Pigmented	a^+a Pigmented
Adults	aa Nonpigmented	a^+a Pigmented	aa Nonpigmented	a^+a Pigmented

Figure 17.3

Electron micrograph of the circular DNA from within a mitochondrion of a mouse cell. Magnification 48,000×.

M. M. K. Nass, "The Circularity of Mitochondrial DNA," Proceedings of the National Academy of Sciences USA 56 (1966):1215–1222. Reproduced by permission of the author.

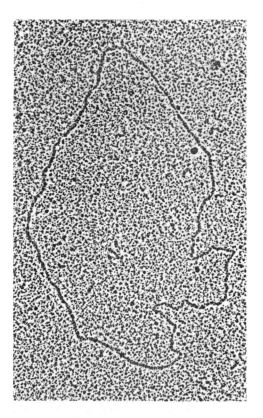

to pass through a female parent before it manifested the symptoms; in chapter 15 we mentioned a similar parental discrepancy in Wilm's tumor. This phenomenon is called *imprinting* (or *molecular imprinting* or *parental imprinting*).

Several other examples are striking. In human beings, two medical syndromes result in mental retardation. In Prader-Willi syndrome, the patients are extremely obese; in Angelman syndrome the patients are sometimes referred to as happy puppets, exhibiting erratic, jerky movements. It now turns out that both syndromes are caused by the same gene located on chromosome 15. If the gene is of paternal origin (e.g., a deletion of the maternal gene), the offspring will have Angelman syndrome; if the gene is of maternal origin (e.g., two maternal copies by a chromosomal error), the offspring will have Prader-Willi syndrome. This unusual situation indicates that the phenotype is not just dependent on what alleles are present, but from which parent the alleles have come. Additional examples are found in other human diseases such as Huntington's disease and several cancers and in traits in mice and fruit flies.

Initial research indicated that the mechanism for the imprinting might be the pattern of methylation of the genes. As you remember from chapter 15, methylation of genes is related to their expression, and differences in methylation of imprinted genes were found, depending on the sex of the bearer of the genes. However, some interest has turned to the possibility that methylation is a symptom, not a cause, of the phenomenon. In some cases of imprinting, changes in chromatin were noted. Imprinting may be similar to variegation position effects in which the expression of a gene is inhibited if the gene is located adjacent to heterochromatin (chapter 8). Currently, the mechanism is unclear, but the phenomenon of imprinting is well established.

CYTOPLASMIC INHERITANCE

Mitochondria

The **mitochondrion** is a cellular organelle of eukaryotes in which the Krebs (citric acid or tricarboxylic acid) cycle and electron transport reactions take place. The actual number of mitochondria per cell can be determined by serial sectioning of whole cells and examination under the electron microscope. This is a tedious and difficult procedure. Estimates range between ten and ten thousand per cell, depending on the organism and cell type. As far as we are concerned, the most interesting aspect of the mitochondrion is that it has its own DNA. In most animal

Gene map of the human mitochondrial chromosome. All but nine loci
are on the heavy (H) strand. The light-strand (L) loci are labeled
inside the circle; the H-strand loci arc labeled on the outside. Also
shown are the origins of H- and L-strand replication and the
directions of transcription. The twenty-two tRNA genes are colored
red. NADH refers to NADH dehydrogenase (subunits 1–4, 4L, 5, and
6); CCO refers to cytochrome-c oxidase (subunits I–III).
Source: Data from V. McKusick, Mendelian Inheritance in Man, *7th ed., 1986.*

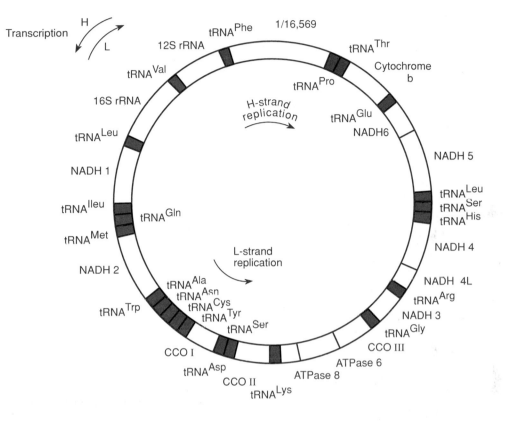

cells, the mitochondrial DNA (mtDNA) is a circle of about
sixteen thousand base pairs (fig. 17.3). However, some organisms (yeast, higher plants) have mitochondrial DNAs
five to twenty-five or more times larger than in animals.

Two general patterns are found in mitochondrial inheritance in animals. First, the mitochondria arc generally inherited in a maternal fashion; that is, the male
gamete usually does not contribute mitochondria to the
zygote. However, there is a small amount of "leakiness"
to this process. For example, it has recently been shown
that about one mitochondrion per thousand is of paternal
origin in mice. In some species, such as mussels, it appears
that mitochondrial inheritance is biparental. That is, the
population of mitochondria in an offspring derives almost
equally from both male and female parent. In some gym-
nosperm plants, such as coast redwoods, mitochondria are
inherited paternally—only paternal mitochondria are
passed into the zygote. However, these are all exceptions
to the general rule of maternal inheritance of mitochondria.

The second pattern of mitochondrial inheritance is
homoplasmy, the existence of a uniform population of mitochondria within an organism. That is, in general, all the
mitochondria within an individual are genetically identical. Certainly biparental inheritance and leakiness of
paternal mitochondria violate that principle, resulting in
heteroplasmy, a heterogeneity of mitochondria within an
organism.

Mitochondrial Genomes

Several mtDNAs have been sequenced, including the
human mtDNA, which is 16,569 bp long. It is a model of
economy, with very few noncoding regions and no introns
(fig. 17.4). Each strand of the duplex is transcribed into
a single RNA product that is then cut into smaller pieces,
primarily by the freeing of the twenty-two tRNAs, which
are interspersed throughout the genome. Also formed are
a 16S and a 12S ribosomal RNA. Although proteins and
small molecules, such as ATP, can move in and out of the
mitochondrion, RNAs cannot, with the possible exception

Figure 17.5

The amino acid sequence of mouse dihydrofolate reductase. The first eighty-five amino acids serve as the signal sequence for transport into mitochondria. Numbers refer to sequential amino acids. Five alpha-helical regions exist in the protein (A–E). Positively and negatively charged amino acids are marked with (+) and (−) signs.

Reprinted by permission from Nature, Vol. 325, p. 499. Copyright © 1987 Macmillan Magazines Ltd.

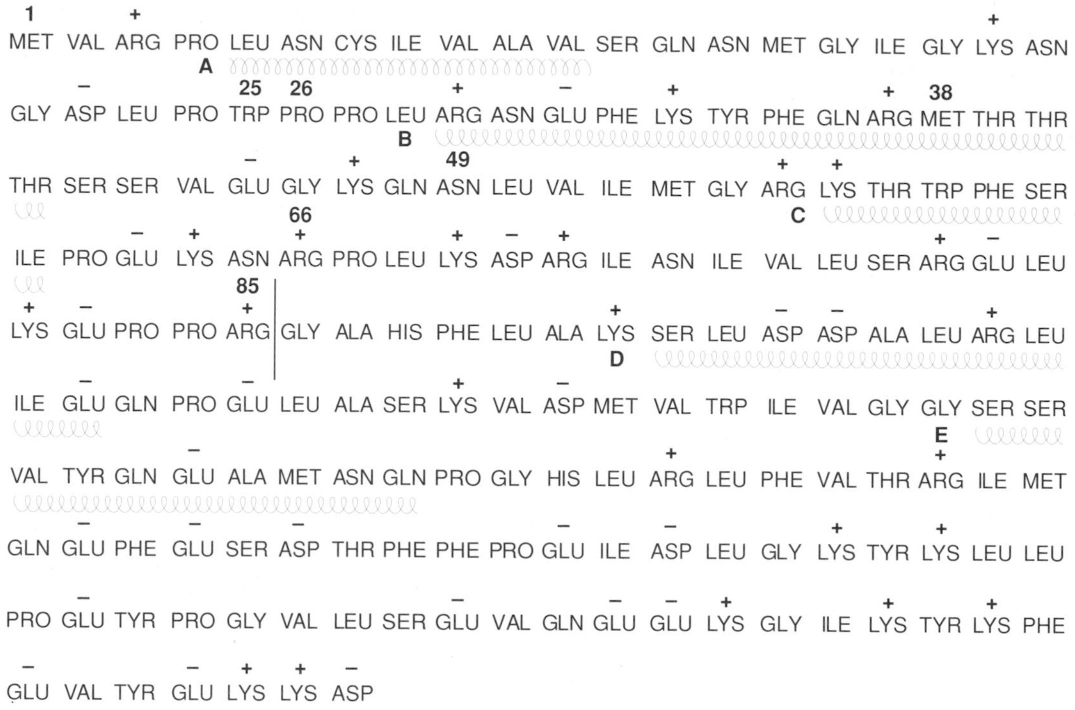

```
    1       +                                                              +
   MET VAL ARG PRO LEU ASN CYS ILE VAL ALA VAL SER GLN ASN MET GLY ILE GLY LYS ASN
            A  ))))))))))))))))))))))))))
    -          25  26              +       -       +           +   38
   GLY ASP LEU PRO TRP PRO PRO LEU ARG ASN GLU PHE LYS TYR PHE GLN ARG MET THR THR
                    B  )))))))))))))))))))))))))))))))))))))))))))))))))))))
                -      +   49                              +   +
   THR SER SER VAL GLU GLY LYS GLN ASN LEU VAL ILE MET GLY ARG LYS THR TRP PHE SER
   )))                    66                          C  ))))))))))))))))))))))
             -   +   +       +   -   +                                  +   -
   ILE PRO GLU LYS ASN ARG PRO LEU LYS ASP ARG ILE ASN ILE VAL LEU SER ARG GLU LEU
   )))       85  |
    +   -       +   |
   LYS GLU PRO PRO ARG|GLY ALA HIS PHE LEU ALA LYS SER LEU ASP ASP ALA LEU ARG LEU
                      |             D  )))))))))))))))))))))))))))))))))))))))))))
        -       -         +       -
   ILE GLU GLN PRO GLU LEU ALA SER LYS VAL ASP MET VAL TRP ILE VAL GLY GLY SER SER
   )))))))                                                            E  ))))))
        -                        +
   VAL TYR GLN GLU ALA MET ASN GLN PRO GLY HIS LEU ARG LEU PHE VAL THR ARG ILE MET
   ))))))))))))))))))))))))))))))))
        -       -       -           -   -           +       +
   GLN GLU PHE GLU SER ASP THR PHE PHE PRO GLU ILE ASP LEU GLY LYS TYR LYS LEU LEU
        -               -           -   -   +       +       +
   PRO GLU TYR PRO GLY VAL LEU SER GLU VAL GLN GLU GLU LYS GLY ILE LYS TYR LYS PHE
    -           -   +   +   -
   GLU VAL TYR GLU LYS LYS ASP
```

of small (135-bp) RNAs. Thus the mitochondrion must be relatively self-sufficient in terms of the RNAs needed for protein synthesis. We previously discussed mitochondrial protein synthesis when we looked at unique attributes of the mitochondrial genetic code in chapter 11.

Oxidative phosphorylation, the process that occurs within the mitochondrion, requires at least sixty-nine polypeptides. The human mitochondrion has the genes for thirteen of these: cytochrome b, two subunits of ATPase, three subunits of cytochrome-c oxidase, and seven subunits of NADH dehydrogenase. The remaining polypeptides needed for oxidative phosphorylation are transported into the mitochondrion after having been synthesized in the cytoplasm under the control of nuclear genes. Proteins that are targeted for entry into the mitochondrion have special signal sequences (see chapter 11).

The signal sequences range up to eighty-five amino acids long. Signal sequences examined so far do not have consensus amino acids, but do have certain attributes (fig. 17.5), including a somewhat regular alternation of basic (positively charged) and hydrophobic (negatively charged) residues. In addition, they form α helices with opposite hydrophobic and hydrophilic faces that must somehow be important in the ability of the protein to enter the mitochondrion. When a signal sequence (such as that in fig. 17.5) is attached to nonmitochondrial proteins by DNA manipulations, those proteins are transported into the mitochondrion.

The mitochondrial ribosomal RNA is more similar to prokaryotic rRNA than to eukaryotic rRNA: The mitochondrial ribosome, although constructed of imported cellular proteins, is sensitive to prokaryotic antibiotics; for example, streptomycin and chloramphenicol inhibit their function. This affinity (close resemblance) of mitochondria and prokaryotes is strong support for the symbiotic origin of mitochondria. That is, according to the model primarily attributed to L. Margulis, it is now generally believed that organelles such as mitochondria and chloroplasts were originally free-living respiring bacteria and free-living cyanobacteria, respectively. These prokaryotes invaded or were eaten by early cells and, over evolutionary time, became the organelles that we see today. Since they arose as prokaryotes, these organelles retain certain evolutionary similarities to other prokaryotes.

Among the mtDNAs from different organisms that have been sequenced, there is great variation in content and organization. Yeast mtDNA, for example, is not as economical as human mtDNA. Yeast mtDNA, about five times larger than human mtDNA, has noncoding regions as well as introns. Because mitochondria are similar in

Lynn Margulis (1938–   )
Courtesy of Lynn Margulis, Boston University Photo Services.

Figure 17.6
Petite yeasts categorized on the basis of segregation patterns. Three types of petites are recognized (segregational, neutral, and suppressive), depending on the meiotic segregation pattern of petite × wild-type diploids. Segregational petite heterozygotes segregate a 1:1 ratio of spores; neutral petites are lost when heterozygous, and suppressive petites act in a dominant fashion under the same circumstances.

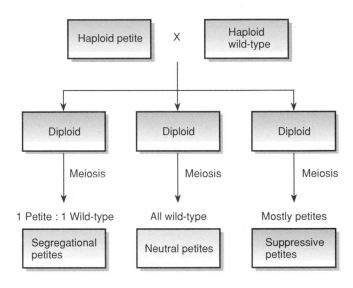

structure and biochemistry to prokaryotic cells, given the general lack of introns in prokaryotic genes, finding introns in yeast mtDNA was surprising. It is indicative of later origins of these genes, which most probably arose as nuclear genes that were then "captured" by the mitochondria, possibly by recombination with nuclear DNA.

There does not seem to be an absolute minimal size or genetic content requirement for mtDNA. A comparison of yeast and human mitochondria makes it clear that introns are not a universal requirement. In fact, many yeast strains differ in the number of introns present in their mtDNA.

There are also differences among species in the enzymes coded by mtDNA. In yeast, a subunit of the mitochondrial ATPase is coded by mtDNA, yet the same subunit is coded by a nuclear gene in human beings and by both a nuclear and a mitochondrial gene in *Neurospora*. If no universal rules exist regarding what a mtDNA must code for, why do mitochondria have DNA? Although this topic is clearly open to speculation, P. Borst of the University of Amsterdam has suggested that mtDNA is simply a relic of the DNA originally belonging to the prokaryotic ancestor of the mitochondrion.

Once the interaction within the mitochondrial-nuclear genetic system is clearly understood, we might expect to see several different inheritance patterns—following either cytoplasmic or nuclear lines—for the genetic defects that lead to interruption of cellular respiration. Among the best-studied phenotypes with such inheritance patterns are the *petite* mutations of yeast.

Petites

Under aerobic conditions, yeast grows with a distinctive colony morphology. Under anaerobic conditions, the colonies are smaller and the structure of the mitochondria become reduced. Occasionally, when growing aerobically, small, anaerobic-like colonies will appear; but in these colonies the mitochondria appear perfectly normal. These colonies are caused by what have been termed **petite mutations** (French for little). When petites are crossed with the wild-type, three modes of inheritance are observed (fig.

17.6). The *segregational petite,* caused by mutation of a chromosomal gene, exhibits Mendelian inheritance. The *neutral petite* is lost immediately upon crossing to the wild-type. The *suppressive petite* shows variability in expression from one strain to the next but is able to convert the wild-type mitochondria to the petite form. All petites represent failures of mitochondrial function, whether the function is controlled by the mitochondria themselves or by the cell's nucleus; they usually lack one or another cytochrome.

Although the mechanisms of neutral and suppressive petites are not known with certainty, observation of their DNA has supplied some interesting information. In some petites, no change in the buoyant density of the DNA is found. (**Buoyant density,** a term that describes the position at which the DNA equilibrates during density-gradient centrifugation, is a measure of the composition of the molecule.) In other petites, changes in buoyant density range from very small to the complete absence of DNA.

Petites, therefore, can be the result of an approximation to a point mutation (in which there is no measurable change in the buoyant density of the DNA), marked changes in the DNA, or the total absence of DNA. In most petites, protein synthesis within the mitochondrion is lacking. Any and all of these changes produce the petite (anaerobic-like) phenotype.

Neutral petites seem to have mitochondria that entirely lack DNA. When neutral petites are crossed with the wild-type to form diploid cells, the normal mitochondria dominate. During meiosis, virtually every spore receives large numbers of normal mitochondria; the progeny are, therefore, all normal.

Suppressive petites could exert their influence over normal mitochondria in one of two ways. The suppressive mitochondria might simply outcompete the normal mitochondria and take over; they might simply reproduce faster within a cell. Alternatively, crossing over between the DNA of the suppressive petite and the wild-type might affect the normal DNA if the suppressive petite's DNA were severely damaged. Presumably, recombination in mtDNAs occurs when two or more mitochondria fuse, bringing the two different DNAs in contact within the same organelle. Recombination would presumably take place by normal crossover mechanisms.

If large portions of the DNA from the suppressive mitochondria were missing or altered, recombination with the normal mitochondria's DNA might exchange some of this damaged DNA. Several experiments have been done in which a suppressive petite and a wild-type, each with mitochondrial DNA of known buoyant density, were crossed. The DNAs of the offspring colonies, which were petites, were of various buoyant densities. For example, when a normal strain having mitochondrial DNA with a buoyant density of 1.684 g/cm³ was crossed with a suppressive petite with a buoyant density of 1.677 g/cm³, there were offspring colonies whose mitochondrial DNA had buoyant densities of 1.671, 1.674, and 1.683 g/cm³. Such information supports the notions that mitochondrial DNA is open to recombination and that the suppressive character takes over a colony by way of recombination.

Human Mitochondrial Inheritance

In human beings, there are about thirty diseases known that trace their dysfunction to mitochondrial pathologies. Only recently, however, have diseases been discovered that are caused by mitochondrial pathologies inherited in a maternal fashion. In 1988, Douglas Wallace and his colleagues showed that *Leber optic atrophy* is a cytoplasmically inherited disease. This disease causes blindness with a median age of onset of twenty to twenty-four years. Apparently, defects in mitochondria are not tolerable in the optic nerve, which has a very great energy demand. The disease also does some damage to the heart.

It was determined from pedigrees that the disease was transmitted only maternally. DNA sequencing of mitochondrial DNAs in affected families resulted in pinning down the disease to a point mutation, a change of nucleotide 11,778, which is in the gene for NADH dehydrogenase subunit 4. A guanine is changed to an adenine at codon 340, which converts an arginine to a histidine. This is the first human disease traced to a mtDNA defect.

Antibiotic Influences

Since the machinery of mitochondrial protein synthesis is prokaryotic in nature, mitochondrial protein synthesis can be inhibited by antibiotics such as chloramphenicol and erythromycin. These antibiotics elicit a petite-type growth response in yeast. Antibiotic-resistant strains can be ob-

Figure 17.7

Inheritance of antibiotic (chloramphenicol) resistance in yeast. Resistant and sensitive cells are produced by a diploid cell that resulted from a cross of resistant and sensitive haploids. The segregation is not in a simple Mendelian ratio, but depends on the random assortment of mitochondria. Sensitive cells have no resistant mitochondria. Resistant cells have resistant mitochondria.

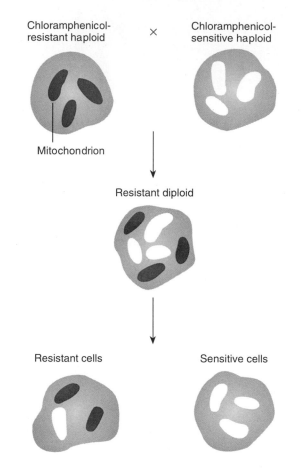

tained by growing yeast on the antibiotic; only resistant mutants will grow. The resistance appears to be inherited in the mitochondrial, not the cellular, DNA. A mitochondrial inheritance pattern results, with crosses between a resistant and a sensitive (wild-type) yeast, shown in figure 17.7. The resulting diploid colonies segregate both resistant and sensitive cells. Although not expected on the basis of a chromosomal gene, the random sorting of diploid mitochondria through cell division could result in a wild-type cell containing only sensitive mitochondria. Since yeast have only one to ten mitochondria per cell, this random assortment of sensitive mitochondria can be expected to occur at a relatively high rate.

Chloroplasts

The **chloroplast** is the chlorophyll-containing organelle that carries out photosynthesis and starch-grain formation in plants (fig. 17.8). Chloroplasts are referred to as **plastids** before chlorophyll develops. However, when grown

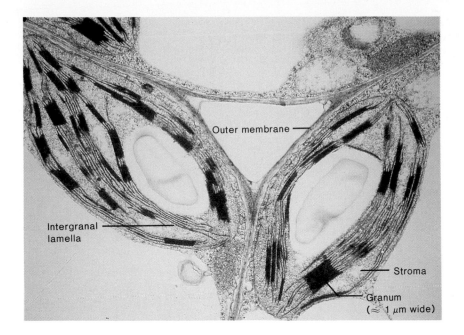

Figure 17.8
Electron micrograph of a lettuce chloroplast. The chloroplast consists of an outer membrane, stacks of grana, lamellae, and stroma. Magnification 3,570×.
© *Dr. J. Burgess/Science Photo Library/Photo Researchers, Inc.*

in the dark (and under some other circumstances), plastids do not develop into chloroplasts, but remain reduced in size and complexity. These undeveloped plastids, referred to as **proplastids,** are each about the size and shape of a mitochondrion.

Like mitochondria, chloroplasts contain DNA and ribosomes, both with prokaryotic affinities. The DNA of chloroplasts (cpDNA) is a circle that ranges in size from 85 kb (kilobases) in the green alga, *Codium,* to as large as 2,000 kb in the green alga, *Acetabularia.* Thus cpDNA is minimally about five times the size of an animal mitochondrial DNA. The cpDNA, like mtDNA, controls the production of tRNAs, rRNAs, and some of the proteins found within the organelle. Scientists believe that the chloroplast evolved from symbiotic cyanobacteria (blue-green algae), which have many affinities with the chloroplast: The ribosomal RNA of cyanobacteria will hybridize with the DNA of chloroplasts.

The similarities between mitochondria and chloroplasts make it possible to predict the inheritance patterns of chloroplast mutations on the basis of existing knowledge of mitochondrial genetics: We should find both chromosomal and plastid mutants of chloroplast functions. Simple segregation should occur in the chromosomal mutations, and cytoplasmic patterns in the cpDNA mutations. Investigation of these inheritance patterns is complicated by the fact that plant cells have both mitochondria and chloroplasts. Since both have prokaryotic affinities, it is sometimes difficult to determine whether a genetic trait is due to a defect in the genetic system of the chloroplast or the mitochondrion. Like mitochondria, chloroplasts generally show homoplasmy and maternal inheritance, although, like mitochondria there are exceptions. For example, gymnosperms usually have paternal inheritance of chloroplasts.

Marcus M. Rhoades (1903–1991)
Courtesy of Dr. Marcus M. Rhoades.

Lesions in the photosystems of the chloroplast result in proplastid formation, with a loss of green color. When proplastid formation occurs in a particular tissue of a plant, variegation will result. That is, there will be both green and white parts, often as stripes. Some interesting genetic studies have been done on the inheritance of variegation, especially in the area of the interaction of chloroplast and chromosomal genes.

Zea mays

M. Rhoades worked on the variegation in corn (*Zea mays*) controlled by the *iojap* chromosomal locus, which, when homozygous, prevents proplastids from developing as chloroplasts and thus results in variegation. The *iojap*-affected plastids do not contain ribosomes or ribosomal RNA; they therefore lack protein synthesis.

The interaction of chromosomal and extrachromosomal inheritance is shown in the reciprocal crosses depicted in figure 17.9, in which one cross produces results exactly as would be predicted on the basis of simple Men-

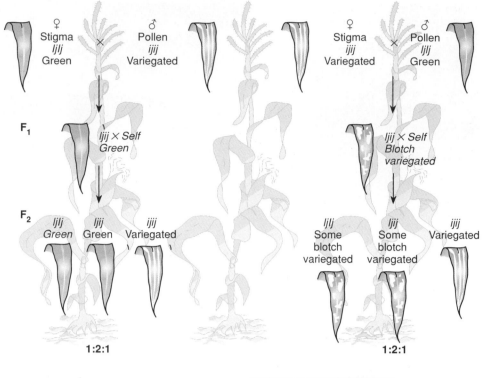

Figure 17.9
Reciprocal crosses involving the chromosomal gene *iojap* in corn. The homozygous recessive condition (*ijij*) induces variegation (representative corn leaves are shown). (Blotch variegation consists of irregularly shaped *white* areas rather than striping.) However, plants with the dominant allele (*IjIj, Ijij*) can still be variegated if their mothers were variegated, since mothers pass on their chloroplasts to their offspring; males (pollen parents) do not pass on their chloroplasts. *Iojap* homozygotes induce variegation. The defective chloroplasts are then inherited in a cytoplasmic fashion.

delian inheritance, with the homozygous recessive genotype (*ijij*) inducing variegation. When the reciprocal cross is carried out, blotch variegation is seen in both the F₁ and F₂ that carry the dominant *Ij* allele.

This inheritance pattern is caused by the fact that the pollen grain in corn does not carry any chloroplasts, whereas the ovule does. Thus the first cross in figure 17.9 (*left*) deals with the passage into the F₂ of normal chloroplasts only. In the F₂ the *ijij* genotype then induces variegation. The chloroplasts of the pollen parent are unimportant because they do not enter the F₁. In the reciprocal cross, however, because the stigma parent is variegated, the F₁ is heterozygous but carries proplastids from the ovule that will remain proplastids even under the dominant normal (*Ij*) allele. Therefore, regions of colorless cells will produce white spots (blotchy variegation). Once the *ij* allele induces chloroplasts to become proplastids, they do not revert to the normal type even under the *Ij* allele. Thus we see the interaction of a chromosomal gene and the chloroplast itself, which "inherits" a changed condition.

There is some evidence that *iojap* may suppress the chloroplast rather than cause a mutation of some function. There are loci in corn and in other species that can induce back mutation in the chloroplasts. Removal of suppression rather than an actual reversion is more likely to occur because the reversion rate is too high to be due to simple back mutation.

Four-O'clocks

The first work with corn variegation was done by Correns, one of Mendel's rediscoverers. He also found maternal inheritance of variegation in the four-o'clock plant, *Mira-*

Ruth Sager (1918–   )
Courtesy of Dr. Ruth Sager.

bilis jalapa. Correns could predict color and variegation of offspring solely on the basis of the region of the plant on which the stigma parent was located. A flower from a white sector, when pollinated by any pollen, would produce white plants; a flower on a green sector or a variegated sector produced green or variegated plants, respectively, when pollinated by pollen from any region of a plant. We thus see the simple maternal nature of the inheritance of the variegation. A chromosomal gene, like *iojap,* induces the occurrence of variegation. Inheritance of this induced variegation follows the "maternal" pattern of chloroplast inheritance.

Chlamydomonas

The single-celled green alga, *Chlamydomonas reinhardi,* has been used in the study of extrachromosomal inheritance for several reasons. It has a single, large chloroplast; it can survive by culture technique even when the chlo-

Figure 17.10

Inheritance pattern of streptomycin resistance in *Chlamydomonas* is dependent on the genotype of the *mt*⁺ parent. (The *n* and *2n* refer to the ploidy of the cells.) If the *mt*⁺ parent is streptomycin resistant (*red*), then the diploid heterozygote as well as the meiotic products will be streptomycin resistant. If, however, the *mt*⁺ parent is streptomycin sensitive (*green*), the diploid heterozygote as well as the meiotic products will be streptomycin sensitive.

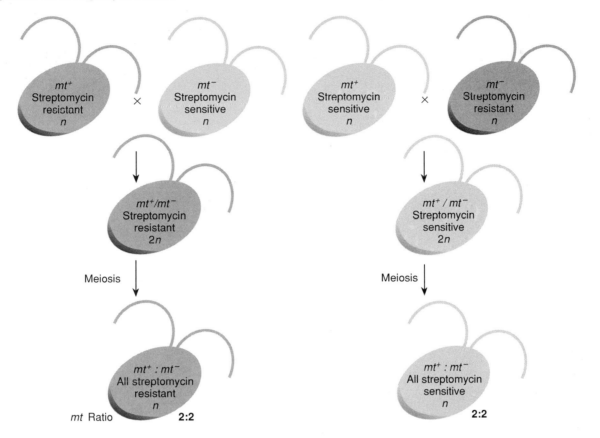

roplast is not functioning; and it shows some interesting non-Mendelian inheritance patterns related to the mating type. R. Sager has done extensive work on inheritance of streptomycin resistance in *Chlamydomonas*.

Streptomycin resistance can be selected for in *Chlamydomonas* in several ways. Normal cells, sensitive to the antibiotic, are killed in its presence. If cells are grown in low levels of the antibiotic (100 μg/ml), some cells will show resistance to it. When these cells are crossed to the wild-type, the resistance will segregate in a 1:1 ratio, indicating that streptomycin resistance is controlled by a chromosomal locus. The same experiment can be repeated using high levels of the antibiotic in the medium (500–1600 μg/ml). Again resistant colonies are found. If they are crossed to the wild-type, a 1:1 ratio does not ensue.

Chlamydomonas does not have sexes but does have mating types, *mt*⁺ and *mt*⁻. Only individuals of opposite type can mate. Mating type is inherited as a single locus with two alleles. When two haploid cells of opposite mating type fuse, they form a diploid zygote, which then undergoes meiosis to produce four haploid cells, two of *mt*⁺ and

two of *mt*⁻. The high-level resistance always segregates with the *mt*⁺ parent (fig. 17.10). It is as if the *mt*⁺ parent were contributing the cytoplasm to the zygote in a manner similar to that of maternal plastid inheritance in plants. The *mt*⁻ parent is acting like a pollen parent by making a chromosomal contribution but not a cytoplasmic one.

The mechanism of the extrachromosomal inheritance of *Chlamydomonas* is not known. It is especially mystifying in that both parent cells fuse and appear to contribute equally to the zygote. Currently, we believe that the target of streptomycin is the chloroplast.

More recent work has shown that the *mt*⁺ inheritance is only 99.98% effective—that is, 0.02% of the offspring in crosses of the type shown in figure 17.10 have the streptomycin phenotype of the *mt*⁻ parent. Thus we have the possibility of studying recombination in chloroplast genes. Although most of the evidence is only indirect and plagued by the previously mentioned problems of separating chloroplast and mitochondrial effects, some initial mapping studies have been done.

Figure 17.11

Conjugation in *Paramecium*. The letters *K* and *k* represent alleles of a gene in each micronucleus. When a *KK* and a *kk* individual conjugate, the exconjugants have the identical *Kk* genotype.

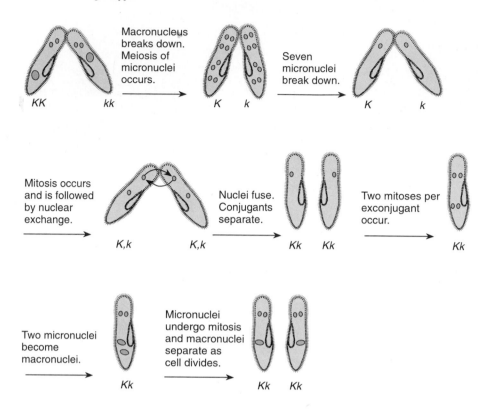

Tracy M. Sonneborn (1905–1981)

Photograph by William Dellenback Photography.

Infective Particles

Paramecium

Tracy Sonneborn discovered the killer trait in *Paramecium*. Before analyzing this trait, we must digress a moment to look at the life cycle of *Paramecium*, a ciliated protozoan familiar to most biologists. Ciliates have two types of nuclei: macronuclei and micronuclei. In *Paramecium* there are two micronuclei, which are primarily reproductive nuclei, and one macronucleus, which is a polyploid nucleus concerned with the vegetative functions of the cell. During cell division, termed **binary fission,** the micronuclei divide by mitosis and the macronucleus constricts and is pulled in half.

Paramecium undergoes two types of nuclear rearrangements, conjugation and **autogamy.** In conjugation, individuals of two mating types come together and form a bridge between themselves. The nuclear events are shown in figure 17.11. Briefly, the macronucleus of each cell disintegrates while the micronuclei undergo meiosis. Of the resulting eight micronuclei per cell, seven disintegrate and one remains; this one undergoes mitosis to form two haploid nuclei per cell. A reciprocal exchange of nuclei across the bridge then occurs. Each cell now has two haploid nuclei, one original and one migrant. The two nuclei fuse to form a diploid nucleus. The diploid nuclei in the two conjugating cells are genetically identical because of the reciprocity of the process. These nuclei then undergo two mitoses each to form four diploid nuclei per cell. Two nuclei become macronuclei, which separate at the next cell division; two remain as micronuclei that divide by mitosis at the next cell division. The two cells that separate are known as **exconjugants.** Depending primarily on the amount of time conjugating cells remain united, an exchange of cytoplasm may occur along with the exchange of nuclei.

Figure 17.12

Autogamy in *Paramecium*. The letters *K* and *k* represent alleles of a gene in each micronucleus. If a heterozygote undergoes autogamy, it becomes homozygous for one of the alleles (*KK* or *kk*).

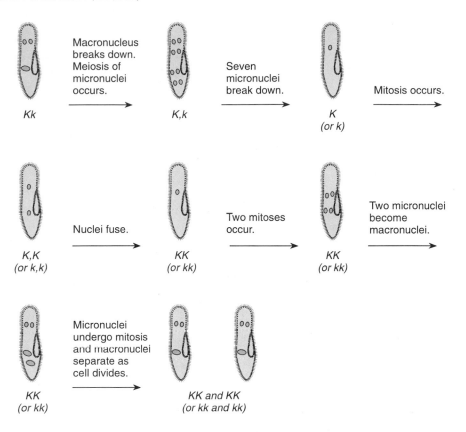

In the second type of nuclear rearrangement, autogamy, only one *Paramecium* is involved (fig. 17.12). The nuclear events are the same as in conjugation except that, at the point where a reciprocal exchange of nuclei would take place, the two haploid nuclei within the cell fuse. All cells after autogamy are homozygous.

Killer Paramecium and Kappa Particles

Sonneborn and his colleagues found that when certain stocks of *Paramecium* were mixed together, one stock had the ability to cause the death of individuals of the other stock. Those individuals causing death were called "killers" and those dying were referred to as "sensitives." During conjugation, the sensitives are temporarily resistant to the killers. If cytoplasm is not exchanged during the conjugation, the exconjugants retain their original phenotypes so that killers stay killers and sensitives stay sensitives. When there is an exchange of cytoplasm between sensitive and killer cells, both exconjugants are killers. The transfer of some cytoplasmic particle seems to be implied. Indeed, Sonneborn observed such particles in the cytoplasm of killers and called them **kappa particles** (fig. 17.13).

Although the occurrence of killer *Paramecium* does not appear to involve chromosomal genes, Sonneborn reported one case in which exconjugant killer paramecia of hybrid origin underwent autogamy. He found that half of the resulting cells had no kappa particles and had become sensitives. He concluded that a gene is required for the presence of kappa particles, which has subsequently been verified by numerous crosses. Figure 17.14 illustrates the sequence of genetic events that would produce a heterozygous killer *Paramecium* that, upon autogamy, would have a 50% chance of becoming sensitive.

Although not yet cultured outside of a *Paramecium*, kappa is presumably a bacterium because it has many bacterial attributes including size, cell wall, presence of DNA, and presence of certain prokaryotic reactions (fig. 17.15). J. Preer and his colleagues, who studied kappa itself, named it *Caedobacter taeniospiralis*. Kappa occurs in at least two forms. The *N* form, which is the infective form that is passed from one *Paramecium* to another, does not confer killer specificity on the host cell. The *N* form is attacked by bacteriophages that induce formation of inclusions, called *R* bodies, inside the kappa particle and thus convert it to the *B* form. These *R* bodies are visible under the light microscope as refractile bodies (fig. 17.15).

Figure 17.13

(a) Normal (sensitive) *Paramecium*. (b) Kappa-containing (killer) *Paramecium*. A *Paramecium* is about 200 μm long.

Source: T. M. Sonneborn, figure 29.3, p. 373 in I. H. Herskowitz, Genetics, 2nd ed. (Boston: Little, Brown, 1965). Reproduced by permission.

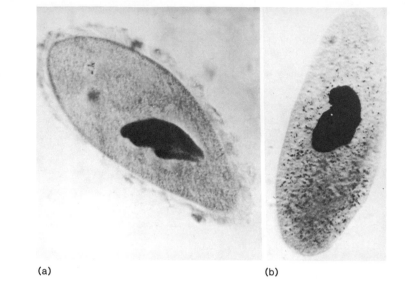

(a)   (b)

Figure 17.14

Autogamy in a heterozygous (*Kk*) killer *Paramecium* (formed by conjugation, with cytoplasmic exchange, of a *KK* killer and a *kk* sensitive cell). Upon autogamy, the heterozygote has a 50 percent chance of becoming a homozygous (*KK*) killer or a homozygous (*kk*) sensitive that loses its kappa particles.

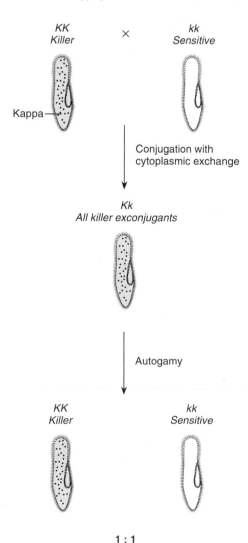

In the *B* form, kappa can no longer replicate; it is often lysed within the cell. It confers killer specificity on the host cell, however. The sensitives are killed by the toxin **paramecin** that is released by the killer *Paramecium* into the environment. Precisely what steps are involved in its formation are not known, although it is plain that the virus plays an integral role. Whether the viral DNA or the kappa DNA codes the toxin is also not known at present.

Mate-Killer Infection and Mu Particles

Kappa is not the only infective agent known in *Paramecium*. Another interesting agent is seen in the **mate-killer** infection. Here again, killer cells have visible bacteria-like particles, called **mu particles,** in the cytoplasm. Preer and his colleagues have named them *Caedobacter conjugatus*. Mate-killers do not release a toxin into the environment but instead kill their mates during conjugation. One of two unlinked dominant genes, M_1 and M_2, is required for the presence of mu particles. An interesting phenomenon occurs when a mate-killer becomes homozygous $m_1m_1m_2m_2$ by autogamy. Although the offspring will eventually lose their mu particles, virtually no loss of particles occurs until about the eighth generation, when some offspring lose all their mu. Up to this generation, all the cells had maintained a full complement of mu. In the fifteenth generation, only about 7% of the cells still have mu particles.

This phenomenon is explained as the diluting out, not of the mu themselves, but of a factor called **metagon,** which is necessary for the maintenance of mu in the cell. Once the cell becomes homozygous recessive, no further metagon production occurs. The verification that metagon is subsequently diluted out is seen in fifteenth-generation cells that still have their mu. We would expect that after fission one daughter cell would have a metagon and the other cell would not. What we expect in fact happens. The rate of dilution is consistent with an original number of

Figure 17.15
Electron micrograph of a sectioned kappa particle (*Caedobacter taeniospiralis*). Phage particles are seen as dark inclusions. The plane of the section cuts through a rolled-up *R* body. Magnification 61,200×.
Reproduced by permission of J. R. Preer, Jr.

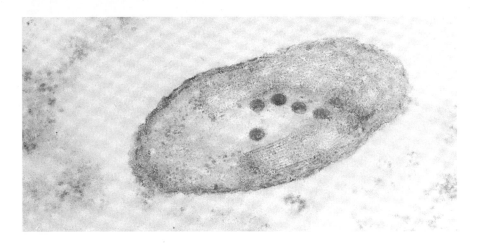

Figure 17.15
Electron micrograph of a sectioned kappa particle (*Caedobacter taeniospiralis*). Phage particles are seen as dark inclusions. The plane of the section cuts through a rolled-up *R* body. Magnification 61,200×.
Reproduced by permission of J. R. Preer, Jr.

Figure 17.16
Electron micrograph of the spirochete associated with the extrachromosomal sex-ratio trait in *Drosophila*. Magnification 22,700×.

K. Oishi and D. F. Poulson, "A Virus Associated with SR-spirochetes of Drosophila nebulosa," Proceedings of The National Academy of Sciences, USA 67 (1970):1565–1572. Reproduced by permission of the authors.

about one thousand metagons per cell. The metagon appears to be mRNA because it is destroyed by RNase. Its protein product is presently unknown.

We thus see several instances of infective particles that interact with the genome of *Paramecium* with interesting phenotypic results. Similar interactions are known in other organisms—for example, the killer trait in yeast.

Drosophila

Several instances occur in insects in which infective particles mimic patterns of inheritance. In *Drosophila,* we find forms of the **sex-ratio phenotype** in which females produce mostly, if not only, daughters. One form is inherited as a chromosomal gene; another form, however, is not chromosomal. In the nonchromosomal form, females usually produce a few sons. These sons do not pass on the sex-ratio trait. The daughters of sex-ratio females do pass on the trait, which was shown to be extrachromosomal by the fact that it persisted even after all the chromosomes had been substituted out of the stock by appropriate crosses.

About half the eggs of a sex-ratio female fail to develop. Cytoplasm can be withdrawn from the undeveloped eggs and used to infect other females. The trait, then, is caused by some cytoplasmic factor that could infect other females and is not passed on by sperm. Detailed cytological examination of the cytoplasm of sex-ratio females has revealed a spirochete (fig. 17.16) that has been isolated and used to infect other female *Drosophila* with the sex-ratio trait; it is, therefore, the causal agent of this phenotype.

Prokaryotic Plasmids

In chapters 7 and 12 we discussed plasmids in regard to their role in the study of prokaryotic genetics and their use in recombinant DNA work. They are mentioned again here because they represent extrachromosomal genetic systems, primarily in prokaryotes. The autonomous segments of DNA known as plasmids are, for the most part, known from bacteria, in which they occur as circles of DNA within the host cell (noncircular DNA is soon degraded). When plasmids become integrated into the chromosomes, they become indistinguishable from chromosomal material.

R and Col Plasmids

In addition to the F factor found frequently in bacteria, there are a variety of other plasmids, including the R and Col plasmids. The **R plasmids** carry genes for resistance to various antibiotics, and the **Col plasmids** have genes that are responsible for producing proteins called *colicins,* which are toxic to strains of *E. coli* (fig. 17.17). Plasmids containing genes for Col-like toxins specific for other bacterial species are also known. Col and R plasmids can exist

Figure 17.17

Electron micrograph of replication of Col E1 circular plasmid. The *arrows* mark the branch points of the theta structure. Magnification 90,000×.

Source: J. I. Tomizawa, Y. Sakakibara, and T. Kakefuda, "Replication of Colicin E1 Plasmid DNA in Cell Extracts: Origin and Direction of Replication," Proceedings of The National Academy of Sciences, USA 71 (1974):2260–2264.

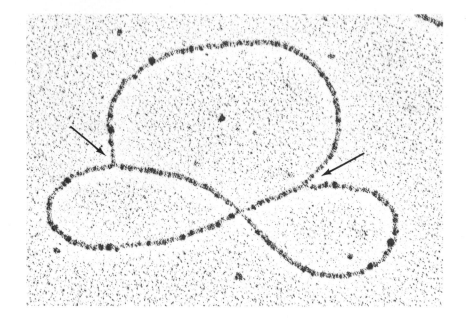

in two states. In one state the plasmid has a sequence of genes called the **transfer operon** (*tra*), which makes the plasmids similar to F factors in that they can transfer their genes from one bacterium to the next. In the other state the plasmids lack this operon and cannot transfer their loci to another cell. Thus Col and R plasmids are actually made of two parts: the loci for antibiotic resistance or colicin production and the part responsible for infectious transfer. In R plasmids the infectious transfer part is abbreviated as RTF (**resistance transfer factor**).

The occurrence of resistance plasmids was observed in Japan in the late 1950s, when it was discovered that bacteria were simultaneously acquiring resistance to several antibacterial agents. When cultures of *Shigella,* a dysentery-causing bacterium, were exposed to streptomycin, sulfonamide, chloramphenicol, or tetracycline, the bacteria exhibited resistance not only to the one particular agent that they were exposed to but to one or more of the others as well. The plasmid responsible for this multiple resistance was named R222.

The Col plasmids contain loci that produce proteins that are toxic, for various reasons, to strains of bacteria not carrying the plasmids. Colicins attack sensitive bacterial cells at bacterial surface receptors. On the basis of the types of receptors they attack, colicins have been classified into twenty or more categories. Some colicins may enter the cell directly, but others do not. For example, colicin K appears to kill sensitive cells by inhibiting DNA, RNA, and protein synthesis although not directly entering the cell. Colicin E3, however, acts as an intracellular ribonuclease that cleaves off about fifty nucleotides from the 3′ end of the 16S rRNA within the ribosome. The cleavage inactivates the sensitive cell's ribosomes and is, of course, lethal.

Since many R plasmids, Col plasmids, and F factors, as well as host chromosomes, have insertion sequences

(chapter 13), a good deal of exchange occurs among the plasmids, and many are able to integrate into the host chromosome. Although their mobility makes it easier to map and study plasmids, it also poses a human health problem. Resistance to various antibacterial agents is easily transferred among enterobacteria worldwide. Transfer of resistance can even occur outside of host organisms (people) where pollution or sewage is found. In addition, resistance found in relatively harmless enterobacteria, such as *E. coli,* can easily be passed to more pathogenic bacteria, such as *Shigella* and *Salmonella.* Since every time we use antibacterial drugs we are selecting for resistance, we should not use these drugs indiscriminately. For some time health workers have been concerned over excessive medical use of antibacterial drugs as well as the use of large quantities of antibiotics in animal feed.

Uncovering Plasmids

How do we know when the phenotype is controlled by a plasmid rather than by the chromosomal genes of a bacterium? Plasmids can be seen with an electron microscope or by density-gradient centrifugation of the cell's DNA. But several less direct lines of evidence will also supply the answer. To begin with, multiple aspects of the phenotype (e.g., resistance to several antibacterial agents) change simultaneously, as with plasmid R222. Another clue is that the phenotypic change is infectious: Japanese workers found that with R222, resistant cells converted nonresistant cells. As B. Lewin stated, "Resistance is infectious. . . ."

There are several other clues to the presence of a plasmid. In linkage studies, using transduction for example, plasmid loci show no linkage to host loci; plasmids themselves can be mapped because their loci are linked to each other. Since the plasmid DNA replicates at its own speed, it can miss being incorporated into a daughter cell. Thus many spontaneous losses of the plasmid occur. And finally, certain treatments—with acridine dyes, for example—have little effect on the replication of the host chromosome but selectively prevent the plasmid from replicating; thus the plasmid can be eliminated from the cell population. The existence of plasmids in a bacterial population can, therefore, be verified with morphological, physiological, and analytical evidence.

SUMMARY

Patterns of non-Mendelian inheritance fall into two categories: maternal effects and cytoplasmic inheritance. Maternal effects are illustrated by snail-shell coiling. The direction of coiling is determined by the genotype of the maternal parent, in which dextral coiling is dominant to sinistral coiling.

Cytoplasmic inheritance is usually seen in organelles, symbionts, or parasites that have their own genetic material. Chloroplasts and mitochondria have relatively small circular chromosomes with prokaryotic affinities. An interaction exists between organelles and nuclei; the organelles do not encode all their own proteins and enzymes. Mitochondrial defects can be inherited through nuclear genes or through the mitochondrion itself. A similar pattern is seen in chloroplasts. The processes of cytoplasmic inheritance are exemplified by symbiotic bacteria in *Paramecium*.

Plasmids are autonomous segments of DNA. In prokaryotes, R and Col plasmids, as well as the F factor, have been well studied. Plasmids usually carry an operon for transfer and insertion sequences for attachment to cell chromosomes and to each other. Hence they represent highly mobile segments of genetic material.

SOLVED PROBLEMS

1. What possible phenotypes and genotypes could the female parent of a sinistrally coiled snail have?

   ANSWER: If a snail is sinistrally coiled, its mother must have had the *dd* genotype, since sinistrality is recessive. If the female parent is a recessive homozygote, its mother must have contributed a recessive *d* allele. Therefore *its* mother (the grandmother) could have had either a *Dd* or *dd* genotype. Its daughter could therefore be either dextrally or sinistrally coiled (respectively). Thus, to answer the question, a sinistrally coiled snail could have had a mother that was either dextrally or sinistrally coiled but only of the *dd* genotype.

2. What is the evidence that mitochondrial DNA might be superfluous?

   ANSWER: By superfluous we mean that there is not one specific function that the mitochondrial DNA *must* control. The evidence for this is simple. When many organisms have their mtDNA sequenced, there is no one gene that is found in all of these mtDNAs. Therefore, any function of a mitochondrion can be controlled by nuclear genes in one organism or another.

3. You have just noticed a petite yeast colony growing in a petri plate under aerobic conditions. What type of petite is it?

   ANSWER: The simplest way to determine the nature of the lesion resulting in the petite phenotype is to make a cross of the petite strain with a wild-type strain. After meiosis, the four products (spores) are isolated and allowed to grow separately under normal, aerobic conditions. If the ratio of petite to wild-type is 1:1, the mutation is of a nuclear gene. If progeny are wild-type, the mutation is of the mitochondrial genome and is of the neutral type. If progeny are mostly petites, the mutation is also of the mitochondrial genome, but is of the suppressive type.

4. *Paramecium* with the genotype *KK* are mated with cells that are *kk* under a situation that allows cytoplasmic exchange. If the exconjugants undergo autogamy, what types of progeny do you expect?

   ANSWER: Both exconjugants will be *Kk,* and since cytoplasmic exchange occurred, both cytoplasms will contain kappa. Autogamy will produce either *KK* or *kk* cells. Since at least one *K* gene is needed for the maintenance of kappa, the *kk* cells will eventually lose the kappas and become sensitive. Thus we expect 1/2 sensitive:1/2 killers

EXERCISES AND PROBLEMS

1. J. Christian and C. Lemunyan have shown that mice raised under crowded conditions produce two generations with reduced growth rates. What sort of genetic control might exist, and how could this control be demonstrated?

2. Snail coiling is called a maternal trait. Is it possible that it is caused by an allele at a sex-linked locus?

3. How would you rule out a viral origin for snail-shell coiling?

4. Give the genotypes such that a sinistral female snail can produce dextral offspring. What genotypes could the male parent of the sinistral female have?

5. What evidence indicates that it is not absolutely essential, in an evolutionary sense, for mitochondria to have genes for specific components of oxidative phosphorylation?

6. How would you determine that a segregative petite mutant in yeast is controlled by a chromosomal gene?

7. What results would be obtained by making all possible pairwise crosses of the three types of yeast petites?

8. An ornamental spider plant has green and white striped leaves. How can you determine whether cytoplasmic inheritance is responsible for the striping and whether there is interaction with an *iojap*-type chromosomal gene?

9. In *Chlamydomonas*, 0.02% of the meiotic products are of the mt^- parental type. How can you use this information in mapping? (Use streptomycin sensitivity, str^s, and resistance, str^r, as an example.)

10. What similarities are shared by mitochondria and plastids?

11. What evidence is there for a prokaryotic origin of mitochondria and chloroplasts?

12. Individuals from killer and nonkiller strains of *Paramecium* are mixed together. Cytoplasmic exchange occurs during conjugation. Approximately 25% of the exconjugants are sensitive and the remaining 75% are killers. What are the genotypes of the individuals of the two strains and what ratios of sensitives and killers would result if the various exconjugants underwent autogamy?

13. What genetic tests could you conduct to show that the mate-killer phenotype in *Paramecium* requires a dominant allele at any one of *two* loci?

14. Resistant and sensitive strains of *Drosophila melanogaster* differ in their ability to tolerate CO_2—anesthetization with it kills sensitive flies. What genetic experiments would you perform to determine that the trait is caused by a virus? How would you rule out chromosomal genes?

15. Describe the types of evidence that could be gathered to determine whether a trait in *E. coli* is controlled by chromosomal or plasmid genes.

16. Suppose you have identified a person who has introns in his or her mitochondrial DNA. What would you deduce about the origin of this DNA?

17. A mitochondrial mutant in people causes blindness. If reciprocal matings are found in a family pedigree between affected and normal individuals, what types of children do you expect for each cross?

18. When chloroplast DNA from *Chlamydomonas* is digested with a particular restriction enzyme and then hybridized with a particular probe, two bands are detected. Some strains (type 1) yield bands of 1.5 and 3.7 kb; other strains (type 2) yield bands of 2.5 and 6.0 kb. For the following crosses, predict the progeny:
   a. mt^+, strain 1 $\times$ mt^-, strain 2
   b. mt^+, strain 2 $\times$ mt^-, strain 1

19. What type of asci do you expect if you cross a yeast strain carrying an antibiotic resistance gene in its mitochondria with a strain that has normal mitochondria?

20. In *Paramecium*, maintenance of kappa particles requires the dominant nuclear gene *K*. A *Kk* killer cell conjugates with a sensitive cell of the same genotype without cytoplasmic exchange. Predict the genotypes and phenotypes that will result if each exconjugant then undergoes autogamy.

21. In *Neurospora*, the slow-growing trait *poky* is inherited maternally and is due to an abnormal respiratory protein. A nuclear gene *F* makes *poky* individuals grow faster, even though the protein is still defective. Such strains are called *fast-poky* (*F'* is normal *poky*). *Poky* cytoplasm is not altered by *F* in a zygote, and *F* has no effect on normal cytoplasm. What genotypes and phenotypes do you expect if the maternal parent is *fast-poky* and the paternal parent is normal?

22. A dextral snail (A) that resulted from a cross is self-fertilized and produces only sinistral progeny. What is the probable genotype of (A) and its parents?

23. In corn, two independent, recessive nuclear genes, *japonica* (*j*) and *iojap* (*ij*), produce green and white leaves. Matings between individuals heterozygous for *japonica* always produce 3 green:1 striped individuals regardless of how the cross is performed. The behavior of *iojap* was described in figure 17.9. You have a variegated plant that could be either *jj* or *ijij*. What cross can you make to determine the genotype of this plant, and what results in the F_1 generation do you expect in each case?

24. If *Paramecium* cells heterozygous for both genes involved in the maintenance of the mate-killer trait are forced to undergo autogamy, what phenotypic ratios do you expect?

25. A petite yeast strain is crossed to a wild-type strain. What phenotypic ratio do you expect after meiosis if
   a. the petite is nuclear?
   b. the petite is suppressive?
   c. the petite is neutral?

26. The maroon-like (*ma-l*) locus in *Drosophila* is inherited as an X-linked recessive trait. If you cross a heterozygous female with a maroon-like male, all the progeny are wild-type. If the female progeny from this cross are mated again with maroon-like males, half of the females produce all maroon-like progeny, and the other half produce all wild-type progeny. Explain these results.

27. In corn, male sterility is controlled by a maternal cytoplasmic element. A dominant nuclear gene, Restorer (*Rf*) restores fertility to male sterile lines. If pollen from a homozygous *RfRf* plant is used to pollinate a male sterile plant, what genotypes and phenotypes do you expect in the progeny?

Suggested Readings for chapter 17 are on page 657.

Part IV

Quantitative and Evolutionary Genetics

Giant land tortoise (*Geochelone elephantopus vanderburghi*) from
Santa Cruz Island, Galápagos.
© *George H. Harrison/Grant Heilman Photography*

18

Quantitative Inheritance

OBJECTIVES

◆ To understand the patterns of inheritance of phenotypic traits controlled by many loci

◆ To investigate the way that geneticists and statisticians describe and analyze normal distributions of phenotypes

◆ To define and measure heritability, the unit of inheritance of variation in traits controlled by many loci

African violets (*Saintpaulia* sp.) demonstrating a range of flower colors, a trait frequently controlled by quantitative inheritance.
© *Barry L. Runk/Grant Heilman Photography*

When we talked previously of genetic traits, we were usually discussing traits controlled by single genes whose inheritance patterns led to simple ratios. However, many traits, including some of economic importance—such as yields of milk, corn, and beef—exhibit what is called continuous variation.

Although there was some variation in height of Mendel's pea plants, all of them could be scored as either tall or dwarf; there was no overlap. Using the same methods that Mendel used, we can look at ear length in corn (fig. 18.1). With Mendel's peas, all of the F_1 were tall. In a cross between corn plants with long and short ears, all of the F_1 plants have ears intermediate in length. When both these F_1's are self-fertilized, the results are again different. In the F_2 generation, Mendel obtained exactly the same height categories (tall and dwarf) as in the parental generation, only the ratio was different—3:1.

In corn, however, ears of every length, from the shortest to the longest, will be found in the F_2; there will be no discrete categories. A genetically controlled trait exhibiting this type of variation is usually controlled by many loci. In this chapter we will study this type of variation by looking at traits that are controlled by progressively more loci. We will then turn to the concept of heritability, which is used as a statistical tool to evaluate the genetic control of traits determined by many loci.

TRAITS CONTROLLED BY MANY LOCI

Let us begin by considering grain color in wheat. When a particular strain of wheat having red grain is crossed with another strain having white grain, all the F_1 plants have kernels that are intermediate in color. When these plants are self-fertilized, the ratio of kernels in the F_2 is 1 red:2 intermediate:1 white (fig. 18.2). This is simple inheritance involving one locus with two alleles. The white allele, *a*, produces no pigment (which results in the background color, white); the red allele, *A*, produces red pigment. The F_1 heterozygote, *Aa*, is intermediate (incomplete dominance). When this monohybrid is self-fertilized, the typical 1:2:1 ratio results. (For simplicity, we use dominant-recessive allele designations, *A* and *a*. Keep in mind, however, that the heterozygote is intermediate in color.)

Two-Locus Control

Now let us examine the same kind of cross using two other stocks of wheat with red and white kernels. Here, when the resulting intermediate (medium-red) F_1 are self-fertilized, five color classes of kernels emerge in a ratio of 1 dark red:4 medium dark red:6 medium red:4 light red:1 white (fig. 18.3). The offspring ratio, in sixteenths, comes from the self-fertilization of a dihybrid in which the two loci are unlinked. In this case, both loci affect the same trait in the same way. In figure 18.3, each capital letter

Figure 18.1

Comparison of continuous variation (ear length in corn) with discontinuous variation (height in peas)

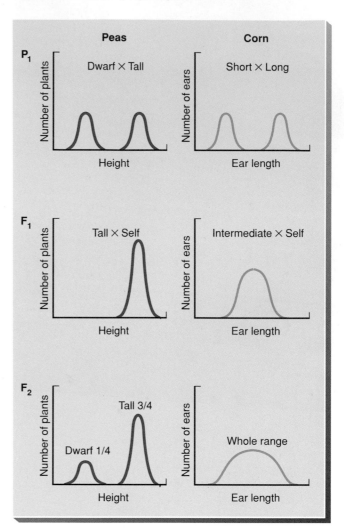

Figure 18.2

Cross involving the grain color of wheat

P_1	Red	×	White
	AA		*aa*

F_1	Intermediate color
	Aa

F_2	Red	: Intermediate :	White
	AA	*Aa*	*aa*

1 : 2 : 1

Figure 18.3
Another cross involving grain color of wheat

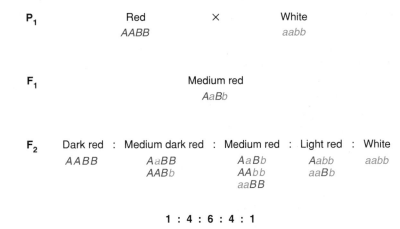

P₁	Red	×	White
	AABB		aabb

F₁ — Medium red — AaBb

F₂	Dark red	:	Medium dark red	:	Medium red	:	Light red	:	White
	AABB		AaBB AABb		AaBb AAbb aaBB		Aabb aaBb		aabb

1 : 4 : 6 : 4 : 1

represents an allele that produces one unit of color, and each lowercase letter represents an allele that produces no color. Thus the genotype *AaBb* has two units of color just like the genotypes *AAbb* and *aaBB*. All produce the same intermediate grain color. Recall from chapter 2 that a cross such as this produces nine genotypes in a ratio of 1:2:1:2:4:2:1:2:1. If these classes are grouped according to numbers of color-producing alleles as shown in figure 18.3, the 1:4:6:4:1 ratio appears. This ratio is a product of the binomial expansion.

Three-Locus Control

In yet another cross of this nature, Nilsson-Ehle in 1909 crossed two wheat strains, one with red and the other with white grain, that yielded plants in the F₁ generation with grain of intermediate color. When these plants were self-fertilized, at least seven color classes, from red to white, were distinguishable in a ratio of 1:6:15:20:15:6:1 (fig. 18.4). This result is explained by assuming that three loci are assorting independently, each with two alleles, such that one allele produces a unit of red color whereas the other allele does not. We then see seven color classes, from red to white, in the 1:6:15:20:15:6:1 ratio. This ratio is in sixty-fourths, directly from the 8 X 8 (trihybrid) Punnett square, and comes from grouping genotypes in accordance with the number of color-producing alleles they contain. Again the ratio is one that is generated in a binomial distribution.

Multilocus Control

From here we need not go on to an example with four loci and then one with five, and so on. We have enough information to draw generalities. It should not be hard to see how discrete loci can generate a continuous distribution (fig. 18.5). Theoretically, it should be possible to distinguish different color classes down to the level of the eye's

Figure 18.4
One of Nilsson-Ehle's crosses involving three loci controlling grain color in wheat. Within the Punnett square, only the number of color-producing alleles are shown to emphasize color production.

P₁	Red	×	White
	AABBCC		aabbcc

F₁ — Intermediate color × Self — AaBbCc

	ABC	ABc	AbC	aBC	Abc	aBc	abC	abc
ABC	6	5	5	5	4	4	4	3
ABc	5	4	4	4	3	3	3	2
AbC	5	4	4	4	3	3	3	2
aBC	5	4	4	4	3	3	3	2
Abc	4	3	3	3	2	2	2	1
aBc	4	3	3	3	2	2	2	1
abC	4	3	3	3	2	2	2	1
abc	3	2	2	2	1	1	1	0

Phenotype  Red ⟶ White

Number of color-producing alleles   6 : 5 : 4 : 3 : 2 : 1 : 0

Ratio   1 : 6 : 15 : 20 : 15 : 6 : 1

Figure 18.5

The change in shape of the distribution as increasing numbers of loci control grain color in wheat. If each locus is segregating two alleles with each affecting the same trait, eventually a continuous distribution will be generated in the F_2 generation.

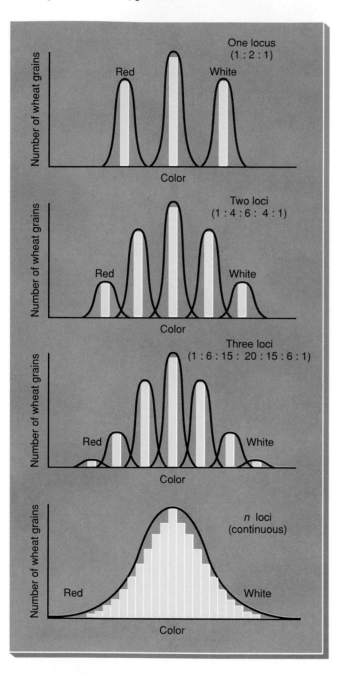

The variation within each genotype is due to the environment—that is, two organisms with the same genotype may not necessarily be identical in color because nutrition, physiological state, and many other variables will influence the phenotype. Figure 18.6 shows that it is possible for the environment to obscure genotypes even in a one-locus, two-allele system. That is, a height of 17 cm could result in the F_2 from either the *aa* or *Aa* genotype (fig. 18.6, column 3) when there is excessive variation. In the other two cases of figure 18.6, there would be virtually no organisms 17 cm tall. Systems such as we are considering, in which each allele contributes a small unit to the phenotype, are easily influenced by the environment with the result that the distribution of phenotypes approaches the bell-shaped curve seen at the bottom of figure 18.5.

Thus phenotypes that are determined by loci with alleles that contribute identical dosages to the phenotype will approach a continuous distribution in the F_2 generation. This type of trait is said to exhibit **continuous, quantitative,** or **metrical variation.** The inheritance pattern is called **polygenic** or **quantitative.** The system is termed an **additive model** because each allele adds a certain amount to the phenotype.

From the three wheat examples just discussed, we can generalize to systems with more than three polygenic loci, each segregating two alleles. From table 18.1 we can predict the distribution of genotypes and phenotypes expected from an additive model with any number of loci segregating two alleles each. This table is useful when we seek to estimate how many loci are producing a quantitative trait, assuming it is possible to distinguish the various phenotypic classes. For example, when a strain of heavy mice was crossed with a lighter strain, the F_1 were of intermediate weight. When these F_1 were interbred, a continuous distribution of adult weights was obtained in the F_2. Since only about 1 mouse in 250 was as heavy as the heavy parent stock, we could guess that if an additive model holds, then four loci are segregating. This is because we expect $1/(4)^n$ to be as extreme as either parent; 1 in 250 is roughly $1/(4)^4 = 1/256$.

Location of Polygenes

The fact that traits with continuous variation can be controlled by genes dispersed over the whole genome was shown by James Crow, who studied DDT resistance in *Drosophila.* A DDT-resistant strain of flies was created by growing them on increasing concentrations of the insecticide. Crow then systematically tested each chromosome for the amount of resistance it conferred. Susceptible flies were mated to resistant flies and the sons from this cross were backcrossed. Offspring were then scored for the particular resistant chromosomes they contained (each chromosome had a visible marker) and were tested for their

ability to perceive differences in wavelengths of light. In fact, we rapidly lose the ability to assign unique color classes to genotypes because the variation within each genotype is such that before long the phenotypes overlap. For example, with three loci, a color somewhat lighter than medium dark red may belong to the medium-dark-red class with three color alleles, or it may belong to the medium-red class with only two color alleles (fig. 18.5).

Figure 18.6
Influence of environment on phenotype distribution

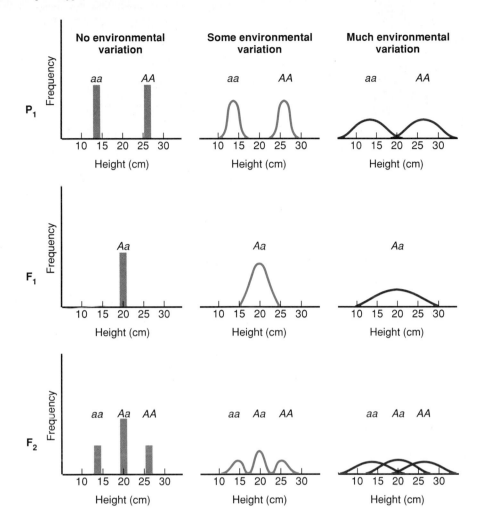

TABLE 18.1 Generalities from an Additive Model of Polygenic Inheritance

	1 Locus	2 Loci	3 Loci	*n* Loci
Number of Gamete Types Produced by an F₁ Multihybrid	2 (A, a)	4 (AB, Ab, aB, ab)	8 (ABC, ABc, AbC, Abc, aBC, aBc, abC, abc)	2^n
Number of Different F₂ Genotypes	3 (AA, Aa, aa)	9 (AABB, AABb, AAbb, AaBB, AaBb, Aabb, aaBB, aaBb, aabb)	27 (AABBCC, AABBCc, AABBcc, AABbCC, AABbCc, AABbcc, AAbbCC, AAbbCc, AAbbcc, AaBBCC, AaBBCc, AaBBcc, AaBbCC, AaBbCc, AaBbcc, AabbCC, AabbCc, Aabbcc, aaBBCC, aaBBCc, aaBBcc, aaBbCC, aaBbCc, aaBbcc, aabbCC, aabbCc, aabbcc)	3^n
Number of Different F₂ Phenotypes	3	5	7	$2n + 1$
Number of F₂ as Extreme as One Parent or the Other	1/4 (AA or aa)	1/16 (AABB or aabb)	1/64 (AABBCC or aabbcc)	$1/4^n$
Distribution Pattern of F₂ Phenotypes	1:2:1	1:4:6:4:1	1:6:15:20:15:6:1	$(A + a)^{2n}$

James F. Crow (1916–   )
Courtesy of Dr. James F. Crow.

resistance to DDT. Sons were used in the backcross because there is no crossing over in males. Therefore, resistant and susceptible chromosomes were passed on intact by the sons. Crow's results are shown in figure 18.7. As you can see, each chromosome has the potential to increase the fly's resistance to DDT. Thus each chromosome contains loci (polygenes) that contribute to the phenotype of this additive trait.

Significance of Polygenic Inheritance

The concept of additive traits is of great importance to genetic theory because it demonstrates that Mendelian rules of inheritance can explain traits that have a continuous distribution—that is, Mendel's rules for discrete characteristics also hold for quantitative traits. Additive traits are also of practical interest. Many agricultural products, both plant and animal, follow polygenic inheritance, including milk production and fruit and vegetable yield. In addition, many human traits, such as height and IQ, appear to be polygenic, although with substantial environmental components.

Historically, the study of quantitative traits began before the rediscovery of Mendel's work at the turn of the century. In fact, there was debate among biologists in the early part of this century as to whether "Mendelians" were correct or whether the "biometricians" were correct in regard to the rules of inheritance. Biometricians used statistical techniques to study traits characterized by continuous variation and claimed that single discrete genes were not responsible for the observed inheritance patterns. They were interested in evolutionarily important facets of the phenotype, traits that can change slowly over time. Mendelians claimed that the phenotype was controlled by discrete "genes." Eventually the Mendelians were proven correct, but the biometricians' tools were the only ones suitable for studying quantitative traits.

Figure 18.7
Survival of *Drosophila* with varying numbers and arrangements of DDT-resistant and susceptible chromosomes in the presence of DDT.
Reproduced with permission, from the Annual Review of Entomology, *Volume 2, © 1957 by Annual Reviews, Inc.*

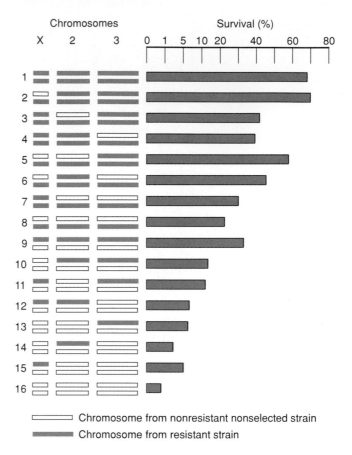

The biometric school was founded by Galton and Pearson, who showed that many quantitative traits, such as height, were inherited. They invented the statistical tools of correlation and regression analysis in order to study the inheritance of traits that fall into smooth distributions.

POPULATION STATISTICS

A distribution (see fig. 18.5, *bottom*) can be described in several ways. One way is the formula for the shape of the curve formed by the frequencies within the distribution. A more functional description of a distribution starts by defining its center, or **mean** (fig. 18.8). As can be seen from the figure, the mean is not itself enough to describe the distribution. Variation about this mean will determine the actual shape of the curve. (We will confine our discussion to symmetrical, bell-shaped curves called **normal distributions.** Many distributions approach a normal distribution.)

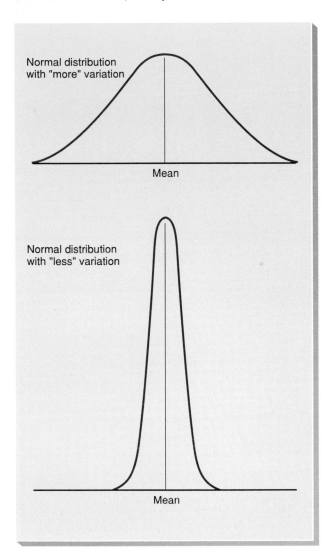

Normal distribution
with "more" variation

Mean

Normal distribution
with "less" variation

Mean

Mean, Variance, and Standard Deviation

The mean of a set of numbers is the arithmetic average of the numbers and is defined as

$$\bar{x} = \Sigma x / n \qquad (18.1)$$

in which

$\bar{x}$ = the mean

Σx = the summation of all values

n = the number of values summed

In table 18.2, the mean is calculated for the distribution shown in figure 18.9. The variation about the mean is calculated as the average squared deviation from the mean:

$$s^2 = V = \frac{\Sigma(x - \bar{x})^2}{n - 1} \qquad (18.2)$$

This value (V or s^2) is called the **variance.** Observe that the flatter the distribution is, the greater the variance will be.

The variance is one of the simplest measures that we can calculate of variation about the mean. You might wonder why we simply don't calculate an average deviation from the mean rather than an average squared deviation. For example, we could calculate a measure of variation as

$$\frac{\Sigma(x - \bar{x})}{n - 1}$$

(We will get to why we use $n - 1$ rather than n in the denominator in a moment.) Note, however, that the above measure is zero. By the definition of the mean, the absolute value of the sum of deviations above it is equal to the absolute value of the sum of deviations below it—one is negative and the other is positive. However, by squaring each deviation, as in equation (18.2), we create a relatively simple index, the variance, which is not zero.

The ear lengths measured in table 18.2 are a sample of all ear lengths in the theoretically infinite population of ears in that variety of corn. Statisticians call sample values **statistics** (and use letters from the Roman alphabet), whereas they call population values **parameters** (and use Greek letters). The sample value is an estimate of the true value for the population. Thus, in the variance formula (equation 18.2), the sample value, V or s^2, is an estimate of the population variance, σ^2. When sample values are used to estimate parameters, one degree of freedom is lost for each parameter estimated. To determine the sample variance, we divide not by the sample size, but by the degrees of freedom ($n - 1$ in this case, as defined in chapter 4). The variance for the entire population (assuming we know the population mean, μ, and all the data values) would be calculated by dividing by n. The sample variance is calculated in table 18.2.

The variance has several interesting properties, not the least of which is the fact that it is additive. That is, if we can determine how much a given variable contributes to the total variance, we can subtract that amount of variance from the total and the remainder is caused by whatever other variables affect the trait. This property makes the variance extremely important in quantitative genetic theory.

TABLE 18.2 Hypothetical Data Set of Ear Lengths (x) Obtained When Corn Is Grown from an Ear of Length 11 cm

x	$(x - \bar{x})$	$(x - \bar{x})^2$
7	−4.12	16.97
8	−3.12	9.73
9	−2.12	4.49
9	−2.12	4.49
10	−1.12	1.25
10	−1.12	1.25
10	−1.12	1.25
10	−1.12	1.25
10	−1.12	1.25
10	−1.12	1.25
11	−0.12	0.01
11	−0.12	0.01
11	−0.12	0.01
11	−0.12	0.01
11	−0.12	0.01
11	−0.12	0.01
12	0.88	0.77
12	0.88	0.77
12	0.88	0.77
13	1.88	3.53
13	1.88	3.53
13	1.88	3.53
14	2.88	8.29
14	2.88	8.29
16	4.88	23.81

$\Sigma x = 278$ $\Sigma (x - \bar{x})^2 = 96.53$

$n = 25$

$$\bar{x} = \frac{\Sigma x}{n} = \frac{278}{25} = 11.12$$

$$s^2 = V = \frac{\Sigma(x - \bar{x})^2}{n - 1} = \frac{96.53}{24} = 4.02$$

$$s = \sqrt{s^2} = \sqrt{4.02} = 2.0$$

The data are graphed in figure 18.9.

The **standard deviation** is also a measure of variation of a distribution. It is the square root of the variance:

$$s = \sqrt{V} \qquad (18.3)$$

In a normal distribution, approximately 67% of the area of the curve lies within one standard deviation on either side of the mean, 96% lies within two standard deviations, and 99% lies within three standard deviations (fig. 18.10). Thus, for the data in table 18.2, about 2/3 of the population would have ear lengths between 9.12 and 13.12 cm (mean ± standard deviation).

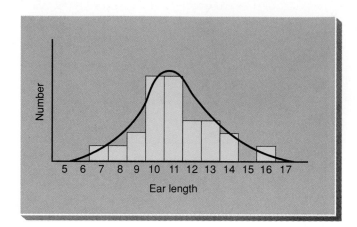

Figure 18.9
Normal distribution of ear lengths in corn. Data are given in table 18.2.

Figure 18.10
Area under the bell-shaped curve. The abscissa is in units of standard deviation (s) around the mean (x̄).

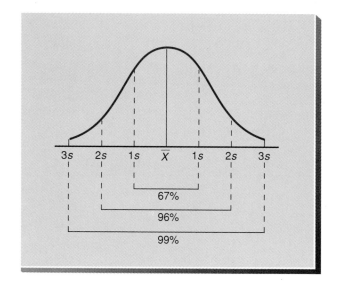

One final measure of variation about the mean is the **standard error of the mean** (SE):

$$\text{SE} = s/\sqrt{n}$$

The standard error (of the mean) is the standard deviation about the mean of a distribution of sample means. In other words, if we repeated the experiment many times, each time we would generate a mean value. We could then use these mean values as our data points. We would expect the variation among a population of means to be less than among individual values, and it is. Data are often summarized as "the mean ± SE." In our example of table 18.2, SE $= 2.0/\sqrt{25} = 2.0/5.0 = 0.4$. We can summarize the data set of table 18.2 as 11.1 ± 0.4 (mean ± SE).

TABLE 18.3 The Relationship between Two Variables, x (Aptitude Test Score) and y (Grade-Point Average)

x	y	x	y	x	y	x	y	x	y
14	4.0	11	2.9	10	2.9	10	1.4	8	2.6
14	3.4	11	2.8	10	2.8	9	2.8	8	2.4
14	3.2	11	2.7	10	2.7	9	2.7	8	2.3
13	3.7	11	2.6	10	2.6	9	2.6	8	1.8
13	2.7	11	2.5	10	2.2	9	2.4	8	1.4
12	2.7	11	2.4	10	2.1	9	2.1	8	1.1
12	2.4	11	2.2	10	1.9	9	1.7	8	0.9
12	2.2	11	2.0	10	1.8	9	1.5	8	0.8
12	2.1	11	1.9	10	1.7	9	1.1	7	1.7
11	3.5	10	3.2	10	1.6	9	0.9	7	0.8

$\Sigma x = 505 \qquad n = 50 \qquad \Sigma y = 112.4$

$$\bar{x} = \frac{\Sigma x}{n} = 10.1 \qquad\qquad \bar{y} = \frac{\Sigma y}{n} = 2.25$$

$$s_x^2 = \frac{\Sigma(x - \bar{x})^2}{n - 1} = 3.07 \qquad s_y^2 = \frac{\Sigma(y - \bar{y})^2}{n - 1} = 0.57$$

$$s_x = \sqrt{s_x^2} = 1.75 \qquad\qquad s_y = \sqrt{s_y^2} = 0.75$$

$$\text{cov}(x, y) = \frac{\Sigma(x - \bar{x})(y - \bar{y})}{n - 1} = 0.87$$

$$r = \frac{\text{cov}(x, y)}{s_x s_y} = \frac{0.87}{(1.75)(0.75)} = 0.66$$

From Paul Blommers and Robert A. Forsyth, *Statistical Methods in Psychology and Education,* Second edition. Copyright © 1977 by Houghton Mifflin Company, Boston, MA.

Covariance, Correlation, and Regression

It is often desirable in genetic studies to know whether there is a relationship between two given characteristics in a series of individuals. For example, is there a relationship between height of a plant and its weight, or between scholastic aptitude and grades? If one increases, does the other also? An example is given in table 18.3; the same data set is graphed in figure 18.11, which is referred to as a scatter plot. There does appear to be a relation between the two variables. With increasing aptitude (x axis), there is an increase in grade point (y axis). We can determine how closely the two variables are related by calculating a **correlation coefficient**—an index that goes from -1.0 to $+1.0$ depending on the degree of relationship between the variables. If there is no relation (if the variables are independent), then the correlation coefficient will be zero. If there is perfect correlation, where an increase in one variable is associated with a proportional increase in the other, the coefficient will be $+1.0$. If an increase in one is

associated with a proportional decrease in the other, the coefficient will be -1.0 (fig. 18.12). The formula for the correlation coefficient (r) is

$$r = \frac{\text{covariance of } x \text{ and } y}{s_x \cdot s_y} \qquad (18.4)$$

where s_x and s_y are the standard deviations of x and y, respectively.

To calculate the correlation coefficient, we need to define and calculate the **covariance** of the two variables, $\text{cov}(x, y)$. The covariance is analogous to the variance, but involves the simultaneous deviations from the means of both the x and y variables:

$$\text{cov}(x, y) = \frac{\Sigma(x - \bar{x})(y - \bar{y})}{n - 1} \qquad (18.5)$$

The analogy between variance and covariance can be seen by comparing equations (18.5) and (18.2). The covariance, variances, and standard deviations are calculated in table 18.3, in which the correlation coefficient, r, is 0.66. (There are computational formulas available that substantially cut down on the difficulty of calculating these statistics. If a computer or calculator is used, only the individual data points need be entered—most computers and calculators can be programmed to do all the computations.)

Many experiments deal with a situation in which we assume that one variable is dependent on the other (cause and effect). For example, we may ask, what is the relationship of DDT resistance in *Drosophila* to an increased number of DDT-resistant alleles? With more of these alleles (fig. 18.7), the DDT resistance of the flies should increase. Number of DDT-resistant alleles is the independent variable and resistance of the flies is the dependent variable. That is, a fly's resistance is dependent on the number of DDT-resistant alleles it has, not the other way around. Going back to figure 18.11, we could make the assumption that grade-point average is dependent on scholastic aptitude. If this were so, an analysis called *regression analysis* could be used. This analysis allows us to predict a grade-point average (y variable) given a particular aptitude (x variable). (It is important to note here that once a cause-and-effect relationship is assumed, strictly speaking, a correlation analysis is not valid. Correlation or regression may be used on a given data set, but not both. Regression analysis assumes a cause-and-effect relationship, whereas correlation analysis does not.)

The formula for the straight-line relationship (regression line) between the two variables is $y = a + bx$, where b is the slope of the line (change in y divided by change

Figure 18.11

Relationship between two variables, grade point average (GPA) and aptitude test score. The raw data are given in table 18.3.

From Paul Blommers and Robert A. Forsyth, Elementary Statistical Methods in Psychology and Education, *2d ed. Copyright © 1977 Houghton Mifflin Company, Boston, MA.*

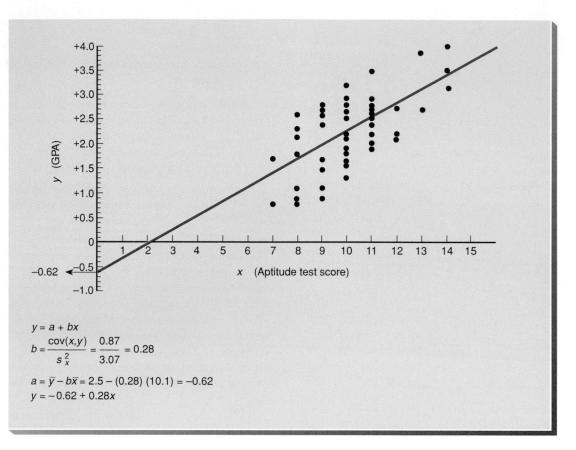

$y = a + bx$

$b = \dfrac{cov(x,y)}{s_x^2} = \dfrac{0.87}{3.07} = 0.28$

$a = \bar{y} - b\bar{x} = 2.5 - (0.28)(10.1) = -0.62$

$y = -0.62 + 0.28x$

in x, or $\Delta y/\Delta x$) and a is the y-intercept of the line (see fig. 18.11). To define any line we need only to calculate the slope, b, and the y intercept, a:

$$b = cov(x, y)/s_x^2 \qquad (18.6)$$
$$a = \bar{y} - b\bar{x} \qquad (18.7)$$

Thus equipped, if a cause-and-effect relationship does exist between the two variables, we can predict any y value given an x value. We can either use the formula $y = a + bx$ or graph the regression line and determine directly the y value for any x. We now continue our examination of the genetics of quantitative traits.

POLYGENIC INHERITANCE IN BEANS

In 1909, W. Johannsen, who studied seed weight of the dwarf bean plant (*Phaseolus vulgaris*), demonstrated that polygenic traits are controlled by many genes. The parent population was made up of seeds (beans) with a continuous distribution of weights. Johannsen divided this parental group into classes according to weight, planted them, self-fertilized the plants that grew, and weighed the F_1 beans. He found that the parents with the heaviest beans produced the progeny with the heaviest beans, and the parents with the lightest beans produced the progeny with the lightest beans (table 18.4). There was a significant correlation coefficient between parent and progeny bean weight ($r = 0.34 \pm 0.01$). He continued this work by beginning nineteen lines (populations) with beans from various points on the original distribution and selfing each successive generation for the next several years. After a few generations, the means and variances stabilized within each line. That is, when Johannsen chose, within each line, parent plants with heavier-than-average or lighter-than-average seeds, the offspring had the parental mean with the parental variance for seed size. For example, in one line, plants with both the lightest average bean weights (24 centigrams) and plants with the heaviest average bean weights (47 cg) produced offspring with average bean weights of 37 cg. By selfing the plants each generation, Johannsen had made them more and more homozygous, thus lowering the number of segregating polygenes. Therefore, the lines became homozygous for certain of the polygenes (different in each line) and any variation in bean weight was then caused only by the environment. Johannsen thus showed that quantitative traits were controlled by many segregating loci.

Figure 18.12
Plots showing varying degrees of correlation within data sets

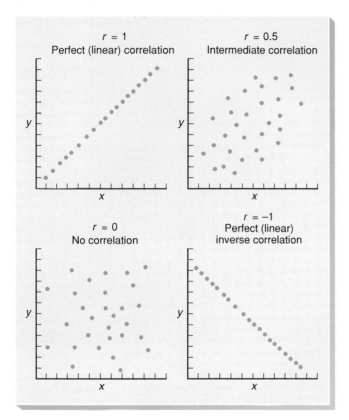

SELECTION EXPERIMENTS

Selection experiments are done for several reasons. Plant and animal breeders select the most desirable individuals as parents in order to improve their stock. Population geneticists select specific characteristics for study in order to understand the nature of quantitative genetic control.

For example, *Drosophila* was tested in a fifteen-choice maze for geotactic response (fig. 18.13). The maze was on its side so that at every intersection a fly had to make a choice between going up or going down. The flies with the highest score were chosen as parents for the "high" line (positive geotaxis; favored downward direction) and the flies with the lowest score were chosen as parents for the "low" line (negative geotaxis; favored upward direction). The same selection was made each generation. As time progressed, the two lines diverged quite significantly. This tells us that there is a large genetic component to the response; the experimenters are successfully amassing more of the "downward" alleles in the high line and the "upward" alleles in the low line. Several other points emerge from this graph. First, the high and low responses are slightly different, or asymmetrical. The high line responded more quickly, leveled out more quickly, and tended toward the original state more slowly after selection was relaxed. (The relaxation of selection occurred when the parents were a random sample of the adults rather than the extremes for geotactic scores.) The low line responded more slowly and erratically. In addition, the low line returned toward the original state more quickly when selection was relaxed.

The nature of these responses (fig. 18.13) indicates that the high line became more homozygous than the low line. This is shown by the former's slight response when selection is relaxed. It has exhausted a good deal of its variability for the polygenes responsible for geotaxis. The low line, however, seems to have much of its original genetic variability because the relaxation of selection caused the mean score of this line to increase rapidly. It still had

TABLE 18.4 Johannsen's Findings of Relationship between Bean Weights of Parents and Their Progeny

Weight of Parent Beans	Weight of Progeny Beans (centigrams)																n	Mean ± SE
	15	20	25	30	35	40	45	50	55	60	65	70	75	80	85	90		
65–75				2	3	16	37	71	104	105	75	45	19	12	3	2	494	58.47 ± 0.43
55–65			1	9	14	51	79	103	127	102	66	34	12	6	5		609	54.37 ± 0.41
45–55			4	20	37	101	204	281	234	120	76	34	17	3	1		1,138	51.45 ± 0.27
35–45	5	6	11	36	139	278	498	584	372	213	69	20	4	3			2,238	48.62 ± 0.18
25–35		2	13	37	58	133	189	195	115	71	20	2					835	46.83 ± 0.30
15–25			1	3	12	29	61	38	25	11							180	46.53 ± 0.52
Totals	5	8	30	107	263	608	1,068	1,278	977	622	306	135	52	24	9	2	5,491	50.39 ± 0.13

Figure 18.13

Selection for geotaxis. The *dotted lines* represent relaxed selection.

Source: Data from T. Dobzhansky and B. Spassky, "Artificial and Natural Selection for Two Behavioral Traits in Drosophila pseudoobscura" in Proceedings of the National Academy of Sciences, *62:75–80, 1969.*

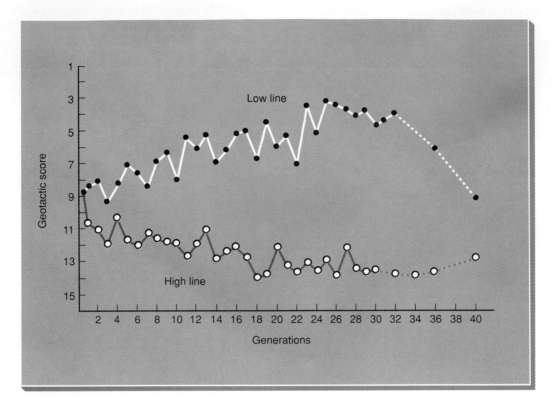

HERITABILITY

Plant and animal breeders want to improve the yields of their crops to the greatest degree they can. They must choose the parents of the next generation on the basis of this generation's yields; thus they are performing selection experiments. Breeders run into two economic problems. They cannot pick only the very best to be the next generations' parents because (1) they cannot afford to decrease the size of a crop by using only a very few select parents and (2) they must avoid **inbreeding depression,** which occurs when plants are self-fertilized or animals are bred with close relatives for many generations. After frequent inbreeding, too much homozygosity occurs, and many genes that are slightly or partially deleterious begin to show themselves, depressing vigor and yield. (More on inbreeding will be presented in chapter 19.) Thus breeders need some index of the potential response to selection so that they can then get the greatest amount of selection for the lowest risk of inbreeding depression.

enough genetic variability to head back to the original population mean. The response to a selection experiment is one way that plant and animal breeders can predict future response.

Realized Heritability

Breeders often calculate a **heritability** estimate, a value that will predict to what extent their selection effort will be successful. Heritability is defined in the following equation:

$$H = \frac{Y_O - \overline{Y}}{Y_P - \overline{Y}} = \frac{\text{gain}}{\text{selection differential}} \qquad (18.8)$$

in which

H = heritability

Y_O = offspring yield

$\overline{Y}$ = mean yield of the population

Y_P = parental yield

From this equation we can see that heritability is the gain in yield divided by the amount of selection that has been practiced (fig. 18.14). $Y_O - \overline{Y}$ is the improvement over the population average due to $Y_P - \overline{Y}$, which is the amount of difference between the parents and the population av-

Figure 18.14

Realized heritability is the gain divided by the selection differential.

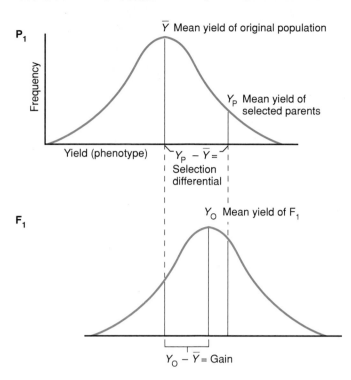

TABLE 18.5 Some Realized Heritabilities

Animal	Trait	Heritability
Cattle	Birth weight	0.49
	Milk yield	0.30
Poultry	Body weight	0.31
	Egg production	0.30
	Egg weight	0.60
Swine	Birth weight	0.06
	Growth rate	0.30
	Litter size	0.15
Sheep	Wool length	0.55
	Fleece weight	0.40

erage (fig. 18.14). If there is no gain ($Y_O = \overline{Y}$), then the heritability will be zero, and breeders will know that no matter how much selection they practice, they will not improve their crops and might as well not waste their time. Since this value is calculated after the breeding has been done, it is referred to as a **realized heritability.** Some typical values for realized heritabilities are shown in table 18.5.

The following example may help to clarify the calculation of realized heritability. The number of bristles on the sternopleurite, a thoracic plate in *Drosophila,* is under polygenic control. In a population of flies, the mean bristle number was 6.4. Three pairs of flies were used as parents; they had a mean of 7.2 bristles. Their offspring had a mean of 6.6 bristles. Hence $Y_O = 6.6$, $\overline{Y} = 6.4$, and $Y_P = 7.2$. Then dividing the gain by the selection differential—that is, substituting in equation 18.8—gives us:

$$H = \frac{6.6 - 6.4}{7.2 - 6.4} = \frac{0.2}{0.8} = 0.25$$

If both a low line and a high line were begun, and if both were carried over several generations, the heritability would be measured by the final difference in means of the high and low lines (gain) divided by the cumulative selection differentials summed for both the high and low lines.

Note from figure 18.13 that the response to selection declines with time as the selected population becomes homozygous for various alleles controlling the trait. As the response declines, the calculated heritability value itself declines. After intense or prolonged selection, heritability may be zero. It does not mean that the trait is not controlled by genes, only that there is no longer a response to selection. Hence heritability is specific for a particular population at a particular time. Intense selection will exhaust the genetic variability, rendering the response to selection, and thus heritability itself, zero.

Quantitative geneticists treat the realized heritability as an estimate of the **true heritability.** True heritability is actually viewed two different ways: as *heritability in the narrow sense* and *heritability in the broad sense.* We will define these on the basis of partitioning of the variance of the quantitative character under study.

Partitioning of the Variance

Given that the variance of a distribution has genetic and environmental causes, and given that the variance is additive, we can construct the following formula:

$$V_{Ph} = V_G + V_E \tag{18.9}$$

in which

V_{Ph} = total phenotypic variance

V_G = variance due to genotype

V_E = variance due to environment

Throughout the rest of this discussion we will stay with this model. A more complex variance model can be constructed if there are interactions between variables. For example, if one genotype responded better in one soil condition than in another soil condition, there would be an environmental-genotype interaction that would require a separate variance term (V_{GE}).

The variance due to the genotype (V_G) can be further broken down according to the effects of additive polygenes (V_A), dominance (V_D), and epistasis (V_I) to give us the final formula of

$$V_{Ph} = V_A + V_D + V_I + V_E \qquad (18.10)$$

We can now define the two commonly used—and often confused—measures of heritability. *Heritability in the narrow sense* is

$$H_N = V_A/V_{Ph} \qquad (18.11)$$

This heritability is the proportion of the total phenotypic variance that is caused by additive genetic effects. It is the heritability of most interest to plant and animal breeders because it predicts how much of a response will occur under selection.

Heritability in the broad sense is

$$H_B = V_G/V_{Ph} \qquad (18.12)$$

This heritability is the proportion of the total phenotypic variance that is caused by all genetic factors, not just additive factors. It measures the extent to which individual differences in a population are caused by genetic differences. It is the measure most often used by psychologists. We will be concerned primarily with H_N, heritability in the narrow sense.

Measurement of Heritability

Three general methods are used to estimate heritability. First, as discussed earlier, we can measure heritability by the response of a population to selection. Second, we can estimate directly the components of variance by eliminating one component; the remaining variance can then be attributed to other causes. For example, by eliminating environmental causes of variance we can estimate the genetic component directly. Or, by eliminating the genetic causes of variance, we can estimate the environmental component directly. Third, we can measure the similarity between relatives. We will look now at the latter two methods.

Variance components can be eliminated several different ways. If we use genetically identical organisms, then the additive, dominance, and interaction variances are zero and all that is left is the environmental variance. For example, Robertson determined the variance components for the length of the thorax in *Drosophila*. The total variance (V_{Ph}) in a genetically heterogeneous population was 0.366 (measured directly from the distribution of the trait as in

tables 18.2 and 18.3). He then looked at the variance in flies that were genetically homogeneous. These were from isolated lines that had been inbred in the laboratory over many generations to become virtually homozygous. He studied the F_1 in several different matings of inbred lines and found the variance in thorax length to be 0.186 (V_E). By subtraction (0.366 − 0.186), the total genetic variance (V_G) was 0.180. From this we can calculate heritability in the broad sense as

$$H_B = V_G/V_{Ph} = 0.180/0.366 = 0.49$$

In order to calculate a heritability in the narrow sense, it is necessary to extract the components of the genetic variance, V_G.

Genetic variance can be measured directly by eliminating the influence of the environment. This is most easily done with plants grown in a greenhouse. Under that circumstance, environmental variables, such as soil quality, water, and sunlight, can be controlled to a very high degree. Hence the variance among individuals grown under these circumstances is exclusively genetic variance. The total phenotypic variance can be obtained from the plants grown under natural circumstances. Here again we can calculate heritability in the broad sense.

Several methods exist to sort out the additive from the dominant and epistatic portions of the genetic variance. The methods mostly rely on correlations between relatives. That is, the expected amount of genetic similarity between certain relatives can be compared with the actual similarity. The expected amount of genetic similarity is the proportion of genes shared; it is a known quantity for any form of relatedness. For example, parents and offspring have half their genes in common. The relation of observed and expected correlations between relatives is a direct measure of heritability in the narrow sense. We can thus define

$$H_N = r_{obs}/r_{exp} \qquad (18.13)$$

where r_{obs} is the observed correlation between the relatives and r_{exp} is the expected correlation. The expected correlation is simply the proportion of the genes in common.

It should be pointed out here that the observed correlation between relatives can be artificially inflated if the environments are not random. Since we know that relatives frequently share similar (or correlated) environments, they may show a phenotypic similarity irrespective of genetic causes. It is important to keep that in mind, especially when we analyze human traits where it may be almost impossible to rule out environmental similarity or to quantify it. Hence r_{obs} may be inflated, which will inflate H_N.

In human beings, finger-ridge counts (fingerprints, fig. 18.15) have a very high heritability; there seems to be very little environmental interference in the embryonic devel-

Figure 18.15

The three basic fingerprint patterns. Ridges are counted where they intersect the line connecting a triradius with a loop or whorl center. (*a*) An arch; there is no triradius; the ridge count is zero. (*b*) A loop; thirteen ridges. (*c*) A whorl; there are two triradii and counts of seventeen and eight (the higher one is routinely used).

From Sarah B. Holt, "Quantitative Genetics of Finger-Print Patterns" in British Medical Bulletin, *17. Copyright © 1961 Churchill Livingstone Medical Journals, Edinburgh, Scotland. Reprinted by permission.*

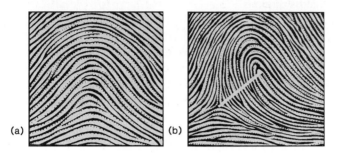

(a)   (b)

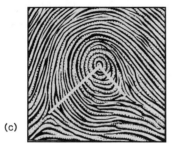

(c)

TABLE 18.6 Correlations Between Relatives, and Heritabilities, for Finger-Ridge Counts

Relationship	r_{obs}	r_{exp}	H_N
Mother–Child	0.48	0.50	0.96
Father–Child	0.49	0.50	0.98
Siblings	0.50	0.50	1.0
Dizygotic Twins	0.49	0.50	0.98
Monozygotic Twins	0.95	1.00	0.95

From Sarah B. Holt, "Quantitative Genetics of Finger-Print Patterns" in *British Medical Bulletin,* 17. Copyright © 1961 Churchill Livingstone Medical Journals, Edinburgh, Scotland. Reprinted by permission.

TABLE 18.7 Some Estimates of Heritabilities (H_N) for Human Traits and Disorders

Trait	Heritability
Schizophrenia	0.85
Diabetes mellitus	
Early onset	0.35
Late onset	0.70
Asthma	0.80
Cleft lip	0.76
Heart disease, congenital	0.35
Peptic ulcer	0.37
Depression	0.45
Stature*	1.00+

*A heritability higher than 1 can be obtained when the correlation among relatives is higher than expected. This is usually the result of dominant alleles.

opment of the ridges (table 18.6). Monozygotic twins are from the same egg, which divides into two embryos at a very early stage. They have an identical genotype. Dizygotic twins result from the simultaneous fertilization of two eggs. They have the same relationship to each other as siblings. (However, environmental influences may be different on these three relationships; they may be treated differently by relatives and friends.) The data therefore suggest that human finger ridges are almost completely controlled by additive genes with a negligible input of environmental and dominance variation. Few human traits are controlled this simply (table 18.7).

This brief discussion should make it clear that the components of the total variance can be estimated. For a given quantitative trait, the total variance can be measured directly. If identical genotypes can be used, then the environmental component of variance can be discovered. By correlation of various relatives, it is possible to measure heritability in the narrow sense directly. If heritability is known and if the total phenotypic variance is known, then all that are left, assuming no interaction, are the dominance and epistatic components. In practice, the epistatic components are usually ignored. Thus, operationally, all that is left is the dominance variance, obtained by subtracting the additive from the total genetic vari-

ance. In addition, plant and animal breeders use sophisticated statistical techniques of covariance and variance analysis, techniques that are beyond our scope here.

QUANTITATIVE INHERITANCE IN HUMAN BEINGS

As with most human studies, the measurement of heritability is limited by a lack of certain types of information. We cannot develop pure lines, nor can we manipulate human beings into various kinds of environments or do selection experiments. However, there are certain kinds of information available that allow some estimation of heritabilities.

Skin Color

Skin color is a quantitative human trait for which a simple analysis can be done on naturally occurring matings. Certain groups of people have black skin; other groups do not. Many of these groups breed true in the sense that skin

Figure 18.16
Inheritance of skin color in human beings

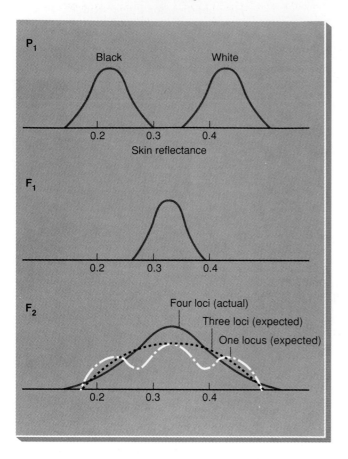

colors stay the same generation after generation within a group; when intermarriage between groups occurs, the F_1 are intermediate in skin color. When F_1 individuals intermarry and produce offspring, the skin color of the F_2 is, on the average, about the same as the F_1 but with more variation (fig. 18.16). The data are consistent with a model of four loci, each segregating two alleles. At each locus, one allele adds a measure of color whereas the other adds none.

Twin Studies

In human beings, the use of twin studies has been helpful in estimating the heritability of quantitative traits. Some monozygotic twins (MZ) have been reared apart. The same is true for dizygotic twins (DZ) and nontwin siblings. Twin studies provide information about environmental influences on many quantitative traits as well as estimates of their heritabilities. Figure 18.17 shows the range of data available for the quantitative trait intelligence (IQ). It shows that there is a high correlation among relatives. However, there is an environmental effect (monozygotic twins raised apart have a median correlation coefficient of 0.67). According to T. Bouchard and M. McGue (fig. 18.17), "That the data support the inference of partial genetic determination for IQ is indisputable; that they are informative about the precise strength of this effect is dubious."

Another way of looking at quantitative traits is by the **concordance** of twins. Concordance means that if one twin has the trait, the other does also. Discordance is the case in which one has the trait and the other does not. Table 18.8 shows some values. High concordance of monozygotic as compared with dizygotic twins is another indicator of the heritability of a trait. Concordance values for measles susceptibility and handedness demonstrate the high environmental influence on some traits.

At present, twin studies are emerging from the shadow of a well-known scandal involving a knighted British psychologist, Cyril Burt (1883–1971), who did classical twin research on the inheritance of IQ. He was posthumously accused of fraud, which was almost universally accepted, and which cast doubt on all of his data and conclusions. More recently, new information seems to cast doubt on the charges of fraud. These on-again, off-again charges of fraud have been a focus of scientific and public interest.

SUMMARY

Some genetically controlled phenotypes do not fall into discrete categories. This type of variation is referred to as quantitative, continuous, or metrical. The genetic control of this variation is referred to as polygenic. If the number of controlling loci is small, and offspring fall into recognizable classes, it is possible to analyze the genetic control of the phenotypes with standard methods. Polygenes controlling DDT resistance are located on all chromosomes in *Drosophila*.

When phenotypes fall into a continuous distribution, the methods of analysis change. We must describe a distribution using means, variances, and standard deviations. Then we must describe the relationship between two variables using variances and correlation coefficients.

Equipped with these tools, we analyze the genetic control of continuous traits. The heritability estimate tells us how much of the variation in the distribution of a trait can be attributed to genetic causes. Heritability in the narrow sense is the relative amount of variance due to additive loci. Heritability in the broad sense is the relative amount of variance due to all genetic components, including dominance and epistasis. In practice, heritability can be calculated as realized heritability—gain divided by selection differential. Estimates of human heritabilities can be obtained from correlations among relatives, concordance and discordance between twins, and studies in which monozygotic twins are reared apart.

Figure 18.17

Correlations for IQ in various degrees of relatedness pairings in 111 studies. *Vertical lines* represent median correlations, measures of the "middle" value of a distribution. *Arrowheads* represent the values obtained if the trait were under simple polygenic control without environmental effects. The term "midparent" refers to the average between the two parents. The chi-square values test for heterogeneity among the studies. A significant chi-square value indicates that the studies in that category differ significantly in their correlations. Assortative mating refers to unrelated husband-wife pairs.

From T. J. Bouchard and M. McGue, "Familial Studies of Intelligence: A Review" in Science, *212:1056–1059. Copyright 1981 by the AAAS.*

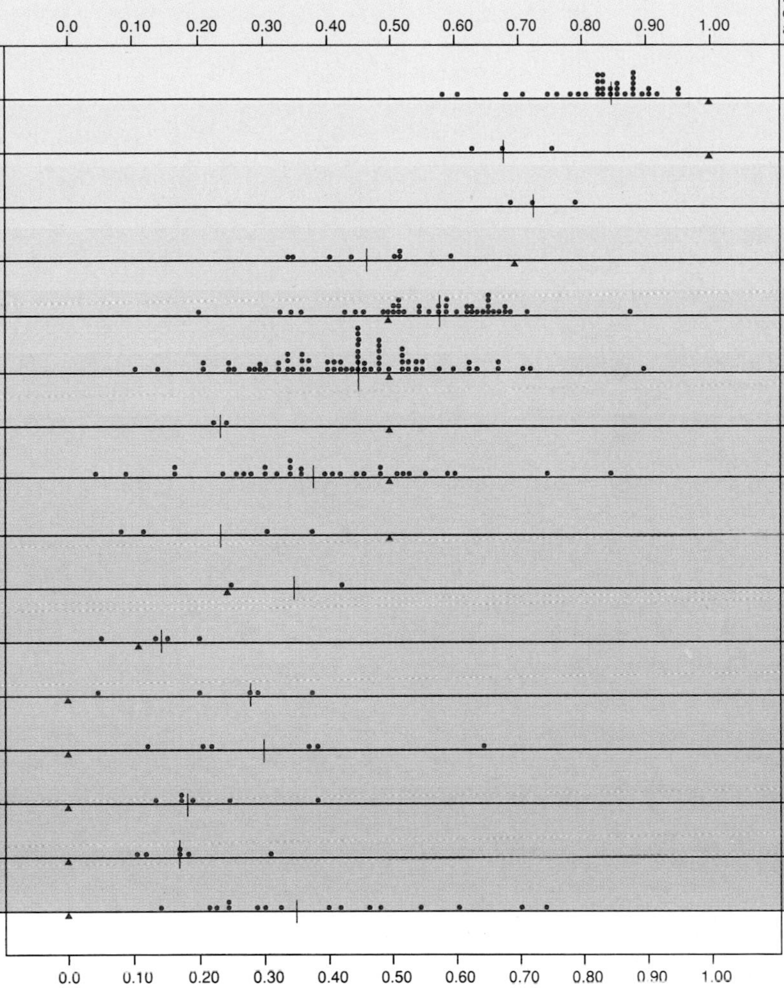

	No. of correlations	No. of pairings	Median correlation	Weighted average	χ^2 (d.f.)	χ^2_+ (d.f.)
Monozygotic twins reared together	34	4672	.85	.86	81.29 (33)	2.46
Monozygotic twins reared apart	3	65	.67	.72	0.92 (2)	0.46
Midparent-midoffspring reared together	3	410	.73	.72	2.66 (2)	1.33
Midparent-offspring reared together	8	992	.475	.50	8.11 (7)	1.16
Dizygotic twins reared together	41	5546	.58	.60	94.5 (40)	2.36
Siblings reared together	69	26,473	.45	.47	403.6 (64)	6.31
Siblings reared apart	2	203	.24	.24	.02 (1)	.02
Single parent-offspring reared together	32	8433	.385	.42	211.0 (31)	6.81
Single parent-offspring reared apart	4	814	.22	.22	9.61 (3)	3.20
Half-siblings	2	200	.35	.31	1.55 (1)	1.55
Cousins	4	1,176	.145	.15	1.02 (2)	0.51
Non-biological sibling pairs (adopted/natural pairings)	5	345	.29	.29	1.93 (4)	0.48
Non-biological sibling pairs (adopted/adopted pairings)	6	369	.31	.34	10.5 (5)	2.10
Adopting midparent-offspring	6	758	.19	.24	6.8 (5)	1.36
Adopting parent-offspring	6	1397	.18	.19	6.64 (5)	1.33
Assortative mating	16	3817	.365	.33	96.1 (15)	6.41

TABLE 18.8 Concordance of Traits between Identical and Fraternal Twins

	Identical (MZ) Twins (%)	Fraternal (DZ) Twins (%)
Hair Color	89	22
Eye Color	99.6	28
Blood Pressure	63	36
Handedness (Left or Right)	79	77
Measles	95	87
Clubfoot	23	2
Tuberculosis	53	22
Mammary Cancer	6	3
Schizophrenia	80	13
Down Syndrome	89	7
Spina Bifida	72	33
Manic-depression	80	20

SOLVED PROBLEMS

1. In a certain stock of wheat, grain color is controlled by four loci following an additive model. How many different gametes can a tetrahybrid produce? How many different genotypes will result if tetrahybrids are self-fertilized? What will be the phenotypic distribution of these genotypes?

   ANSWER: Assume the A, B, C, and D loci with A and a, B and b, C and c, and D and d alleles, respectively. A tetrahybrid will have the genotype *AaBbCcDd*. A gamete can get either allele at each of four independently assorting loci, or $2^4 = 16$ different gametes. Three genotypes are possible for each locus, two homozygotes and a heterozygote. Therefore, for four independent loci, there are $3^4 = 81$ different genotypes. Phenotypes are distributed according to the binomial distribution. Thus there will be a pattern of $(A + a)^{2n} = (A + a)^8 =$ a ratio of 1:8:28:56:70:56:28:8:1 of phenotypes of decreasing red color from left to right, of eight red colors and white.

2. In horses, white facial markings are inherited in an additive fashion. These markings are scored on a scale that begins at zero. In a particular population, the average score is 2.2. A group of horses with an average score of 3.4 is selected to be parents of the next generation. The offspring of this group of selected parents have a mean score of 3.1. What is the realized heritability of white facial markings in this herd of horses?

   ANSWER: This is a simple selection experiment; the data fit our equation for realized heritability (equation 18.8). In this case,

   Y_O = offspring yield = 3.1

   $\overline{Y}$ = mean yield of the population = 2.2

   Y_P = parental yield = 3.4

   Substituting into equation 18.8:

   $$H = \frac{Y_O - \overline{Y}}{Y_P - \overline{Y}} = \frac{3.1 - 2.2}{3.4 - 2.2} = \frac{0.9}{1.2} = 0.75$$

3. Corn growing in a field in Indiana had a lysine (amino acid) content of 2.0% with a variance of 0.16. When grown in the greenhouse under controlled and uniform conditions, the mean lysine content was again 2.0% but the variance was 0.09. What measure of heritability can you calculate?

   ANSWER: We use formula 18.9 for the calculation of heritability by partitioning of the variance ($V_{Ph} = V_G + V_E$). In this case:

   V_{Ph} = total phenotypic variance = 0.16

   V_G = variance due to genotype = 0.09

   V_E = variance due to environment = ?

   Here we have eliminated environmental variance, leaving the total genotypic variance (0.09). If we subtract this from the total variance, we get the environmental variance: $0.16 - 0.09 = 0.7$. Heritability in the broad sense is the genetic variance divided by the total phenotypic variance, or $0.09/0.16 = 0.56$.

1. A geneticist wished to know if variation in the number of egg follicles produced by chickens was inherited. As a first step in his experiments, he wished to determine if the number of eggs laid could be used to predict the number of follicles. If this were true, he could then avoid the killing of chickens. He obtained the following data from fourteen chickens.

Chicken Number	Eggs Laid	Ovulated Follicles
1	39	37
2	29	34
3	46	52
4	28	26
5	31	32
6	25	25
7	49	55
8	57	65
9	51	44
10	21	25
11	42	45
12	38	26
13	34	29
14	47	30

Calculate a correlation coefficient. Graph the data and then calculate the slope and y-intercept of the regression line. Draw the regression line on the same graph.

2. A variety of squash has fruits that weigh about 5 pounds. In a second variety, the average weight is 2 pounds. When the two varieties are crossed, the F_1 produce fruit with an average weight of 3.5 pounds. When two of these are crossed, a range of fruit weights is found, from 2 to 5 pounds. Of two hundred offspring, three produce fruits weighing about 5 pounds and three produce fruits about 2 pounds in weight. How many allelic pairs are involved in the weight difference between the varieties, and approximately how much does each effective gene contribute to the difference?

3. In rabbit variety 1, ear length averages 4 inches. In a second variety, it is 2 inches. Hybrids between the varieties average 3 inches in ear length. When these hybrids are crossed to each other, the offspring exhibit a much greater variation in ear length, ranging from 2 to 4 inches. Of five hundred F_2 animals, two have ears about 4 inches long and two have ears about 2 inches long. Approximately how many allelic pairs are involved in determining ear length, and how much does each effective gene seem to contribute to the length of the ear? What do the distributions of P_1, F_1, and F_2 probably look like?

4. Assume that height in people depends on four pairs of alleles. How can two persons of moderate height produce children who are much taller than they are? Assume that the environment is exerting a negligible effect.

5. How are polygenes different from traditional Mendelian genes?

6. Outstanding athletic ability is often found in several members of a family. Devise a study to determine to what extent athletic ability is inherited. (What is outstanding athletic ability?)

7. Variations in stature are almost entirely due to heredity. Yet average height has increased substantially since the Middle Ages, and the increase in height of children of immigrants to the United States, as compared with height of the immigrants themselves, is especially noteworthy. How can these observations be reconciled?

8. Would you expect good nutrition to increase or decrease the heritability of height?

9. Two adult plants of a particular species have extreme phenotypes for height (1 foot tall and 5 feet tall), a quantitative trait. If you had only one uniformly lighted greenhouse, how would you determine whether the variation in plant height is environmentally or genetically determined? How would you attempt to estimate the number of allelic pairs that may be involved in controlling this trait?

10. If skin color is caused by additive genes, can marriages between individuals with intermediate-colored skin produce light-skinned offspring? Can such marriages produce dark-skinned offspring? Can marriages between individuals with light skin produce dark-skinned offspring?

11. The tabulated data from Emerson and East (The inheritance of quantitative characters in maize. 1913. *Univ. Nebraska Agric. Exp. Sta. Bull.* no. 2) show the results of crosses between two varieties of corn and their F_2 offspring (see top of next page). Provide an explanation for these data in terms of number of allelic pairs controlling ear length. Do all the genes involved affect length additively? Explain.

	5	6	7	8	9	10	11	12	13	14	15	16	17	18	19	20	21
Ear length in corn (cm)																	
Variety P_{60}	4	21	24	8													
Variety P_{54}								3	11	12	15	26	15	10	7	2	
F_1				1	12	12	14	17	9	4							
F_2 ($F_1 \times F_1$)			1	10	19	26	47	73	68	68	39	25	15	9	1		

12. In *Drosophila,* a marker strain exists containing dominant alleles that are lethal in the homozygous condition on both chromosome 2, 3, and 4 homologues. These six lethal alleles are within inversions so there is virtually no crossing over. The strain thus remains perpetually heterozygous for all six loci and therefore all three chromosome pairs. (Geneticists use a shorthand notation in these "balanced-lethal" systems in which only the dominant alleles on a chromosome are shown, with a slash separating the two homologous chromosomes.) The markers are: chromosome 2, Curly and Plum (*Cy/Pm,* shorthand for $CyPm^+/Cy^+Pm$); chromosome 3, Hairless and Stubble (*H/S*); and chromosome 4, Cell and Minute(4) (*Ce/M*[4]). With this strain, which allows you to follow particular chromosomes by the presence or absence of phenotypic markers, construct crosses to give the strains used by Crow (fig. 18.7) to determine the location of polygenes for DDT resistance.

13. The components of variance for two characters of *D. melanogaster* are shown in the following table (data from A. Robertson, Optimum group size in progeny testing. . . .1957. *Biometrics* 13:442–50). Estimate the dominance and epistatic components and calculate heritabilities in the narrow and broad sense.

Variance components	Thorax length	Eggs laid in 4 days
V_{Ph}	100	100
V_A	43	18
V_E	51	38
$V_D + V_I$	?	?

14. Defecation rate in rats is termed "emotionality." The chart at the top of the next column (data modified from Broadhurst. 1960. *Experiments in Personality.* Vol. 1. London: Eysenck) shows mean emotionality scores during five generations in high and low selection lines. In the final generation, the parental mean was 4 for the high line and 0.9 for the low line. The cumulative selection differential is five for each line. Calculate realized heritability overall and separate heritabilities for each line. Do these differ? Why? Why was the response to selection asymmetrical?

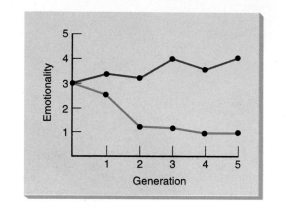

15. The table below (data from Ehrman and Parsons, 1976. *The Genetics of Behavior,* 121. Sunderland, Mass.: Sinauer Associates) gives heights in centimeters of eleven pairs of brothers and sisters. Calculate a correlation coefficient and a heritability. Is this realized heritability, heritability in the broad sense, or heritability in the narrow sense?

Pair	Brother	Sister	Pair	Brother	Sister
1	180	175	7	178	165
2	173	162	8	186	163
3	168	165	9	183	168
4	170	160	10	165	160
5	178	165	11	168	157
6	180	157			

How can environmental factors influence this heritability value?

16. Data were gathered during a selection experiment for six-week body weight in mice. Graph these data and calculate a realized heritability.

	High Line			Low Line		
Generation	$\overline{Y}$	Y_P	Y_O	$\overline{Y}$	Y_P	Y_O
0	21			21		
1		24	22		18	20
2		24	23		18	20
3		26	23		18	20
4		26	24		16	19
5		26	23		16	18

17. A red-flowered plant is crossed with a yellow-flowered plant to produce F_1 plants with orange flowers. The F_1 offspring are selfed and produce plants with flowers of seven different colors. How many genes are probably involved in color production?

18. A plant with a genotype of *aabb* and a height of 40 cm is crossed to a plant with a genotype of *AABB* and a height of 60 cm. If each dominant allele contributes to height additively, what is the expected height of the F_1 progeny?

19. If the F_1 generation in the cross in problem 18 is selfed, what proportion of the F_2 offspring are expected to be 50 cm tall?

20. You determine the following variance components for leaf width in a particular species of plant:

Additive genetic variance (V_A)	4.0
Dominance genetic variance (V_D)	1.8
Epistatic variance (V_I)	0.5
Environmental variance (V_E)	2.5

Calculate the broad sense and narrow sense heritability.

21. In a mouse population, the average tail length is 10 cm. Six mice with an average tail length of 15 cm are interbred. The mean tail length in their progeny is 13.5 cm. What is the realized heritability?

22. The narrow sense heritability of egg weight in chickens in one coop is 0.5. A farmer selects for heavier eggs by breeding a few chickens with heavier eggs. He finds a difference in the mean egg weight of 9 g between selected and unselected chickens. By how much can he expect egg weight to increase in the selected chickens?

23. Does schizophrenia seem to have a strong genetic component (table 18.8)? Explain.

24. Two strains of wheat were compared for the time required to mature. Strain X required 14 days and strain Y required 28 days. The strains were crossed, and the F_1 generation was selfed. One hundred F_2 progeny out of 6,200,000 matured in 14 days or less. How many genes may be involved in maturation?

25. If, in a population of swine, the narrow sense heritability of maturation weight is 0.5, the phenotypic variance is 100 lb², the total genetic variance is 50 lb², and the epistatic variance is 0, calculate the dominance genetic variance and the environmental variance.

26. A group of four-month-old hogs has an average weight of 170 pounds. The average weight of selected breeders is 185 lbs. If the heritability of weight is 40%, what is the expected average weight of the first generation progeny?

Suggested Readings for chapter 18 are on page 658.

19

Population Genetics: The Hardy-Weinberg Equilibrium and Mating Systems

The cheetah (*Acinonyx jubatus*) is in peril of extinction because it has very low levels of genetic variability presumably caused by periods of relatively low numbers (bottlenecks).

© Stephenie S. Ferguson/William E. Ferguson

Population genetics is the algebraic description of the genetic makeup of a population and the way in which allelic frequencies change in populations over time. This chapter is the first of three in which we will look at what population genetics can tell us about the way evolution proceeds.

Almost all of the mathematical foundations of genetic changes in populations were developed in a short period of time during the 1920s and 1930s by three men: R. A. Fisher, J. B. S. Haldane, and S. Wright. Some measure of disagreement emerged among these men, but they disagreed on which evolutionary processes were more important than others—not on how the processes worked. Since the 1960s, excitement has arisen in the field of population genetics, primarily on three fronts. First, the high-speed computer has made it possible to do a large amount of arithmetic in a very short period of time so that complex simulations of real populations can be added to the repertoire of the experimental geneticist. Second, the technique of electrophoresis has provided a means of gathering the large amount of empirical data necessary to check some of the assumptions used for many of the mathematical models. The information and interpretation of the electrophoretic data have generated some controversy about the role of "neutral" evolutionary changes in natural populations. Last, newer techniques of molecular genetics are being used to analyze the relationships among species and the rate of evolutionary processes. We consider these studies later.

HARDY-WEINBERG EQUILIBRIUM

Let's begin with a few definitions. For the most part, we define a species as a group of organisms potentially capable of interbreeding. Most species are made up of **populations,** interbreeding groups of organisms that themselves are usually subdivided into partially isolated breeding groups called **demes.** As we will see, it is these demes, or local populations, that can evolve.

In 1908, G. H. Hardy, a British mathematician, and W. Weinberg, a German physician, independently discovered a rule that relates allelic and genotypic frequencies in a population of diploid, sexually reproducing individuals if that population has random mating, large size, no mutation or migration, and no selection. The rule has three aspects:

1. The allelic frequencies at an autosomal locus in a population will not change from one generation to the next (allelic-frequency equilibrium).

2. The genotypic frequencies of the population are determined in a predictable way by the allelic frequencies (genotypic-frequency equilibrium).

Sir Ronald A. Fisher (1890–1962)
Courtesy of The National Portrait Gallery, England.

Sewall Wright (1889–1988)
Courtesy of Dr. Sewall Wright.

3. The equilibrium is neutral—that is, if it is perturbed, it will be reestablished within one generation of random mating, although at the new allelic frequency (if all the other requirements are maintained).

Calculating Allelic Frequencies

If we consider an autosomal locus in a diploid, sexually reproducing species, allelic frequencies can be measured in either of two ways. The first way is simply by counting genes:

frequency of the *a* allele, *q,* =

$$\frac{\text{number of } a \text{ alleles}}{\text{total number of alleles}}$$

The expression "frequency of" can be shortened to $f()$. For example, the frequency of the *a* allele will be written as $f(a)$. Since the homozygotes have two of a given allele and heterozygotes have only one, and since the total number of alleles is twice the number of individuals (each individual carries two alleles), we can calculate allelic frequencies in the following manner. Consider, for example, the phenotypic distribution of MN blood types (controlled by the codominant *M* and *N* alleles) among two hundred persons chosen randomly in Columbus, Ohio:

$$\begin{aligned} \text{type M } (MM \text{ genotype}) &= 114 \\ \text{type MN } (MN \text{ genotype}) &= 76 \\ \text{type N } (NN \text{ genotype}) &= \underline{10} \\ &\quad\ 200 \end{aligned}$$

Then,

$$p = f(M) = \frac{2(114) + 76}{2(200)} = \frac{304}{400} = 0.76$$

Similarly,

$$q = f(N) = \frac{2(10) + 76}{2(200)} = \frac{96}{400} = 0.24$$

Alternatively, because the frequencies of the two alleles M and N must add up to unity ($p + q = 1$), $q = 1 - p$ (and $p = 1 - q$).

Another way of calculating allelic frequencies is based on knowledge of the genotypic frequencies, which in this example are:

$$f(MM) = \frac{114}{200} = 0.57$$

$$f(MN) = \frac{76}{200} = 0.38$$

$$f(NN) = \frac{10}{200} = 0.05$$

We derive an expression for calculating p and q based on genotypic frequencies as follows:

$$p = f(M) = \frac{2 \times \text{number of } MM + \text{number of } MN}{2 \times \text{total number}}$$

$$= \frac{2 \times \text{number of } MM}{2 \times \text{total number}} + \frac{\text{number of } MN}{2 \times \text{total number}}$$

$$= f(MM) + (1/2)f(MN)$$

and,

$$q = f(N) = \frac{2 \times \text{number of } NN + \text{number of } MN}{2 \times \text{total number}}$$

$$= \frac{2 \times \text{number of } NN}{2 \times \text{total number}} + \frac{\text{number of } MN}{2 \times \text{total number}}$$

$$= f(NN) + (1/2)f(MN)$$

Thus allelic frequencies can be calculated as the frequency of homozygotes plus half the frequency of heterozygotes as follows:

$$p = f(M) = f(MM) + (1/2)f(MN)$$

$$= 0.57 + (1/2)0.38 = 0.76$$

$$q = f(N) = f(NN) + (1/2)f(MN)$$

$$= 0.05 + (1/2)0.38 = 0.24$$

or,

$$q = 1 - p = 1 - 0.76 = 0.24.$$

Assumptions of Hardy-Weinberg Equilibrium

We will consider a population of diploid, sexually reproducing organisms with a single autosomal locus segregating two alleles (i.e., every individual is one of three genotypes—MM, MN, or NN). Later on we will generalize the discussion to include multiple alleles and multiple loci. For the moment, the focus is on a genetic system such as the MN locus in human beings. The following five major assumptions are necessary for the Hardy-Weinberg equilibrium to hold.

Random Mating

The first of these assumptions is **random mating,** which means that the probability of two genotypes mating is the product of the frequencies (or probabilities) of the genotypes in the population. If the MM genotype makes up 90% of a population, then any individual has a 90% chance (probability = 0.9) of mating with a person with an MM genotype. The probability of an MM by MM mating is (0.9)(0.9), or 0.81.

Deviations from random mating come about for two reasons, choice or circumstance. If members of a population choose individuals of a particular phenotype as mates more or less often than at random, the population is engaged in **assortative mating.** If individuals with similar phenotypes are mating more often than at random, *positive assortative mating* is in force; if matings occur between individuals with dissimilar phenotypes more often than at random, *negative assortative mating,* or **disassortative mating,** is at work.

Deviations from random mating also arise when mating individuals are either more closely related genetically or more distantly related than individuals chosen at random from the population. **Inbreeding** is the mating of individuals who are related and **outbreeding** is the mating of genetically unrelated individuals. Inbreeding is a consequence of either pedigree relatedness (e.g., cousins) or small population size.

One of the first counterintuitive observations of population genetics is that deviations from random mating alter genotypic frequencies but not allelic frequencies. Envision a population in which every individual is the parent of two children. On the average, each individual will pass on one copy of each of his or her alleles. Assortative mating and inbreeding will change the zygotic (genotypic) combinations from one generation to the next but will not change which alleles are passed into the next generation. Thus genotypic, but not allelic, frequencies change under nonrandom mating.

Large Population Size

Although an extremely large number of gametes is produced each generation, each successive generation is the result of a sampling of a relatively small portion of the gametes of the previous generation. A sample may not be an accurate representation of a population, especially if the sample is small. Thus the second assumption of the Hardy-Weinberg equilibrium is that the population is infinitely large. A large population produces a large sample of successful gametes. The larger the sample, the greater the probability that the allelic frequencies of the offspring will accurately represent allelic frequencies in the parent population. When populations are small or when alleles are rare, changes in allelic frequencies take place due to chance alone. These changes are referred to as **random genetic drift,** or just *genetic drift.*

No Mutation or Migration
Allelic and genotypic frequencies may be changed by the loss or addition of alleles through mutation, or through migration (immigration or emigration) of individuals from or into a population. The third and fourth Hardy-Weinberg assumptions are that there is no such allelic loss or addition in the population due to mutation or migration.

No Natural Selection
The final assumption necessary to the Hardy-Weinberg equilibrium is that no individual will have a reproductive advantage over another individual because of its genotype. In other words, there is no natural selection occurring. (Artificial selection, as practiced by animal and plant breeders, will also perturb the Hardy-Weinberg equilibrium.)

In summary, the Hardy-Weinberg equilibrium holds (is exactly true) for an infinitely large, randomly mating population in which mutation, migration, or selection does not occur. In view of the assumptions, it seems that such an equilibrium would not be characteristic of natural populations. However, this is not the case. Hardy-Weinberg equilibrium is approximated in natural populations for two major reasons. First, the consequences of violating some of the assumptions, such as no mutation or infinitely large population size, are small. Mutation rates, for example, are on the order of one change per locus per generation per 10^5 gametes. Thus there is virtually no measurable effect of mutation in a single generation. In addition, populations do not have to be infinitely large to act as if they were. As we will see, a relatively small population can still closely approximate Hardy-Weinberg equilibrium. In other words, minor deviations from the other assumptions can still result in a good fit to the equilibrium; only major deviations can be detected statistically. Second, the Hardy-Weinberg equilibrium is extremely resilient to change because, regardless of the perturbation, the equilibrium is usually reestablished after only one generation of random mating. The new equilibrium will be, however, at the new allelic frequencies—the Hardy-Weinberg equilibrium does not "return" to previous allelic values.

Figure 19.1
Gene pool concept of zygote formation. Males and females have the same frequencies of the two alleles: $p = f(A)$ and $q = f(a)$. After one generation of random mating, the three genotypes, *AA, Aa,* and *aa,* have the frequencies of p^2, $2pq$, and q^2, respectively.

Proof of Hardy-Weinberg Equilibrium

The three properties of the Hardy-Weinberg equilibrium are that (1) allelic frequencies do not change from generation to generation, (2) allelic frequencies determine genotypic frequencies, and (3) the equilibrium is achieved in one generation of random mating. We will concentrate for a moment on the second property. In a population of individuals segregating the *A* and *a* alleles at the A locus, each individual will be one of three genotypes: *AA, Aa,* or *aa.* If $p = f(A)$ and $q = f(a)$, then the genotypic frequencies in the next generation can be predicted. If all the assumptions of the Hardy-Weinberg equilibrium are met, the three genotypes should occur in the population in the frequencies at which gametes would be randomly drawn in pairs from a **gene pool,** defined as the complement of alleles available among the reproductive members of a population from which gametes can be drawn. Thus

$$f(AA) = (p \times p) = p^2$$

$$f(Aa) = (p \times q) + (q \times p) = 2pq$$

$$f(aa) = (q \times q) = q^2$$

demonstrates the second property of the Hardy-Weinberg equilibrium (fig. 19.1).

All three properties of the Hardy-Weinberg equilibrium can be proven for the one-locus, two-allele case in sexually reproducing diploids by simply observing the offspring of a randomly mating, infinitely large population. We let the initial frequencies of the three genotypes be any values that sum to 1; for example, let *X, Y,* and *Z* be the proportions of the *AA, Aa,* and *aa* genotypes, respectively. Then the proportions of offspring after one generation of random mating are as shown in table 19.1. For example, the probability of an *AA* individual mating with an *AA* individual is $X \times X$, or X^2. Since all the offspring of this mating are *AA,* they are counted only under the

TABLE 19.1 Proportions of Offspring in a Randomly Mating Population Segregating the A and a Alleles at the A locus: $X = f(AA)$, $Y = f(Aa)$, and $Z = f(aa)$

		Offspring		
Mating	Proportion	AA	Aa	aa
$AA \times AA$	X^2	X^2		
$AA \times Aa$	XY	$\frac{1}{2}XY$	$\frac{1}{2}XY$	
$AA \times aa$	XZ		XZ	
$Aa \times AA$	XY	$\frac{1}{2}XY$	$\frac{1}{2}XY$	
$Aa \times Aa$	Y^2	$\frac{1}{4}Y^2$	$\frac{1}{2}Y^2$	$\frac{1}{4}Y^2$
$Aa \times aa$	YZ		$\frac{1}{2}YZ$	$\frac{1}{2}YZ$
$aa \times AA$	XZ		XZ	
$aa \times Aa$	YZ		$\frac{1}{2}YZ$	$\frac{1}{2}YZ$
$aa \times aa$	Z^2			Z^2
Sum	$(X + Y + Z)^2$	$(X + \frac{1}{2}Y)^2$	$2(X + \frac{1}{2}Y)(Z + \frac{1}{2}Y)$	$(Z + \frac{1}{2}Y)^2$

AA column of offspring in table 19.1. When all possible matings are counted, the offspring with each genotype are summed. The proportion of AA offspring is $X^2 + XY + (1/4)Y^2$, which factors to $(X + [1/2]Y)^2$. Recall that the frequency of an allele is the frequency of its homozygote plus half the frequency of the heterozygote. Hence $X + (1/2)Y$ is the frequency of A since $X = f(AA)$ and $Y = f(Aa)$. If $p = f(A)$, then $(X + [1/2]Y)^2$ is p^2. Thus, after one generation of random mating, the proportion of AA homozygotes is p^2. Similarly, the frequency of aa homozygotes after one generation of random mating is $Z^2 + YZ + (1/4)Y^2$, which factors to $(Z + [1/2]Y)^2$, or q^2. The frequency of heterozygotes when summed and factored (table 19.1) is $2(X + [1/2]Y)(Z + [1/2]Y)$, or $2pq$. Therefore, after one generation of random mating, the three genotypes (AA, Aa, and aa) occur as p^2, $2pq$, and q^2.

Finally, we concentrate on the first property of the Hardy-Weinberg equilibrium, that allelic frequencies do not change generation after generation. Have the allelic frequencies changed from one generation to the next (parents to offspring)? Before random mating, the frequency of the A allele is, by definition, p:

$$f(A) = p = f(AA) + (1/2)f(Aa) = X + (1/2)Y$$

After random mating, the frequency of the A homozygote is p^2 and the frequency of the heterozygote is $2pq$. Thus

the frequency of the A allele, the frequency of its homozygote plus half the frequency of the heterozygotes, is

$$f(A) = f(AA) + (1/2)f(Aa)$$

$$= p^2 + (1/2)(2pq)$$

$$= p^2 + pq = p(p + q)$$

$$= p \text{ (remember, } p + q = 1)$$

Thus, in a randomly mating population of sexually reproducing diploid individuals, the allelic frequency, p, does not change from generation to generation. Here, by observing the offspring of a randomly mating population, we have proven all three properties of the Hardy-Weinberg equilibrium.

Generation Time

Although generation interval is commonly thought of as the average age of the parents when their offspring are born, the statistical concept of generations is more complex. Demographers use formulas relating generation time to the age of reproducing females, the reproductive level of each age group, and the probability of survival in each age group. Here, to avoid these complexities, we will use **discrete generations,** unless otherwise noted. That is, we

BOX 19.1

DE FINETTI DIAGRAM

The triangle devised by B. De Finetti in 1926 and known as the De Finetti diagram is an interesting way to look at the frequencies of three genotypes in a population. Any point within the triangle (Box fig. 1) represents a population, with the lengths of perpendiculars extended to the three sides from such a point representing the relative proportions of the genotypes in this population. For example, in population 1, the lines D, R, and H represent the relative frequencies of *AA, aa,* and *Aa,* respectively. If the triangle is drawn with an altitude of 1, the three perpendiculars can be measured directly because their sum will also be 1. There are two other interesting properties of this triangle. First, the populations in Hardy-Weinberg equilibrium trace a parabola within the triangle—population 2 is in Hardy-Weinberg equilibrium, population 1 is not. Second, the perpendicular heterozygote line (H) divides the base of the triangle into the proportions of the allelic frequencies, p and q.

The shape of the parabola within the triangle shows several things. Its peak is directly over the center of the base. Since the perpendicular to the base, H, is the proportion of heterozygotes, it is apparent that heterozygotes are most numerous when the base is bisected at $p = q = 0.5$. As allelic frequencies increase or decrease, the parabola approaches the sides of the triangle—that is, as allelic frequencies increase or decrease, the relative proportion of heterozygotes to rarer homozygotes increases. Thus, in populations in Hardy-Weinberg equilibrium with allelic frequencies far from 0.5, the less-common allele is more often found in heterozygotes than in homozygotes. (See the relative lengths of R and H in populations 2 and 3.)

Box fig. 2 presents another way of looking at populations in Hardy-Weinberg equilibrium. Here the frequency of genotypes (y axis) is plotted against allelic frequency [x axis: $q = f(a)$]. As in the De Finetti diagram, you can see that heterozygotes are maximal (50%) when $p = q = 0.5$. Also, as an allele becomes rare, it is found predominantly in heterozygotes rather than in homozygotes.

Box figure 1

The De Finetti diagram. The perpendiculars from any point are in the proportions of the three genotypes. Any points (populations) on the colored parabola are in Hardy-Weinberg proportions. The vertical H line divides the base in the ratio of $p:q$.

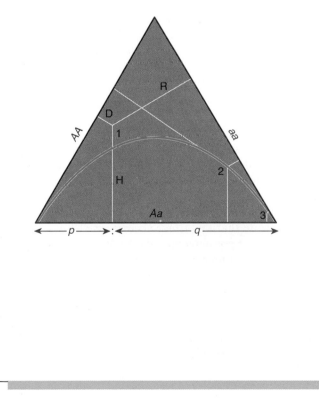

Box figure 2

The distribution of genotypes in populations in Hardy-Weinberg equilibrium. Note that heterozygote proportions are maximal when $q = p = 0.5$.

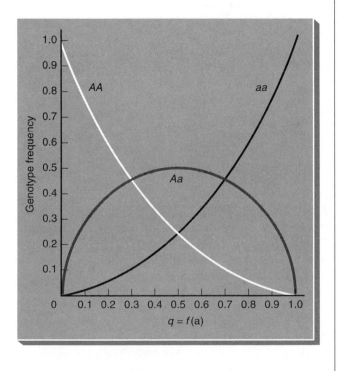

TABLE 19.2 Chi-Square Test of Goodness-of-Fit to the Hardy-Weinberg Proportions of a Sample of 200 Persons for MN Blood Types for Which $p = 0.76$ and $q = 0.24$

	MM	MN	NN	Total
Observed Numbers	114	76	10	200
Expected Proportions	p^2	$2pq$	q^2	1.0
	(0.58)	(0.36)	(0.06)	1.0
Expected Numbers	116.0	72.0	12.0	200.0
$\chi^2 = (O - E)^2/E$	0.03	0.22	0.33	0.58

will assume that all the individuals drawn in a sample, for purposes of determining allelic and genotypic frequencies, are drawn from the same generation and that, in resampling the population, the second sample represents the next generation, offspring of the first generation. The discrete-generation model holds for organisms such as annual plants and fruit flies maintained under laboratory conditions, in which there is no breeding among individuals of different generations. Generations that overlap, such as in populations of human beings and many other organisms, usually are described properly by more complex mathematical models employing integral calculus.

Testing for Fit to Hardy-Weinberg Equilibrium

There are several ways to determine whether a given population conforms to the Hardy-Weinberg equilibrium at a particular locus. However, the question usually arises when there is just a single sample from a population, representing only one generation. Can existence of the Hardy-Weinberg equilibrium be determined with just one sample? The answer is that we can determine whether the three genotypes (*AA, Aa,* and *aa*) occur with the frequencies p^2, $2pq$, and q^2. If they do, then the population is considered to be in Hardy-Weinberg proportions; if they do not, then the population is considered not to be in Hardy-Weinberg proportions.

MN Blood Types
In order to determine whether observed and expected allelic frequencies are the same, the chi-square statistical test can be used. In a chi-square test, we compare an observed number with an expected number. In this case, the observed values are the actual numbers of the three genotypes in the sample, and the expected values come from the prediction that the genotypes occur in the proportions of p^2, $2pq$, and q^2. An analysis for the Ohio MN blood-type data is presented in table 19.2. The agreement between observed and expected numbers is very good, obvious even before calculation of the chi-square value. Since

the critical chi-square for one degree of freedom at the 0.05 level is 3.841 (see table 4.4), we find that the Ohio population does not deviate from Hardy-Weinberg proportions at the MN locus.

Earlier (chapter 4), we used the chi-square statistic to test for fit of real data to an expected data set based on a ratio predicted before doing the test. For example, we tested the fit of data to a 3:1 ratio in table 4.2. In that case, the number of degrees of freedom was simply the number of independent categories: the total number of categories minus one. Here, however, our expected ratio is derived from the data set itself. The values p^2, $2pq$, and q^2 came from p and q, which were estimated from the data. In this case, we lose one additional degree of freedom for every independent value we estimate from the data. If we calculate p from a sample, we lose one degree of freedom. However, we do not lose a degree of freedom for estimating q since q is no longer an independent variable: $q = 1 - p$. So in the case above, we lose two degrees of freedom—one for estimating p and one for independent categories. The general rule of thumb in using chi-square analysis to test for fit of data to Hardy-Weinberg proportions is that the number of degrees of freedom equals the number of phenotypes minus the number of alleles (in this case, $3 - 2 = 1$).

The chi-square analysis in table 19.2 may seem paradoxical. Because the observed allelic frequencies calculated from the original genotypic data are used to calculate the expected genotypic frequencies, it may appear to some individuals that the analysis must, by its very nature, show that the population is in Hardy-Weinberg proportions. To demonstrate that this is not necessarily the case, a counterexample is presented in table 19.3. We use data similar to the Ohio sample, except that the original number of heterozygotes has been distributed equally among the two homozygote classes. The same allelic frequencies are maintained, yet a different genotypic distribution is created. The chi-square value of 191.17 for these data demonstrates that the population represented in table 19.3 is not in Hardy-Weinberg proportions. Thus a chi-square analysis of fit to the Hardy-Weinberg proportions by no means represents circular reasoning.

TABLE 19.3 Chi-Square Test of Goodness-of-Fit to the Hardy-Weinberg Proportions of a Second Sample of 200 Persons for MN Blood Types for Which $p = 0.76$ and $q = 0.24$ and Heterozygotes Are Absent

	MM	MN	NN	Total
Observed Numbers	152	0	48	200
Expected Proportions	p^2	$2pq$	q^2	1.0
	(0.58)	(0.36)	(0.06)	1.0
Expected Numbers	116.0	72.0	12.0	200.0
$\chi^2 = (O - E)^2/E$	11.17	72.00	108.00	191.17

PKU

Circumstances do exist in which Hardy-Weinberg proportions cannot be tested. In the case of a dominant trait, for example, allelic frequencies cannot be calculated from the genotypic classes because the homozygous dominant individuals cannot be distinguished from the heterozygotes. However, allelic frequencies can be estimated by assuming that the Hardy-Weinberg equilibrium exists and, thereby, assuming that the frequency of the recessive homozygote is q^2, from which q and then p can be determined.

If, for example, Hardy-Weinberg equilibrium is assumed for a disease such as phenylketonuria (PKU), which is expressed only in the homozygous recessive state, it is possible to calculate the proportion of the population that is heterozygous (carriers of the PKU allele). But is it fair to assume Hardy-Weinberg equilibrium here? There was, until recent medical practices intervened, a good deal of selection against individuals with PKU, who were usually mentally retarded. Thus the assumption of no selection required for equilibrium is violated. However, only one child in ten thousand live births has PKU. When a genotype is as rare as one in ten thousand, selection is having a negligible effect on allelic frequencies. Therefore, because of the rarity of the trait, we can assume Hardy-Weinberg equilibrium here and calculate

frequency of recessive homozygote = $q^2 =$
$1/10,000 = 0.0001$

so,

$q = \sqrt{0.0001} = 0.01$

and

$p = 1 - q = 0.99$

Therefore,

frequency of normal homozygote = $p^2 = (0.99)^2$
$= 0.98$ or 98 in 100

frequency of heterozygote = $2pq = 2(0.01)(0.99)$
$= 0.02$ or 2 in 100.

By assuming the Hardy-Weinberg equilibrium, we have discovered something not intuitively obvious: A recessive gene causing a trait as rare as one in ten thousand is carried in the heterozygous state by one in fifty individuals. Obviously, the chi-square test cannot be used here to verify Hardy-Weinberg proportions since the allelic frequencies were derived by assuming Hardy-Weinberg proportions to begin with. In statistical terms, the number of phenotypes minus the number of alleles = $2 - 2 = 0$ degrees of freedom, which precludes doing a chi-square test.

EXTENSIONS OF HARDY-WEINBERG EQUILIBRIUM

The Hardy-Weinberg equilibrium can be extended to include, among other cases, multiple alleles and multiple loci.

Multiple Alleles

Multinomial Expansion

The expected genotypic array under Hardy-Weinberg equilibrium is p^2, $2pq$, and q^2, which form the terms of the binomial expansion $(p + q)^2$. If males and females each have the same two alleles in the proportions of p and q, then genotypes will be distributed as a binomial expansion in the frequencies p^2, $2pq$, and q^2 (see fig. 19.1). To generalize to more than two alleles, one need only add terms to the binomial expansion and thus create a multinomial expansion. For example, with alleles a, b, and c with frequencies p, q, and r, the genotypic distribution should be $(p + q + r)^2$ or

$$p^2 + 2pq + 2pr + q^2 + 2qr + r^2$$

Homozygotes will occur as the terms p^2, q^2, and r^2, and heterozygotes will occur with frequencies $2pq$, $2pr$, and $2qr$. The ABO blood-type locus in human beings is an interesting example because it has multiple alleles and dominance.

ABO Blood Groups

The ABO locus has three alleles, A, B, and O, in which the A and B alleles are codominant, both dominant to the O allele. These alleles control the production of a surface

TABLE 19.4 ABO Blood-Type Distribution in 500 Persons from Massachusetts

Blood Type	Genotype	Number
A	*AA* or *AO*	199
B	*BB* or *BO*	53
AB	*AB*	17
O	*OO*	231
Total		500

antigen on red blood cells (see fig. 2.13). Table 19.4 contains blood-type data from a sample of five hundred persons from Massachusetts. Is the population in Hardy-Weinberg proportions? The answer is not apparent from the data in table 19.4 alone, since there are two possible genotypes for both the A and the B phenotypes. No estimate of the allelic frequencies is possible without making assumptions about the number of each genotype within these two phenotypic classes. Is it possible to estimate the allelic frequencies? The answer is yes, if we assume that Hardy-Weinberg equilibrium exists.

One procedure is as follows. Let us assume that $p = f(A)$, $q = f(B)$, and $r = f(O)$. Blood type O has the *OO* genotype and, if the population is in Hardy-Weinberg equilibrium, this genotype should occur at a frequency of r^2. Thus

$$f(OO) = 231/500 = 0.462 = r^2$$

and

$$r = f(O) = \sqrt{0.462} = 0.680$$

From table 19.4 we see that blood type A plus blood type O include only the genotypes *AA, AO,* and *OO.* If the population is in Hardy-Weinberg equilibrium, these together should be $(p + r)^2$, in which $p^2 = f(AA)$, $2pr = f(AO)$, and $r^2 = f(OO)$:

$$(p + r)^2 = (199 + 231)/500 = 0.860$$

Then, taking the square root of each side:

$$p + r = \sqrt{0.860} = 0.927$$

and

$$p = 0.927 - r = 0.927 - 0.680 = 0.247$$

The frequency of allele *B, q,* can be obtained by similar logic with blood types B and O, or simply by subtraction:

$$q = 1 - (p + r) = 1 - 0.927 = 0.073$$

Thus the Hardy-Weinberg equilibrium can be extended to include multiple alleles and can be used to obtain estimates of the allelic frequencies in the ABO blood groups. With ABO, it is statistically feasible to do a chi-square test because there is one degree of freedom (number of phenotypes − number of alleles = 4 − 3 = 1). We are really testing only the AB and B categories: If we did our calculations as above, the observed and expected values of phenotypes A and O must be equal.

Multiple Loci

The Hardy-Weinberg equilibrium can also be extended for consideration of several loci at the same time in the same population. This situation deserves mention because the whole genome is likely involved in evolutionary processes and we must, eventually, consider simultaneous allelic changes in all loci segregating alleles in an organism. (Even with a high-speed computer, simultaneous consideration of all loci is a bit far off in the future.) When two loci, A and B, on the same chromosome are in equilibrium with each other, the combinations of alleles on a chromosome in a gamete follow the product rule of probability. Consider the A locus with alleles *A* and *a* and the B locus with alleles *B* and *b,* respectively, with allelic frequencies p_A and q_A for *A* and *a,* respectively, and p_B and q_B for *B* and *b,* respectively. Given completely random circumstances, the chromosome with the *A* and *B* alleles should occur at the frequency $p_A p_B$. This condition is referred to as **linkage equilibrium.** When alleles of different loci are not in equilibrium (i.e., not randomly distributed in gametes), the condition is referred to as **linkage disequilibrium.** The approach to linkage equilibrium is gradual and is a function of the recombination distance between the two loci.

For example, let's start with a population out of equilibrium such that all chromosomes are *AB* (70%) or *ab* (30%). Then $p_A = 0.7$, $q_A = 0.3$, $p_B = 0.7$, and $q_B = 0.3$. We expect the *Ab* chromosome to occur $0.7 \times 0.3 = 0.21$, or 21% of the time. The frequency of the *Ab* chromosome is zero. Assume the map distance between the two loci is 0.1; in other words, 10% of chromatids in gametes are recombinant. Initially, we consider that each locus is in Hardy-Weinberg proportions, or the frequency of *AB/AB* individuals = 0.49 (0.7×0.7); the frequency of *ab/ab* individuals is 0.09 (0.3×0.3); and the frequency of *AB/ab* individuals is 0.42 ($2 \times 0.7 \times 0.3$). After one generation of random mating, gametes will be as follows:

from *AB/AB* individuals (49%): only *AB* gametes, 49% of total

from *ab/ab* individuals (9%): only *ab* gametes, 9% of total from *AB/ab* individuals (42%):

AB gametes, 18.9% of total (0.45 × 0.42)

ab gametes, 18.9% of total (0.45 × 0.42)

Ab gametes, 2.1% of total (0.05 × 0.42)

aB gametes, 2.1% of total (0.05 × 0.42)

(The values of 18.9% and 2.1% from the dihybrids result from the fact that since map distance is 0.1, 10% of gametes will be recombinant, split equally between the two recombinant classes—5% and 5%. Ninety percent will be parental, split equally between the two parental classes—45% and 45%. Each of these numbers must be multiplied by 0.42 because the dihybrid makes up 42% of the total number of individuals.)

Although we expect 21% of the chromosomes to be of the *Ab* type, only 2.1%, 10% of the expected, appear in the gene pool after one generation of random mating. You can see that linkage equilibrium is achieved at a rate dependent on the map distance between loci. Unlinked genes, appearing 50 map units apart, also approach linkage equilibrium gradually.

Although we will not derive these extensions here, we note two others. If the frequencies of alleles at an autosomal locus differs in the two sexes, it takes two generations of random mating to achieve equilibrium. In the first generation, the allelic frequencies in the two sexes are averaged such that each sex now has the same allelic frequencies. Genotype frequencies are then brought into Hardy-Weinberg proportions in the second generation. However, if the allelic frequencies differ in the two sexes for a sex-linked locus, Hardy-Weinberg proportions are established only gradually. The reasoning is straightforward. Females, with an X chromosome from each parent, average the allelic frequencies from the previous generation. However, males, who get their X chromosomes from their mothers, have the allelic frequencies of the females in the previous generation. Hence allelic frequencies are not the same in the two sexes after one generation of random mating, and equilibrium is achieved slowly.

NONRANDOM MATING

The Hardy-Weinberg equilibrium is based on the assumption of random mating. Deviations from random mating come about when phenotypic resemblance or relatedness influence mate choice. When phenotypic resemblance influences mate choice, either *assortative* or *disassortative* mating occurs, depending on whether individuals choose mates on the basis of similarity or dissimilarity, respectively. For example, in human beings, there is assortative mating for height—short men tend to marry short women and tall men tend to marry tall women. When relatedness influences mate choice, either *inbreeding* or *outbreeding* occur, depending on

whether mates are more or less related than two randomly chosen individuals from the population. An example of inbreeding in human beings is the marriage of first cousins. Both types of nonrandom mating (assortative-disassortative mating and inbreeding-outbreeding) have the same qualitative effects on the Hardy-Weinberg equilibrium: Assortative mating and inbreeding increase homozygosity without changing allelic frequencies, whereas disassortative mating and outbreeding increase heterozygosity without changing allelic frequencies.

Two differences are apparent, however, between the effects of phenotypic resemblance and relatedness on mate choice. First, assortative or disassortative mating will disturb the Hardy-Weinberg equilibrium only when the phenotype and genotype are closely related. That is, if assortative mating occurs for a nongenetic trait, then there will be no distortions of the Hardy-Weinberg equilibrium. Inbreeding and outbreeding affect the genome directly. A second difference between the two types of mating is that the effects of inbreeding or outbreeding are felt across the whole genome, whereas the disturbances to the Hardy-Weinberg equilibrium by assortative and disassortative mating occur only for the particular trait being considered (and closely linked loci). Given the similarities in the consequences of the two types of matings, we will concentrate our discussion on inbreeding.

Inbreeding

Inbreeding comes about in two ways: (1) the systematic choice of relatives as mates and (2) the subdivision of a population into small subunits within which individuals have little choice but to mate with relatives. We will concentrate on inbreeding as the systematic choice of relatives as mates. The consequences of both are similar.

Common Ancestry

An inbred individual is one whose parents are related—that is, there is **common ancestry** in the family tree. The extent of inbreeding is thus a function of the degree of common ancestry shared by the parents of an inbred individual. When mates share ancestral genes, each may pass on copies of the same ancestral allele to their offspring. An inbred individual can then carry identical copies of a single ancestral allele. In other words, an individual of *aa* genotype is homozygous and, if it is possible that the *a* allele from each parent is a length of DNA originally copied from the same DNA of a common ancestor, the *aa* individual is said to be inbred.

The first observable effect of inbreeding is the expression of hidden recessives. In human beings, each individual carries, on the average, about four **lethal-equivalent alleles,** alleles that kill when paired to form a homozygous genotype (see Box 19.2). In many, probably most, human societies, zygotes are protected from these lethal alleles

THE DETERMINATION OF LETHAL EQUIVALENTS

The average person carries about four lethal-equivalent alleles that are hidden as recessive alleles. Four lethal equivalents means four alleles that are lethal when homozygous, or eight alleles conferring a 50% chance of mortality when homozygous, or any similar combination of lethal and semilethal alleles. The exact arrangement cannot be determined with current analytical methods. The estimate of hidden defective and lethal alleles is arrived at by using inbreeding data.

Crow and Kimura, in 1970, analyzed data showing that in Swedish families in which marriages occurred between first cousins, between 16 and 28% of the offspring had genetic diseases. For unrelated parents the comparable figure is between 4 and 6%. It is, therefore, estimated that the offspring of first cousins have an added risk of 12 to 22% of having a genetic defect. The children of first cousins have an inbreeding coefficient of 1/16. Hence a theoretical individual who is completely inbred has the risk of genetic defect increased sixteenfold over an individual whose parents are first cousins. If 100% risk is considered 1 lethal equivalent, then a completely inbred individual would carry 2 to 3.5 lethal equivalents (16 × 12%—16 × 22%). However, a completely inbred individual is, in essence, a doubled gamete. Since our interest is in the number of deleterious alleles carried by a normal person, it is necessary to further multiply the risk by a factor of 2 to determine the number of lethal-equivalent alleles carried by a normal individual. The conclusion is that the average person carries the equivalent of four to seven alleles that would, in the homozygous state, cause a genetic defect.

A similar calculation can be made using viability data rather than genetic defects to determine the occurrence of lethal equivalents. A study from rural France (see Crow and Kimura, 1970) showed that the mortality rate of offspring of first cousins was 25%, whereas the analogous figure for the offspring of unrelated parents was about 12%, an increased risk of 13% for the offspring of cousins. Multiplying this risk figure of 0.13 by 32 (16 × 2) presents a figure of four lethal equivalents per average person in the population. In 1971, Cavalli-Sforza and Bodmer, using data primarily from Japanese populations, reported an estimate of about two lethal equivalents per average person. Despite some interpopulation differences in these estimates, they are about the same order of magnitude—two to seven lethal equivalents per person.

by a cultural pattern of outbreeding, mating with non-relatives. Rarely does an outbred zygote receive the same recessive lethal from each parent. Dominance acts to mask the expression of deleterious recessive alleles. But, in the process of inbreeding, during which the zygote may receive copies of the same ancestral allele from each parent, there is a substantial increase in the probability that a deleterious allele will pair to form a homozygous genotype (fig. 19.2). Inbreeding can result in spontaneous abortions (miscarriages), fetal deaths, and congenital deformities. In many species, however, inbreeding—even self-fertilization—occurs normally. These species usually do not have the problem with lethal equivalents that normally outbreeding species do. Through time, species that normally have inbreeding have had these deleterious alleles mostly eliminated, presumably by natural selection. Inbreeding has even been used successfully in artificial selection regimes in livestock and crop plants.

From our previous discussion, you can see that there are two types of homozygosity—**allozygosity,** in which two alleles are alike but unrelated (not copies of the same ancestral allele) and **autozygosity,** in which two alleles have **identity by descent** (i.e., are copies of the same ancestral allele). An **inbreeding coefficient, F,** can be defined as the

Figure 19.2

Homozygosity by descent of copies of the same ancestral allele, *a.* The individual at the bottom of the pedigree is inbred with the *aa* genotype.

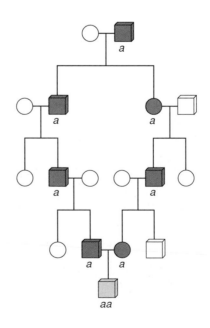

TABLE 19.5 Genotypic Proportions in a Population with Inbreeding

Genotype	Due to Random Mating $(1 - F)$		Due to Inbreeding (F)		Observed Proportions
AA	$p^2(1 - F)$	+	pF	=	$p^2 + Fpq$
Aa	$2pq(1 - F)$			=	$2pq(1 - F)$
aa	$q^2(1 - F)$	+	qF	=	$q^2 + Fpq$
TOTAL	$(p^2 + 2pq + q^2)(1 - F)$	+	$(p + q)F$	=	
	$(1 - F)$	+	F	=	1

probability of autozygosity, the probability that the two alleles in an individual at a given locus are identical by descent. This coefficient can range from zero, at which point there is no inbreeding, to 1, at which point an individual is autozygous with certainty.

Increased Homozygosity from Inbreeding

What are the effects of inbreeding on the Hardy-Weinberg equilibrium? Let us for a moment return to the gene pool concept to produce zygotes. Assume that an allele drawn from this gene pool is of the A type, drawn with a probability of p. On the second draw, the probability of autozygosity, that is, drawing a copy of the same allele A, is F, the inbreeding coefficient. Thus the probability of an autozygous AA individual is pF. On the second draw, however, with probability $(1 - F)$, either the A or a allele can be drawn, with probabilities of $p^2(1 - F)$ and $pq(1 - F)$, respectively. Note that a second A allele here produces a homozygote that is not inbred (allozygous). If the first allele drawn was an a allele, with probability q, then the probability of drawing the same allele (copy of the same ancestral allele) is F and thus the probability of autozygosity is qF. However, the probability of drawing an a or A allele that does not contribute to inbreeding is $(1 - F)$ and, therefore, the probability of an aa or Aa genotype is $q^2(1 - F)$ and $pq(1 - F)$, respectively. These calculations are summarized in table 19.5, a summary of the genotypic proportions in a population with inbreeding.

Several points emerge from table 19.5. First, when the inbreeding coefficient is zero (complete random mating), the table reduces to Hardy-Weinberg proportions. Second, compared with Hardy-Weinberg proportions, inbreeding increases the proportion of homozygotes in the population (identity by descent implies homozygosity). With complete inbreeding ($F = 1$), only homozygotes will occur in the population.

How does inbreeding affect allelic frequencies? Recall that an allelic frequency is calculated as the frequency of homozygotes for one allele plus half the frequency of the

heterozygotes. Here we let p_{n+1} be the frequency of the A allele after one generation of inbreeding:

$$p_{n+1} = p^2(1 - F) + pF + (1/2)(2pq)(1 - F)$$
$$= p^2(1 - F) + pF + pq(1 - F)$$
$$= p^2 + pq + F(p - p^2 - pq)$$
$$= p(p + q) + pF(1 - p - q)$$
$$= p(1) + pF(0)$$
$$= p$$

Thus inbreeding does not change allelic frequencies. We can also see intuitively that inbreeding affects zygotic combinations (genotypes) but not allelic frequencies: Although inbreeding may determine the genotypes of offspring, inbreeding does not change the number of each allele that an individual transmits into the next generation.

In summary, inbreeding causes an increase in homozygosity, affects all loci in a population equally, and, by itself, has no effect on allelic frequencies, although it can expose deleterious alleles to selection. The results of inbreeding can be seen by the appearance of recessive traits that are often deleterious. Inbreeding increases the rate of fetal deaths and congenital malformations in human beings and in other species that are normally outbred. In outbred agricultural crops and farm animals, decreases in size, fertility, vigor, and yield often result from inbreeding. Once deleterious traits appear due to inbreeding, natural selection can cause their removal from the population. However, in species adapted to inbreeding, including many crop plants and farm animals, inbreeding does not expose deleterious alleles because those alleles have generally been eliminated already.

Figure 19.3

Conversion of a pedigree of the mating of first cousins to a path diagram. All extraneous individuals are removed, leaving only those who could contribute to the inbreeding of individual I. Individuals in the line of descent are connected directly with *straight lines,* indicative of the paths in which gametes are passed.

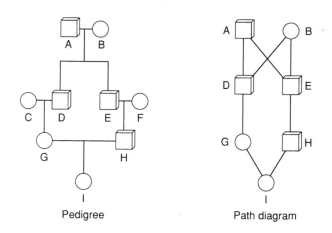

Pedigree   Path diagram

Figure 19.4

The path diagram of the mating of first cousins with gametes labeled in *lowercase letters*

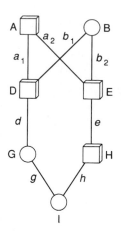

Pedigree Analysis

Path Diagram Construction

The inbreeding coefficient, *F*, of an individual (the probability of autozygosity) can be determined by pedigree analysis. It is determined by converting a pedigree to a **path diagram** by eliminating all extraneous individuals, those who cannot contribute to the inbreeding coefficient of the individual in question. A path diagram shows the direct line of descent from common ancestors. An example of the conversion of a pedigree to a path diagram is shown in figure 19.3, in which individuals C and F are omitted from the path of descent because they are not related to anyone on the other side of the family tree and, therefore, do not contribute to the "common ancestry" of I. The pedigree in figure 19.3 shows an offspring who is the daughter of first cousins. Since first cousins are the offspring of siblings, they share a set of common grandparents. Thus individual I can be autozygous for alleles from either ancestor A or B, her great-grandparents. The path diagram shows the only routes by which autozygosity can occur.

The inbreeding coefficient of the offspring of first cousins can be calculated as follows. The path diagram of figure 19.3 is shown again in figure 19.4, in which the lowercase letters designate gametes. There are two paths of autozygosity in this diagram, one path for each grandparent as a common ancestor: A to D and E, then to G and H, and finally to I; or B to D and E, then to G and H, and finally to I.

In the path with A as the common ancestor, A contributes a gamete to D and a gamete to E. The probability is 1/2 that D and E each carry a copy of the same allele. That is, there are four possible allelic combinations for the two gametes, a_1 and a_2: *A-A; A-a; a-A;* and *a-a*. Of these combinations, the first and last (*A-A* and *a-a*) give a copy

of the same allele to the two offspring, D and E, and can thus contribute to autozygosity. The probability that gametes a_1 and *d* carry copies of the same allele is 1/2, and the probability that *d* and *g* carry copies of the same allele is also 1/2. Similarly, on the other side of the pedigree, the probability is 1/2 that a_2 and *e* carry copies of the same allele and is 1/2 that *e* and *h* carry copies of the same allele. Thus the overall probability that the alleles carried by *g* and *h* are identical by descent (autozygous) is $(1/2)^5$. In general it would be $(1/2)^n$, where *n* is the number of ancestors in the path.

The reader may have spotted an additional factor here. Of the possible combinations of allelic copies passed on to D and E, one half (*A-A* and *a-a*) are autozygous combinations. However, the other half of combinations, *A-a* and *a-A* can lead to autozygosity if A is itself inbred. If we let F_A be the inbreeding coefficient of A (the probability that any two alleles at a locus in A are identical by descent), then F_A is the probability that the *A-a* and *a-A* combinations are also autozygous. Thus the probability that a common ancestor, A, passes on copies of an identical ancestral allele is $1/2 + (1/2)F_A$, or $(1/2)(1 + F_A)$. In other words, there is a 1/2 probability that the alleles transmitted from A to D and E are copies of the same allele. In the other half of the cases, these alleles can be identical if A is inbred. The probability of identity of A's two alleles is F_A. The expression for the inbreeding coefficient of I, F_I, can now be changed from $(1/2)^n$ by substituting $(1/2)(1 + F_A)$ for one of the $(1/2)$s to

$$F_I = (1/2)^n(1 + F_A)$$

This equation accounts only for the inbreeding of I by the path involving the common ancestor, A, and does not account for the symmetrical path with B as the common ancestor. To obtain the total probability of inbreeding, the values from each path must be added (mutually exclusive events, see chapter 4). Thus the complete formula for the inbreeding coefficient of the offspring of first cousins is

$$F_I = \Sigma[(1/2)^n(1 + F_J)] \qquad (19.1)$$

in which F_I is the probability that the two alleles in I are identical by descent, n is the number of ancestors in a given path, F_J is the inbreeding coefficient of the common ancestor of that path, and all paths are summed.

In the example of the mating of first cousins (fig. 19.4):

$$F_I = (1/2)^5(1 + F_A) + (1/2)^5(1 + F_B)$$

If we assume that F_A and F_B are zero (which we must assume when the pedigrees of A and B are unknown), then:

$$F_I = 2(1/2)^5 = (1/2)^4 = 0.0625$$

which can be interpreted to mean that about 6.25% of individual I's loci are autozygous or that there is a 6.25% chance of autozygosity at any one of I's loci.

The inbreeding coefficient of the offspring of siblings (fig. 19.5) can be calculated, assuming that A and B are not themselves inbred (F_A and F_B are zero), as

$$F_I = 2(1/2)^3 = 0.25$$

Thus, about 25% of the loci in an offspring of siblings are autozygous.

Path Diagram Rules
The following points should be kept in mind when an inbreeding coefficient is calculated:

1. All possible paths must be counted. A path is possible if gametes can actually pass in that direction. Paths that violate the rules of inheritance cannot be used. For example, in figure 19.4, the following path is unacceptable: I G E A D H I.

2. In any path, an individual can be counted only once.

3. Every path must have one and only one common ancestor. The inbreeding coefficient of any other individual in the path is immaterial.

In figure 19.6, we present a complex pedigree produced from repeated sib mating, a pattern found in livestock and laboratory animals. This pedigree has several interesting points. First, there are common ancestors in several different generations. Second, some of the paths are complex. Thus we must be sure to count all paths (paths 5 and 6 might not be immediately obvious). Third,

Figure 19.5

Conversion of a sib-mating pedigree to a path diagram. Individual I is inbred.

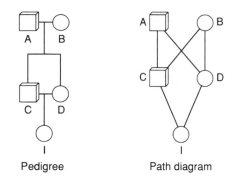

Pedigree   Path diagram

although not shown in figure 19.6, one of the common ancestors, A, is also inbred ($F_A = 0.05$)—a fact that we must take into consideration in paths 3 and 5. Thus, F_I is as follows:

From path 1 $(1/2)^3$	$= 0.1250$
From path 2 $(1/2)^3$	$= 0.1250$
From path 3 $(1/2)^5(1 + 0.05)$	$= 0.0328$
From path 4 $(1/2)^5$	$= 0.0313$
From path 5 $(1/2)^5(1 + 0.05)$	$= 0.0328$
From path 6 $(1/2)^5$	$= 0.0313$

$$F_I = 0.3781$$

Population Analysis

It is also possible to define the inbreeding coefficient, F, of a population as the relative reduction in heterozygosity in the population due to inbreeding. In an individual, F is the probability of autozygosity; it is an increase in homozygosity, which is therefore a decrease in heterozygosity. In a population, it is also the reduction in heterozygosity. From the definition, we can calculate the population F as follows:

$$F = \frac{(2pq - H)}{2pq}$$

where H is the actual proportion of heterozygotes in a population and $2pq$ is the expected proportion of heterozygotes based on Hardy-Weinberg proportions. This equation reduces to:

$$F = 1 - \frac{H}{2pq} \qquad (19.2)$$

Figure 19.6

Pedigree and path diagram of two generations of sib matings. The six paths involving the potential for autozygosity are shown. $F_A = 0.05$. The paths involve common ancestors in two generations.

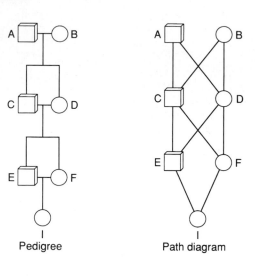

Pedigree   Path diagram

Paths

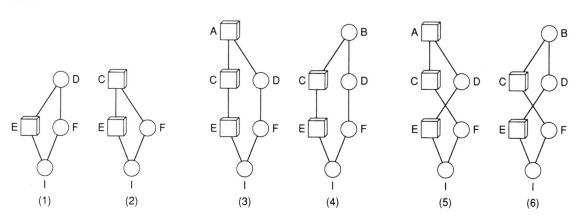

(1)   (2)   (3)   (4)   (5)   (6)

This equation shows that when $H = 2pq$, F is zero, the case when there is no decrease in heterozygotes and therefore, apparently, no inbreeding. When there are no heterozygotes, $F = 1$. This could be the case of a completely inbred population, for example, a self-fertilizing plant species.

As an example of an intermediate case, take the sample of one hundred individuals segregating the A_1 and A_2 alleles at the A locus: A_1A_1, 54; A_1A_2, 32; and A_2A_2, 14. In this example, $p = 0.7$, $q = 0.3$, and $H = 0.32$. Since $2pq = 0.42$, $H/2pq = 0.32/0.42 = 0.76$, and $F = 1 - 0.76$, or 0.24. Thus the inbreeding coefficient of this population is 0.24; there is a 24% reduction in heterozygotes, due presumably to inbreeding.

SUMMARY

In a large, randomly mating population of sexually reproducing diploid organisms, without the influence of mutation, migration, or selection, an equilibrium will be achieved for an autosomal locus with two alleles. This Hardy-Weinberg equilibrium predicts that (1) allelic frequencies (p, q) will not change from generation to generation; (2) genotypes will occur according to the binomial distribution $p^2 = f(AA)$, $2pq = f(Aa)$, and $q^2 = f(aa)$; and (3) if perturbed, equilibrium will be reestablished in just one generation of random mating.

To determine whether a population is in Hardy-Weinberg proportions, the observed and expected distribution of genotypes can be compared by the chi-square statistical test. In some circumstances, in which it is reasonable to assume equilibrium, allelic and genotypic frequencies can be estimated, even when dominance occurs. The Hardy-Weinberg equilibrium is easily extended to prediction of the frequencies of multiple alleles, multiple loci, and different frequencies of alleles in the two sexes, for both sex-linked and autosomal loci.

Random mating is required for the Hardy-Weinberg equilibrium to hold. Deviations from random mating fall into two categories, depending on whether phenotypic resemblance or relatedness is involved in mate choice. Phenotypic resemblance is the basis for assortative and disassortative mating, in which individuals choose similar or dissimilar mates, respectively. Assortative mating causes increased homozygosity only among loci controlling the traits for which mate choice is made. There are no changes in allelic frequencies. Similarly, disassortative mating causes increased heterozygosity without changing allelic frequencies.

Mating among relatives is termed inbreeding and is represented by F, the inbreeding coefficient, which measures the probability of autozygosity (homozygosity by descent). It can be calculated from pedigrees by using the formula:

$$F = \Sigma[(1/2)^n(1 + F_J)]$$

where n is the number of ancestors in a given path and F_J is the inbreeding coefficient of the common ancestor of that path. Inbreeding exposes recessive deleterious traits already present in the population and causes homozygosity throughout the genome. It does not, by itself, change allelic frequencies. F can also be calculated from the reduction in heterozygosity in a population.

SOLVED PROBLEMS

1. One hundred fruit flies (*Drosophila melanogaster*) from California were tested for their genotype at the transferrin (blood protein) locus using starch-gel electrophoresis. There were two alleles present, S and F, for slow and fast migration, respectively. The following results were noted: SS, 66; SF, 20; FF, 14. What are allelic and genotypic frequencies in this population?

   ANSWER: Since the sample size is 100, the proportions of the three genotypes, SS, SF, and FF, are: 0.66, 0.20, and 0.14, respectively. We can calculate allelic frequencies directly from these genotypes, remembering that the frequency of an allele is the frequency of its homozygote plus half the frequency of the heterozygote, or:

   $$p = f(S) = f(SS) + (1/2)f(SF) =$$
   $$0.66 + (1/2)(0.20) = 0.76$$
   $$q = f(F) = f(FF) + (1/2)f(FS) =$$
   $$0.14 + (1/2)(0.20) = 0.24$$

   Alternatively, we could get allelic frequencies by counting alleles. Thus,

   $$p = \frac{2 \times \text{number of } SS + \text{number of } SF}{2 \times \text{total number}} = \frac{2(66) + 20}{2(100)} = \frac{152}{200} = 0.76$$
   $$q = \frac{2 \times \text{number of } FF + \text{number of } SF}{2 \times \text{total number}} = \frac{2(14) + 20}{2(100)} = \frac{48}{200} = 0.24$$

2. Is the population described in question 1 in Hardy-Weinberg equilibrium?

   ANSWER: We can determine whether the numbers of the three genotypes (SS, SF, and FF) are in Hardy-Weinberg proportions with the chi-square statistical test. The observed numbers of the three genotypes are 66, 20, and 14, respectively. Using allelic frequencies of $p = f(S) = 0.76$ and $q = f(F) = 0.24$, we expect p^2, $2pq$, and q^2, respectively, of the three genotypes. That is:

   $$p^2 = 0.76^2 = 0.578, \text{ or } 57.8 \text{ in } 100$$
   $$2pq = 2(0.76)(0.24) = 0.365, \text{ or } 36.5 \text{ in } 100$$
   $$q^2 = 0.24^2 = 0.058, \text{ or } 5.8 \text{ in } 100$$

   We can now set up a chi-square table as follows (see next page):

	SS	SF	FF	Total
Observed Numbers	66	20	14	100
Expected Proportions	p^2	$2pq$	q^2	1.0
	(0.578)	(0.365)	(0.058)	1.0
Expected Numbers	57.8	36.5	5.8	100
$\chi^2 = (O - E)^2/E$	1.163	7.459	11.593	20.215

The critical chi-square value (0.05 at 1 degree of freedom) is 3.841, so we reject the hypothesis that this population is in Hardy-Weinberg proportions. From inspection of the table, it appears that there are too few heterozygotes and too many homozygotes, indicating that inbreeding could be the cause of the discrepancy.

3. Convert the pedigree in figure 19.2 into a path diagram and determine the inbreeding coefficient of the inbred individual, assuming that the common ancestors are not themselves inbred.

ANSWER: There are two paths, each with seven ancestors. Thus the inbreeding coefficient is:

$$F = \Sigma[(1/2)^n(1 + F_J)] = 2(1/2)^7 = 0.016$$

Hence, the inbreeding coefficient is 0.016, or about 1.6% of the loci of the inbred individual are autozygous.

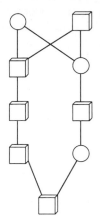

EXERCISES AND PROBLEMS

1. One hundred persons from a small town in Pennsylvania were tested for their MN blood types. Is the population they represent in Hardy-Weinberg proportions? The genotypic data are: *MM*, 41; *MN*, 38; and *NN*, 21.

2. In the following two sets of data, calculate allelic and genotypic frequencies and determine whether the populations are in Hardy-Weinberg proportions. Do a statistical test if one is appropriate.
   a. Allele *A* is dominant to *a: A−*, 91; *aa*, 9.
   b. Electrophoretic alleles *F* and *S* are codominant at the alcohol dehydrogenase locus in *Drosophila*: *FF*, 137; *FS*, 196; *SS*, 87.

3. The dominant ability to taste PTC is due to the allele *T*. Among a sample of 215 individuals from a population, 150 could detect the taste of PTC and 65 could not. Calculate the allelic frequencies of *T* and *t*. Is the population in Hardy-Weinberg proportions?

4. The frequency of children homozygous for the recessive allele for cystic fibrosis is about one in 2,500. What is the percentage of heterozygotes in the population?

5. The following data are ABO phenotypes from a population sample of one hundred persons. Determine the frequencies of the three alleles: type A, 7; type B, 72; type AB, 12; type O, 9. What did you have to assume? Is the population in Hardy-Weinberg proportions?

6. How quickly and in what manner is Hardy-Weinberg equilibrium achieved under the following initial conditions (assuming a diploid, sexually reproducing population)?
   a. One locus, five alleles
   b. Two loci, two alleles each, not linked

7. A group of fruit flies was testcrossed to determine allelic arrangements of two linked loci in the gametes of that generation. With the following data, can you determine whether linkage equilibrium holds? Gametic arrangements are: *AB*, 58; *ab*, 8; *Ab*, 12; and *aB*, 22.

8. PTC tasting is dominant in human beings.
   a. Should most human populations be heading toward a 3:1 ratio of tasters to nontasters? Explain.
   b. Confronted with a population sample of human beings of unknown origin, would you expect more or less than half the sample to be tasters?

9. Under what circumstances is inbreeding deleterious?

10. What is the inbreeding coefficient of I in the following pedigree? Assume that the inbreeding coefficients of other members of the pedigree are zero unless other information tells you differently.

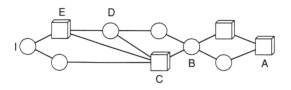

11. What is the inbreeding coefficient of individual I?
$F_A = 0.01$; $F_B = 0.02$; $F_C = 0.02$.

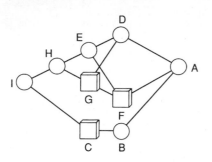

12. The following is the pedigree of a child produced by a mating of half siblings. Individuals A and C have inbreeding coefficients of 0.2; all others are zero. Convert the pedigree to a path diagram and determine the inbreeding coefficient of individual G.

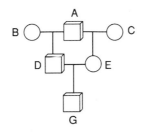

13. Given the population in problem 1 of this chapter, what is its inbreeding coefficient?

14. a. Prove that the perpendicular to the base divides the base of the De Finetti diagram in the ratio of p:q.
    b. Prove that populations in Hardy-Weinberg equilibrium form a parabola on the De Finetti diagram.

15. A particular recessive disorder is present in one in 10,000 individuals. If the population is in equilibrium, what are the frequencies of the two alleles?

16. What allelic frequency will generate twice as many recessive homozygotes as heterozygotes?

17. Attack or defend mathematically the following statement: "Assume brown eye color is the result of a dominant allele at one locus. With time, the frequency of brown-eyed individuals will increase, until about three out of four individuals are brown-eyed."

18. A particular human population has 500 *MM* individuals, 300 *MN*, and 700 *NN*. Calculate the allelic frequencies and determine whether the population is in equilibrium.

19. Assume random mating occurs among the individuals of the population described in problem 18. What will be the frequencies of each type of individual in the next generation?

20. On a small island, 235 mating individuals are all true-breeding for brown eyes. An epidemic eliminates all the population except 10 young women, two young men, and four elderly women. A boatload of foreigners arrive; the foreign population consists of six heterozygous brown-eyed females, four homozygous brown-eyed males, and 10 blue-eyed males. Assuming that eye color is controlled by one locus, that mating is random with respect to eye color, and that each male and female capable of breeding does so, calculate the allelic frequencies of their offspring.

21. In a large, randomly mating human population, the frequencies of the *A, B,* and *O* alleles are 0.7, 0.2 and 0.1, respectively. Calculate the expected frequencies for each blood type.

22. In a given population, only the *A* and *B* alleles are present in the ABO system; there are no individuals with type O blood or with *O* alleles. If 200 people have type A blood, 75 have type AB blood, and 25 have type B blood, what are the allelic frequencies in this population?

23. In a human population of 100 people, 17 have type A blood, 15 have type B, 4 have type AB, and 64 have type O. If this population is in equilibrium, what are the allelic frequencies?

24. In a sample of 100 people, there are 21 *MM*, 38 *MN*, and 41 *NN* individuals. Calculate the inbreeding coefficient.

25. If, in a population with two alleles at an autosomal locus, $p = 0.7$, $q = 0.3$, and the frequency of heterozygotes is 0.32, what is the inbreeding coefficient?

Suggested Readings for chapter 19 are on page 659.

20

Population Genetics: Processes that Change Allelic Frequencies

Natural selection works on the variation found in nature, here shown by differences in *spot patterns* on the wing covers of ladybird beetles, *Hippodamia convergens.*
© William E. Ferguson

This chapter is devoted to a discussion of some of the effects of violating, or relaxing, the assumptions of the Hardy-Weinberg equilibrium other than random mating, which we discussed in chapter 19. Here we consider the effects of mutation, migration, small population size, and natural selection on the Hardy-Weinberg equilibrium. These processes usually result in changes in allelic frequencies.

MODELS FOR POPULATION GENETICS

The steps to be taken in solving for equilibrium in population genetic models follows the same general pattern regardless of what model we are analyzing. We emphasize that these models are developed to help us understand the genetic changes taking place in a population. The models help us to understand nonintuitive processes and to quantify intuitive processes. The models can be outlined as follows:

1. Set up an algebraic model.

2. Calculate allelic frequency in the next generation, q_{n+1}.

3. Calculate change in allelic frequency between generations, Δq.

4. Calculate the equilibrium condition, $\hat{q}$ (q-hat), at $\Delta q = 0$.

5. Determine, when feasible, if the equilibrium is stable.

MUTATION

Mutational Equilibrium

Mutation affects the Hardy-Weinberg equilibrium by changing one allele to another and thus changing allelic and genotypic frequencies. Consider a simple model in which two alleles A and a exist. A mutates to a at a rate μ (mu), and a mutates back to A at a rate of ν (nu):

$$A \underset{\nu}{\overset{\mu}{\rightleftharpoons}} a$$

If p_n is the frequency of A in generation n and q_n is the frequency of a in generation n, then the new frequency of a, q_{n+1}, is the old frequency of a plus the addition of a alleles from forward mutation and the loss of a alleles by back mutation. That is,

$$q_{n+1} = q_n + \mu p_n - \nu q_n \qquad (20.1)$$

in which μp_n is the increment of a alleles added by forward mutation and νq_n is the loss of a alleles due to back mutation. Equation 20.1 takes into account not only the rate of forward mutation, μ, but also p_n, the frequency of A alleles available to mutate. Similarly, the loss of a alleles to A alleles is the product of both the rate of back mutation, ν, and the frequency of the a allele, q_n. Equation 20.1 completes the second modeling step, derivation of an expression for q_{n+1}, allelic frequency after one generation of mutation pressure. The third step is to derive an expression for change in allelic frequency between two generations. This change (Δq) is simply the difference between the allelic frequency at generation $n + 1$ and the allelic frequency at generation n. Thus, for the a allele:

$$\Delta q = q_{n+1} - q_n = (q_n + \mu p_n - \nu q_n) - q_n \quad (20.2)$$

which simplifies to

$$\Delta q = \mu p_n - \nu q_n \qquad (20.3)$$

The next step in the model is to calculate the equilibrium condition $\hat{q}$, which is the allelic frequency that occurs when there is no change in allelic frequency from one generation to the next— that is, when Δq (equation 20.3) is equal to zero:

$$\Delta q = \mu p_n - \nu q_n = 0 \qquad (20.4)$$

Thus

$$\mu p_n = \nu q_n \qquad (20.5)$$

Then, substituting $(1 - q_n)$ for p_n (since $p = 1 - q$), gives

$$\mu(1 - q_n) = \nu q_n$$

or by rearranging

$$\hat{q} = \frac{\mu}{\mu + \nu} \qquad (20.6)$$

And, since $p + q = 1$,

$$\hat{p} = \frac{\nu}{\mu + \nu} \qquad (20.7)$$

We can see from equations 20.6 and 20.7 that an equilibrium of allelic frequencies does exist. Also, the equilibrium value of allele a ($\hat{q}$) is directly proportional to the relative size of μ, the rate of forward mutation toward a.

Figure 20.1

Types of equilibria: stable, unstable, and neutral

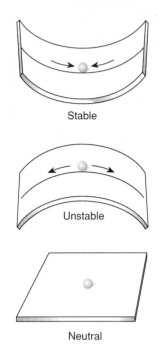

Stable

Unstable

Neutral

If $\mu = \nu$, the equilibrium frequency of the a allele ($\hat{q}$) will be 0.5. As μ gets larger, the equilibrium value shifts toward higher frequencies of the a allele.

Stability of Mutational Equilibrium

Having demonstrated that allelic frequencies can reach an equilibrium due to mutation, we can ask whether the mutational equilibrium is stable. A stable equilibrium is one that returns to the original equilibrium point after it has been perturbed. An unstable equilibrium is one that will not return after being perturbed but, rather, will continue away from the equilibrium point. As we mentioned in the last chapter, the Hardy-Weinberg equilibrium is a neutral equilibrium; it remains at the allelic frequency to which it is moved when perturbed.

Stable, unstable, and neutral equilibrium points can be visualized as marbles in the bottom of a concave surface (stable), on the top of a convex surface (unstable), or on a level plane (neutral; fig. 20.1). Although more sophisticated mathematical ways exist for determining whether an equilibrium is stable, unstable, or neutral, we will use graphical analysis for this purpose.

Figure 20.2 introduces the process of graphical analysis, whereby an understanding of the dynamics of an event or process can be obtained by the representation of the event in graphical form. In figure 20.2, we have graphed equation 20.3, the Δq equation of mutational dynamics. The ordinate, or y axis, is Δq, change in allelic frequency. The abscissa, or x axis, is q, or allelic frequency. The diagonal line is the Δq equation, the relationship of Δq and

Figure 20.2

Graphical analysis of mutational equilibrium. The graph of the mutational Δq equation showing that when the population is perturbed from the equilibrium point ($\hat{q} = 0.167$), it returns to that equilibrium point. At q values above equilibrium, change is negative, tending to return the population to equilibrium. At q values below equilibrium, change is positive, also tending to return the population to equilibrium.

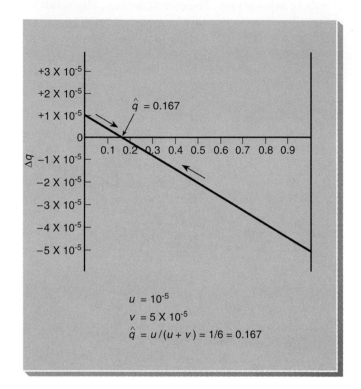

q. Note that Δq can be positive (q increasing) or negative (q decreasing) whereas q is always positive (0–1.0). Graphical analysis has the potential to provide insights into the dynamics of many processes in population genetics.

The diagonal line in figure 20.2 crosses the $\Delta q = 0$ line at the equilibrium value ($\hat{q}$) of 0.167. This line also shows us the changes in allelic frequency that will occur in a population not at the equilibrium point. We will look at two examples of populations under the influence of mutation pressure but not at equilibrium: one at $q = 0.1$ (below equilibrium) and one at $q = 0.9$ (above equilibrium).

If we substitute $q = 0.1$ into equation 20.3, we get a Δq value of 4×10^{-6}. If we substitute $q = 0.9$ into the equation, we get a Δq value of -4.4×10^{-5}. In other words, when the population is below equilibrium, q will increase ($\Delta q = +4 \times 10^{-6}$); if the population is above equilibrium, q will decrease ($\Delta q = -4.4 \times 10^{-5}$). These same conclusions can be read directly from the graph in figure 20.2.

We can see that the mutational equilibrium is a stable one. Any population whose allelic frequency is not at the equilibrium value will tend to return to that equilibrium value. A shortcoming of this model is that there is no obvious information revealing the time frame for reaching equilibrium. To derive equations to determine this parameter is beyond our scope here. (We could use computer simulation or integrate equation 20.3 with respect to time.) In a large population, any great change in allelic frequency by mutation pressure alone takes an extremely long time. Most mutation rates are on the order of 10^{-5}, and equation 20.3 shows that change will be very slow with values of this magnitude. For example, if $\mu = 10^{-5}$, $\nu = 10^{-6}$, and $p = q = 0.5$, $\Delta q = (0.5 \times 10^{-5}) - (0.5 \times 10^{-6}) = 4.5 \times 10^{-6}$, or 0.0000045. It takes, usually, thousands of generations to get near equilibrium, which is approached asymptotically.

As you can see from the low values of mutation rates, perturbations to the Hardy-Weinberg equilibrium by mutation in any one generation usually cannot be detected. Mutation rate can, however, determine the eventual allelic frequencies at equilibrium if no other factors act to perturb the gradual changes that mutation rates cause. Mutation can also affect final allelic frequencies when it restores alleles that natural selection is removing, a situation we will discuss at the end of the chapter. More importantly, mutation provides the alternative alleles upon which natural selection acts.

MIGRATION

Migration is similar to mutation in the sense that allelic frequencies are changed by adding or removing alleles. Human populations are frequently influenced by migration.

Assume two populations, both containing alleles A and a at the A locus, but at different frequencies (p_1 and q_1 versus p_2 and q_2), as shown in figure 20.3. Assume that a group of individuals moves from population 2 and joins population 1 and further that this group of migrants makes up a fraction m (e.g., 0.2) of the new conglomerate population. Thus the old residents, or natives, will make up a fraction $(1 - m;$ e.g., 0.8) of the combined population. The conglomerate a-allele frequency, q_c, will be the weighted average of the allelic frequencies of the natives and migrants (the allelic frequencies weighted by—multiplied by—their proportions):

$$q_c = mq_2 + (1 - m)q_1 \tag{20.8}$$

$$q_c = q_1 + m(q_2 - q_1) \tag{20.9}$$

The change in allelic frequency, a, from before to after the migration event is

$$\Delta q = q_c - q_1 = [q_1 + m(q_2 - q_1)] - q_1 \tag{20.10}$$

$$\Delta q = m(q_2 - q_1) \tag{20.11}$$

Figure 20.3

Diagrammatic view of migration. A group of migrants enters a native population, making up a proportion, m, of the final conglomerate population.

From Wallace, Population Genetics, *BSCS pamphlet 12. Copyright © 1968 BSCS, Colorado Springs, CO.*

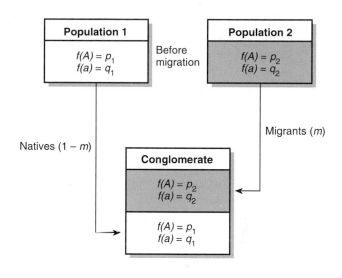

We then find the equilibrium value, $\hat{q}$ (at $\Delta q = 0$). Remembering that in a product series, any multiplier of zero makes the whole expression zero, Δq will be zero when either

$$m = 0 \quad \text{or} \quad q_2 - q_1 = 0; q_2 = q_1$$

The conclusions to be drawn from this model are intuitive. Migration can upset the Hardy-Weinberg equilibrium. Allelic frequencies in a population under the influence of migration will not change if either the size of the migrant group drops to zero (m, the proportion of the conglomerate made up of migrants, drops to zero) or the allelic frequencies in the migrant and resident groups are identical.

This migration model can be used to determine the degree to which alleles from one population have entered another population. It can be applied for analysis of any two populations. We can, for example, analyze the amount of admixture of alleles from Mongol populations with eastern European populations to explain the relatively high levels of blood type B in eastern European populations (if we make the unrealistic assumption that these groups are homogeneous populations). The calculations are also based on a change all in one generation, which did not happen. Blood-type and other loci can be used to determine allelic frequencies in western European, eastern European, and Mongol populations. We can use equation 20.9 to solve for m, the proportion of migrants:

$$m = \frac{q_c - q_1}{q_2 - q_1} \tag{20.12}$$

From one sample, we find that the B allele is 0.10 in western Europe, taken as the resident or native population (q_1); 0.12 in eastern Europe, the conglomerate population (q_c); and 0.21 in Mongols, the migrants (q_2). Substituting into equation 20.12 gives a value of m of 0.18. That is, given the above assumptions, 18% of the alleles in the eastern European population have been brought in by genetic mixture with Mongols.

When a migrant group first joins a native group, before genetic mixing takes place, the Hardy-Weinberg equilibrium of the conglomerate population is perturbed even though both subgroups are themselves in Hardy-Weinberg proportions. There will be a decrease in heterozygotes in the conglomerate population from what is predicted from the allelic frequencies of the conglomerate (the average allelic frequencies of the two groups). This is a phenomenon of subdivision referred to as the **Wahlund effect.** The reason this happens stems from the fact that the relative proportions of heterozygotes increase at intermediate allelic frequencies, as was seen in the boxed material in chapter 19 on the De Finetti diagram. As allelic frequencies rise above 0.5 or fall below 0.5, the relative proportion of heterozygotes decreases.

In a conglomerate population, the allelic frequencies will be intermediate between the values of the two subgroups (because of averaging). This in general will cause the expectation of heterozygotes to be higher than the average proportion of heterozygotes in the two subgroups that are actually there. An example is worked out in table 20.1. Assume that the two subgroups each make up 50% of the conglomerate population. In subgroup 1, $p = 0.1$ and $q = 0.9$; in subgroup 2, $p = 0.9$ and $q = 0.1$. Each subgroup will have 18% heterozygotes. The average, $(0.18 + 0.18)/2 = 0.18$, is the proportion of heterozygotes actually in the population. However, the conglomerate allelic frequencies are $p = 0.5$ and $q = 0.5$, leading to an expectation of 50% heterozygotes. Hence the observed frequency of heterozygotes is lower than expected (i.e., the Wahlund effect).

It should be noted that the same logic holds even if both populations have allelic frequencies above or below 0.5. Also, this effect happens when an observer samples what he or she thinks is a single population but is actually subdivided into several demes. When most population geneticists sample a population and find a deficiency of heterozygotes, they first think of inbreeding and then of subdivision, the Wahlund effect. (A further complication that we will not deal with here is that inbreeding leads to subdivision and subdivision leads to inbreeding. Statistics have been developed to try to separate these two phenomena.) As soon as random mating occurs in a subdivided population, Hardy-Weinberg equilibrium is established in one generation. We refer to a population in which the individuals are mating at random as unstructured or **panmictic.**

TABLE 20.1 The Wahlund Effect, a Deficiency of Heterozygotes, Is Seen When a Population Is Made Up of Subpopulations, Each of Which Is Composed of Individuals Mating at Random

	Subgroup 1	Subgroup 2	Conglomerate	
p	0.1	0.9	0.5	
q	0.9	0.1	0.5	
			Expected	*Observed*
p^2	0.01	0.81	0.25	0.41
$2pq$	0.18	0.18	0.50	0.18
q^2	0.81	0.01	0.25	0.41

SMALL POPULATION SIZE

Another variable that can upset the Hardy-Weinberg equilibrium is small population size. The Hardy-Weinberg equilibrium assumes an infinitely large population because, as defined, it is **deterministic,** not stochastic. That is, the Hardy-Weinberg equilibrium predicts exactly what the allelic and genotypic frequencies should be after one generation; it ignores variation due to sampling error. To some extent, every population of organisms on earth violates the Hardy-Weinberg assumption of infinite population size.

Sampling Error

The zygotes of every generation are a sample of gametes from the parent generation. The changes in allelic frequencies from one generation to the next that are due to inexact sampling of the alleles of the parent generation are sampling errors. Toss a coin one hundred times and chances are it will not land heads exactly fifty times. However, as the number of coin tosses increases, the percentage of heads will approach 50%, a percentage that is reached with certainty only after an infinite number of tosses. The same applies to any sampling problem, from drawing cards from a deck to drawing gametes from a gene pool.

If small population size is the only factor causing deviation from Hardy-Weinberg equilibrium, it will cause the allelic frequencies of a given population to fluctuate from generation to generation in the process known as random genetic drift. In other words, an Aa heterozygote will sometimes produce several offspring that have only its A allele, or sometimes random mortality will kill a disproportionate number of aa homozygotes. In either case, the next generation may not have the same allelic frequencies as the present generation. The end result will be either fixation or loss of any given allele ($q = 1$ or

Figure 20.4

Random genetic drift. Ten populations, each consisting of two individuals with initial $q = 0.5$, all go to fixation or loss of the a allele (4 or 0 copies) within ten generations due only to the sampling error of gametes. Once the a allele has been fixed or lost, no further change in allelic frequency will occur (barring mutation or migration). We show a population of only two individuals to exaggerate the effects of random genetic drift.

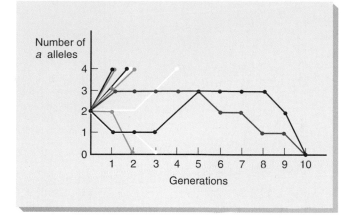

Figure 20.5

Initial conditions of random drift model. One thousand populations, each of size one hundred, and each with an allelic frequency (q) of 0.5.

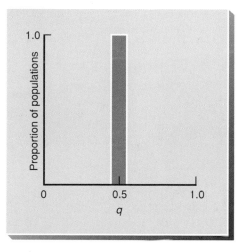

Figure 20.6

Genetic drift in small populations: $q = 0.5$. After time passes, the populations of figure 20.5 begin to diverge in their allelic frequencies. Time is measured in population size (N), showing that the effects of random genetic drift are qualitatively similar in populations of all sizes; the only difference is the time scale.

From M. Kimura, "Solution of a Process of Random Genetic Drift with a Continuous Model" in Proceedings of the National Academy of Sciences, 41:144–150, 1955. Reprinted by permission.

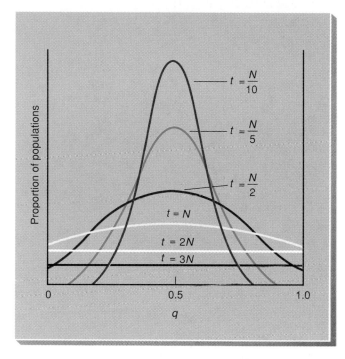

$q = 0$; fig. 20.4), although which will be fixed or lost is a function of the original allelic frequencies. The amount of time it will take to reach the fixation-loss endpoint depends on the size of the population.

Simulation of Random Genetic Drift

The process of random genetic drift can be investigated mathematically by starting with a large number of populations of the same finite size and observing how the distribution of allelic frequencies among the populations changes in time due only to random genetic drift. For example, we start with one thousand hypothetical popula-

tions, each containing one hundred individuals, and in each the frequency of the a allele, q, is 0.5 (fig. 20.5). We measure time in generations, t, as a function of the population size, N (100 in this example). For instance, $t = N$ is generation 100, $t = N/5$ is generation 20, and $t = 3N$ is generation 300. Then, by using computer simulation (or the **Fokker-Planck equation,** which physicists use to describe diffusion processes such as Brownian motion), we generate a series of curves shown in figure 20.6. These curves show that as the number of generations increases, the populations begin to diverge from $q = 0.5$. Approximately the same number of populations go to q values above 0.5 as go to q values below 0.5. Therefore, the distribution spreads symmetrically. When the distribution of allelic frequencies reaches the sides of the graph, some populations become fixed for the a allele and some lose the a allele. In a sense, the sides act as sinks: Any population that has the a allele lost or fixed will be permanently removed from the process of random genetic drift. Without mutation to bring one or the other allele back into the gene pool, these populations will maintain a constant allelic frequency of zero or 1.0.

After a point between N (100) and $2N$ (200) generations, the distribution of allelic frequencies flattens out and begins to lose populations to the edges (fixation or loss)

Figure 20.7

Continued genetic drift in the one thousand populations, one hundred each in size, shown in figures 20.5 and 20.6. After approximately $2N$ generations, the distribution is flat and populations are going to loss or fixation of that a allele at a rate of $1/2N$ populations per generation.

From S. Wright, "Evolution in Mendelian Population" in Genetics, *97:114. Copyright © 1931 Genetics Society of America, Chapel Hill, NC. Reprinted by permission.*

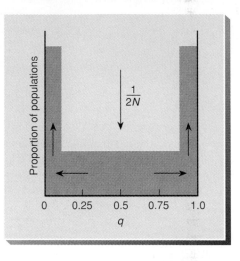

Figure 20.8

Random genetic drift in small populations. $q = 0.1$. Compare this figure with figure 20.6. In this case, the probability of fixation of the a allele is 0.1 and the probability of its loss is 0.9.

From M. Kimura, "Solution of a Process of Random Genetic Drift with a Continuous Model" in Proceedings of the National Academy of Sciences, *41:144–150, 1955. Reprinted by permission.*

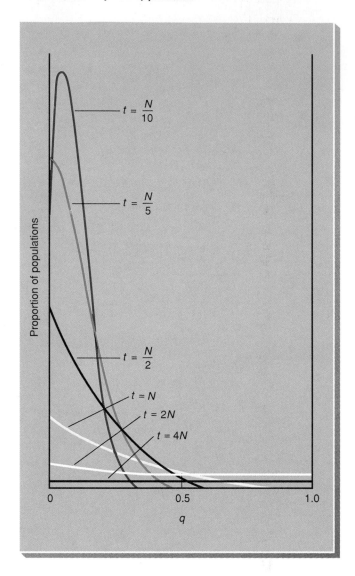

at a constant rate as shown in figure 20.7. The rate of loss of populations to the edges is about $1/2N$ ($1/200$), or 0.5% of the populations per generation. If the initial allelic frequency was not 0.5, everything is shifted in the distribution (fig. 20.8), but the basic process is the same—in all populations, sampling error will cause allelic frequencies to drift toward fixation or elimination. If no other factor counteracts this drift, every population is destined eventually to be either fixed for or deficient in any given allele.

The amount of time the process takes depends on the size of the population. The example used here was based on small populations of one hundred. If one million is substituted in figure 20.6 for the population size of one hundred, a flat distribution of populations would not be reached until two million generations, rather than two hundred generations, and so on. Thus a population experiences the effect of random genetic drift in inverse proportion to its size: Small populations rapidly reach fixation or loss of a given allele, whereas large populations take longer to show the same effects. Genetic drift also shows itself in several other ways.

Founder Effects and Bottlenecks

Several well-known genetic phenomena are caused by populations starting at or proceeding through small numbers. When a population is initiated by a small, and therefore genetically unrepresentative, sample of the parent population, the genetic drift observed in the subpopulation is referred to as a **founder effect.** A classic human example is the population founded on Pitcairn Island by several of the *Bounty* mutineers and some Polynesians. The unique combination of Caucasian and Polynesian traits

that characterizes today's Pitcairn Island population resulted from the small number of founders of the population.

Sometimes populations go through **bottlenecks,** periods of very small population size, with predictable genetic results. After the bottleneck, the parents of the next generation have been reduced to a small number and may not be genetically representative of the original population. The field mice on Muskeget Island, Massachusetts have a white forehead blaze of hair not commonly found in nearby mainland populations. Presumably, the island population went through a bottleneck at the turn of the century when cats existed on the island and reduced the number of mice to near zero. The population was reestablished by a small group of mice that happened by chance to contain several animals with this forehead blaze.

NATURAL SELECTION

Although mutation, migration, and random genetic drift all influence allelic frequencies, they do not of necessity produce populations of individuals that are better adapted to their environments. Natural selection, however, tends to that end. The consequence of natural selection, Darwinian evolution, is considered in detail in the next chapter. We discuss here the algebra of the process of natural selection. Artificial selection, as practiced by animal and plant breeders, follows the same rules.

Ways in Which Natural Selection Acts

Selection, or **natural selection** (or artificial selection), is a process whereby one phenotype and, therefore, one genotype leaves relatively more offspring than another genotype. (Natural selection occurs in nature, whereas artificial selection is imposed by animal or plant breeders.) Selection is a matter of **reproductive success,** the relative contribution of that genotype to the next generation. It is important to note that selection acts on whole organisms and thus on phenotypes. However, we will analyze the process by looking directly at the genotype, usually only at one locus.

Fitness

A measure of reproductive success is the **fitness,** or **adaptive value,** of a genotype. The genotype that leaves relatively more offspring than another has the higher fitness.

Fitness usually is defined to vary from zero to 1 and is always relative to a given population at a given time. For example, in a normal environment, fruit flies with long wings may be more fit than fruit flies with short wings. But in a very windy environment, a fruit fly with limited flying ability may do better than the long-winged genotype that will be blown around by the wind. Thus fitness is relative to a given circumstance. In a given environment the genotype that leaves the most offspring is usually assigned a fitness of 1 and a lethal genotype has a fitness of zero. Any other genotype has a fitness value between zero and 1. A number of factors can decrease this fitness value, W, below 1. A **selection coefficient** measures the sum of forces acting to prevent reproductive success. It is usually given the letter s or t and is defined by the fitness equation:

$$W = 1 - s \qquad (20.13)$$

and

$$s = 1 - W \qquad (20.14)$$

Thus, as the selection coefficient increases, fitness decreases, and vice versa.

Components of Fitness

Natural selection can act at any stage of the life cycle of an organism. It usually acts in one of four ways. (1) The reproductive success of a genotype can be affected by its prenatal, juvenile, or adult survival. Differential survival of genotypes is referred to as viability selection or **zygotic selection.** (2) A heterozygote can have differential success of its gametes when one of its alleles fertilizes more often than the other. This phenomenon is termed **gametic selection.** A well-studied case is the t-allele (tailless) locus in house mice, in which heterozygous males of the Tt genotype have as many as 95% of their gametes carry the t allele. Selection can also take place in two areas of the reproductive segment of an organism's life cycle. (3) Some genotypes may mate more often than others (have greater mating success) resulting in **sexual selection.** Specifically, sexual selection usually refers to situations in which there is competition among members of the same sex for mates. Adaptations for fighting, such as antlers in male elk, or displaying, such as the peacock's tail, are examples of the results of sexual selection. (4) Or, some genotypes may be more fertile than other genotypes resulting in **fecundity selection.** The particular variable of the life cycle upon which selection acts is termed a **component of fitness.**

Types of Selection

Figure 20.9 shows the three main ways that the sum total of selection can act. **Directional selection** works by continuously removing individuals from one end of the phenotypic (and therefore, presumably, genotypic) distribution (e.g., short individuals are removed). By removal we mean through death or failure to reproduce (genetic death). Thus the mean is constantly shifted toward the other end of the phenotypic distribution; in our example, the mean shifts toward the tall end. An example from the geologic record would be the evolution of neck length in giraffes by directional selection.

Stabilizing selection (fig. 20.9) works by constantly removing individuals from both ends of a phenotypic distribution, thus maintaining the same mean over time. Stabilizing selection now works on the neck length of giraffes—it is neither increasing nor decreasing. **Disruptive selection** works by favoring individuals at both ends of a phenotypic distribution at the expense of individuals in the middle. It, like stabilizing selection, should maintain the same mean value of the phenotypic distribution. Disruptive selection has been carried out successfully in the laboratory for bristle number in *Drosophila*. Starting with a population with a mean sternopleural chaeta (bristles on one of the body plates) number of about eighteen, investigators succeeded after twelve generations of getting a fly population with one peak of bristle numbers at about sixteen and another at about twenty-three (fig. 20.10).

Figure 20.9

Directional, stabilizing, and disruptive selection. *Colored areas* show the groups being selected against. At the *top* is the original distribution of individuals. After selection, the final distributions are seen in the bottom *row*.

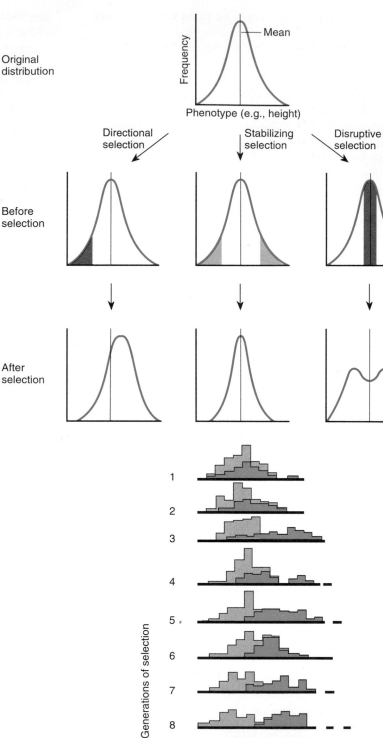

Figure 20.10

Disruptive selection in *Drosophila melanogaster*. After twelve generations of selection by the investigators for flies with either many or few bristles (chaetae) on the sternopleural plate, the population was bimodal. In other words, there were many flies with few or many bristles in the population, but few flies with intermediate bristle number.

Reprinted by permission from Nature, *Vol. 193, p. 1165. Copyright © 1962 Macmillan Magazines Ltd.*

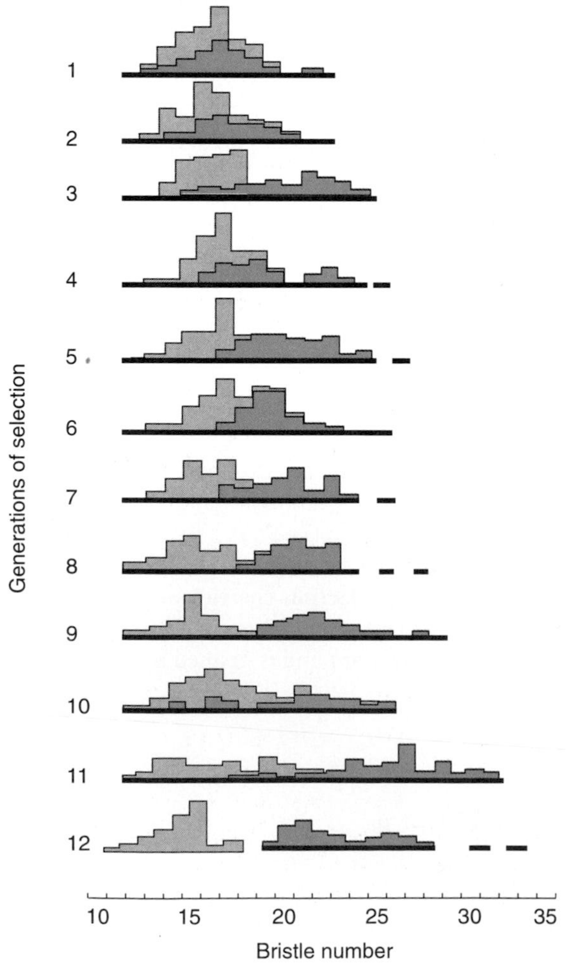

Selection against the Recessive Homozygote

Selection can be analyzed by our standard model-building protocol of population genetics—namely, define the initial conditions; allow selection to act; calculate the allelic frequency after selection (q_{n+1}); calculate Δq (change in allelic frequency from one generation to the next); then calculate equilibrium frequency, $\hat{q}$, when Δq becomes zero; and examine the stability of the equilibrium. In the analysis that follows, we consider a single autosomal locus in a diploid, sexually reproducing species with two alleles and assume that selection acts directly on the genotypes in a simple fashion (i.e., it occurs at a single stage in the life of the organism, such as larval mortality in *Drosophila*). After selection, the individuals remaining within the population mate at random to form a new generation in Hardy-Weinberg equilibrium.

Selection Model

In table 20.2, we outline the model for the case of selection against the homozygous recessive genotype. The initial population is in Hardy-Weinberg equilibrium. Even with selection acting during the life cycle of the organism, the Hardy-Weinberg equilibrium will be reestablished anew after each round of random mating, although possibly at new allelic frequencies. All selection models start out the same way. They diverge at the point of assigning fitnesses, which depend on the way in which natural selection is acting. In the model of table 20.2, the dominant homozygote and the heterozygote have the same fitness ($W = 1$). Natural selection cannot differentiate between the two genotypes because they both have the same phenotype. The recessive homozygote, however, is being selected against, which means that it has a lower fitness than the two other genotypes ($W = 1 - s$).

After selection, the ratio of the different genotypes is determined by multiplying their frequencies (Hardy-Weinberg proportions) by their fitnesses. The procedure follows from the definition of fitness, which in this case is a relative survival value. Thus only $1 - s$ of the *aa* genotype survive for every one of the other two genotypes. For example, if *s* were 0.4, then the fitness of the *aa* type would be $1 - s$, or 0.6. For every ten *AA* and *Aa* individuals that survived to reproduce, only six *aa* individuals would have survived to reproduce. The total of the three genotypes after selection is $1 - sq^2$. That is,

$$p^2 + 2pq + q^2(1 - s) = p^2 + 2pq + q^2 - sq^2$$
$$= 1 - sq^2$$

Mean Fitness of a Population

The value ($1 - sq^2$) is referred to as the **mean fitness of the population,** $\overline{W}$, because it is the sum of the fitnesses of the genotypes multiplied (weighted) by the frequency at which they occur. Thus it is a mean of the fitnesses, weighted by their frequencies. The new ratios of the three

TABLE 20.2   Selection Against the Recessive Homozygote: One Locus with Two Alleles, *A* and *a*

	Genotype			
	AA	*Aa*	*aa*	Total
Initial genotypic frequencies	p^2	$2pq$	q^2	1
Fitness (W)	1	1	$1 - s$	
Ratio after selection	p^2	$2pq$	$q^2(1 - s)$	$1 - sq^2 = \overline{W}$
Genotypic frequencies after selection	$\dfrac{p^2}{\overline{W}}$	$\dfrac{2pq}{\overline{W}}$	$\dfrac{q^2(1 - s)}{\overline{W}}$	1

genotypes can be returned to genotypic frequencies by simply dividing by the mean fitness of the population, $\overline{W}$, as in the last line of table 20.2. (Remember that a set of numbers can be converted to proportions of unity by dividing them by their sum.) The new genotypic frequencies are thus the products of their original frequencies times their fitnesses divided by the mean fitness of the population.

After selection, the new allelic frequency (q_{n+1}) is the proportion of *aa* homozygotes plus half the proportion of heterozygotes, or

$$q_{n+1} = \frac{q^2(1 - s)}{1 - sq^2} + \frac{pq}{1 - sq^2}$$

$$= \frac{q(q - sq + p)}{1 - sq^2}$$

$$= \frac{q(1 - sq)}{1 - sq^2} \tag{20.15}$$

This model can be simplified somewhat if we assume that the *aa* genotype is lethal. Its fitness would be 0, and *s*, the selection coefficient, would be 1. Equation 20.15 would then change to:

$$q_{n+1} = \frac{q(1 - q)}{1 - q^2} \tag{20.16}$$

Since $(1 - q^2)$ is factorable into $(1 - q)(1 + q)$, equation 20.16 becomes

$$q_{n+1} = \frac{q(1 - q)}{(1 - q)(1 + q)}$$

$$= \frac{q}{1 + q} \tag{20.17}$$

The change in allelic frequency is then calculated as

$$\Delta q = q_{n+1} - q = \frac{q}{1 + q} - q.$$

To solve this equation, q is multiplied by $(1 + q)/(1 + q)$ so that both parts of the expression are over the common denominator $(1 + q)$:

$$\Delta q = \frac{q - q(1 + q)}{1 + q}$$

$$= \frac{-q^2}{1 + q} \qquad (20.18)$$

which is the expression for the change in allelic frequency caused by selection. Since selection will not act again until the same stage in the life cycle next generation, equation 20.18 is also an expression for the change in allelic frequency between generations.

Two facts should be apparent from equation 20.18. First, the frequency of the recessive allele (q) is declining, as indicated by the negative sign of the fraction. This fact should be intuitive from the way that selection has been defined in the model (eliminating aa homozygotes). Second, the change in allelic frequency is proportional to q^2, which appears in the numerator of the expression. In other words, allelic frequency is declining as a relative function of the occurrence of homozygous recessive individuals in the population. This fact is consistent with the method in which the selection model was set up (in which selection is against the homozygous recessive genotype). The final formula supports the methodology of the model making.

Equilibrium Conditions

Next we calculate the equilibrium q by setting the Δq equation equal to zero, since a population that is in equilibrium will show no change in allelic frequencies from one generation to the next:

$$\frac{-q^2}{1 + q} = 0 \qquad (20.19)$$

For a fraction to be zero, the numerator must equal zero. Thus $-q^2 = 0$, and $\hat{q} = 0$. At equilibrium, the a allele should be entirely removed from the population. If the aa homozygotes are being removed, and if there is no mutation to return a alleles to the population, then eventually the a allele will disappear from the population.

Time Frame for Equilibrium

One shortcoming of this selection model is that it is not immediately apparent how many generations will be required to remove the a allele. The deficiency can be compensated for by the use of a computer simulation or by

Figure 20.11

Decline in q (the frequency of the a allele) with selection against the aa homozygote under different intensities of selection. Note that the loss of the a allele is asymptotic in both cases, but the drop in allelic frequency is more rapid with the larger selection coefficient.

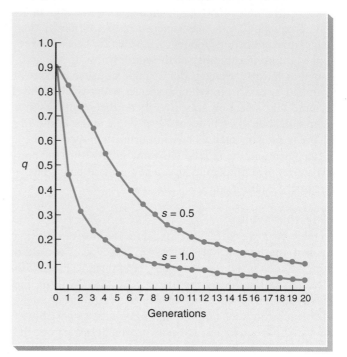

the introduction of a calculus differential into the model. Either method would produce the frequency-time graph of figure 20.11. This figure clearly shows that the a allele is removed more quickly when selection is stronger (when s is larger) and that the curves appear to be asymptotic— the a allele is not immediately eliminated and would not be entirely removed until an infinitely large number of generations had passed. The reason for the asymptotic behavior of the graph is that as the a allele becomes rarer and rarer, it tends to be found in heterozygotes (table 20.3). Since selection can remove only aa homozygotes, an a allele hidden in an Aa heterozygote will not be selected against. When $q = 0.5$, there are two heterozygotes for every aa homozygote. When $q = 0.001$, there are almost two thousand heterozygotes per aa homozygote. Remember, only the recessive homozygote is selected against. Natural selection cannot distinguish the dominant homozygote from the heterozygote.

Selection-Mutation Equilibrium

Although a deleterious allele is eliminated slowly from a population, the time frame is so great that there is opportunity for mutation to bring the allele back. Given such a population in which alleles are removed by selection while being added by mutation, the point at which there is no change in allelic frequency, the **selection-mutation equilibrium,** may be determined as follows. The new fre-

TABLE 20.3 Relative Occurrence of Heterozygotes and Homozygotes as Allelic Frequency Declines: $q = f(a)$; $p = f(A)$

q	$f(Aa)$ $(2pq)$	$f(aa)$ (q^2)	$f(Aa)/f(aa)$
0.5	0.50	0.25	2
0.2	0.32	0.04	8
0.1	0.18	0.01	18
0.01	0.0198	0.0001	198
0.001	0.001998	0.000001	1,998

TABLE 20.4 All Possible One-Locus, Two-Allele Selection Models (Assuming All Selection Coefficients Are Constants)

Type of Selection	Genotypic Fitness		
	A_1A_1	A_1A_2	A_2A_2
1. Against recessive homozygotes	1	1	$1 - s$
2. Against heterozygotes	1	$1 - s$	1
3. Against one allele	1	$1 - s_1$	$1 - s_2$
4. Against homozygotes	$1 - s_1$	1	$1 - s_2$

quency (q_{n+1}) of the recessive a allele after nonlethal selection ($s > 1$) against the recessive homozygote is given by equation 20.15:

$$q_{n+1} = \frac{q(1 - sq)}{1 - sq^2}$$

Change in allelic frequency under this circumstance will thus be:

$$\Delta q = q_{n+1} - q = \frac{q(1 - sq)}{(1 - sq^2)} - \frac{q(1 - sq^2)}{(1 - sq^2)}$$

$$= \frac{q - sq^2 - q + sq^3}{1 - sq^2}$$

$$= \frac{-sq^2(1 - q)}{1 - sq^2} \quad (20.20)$$

Equation 20.20 is the general form of equation 20.18 for any value of s. The change in allelic frequency due to mutation can be found by using equation 20.4:

$$\Delta q = \mu p - \nu q$$

where μ and ν are the rate of forward and back mutation, respectively. When equilibrium exists, the change from selection will just balance the change from mutation. Thus:

$$\mu p - \nu q + \left(-\frac{sq^2(1 - q)}{1 - sq^2} \right) = 0$$

and,

$$\mu p - \nu q = \frac{sq^2(1 - q)}{1 - sq^2} \quad (20.21)$$

Now, some judicious simplifying is justified since, in a real situation, q will be very small because the a allele

is being selected against. Thus νq will be close to zero and $1 - sq^2$ will be close to unity. Equation 20.21, therefore, becomes:

$$\mu p \approx sq^2(1 - q)$$

$$\mu(1 - q) \approx sq^2(1 - q)$$

$$q^2 \approx \mu/s$$

$$\hat{q} \approx \sqrt{\mu/s} \quad (20.22)$$

In the case of a recessive lethal, s would be unity, so:

$$q^2 \approx \mu \text{ and } \hat{q} \approx \sqrt{\mu}$$

If a recessive homozygote has a fitness of 0.5 ($s = 0.5$) and a mutation rate, μ, of 1×10^{-5}, the allelic frequency at selection-mutation equilibrium will be:

$$\hat{q} \approx \sqrt{\mu/s} \approx \sqrt{(1 \times 10^{-5})/0.5} \approx \sqrt{2 \times 10^{-5}}$$

$$\approx 0.004$$

If the recessive phenotype were lethal, then:

$$\hat{q} \approx \sqrt{\mu/s} \approx \sqrt{(1 \times 10^{-5})/1}$$

$$\approx 0.003$$

These are very low equilibrium values of the a allele.

Types of Selection Models

In view of the limited ways that fitnesses can be assigned, only a limited number of selection models are possible. Table 20.4 is a list of all possible selection models if we assume that fitnesses are constants and the highest fitness is 1. (The reader might now go through the list of models and determine the equilibrium conditions for each.) Some readers might note that two possible fitness distributions are missing. There is no model in which fitnesses are

TABLE 20.5 Selection Model of Heterozygote Advantage: The A Locus with A_1 and A_2 Alleles

| | Genotype | | | |
	A_1A_1	A_1A_2	A_2A_2	Total
Initial genotypic frequencies	p^2	$2pq$	q^2	1
Fitness (W)	$1 - s_1$	1	$1 - s_2$	
Ratio after selection	$p^2(1 - s_1)$	$2pq$	$q^2(1 - s_2)$	$1 - s_1p^2 - s_2q^2 = \overline{W}$
Genotypic frequencies after selection	$\dfrac{p^2(1 - s_1)}{\overline{W}}$	$\dfrac{2pq}{\overline{W}}$	$\dfrac{q^2(1 - s_2)}{\overline{W}}$	1

$1 - s$, 1, and 1 for the A_1A_1, A_1A_2, and A_2A_2 genotypes, respectively (remembering that $p = f[A_1]$ and $q = f[A_2]$). That model is for selection against the A_1A_1 homozygote. Some reflection should show that this is the same model as model 1 of table 20.4, except that the A_1 allele is acting like a recessive allele. In other words, natural selection acts against A_1A_1 homozygotes but not against the A_1A_2 and A_2A_2 genotypes. Thus the model reduces to model 1 if we treat A_1 as the recessive allele and A_2 as the dominant allele. Similarly, the $(1 - s_1, 1 - s_2, 1)$ model is eliminated for the same reason (allele A_2 is acting like the dominant allele and A_1 like the recessive allele), making the list in table 20.4 inclusive. We now describe the outcome of each of the models in the table.

In both models 1 and 3 (table 20.4), selection is against genotypes containing the A_2 allele. Model 1, which was just derived in detail, is the case of a deleterious recessive allele. Almost any enzyme defect in a metabolic pathway fits this model, such as PKU, alkaptonuria, Tay-Sachs disease, and so on. In model 3, however, natural selection can detect the heterozygote, which is the case with deleterious alleles that are not completely recessive. An example would be the hemoglobin anomaly called thalassemia, which produces a severe anemia in homozygotes and a milder anemia in heterozygotes. The disorder is common in some European and Oriental populations. It should be clear that selection can eliminate more quickly a partially recessive allele than a completely recessive allele because the allele can no longer "hide" in the heterozygote.

Dominant or semidominant alleles (model 3) are usually more quickly removed from a population because they are completely open to selection. It takes an infinite number of generations to remove a recessive lethal allele but only one generation for natural selection to remove a completely dominant lethal allele (see model 3, where $s_1 = s_2 = 1$). Examples of dominant deleterious traits in people are Huntington's chorea, facioscapular muscular dystrophy, and chondrodystrophy.

Model 2 is interesting because selection against the heterozygote leads to an unstable equilibrium at $q = 0.5$. If one heterozygote is removed by selection, one each of the two alleles is eliminated. However, if p and q are not equal (and thus not equal to 0.5), then one A_1 allele is not the same proportion of the A_1 alleles as one A_2 allele is of

the A_2 alleles. In other words, in a population of fifty individuals with $q = 0.1$ and $p = 0.9$, one A_2 allele is 10% (1/10) of the A_2 alleles, whereas one A_1 allele is only 1.1% (1/90) of the A_1 alleles. Removing one each of the two alleles causes a decrease in q. Therefore, a population following model 2 is at equilibrium at $p = q = 0.5$. However, this is an unstable equilibrium. Any perturbation that changes the allelic frequencies will cause the rarer allele to be selected against and eventually removed from the population. An example is the maternal-fetal incompatibility at the Rh locus in human beings. The disease erythroblastosis occurs only in heterozygous fetuses (Rh^+Rh^-) in Rh-negative (Rh^-Rh^-) mothers. Heterozygotes are, therefore, selected against.

In model 4, selection is against homozygotes. This model is called **heterozygote advantage** and the equilibrium condition will be derived because the results are important to evolutionary theory (table 20.5). At equilibrium:

$$\Delta q = \frac{pq(s_1p - s_2q)}{\overline{W}} = 0 \tag{20.23}$$

For this expression to be zero, either:

$$p = 0, \quad q = 0, \quad \text{or} \quad (s_1p - s_2q) = 0$$

If $p = 0$ or $q = 0$, the result is trivial in that the equilibrium exists only because of the absence of one of the alleles. The more meaningful equilibrium occurs when $s_1p - s_2q = 0$. In that case:

$$s_1p = s_2q \quad \text{or} \quad s_1(1 - q) = s_2q$$

and,

$$\hat{q} = \frac{s_1}{s_1 + s_2} \tag{20.24}$$

Since $p + q = 1$,

$$\hat{p} = \frac{s_2}{s_1 + s_2} \tag{20.25}$$

BOX 20.1

A COMPUTER PROGRAM TO SIMULATE THE APPROACH TO ALLELIC EQUILIBRIUM UNDER HETEROZYGOUS ADVANTAGE

It is surprising how much insight into the processes of population genetics can be gained by modeling them on the computer. The simple computer program presented here calculates changing allelic frequencies due to random mating when alleles at a locus are under a heterozygous-advantage selection regime. The program is written in the BASIC language. The interested student can simulate any of the processes described in chapters 19 and 20 by using this program as a model. Usually, only a few lines need to be changed to look at an entirely different process. Output should be graphed. The more computer-literate student can direct the output of the program to a file that can then be used by a graphics program to do the graphing directly. The program should be rerun several times with various sets of values for the variables p, q, s_1, s_2, and number of generations.

In the computer program (Box fig. 1), p is set to 0.9, q is $1 - p$ (0.1), s_1 is 0.6, s_2 is 0.4, and the number of generations is 25. The program prints the initial conditions and then calculates Δq using equation 20.23. It determines the new q by adding Δq to the previous q. The program then prints allelic frequencies along with the generation number. It repeats this process twenty-five times. After the command to run, the following output results (Box fig. 2). As you can see, p is heading for 0.4 with q approaching 0.6. (It takes forty-two generations to reach equilibrium to a seven-place accuracy.)

Box figure 1

A BASIC computer program to simulate heterozygous advantage. Initial conditions are given in lines 10–40. The equation being iterated is in line 120.

```
10  P=.9
20  Q=1–P
30  S1=.6
40  S2=.4
50  PRINT "Initial  Conditions: "
60  PRINT "p = ", P; " q = ",Q
70  PRINT "s1 = ",  S1; " s2 = ", S2
80  PRINT "............."
90  PRINT "Generations        p            q"
100 FOR I=1 TO 25
110 WBAR = 1–S1*P*P–S2*Q*Q
120 DELTAQ= P*Q* (S1*P–S2*Q) / WBAR
130 Q=Q+DELTAQ
140 P=1–Q
150 PRINT I, P , Q
160 NEXT I
170 STOP
```

Box figure 2

Output of the computer program of box figure 1.

```
    Initial  Conditions:
    p =          .9      q =           .1
    s1 =         .6      s2 =          .4
    . . . . . . . . . . . . . .
    Generations       p               q
    1          . 8117647       . 1882353
    2          . 7052039       . 2947962
    3          . 6100563       . 3899438
    4          . 540254        . 459746
    5          . 493199        . 5068011
    6          . 4621927       . 5378073
    7          . 4417475       . 5582525
    8          . 4281701       . 57183
    9          . 4190853       . 5809147
    10         . 4129688       . 5870313
    11         . 4088311       . 591169
    12         . 4060224       . 5939776
    13         . 4041113       . 5958888
    14         . 4028086       . 5971914
    15         . 4019196       . 5980804
    16         . 4013125       . 5986875
    17         . 4008976       . 5991025
    18         . 4006139       . 5993861
    19         . 40042         . 5995801
    20         . 4002873       . 5997128
    21         . 4001966       . 5998035
    22         . 4001345       . 5998656
    23         . 400092        . 599908
    24         . 4000629       . 5999371
    25         . 4000431       . 599957
```

Figure 20.12

Plot of allelic frequency (q) versus change in allelic frequency (Δq) for a polymorphism maintained by heterozygote advantage. In this case $s_1 = 0.2$ and $s_2 = 0.3$; the equilibrium value, $\hat{q}$, is 0.4. When perturbed, the population tends to return to this value unless the perturbation brings q to either 1.0 or 0.0, in which case the population is either fixed for or has lost, respectively, the a allele and no further change in allelic frequency will take place, barring mutation or migration.

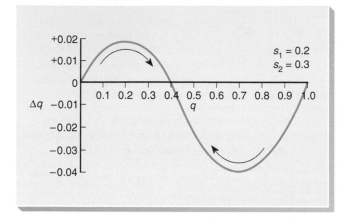

Several interesting conclusions follow. First, unlike the other models of selection, this model allows a population to maintain both alleles. We can demonstrate that this equilibrium is stable by graphing the Δq value against q. Such a graph is shown in figure 20.12, in which q is the frequency of allele A_2 and the fitness of genotypes A_1A_1, A_1A_2, and A_2A_2 are assumed to be 0.8, 1, and 0.7, respectively. Note that if the equilibrium is perturbed by an increase or decrease of q, the population will return to the

point of equilibrium. Second, the equilibrium is independent of the original allelic frequencies since it involves only the selection coefficients, s_1 and s_2. Last, the equilibrium for each allele (equations 20.24 and 20.25) is dependent on the coefficient of selection against the other allele. As the selection against A_1 increases (s_1 increases), the equilibrium shifts toward a higher value of q (more A_2 alleles).

SUMMARY

The effects of relaxing some of the assumptions of the Hardy-Weinberg equilibrium are analyzed. Both mutation and migration transport alleles in and out of a population. Mutation provides the variability on which natural selection acts, but it usually does not directly affect the equilibrium because mutation rates are usually very low. If two randomly mating populations merge, or if two randomly mating demes are mistakenly treated as a single deme, the conglomerate will be deficient in heterozygotes. This deviation is called the Wahlund effect.

Finite population size is a source of sampling error. It results in changes in allelic frequencies known as random genetic drift. The smaller the population, the more rapidly allelic frequencies change. The dynamics of random genetic drift were studied graphically.

Natural selection is defined by differential reproductive success. Depending upon which phenotypes are most fit, natural selection can act in several ways to change allelic and genotypic frequencies. Selection against the recessive homozygote acts to remove the allele from the population. Mutation will bring the allele back into the population. Thus there exists a selection-mutation equilibrium that will maintain the unfavorable allele at a relatively low frequency. Heterozygous advantage will maintain both alleles in a population.

SOLVED PROBLEMS

1. At a particular locus, there are two alleles, B and b. The mutation rate of B to b is 3.5×10^{-4} whereas the mutation rate of b to B is 6×10^{-8}. What is the equilibrium frequency of the b allele assuming no other factor is operating in this population to disturb the Hardy-Weinberg equilibrium?

   ANSWER: We let $q = f(b)$, $\mu = 3.5 \times 10^{-4}$, and $\nu = 6 \times 10^{-8}$. We then simply substitute μ and ν into equation 20.6:

   $$\hat{q} = \mu/(\mu + \nu) = 3.5 \times 10^{-4}/(3.5 \times 10^{-4} + 6 \times 10^{-8})$$
   $$= 0.9998$$

2. Given a population of about one million cicadas with a frequency of the a allele at the a locus of 0.75, what is the probability of the loss of the a allele by random genetic drift? How much longer will the possible loss of the allele take as compared to the loss of the allele in a population of one thousand?

   ANSWER: Regardless of the size of a finite population, random genetic drift takes place. The probability of the loss of an allele with a frequency of 0.75 is 0.25; the probability of its fixation is 0.75 (see fig. 20.8). Since it is convenient to measure time (number of generations) within populations of finite size in units of population size, we can see that an event that will take N generations will be one thousand generations in the small population but one million generations in the large population. Drift phenomena will occur in the larger population at about one thousandth the rate as in the small population.

3. At an electrophoretic loci in fruit flies, two alleles seem to be maintained indefinitely in a population. In a laboratory colony, the fitnesses of the three genotypes, FF, FS, and SS are 0.85, 1.0, and 0.6, respectively. What is the equilibrium frequency of the slow allele?

EXERCISES AND PROBLEMS

1. Consider a locus with alleles A and a in a large, randomly mating population under the influence of mutation.
   a. If the mutation rate of A to a is 6×10^{-5} and the back-mutation rate to A is 7×10^{-7}, what is the equilibrium frequency of a?
   b. If $q = 0.9$ in generation n, what would it be one generation later, only under the influence of mutation?

2. The following data refer to the R^O allele in the Rh blood system:

   frequency in western Europeans = 0.62

   frequency in eastern Europeans = 0.45

   frequency in Mongols = 0.03

   What is the total proportion of alleles that have entered the eastern European population?

3. Given the data from problem 1 of chapter 19, what factors could have led to the population not being in Hardy-Weinberg equilibrium?

4. In a population of five hundred individuals with a frequency of allele A of 0.7, what is the ultimate fate of the A allele? What is the probability that the population will eventually lose the A allele? How many are $N/5$ generations? $4N$ generations?

5. Derive an expression for mutation equilibrium when there is no back mutation.

6. Differentiate among stabilizing, directional, and disruptive selection.

7. Derive a model of selection in which the fitness of the heterozygote is half that of one of the homozygotes and twice the fitness of the other. Give expressions for the following:
   a. Mean population fitness
   b. Equilibrium allelic frequency (stable?)

8. Derive an expression for the equilibrium allelic frequencies under a model in which selection is against the heterozygotes. Is the equilibrium stable?

9. Table 20.6 describes selection at the A locus in a given diploid species in which $p = f(A)$ and $q = f(a)$.
   a. Describe the type of selection that is occurring here. Why does the total equal 1 before selection but $\overline{W}$ after?
   b. Derive an equation for q after one generation of selection (q_{n+1}).
   c. This system will reach equilibrium, with $\hat{p} = s_2/(s_1 + s_2)$. If selection is twice as strong against aa as against AA, what will the equilibrium allelic frequencies be? If $s_1 = 0.1$ and $s_2 = 0.3$, what will the percentage of heterozygotes be at equilibrium?

10. Given a locus with alleles A and a in a sexually reproducing, diploid population in Hardy-Weinberg equilibrium, set up a model and the initial formula for the frequency of the dominant allele after one generation (p_{n+1}) if selection acts against the dominant phenotype.

11. There is a locus with alleles A and a in a large, randomly mating, diploid, sexually reproducing population. Allele A mutates to a at a rate of μ and there is no back mutation. However, the aa homozygote is selected against with a fitness of $1 - s$. Give a formula for the equilibrium condition. If $\mu = 5 \times 10^{-5}$ and $s = 0.15$, what are the equilibrium allelic frequencies?

12. If a locus has alleles A_1 and A_2, what will be the equilibrium frequency of A_1 if both homozygotes are lethal?

13. The following data were collected from a population of *Drosophila* segregating sepia (s) and wild-type (s^+) eye colors. A sample was taken when the eggs were deposited and later among adults. Reconstruct the model of selection.

	s^+/s^+	s^+/s	s/s
Egg	25	50	25
Adult	30	60	10

TABLE 20.6

	Genotypes			Total
	AA	*Aa*	*aa*	
Before selection	p^2	$2pq$	q^2	1
Fitness (*W*)	$1 - s_1$	1	$1 - s_2$	
After selection	$p^2(1 - s_1)$	$2pq$	$q^2(1 - s_2)$	$\overline{W} = 1 - s_1p^2 - s_2q^2$

TABLE 20.7

Elevation of Sample	Inversion			
	ST	AR	CH	Others
6,800 ft	26	44	16	14
4,600 ft	32	37	19	12
3,000 ft	41	35	14	10
800 ft	46	25	16	13

(*ST* = Standard; *AR* = Arrowhead; *CH* = Chiricahua)

14. The data in table 20.7 are taken from Dobzhansky's work with chromosomal inversions in *Drosophila pseudoobscura* and represent four samples from various altitudes in the Sierra Nevada Mountains in California. What would you say about, and what would you do in the lab to determine, the fitnesses of the inversions? What factors could cause the changes in fitness?

15. In a population of 900 butterflies, the frequency (p) of the fast allele of the enzyme phosphoenol pyruvate is 0.6, and the frequency of the slow form (q) is 0.4. Ninety butterflies migrate to this population and, among the migrants, the frequency of the slow allele is 0.8. Calculate the allelic frequencies of the new population.

16. In a particular population with two alleles at a locus, the frequency of *AA* individuals = 0.25, *Aa* = 0.5, and *aa* = 0.25. If the *AA* genotype has a fitness = 1, *Aa* = 0.8, and *aa* = 0.6, what will the frequencies of *A* and *a* be in the next generation? Assume mutations are nonexistent.

17. If the frequency of the *N* allele in a native population is 0.25, 0.32 in a conglomerate population, and 0.4 in a migrant population, what percent of the *N* alleles in the conglomerate population were derived from the migrant population?

18. Consider a population in which $p = 0.9$ and $q = 0.1$. If the forward mutation rate, $A \rightarrow a$, is 5×10^{-5}, and the reverse mutation rate, $a \rightarrow A$, is 2×10^{-5}, calculate the equilibrium frequency $\hat{q}$, of the *a* allele.

19. If the forward mutation rate, $A \rightarrow a$, is 5 times the reverse mutation rate, what will the equilibrium frequency of the *a* allele be?

20. Calculate the frequency of the recessive *b* allele in a population one generation after selection if in the original population $q = f(b) = 0.7$ and the relative fitness of *bb* homozygotes is 0.4.

21. A type of dwarfism in people is caused by a dominant allele. The mutation rate from the normal to the mutant allele has been estimated at 5×10^{-5}, and the fitness of the dwarf is 0.2 compared to normal individuals. Calculate the equilibrium frequency of the dwarf allele.

22. In a particular population, the frequency of an allele *t* was 0.25 in a migrant population and 0.45 in the conglomerate population. If the migration rate was 0.1, calculate the frequency of *t* in the original population.

23. A recessive allele ($q = 0.5$) was initially neutral, but suddenly the environment changed and the recessive homozygote became lethal. What is q one generation after selection begins? What is the expected frequency of the recessive allele two generations after selection?

Suggested Readings for chapter 20 are on page 659.

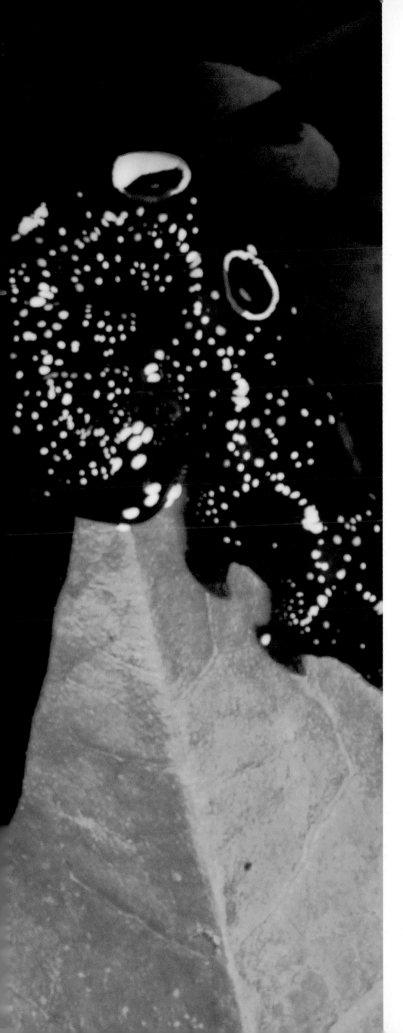

21

Genetics of the Evolutionary Process

Eyespots on a caterpillar (*Orthreis fullonia*) can frighten and disorient predators.
© *William E. Ferguson*

In the two previous chapters, we laid the theoretical groundwork for an understanding of the process of evolution in natural populations. Populations change, or evolve, through natural selection and the other forces that perturb the Hardy-Weinberg equilibrium. The merger of population genetics theory with classical Darwinian views of evolution is known as **neo-Darwinism,** or the "new synthesis." In this chapter we will concern ourselves with evolution and speciation.

DARWINIAN EVOLUTION

Charles Darwin (fig. 21.1) was a British naturalist who published his theory of evolution in 1859 in a book entitled *The Origin of Species by Means of Natural Selection or the Preservation of Favored Races in the Struggle for Life.* This book provided overwhelming support for the reality of evolution as well as a mechanism for evolution. Darwin had been greatly influenced by the writings of the Reverend Thomas Malthus, who is best known for his theory that populations increase exponentially, whereas their food supplies increase arithmetically. Malthus, in *An Essay on the Principle of Population* in 1798, was referring specifically to human populations and was trying to encourage people to reduce their birthrate rather than have their offspring starve to death. Malthus's writings impressed upon Darwin the realization that under limited resources—the usual circumstance in nature—not all organisms survive. In nature, organisms compete for the resources needed to survive.

Darwin was the naturalist aboard the *Beagle,* a ship that sailed around the world from 1831 to 1836 with the primary purpose of charting the coast of South America. During the travels of the *Beagle,* Darwin amassed great quantities of observations (especially on South America and the Galápagos Islands) that led him to suggest a theory wherein organisms become adapted to their environment by the process of natural selection. In outline, the process, as proposed by Darwin, is as follows:

1. *Variation is a characteristic of virtually every group of animals and plants.* Darwin saw variation as an inherent property among individuals of all populations.

2. *Every group of organisms overproduces offspring.* Most populations maintain a relatively constant density over time. Thus every parent, on average, just replaces itself. Therefore, most offspring produced by the individuals of a population will die before they reproduce. Hence, in every group of organisms, there is an overabundance of young.

3. *The most fit will survive.* This step is the cornerstone and the best-known part of Darwin's theory. Among all the organisms competing for a limited array of resources, only the organisms best

Figure 21.1
Charles Darwin (1809–1882). Darwin was an English naturalist who first established the theory of organic evolution ("Origin of Species"). *Painting by George Richmond, 1840. Downe House, Downe, Kent. © Archiv/ Photo Researchers, Inc.*

able to obtain and utilize these resources will survive (**survival of the fittest**). If the favorable characteristics of the most fit individuals are inherited, these traits will be passed on to the next generation. The most fit organisms have the greatest reproductive success (see Box 21.1).

Thus, over time, if advantageous mutations arise or if the environment changes, the characteristics of a population should change through the process of natural selection (directional or disruptive selection). A particularly well-adapted population in a stable environment may remain the same through the forces of stabilizing selection (see fig. 20.9). Nonrandom mating, genetic drift, and migration, may also play a role in population differentiation.

EVOLUTION AND SPECIATION

The term **evolution** describes a change in phenotypic frequencies, which results in a population of individuals that are better adapted to the environment than their ancestors were. **Speciation** comes in two different forms. (1) It may be the evolution of a population over time until a point is reached at which the current population cannot be classified as belonging to the same species as the original population. This process is known as **anagenesis,** or **phyletic evolution** (*an* is Latin for without; *genesis* is Latin for birth or creation). (2) Speciation may also be the divergence of

BOX 21.1

ATTACKS ON DARWINISM

From time to time, attacks on neo-Darwinism are mounted, usually by persons who either see evolutionary theory as antireligious or who basically misunderstand Darwin's theory. One attack, entitled "Darwin's Mistake," by Tom Bethell, was published in *Harper's* magazine.

Bethell began by pointing out that Darwinian theory is a tautology rather than a predictive theory. (The term *tautology* means a statement that is true by definition.) That is, evolution is the survival of the fittest. But who are the fittest? Obviously, the individuals who survive. Thus, without an independent criterion for fitness, other than survival, we are left with the statement that evolution is the survival of the survivors. This indeed is a tautology. But it is possible to assign independent criteria for fitness. Darwin wrote extensively about artificial selection in pigeons, in which the breeders' choice was the criterion for fitness. (Many novel breeds of pigeon have been created this way.) Artificial selection has been practiced extensively by plant and animal breeders. Here too, survival is not the criterion for fitness; productivity is.

It is more difficult to establish, a priori, independent criteria of fitness in nature. Often, uncontrolled or unseen vagaries have major impacts on the course of events. Surely the temperature became colder before the mammoths became woolly. Is it then reasonable or safe to predict that elephants would get woolly if the climate became colder in Africa today? The answer is no, for several reasons. First, the elephants might adapt to colder weather in any of a large number of different ways—they could get fatter, they could migrate, and so on. To some extent, adaptation depends not only on the changing environment but also on the reserve variation within the gene pool of the species from which the adaptive traits must come. Second, the elephants could become extinct; they might not be able to adapt at all. And third, if the climatic changes were not severe, the elephants might not change.

Predicting the exact course of evolution is nearly impossible. To provide independent criteria for fitness in nature is, therefore, very difficult. Some modern evolutionary biologists, although not doubting neo-Darwinism, do worry to some extent about the difficulties in testing modern evolutionary theory. However, the support for Darwinism (the fossil record, embryology, comparative anatomy, geographic distributions, etc.) is so overwhelming that the general nature of evolution is not in doubt. We can clearly trace its path although we cannot make exact predictions.

From a philosophical point of view, neo-Darwinism is the general paradigm (broad concept) defining "normal" science in biology. Every scientific endeavor works under the umbrella of a paradigm. When enough inconsistencies appear, a new paradigm is sought that replaces the old in what Thomas Kuhn called a "scientific revolution." In physics, relativity overthrew Newtonian principles. In biology, Darwinism overthrew the concept of a recent, biblically described origin of animals and plants. Darwinism became the paradigm because it explained many things in a consistent fashion that a recent origin of all forms of life could not. Neo-Darwinism will remain the current paradigm unless it is overthrown by a better theory that explains previous inconsistencies. To date there are no major inconsistencies that suggest that neo-Darwinism is not correct.

Bethell then went on to try to refute neo-Darwinism by using the following argument: Survival of the fittest can be redefined to mean that some organisms have more offspring than others. Thus natural selection cannot be a creative force because the only thing it works on is organisms alive now, some having more offspring than others. How, asks Bethell, can this possibly give us tigers and horses from ancestors that did not look like tigers and horses? The answer is that mutation produces variants in the population. The organism best able to compete will leave the most offspring. Given an array of different genotypes in a population, natural selection is the process that determines which genotypes will increase in future generations. Traits that give the bearer an advantage increase in the population and evolution takes place. Natural selection was the force behind the evolution from the small Eocene horse to the modern *Equus*.

Misinterpretation of mutation is the basis for other attacks on Darwinism. For example, Darwinian evolution has been attacked as not feasible since most mutations are deleterious. How, the argument goes, can evolution proceed by a combination of deleterious events? The answer is that although most mutations are deleterious, some are not. This is especially true in changing environments in which yesterday's deleterious mutant may be today's favored mutant.

The most recent attacks on Darwinism have been by creationists who have attempted to get laws passed in many states requiring the biblical version of creation to be taught as an alternative to Darwinism. This position has been denied in the courts because creationism is not a scientific theory. It does not follow the rules of the scientific method wherein empirical evidence can refute it.

Figure 21.2

Forms of speciation. In anagenesis (*a*), a species changes over time until it is so different from the progenitor that it is classified as a new species. In cladogenesis (*b*), speciation takes place as a branching process wherein one species becomes two or more.

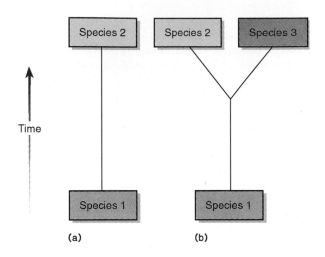

(a)   (b)

a population into two distinct forms (**species**) that exist simultaneously. This branching process is known as **cladogenesis,** or **true speciation** (*clado* is Greek for branch; fig. 21.2). What do we mean by the term species?

Before Darwin's time, **typological thinking** prevailed in which a species was defined as a group of organisms that were morphologically similar. All variants were considered to be imperfections. One of Darwin's greatest contributions to modern biological theory was to treat variation as a normal part of the description of a group of organisms. The modern **biological species concept** groups together as members of the same species organisms that can potentially interbreed. A species, therefore, is a group of organisms that can mate among themselves to produce fertile offspring.

The definition of species on the basis of interbreeding unfortunately cannot be used in many places, mostly due to technical problems of applying it. Taxonomists and paleontologists, who often use nonliving specimens (preserved or fossilized), use the **morphological species concept** as a working definition. In it, two organisms are classified as belonging to the same species if they are morphologically similar. They are classified as belonging to two different species if they are as different as two organisms belonging to two recognized species. Other problems for taxonomists arise since speciation is a dynamic process. For example, isolated subgroups of a population may be in various stages of becoming new species; the rate of successful interbreeding among individuals from these subgroups may range from zero to 100%. How should the in-betweens be classified? There is no correct answer. It depends on the circumstances.

Still other problems make it necessary to turn to the morphological species concept. Haploid and asexual species are hard to classify. Also, two organisms that will not interbreed in nature may do so in a laboratory setting. Thus the interbreeding test carried out in the laboratory (as is done frequently) is not necessarily an adequate criterion of speciation. Other problems arise in classifying groups that are geographically isolated from each other, such as populations on islands. There individuals are physically isolated, but in many cases can interbreed freely when brought together with their mainland counterparts. So, although there is a good theoretical definition of a species (potentially interbreeding individuals), more often than not it is necessary for biologists to apply the morphological species concept to determine whether two populations belong to the same species. In some cases, no decision can be made about the species status of a population. It is clear that a population has evolved but it is not clear whether it has evolved enough to be called a new species. Again, this is a problem more for taxonomists and evolutionary biologists than for the organisms themselves.

Mechanisms of Cladogenesis

Reproductive Isolation

How does one species become two? Basically, **reproductive isolating mechanisms** must evolve to prevent two subpopulations from interbreeding when they are in contact. Reproductive isolating mechanisms are environmental, behavioral, mechanical, and physiological barriers that prevent individuals of two species from producing viable offspring. Following is a modification of the classification system of isolating mechanisms suggested by the evolutionary biologist, G. L. Stebbins:

A. Prezygotic mechanisms prevent fertilization and zygote formation.
1. Residential—The populations live in the same region but occupy different habitats.
2. Seasonal or temporal—The populations exist in the same region but are sexually mature at different times.
3. Ethological (in animals only)—The populations are isolated by incompatible premating behavior.
4. Mechanical—Cross-fertilization is prevented or restricted by incompatible differences in reproductive structures (genitalia in animals, flowers in plants).
B. Postzygotic mechanisms affect the hybrid zygotes after fertilization has taken place.
1. F_1 hybrid breakdown—F_1 hybrids are inviable or weak.
2. Developmental hybrid sterility—Hybrids are sterile because gonads develop abnormally or because meiosis breaks down before it is completed.
3. Segregational hybrid sterility—Hybrids are sterile because of abnormal distribution to the gametes of whole chromosomes, chromosome segments, or combinations of genes.
4. F_2 breakdown—F_1 hybrids are normal, vigorous, and fertile, but the F_2 generation contains many weak or sterile individuals.

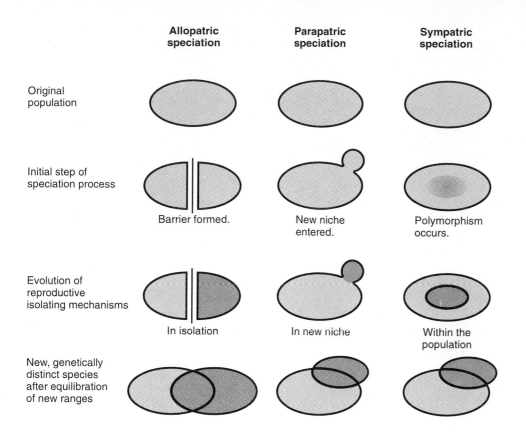

Allopatric speciation	Parapatric speciation	Sympatric speciation

Original population

Initial step of speciation process

Barrier formed.   New niche entered.   Polymorphism occurs.

Evolution of reproductive isolating mechanisms

In isolation   In new niche   Within the population

New, genetically distinct species after equilibration of new ranges

Figure 21.3
The three general mechanisms of speciation. In allopatric speciation, reproductive isolation evolves after the population has been geographically divided. In parapatric speciation, reproductive isolation evolves when a segment of the population enters a new niche. In sympatric speciation, reproductive isolation evolves while the incipient group is still in the vicinity of the parent population.

Allopatric, Parapatric, and Sympatric Speciation

Barriers to **gene flow,** the spread of genes between populations, can arise in three different ways, each of which defines a different mechanism of speciation. Usually, the mode of speciation is dictated by both the properties of the genetic systems of the organisms and stochastic (random) or accidental events. For example, vertebrates tend to have different speciation modes than phytophagous (plant-feeding) insects.

The appearance of a geographic barrier, such as a river or mountain, through the range of a species will physically isolate populations of the species. Physical isolation can also occur if migrants cross a particular barrier and begin a new population (founder effect). The physically isolated populations can then evolve independently. If reproductive isolating mechanisms evolve, then two distinct species will be formed, which, if they come together in the future, will remain distinct species. Speciation in which the evolution of reproductive isolating mechanisms occurs during physical separation of the populations is called **allopatric speciation** (fig. 21.3). As the evolutionary biologist Guy Bush pointed out, "Although examples in nature are difficult to substantiate . . . it [allopatric speciation] has been convincingly demonstrated in frogs . . . and lizards. . . ."

It should be noted that reproductive isolating mechanisms usually originate incidentally to the speciation process. That is, they arise during the process of evolution in isolated populations rather than necessarily being selected for. When isolated populations come together again, incomplete isolating mechanisms may allow hybrids to form. If the hybrids are normal and viable and can freely interbreed with individuals of each parent population, then no speciation has taken place. However, if the hybrids are at a disadvantage, natural selection may favor stronger isolating mechanisms. In this case, organisms that mate with individuals from the other population will leave fewer offspring. The result will be a more effective barrier to hybridization. Regions in which previously isolated populations come into contact and produce hybrids are called **hybrid zones.**

Until recently, evolutionary biologists believed that allopatric speciation was the general rule. Many now believe that two other modes of speciation may occur frequently in certain groups of organisms. **Parapatric speciation** occurs when a population of a species that occupies a large range enters a new niche, or habitat (fig. 21.3). Although no physical barrier arises, occupancy of the new niche will result in a barrier to gene flow between the population in the new niche and the rest of the species. Here again, the evolution of reproductive isolating mechanisms will produce two species where there was only one before. Parapatric speciation is believed to have occurred often in relatively nonvagile animals such as snails, flightless grasshoppers, and annual plants.

Sympatric speciation occurs when a polymorphism, which is the occurrence of alternative phenotypes in the same population, arises within an interbreeding population before a shift to a new niche. This mode of speciation may be common in parasites and phytophagous insects.

Figure 21.4

The apple maggot fly, *Rhagoletis pomonella*. This species has exhibited host range expansion since the last century, from hawthorn to apple, cherry, and roses. Host races are presumably the initial step in sympatric speciation. Magnification 10×.

Source: Jeffrey L. Feder and Guy L. Bush, Zoology Department, Michigan State University.

Figure 21.5

The Galápagos Archipelago located about seven hundred miles west of Ecuador. This isolated chain of islands is a natural laboratory for the evolutionary machinery.

From David Lack, Darwin's Finches. Copyright © 1947 by Cambridge University Press, New York, NY. Reprinted by permission.

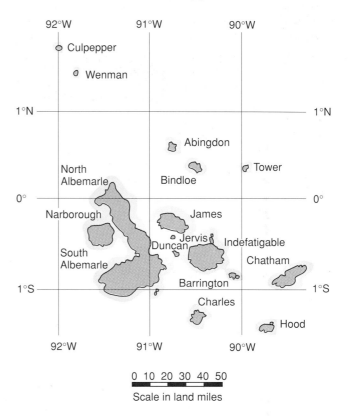

For example, if a polymorphism arises within a parasite species such that an individual with a certain genotype can adapt to a new host, this genotype may be the forerunner of a new species. If the parasite not only feeds on the new host but also mates on the new host, a barrier to gene flow arises, although the parasite may be surrounded by other members of its species with the original genotype. Sympatric speciation can thus occur in the middle of a species range rather than at the edges (fig. 21.3).

An example of incipient sympatric speciation has been seen recently in host races of the apple maggot fly (*Rhagoletis pomonella*) in North America (fig. 21.4). This fly was found originally only on hawthorn plants. However, in the nineteenth century it spread as a pest to apple trees, which had been introduced. In fact, races are now known on pear and cherry trees and rose bushes. These races have developed genetic, behavioral, and ecological differences from the original hawthorn-dwelling parent. Evolutionary biologists view this as an opportunity to observe parapatric speciation as it occurs.

Another form of sympatric speciation occurs when cytogenetic changes take place that result in "instantaneous speciation." These cytogenetic changes include polyploidy and translocations. For example, if polyploid offspring cannot produce fertile hybrids with individuals from a parent population, then the polyploid is reproductively isolated. This mechanism is much more common in plants because they can exist vegetatively despite odd

ploidy and they usually do not have chromosomal sex-determining mechanisms, which are especially vulnerable to ploidy problems (see chapter 8).

The end result of the true speciation process (cladogenesis) is the divergence of a homogeneous population into two or more species. One of the classic examples of cladogenesis is seen in the ground finches of the Galápagos Islands. The birds are very well studied not only because they present a striking case of speciation but also because they were studied by Darwin and were a strong influence on his views. Figure 21.5 is a map of the Galápagos Islands, and figure 21.6 is a diagram of the species of "Darwin's Finches."

An original finch somehow reached the Galápagos Archipelago from South America and with time spread to its various islands. Given the limited ability of the birds to get from island to island, allopatric speciation took place. On each island the finch population evolved reproductive isolating mechanisms while evolving to fill certain niches not being filled on the islands. For example, in South America no finches have evolved to be like woodpeckers because there are many woodpecker species already there. But the Galápagos, being isolated from South America, have what is called a **depauperate fauna,** a fauna lacking

Figure 21.6

Species of Darwin's finches. These birds apparently evolved from a single group of migrants from the South American mainland. Isolated on the different islands, the birds evolved to fill many vacant niches.

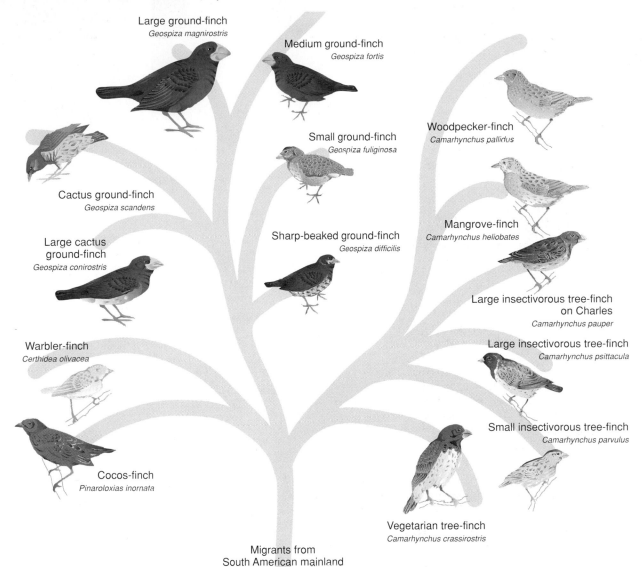

Large ground-finch
Geospiza magnirostris

Medium ground-finch
Geospiza fortis

Small ground-finch
Geospiza fuliginosa

Woodpecker-finch
Camarhynchus pallidus

Cactus ground-finch
Geospiza scandens

Sharp-beaked ground-finch
Geospiza difficilis

Mangrove-finch
Camarhynchus heliobates

Large cactus
ground-finch
Geospiza conirostris

Large insectivorous tree-finch
on Charles
Camarhynchus pauper

Large insectivorous tree-finch
Camarhynchus psittacula

Warbler-finch
Certhidea olivacea

Small insectivorous tree-finch
Camarhynchus parvulus

Cocos-finch
Pinaroloxias inornata

Vegetarian tree-finch
Camarhynchus crassirostris

Migrants from
South American mainland

many species found on the mainland. The islands lacked woodpeckers, and a very useful food resource for birds—insects beneath the bark of trees—was going unused. Finches that could make use of this resource would be at an advantage and thus favored by natural selection. On one island a finch did evolve to use this food resource. The woodpecker finch acts like a woodpecker by inserting cactus needles into holes in dead trees to extract insects. Darwin wrote: "Seeing this gradation and diversity of structure in one small, intimately related group of birds, one might really fancy that from an original paucity of birds in this archipelago, one species had been taken and modified for different ends."

Phyletic Gradualism versus Punctuated Equilibrium

Darwin visualized the process of cladogenesis as a gradual one, which we refer to as **phyletic gradualism.** However, an alternative view arose in 1972 when N. Eldredge and S. J. Gould suggested that speciation itself, and morphological changes accompanying speciation, occur rapidly, accompanied by long periods of time when little change occurs (*stasis*). They called their model **punctuated equilibrium** (periods of stasis punctuated by rapid evolutionary change). Although figure 21.7 presents what appears to be two clear alternatives, in practice the models

Figure 21.7

Diagrammatic interpretation of cladogenesis. (*a*) Phyletic gradualism is depicted as a gradual divergence over time. (*b*) Punctuated equilibrium is depicted as a rapid divergence of the two groups. The horizontal axis is some arbitrary measure of species differences.

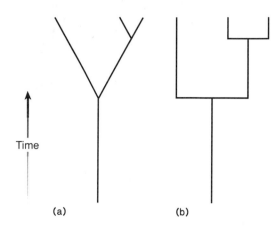

Time

(a)   (b)

Stephen J. Gould (1943–)
Courtesy of Dr. Stephen J. Gould and the Harvard University News Office.

Edmund Brisco Ford (1901–1988)
Courtesy of Professor Edmund Brisco Ford.

Richard C. Lewontin (1929–)
Courtesy of Dr. Richard C. Lewontin.

are very hard to tell apart. They both start with the same ancestral species and predict the same number of modern species. Allopatric, parapatric, and sympatric speciation mechanisms hold for both punctuated equilibrium and phyletic gradualism. The only major difference between the models is rate of change, and this can only be discovered from an almost complete fossil record. The punctuated-equilibrium model has brought much excitement to modern evolutionary biology. We await a time in the near future when we can decide which model has predominated in evolutionary history.

GENETIC VARIATION

Darwinian evolution depends on the occurrence of variation within a population. E. B. Ford, a British evolutionary biologist, applied the term **genetic polymorphism** to the occurrence of more than one allele at a given locus. Usually we consider a locus polymorphic if a second allele occurs in the population at a frequency of 5% or more. Before the mid-1960s, the general belief was that only a few loci were polymorphic in any individual or any population.

In 1966, two researchers found a way to randomly sample the genome. R. C. Lewontin and J. L. Hubby used acrylamide-gel electrophoresis (see chapter 5) to investigate variability in a fruit fly species, *Drosophila pseudoobscura*. (H. Harris reported independent, similar work in human beings.) Lewontin and Hubby reasoned that choosing enzymes and general proteins that are amenable to separation by electrophoresis is, in fact, choosing a random sample of the genome of the fruit fly. If this is the case, then the degree of polymorphism found by electrophoretic sampling would provide an estimate of the amount of variability occurring in the individual organism and in the population. Their results were startling.

Lewontin and Hubby found that the species was polymorphic at 39% of eighteen loci examined, the average population was polymorphic at 30% of its loci, and the average individual was heterozygous at 12% of its loci. The high rate of polymorphism sparked two interrelated controversies. The first was whether electrophoresis does, in fact, randomly sample the genome. The second was whether most electrophoretic alleles are maintained in the population by natural selection. Let us return to the arguments after looking at ways in which genetic polymorphisms could be maintained in natural populations.

Maintaining Polymorphisms

Heterozygote Advantage

When selection acts against both homozygotes, an equilibrium is achieved, dependent solely on the selection coefficients, that will maintain both alleles (see chapter 20). The classic example of heterozygote advantage in human beings is sickle-cell anemia. Sickle-cell hemoglobin (Hb^S) differs from normal hemoglobin (Hb^A) by having a valine

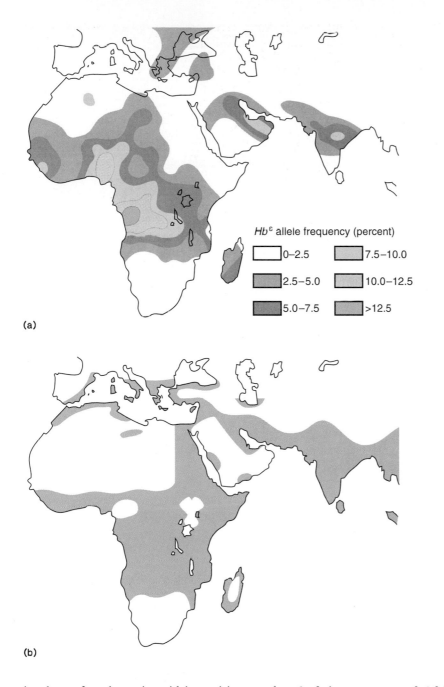

Hb^S allele frequency (percent)

☐ 0–2.5	☐ 7.5–10.0
☐ 2.5–5.0	☐ 10.0–12.5
☐ 5.0–7.5	☐ >12.5

(a)

(b)

in place of a glutamic acid in position number 6 of the beta chain of the globin molecule. Under reduced oxygen, the erythrocytes containing sickle-cell hemoglobin change from round to sickle-shaped cells (see fig. 2.27). There are two unfortunate consequences of this change of cell shape: (1) sickle-shaped cells are rapidly broken down, which causes anemia as well as hypertrophy of the bone marrow, and (2) the sickle cells clump, which blocks capillaries and produces local losses of blood flow resulting in tissue damage.

Such a condition of reduced fitness leads to the prediction that the sickle-cell allele would be selected against in all populations and, therefore, would be rare. But this is not the case. The sickle-cell allele is common in many parts of Africa, India, and southern Asia. What could possibly maintain this detrimental allele? In the search for an answer to this question, it was discovered that the distribution of the sickle-cell allele coincided well with the distribution of malaria (fig. 21.8). The following facts have now been uncovered. The sickle-cell homozygote (*Hb^S Hb^S*) almost always dies of anemia. The sickle-cell heterozygote (*Hb^A Hb^S*) is only slightly anemic and has resistance to malaria. The normal homozygote (*Hb^A Hb^A*) is not anemic and has no resistance to malaria. Thus, in areas of malaria, the most fit genotype of the three appears to be the sickle-cell heterozygote, which has resistance to malaria and only a minor anemia.

TABLE 21.1 Sickle-Cell Anemia Frequencies in African Blacks and African Americans

	Percent of Homozygotes $Hb^A Hb^A$	Percent of Heterozygotes $Hb^A Hb^S$	Frequency of Hb^S (q)
African Blacks (Midcentral Africa)	82	18	0.09
African Americans	92	8	0.04

This conclusion is supported by the changes in allelic frequencies that occur when a population from a malarial area moves to a nonmalarial area. Since the normal homozygote is no longer at risk for malaria, selection acts mainly on the sickle-cell homozygote and to a slight extent on the heterozygote. Table 21.1 shows data for African blacks versus African Americans. The African population is, of course, under malarial risk, whereas the American population is not. The sickle-cell hemoglobin allele (Hb^S) is reduced in African Americans. (Very few other examples of heterozygote advantage have been documented.)

Heterozygote advantage is an expensive mechanism for maintaining a polymorphism. Losses must occur in both homozygous groups in order for the polymorphism to exist. Thus part of the reproductive output of a population is lost each generation to maintain each polymorphism under heterozygote advantage. In the case of sickle-cell anemia, a tragic loss of human life due to either anemia or malaria results. The loss of individuals to maintain genetic variation at a particular locus is called **genetic load**. In the sickle-cell case, it is due to the segregation of individuals with lowered fitness and is therefore called **segregational load.**

Frequency-Dependent Selection

All the selection models discussed so far have had selection coefficients that were constants. This is not always the case. For example, L. Ehrman has shown that when a female fruit fly has a choice of mates of different genotypes, the female fly will choose to mate with a male with a rare genotype. **Frequency-dependent selection** is selection in which the fitnesses of genotypes change according to their frequencies in the population. In the previous chapter, we looked at heterozygote disadvantage in which an unstable equilibrium exists at $p = q = 0.5$. At allelic frequencies above or below 0.5, the rarer allele was eliminated. However, that was not frequency-dependent selection because fitnesses did not change as allelic frequencies change. Here we will be looking at models that allow the maintenance of polymorphisms. That is, we will look at frequency-dependent models in which a genotype is at an advantage when rare and at a disadvantage when common. Frequency-dependent models can also be built with results similar to heterozygote disadvantage in which a genotype is at an advantage when common and at a disadvantage when rare.

Lee Ehrman (1935–   )
Courtesy of Dr. Lee Ehrman. Photo by Jan Robert Factor.

The population geneticist Bruce Wallace has coined the terms *hard selection* and *soft selection* to deal with cases of frequency and density dependence. (Density-dependent selection exists when the fitness of a genotype changes as population density changes. We will not deal with that here.) Wallace defined soft selection as selection in which the selection coefficients are dependent on frequency and density of genotypes. Hard selection is selection that is independent of both frequency and density. For example, the low fitness of sickle-cell anemia homozygotes involves hard selection because of the objectively deleterious effects of the anemia. Soft selection could be envisioned as selection that acts perhaps on aggressive behavioral genotypes in some field mouse species. When population density and frequency of the genotypes are low, they will survive and reproduce. As population density and frequency of the genotypes increase, they may be selected against because of the preoccupation of these aggressive individuals with territory defense under crowded conditions. This has been suggested as a mechanism of wildlife's "lemming cycle," rapid declines in density of lemming and field mouse populations every three to five years.

A model for frequency-dependent selection can be constructed by assigning fitnesses that are not constants. One way to do this is to assign fitnesses that are a function of allelic frequencies. Thus the assigned fitnesses for one locus with two alleles could be $(1.5 - p)$, 1, and

TABLE 21.2 Selection Model of Frequency-Dependent Selection: The A Locus with the *A* and *a* Alleles.

| | Genotype | | | |
	AA	*Aa*	*aa*	Total
Initial genotypic frequencies	p^2	$2pq$	q^2	1
Fitness (W)	$1.5 - p$	1	$1.5 - q$	
Ratio after selection	$p^2(1.5 - p)$	$2pq$	$q^2(1.5 - q)$	$\overline{W} = 0.5 + 2pq$
Genotypic frequencies after selection	$\dfrac{p^2(1.5 - p)}{\overline{W}}$	$\dfrac{2pq}{\overline{W}}$	$\dfrac{q^2(1.5 - q)}{\overline{W}}$	1

$(1.5 - q)$ for the *AA, Aa,* and *aa* genotypes, respectively (table 21.2). An interesting outcome of this model is that at $p = q = 0.5$, the system is in equilibrium and there is no selection because all the fitnesses are equal to 1.

Another way of looking at frequency-dependent selection is to look at the situation in which each genotype exploits a slightly different resource. As a genotype becomes rare, there most likely will be less competition for the resource used by that genotype and therefore the genotype will be at an advantage compared with the common genotypes, which are competing for resources. This type of selection is probably very common.

Transient Polymorphism

A genetic polymorphism can result when an allele is being eliminated either by random or selective mechanisms. If a population starts out homozygous for the *a* allele, for example, and a mutation brings in a more-favored *A* allele, the population will gradually become all *A* through directional selection. However, during the process of replacement, both alleles will be present.

Other Systems

Selection at one stage in the life cycle of an organism can be balanced by a different form of selection at another stage in the life cycle. For example, an allele can be favored in a larva but selected against in an adult. There can also be a balance of selection in different parts of the habitat in a heterogeneous environment. For instance, an allele can be favored in a wet part of the habitat but selected against in a dry part of the habitat.

Maintaining Many Polymorphisms

In summary, allelic polymorphisms in a population were accounted for classically by heterozygote advantage, frequency-dependent selection, or, infrequently, some other mechanism. Until the work of Lewontin and Hubby, heterozygote advantage was believed to be the most common method of maintaining a polymorphism at a given locus. The maintenance of an allele by heterozygote advantage costs the population a certain number of its offspring due to the mortality (or sterility) of the homozygotes. Most populations can afford the loss if polymorphisms are maintained at only a few loci. After Lewontin and Hubby reported that polymorphisms seemed to exist at a large proportion of loci, new explanations were needed to account for them. Three explanations were considered:

1. Electrophoresis (the technique used in the research of Lewontin and Hubby) does not randomly sample the genome, and thus a large amount of variability does not really exist.

2. New population genetic models can be derived that explain how this large amount of variability is maintained.

3. Electrophoretic alleles are not under selective pressure. That is, allozymic forms of an enzyme all perform the function of the enzyme equally well. This idea is called the **neutral gene hypothesis.**

Sampling the Genome

Does electrophoresis randomly sample the genome? Since, on the basis of DNA content, the genome of higher organisms has the potential to contain half a million genes, there may always be a question as to whether it is really being sampled randomly by electrophoresis. Since the original reports of Lewontin and Hubby and Harris, numerous studies on many different organisms agree, for the most part, on the high amount of polymorphism in natural populations (table 21.3). However, several lines of evidence suggest that the results from electrophoresis are actually underestimates of the true amount of genetic variability present in a population.

The majority of amino acid substitutions, for example, do not change the charge of the protein. Thus, what appear as single bands on an electrophoretic gel could actually be heterogeneous mixtures of the products of several alleles. Also, it is now known that glycolytic enzymes are less polymorphic than other enzymes. And, since glycolysis is a limited process in which most enzymes are not involved, it follows that the average heterozygosity over all loci should be slightly higher than the original estimates that included glycolytic enzymes. Recent technical

TABLE 21.3 Survey of Genic Heterozygosity

Species	Number of Populations	Number of Loci	Proportion of Loci Polymorphic per Population	Heterozygosity per Locus	Standard Error of Heterozygosity
Homo sapiens	1	71	0.28	0.067	0.018
Mus musculus musculus	4	41	0.29	0.091	0.023
M. m. brevirostris	1	40	0.30	0.110	—
M. m. domesticus	2	41	0.20	0.056	0.022
Peromyscus polionotus	7 (regions)	32	0.23	0.057	0.014
Drosophila pseudoobscura	10	24	0.43	0.128	0.041
D. persimilis	1	24	0.25	0.106	0.040
D. obscura	3 (regions)	30	0.53	0.108	0.030
D. subobscura	6	31	0.47	0.076	0.024
D. willistoni	2–21	28	0.86	0.184	0.032
	10	20	0.81	0.175	0.039
D. melanogaster	1	19	0.42	0.119	0.037
D. simulans	1	18	0.61	0.160	0.052
Limulus polyphemus	4	25	0.25	0.061	0.024

Note: See source (Lewontin, 1974) for individual references.
From R. C. Lewontin, *The Genetic Basis of Evolutionary Change.* Copyright © 1974 Columbia University Press, New York, NY. Reprinted by permission.

advances of multidimensional electrophoresis and DNA sequencing also support the hypothesis that electrophoresis does randomly sample the genome. However, DNA sequencing studies have shown that abundant variation exists, especially in the third (wobble) position of codons, and in parts of introns. Heterozygosity at the DNA sequence level seems to approach 100%.

Multilocus Selection Models

Can the high degree of variability in natural populations be accounted for by standard genetic models? If each locus is considered independently, then for each polymorphic locus, offspring in a population lost in order to maintain that polymorphism by heterozygote advantage are independent of offspring lost due to selection at other loci. The losses would soon outstrip the reproductive capacity of any species. Models proposed since Lewontin and Hubby's report have suggested that natural selection favors the individuals that are the most heterozygous overall. Individuals selected against because of their homozygosity would be the individuals with many homozygous loci. In other words, natural selection acts on the entire genome, not on each locus separately. We can show algebraically that the large number of polymorphisms that exist in natural populations could be maintained according to these models.

Neutral Alleles

The high incidence of polymorphism revealed by electrophoresis may not be important from an evolutionary point of view. If all or most electrophoretic alleles are neutral (i.e., if no allele is more fit than its alternative) or only very slightly deleterious, there will be virtually no selec-

Motoo Kimura (1924–   )
Courtesy of Dr. Motoo Kimura.

tion at these loci and the variation observed in the population is merely a chance accumulation of mutations, a combination of mutation and genetic drift. This model, proposed by M. Kimura of Japan, is an alternative to the natural-selection model.

Which Hypothesis Is Correct?

Researchers favoring the concept that most electrophoretic alleles are neutral do not deny that selection exists. They do not hold that evolution is nonadaptive but say merely that most of the molecular variation (electrophoretic) found in nature is not related to fitness—it is neutral. Thus the demonstration that selection actually exists, in electrophoretic systems or otherwise, is not proof against

Figure 21.9

Relation of latitude and frequency of the warm-adapted esterase allele *Es-I^a* in populations of the fish *Catostomus clarki*. Note how the frequency of the allele increases as latitude decreases (warmer water).

From Richard Koehn, "Functional and Evolutionary Dynamics of Polymorphic Esterases in Catostomid Fishes" in Transactions of the American Fisheries Society, 99:223. Copyright © 1970 American Fisheries Society, Bethesda, MD. Reprinted by permission.

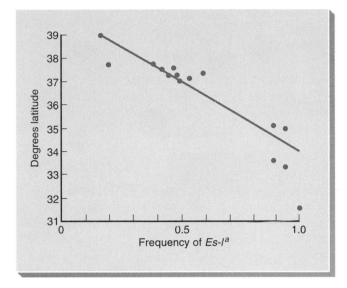

Grand Patterns of Variation

Clinal Selection

Data on the geographic distribution of alleles also fail to support adequately either theory. Often a single allele predominates over the range of a species (fig. 21.10). Changes in allelic frequency from one geographic area to another can often be attributed to **clinal selection,** selection along a geographic gradient, in which allelic frequencies change as altitude, latitude, or some other geographic attribute changes. Note the general increase in the *Es-5^b* frequency from west to east in the southern United States (fig. 21.10). But, in line with the neutralist view, geographic patterns similar to those in figure 21.10 can also be produced by neutral alleles with a very low level of migration, as little as one individual per one thousand per generation.

Molecular Evolutionary Clock

The advancing technology that made it possible to detect the sequence of amino acids in a protein also made it possible to discover by how much the proteins and DNA of various species differ. In chapter 17 we discussed the use of mitochondrial DNA (mtDNA) to determine evolutionary relationships. Currently protein, nuclear DNA, and mtDNA clocks are being studied.

Knowledge of the changes of amino acid sequences can be used to estimate the rate of evolutionary change. That is, the data show how many amino acid substitutions have occurred between two known groups of organisms. The genetic code dictionary allows us to estimate the minimum number of nucleotide substitutions required for this change. For example, if one protein contains a phenylalanine in position 7 (codons UUU, UUC) and the same protein in a different species has an isoleucine in the same position (AUU, AUC, AUA), we can see that the minimum number of substitutions to convert a phenylalanine codon to an isoleucine codon is one (UUU → AUU). With the minimum number of substitutions known, we can calculate molecular **evolutionary rates,** nucleotide substitutions per million years. In a sense these rates provide us with a **molecular evolutionary clock,** measuring evolutionary time in nucleotide substitutions.

Many studies of the rate of amino acid and nucleotide substitutions have been done on hemoglobin, cytochrome *c,* a class of proteins involved in blood clotting called fibrinopeptides, and many others. Figure 21.11 shows the way in which an amino acid sequence differs among species. From comparisons of this type, we can calculate the actual number of amino acid differences as well as percentage differences. Table 21.4 is a compilation of percentage differences between various species based on the cytochrome *c* protein. This type of information can be used two ways.

the neutralist view. No one denies the explanation for the maintenance of sickle-cell anemia. Selection at several other electrophoretic systems is also known.

For example, Koehn showed that different alleles of an esterase locus in a freshwater fish in Colorado produced proteins with different enzyme activities at different water temperatures. Koehn then showed that the alleles were distributed as would be predicted on the basis of the temperature of the water. In other words, the distribution of alleles correlated with the distribution of water temperature. The enzyme produced by the *ES-I^a* allele functioned best at warm temperatures, whereas the enzyme produced by the *ES-I^b* allele functioned best at cold temperatures. The cold-adapted enzyme was prevalent in the fish in colder waters (higher latitudes) and the warm-adapted enzyme was prevalent in the fish in warmer waters (lower latitudes; fig. 21.9).

Isolated instances of selection, however, are not adequate to prove the case for the maintenance of variation by means of natural selection or disprove the case for neutral alleles. Both theories recognize natural selection as the guiding force in producing adapted organisms. What is needed is proof that the majority of polymorphic loci are either being selected or are neutral. For this proof, many loci must be examined independently—a very difficult undertaking—or some grand pattern must emerge supporting one hypothesis or the other.

Figure 21.10
Esterase allele frequency distribution of the *Es-5* locus in house mice. Each *circle* represents allelic frequencies at that geographic location. Note the general tendency for the *Es-5ᵇ* allele to increase in the continental United States from west to east.

From Linda L. Wheeler and Robert K. Selander, "Genetics Variation in Populations of the House Mouse, Mus musculus, *in the Hawaiian Islands" in* Studies in Genetics, *VII. Copyright © 1972 University of Texas Publications, Austin, TX. Reprinted by permission of the authors.*

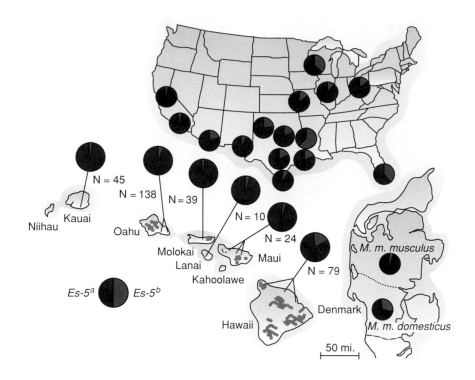

Figure 21.11
The amino acids making up the terminal portion of cytochrome *c* in three species. Note the similarities and differences.

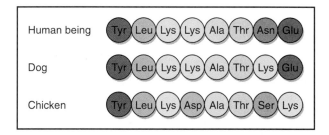

First, a **phylogenetic tree** can be constructed that tells us the evolutionary history of the species under consideration (fig. 21.12). This tree can be compared with phylogenetic trees constructed by more classical means using fossil evidence and evidence from morphology, physiology, and development. From the comparisons, we can look at areas of disagreement in an attempt to find out the best way of creating phylogenetic trees. In addition, molecular phylogenies can give us information unattainable any other way, as, for example, when the fossil record is incomplete or ambiguous.

A second use of DNA or amino acid difference data is to determine average rates of substitution. Knowing the current amino acid differences in a protein between species, it is possible to estimate the actual number of nucleotide substitutions that have taken place over evolutionary time using the statistical Poisson distribution, which deals with rare events. The index, K, is the average number of amino acid substitutions, per site, between two proteins:

$$K = -\ln(1 - p)$$

in which ln is the natural logarithm (to the base e) and $p = d/n$ in which d is the number of amino acid differences and n is the total number of amino acid sites being compared. For example, in figure 21.11, $n = 8$ and $d = 3$ between the dog and chicken. Thus

$$K = -\ln(1 - 0.375) = 0.47$$

Therefore the average number of amino acid substitutions, per site, between dog and chicken is 0.47.

We can take this calculation one step further by determining the per-year rate:

$$k = K/2T$$

in which k is the amino acid substitution rate per site per year and T is the number of years since the two species diverged from a common ancestor. We divide by $2T$ because each side of the tree has evolved independently for T years. When ks are calculated for many proteins over many species, they cluster around 10^{-9} (table 21.5). In fact, Kimura has suggested the unit of a *pauling* to be equal to 10^{-9} amino acid substitutions per year per site in honor of Linus Pauling, who, along with E. Zuckerkandl, first proposed the concept of a molecular clock in 1963. If

Table 21.4 Amino Acid Differences (percent) in Cytochrome c between Different Organisms

	Human Being	Pig	Horse	Chicken	Turtle	Bullfrog	Tuna	Carp	Lamprey	Fruit Fly	Screw-Worm	Silkworm	Sesame	Sunflower	Wheat	C. krusei	Yeast	N. crassa	R. rubrum
Human Being	0	10	12	13	14	17	20	17	19	27	25	29	35	38	38	46	41	44	65
Pig, Bovine, Sheep		0	3	9	9	11	16	11	13	22	20	25	38	40	40	45	41	43	64
Horse			0	11	11	13	18	13	15	22	20	27	39	41	41	46	42	43	64
Chicken, Turkey				0	8	11	16	14	17	23	21	26	40	41	41	45	41	44	64
Snapping Turtle					0	10	17	13	18	22	22	26	38	39	41	47	44	45	64
Bullfrog						0	14	13	20	20	20	27	41	42	43	46	43	45	65
Tuna Fish							0	8	18	23	22	30	42	43	44	43	43	45	65
Carp								0	12	21	20	25	40	41	42	45	42	43	64
Lamprey									0	27	26	30	44	44	46	50	45	47	66
Fruit Fly										0	2	14	42	41	42	43	42	38	65
Screw-Worm Fly											0	13	41	40	40	43	42	38	64
Silkworm Moth												0	39	40	40	43	44	44	65
Sesame													0	10	13	47	44	48	65
Sunflower														0	13	47	43	49	67
Wheat															0	45	42	48	66
Candida krusei																0	25	39	72
Baker's Yeast																	0	38	69
Neurospora crassa																		0	69
Rhodospirillum rubrum																			0

Source: *Atlas of Protein Sequence and Structure*, M. O. Dayhoff (ed.). Copyright © 1972 National Biomedical Research Foundation, Washington, DC.

Figure 21.12

Composite evolution of hemoglobin, cytochrome c, and fibrinopeptide A. The total number of nucleotide substitutions is given on the horizontal axis. Note how the tree groups similar organisms and generally agrees with classical systematics.

From C. H. Langley and W. M. Fitch, "An Examination of the Constancy of the Rate of Molecular Evolution" in Journal of Molecular Evolution, *3:168. Copyright © 1974 Springer-Verlag, Heidelberg. Reprinted by permission.*

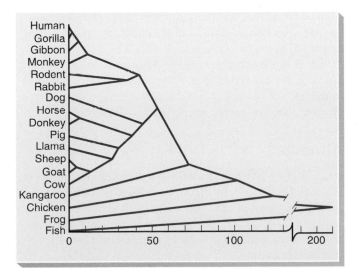

Table 21.5 Evolutionary Rates (k) as 10^{-9} Substitutions per Amino Acid Site per Year for Various Proteins

Protein	k
Fibrinopeptide	8.3
Pancreatic ribonuclease	2.1
Lysozyme	2.0
Hemoglobin alpha	1.2
Myoglobin	0.89
Insulin	0.44
Cytochrome c	0.3
Histone H4	0.01

Source: Data from Kimura, 1983.

the values of k (such as those in table 21.5) form a normal distribution around 10^{-9}, then 10^{-9} would be the rate of "the" molecular evolutionary clock. So far the data have been too limited to determine the distribution.

Although controversy still exists, the neutralists have interpreted the relative constancy of the molecular evolutionary clock as strong evidence in support of the neutral gene hypothesis. A constant rate of molecular evolution over many groups of organisms over many different time intervals implies that the substitution rate is a stochastic or random process rather than a directed or selectional process. This is not to say that there are no adapted changes in proteins or that there are no constraints.

It would be unreasonable to think that all areas of a protein are under the same constraints in terms of amino acid changes that could be neutral. In fact, the evidence suggests that there are three classes of amino acids in terms of substitution rate: invariant, moderately variant, and hypervariant. It seems possible that there will be virtually no substitutions of amino acids in and around the active site of the enzyme since any amino acid change in that area might be deleterious or lethal. For example, there is a segment of cytochrome c, from amino acids 70 to 80, that is invariant in all organisms tested. This area includes a binding site of the protein.

DNA Variation

If the neutralist view of molecular evolution is correct, we should be able to make some predictions about rates of change in DNA. For example, we predict that DNA under greater constraint should amass fewer base changes than DNA under lesser constraint. We could test this by looking at the accumulation of mutations in the three positions of the codon, or we could look at DNA that is not directly translated such as pseudogenes (chapter 14) or introns which are probably under lesser constraint. Let us first look at the three positions of the codon.

A reexamination of the codon dictionary (see table 11.4) shows that the third, or wobble, position of the codon should be under less constraint. Eight amino acids belong to unmixed families in which the amino acid is defined by the first and second positions coupled with any of the four bases in the third position of the codon. The remaining amino acids belong to mixed families in which the first two positions and the purine or pyrimidine nature of the third position is important. Hence the wobble (third) position of the codon is under the least constraint and should build up the most neutral or near-neutral mutations.

In addition, analysis of changes in the first and second positions indicate that more drastic change in the physical chemistry of the amino acids takes place by mutation of the second rather than the first position of the codon. Thus we predict that evolutionary distance, as measured by base substitutions, should be greatest for the third position and least for the second position of codons. This turns out to be generally true (table 21.6).

It should be clear here that a major problem facing those studying evolutionary clocks is calibrating them. Are average changes uniform throughout lineages? Do clocks speed up, slow down, or show other unpredictable changes through time? There is evidence, for example, that both the nuclear and mitochondrial DNA clocks have slowed down in the hominid lineage as compared with old-world monkeys. If the clocks change speed in different lineages,

Table 21.6 Evolutionary Distance of Codons, Measured in Base Substitutions per Nucleotide Site

	Codon Site		
	2	1	3
Beta globin, human being vs. mouse	0.13	0.17	0.34
Beta globin, chicken vs. rabbit	0.19	0.30	0.64
Rabbit, alpha vs. beta globin	0.44	0.54	0.90

Source: Data summary from Kimura, 1983.

at different times, and for different parts of the genome, then there will be errors interpreting lineages and errors in using averages to understand the general patterns of change.

At this point in time, the neutral mutation theory is still not universally accepted as the explanation for most of the protein and DNA variations within and between species. Although, it is probably safe to say that natural selection acts to create organisms that are adapted to their environments (see Boxes 21.2 and 21.3 on mimicry and industrial melanism), many nucleotide and amino acid changes may not have measurable effects on the fitness of the organism, and hence their frequencies may be determined by the stochastic processes of mutation and genetic drift. Adaptation is by natural selection, but there is most certainly neutral variation in organisms.

SOCIOBIOLOGY

We close this chapter by looking at a level of evolution that has only recently been addressed. In 1975, E. O. Wilson published a mammoth tome entitled *Sociobiology: The New Synthesis*. This book has been the center of major controversies that have spread to the fields of sociology, psychology, anthropology, ethology, and political science. The basic premise of the book is that social behavior is under genetic control. Although Wilson's book contains twenty-six chapters concerned with the animal kingdom, controversies have arisen because of the one chapter that applies the theory to human beings.

Altruism

V. C. Wynne-Edwards published a book in 1962 entitled *Animal Dispersion in Relation to Social Behavior*, in which he suggested that animals regulated their own population density through altruistic behavior. For example, under crowded conditions many birds cease reproducing. The interpretation of this phenomenon was that these birds were being altruistic: Their failure to breed was for the

BOX 21.2
MIMICRY

Mimicry is a phenomenon whereby an individual of one species gains an advantage by resembling an individual of a different species. There are at least two types of mimicry.

In **Müllerian mimicry,** named after F. Müller, several groups of organisms gain an advantage by looking like one another. This mimicry occurs among organisms in which all the mimetic species are offensive and obnoxious. The classical example is the general similarity of bees, wasps, and hornets.

In **Batesian mimicry,** named after H. W. Bates, a vulnerable organism (mimic) gains a selective advantage by looking like a dangerous or distasteful organism (model). The classical example of Batesian mimicry was, until 1991, the monarch (*Danaus plexippus*) and viceroy (*Limenitis archippus*) butterflies (Box fig. 1). Although the viceroy is smaller and, on close examination, looks quite different from the monarch, the resemblance is striking at first glance. Monarch butterflies feed on milkweed plants and from them obtain noxious chemicals called cardiac glycosides that the monarchs store in their bodies. When a bird tries to eat a monarch, it becomes sick and regurgitates what it has eaten. Thereafter, the bird will not only avoid eating monarchs but also will avoid eating any butterflies that look anything like monarchs. Previously it had been believed that the mimetic viceroy butterfly gained a selective advantage by looking like the monarch and fooling bird predators into thinking that the viceroy was bad to eat. However, Ritland and Brower demonstrated a previously unrealized fact: the viceroys taste as bad as the monarchs to birds. This fact changes the mimicry situation of these two species from one of Batesian to one of Müllerian mimicry.

Examples of Batesian mimicry occur in numerous butterfly species. For example, in West Africa, *Pseudacraea* species mimic species of the genus *Bematistes* (Box fig. 2). These species are primarily black and white or black and orange, and in some the sexes differ, each having a different mimic. Upwards of 20 species can be involved in these mimicry complexes in one area.

Both forms of mimicry depend on a selective pressure generated by predation. Certain requirements must be met in order for each system to work properly. Batesian mimicry has the following requirements:

1. The model species must be conspicuous and inedible or dangerous.
2. Both model and mimic species must occur in the same area with the model being very abundant. If the model is rare, predators do not have sufficient opportunity to learn that its pattern is associated with a bad taste. In fact, the reverse can happen in which the model can be at a selective disadvantage if it is rare because the predators will learn from the mimic that the pattern is associated with something good to eat.

Box figure 1

Müllerian mimicry. (*a*) Monarch butterfly and (*b*) viceroy butterfly. Both have a similar color (*orange* and *black*) and a generally similar pattern.

(a) © Robert Finke/Photo Researchers, Inc. (b) © Richard Parker/Photo Researchers, Inc.

(a)

(b)

3. The mimic should be very similar to the model in morphological characteristics easily perceived by predators but not necessarily similar in other traits. The mimic is not evolving to *be* the model, only to look like it.

Müllerian mimicry requires that all the species be similar in appearance and distinctively colored. They can, however, be equally numerous. And, as the British geneticist P. M. Sheppard pointed out, the resemblance among Müllerian mimics need not be as good as between the mimic and model of a Batesian pair because Müllerian mimics are not trying to deceive a predator, only to remind the predator of the relationship.

Continued on next page

BOX 21.2—*continued*

Box figure 2

Batesian mimicry seen in West African butterflies occurring in the same places. Those on the left are different model species belonging to the genus *Bematistes*. Those on the right that mimic them are different species belonging to the genus *Pseudocraea*.

© *J. A. L. Cooke/Oxford Scientific Films/Animals, Animals.*

Although there have been some critics of mimicry theory, especially critics of the way in which the system could evolve, the general model put forth by the population geneticist and mathematician R. A. Fisher is generally accepted. According to Fisher, any new mutation that gave a mimic any slight advantage would be selected for. As time proceeded, other loci that might favorably modify expression of mimetic genes would also be selected for in order to improve the similarity of mimic and model. This mechanism surmounts the criticism that a single mutation could not produce a mimic that so closely resembled its model.

ultimate good of the species. (**Altruism** means risking loss of fitness in an act that could improve the fitness of another individual.) Wynne-Edwards suggested a mechanism called **group selection:** groups that had altruistic behavior would have a survival advantage over groups that did not.

Then in 1966, G. Williams, in his book *Adaptation and Natural Selection: A Critique of Some Current Evolutionary Thought,* refuted the altruistic view with the charge that individuals that performed altruistic acts would be selected against. In other words, organisms not performing altruistic acts would have a higher fitness. Williams held that apparent altruism had to be interpreted on the basis of benefits accruing to the individual performing the altruistic act. After his book, the idea of doing something for the good of the species became passé. How, then, can apparent altruism be accounted for? How

BOX 21.3

INDUSTRIAL MELANISM

Industrial melanism is the darkening of moths during the Industrial Revolution and is an interesting case study of natural selection. It illustrates the type of selection that can be caused by human intervention in natural systems: It is a spectacular example of very rapid evolution in nature, and it shows the importance to natural populations of having a reserve of genetic variability. The phenomenon was first observed in England and gets its name from the increase in frequency of dark-colored (melanic) phenotypes of several species of moths concurrent with a change in the environment due to industrialization. Industrial melanism has occurred in more than fifty species, of which the best studied is the peppered moth, *Biston betularia*. Until the middle of the nineteenth century, there was little industrialization in England, and the only known form of the peppered moth was the *typical* form, which is white with a peppering of black (Box fig. 1). A black form, known as *carbonaria,* was discovered in 1848 and, as industrial pollution increased, so did the frequency of *carbonaria,* until it often made up 95% of a population. The black pigmentation is controlled by a single dominant allele.

Before industrialization, trees were covered by lichens and the moths had a pattern remarkably similar to the pattern of the lichens. On the lichen, these moths were hidden from predators. With predators present, this **cryptic coloration** was selected for. With the advent of industrialization, pollution killed the lichens and blackened the trunks of the trees; the typical moths were then obvious to predators, whereas the black *carbonaria* form had become nearly invisible. The *carbonaria* form increased in frequency as a result of predators selectively eating the typical form.

H. B. D. Kettlewell, a British biologist, demonstrated that predation by birds was the selective force here. He released both forms of the moth and observed the feeding of birds on these moths as well as the recapture rates of the two forms of moths. He found that in an industrial area, 43 typical moths were taken to 15 *carbonaria;* in a pollution-free area, 164 *carbonaria* were taken to 26 typicals. In recent years, industrial cities have lowered pollution levels and, as expected, the frequency of the *carbonaria* form has decreased.

Box figure 1

Peppered moth (*a*) on soot-covered tree trunk from a polluted industrial area, melanic moth (*b*) on lichen-covered tree trunks from unpolluted rural area, and (*c*) both on a soot-covered tree trunk from a polluted industrial area. Note how the peppered form on soot-covered trees and the melanic form on lichen-covered trees are highly visible.

(a,b) *Courtesy © J. A. Bishop and L. M. Cook.* (c) © *John D. Cunningham/ Visuals Unlimited.*

(a)

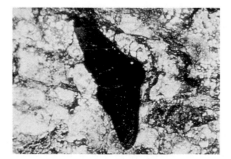

(b)

(c)

Edward O. Wilson (1929–   )
Courtesy of Dr. Edward O. Wilson. Photo by Pat Hill/OMNI Publications Int'l Ltd.

Figure 21.13

Haplodiploidy in eusocial hymenoptera produces sisters with an average $r = 0.75$. Because drones (males) are haploid, queens produce daughters of only two genotypes. A given daughter has an $r = 1.0$ with sisters of identical genotype and an $r = 0.5$ with sisters having the other genotype, for an average $r = 0.75$. In other words, females have a 75 percent genetic similarity with their sisters while having only a 50 percent similarity with their own offspring.

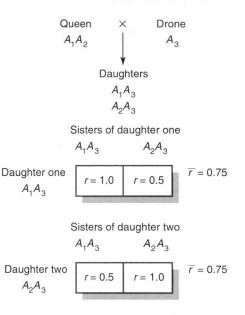

can we explain why ground squirrels appear to put themselves at risk to predators by giving alarm calls and why female workers in ant, wasp, and bee colonies forsake reproduction in order to work for the colony? **Sociobiology,** the study of the evolution of social behavior, attempts to answer these questions.

Kin Selection and Inclusive Fitness

In 1964, W. D. Hamilton developed concepts that explained altruistic acts on the basis of improving individual fitness. Starting with the known fact that relatives have alleles in common, Hamilton suggested that natural selection would favor an allele that prompted altruistic behavior toward relatives because the result of this behavior might be an increase in copies of that allele in the next generation. The proportion of alleles shared by two individuals can be defined as a **coefficient of relationship, r.** If an individual has a certain allele, the probability that a particular relative also has that allele is r. Siblings have an $r = 1/2$. A squirrel is likely to have all its alleles still viable if it sacrifices itself for two or more siblings. In fact, natural selection should definitely favor altruism of an individual toward three siblings because, in a sense, natural selection is weighing one copy of an individual's alleles (the individual itself) versus 1.5 copies (three siblings).

This sort of reasoning has been termed the **calculus of the genes.** It does not imply that individuals actually think these things out; rather, natural selection has favored those individuals that do behave this way. Hamilton referred to the sum of an individual's fitness plus fitness effects of alleles shared by relatives as **inclusive fitness.** He referred to the way natural selection acts on inclusive fitness as **kin selection.**

Hamilton applied his ideas of inclusive fitness and kin selection to explain sterile castes in the eusocial (truly social) hymenoptera (bees, ants, and wasps). The workers in these colonies are sterile females. Why do they forsake their ability to reproduce in order to help with the chores of maintaining the hive or colony? The answer seems to come from **haplodiploidy,** the unusual sex-determining mechanism of these species. In the eusocial hymenoptera with sterile castes, fertilized eggs produce diploid females whereas unfertilized eggs produce haploid males (drones). The difference between a reproductive queen and a sterile worker in bees is larval nutrition: Larvae fed "royal jelly" can become queens. Hamilton showed that since a worker is more closely related to her sisters than to her own potential offspring, kin selection could favor a worker helping her sisters at the expense of her own reproduction.

Figure 21.13 shows the case of a queen (female) with alleles A_1 and A_2 at the A locus and a haploid drone (male) with the A_3 allele. A daughter will have either the A_1A_3

or A_2A_3 genotype. If we compare one of these daughters with her sisters, we see that the average $r = 0.75$—half of the time $r = 1.0$ and the other half of the time $r = 0.5$. A queen and her daughters have an $r = 0.5$. Thus we see that workers (females) are more closely related to their sisters and hence are at a reproductive advantage by raising them rather than their own young. Wilson has pointed out that sterile caste systems have evolved among insects in only one other group beside the eusocial hymenoptera, the termites. Although eusocial hymenoptera make up only 6% of insects, sterile castes have independently evolved at least eleven times. This is compelling evidence for the validity of Hamilton's analysis. Only one noninsect example of castes has been discovered: The naked mole rat, a small subterranean rodent living in Africa has this type of social system.

Many studies concerned with apparently altruistic acts have provided a large body of support for Hamilton's theory of kin selection and inclusive fitness. P. Sherman, working with ground squirrels, for example, has observed that alarm calls are made primarily by those individuals that have the most to gain from the standpoint of inclusive fitness, namely resident females surrounded by kin.

One other explanation for altruism is also consistent with benefits to individual fitness. It is that many apparently altruistic acts are in reality selfish—they just look altruistic. To be altruistic, an individual must risk reducing its fitness for the potential benefit of the fitness of others. We may, in fact, misinterpret some acts as altruistic that simply are not.

This turnaround in thought, from group selection to individual selection, has been an intellectual revolution in modern evolutionary biology. Before this revolution, many of the behaviors in nature that involved apparent altruism were difficult to explain. Now sociobiological reasoning provides an explanation.

The reason that so much controversy has sprung up over the theory of genetic control of social behavior is in the implications the theory has for human social, political, and legal issues. Human husband–wife, parent–child, and child–child conflicts, for example, may be built into the genes. Altruism, our highest nobility, may be mere selfishness. Many critics fear that sociobiological concepts can be used to support sexism and racism. For human beings, the alternative to the theory of sociobiology is the theory that most human behavior is a result of the environment, including cultural learning. At present, although much evidence remains to be gathered, the sociobiology concept is very attractive to many evolutionists.

SUMMARY

The theory of evolution by natural selection was put forward by Charles Darwin, who recognized the natural variation among individuals within a population of similar organisms. He noted also that offspring are overproduced in nature, and this overproduction inevitably leads to competition for scarce resources. Darwin assumed that, when competition occurs, the most fit will survive and, through time, a population would become better adapted to its environment through the process of natural selection. Applying the algebra of population genetics to this theory leads to the modern concept of evolution, neo-Darwinism.

Cladogenic speciation occurs when reproductive isolating mechanisms arise, usually after gene flow in a population is blocked. Different populations of a species can then evolve independently. When the point is reached at which individuals from the isolates can no longer interbreed, speciation has taken place. If the isolates then come in contact again, they will remain as separate species. Speciation may occur gradually or in a punctuated manner; it can be by allopatric, parapatric, or sympatric mechanisms.

Evolution depends on variation. In 1966, Lewontin and Hubby, using electrophoresis, showed that a tremendous amount of heterozygosity occurred in natural populations. Attempts to explain this variation have led to two major competing theories: (1) variation is being maintained selectively and (2) variation is not under selective pressure but is instead neutral. Two areas of evidence support the neutralist view.

First, the molecular evolutionary clock (the per-year, per-amino acid, substitution rate) appears to be fairly constant at 10^{-9}. This constancy implies that the majority of amino acid changes are the result of stochastic processes. Second, there have been greater numbers of nucleotide substitutions in DNA under lesser constraint than DNA under greater constraint. For example, the third, or wobble, position of the codon has had an accumulation of more mutations than the other two positions. We conclude that natural selection creates adapted organisms, but the majority of base and amino acid changes may be neutral.

Sociobiology is another term for evolutionary behavioral ecology. It attempts to provide evolutionary explanations for social behaviors. Apparent altruistic behavior can be explained either by kin selection or selfishness. Sterile castes in insects have come about because of the unusual haplodiploid sex-determining mechanism in the eusocial hymenoptera. There is much controversy about and little information for applying sociobiological principles to human behavior.

1. What are the relationships between evolution and reproductive isolating mechanisms?

   ANSWER: Reproductive isolating mechanisms prevent individuals in two populations from mating with each other or producing viable offspring. These mechanisms can be prezygotic or postzygotic. They usually evolve during the time that populations are isolated from each other, either physically or during parapatric or sympatric speciation. For example, if a species is split by a new river, the populations on either side of the river can evolve in isolation from each other. Reproductive isolating mechanisms usually evolve irrespective of the other facets of evolution that are taking place. Thus if, after time, the two populations come into contact (the river might dry up), reproductive isolating mechanisms may have evolved to prevent mating. If weak reproductive isolating mechanisms have evolved, natural selection will usually favor a strengthening of them by selecting against hybrids and any mating behavior that leads to the formation of hybrids.

2. What is our modern evolutionary concept of altruism?

   ANSWER: An altruistic act is one in which an individual risks the loss of fitness in order to benefit another individual. Human beings value these "selfless" acts; they are not favored in natural animal populations, however, except under very specific circumstances, because altruistic acts should be selected against. In other words, all other things being equal, an individual that did not do altruistic acts would have a higher fitness than one that did do these acts, and therefore fitness is higher for "selfish" individuals. Altruistic acts, however, are expected if the recipient of the acts shares genes in common with the individual performing the altruistic acts. Generally, altruism can be expected among relatives, following the rules of kin selection.

1. The following electrophoretic data are from a sample of one hundred field mice for their salivary amylase-1 genotypes. The two alleles are F and S, for fast and slow migration in an electric field: FF, 43; FS, 54; and SS, 3. Is selection acting? What would you look for in data to determine among frequency-dependent selection, heterozygote advantage, and transient polymorphism?

2. What mechanisms permit the maintenance of genetic variability in natural populations? Give examples where possible.

3. Outline the Darwinian mechanism of the process of evolution. What is meant by neo-Darwinism?

4. The population geneticist Hampton Carson has defined a "population flush" as a period of reduced selection during population increase. Why should there be reduced selection during a flush?

5. Describe how the processes of allopatric, parapatric, and sympatric speciation could take place.

6. Discuss the "neutral gene hypothesis." What are its alternatives? What data are needed to distinguish among these views?

7. Koehn showed that different functioning alleles of an esterase system in fish were correlated with water temperature. What sorts of selection can you imagine could affect the same type of alleles in mammals, which are homeothermic (warm-blooded) and hence maintain a relatively constant internal temperature?

8. Niemälä and Tuomi have suggested that the irregular leaf outlines seen in some plant species is a form of mimicry. What would the leaves be mimicking? What form of mimicry might this be?

9. Can information on evolutionary rates gained from molecular techniques shed light on the punctuated equilibrium-phyletic gradualism controversy? What additional data are needed to decide this controversy?

10. From figure 21.11, what is K (the average number of amino acid substitutions per site) between human beings and chickens? Between dogs and human beings? Do all three possible comparisons support known evolutionary relationships?

11. What is meant by "constraint" when applied to the molecular evolution of DNA and proteins?

12. How does the acceptance of the neutral mutation theory change our basic views of neo-Darwinism?

13. What are the differences among individual selection, group selection, and kin selection? How are altruistic acts explained by each?

14. If the "calculus of the genes" suggests sacrificing oneself for two siblings, how many first cousins should one sacrifice oneself for?

15. Recently, a vial of bull semen was stolen from an artificial insemination facility. Your friend is about to undergo artificial insemination and is concerned that she may give birth to a minotaur, or cow–human hybrid. Provide two explanations for why she should not worry about this possibility.

16. In *Drosophila*, females in population A produce an average of 250 offspring, and those in population B produce an average of 200 offspring. When the two populations are crossed, A/B females produce 100 offspring. Are populations A and B in the process of becoming different species?

17. A few plants of species Q ($2n = 14$) suddenly double their chromosomes ($2n = 28$) and immediately become a new species, R. Why are Q/R hybrids sterile?

18. One of the arguments creationists use to refute evolution is the gaps in the fossil record. How can you explain the gaps from an evolutionary standpoint?

19. In a given population, the frequencies of *AA, Aa,* and *aa* genotypes are 0.36, 0.48, and 0.16, respectively. If the assigned fitnesses are $1.5 - p$, 1.0, and $1.5 - q$, what will be the genotypic frequencies after selection?

20. If the rate of amino acid substitution per site per year is 2×10^{-9}, and the average number of amino acid substitutions per site is 0.2, how long has it been since the two species diverged?

21. Scientists have examined 1,000 amino acids in proteins of human beings and chimps and have found a difference of 23. Calculate the average number of amino acid substitutions per site.

22. Scientists are now using DNA sequences to show phylogenetic relationships between or among species. In many cases, cDNA is made from isolated mRNA, and then sequenced. Is the method a reasonable approach to show evolutionary relationships?

23. In which codon position should the greatest abundance of variation occur?

24. In certain animal populations, infanticide is practiced by one or more males in the population. Do you think this infanticide is random, or would you expect specific individuals to be eliminated?

Suggested Readings for chapter 21 are on page 660.

Appendix
Brief Answers to Selected
Exercises and Problems

Chapter 2
Mendel's Principles

1. Dwarf F_2 (1/4 of total F_2), when selfed, produce all dwarf progeny. Tall F_2 (3/4 of total F_2), when selfed, produce progeny that fall into two categories: one-third (TT, 1/4 of total F_2) produce all tall, and two-thirds (Tt, 1/2 of total F_2) produce tall and dwarf progeny in a 3:1 ratio. (The 3:1 ratio is from 1/2 the F_2 so the tall component is 3/8 of the total F_3 [3/4 × 1/2] and the dwarf is 1/8 of the total F_3 [1/4 × 1/2].) Overall, the F_3 are 3/8 TT (tall), 2/8 Tt (tall), and 3/8 tt dwarf (see fig. 2.7).

2. The round, yellow F_2 plants are made up of four genotypes; the round, green of two genotypes; the wrinkled, yellow of two genotypes; and the wrinkled, green of one genotype. Testcrossing all these genotypes produces the following results:

   R-Y-:   1/16 $RRYY$ × $rryy$ → all $RrYy$
              2/16 $RrYY$ × $rryy$ → 1/2 $RrYy$, 1/2 $rrYy$
              2/16 $RRYy$ × $rryy$ → 1/2 $RrYy$, 1/2 $Rryy$
              4/16 $RrYy$ × $rryy$ → 1/4 $RrYy$, 1/4 $Rryy$,
              1/4 $rrYy$, 1/4 $rryy$
   R-yy: 1/16 $RRyy$ × $rryy$ → all $Rryy$
              2/16 $Rryy$ × $rryy$ → 1/2 $Rryy$, 1/2 $rryy$
   rrY-:   1/16 $rrYY$ × $rryy$ → all $rrYy$
              2/16 $rrYy$ × $rryy$ → 1/2 $rrYy$, 1/2 $rryy$
   $rryy$:   1/16 $rryy$ × $rryy$ → all $rryy$.

3. 119:32:9 is very close to a 12:3:1 ratio indicating two loci with epistasis. For example:

   P_1 red ($RRYY$) × white ($rryy$)
   F_1 red ($RrYy$) × self
   F_2 119 (R-Y- + R-yy):32 (rrY-):9($rryy$)

Note: For chapters containing a large number of questions, only the even-numbered answers are given for the later questions.

4. $RrTt$ × self yields:

   1/16 $RRTT$ red, tall
   2/16 $RRTt$ red, medium
   1/16 $RRtt$ red, dwarf
   2/16 $RrTT$ pink, tall
   4/16 $RrTt$ pink, medium
   2/16 $Rrtt$ pink, dwarf
   1/16 $rrTT$ white, tall
   2/16 $rrTt$ white, medium
   1/16 $rrtt$ white, dwarf

5. $CcTt$ × self yields:

   3/16 C-TT purple, tall
   6/16 C-Tt purple, medium
   3/16 C-tt purple, dwarf
   1/16 $ccTT$ white, tall
   2/16 $ccTt$ white, medium
   1/16 $cctt$ white, dwarf

6. Two loci with epistasis. $AaBb$ × self yields 9/16 A-B-:6/16 A-bb + aaB-:1/16 $aabb$

   Verify by testcrossing the various classes.

7. Choice (b) is preferred because although each will give the correct genotype, generally, testcrossing has the greatest probability of exposing the recessive allele in a heterozygote. For example, an Aa genotype when selfed will produce aa offspring 1/4 of the time; when testcrossed, aa offspring will appear 1/2 of the time; and when crossed with the Aa type (backcross), aa offspring will occur 1/4 of the time. Thus, with a limited number of offspring examined per cross, testcrossing most reliably exposes the recessive allele.

8. The disease is recessive at the individual level, but incompletely dominant at the enzymatic level. Check the glossary for definitions.

9. All. Since the child was type A, it must have gotten the *A* allele from its mother. The other allele in the child is either *A* or *O*. A type A (*AO*), type B (*BO*), type O, or type AB man could have supplied either an *A* or *O* allele.

10. The following table shows female and child phenotypes in the ABO system and the phenotypes of men who cannot be the father of the child.

Mother	Child	Nonfathers	Mother	Child	Nonfathers
A	A	—	AB	A	—
	B	A, O		B	—
	AB	A, O		AB	O
	O	AB			
B	A	B, O	O	A	B, O
	B	—		B	A, O
	AB	B, O		O	AB
	O	AB			

11. Universal donor, type O (no red-cell antigens); universal recipient, type AB (no serum antibodies).

12. A:AB:B is 4:8:4 or 1:2:1; Rh⁺:Rh⁻ is 12:4 or 3:1. Both loci are segregating properly. The lack of independent assortment would result in the lack of simple ratios (see chapter 6).

13. The F_1 are tetrahybrids (*AaBbCcDd*). If selfed, an F_1 would form $2^4 = 16$ different types of gametes; 2^4 different phenotypes would appear in the F_2, which would be made up of $3^4 = 81$ different genotypes; $1/(16)^2 = 1/256$ of the F_2 would be of the *aabbccdd* genotype.

14. When a decahybrid is selfed, it would produce $2^{10} = 1,024$ different gametes; $1/(2^{10})^2 = 9 \times 10^{-7}$ of the F_2 would be homozygous recessive; $3^{10} = 59,049$ different genotypes yielding 2^{10} different phenotypes would appear. If the decahybrid were testcrossed, it would produce 2^{10} different gametes; $1/2^{10} = 0.00098$ of the F_2 would be homozygous recessive; 2^{10} different genotypes and phenotypes would appear.

15. Properties that make an organism popular among geneticists include a well-understood life cycle; ease of maintenance in the lab or in experimental gardens; a lot of variability in the phenotype that is genetically controlled; ease of manipulation (breeding); and short generation time.

16. Wingless mutant, *Wi*; wild-type, *Wi⁺*. (*W* is already the allele designation for the wrinkled phenotype.)

17. Rule of segregation: adult diploid organisms possess two copies of each gene. Gametes get one copy. Fertilization restores the diploid number to the zygote. Rule of independent assortment: alleles of different genes segregate independently of each other.

18. The animals would be bred for several generations of controlled matings. You would look for offspring types and ratios consistent with a hypothesis of the number of genes controlling the trait and the dominance relationships of alleles. One approach might be to try to establish pure-breeding lines (homozygotes), which can then be bred in a more controlled manner. If both rooster and hen were pure-breeding rose-combed types (all offspring were also rose combed), then it would not be possible to determine the nature of the genetic control of this trait.

19. *A* and *B* each add a substance to the H structure, whereas *O* adds nothing. Thus, for example, the *AO* heterozygote converts all the H structures to the A antigen. *AB* heterozygotes convert the H structures to approximately equal numbers of A and B antigens.

20. In the table: D, dominant; R, recessive; M, mutant; +, wild-type.

Allele	D/R	M/+	Alternative	D/R
y⁺	D	+	*y*	R
Hw	D	M	*Hw⁺*	R
Ax⁺	R	+	*Ax*	D
Co	D	M	*Co⁺*	R
rv⁺	D	+	*rv*	R
dow	R	M	*dow⁺*	D
M(2)e⁺	R	+	*M(2)e*	D
J	D	M	*J⁺*	R
tuf⁺	D	+	*tuf*	R
bur	R	M	*bur⁺*	D

21. Constraints are set by the length of the gene (number of mutable sites) and the different possible phenotypic effects that can result from alteration of the amino acid sequence of the protein product of the gene.

22. See for example figure 2.25. Other ratios are 10:3:3; 10:6; 12:3:1; 12:4.

23. Dominance refers to the expression of an allele (and the masking of the other) when in the heterozygous condition. Epistasis is similar to dominance but is intergenic; that is, an allelic combination at one gene can mask the expression of alleles of another gene. Pleiotropy is suspected when different phenotypic effects are observed together. It is supported when crosses cannot separate these effects and verified when a molecular or physiological mechanism of gene action is determined.

24.

Blockage	Build Up	Never Appears
1	Q	R, S, T, U
2	R	S, T, U
3	S	T, U
4	T	U

25. 2 → indole → tryptophan → 3-hydroxy- → niacin
        1   or   anthranilic
                      kynurenine   acid

Accumulation: 1, indole; 2, ?; 3, tryptophan or kynurenine; 4, 3-hydroxyanthranilic acid. Without serine, the pathway would be blocked at the point of tryptophan production, between indole and kynurenine.

26.
$$? \xrightarrow{4} D \xrightarrow{2} B \xrightarrow{5} C \xrightarrow{3} A \xrightarrow{1} E$$
Accumulation: 1, A; 2, D; 3, C; 4, ?; 5, B

28. $Tt \times tt$. We see two phenotypes in approximately a 1:1 ratio. We don't know which allele is dominant from this cross. The two phenotypes indicate one gene, and the 1:1 ratio indicates a mating of a homozygote and a heterozygote.

30. All AB; or 1/2 AB, 1/2 A; or 1/2 AB, 1/2 B; or 1/4 A:1/4 AB:1/4 B:1/4 O. Crosses can be $AA \times BB$, $AA \times BO$, $AO \times BB$, or $AO \times BO$. Different genotypes give the same phenotype, a result indicative of multiple alleles.

$$AA \times BB$$
$$\downarrow$$
All AB
(AB)

$$AO \times BB$$
$$\downarrow$$
$1/2\ AB : 1/2\ BO$
(AB)   (B)

$$AA \times BO$$
$$\downarrow$$
$1/2\ AB : 1/2\ AO$
(AB)   (A)

$$AO \times BO$$
$$\downarrow$$
$1/4\ AB : 1/4\ AO : 1/4\ BO : 1/4\ OO$
(AB)   (A)   (B)   (O)

32. Red-eyed is dominant and the cross is between two heterozygotes since we see two phenotypes in an approximate 3:1 ratio.

34. In the first cross, look at the yellow-to-green ratio, 120:43—almost exactly 3:1. Therefore yellow is dominant and both parents must be heterozygous. Now look at the tall-to-short ratio, 122:41. Again, we see a 3:1 ratio, which indicates that tall is dominant and each parent is heterozygous. Thus, the cross is probably $YyTt \times YyTt$.

In the second cross, there are no tall progeny. Therefore, either the short phenotypes are homozygous, or short is dominant and at least one parent is homozygous. In the absence of the first cross, we can't determine the mode of inheritance of height. We can, however, conclude that yellow is dominant (we got a 3:1 ratio) and that each parent is heterozygous. Based on the first cross, we can conclude that this cross is $Yytt \times Yytt$.

In the third cross, we see 41 yellow to 46 green, and 45 tall to 42 short. Both of these ratios are 1:1, and all we can conclude is that these ratios result from matings between a heterozygote and a recessive homozygote. With only this cross, we can't determine dominance. However, we can if we use all three crosses. The cross is $yyTt \times Yytt$.

36. Examine each trait separately. There are 104 long:34 short; and 69 brown:69 red. Length appears in a 3:1 ratio; therefore, long is dominant and each parent is heterozygous. Eye color appears in a 1:1 ratio. We can't conclude which allele is dominant; all we can conclude is that one parent is a recessive homozygote and that one

parent is a heterozygote. If L = long, l = short, R = red, and r = brown, one possible way to indicate the cross is $Llrr \times LlRr$.

38. You could set up a Punnet square and count the boxes; but unfortunately, this is an 8×8 matrix that yields 64 squares to count. Very tedious. Set up the cross:

$$TtYySs \times TtYySs$$

and look at one gene at a time.

In the first question, the chance of getting each dominant trait is 3/4. Therefore, the chance of all three together is $3/4 \times 3/4 \times 3/4 = 27/64$.

In the second question, the chance of getting each recessive is 1/4. Therefore, the chance for all three recessives is $1/4 \times 1/4 \times 1/4 = 1/64$.

In the third question, the chance of short is 1/4, the chance of green is 1/4, and the chance of smooth is 3/4. Therefore the total chance is $1/4 \times 1/4 \times 3/4 = 3/64$.

40. Red and crossveins are dominant.
   1. $rrCc \times rrCc$
   2. $RrCc \times Rrcc$
   3. $R\text{-}cc \times RRCc$
   4. $RrCc \times RrCc$

Let R = red, r = orange, C = crossveins, c = crossveinless. Look first at those crosses that yield only two phenotypes and thus indicate that one gene is homozygous. Crosses 1 and 3 are such crosses. In cross 1 we see only orange eyes, and in cross 3 we see only red eyes. The eye color must be homozygous in at least one parent. From these two crosses we can't determine whether red or orange is dominant. In cross 1, both parents have crossveins, and we see a 3:1 ratio in the progeny. Crossveins must be dominant, and this cross must be $Cc \times Cc$. Similar logic applies to cross 3.

In cross 2 we see an approximate 3:3:1:1 ratio, which suggests that the same gene is heterozygous in each parent. We see a 3:1 red:orange ratio. Therefore, red is dominant and heterozygous in each parent: $Rr \times Rr$. We have already determined that the allele for crossveins is dominant, which indicates the cross: $RrCc \times Rrcc$. Thus a 3:3:1:1 ratio indicates this general situation: $AaBb \times Aabb$. In cross 4 we see four phenotypes in an approximate 9:3:3:1. This result indicates a mating between two double heterozygotes $RrCc \times RrCc$.

42. a. 4/64 or 1/16
     b. 33/64
To be colorless, individuals must be at least $aabb$; they could be either $aabbC$- or $aabbcc$. The chance of the first genotype is $(1/4\ aa)(1/4\ bb)(3/4\ C\text{-}) = 3/64$, and the chance of the second is $(1/4\ aa)(1/4\ bb)(1/4\ cc) = 1/64$. Since their genotype is colorless, we add probabilities. Or, $(1/4\ aa)(1/4\ bb)(1\ C\text{-}\ or\ cc) = 1/16$.

For red probabilities, determine genotypes that will be red, calculate probability for each genotype, and add.

A-bbC- $3/4 \times 1/4 \times 3/4 = 9/64$
A-$bbcc$ $3/4 \times 1/4 \times 1/4 = 3/64$
A-B-cc $3/4 \times 3/4 \times 1/4 = 9/64$
aaB-C- $1/4 \times 3/4 \times 3/4 = 9/64$
aaB-cc $1/4 \times 3/4 \times 1/4 = \underline{3/64}$
$33/64$

Alternatively, calculate the frequency of blacks as $3/4$ A- $\times 3/4$ B- $\times 3/4$ C- $= 27/64$. We calculated $4/64$ colorless, so $1 - (27/64 + 4/64) = 33/64 =$ frequency of red.

44. Steve and his fiancé could be related. Both the dean and Steve's father must be AO to produce O children, and each could have contributed M to produce M offspring. If the dean and Steve's father each contributed an S allele, the daughter would be SS. Note that if the daughter had B blood, she and Steve could not be related.

Chapter 3
Mitosis and Meiosis

1. See table 3.1 for a summary answer.

2. Kinetochore: chromosomal attachment point for spindle fibers, located within the centromere. Centromere: constriction in eukaryotic chromosome in which kinetochore lies.

3. When a eukaryotic chromosome replicates during the S-phase of the cell cycle, one chromosome becomes two chromatids, attached at the centromere. These are sister chromatids. Chromatids that are not identical copies of each other are nonsisters. Homologous chromosomes are members of a pair of essentially identical chromosomes. In diploid organisms one member from each pair comes from each parent. Nonhomologous chromosomes do not share this relationship.

4.

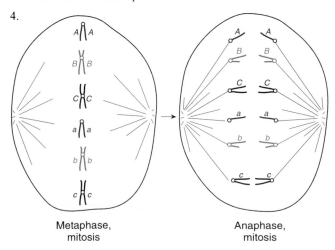

Metaphase, mitosis   Anaphase, mitosis

5. $2^3 = 8$ different gametes can arise. A crossover between the A locus and its centromere does not alter gametic combinations.

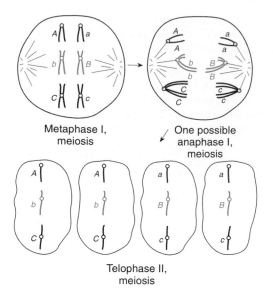

Metaphase I, meiosis

One possible anaphase I, meiosis

Telophase II, meiosis

6. a. Mitosis, prophase, $2n = 6$, or meiosis II, prophase, $2n = 12$
   b. Mitosis, metaphase, $2n = 6$, or meiosis II, metaphase, $2n = 12$
   c. Mitosis, late anaphase, $2n = 6$, or meiosis II, late anaphase, $2n = 12$
   d. Meiosis I, early prophase, $2n = 6$
   e. Meiosis I, late anaphase, $2n = 6$
   f. Meiosis II, anaphase, $2n = 6$

7. a. 46; b. 0; c. 23; d. 23; e. 23

8. a. 20; b. 20; c. 30; d. 10 per nucleus; e. 10

9.

	$2n$ (= Dyads)	Bivalents (= Tetrads)
Human Beings	46	23
Garden Peas	14	7
Drosophila melanogaster	8	4
House Mouse	40	20
Roundworm	2	1
Pigeon	80	40
Boa Constrictor	36	18
Cricket	22	11
Lily	24	12
Indian Fern	1,260	630

10. S-phase

11. The intent of this question is to make you think about the essential steps of meiosis, primarily the necessity to separate members of homologous pairs of chromosomes. Presumably, any method you devise will force you through that process.

12. The terms reductional and equational refer to the segregation of centromeres (chromosomes) during meiosis. The first division is reductional because homologous chromosomes are separated from each other, reducing the number of chromosomes in half. The second

meiotic division is equational because the number of chromosomes remains the same although the number of chromatids is halved.

13. Meiosis apportions homologous chromosomes the same way that Mendel's rules apportion alleles. Each gamete gets one member of a homologous pair of chromosomes. Segregation predicts the same about alleles. The separation of homologues of one chromosome pair at meiosis is independent of the separation of other homologous pairs. Independent assortment makes the same prediction about alleles.

14. Any possible genotype, from *AAABBB* through *aaabbb* can occur in the endosperm. If at a given locus the endosperm is homozygous, so is the sporophyte. If the locus is heterozygous (e.g., *AAa* or *Aaa*), so is the sporophyte. Thus an *AAabbb* endosperm is associated with an *Aabb* sporophyte.

15. 40 sperm, 10 ova

16. Human being: see figure below. Pea plant: With the obvious differences between corn and peas aside, figure 3.25 describes the pea life cycle. *Neurospora:* see figure 6.15.

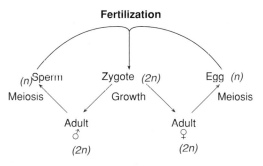

Fertilization

(n) Sperm — Zygote (2n) — Egg (n)

Meiosis — Growth — Meiosis

Adult ♂ (2n) — Adult ♀ (2n)

17. A greater maternal influence in *Drosophila* and corn than in *Neurospora* in which sexes (mating types) do not show the disparity in size between male and female cells as in *Drosophila* and corn.

18. *AaBb* queen × *ab* drone yields

Sons (or gametes): *AB, Ab, aB, ab*
Daughters: *AaBb, Aabb, aaBb, aabb*

19.

	DNA (Number of Chromatids)	Ploidy
Spermatogonium or Oogonium	2	*2n*
Primary Spermatocyte or Primary Oocyte	4	*2n*
Secondary Spermatocyte or Secondary Oocyte	2	*n*
Spermatid or Ovum	1	*n*
Sperm	1	*n*

20. g. Homologous chromosomes will pair during meiosis. Each gamete gets one of each chromosome, A, B, C, D, and E. Fertilization fuses two cells with the above chromosome complement. Since root cells are somatic tissue, these cells will be diploid.

22. A gamete from wheat will have 21 chromosomes, and a gamete from rye will have 7 chromosomes. Even if the 7 rye chromosomes could pair with 7 wheat chromosomes, a highly unlikely possibility, the remaining 14 wheat chromosomes could not pair, and would segregate randomly during meiosis. Each gamete would get an incomplete set, and most gametes would not survive. If fertilization did occur, the zygotes would have extra chromosomes (trisomic) or would be missing some chromosomes (monosomic or nullosomic).

24. a. 200 b. 100 c. 50. The primary spermatocyte is diploid and will undergo meiosis to yield 4 cells; $4 \times 50 = 200$. The secondary spermatocyte has completed one meiotic division, and has one more division to go; $2 \times 50 = 100$. The spermatid is haploid, and each spermatid matures into a sperm cell.

26. a. Since the diploid number is different for each species, the gametes will have 25 and 19 chromosomes, respectively. Even if all 19 of the red fox were identical to 19 of the arctic fox, some chromosomes could not pair, and imbalanced gametes and zygotes would result.
b. There must be some homology between the large chromosomes of the red fox and the small chromosomes of the arctic fox. This homology allows some of the chromosomes to pair.

28. a. *c;* There will be *c* amount of DNA in G_1 because DNA replication has not yet taken place.
b. *2c;* There will be *2c* amount of DNA in G_2 because DNA replication has taken place.
c. *0.5c;* At the end of meiosis I, a cell will have *c* amount of DNA, disregarding differences brought about by differences in sex chromosomes. Since a cell entering meiosis I has *2c* DNA, it will have *c* at the end of the first meiotic division and *0.5c* at the end of the second meiotic division (one tetrad reduced to one chromosome).

Chapter 4
Probability and Statistics

1. a. $(5!/3!2!)(1/2)^3(1/2)^2 = 0.3125$
b. $(1/2)^5 = 0.03125$ [SDSDS, in which S = son, D = daughter]
c. $2(1/2)^5 = 0.0625$ [SDSDS + DSDSD]
d. $(1/2)^5 = 0.03125$
e. $2(1/2)^5 = 0.0625$ [all sons + all daughters]
f. 4 daughters, 1 son + 5 daughters: $(5!/4!1!)(1/2)^4(1/2) + (1/2)^5 = 0.1875$
g. $(1/2)^2 = 0.25$ [DXXXS, in which X is either a daughter or a son, with $p = 1$; $P = (1/2)(1)(1)(1)(1/2)$]

2. Parents are heterozygotes ($Tt \times Tt$)
a. $1/4 = 0.25$
b. $1/8 = 0.125$ [$(1/4)(1/2)$]
c. $(6!/2!2!2!)(1/8)^2(1/8)^2(3/8)^2 = 0.0030899$
d. $3/8 = 0.375$

3. Remember that albinos have blue eyes. Therefore, 7/16 of the offspring will have blue eyes. If we let B = brown, b = blue, C = normal color expression, and c = albinism, the following genotypes are blue eyed: $C\text{-}bb$ and $cc\text{-}\text{-}$.
   a. $(1/4)^5 = 0.0009765$
   b. $(1/8)^5 = 0.0000305$
   c. $(5!/4!1!)(7/32)^4(9/32) = 0.0032199$
   d. $(4!/2!2!)(1/8)^2(1/8)^2 = 0.0014648$

4. a. $1/10{,}000 = 0.0001$
   b. $1/10{,}000 = 0.0001$
   c. $(1/10{,}000)^2 = 0.00000001$

5. a. $2(1/2)^4 = 0.125$
   b. $(1/2)^4(1/2)^4 = 0.0039062$ [Probability that sperm and egg creating the zygote each had only paternal centromeres.]

6. a. Hypothesis: $Rr \times Rr$ produces $R\text{-}{:}rr$ in a 3:1 ratio. Chi-square, 1 degree of freedom (d.f.), = 0.263. Critical chi-square at 0.05, 1 d.f., = 3.841.

	Round	Wrinkled	Sum
Observed	5,474	1,850	7,324
Expected	3/4	1/4	
	5,493	1,831	
$O - E$	−19	19	
$(O - E)^2$	361	361	
$(O - E)^2/E$	0.066	0.197	0.263

Fail to reject (i.e., accept) the null hypothesis at 0.05 level and therefore at the 0.01 level.

   b. Hypothesis: $Rr \times Rr$ produces $R\text{-}{:}rr$ in a 3:1 ratio. Chi-square, 1 d.f., = 0.474. Critical chi-square at 0.05, 1 d.f., = 3.841.

	Round	Wrinkled	Sum
Observed	45	12	57
Expected	3/4	1/4	
	42.75	14.25	
$O - E$	2.25	−2.25	
$(O - E)^2$	5.063	5.063	
$(O - E)^2/E$	0.118	0.355	0.474

Fail to reject (i.e., accept) the null hypothesis at 0.05 level and therefore at the 0.01 level.

   c. Hypothesis: $Rr{:}RR$ are in a 2:1 ratio. Chi-square, 1 d.f., = 0.174. Critical chi-square at 0.05, 1 d.f., = 3.841.

	Rr	*RR*	Sum
Observed	372	193	565
Expected	2/3	1/3	
	376.67	188.33	
$O - E$	−4.67	4.67	
$(O - E)^2$	21.81	21.81	
$O - E)^2/E$	0.058	0.116	0.174

Fail to reject (i.e., accept) the null hypothesis at 0.05 level and therefore at the 0.01 level.
   d. Hypothesis: $VvLl \times vvll$ produces a 1:1:1:1 ratio of offspring. Chi-square, 3 d.f., = 1.084. Critical chi-square at 0.05, 3 d.f., = 7.815.

	VvLl	*vvLl*	*Vvll*	*vvll*	Sum
Observed	47	40	38	41	166
Expected	1/4	1/4	1/4	1/4	
	41.5	41.5	41.5	41.5	
$O - E$	5.5	−1.5	−3.5	−0.5	
$(O - E)^2$	30.25	2.25	12.25	0.25	
$(O - E)^2/E$	0.729	0.054	0.295	0.006	1.084

Fail to reject (i.e, accept) the null hypothesis at 0.05 level and therefore at the 0.01 level.

7. Hypothesis: $RrYy \times RrYy$ produces $R\text{-}Y\text{-}{:}R\text{-}yy{:}rrY\text{-}{:}rryy$ in a 9:3:3:1 ratio. The critical chi-square, 3 degrees of freedom at probability of 0.05, = 7.815.

	R-Y-	*R-yy*	*rrY-*	*rryy*	Total
Observed Numbers (O)	315	108	101	32	556
Expected Ratio	9/16	3/16	3/16	1/16	
Expected Numbers (E)	312.75	104.25	104.25	34.75	556
$O - E$	2.25	3.75	−3.25	−2.75	
$(O - E)^2$	5.06	14.06	10.56	7.56	
$(O - E)^2/E$	0.016	0.135	0.101	0.218	$0.470 = \chi^2$

Since this chi-square, 0.470, is less than the critical chi-square, we fail to reject our hypothesis of two-locus genetic control with dominant alleles at each locus.

8. Hypothesis: $AaCc^a \times AaCc^a$ produces agouti ($A\text{-}C\text{-}$), black ($aaC\text{-}$), albino ($\text{-}\text{-}c^ac^a$) offspring in a 9:3:4 ratio. Critical chi-square, 2 degrees of freedom at probability of 0.05, = 5.991.

	agouti	black	albino	Total
Observed Numbers (O)	28	7	13	48
Expected Ratio	9/16	3/16	4/16	
Expected Numbers (E)	27	9	12	48
$O - E$	1	−2	1	
$(O - E)^2$	1	4	1	
$(O - E)^2/E$	0.037	0.444	0.083	$0.565 = \chi^2$

Since this chi-square, 0.565, is less than the critical chi-square, we fail to reject our hypothesis of two-locus genetic control with epistasis.

9. Assume that Mendel was observing the phenotypes of offspring produced from seeds obtained by testcrossing the plant having a dominant phenotype (A- $\times$ aa). As soon as an offspring with the recessive phenotype appeared ($p = 1/2$), he was completely certain that the dominant parent was heterozygous. If he tested five seeds and all produced plants with the dominant phenotype, there would be a 3% chance that the parent plant was still heterozygous ($p = [1/2]^5$ of getting only A alleles from an Aa genotype). Testing seven seeds would lower the chance of heterozygosity to less than 1%. "Pretty reliable" is arbitrary; complete certainty requires an infinite number of offspring.

10. $1/2$. One half the families would stop at two children, one half of the remaining families would stop at three children, and so on. Thus the average family size is the sum of $(1/2)(2) + (1/4)(3) + (1/8)(4) + (1/16)(5) + \ldots$, or

$$\text{Average family size} = \sum_{n=1}^{\infty} (1/2^n)(n+1)$$

$$= \sum_{n=1}^{\infty} [(n+1)/2^n]$$

This summation reaches a limit of 3.0 at infinity.

12. a. $81/256$ b. $108/256$ c. $9/256$. In (a), since all children have the same phenotype, each child will have the same probability of having no molars. Therefore $(3/4)^4$

$= 81/256$. In (b), $P = \dfrac{4!}{3!1!} (3/4)^3(1/4) = 108/256$.

When order is given, we multiply the chance of each event,
$1/4 \times 1/4 \times 3/4 \times 3/4 = 9/256$.

14. $1/8$. B must be heterozygous (Gg), as must A's father. We assume A's mother is GG, since there is no mention of the disease in her family. Therefore A has $1/2$ chance of getting g from his father. If two heterozygotes mate, the chance of a recessive child is $1/4$, so $P = 1/4 \times 1/2$.

16. We reject the 3:1 ratio as an appropriate null hypothesis. If we calculate the chi-square using 3:1 as the expected ratio, we expect 72 and 24 (3/4 and 1/4 of 96, respectively):

	O	E	$O - E$	$(O - E)^2$	$(O - E)^2/E$
curly-winged flies	61	72	9	81	$81/72 = 1.125$
straight-winged flies	35	24	9	81	$81/24 = 3.375$

Chi-square $= 4.5$. With one degree of freedom $p < 0.05$ (critical chi-square $= 3.841$).

18. $5(1/2)^5 = 5/32$. The cross is $Tt \times tt$. This is a testcross, so there is a $1/2$ chance of either a taster or nontaster offspring. The binomial expression is used for unordered events:

$$P = \frac{5!}{4!1!} (1/2)^4(1/2) = 5(1/2)^5$$

20. $1/512$. The F_1 progeny are $AaBbCcDdEe$. The chance of getting any individual who is completely homozygous is $(1/4)^5$. Since we are looking for two different possibilities, we have $2(1/4)^5 = 2/1024 = 1/512$.

Chapter 5
Sex Determination, Sex Linkage, and Pedigree Analysis

1.

		Cross	Reciprocal
P_1	Female	X^+X^+	$X^{lz}X^{lz}$
	Male	$X^{lz}Y$	X^+Y
F_1	Female	X^+X^{lz}	X^+X^{lz}
	Male	X^+Y	$X^{lz}Y$
F_2	Females	X^+X^+, X^+X^{lz}	$X^+X^{lz}, X^{lz}X^{lz}$
	Males	$X^+Y, X^{lz}Y$	$X^+Y, X^{lx}Y$

2. The differences are in terminology only, not in shape or size of the chromosomes. In species in which females have a homomorphic sex chromosome pair, the members of the pair are called X chromosomes. In species in which males have a homomorphic sex chromosome pair, the members of the pair are called Z chromosomes.

3.

		Cross	Reciprocal
P_1	Female	Z^pW	Z^+W
	Male	Z^+Z^+	Z^pZ^p
F_1	Female	Z^+W	Z^pW
	Male	Z^+Z^p	Z^+Z^p
F_2	Females	Z^+W, Z^pW	Z^+W, Z^pW
	Males	Z^+Z^+, Z^+Z^p	Z^+Z^p, Z^pZ^p

4. Exemptions should be made for hemophilia in brother, sister's son, mother's brother, mother's sister's son, and others, more distantly related.

5. The protein is probably a dimer, which, in the heterozygote, can be of fast-fast, fast-slow, or slow-slow subunit combinations. A female heterozygous for a sex-linked gene controlling a dimeric enzyme should show the pattern of slot 3 in whole blood (mixture of slow-slow and fast-fast dimers) and slots 1 or 2 in individual cells.

6. The pattern should be of fifteen bands made up of combinations of A, B_1, and B_2 subunits combined four at a time. These are both allozymes of B_1 and B_2 subunits and isozymes of A and B subunit interactions.

7. Penetrance is the appearance of the genotype in the phenotype; expressivity is the degree to which a trait is expressed.

8. a. 0; human female, male fly
   b. 1; human female, female fly
   c. 0; human male, male fly
   d. 1; human male, female fly
   e. 2; human female, female fly
   f. 4; human female, female fly
   g. 1/0 mosaic; human male/female mosaic, male/female fly mosaic

9. a. Could be an autosomal recessive, but most likely a sex-linked recessive, mode of inheritance. If complete penetrance is assumed, it could not be a dominant or Y-linked mode of inheritance.
   b. Could be an autosomal recessive, but most likely an autosomal dominant, mode of inheritance. X- and Y-linked inheritance are ruled out.
   c. Could be an autosomal recessive, or an X-linked recessive, mode of inheritance (if the allele causing the trait was brought into the pedigree by the affected's maternal grandmother). Other modes of inheritance are ruled out.

10. a. The phenotype is the propensity to have twin offspring. It could be caused by a recessive or dominant, sex-linked or autosomal allele.
   b. Autosomal dominant or possibly autosomal recessive inheritance.
   c. Autosomal, or sex-linked, recessive inheritance.
   d. Autosomal recessive inheritance.

11. Pseudodominance refers to the hemizygous (one copy) appearance of a recessive trait. A phenocopy is an environmentally induced phenotype similar to a genetically controlled one.

12. Pedigree (*a*) could represent either Y-linkage or autosomal inheritance. Pedigree (*b*) represents an autosomal recessive pattern. Reduced penetrance simply means that any of the affected in these pedigrees could have a normal phenotype.

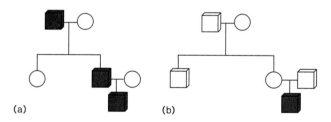

(a)   (b)

13. Apparent ratio of offspring is 3:1, male to female. Fertile males are 50% *tra/tra* and 50% +/*tra*. Fertile daughters are all +/*tra*. If these progeny are mated among themselves, males homozygous for transformer will have offspring in the ratio of 3:1, males to females, whereas males heterozygous for transformer will have offspring in the ratio of 5:3, males to females.

14. 3/8 males, 3/8 females, 2/8 intersexes (*dsx/dsx* homozygotes)

15. Sex switches are genes that determine the developmental pathway of an organism, either toward male or female, depending on the presence or state of the gene. *SRY* is the male sex switch in human beings, whereas *Sxl* is a female switch in *Drosophila*.

16. Early onset in calico cats; tortoiseshell cats have a later onset of Lyonization as seen by the fact that there were more cells present at that time and thus smaller sectors.

17. Assuming 100% penetrance:

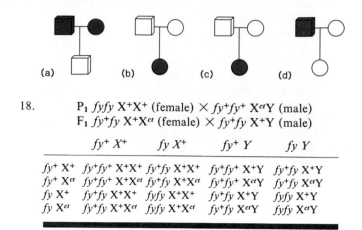

(a)   (b)   (c)   (d)

18.   P₁ *fyfy* X⁺X⁺ (female) × *fy⁺fy⁺* X^{ct}Y (male)
        F₁ *fy⁺fy* X⁺X^{ct} (female) × *fy⁺fy* X⁺Y (male)

fy⁺ X⁺	*fy X⁺*	*fy⁺ Y*	*fy Y*
fy⁺ X⁺	*fy⁺fy⁺* X⁺X⁺	*fy⁺fy* X⁺X⁺	*fy⁺fy⁺* X⁺Y	*fy⁺fy* X⁺Y
fy⁺ X^{ct}	*fy⁺fy⁺* X⁺X^{ct}	*fy⁺fy* X⁺X^{ct}	*fy⁺fy⁺* X^{ct}Y	*fy⁺fy* X^{ct}Y
fy X⁺	*fy⁺fy* X⁺X⁺	*fyfy* X⁺X⁺	*fy⁺fy* X⁺Y	*fyfy* X⁺Y
fy X^{ct}	*fy⁺fy* X⁺X^{ct}	*fyfy* X⁺X^{ct}	*fy⁺fy* X^{ct}Y	*fyfy* X^{ct}Y

F₂: females, 3/4 wild-type, 1/4 fuzzy; males, 3/8 wild-type, 3/8 cut, 1/8 fuzzy, and 1/8 cut and fuzzy

20. a. X-linked.
   b. gray.
   c. 1/2 gray:1/2 yellow in both sexes.
      In both crosses, we see a difference in the phenotypes of the sexes, suggesting some type of sex linkage. The F₁ offspring from cross 1 indicate that gray is dominant to yellow. The F₁ females from this cross must be heterozygous and the two phenotypes in the F₂ males result from each of the X chromosomes in the F₁ female. Cross 1 is therefore (calling gray the wild-type)

            X⁺X⁺   ×   XʸY
                    ↓
            X⁺Xʸ   ×   X⁺Y
            gray   gray
                    ↓
       X⁺X⁺   X⁺Xʸ   X⁺Y   XʸY
       gray   gray   gray   yellow

   Now diagram cross 2:

            XʸXʸ   ×   X⁺Y
                    ↓
            X⁺Xʸ   ×   XʸY
            gray   yellow
                    ↓
       X⁺Xʸ   XʸXʸ   X⁺Y   XʸY
       gray   yellow   gray   yellow

22. Yes. Begin by determining genotypes of the two individuals. The woman must be heterozygous X^CX^c. A man with normal vision must be X^CY, and all his daughters must receive his X chromosome and should be normal, either X^CX^C or X^CX^c. Since color blindness is recessive, the daughter must have two X^c. A *very* rare possibility is that the man is the father and nondisjunction occurred in both parents: at meiosis I in the male and at meiosis II in the female (see chapter 8).

24. a. wild-type females, white-eyed males.

b. 3 wild-type: 3 white-eyed:1 ebony:1 ebony, white-eyed in both sexes.

c. 3 wild-type females:1 ebony female:3 wild-type males:3 white-eyed males:1 ebony male:1 ebony, white-eyed male. Let X^+ = red, X^w = white, e^+ = wild-type, e = ebony.

$$X^w X^w\ e^+ e^+ \quad \times \quad X^w Y\ ee \qquad P_1$$
$$\downarrow$$
$$X^+ X^w\ e^+ e \quad \times \quad X^w Y\ e^+ e \qquad F_1$$
$$\text{(wild-type)} \qquad \text{(white-eyed)}$$

Use probability for the F_2 generation rather than the Punnett square:

$(1/2)X^+ \times (1/2)X^w \times (3/4)e^+$ - = 3/16 wild-type females
$\qquad\qquad\qquad\qquad \times (1/4)ee$ = 1/16 ebony females
$\qquad\qquad \times (1/2)Y \times (3/4)e^+$ - = 3/16 wild-type males
$\qquad\qquad\qquad\qquad \times (1/4)ee$ = 1/16 ebony males
$(1/2)X^w \times (1/2)X^w \times (3/4)e^+$ - = 3/16 white-eyed females
$\qquad\qquad\qquad\qquad \times (1/4)ee$ = 1/16 white-eyed, ebony females
$\qquad\qquad \times (1/2)Y \times (3/4)e^+$ - = 3/16 white-eyed males
$\qquad\qquad\qquad\qquad \times (1/4)ee$ = 1/16 white-eyed, ebony males

For the reciprocal cross,

$$X^+ X^+\ ee \quad \times \quad X^w Y\ e^+ e^+$$
$$\downarrow$$
$$X^+ X^w\ e^+ e \quad \times \quad X^+ Y\ e^+ e$$
$$\text{(wild-type)} \qquad \text{(white-eyed)}$$

All F_2 females will get X^+; e^+:e will be 3:1. The males will be as in the males above.

26. a. autosomal recessive, X-linked recessive

b. X-linked dominant, autosomal dominant, autosomal recessive (unlikely)

c. Y-linked

In (a), the trait appears from normal individuals, indicating that it is recessive. In all generations, 1/2 of sons of normal women have the trait. It could also be autosomal, with parents of affected individuals being heterozygous. In (b), we notice that half of the sons of affected women have the trait, as well as all the daughters of affected men. The trait cannot be an X-linked recessive trait, for all sons of affected women would be affected. An autosomal dominant trait, heterozygous in affected individuals, would yield half the progeny affected. If the trait were autosomal recessive, each normal individual who marries an affected individual must be heterozygous, a very rare possibility. In (c), the trait is passed only from father to son, behavior typical of a Y-linked trait.

28. We see four phenotypes, so we must have at least two genes involved. We see no difference in body color between sexes, so we can conclude that body color is autosomally controlled. Note that if body color were X-linked, males should have had dark bodies. Since all F_1 offspring have tan bodies, tan body must be dominant to dark body.

We see a difference in eye color between the sexes in the F_1 generation, suggesting that eye color is X-linked. Since the F_1 females have red eyes, red must be dominant, and hence the original white-eyed female must be homozygous for the recessive allele.

In the F_2 generation, the ratio of tan to dark is 51:16—very close to the 3:1 ratio expected for one autosomal gene. Red to white is 36:31, very close to the expected 1:1 for an X-linked trait.

Let dk^+ = tan body, dk = dark body, X^+ = red eyes, X^w = white eyes. Parents are: $dk\,dk\ X^w X^w \times dk^+ dk^+$ $X^+ Y$. We can then diagram the cross between the resulting F_1 individuals: $dk^+ dk\ X^+ X^w \times dk^+ dk\ X^w Y$ This cross yields

$(1/2)X^+ \times (1/2)X^w \times 3/4$ tan = 3/16, male or female,
(or Y) red-eye, tan body
$\qquad\qquad\qquad\qquad \times 1/4$ dark = 1/16, male or female, red-eye, dark body
$(1/2)X^w \times (1/2)X^w \times 3/4$ tan = 3/16, male or female,
(or Y) white-eye, tan body
$\qquad\qquad\qquad\qquad \times 1/4$ dark = 1/16, male or female, white eye, dark body

Chapter 6
Linkage and Mapping in Eukaryotes

1. a.
P₁ groucho $\qquad\qquad \times$ rough
$\quad gro/gro\ ro^+/ro^+ \qquad gro^+/gro^+\ ro/ro$
F₁ ♀ $gro^+/gro\ ro^+/ro \times$ ♂ $gro/gro\ ro/ro$
F₂ $gro/gro\ ro^+/ro \qquad\quad$ 518
$\quad gro^+/gro\ ro/ro \qquad\quad$ 471
$\quad gro/gro\ ro/ro \qquad\qquad$ 6
$\quad gro^+/gro\ ro^+/ro \qquad\quad$ 5
$(6 + 5)/1,000 = 0.011 = 1.1\%$ recombination
$= 1.1$ map units apart

b. Given the map units, F_1 gametes are produced on the average by females as follows: $gro\ ro^+$, 49.45% $= 0.4945$; $gro^+\ ro$, 49.45% $= 0.4945$; $gro\ ro$, 0.55% $= 0.0055$; and $gro^+\ ro^+$, 0.55% $= 0.0055$. Males, lacking crossing over, produce only two gamete types: $gro\ ro^+$ and $gro^+\ ro$, each 50% $= 0.50$. Summing from the Punnett square, the phenotypes of the offspring would be as follows: wild-type, 50%; groucho, rough, 0%; groucho, 25%; and rough, 25%.

| | ♂ | |
♀	$gro\ ro^+$ (0.5)	$gro^+\ ro$ (0.5)
$gro\ ro^+$ (0.4945)	groucho 0.24725	wild-type 0.24725
$gro^+\ ro$ (0.4945)	wild-type 0.24725	rough 0.24725
$gro\ ro$ (.0055)	groucho 0.00275	rough 0.00275
$gro^+\ ro^+$ (0.0055)	wild-type 0.00275	wild-type 0.00275

2. The linkage pattern is consistent with X-linkage.

P_1 $X^{abe,bis}$ $X^{abe,bis}$ (female) $\times$ X^+Y (male)
F_1 $X^+X^{abe,bis}$ $\times$ $X^{abe,bix}Y$

F_2: Since crossovers can only occur in F_1 females, this is the same as a testcross: Both sons and daughters will have the phenotype of the X chromosome inherited from the mother. Thus, in the F_2, both sons and daughters have roughly the same distribution of phenotypes. Abnormal and brown are the recombinant classes. The two loci are 16.0 map units apart on the X chromosome.

3. A dihybrid female is testcrossed (with a hemizygous male having both recessive alleles). Each recombinant class will make up about 5% of the offspring. Each parental class will make up about 45% of the offspring. Phenotypic classes will be equally distributed between the two sexes. The same results will be found for an autosomal locus if the dihybrids are females (no crossing over in males). A reciprocal cross cannot be done for X-linked genes because males cannot be dihybrid. Males dihybrid for an autosomal gene produce only two classes of offspring when testcrossed—parentals.

4. This is a simple cross complicated by the fact that these loci are detected by electrophoretic methods. The alleles are codominant and thus, although we can't do a true testcross (no recessive homozygotes exist), we tested the trihybrid females by crossing them with males that were homozygous for the slow alleles. Thus the cross to produce the data above was:

$got^f got^s$ $amy^f amy^s$ $sdh^f sdh^s$ $\times$ $got^s got^s$ $amy^s amy^s$ $sdh^s sdh^s$

In looking at the offspring, we know that each has a slow allele at each locus contributed by the homozygous males. Thus we look at the other allele in each of the offspring to know the alleles contributed by the trihybrid female. By doing that we have the same information we would have in a testcross.

The pattern of numbers among the eight offspring classes is the pattern we are used to for linkage of the three loci. We can tell from the two groups in largest numbers (nonrecombinants—classes 1 and 2) that the alleles are in the coupling (*cis*) arrangement. If we compare either of the nonrecombinant offspring with the double recombinant classes (7 and 8), we can see that the *amy* locus is in the middle. Especially clear should be the comparisons of classes 8 with 2 or 7 with 1. Again, disregard the slow alleles at each locus and look only at the other alleles, the ones contributed by the trihybrid. We can now infer that the trihybrid female had the following chromosomal arrangement:

$$got^f \; amy^f \; sdh^f$$
_____
_____
$$got^s \; amy^s \; sdh^s$$

You should now be able to see that classes 3, 4, 7, and 8 have crossovers in the *got-amy* region and classes 5, 6, 7, and 8 have crossovers in the *amy-sdh* region. Tallying these numbers, we see that the recombinants make up $(11 + 14 + 1 + 1)/1,000 = 0.027$ or 2.7% in the *got-*

amy region and $(58 + 53 + 1 + 1)/1,000 = 0.113$ or 11.3% in the *amy-sdh* region. Thus the map distances between loci are 2.7 and 11.3 centimorgans, respectively.

We expect $0.027 \times 0.113 = 0.00305$ double crossovers, or 3.05 per thousand. We observed only 2. Thus the coefficient of coincidence is $2/3.05 = 0.6557$, or about 66% of the expected double crossovers are actually observed.

5. a. The hotfoot locus is in the middle (compare, e.g., hotfoot, a double crossover, with the wild-type, a parental); it is 16.0 map units from either end locus.
   b. The trihybrid parent was $o \; h \; wa/o^+ h^+ wa^+$
   c. Interference is $1 - (20/25.6) = 0.22$, or 22%.

6. a. The brittle and glossy loci are linked; ragged is assorting independently. The brittle to glossy distance is 8.0 map units. The offspring of this cross form a pattern of four classes in high numbers, all about equal, and four classes in low numbers, all about equal. This is the pattern of simple linkage of two loci with a third locus assorting independently. There are two ways to approach this problem. By inspection you should be able to see that among the four classes in high numbers, the brittle and glossy mutants are together and their wild-type allele are also together, with the ragged phenotype acting independently. Thus the ragged locus is assorting independently and the four classes in high numbers are actually two classes with regard to brittle and glossy—they are the parentals. Similarly, the four classes in low numbers are the recombinants. If you cannot see this, then arbitrarily assign one reciprocal set in high numbers as parentals and one reciprocal set in low numbers as double crossovers. You will then calculate map distances such that brittle and glossy are 8 map units apart and ragged is 50 map units from the "middle" locus. In other words ragged is assorting independently.
   b. The *bt1* and *gl17* alleles are in *cis* (coupling) phase.
   c. Not relevant.

7. a. Work backward from the 0.61% double recombinants $(0.100 \times 0.061 \times 100)$.
   b. With a coefficient of coincidence of 0.60, only 0.366% (0.61×0.60) of the expected double recombinants will occur. For example:

ancon, spiny, arctus oculus	421
wild-type	422
ancon, spiny	28
arctus oculus	29
ancon	48
spiny, arctus oculus	48
ancon, arctus oculus	2
spiny	2

8. Notchy should be known on the X chromosome by the results of reciprocal crosses as well as linkage with other X-linked loci. If a trihybrid female is testcrossed (to an ancon, spiny-legged male fly hemizygous for the *ny* allele), the offspring will consist of, for example:

	Males	Females
Parentals (an sple or an⁺ sple⁺)		
ancon, spiny legs, notchy	45	45
ancon, spiny legs	45	45
notchy	45	45
wild-type	45	45
Recombinants (an sple⁺ or an⁺ sple)		
ancon, notchy	5	5
ancon	5	5
spiny legs, notchy	5	5
spiny legs	5	5

Ancon and spiny legs are linked, separated by 10 map units, whereas notchy segregates independently. Due to the genetic nature of the male parent, the sexes segregate similarly.

9. Construct a pedigree of the Duffy alleles. Arbitrarily assign one allele on a normal chromosome 1 and the other allele to the coiled chromosome. Then, in a drawing accompanying the pedigree, the alleles and their morphologically proper chromosomes would be associated.

10. Single crossover between the loci, involving only two chromatids, should produce a pattern similar to

$ab, ab, ab^+, ab^+, a^+b, a^+b, a^+b^+, a^+b^+$

If crossing over occurred at the two-strand stage, each crossover would involve both chromosomes and therefore all four chromatids, such as

$ab^+, ab^+, ab^+, ab^+, a^+b, a^+b, a^+b, a^+b$

A single crossover involves only two chromatids, but complex crossovers can involve three or four chromatids. For example, a three-strand double crossover between the a and b loci will produce

$ab, ab, ab^+, ab^+, a^+b^+, a^+b^+, a^+b, a^+b$

11. PD, 1, 2, 4, 6, 8–10; NPD, 3; TT, 5, 7. The loci are 20 map units apart.

12. a. PD, 1, 2, 6, 8, 10, 11; NPD, 9; TT, 3, 4, 7; unscorable, 5, 12
    b. Yes, PD >> NPD
    c. 25 map units apart

13. FDS, 3–5, 8, 10; SDS, 1, 2, 6, 7, 9. The distance between the *arg* locus and its centromere is 25 map units.

14. a. FDS, 1, 4–6, 8, 9, 11; SDS, 2, 7, 10; unscorable, 3, 12.
    b. The distance between the fuzzy locus and its centromere is 15 map units.

15. For example, the only variant of the type 2 pattern of table 6.6 is
$a^+b, a^+b, a^+b, a^+b, ab^+, ab^+, ab^+, ab^+$. Other patterns that are variants of the remaining five categories are derived by inverting the eight spores of a pattern or switching spores 1 and 2 with spores 3 and 4 or spores 5 and 6 with spores 7 and 8.

16. $ab \times a^+b^+ \rightarrow ab/a^+b^+$, which undergoes meiosis. Twelve map units means NPD + (1/2) TT = 12% of the total asci.

17. $a \times a^+ \rightarrow a/a^+$, which undergoes meiosis. Twelve map units means that the SDS pattern makes up 24% of the asci.

18. The best first-order estimate is that the a and b loci are on opposite sides of the same centromere. The a locus is 5.5, and the b locus is 10.5, map units from the centromere.

19. For example: 1, 445; 2, 5; 3, 250; 4, 250; 5, 30; 6, 5; 7, 15.

20. Assume the order of centromere, a, b.

    Class 1: No crossovers.
    Class 2: A four-strand double crossover between the a and b loci.
    Class 3: One crossover between the loci.
    Class 4: A two-strand double crossover, one between the centromere and the a locus and the other between the loci.
    Class 5: One crossover between the centromere and the a locus.
    Class 6: Three crossovers. The first is between the centromere and the a locus on chromatids 2 and 3; the second is between the two loci on chromatids 1 and 3; and the third is between the two loci on chromatids 2 and 4.
    Class 7: Two crossovers. The first is between the centromere and the a locus on chromatids 2 and 3; the second is between the loci on chromatids 2 and 4.

21. The +/− patterns seen under the columns headed chromosome 6, 14, X, respectively.

22. 20 map units apart

23. Only the type A, fructose intolerant children are recombinant. Thus, estimate θ at 0.2. Probability of a particular nonrecombinant child = 0.8/2 = 0.4. Probability of a particular recombinant child = 0.2/2 = 0.1. $Z = \log([0.4]^8[0.1]^2/[0.25]^{10}) = \log(6.872) = 0.837$.

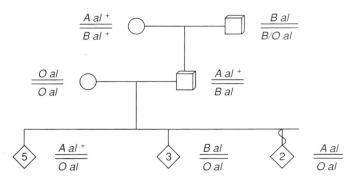

24. a. linked b. *trans* c. 28.7%. The cross is a testcross. If the genes were linked, we would expect a 1:1:1:1 ratio of offspring. The alleles that are linked will appear as the majority classes, and the majority classes are Trembling, long-haired and normal, Rex. Therefore, Trembling and Rex are in the *trans* position. If we let T = Trembling, and R = Rex, the cross is $\dfrac{Tr}{tR} \times \dfrac{tr}{tr}$. Recombinants are Trembling, Rex and normal, long-haired; $\dfrac{42 + 44}{300} \times 100 = 28.7\%$.

26. a. $\dfrac{k\ e^+\ cd}{k^+\ e\ cd^+}$ b. $\underline{k\ 6.9\ e\ 5.1\ cd}$

The initial cross is $\dfrac{k\ cd\ e^+}{k\ cd\ e^+} \times \dfrac{k^+\ cd^+\ e}{k^+\ cd^+\ e}$

And the F$_2$ offspring are $\dfrac{k\ cd\ e^+}{k^+\ cd^+\ e}$, in any order.

The last two classes are double crossovers and allow us to determine order. If the order is $k\ cd\ e$, a double crossover will yield $k\ cd^+\ e^+$ and $k^+\ cd\ e$. If the order is $cd\ k\ e$, a double crossover will yield $cd\ k^+\ e^+$ and $cd^+\ k\ e$. Therefore, the order must be $k\ e\ cd$. After reconstructing the trihybrid $\left(\dfrac{k\ e^+\ cd}{k^+\ e\ cd^+}\right)$ and scoring each of the offspring for crossovers in the k - e and e - cd regions: Map units, k - e = $[(64 + 67 + 7)/2000] \times 100 = 6.9$ and map units, e - cd = $[(49 + 46 + 7)/2000] \times 100 = 5.1$

28. a. yes; PD $>>$ NPD. b. a: 2.5 map units; b: 7.5 map units. Classify each ascus: I: PDT, FDS for both; II: TT, FDS for a, SDS for b; III: TT, FDS for a, SDS for b; IV: TT, FDS for a, SDS for b; V: TT, FDS for a, SDS for b; VI: PDT, SDS for a and b; VII, PDT, SDS for a and b; VIII: PDT, SDS for a and b. We see no NPDs, so genes are linked. For gene to centromere distances, use the formula, $\dfrac{1/2\ (\#SDS)}{100} \times 100$: a to centromere = $(1/2)(3 + 1 + 1)$ = 2.5 map units; b to centromere = $(1/2)(2 + 3 + 2 + 3 + 3 + 1 + 1)$ = 7.5 map units.

30. 33 map units. The woman is heterozygous in *trans* for color blindness and hemophilia: hc^+/h^+c. Recombination between these two markers yields h^+c^+ and hc. We can only detect recombinants in sons, so: $\dfrac{\#\ \text{normal sons} + \#\ \text{double mutant sons}}{\text{Total sons}} \times 100$ = map distance = 2/6 = 0.33.

32. Enzyme A on 11; B on 15; C on 18; D on 3; E on 7. Gene A is present in clones X and Y, and chromosome 11 is common to these two clones. B is present only in X, and 15 is the only chromosome unique to X. Similar logic allows the assignment of the other genes.

34. 0.0125. This problem requires manipulation of equations. We know that interference = 1 − coefficient of coincidence, so coefficient of coincidence = 1 − interference = 1 − (−1.5) = 2.5. Since coefficient of coincidence = $\dfrac{\text{observed double crossovers}}{\text{expected double crossovers}}$, observed double crossovers = (coefficient of coincidence) (expected double crossovers). The expected double crossover frequency is (2.5)(0.005) = 0.0125.

36. Lines A and C. We need to determine the lines that have chromosome 3; these lines are expected to be positive for the enzymes; we see that lines A and C have chromosome 3.

Chapter 7
Linkage and Mapping in Prokaryotes and Viruses

1. The prokaryotic chromosome is a double-stranded DNA circle that is relatively small compared with most eukaryotic chromosomes. Viral chromosomes can be DNA or RNA. Viruses are obligate intracellular parasites. Whether they are alive depends on the definition of alive.

2. a. *pen^r*;
   b. *azi^s*;
   c. *his$^-$*;
   d. *gal$^-$*;
   e. + or *glu$^+$*;
   f. + or *ton$^+$*

3. A heterotroph requires an organic source of carbon; an auxotroph has a particular nutritional requirement. A minimal medium contains only those substances required by the wild-type organism to grow; a complete medium is minimal medium that has been enriched with a complete array of organic compounds including amino acids and nucleic acid subunits. An enriched medium is the minimal medium to which nutrients have been added. A selective medium is a particular case of an enriched medium in which only one or a few items have been added.

4. A colony is a visible mass of cells derived usually from a single progenitor. A plaque is the equivalent growth of phages on a bacterial lawn producing a cleared area lacking intact bacteria.

5. A plasmid is an independent genetic entity within a cell, usually a small auxiliary circle of DNA. It integrates by a single crossover between the plasmid and the bacterial chromosome. It leaves by a reverse process of "looping out" (see figs. 7.20 and 7.21).

6. Far. If the selective locus is near, it will pass into the F$^-$ cell very early during conjugation. Consequently, there will be a great reduction in the recovery of loci distal to the selective marker because both the Hfr and F$^-$ members of a conjugation event can be killed by the selective agent (e.g., antibiotic).

7. When the drug sensitivity locus passes into the F⁻ strain, there will be a general decline in the recovery of recombinants after selection because both the Hfr and F⁻ members of a conjugation event can be killed.

8. See figures 7.10, 7.11, 7.19, and 7.20.

9. The bacterium could have survived and produced a colony if it was on a λ-free area, it became lysogenic, or it was genetically resistant to phage λ.

10. The order is *trp-purB-pyrC* since the class in lowest numbers (4) represents the double recombinants and hence exchanged the outside loci but not the middle one (*purB*). The relative recombination frequency between *purB* and *pyrC* is $(67 + 4)/(67 + 4 + 86) = 0.45$. Since *trp* is a selected locus (there are no *trp⁻* transformants), we cannot get accurate estimates of the other distances.

11.

	9 min		1 min		8 min		7 min	
Origin		*az*		*ton*		*lac*		*galB*

12. *lac* to *gal*, 9 minutes; *gal* to *his*, 27 minutes; *his* to *argG*, 24 minutes; *argG* to *xyl*, 11 minutes; *xyl* to *ilv*, 4 minutes; *ilv* to *thr*, 17 minutes; *thr* to *lac*, 8 minutes. For every interruption, a complete medium plate and one each of the seven selected plates are needed.

13. For example, use an Hfr that is wild-type but *str^s* with the F factor integrated at minute 20. Use an F⁻ strain that is *pyrD⁻*, *purB⁻*, *man⁻*, *uvrC⁻*, *his⁻*, and *str^r*. Interrupt mating at one-minute intervals and plate cells on complete medium with streptomycin to kill Hfr cells and grow up recombinant and nonrecombinant F⁻ cells. The next day, after colonies have grown up, replica plate onto selective media. The following data would be generated:

	Colony Growth on Media Selective for				
	purD⁺	*purB⁺*	*man⁺*	*uvrC⁺*	*his⁺*
Minute 0	−	−	−	−	−
1	+	−	−	−	−
5	+	+	−	−	−
16	+	+	+	−	−
22	+	+	+	+	−
24	+	+	+	+	+

14. The original strain was *arg⁻ his⁻*. Colonies on minimal medium + histidine are *arg⁺ his⁻*. Colonies on minimal medium + arginine are *arg⁻ his⁺*. Colonies on both are wild-type (*arg⁺ his⁺*).

15. 1, *his⁻ arg⁻*; 2, *leu⁻*; 3, *lys⁻*; 4, *his⁻ met⁻* or *his⁺ met⁻*; 5, *arg⁻*

16. Colony 1, *glu⁻ xyl⁺ arg⁻*; colony 2, *arg⁻ his⁻*; colony 3, + or at least *glu⁺* (may be unable to use other sugars); colony 4, *arg⁻*; colony 5, *iso⁻* (may also be *thr⁻* and/or *val⁻* or *ilv⁻*; may also be *thr⁻*); colony 6, *glu⁻ gal⁺ his⁻* or *glu⁻ gal⁺ his⁺*.

17. The class in lowest frequency (*trp⁺ his⁻ tyr⁺*, 107) identifies *his* as the middle locus (*trp-his-tyr*); relative recombination frequencies:

$$trp \text{ to } his = \frac{2600 + 418 + 107 + 3660}{2600 + 418 + 107 + 3660 + 1180 + 11940}$$

$$= 0.34$$

$$his \text{ to } tyr = \frac{418 + 685 + 1180 + 107}{418 + 685 + 1180 + 107 + 3660 + 11940}$$

$$= 0.13$$

18. Where phages cannot grow: *E. coli* Ton^R, phage h⁺. Where phage can grow, *E. coli* Ton^S, phage h⁺ or h, or *E. coli* Ton^R, phage h.

19. Prophage: an integrated phage; lysate: material released from a lysed cell; lysogeny: a bacteria-phage interaction in which the phage has become a prophage that can be induced at a later date; temperate phage: one capable of becoming a prophage.

20. Specialized transduction comes about through incorrect looping out of a prophage (see fig. 7.28). Generalized transduction involves the inclusion of a piece of bacterial DNA in a phage coat.

21. Use replica plating on selective media with arabinose as the sole carbon source, thus selecting for *ara⁺* cells. Although all three loci can be cotransduced, the rarity of *ara⁺ leu⁻ ilvH⁺* indicates *leu* is the middle locus (*ara-leu-ilvH*). Cotransductance frequencies:

$$ara \text{ to } leu = (9 + 340)/(9 + 340 + 32) = 0.92$$

$$ara \text{ to } ilvH = 340/(340 + 32 + 9) = 0.89$$

22. Mix the phages together with bacteria with increasing quantities of the two phages. Knowing the numbers of each in a particular case, it is possible to predict the proportion of cells doubly infected. The recovery of recombinants should increase with that probability. In other words, recombination should occur only in doubly infected cells.

23. *m-r-tu*: *m-r*, 12.8%; *r-tu*, 20.8%; coefficient of coincidence = 334/275.3 = 1.21. The problem is done just like a eukaryotic mapping problem (e.g., *Drosophila*). The data are equivalent to offspring from a testcross. For example, the 3,467 and 3,729 classes are nonrecombinant; the 162 and 172 classes are double crossovers.

24. *thr ara-2 ara-1 ara-3 leu*. Examine those crosses that yield a large difference between *ara⁺ leu⁺* and *ara⁺ thr⁺*: crosses 1, 4, and 6. Arbitrarily choose a gene order for the two mutants; for example, assume the order is *thr ara-1 ara-2 leu*. Cross 1 can then be diagrammed as

thr⁺ ara-1⁺ ara-2⁻ leu⁺	Donor
thr⁻ ara-1⁻ ara-2⁺ leu⁻	Recipient

Draw the crossovers needed to produce *ara*⁺ *leu*⁺ and *ara*⁺ *thr*⁺

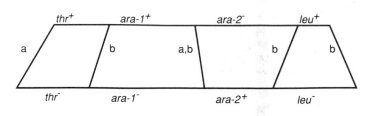

We see that with the above order, *ara*⁺ *leu*⁺ results from two sets of exchanges (b's), and are thus equivalent to a double crossover. Since double crossovers are rare, such transductants should also be rare. This result is not seen; *ara*⁺ *leu*⁺ >> *ara*⁺ *thr*⁺. Therefore, our initial order is wrong, and the correct order is *thr ara-2 ara-1 leu*. In cross 4, by similar logic as above, *ara*⁺ *leu*⁺ > > *ara*⁺ *thr*⁺, so order is *thr ara-1 ara-3 leu*. Since we know *ara-1* is closer to *leu* than *ara-2*, and that *ara-3* is closer to *leu* than *ara-1*, the order becomes *thr ara-2 ara-1 ara-3 leu*. Cross 6 confirms that *ara-3* is closer to *leu* than *ara-2*.

26. | 0.3 | 0.1 | 0.6 |
    a   *b*   *d*   *c*

Recombination frequencies should be approximately additive. The largest distance is between *a* and *c*; therefore *a* and *c* must be at opposite ends. Since *a − b* = 0.3, *b* must be 0.3 units to the right of *a*. This position gives *b − c* as 0.7, the observed distance. We now have the following map

| 0.3 | 0.7 |
  a   *b*   *c*

If *d* is to the left of *b*, then *d − c* should be greater than 0.7, a result not seen. Therefore, *d* is 0.1 unit to the right of *b*.

28. The order is *a c b*, and *c* is close to *a*. Genes *c* and *a* are cotransformed 69% of the time, suggesting that these two genes are very close and *b* is far away. Two orders are possible: *ac____b* and *ca____b*. If the first order is correct, *a*⁺ *b*⁺ *c*⁻ results from a double crossover; this class should be the least frequent. If the second order is correct, a single exchange between *a* and *c* would yield *a*⁺ *b*⁺ *c*⁻, but this frequency should be similar to *a*⁺ *b*⁻ *c*⁻, and it is not.

30. *thr leu pro his*. We see that cells that are *thr*⁺ are the most frequent. The chance of the conjugation being interrupted increases with the length of time for the mating. Therefore, genes far from the origin of transfer will appear less frequently. We can order the genes based merely upon the frequency of genotypes seen. The order must be *thr leu pro his*. Since we see no *his*⁺, and since we stopped the mating at 25 minutes, *his* must be after 25 minutes on the map.

32. *a* and *c* are close; *b* is farther away. The numbers of the first three transformant classes indicate that each gene, by itself, has an equal chance of being incorporated. We notice that classes with *b* and any other gene are quite rare, a situation that indicates *b* is far from *a* and *c*. We notice that *a* and *c* are cotransformed about 30% of the time.

34. a.   *lys*⁺ *his*⁺ *val*⁺   Since there is no lysine in
      lys⁺ *his*⁺ *val*⁻   the medium, *lys*⁺ must be
      lys⁺ *his*⁻ *val*⁺   present to allow growth.
      lys⁺ *his*⁻ *val*⁻

    b.   *lys*⁺ *val*⁺ *his*⁺   *Lys*⁺ and *val*⁺ must be
      lys⁺ *val*⁺ *his*⁻   present to allow growth.

    c.   *lys*⁺ *val*⁺ *his*⁺   *Lys*⁺ and *his*⁺ must be
      lys⁺ *val*⁻ *his*⁺   present to allow growth.

    d.   *lys*⁺ *val*⁺ *his*⁻   We see no *lys*⁺ *val*⁺ *his*⁺
      lys⁺ *val*⁻ *his*⁺   cells.

    e.   *lys*⁺ and *val*⁺ are close together; they are cotransformed 75% of the time. Order could be *lys val his* or *val lys his*.

    f.   *val lys his*. If the order is *val lys his*, *val*⁺ *lys*⁻ *his*⁺ should be rare, since this genotype results from a double exchange, and indeed, this class is the least frequent.

Chapter 8
Cytogenetics

1. All chromosomes form linear bivalents. The cross-shaped figure is seen only in heterozygotes.

2. Yes, if the deletion moves a gene next to heterochromatin.

3. No, there are no inversion loops formed in homozygotes.

4. Any arrangement will suffice that will result in an abnormal chromosomal complement in a gamete, including heterozygous inversions and translocations.

5. A diagram will show that a crossover between a centromere and the center of the cross can change the consequences of the pattern of centromere separation. For example, we diagram a crossover between loci 4 and 5 as in figure 8.11.

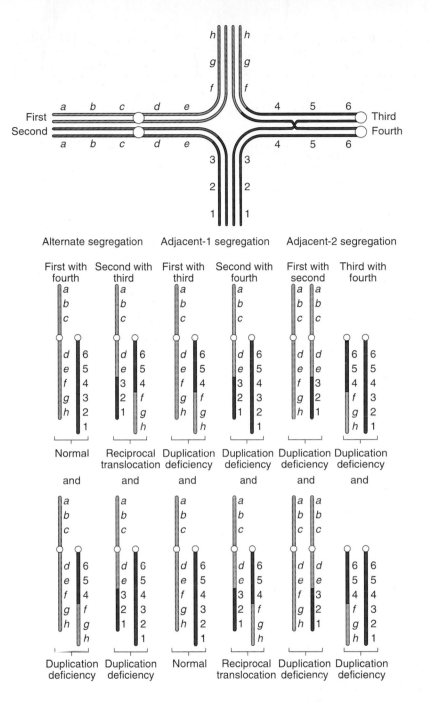

6. Problems occur during meiosis, not mitosis. Problems are worse in the odd-ploid organisms (e.g., triploid).

7. $4n = 92; 2n - 1 = 45$.

8. Usually allopolyploids because of a lack of pairing partners. Amphidiploids should have few or no meiotic problems.

9. $n = 8 + n = 6$ equals $14 \times 2 = 28; 20 + 20 = 40$

10. XX, 0, and X (unaffected) eggs will result.

11. One normal chromosome; one normal inversion chromosome; one dicentric chromosome with duplications and deficiencies; one acentric chromosome with duplications and deficiencies.

12. Reciprocal translocation (some effects occur only in the heterozygous condition). Look for the cross-shaped figure at meiosis or in salivary gland chromosomes.

13. Inversion (some effects occur only in the heterozygous condition). Look for a loop at meiosis or in salivary gland chromosomes.

14. Assume a crossover as shown below.

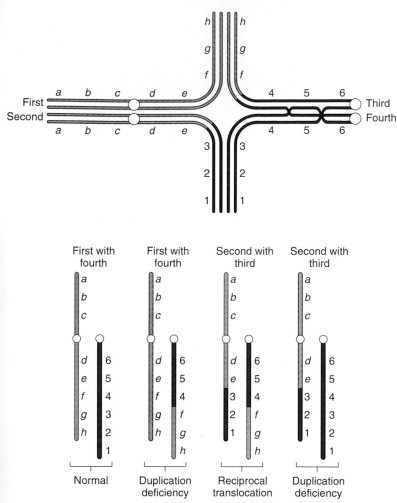

First with fourth	First with fourth	Second with third	Second with third
Normal	Duplication deficiency	Reciprocal translocation	Duplication deficiency

15. An X0/XYY mosaic can occur by nondisjunction of the Y chromosome in a cell during early cleavage in an XY individual. An XX/XXY mosaic can come about if one of the cells during early cleavage in an XX zygote is fertilized again by a Y-bearing sperm. Trisomy-21 usually comes about from an egg with two copies of the chromosome; the egg had two copies because of meiotic nondisjunction.

16. We have five heterozygous genes, so we expect to see 32 genotypes, but we see only 6 genotypes. We see no exchanges between genes B, C, and D. These three genes could be so tightly linked that no recombination occurs between them (unlikely). Genes B, C, and D could be within an inversion that is heterozygous in the heterozygous parent. The recombination that does occur within this inversion results in inviable zygotes.

18. We see about half the enzyme activity in crosses with strains B and C. Therefore the gene must be located in the region that is common to both strains, approximately in the region located 25 to 35 map units from the left end.

20.

$$\begin{array}{cccc} & v & m & s \\ /////////////// \underline{\quad 1.1 \quad} & m^+ & s^+ \end{array}$$

The use of X rays alerts us to chromosomal aberrations. The fact that a vermilion female appears when we expect all wild-type females indicates that we have a deletion. The deletion must end between v and m. If the deletion included miniature, we should have seen a vermilion, miniature female. We can draw the chromosomes of the flies in the second cross as:

$$\begin{array}{cccccc} v & m & s & & v & m & s \\ //////// \underline{\quad} m^+ & s^+ & \times & \underline{\quad\quad\quad} \end{array}$$

The question is, what is the distance between the deletion and the gene for miniature? We must look at the males from the cross. Note that we see only half as many males compared to females; those males that received the deletion X-chromosome must have died. The vermilion male must have resulted from a recombination between the end of the deletion and miniature:

$$\begin{array}{ccc} v & m & s \\ //////// \underline{\quad} | m^+ & s^+ \end{array}$$

yields the following chromosomes:

$$\begin{array}{ccc} v & m^+ & s^+ \end{array} \text{ and } /////////// \underline{\quad} m \quad s$$

Therefore,
map distance = (vermilion males/total males) × 100
= 1/90 × 100
= 1.1 map units

22. The father. The allele for color-blindness can only come from the mother. If meiosis in her is normal, an egg could get the X chromosome carrying the mutant allele. The daughter has only one X chromosome, so the sex chromosomes failed to separate in the man, and a sperm with neither X nor Y fertilized the egg.

24. The first meiotic division in the father is normal, producing cells with either two X or two Y chromatids. During the second meiotic division in the cell with the two Y chromatids, both Y chromatids move to the same pole and end up in the same sperm cell.

26. We expect to see about 32% recombination between these two genes, but we see only 2%. The most likely explanation is that an inversion occurred so that these two genes came to lie close to each other.

28. A translocation from the tip of the normal X in the male to the Y. We expect all males to receive an X chromosome with the white-eye allele from the female. For the male to be wild-type, we still must have part of the wild-type X chromosome. To test, cross this wild-type male to white-eyed females. All the female progeny should be white-eyed and all the male progeny red-eyed. You should draw the chromosomes to convince yourself of this.

1. **Conservative**

Density-gradient
centrifuge tube

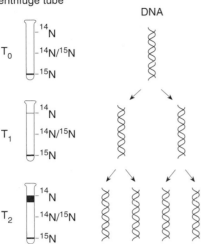

DNA

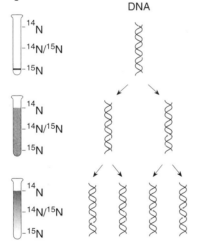

Dispersive

Density-gradient
centrifuge tube

DNA

2. The genetic code would somehow be read in number of tetranucleotide units, in which each unit consists of one each of the four bases (G, C, T, A). For example, one unit might be the amino acid alanine, two units might be the amino acid arginine, and so on.

3. Only proteins and nucleic acids were ever considered seriously.

4. Sugars: DNA has deoxyribose, RNA has ribose; and bases: DNA has thymine in place of uracil, RNA has uracil in place of thymine.

5.

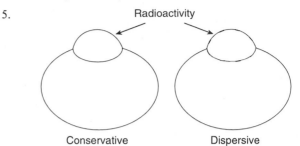

Conservative   Dispersive

6. Original   5'-ATTCTTGGCATTCGC-3'
   DNA complement   3'-TAAGAACCGTAAGCG-5'
   RNA complement   3'-UAAGAACCGUAAGCG-5'
   Replication is $5' \rightarrow 3'$.

7. See figure 9.13*b*.

8. See figures 9.18 and 9.19.

9. Theta structure: prokaryotic DNA replication; D-loop: chloroplasts and mitochondria; rolling-circle: bacterial conjugation; bubbles: eukaryotic DNA replication. The theta structure is an outcome of replicating a closed circle; the rolling-circle is a way to transport a linear copy of a circular chromosome; bubbles are outcomes of multiple origins of replication in a linear DNA; the function of D-loops is not obvious.

10. A primosome is a helicase plus a primase; it initiates DNA replication on lagging strands and is part of the replisome. A replisome includes a primosome plus two copies of DNA polymerase III; it coordinates replication on both the leading and lagging strands at the Y-junction.

11. Continuous replication occurs $5' \rightarrow 3'$ creating a leading strand, discontinuous replication occurs $5' \rightarrow 3'$ backward from the replicating fork on the $5' \rightarrow 3'$ template creating a lagging strand. Discontinuous replication is necessitated because none of the DNA polymerases acts in the $3' \rightarrow 5'$ direction.

12. See figure 9.27.

13. See figure 9.30 for the origin, figure 9.32 for the continuation, and figure 9.35 for the termination of DNA replication in *E. coli*.

14. At one time molecular swivels, located periodically along the DNA, were suggested.

15. Possible mechanism of DNA gyrase involves a double-stranded break in the DNA. Another double-stranded helix is then passed through the break, which is then closed.

DNA gyrase

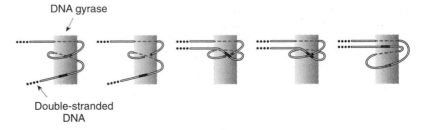

Double-stranded
DNA

16. 1: double-stranded DNA; 2: single-stranded DNA; 3: double-stranded RNA; 4: single-stranded DNA; 5: single-stranded RNA. First, look at what bases are present. If T is present, the molecule is DNA, and if U is present, the molecule is RNA. Remember that for a double-stranded molecule A must equal T (U) and G must equal C. For molecule 1 we see that A = T and G = C. In molecule 2, purines (A + G) > pyrimidines (C + T). In double-stranded molecules, purines = pyrimidines. Molecule 3 must be double-stranded RNA. In 4 and 5, G is not equal to C, therefore, the molecules are single stranded.

18. (lowest) 2, 1, 4, 5, 3 (highest)
Remember that a G-C pair has three hydrogen bonds and thus requires more energy to be broken than an A-T pair. Therefore, the higher the melting temperature, the higher the G-C content. Simply arrange the molecules from lowest to highest melting temperature.

20. You could imagine a situation in which a single molecule is replicated conservatively.

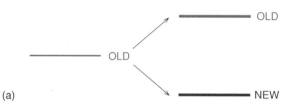

(a)

Unfortunately, all nucleic acids seem to replicate semiconservatively. Therefore, we must propose that a complementary strand (−) is made from the first:

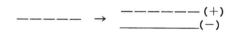

We now need to make more (+) strands from the (−). The (+) strand could separate and allow new (+) strands to be made and peeled off:

(b)

Other possibilities also exist as long as you keep in mind the complementary nature of nucleic acids.

22. Finding small pieces or fragments of DNA suggests the Okazaki pieces are only slowly, if at all, joined, a function of DNA ligase. The fact that not many long DNA molecules are seen also suggests that the DNA is being broken, implicating an endonuclease as well.

24. 1.91×10^4 or 19.1 Kb. One base takes up 3.4×10^{-1} nm. Therefore,

$$\frac{6.5 \text{ m}}{3.4 \times 10^{-1} \text{ nm/bp}} = \frac{6.5 \times 10^3 \text{ nm}}{3.4 \times 10^{-1} \text{ nm/bp}}$$

$$= 1.91 \times 10^4 \text{ bp.}$$

Chapter 10
Gene Expression: Transcription

1. Hybridization of nucleic acids is carried out between species to determine the extent to which it occurs. Presumably, the greater the amount of hybridization, the more similar the genomes of two species and therefore the more closely they are related in an evolutionary sense (see chapter 21).

2. Transcription has higher error rates. Errors of DNA polymerase tend to become permanent, whereas errors of RNA polymerase do not.

3. The transcription start signal is the promoter, recognized with the aid of the sigma factor of RNA polymerase. Transcription stop signals are terminators, some of which require rho factors. Polycistronic transcripts are routine in prokaryotes, virtually unknown in eukaryotes.

4. See figure 10.2. Complementarity is achieved between mRNA and rRNA and between mRNA and tRNA.

5. A consensus sequence is made up of the nucleotides that appear in a significant proportion of cases when similar sequences are aligned. A conserved sequence consists of nucleotides found in all cases when similar sequences are aligned. For example, the Pribnow box (fig. 10.5) is the consensus sequence TATAAT.

6. With no sigma factors, there would be random starts of transcription; with no rho factor, there would be a failure to properly terminate transcription at rho-dependent termination sites.

7. See figure 10.7 for a promoter and figure 10.10 for a terminator. The transcript starting from the promoter would be 5′-CUUAUACGGU···. The transcript from the terminator is shown at the bottom of figure 10.10.

8. Although the overall process is similar in prokaryotes and eukaryotes, differences exist in number of different RNA polymerases and transcription factors in eukaryotes; exact DNA sequences recognized; polycistronic nature of prokaryotic DNA; and posttranscriptional processing in eukaryotes (splicing, capping, polyadenylation).

9. Introns do accumulate more mutations. Presumably, at most sites within an intron, mutations do not affect the phenotype and therefore are not selected against (see chapter 20).

10. Removing one base too many or too few would result in a shift in the reading frame during translation (see chapter 11), thus radically altering the protein product.

11.

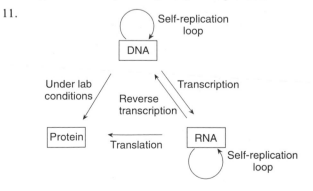

12. Heterogeneous nuclear mRNAs are eukaryotic transcripts before posttranscriptional modifications. Small nuclear ribonucleoproteins are the components of the eukaryotic mRNA-splicing apparatus.

13. Five introns

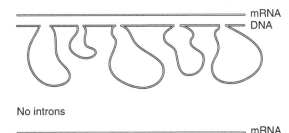

No introns

14. A stem-loop structure, when a single strand of DNA or RNA, forms a short internal double-helical section (see fig. 10.10). An inverted repeat is a sequence read outward on both strands of a double helix from a central point (see fig. 10.10). A tandem repeat is a segment of nucleic acid repeated consecutively; that is, the same sequence repeats in the same direction on the same strand:

5'-TCCGGTCCGGTCCGG-3'
3'-AGGCCAGGCCAGGCC-5'

A DNA sequence with a seven-base inverted repeat is

5'-ATTACCGCGGTAAT-3'
3'-TAATGGCGCCATTA-5'

15. TATAAT (Pribnow box): Part of prokaryotic promoters
TTGACA: −35 sequence in prokaryotic promoters
TATA (Hogness box): Part of eukaryotic promoters
CAAT: −70 sequence in eukaryotic promoters
TACTAAC: Consensus sequence around lariat branch point in nuclear introns

16. A technique in which DNA in contact with a protein is exposed to nucleases; only DNA protected by the protein is undigested. Promoters could be isolated by protection with RNA polymerase and then sequenced.

17. Recognition signals are shown in figures 10.30, 10.32, and 10.34.

18. Rho-dependent terminators require the rho protein for successful termination. Both terminator types have stem-loop structures. Rho-independent terminators have, in addition, a sequence of uracil-containing nucleotides after the inverted repeat (see fig. 10.10).

19. Group I introns are self-splicing and require a guanine-containing nucleotide for splicing. Group II introns are similar but do not require an external nucleotide for splicing. Group I and II introns are released as linear and lariat-shaped molecules, respectively.

20. See figures 10.30 and 10.32.

21. Spliceosomes are composed of small nuclear ribonucleoproteins (see fig. 10.35).

22. The superscripts of the sigma factors refer to their molecular weights (e.g., σ^{70} is 70,000 daltons).

23. A transcriptional factor is a eukaryotic protein that is similar in function to prokaryotic sigma factors. An enhancer is a eukaryotic DNA sequence that increases transcription even if distant from the promoter itself.

24.

leader   coding   trailing

prokaryotic

cap   leader   coding   trailing   Poly A

eukaryotic

Both molecules will have 5' and 3' noncoding regions, the leader and trailing sequences, respectively. Eukaryotic mRNA has a 5' cap and a 3' poly-A tail.

26. Top   3' _____ 5' The top strand is
Bottom   5' _____ 3' transcribed (template strand).

RNA: 5'-G G C A U G C G G A A A G U C C A A-3'

Begin by writing the RNA sequence complementary to each DNA strand. Since the DNA fragment is from the beginning part of the gene, there must be a start signal, AUG, for protein synthesis. Transcription of the top strand yields such an RNA, but the bottom does not. Since transcription proceeds 5' → 3', the left end of the top strand must be the 3' end. Remember that nucleic acids are antiparallel.

28.

There are three introns, so we expect three single-stranded DNA loops. The coding regions (exons) will form RNA-DNA hybrids and will appear thicker.

30. 5' A U A C C G U A C C <u>A U G</u> A A G G C C 3'
The Pribnow box is T A A T G C, beginning six bases from the 3' end of the DNA. Transcription proceeds in a 5' → 3' direction; therefore it moves left to right along the given DNA molecule because all nucleic acids are antiparallel.

32. All transcription could be rho independent, and we would see normal transcription at both temperatures. If transcription is rho dependent, long molecules could

possibly be rapidly degraded, producing short molecules. Neither possibility is too likely since some genes seem to be rho independent and others rho dependent.

34. The double helix must unwind in order for transcription to occur. An A-T pair, because it has only two H-bonds, is more easily disrupted than a G-C pair.

Chapter 11
Gene Expression: Translation

1. The mRNA is: 5'-AUGUUACCGGGAAAAUAG-3'; the anticodons are 3'-UAC-5', 3'-AAU-5', 3'-GGC-5', 3'-CCU-5', 3'-UUU-5'; the amino acids are methionine, leucine, proline, glycine, lysine.

2. If a site on the ribosome was a temporary docking region for the nascent polypeptide, only one tRNA need be present at a time.

3. Five bases are needed for codons composed only of two base types ($2^5 = 32$, a number large enough to include different codons for the twenty amino acids); if double-stranded, the proportion of C and G would be equal.

4. 12/27 phenylalanine; 6/27 serine; 6/27 leucine; 3/27 proline

5. See figure 11.16. Use the mRNA of problem 1 and be sure to include EF-Tu and EF-Ts.

6. In prokaryotes, the Shine-Dalgarno sequence is used for mRNA-ribosomal alignment. In eukaryotes, the scanning hypothesis describes the need for a 5' cap and scanning behavior. Because a 5' cap is needed for ribosomal alignment, only one polypeptide can be synthesized per mRNA (there is only one site of ribosomal attachment per eukaryotic mRNA).

7. There are approximately twenty aminoacyl-tRNA synthetases in an *E. coli* cell, one for each amino acid. Recognition signals can occur at any point on a given tRNA. This system of synthetase-tRNA recognition has been called a second genetic code.

8. The three nonsense codons are 5'-UAA-3', 5'-UAG-3', and 5'-UGA-3'. They all begin with a pyrimidine (uracil) and have purines in the second and third positions. The theoretical anticodons are 3'-AUU/I-5', 3'-AUC/U-5', and 3'-ACU/I-5', respectively.

9. See figure 11.8.

10. An antibiotic would be effective if it interfered with any step of the prokaryotic translation process. Thus we could have antibiotics that interfered with any of the prokaryotic initiation, elongation, translocation, and termination factors, any of the aminoacyl-tRNA synthetases, and any of the exposed proteins of the ribosome.

11. Three; see figure 11.28.

12. 5'-UUA-3' → 5'-UAA-3'. The consequence is that the growing polypeptide will be terminated at an improper point, probably producing a nonfunctioning enzyme or protein.

13. 5'-UAA-3' → 5'-UAA-3'. The consequence is the failure to terminate the particular protein leading to continued chain elongation to the next nonsense codon or to the end of the mRNA. The result is probably a nonfunctioning enzyme or protein.

14. There would be implications to evolutionary theory both in general and in reference to the evolution of the genetic code, indicating a long and isolated evolutionary lineage of the organism in question. Extraterrestrial organisms probably would use a different code system. In other words, there is no obvious structural or functional relationship between a nucleotide base and an amino acid.

15. There would be blockage of further protein synthesis because of the N-terminal formyl group that prevents a peptide bond. The growing peptide would be stopped at that point.

16. EF-Tu brings a tRNA to the A site at the ribosome. EF-Ts is involved in recharging EF-Tu (see fig. 11.15). The eukaryotic equivalent is eEF1.

17. RF1 and RF2 recognize nonsense codons. Their eukaryotic equivalent is eRF.

18. A signal peptide is a sequence of amino acids at the amino-terminal end of a protein that signals that the protein should enter a membrane (see fig. 11.24). The situation in eukaryotes is more complex because there are so many different membrane-bound organelles, each having their own membrane-specific requirements. Signal peptides are usually cleaved off the protein after the protein enters or passes through the membrane.

19. Puromycin does not disrupt eukaryotic translation because it does not bind to the eukaryotic ribosome for reasons that have to do with the general conformation of the eukaryotic translation system.

20. The table could look the same (see table 11.4) except that the position would be left side = first position (5′ end); top = third position (3′ end); right side = second position. For example, the codons for valine (currently 5′-GUU-3′, 5′-GUC-3′, 5′-GUA-3′, and 5′-GUG-3′) would be 5′-GUU-3′, 5′-GCU-3′, 5′-GAU-3′, and 5′-GGU-3′.

21. NH₂-FGKICABHLNOEDJM-COOH

22. GGC. The first polymer produces reading frames of GCG - CGC. . . . Therefore, GCG and CGC represent ala and arg, but we cannot determine which is which. The second polymer produces either CGG - CGG. . . , GGC - GGC. . . , or GCG - GCG. . . . We know that GCG will code for either ala or arg. The new combination, GGC, is not similar to the original codons, and probably codes for gly. CGC and CGG probably code for the same amino acid, either ala or arg.

24. 6. Begin with CUC, then list all possible codons in which the first C is changed; then list all possible codons in which only the U is changed. Finally, list all codons that result in a change of the second C.

List 1	List 2	List 3
AUC: *ile*	CAC: *his*	CUG
UUC: *phe*	CGC: *arg*	CUA all *leu*.
GUC: *val*	CCC: *pro*	CUU

26. a. 5′-AUG AUU GAA UGC GAG CGG AGU-3′
    b. N-met-ile-glu-cys-glu-arg-ser

First determine the sequence of the RNA complementary to the given DNA strand. Don't forget about polarity; as the strand is written, the 5′ end of the RNA will be on the left. Blocking off successive groups of three bases allows the determination of the codons. Use the code to determine the amino acid sequence.

28. Not apparent; either GAA/G or AUA/G if the mutations are single-step changes. GAA/G can change in a single step to GUX (valine) or AAA/G (lysine), as can AUA/G change in a single step to GUX (valine) or AAA/G (lysine).

30. GGX GCU AGC CAU UGC CUC UUC/U. Mutant 1 is shorter than normal, indicating the cysteine codon mutated to a stop codon. Since cysteine is UGU/C, the stop must be UGA in the mutant. In mutant 2, histidine is replaced by leucine; CAU/C → CUX. Changing the A to a U will give a leucine codon. Only the third mutant allows us to determine the normal sequence. All amino acids after the first are changed, indicating a frame shift,

an addition or deletion of one base. We must therefore line up the two possible sequences. Write down the possible sequences for the normal:

GGX GCX UCX   CAU/C UGU/C UUA/G UUC/U

   AGU/C   CUX

Then write down the possible sequences for mutant 3:

GGX GUX GCX AUU/C/A GUX UCX

   AGU/C

By aligning these sequences, the change can be seen as a deletion of the fifth base in the normal sequence (C). Therefore, only certain of the original possibilities can exist.

Chapter 12
DNA Cloning and Sequencing

1. Type II endonucleases are valuable because they cut DNA at specific points and many leave overlapping or "sticky" ends.

2. At least three sequences of twofold symmetry (palindromes) exist (with shorter embedded sequences): GAATTC, GGATCC, GCATGC.

3. In DNA with a random sequence, a four-cutter will find sites approximately once in 4^4 bases ($= 1/256 = 0.0039$). A six-cutter will find sites approximately once in 4^6 bases ($= 1/4096 = .0002$). An eight-cutter will find sites approximately once in 4^8 bases ($= 1/65,536 = 0.000015$).

4. Restriction endonucleases are unsuited for cloning if they destroy the DNA of interest or if they do not have appropriate sites in the DNA of interest (both vector and insert).

5. DNA can be joined by having compatible ends or by blunt-end ligation (linkers combine these methods). The appropriateness of a method depends on what DNA is to be cloned and how that DNA can be obtained. Having DNA with "sticky" ends created by the same restriction enzyme would be easiest, but sometimes is not available.

6. The heteroduplex would look similar to the one in figure 12.19.

7. A plasmid is a self-replicating circle of DNA found in many cells. Foreign DNA inserted into a vector forms an expression vector if that foreign DNA produces a protein product. Cosmids are plasmids that contain *cos* sites and are useful for cloning large segments of DNA (up to 50 kb).

8. The steps in creating cDNA are shown in figure 12.11. Radioactive cDNA and mRNA can be obtained either by using radioactive triphosphate nucleotides during the synthesis of the cDNA and mRNA or by end-labeling the products.

9. Chromosome walking is a technique for cloning overlapping chromosome regions starting from an arbitrary point. It is useful for determining relative locations of genes in uncharted regions as well as cloning regions too big to fit in a single vector.

10. Use radioactive alanine tRNA as a probe in either a whole digest (Southern blotting) or in a genomic library (dot blotting).

11. Southern and northern blotting are gel transfer techniques used to probe for DNA and RNA sequences, respectively. Western blotting is an antibody-probing technique used for protein determination, similar to Southern blotting. Dot blotting is a probing technique that eliminates the electrophoretic separation step.

12. If an mRNA could not be isolated, a probe could be constructed using the codon dictionary if the amino acid sequences of the protein product were known. A genomic library could be constructed with the intent of looking for the location of the gene by its expression. Expression vectors are those whose foreign DNA is expressed.

13. Plasmids of *E. coli* origin survive in yeast when a yeast centromere (CEN region) is added, allowing them to replicate within the yeast cell. Inactivated viruses can function in the presence of intact "helper" viruses that allow them to complete their life cycles.

14. Foreign DNA can be introduced into eukaryotic cells by infection with animal virus vectors as well as several other techniques including electroporation, liposomes, and biolistics. Transfection is the introduction of foreign DNA into eukaryotic cells. A transgenic mouse has been successfully transfected; it has incorporated the foreign DNA into its genome.

15. Hypervariable DNA is DNA showing a great deal of interindividual variation. A RFLP (restriction fragment length polymorphism) is a polymorphism (variation) that shows up after Southern blotting and probing of restriction digests. A VNTR (variable number of tandem repeats) locus is one that is hypervariable due to unequal crossing over among the tandem repeats. VNTR loci are a hypervariable subset of RFLPs.

16. Investigators using DNA fingerprinting (usually of RFLP, preferably VNTR loci) have demonstrated much variability; people are very different from each other in these systems. Hence people can be identified with very high confidence. These techniques are especially suitable for paternity exclusion. RFLPs of mother, child, and putative father are examined. The father's genome must include (barring mutation) all the offspring bands not found in the mother.

17. Polymerase chain reaction (PCR) is a rapid method to amplify DNA segments. It is particularly useful when only a small amount of the DNA is available and primers flanking the region can be created.

18. Electrophoretic bands of the total digest are (* indicates end label) 50, 100*, 150*, 250, 300 bp. Bands of the partial digest are 50, 100*, 150*, 250, 300(×2), 350, 400*(×2), 450*(×2), 600, 700*, 750*, 850* bp.

19. The original was cut between the following lengths: 50–200–100–400.

20. Mutant A: elimination of site between the 300- and 50-bp segments. Mutant B: elimination of site between 100- and 300-bp segments. Mutant C: creation of a new site within the 300-bp segment, 225 bp to the right of the 100-bp end segment and 75 bp to the left of the 50-bp segment.

21. Alternative II is correct. Note that among the clues, alternative I requires a band of a segment that is 200 bp long.

22. For the steps in the dideoxy sequencing method, see figures 12.36 and 12.37. Use of fluorescent dyes has allowed for the automation of the process and the elimination of radioactive tags.

23. 5'-CAATAGGTCGAGGTTCAATGG-3'

24. The DNA can be inserted into the M13 general sequencing vector.

25. Overlap can work for one, four, or seven bases. Lowercase letters indicate the stop sequence (tga), boldface indicates the **ATG** start sequence, and X indicates any base. One-base overlap, ···tga**TG**···; four-base overlap, ···**Atga** ···; seven-base overlap, ···**ATGX**tga···

26. Benefits of gene cloning are described for medicine, agriculture, and industry. Many people are concerned that toxic, carcinogenic, or generally damaging genes will be introduced into the environment or pass from person to person in an epidemic.

28.   __3__ | __5__ | __4.2__ or

  __3__ | __4.2__ | __5.0__ or

  __5__ | __3__ | __4.2__

A linear molecule with *n* sites will produce *n* + *1* fragments, so there are two sites within the molecule. Any one of the three fragments could be in the middle. Draw the possibilities by putting one of the fragments in the middle.

30.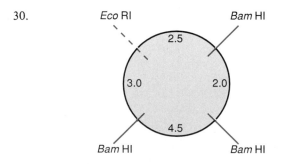

The molecule either has no *Eco* RI sites, or it is a circle. A circle with *n* sites will yield *n* molecules. If there are no *Eco* RI sites, and the molecule is linear, we expect to see the same number of *Bam* HI fragments from both the single *Bam* HI digest and from the double digest. Since we got four fragments in the double digest, but only three in the *Bam* HI digest, the molecule must be a circle. The

2.0 kb and 4.5 kb fragments appear for both *Bam* HI and *Bam* HI + *Eco* RI, so they must not contain an *Eco* RI site. The 5.0 kb fragment must contain an *Eco* RI site 2.5 kb from one end.

32.

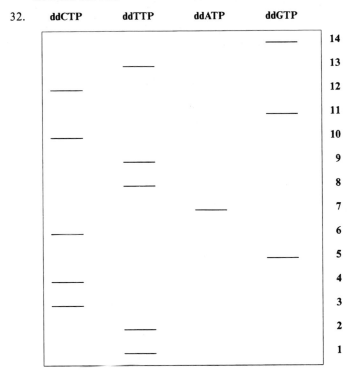

What will appear in the gel are fragments of the newly synthesized strand. Since DNA synthesis proceeds 5′ to 3′, the 5′ base will be T in the new strand. Proceed up the gel by indicating the base complementary to the sequence given.

34. We must insert DNA that has no introns into bacterial plasmids. This DNA can be obtained by isolating mature, cytoplasmic mRNA and then using reverse transcriptase to make double-stranded cDNA. Plasmids with cDNA inserted can then be used to isolate human genomic DNA.

36.

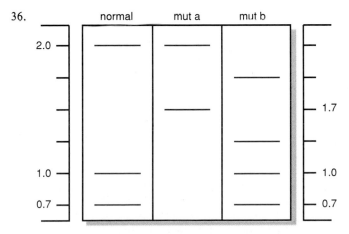

The first mutant will be cleaved only once, yielding bands of 1.7 kb and 2.0 kb. The second mutant will produce the 1.0 kb and the 0.7 kb fragments since this site is conserved. A new site within the 2.0 kb fragment will

generate two new fragments, each smaller than 2.0 kb. Thus the loss of a site produces two bands, and the acquisition of a new site yields four total bands.

38. 0.5, 1.0, 2.0, 3.5 kb. The 1.0 band should have a higher intensity than the other bands. It helps to redraw the *Eco* RI sites:

0.5	1.0	3.0	3.5

2.0

Since *Hind* III does not cut the 3.5 band, it must cut within the 3.0 band. To produce a 5.5 kb fragment, it must cut 2.0 kb to the left of the *Eco* RI site. Thus, two 1.0 kb bands will be produced as well as 0.5, 2.0, and 3.5 kb.

40. Two of the three chromosome 21s present in the child came from the father, not the mother. Since the probe produces different bands in the mother and the father, all of these bands must also be present in the child. The intensity of a band is proportional to the amount of DNA present. The bands that are of paternal origin are more intense than the maternal bands. The father contributed two 21st chromosomes, a situation that occurs in 20% to 30% of Down syndrome patients.

Chapter 13
Gene Expression: Control in Prokaryotes and Phages

1. a. Inducible (wild-type); b–d. constitutive; e. neither, superrepressed; f. inducible

2. a. (a) z^+; (b) no transcription
   b. (a) z^+, z^-; (b) no transcription
   c. (a) z^+, z^-; (b) no transcription
   d. (a) z^+, z^-; (b) z^-
   e. (a) z^+, z^-; (b) z^+

3. $p\ o^c\ e^+ \cdots a^+/F'\ p\ o^+\ e^- \cdots a^+$. Operator constitutive mutants (o^c) are *cis* dominant; that operon cannot be repressed. Regulator mutants (i^-) are recessive in that a wild-type regulator (i^+) can control all operons in a merozygote. This type of control can work for any synthetic pathway in which an end product or derivative can be a corepressor.

4. The *E. coli trp* operon functions as a normal repressible operon. In addition, attenuation, based on secondary structure and stalling of the ribosome on the leader transcript, can further prevent transcription (see fig. 13.16). Attenuator control can be exerted based on other amino acids if their codons appear in the leader transcript, causing ribosome stalling.

5. Repression; lytic response

6. Assuming that the mutants produce inactive proteins: *cI*, lytic response; *cII*, lytic response; *cIII*, lytic response; *N*, neither lytic nor lysogenic responses possible; *cro*, no lytic response possible; *att*, no lysogenic response possible; *Q*, no lytic response possible.

7. Assuming that the operator mutants prevent the operator from recognizing repressor: o_{R1}, should result in lysogeny; o_{R3}, should result in lytic response. Assuming that promoter mutants do not recognize RNA polymerase: p_L, no action by phages possible; p_{RE}, lytic response; p_{RM}, lytic response; p_R, probably lysogeny because no right or late operon production; t_{L1}, read through transcription of left operon even if N product absent, lysogeny possible. Same results for $nutL$ mutant; t_{R1}, read through transcription of right and late operons even without N-gene product; should result in lytic response. Same result for $nutR$ mutant.

8. The λ chromosome has one circular and two linear forms. The circular form is the infective cellular form. A break at one point (*cos* site) takes place during packaging into the phage head and a break at another point forms the linear integrative prophage (see fig. 13.17).

9. UV damage induces DNA repair systems. The RecA protein, involved in repair, also cuts the λ repressor protein, leading to induction of the lytic cycle.

10. The prophage region of the Hfr chromosome may enter the F^- cell with no repressors present. The situation is thus similar to regular phage infection, which can go either way (lysogenic or lytic cycles).

11. Transcription is the level at which most control mechanisms work. These include sigma factors, efficiency of promoter recognition, catabolite repression, operon-repressor systems, attenuation, and transposition. Translational control mechanisms include polarity, antisense RNA, differences in the efficiency of processes due to nucleotide sequence differences, codon preference, and the stringent response. Posttranslational mechanisms include feedback inhibition and differential rates of protein degradation.

12. IS elements have been called selfish DNA because they replicate copies of themselves in the host genome without necessarily providing the host with any obvious benefit.

13. An IS element is a simple transposon. A transposon is a segment of DNA that can make a copy of itself to be inserted at another place in the genome. An intron is an intervening sequence, a region excised from mRNA before translation. A plasmid is an autonomous, self-replicating genetic particle. A cointegrate is an intermediate structure in transposition.

14. See figure 13.32. Transposase is involved in joining nonhomologous single-stranded DNA. Resolvase catalyzes a recombinational event that separates the two elements of a cointegrate. DNA polymerase I and ligase function in processes that repair small patches of single-stranded DNA during transposition.

15. See figure 13.28.

16. See figure 13.33.

17. See figure 13.34.

18. A polar mutation is one, usually within an operon, that prevents the expression of a distal gene. It can be caused by transcription stop signals within a transposon.

19. Transcriptional control is much more efficient.

20. Cyclic AMP, combined with CAP protein, attaches to CAP sites enhancing transcription of nonglucose, sugar-metabolizing operons in *E. coli*.

21. Feedback inhibition: allosteric enzymes in some synthetic pathways can be inhibited by the end product of that pathway. In addition, many repressor proteins are allosteric.

22. An operon with a CAP site (e.g., arabinose in *E. coli*) has enhanced transcription when glucose is absent (positive control). However, these operons are also usually inducible. Lack of an induction (negative control) mechanism is very inefficient.

23. Heat-shock proteins are normally induced by the presence of a specific sigma factor, which itself is induced by heat shock.

24. Protein degradation can be controlled by the nature of the N-terminal end or by PEST regions (regions rich in proline, glutamic acid, serine, and threonine).

25. The stringent response is the response of a prokaryotic cell to amino acid starvation. The idling reaction of the ribosome results in production of 3′–ppGpp–5′, whose appearance is associated with the cessation of transcription, especially of tRNAs and rRNAs, through an unknown mechanism.

26. Antisense RNA is complementary to mRNA. It can have regulatory functions by forming a double helix with (and thus inactivating) its functional complement. An obvious source of antisense RNA is transcription of the complementary DNA strand of the strand normally transcribed. Another source could be from the transcription of a gene that has evolved from the original gene and thus has a similar base sequence. Clinically, it might prevent functioning of RNAs that have disease functions.

28. We must think about how these operons are controlled. Not only do they need inducer, but they also require the catabolite repression-activation system. These mutants could be unable to make cAMP because the adenylcylase gene is defective. Alternatively, they could be making a defective catabolite activating protein (CAP^-).

30. a. No transcription; *cis* dominant. b. yes, transcription; recessive to wild-type. c. no transcription; dominant. If polymerase cannot bind, no transcription or translation can occur, but this effect will only be seen in the one DNA strand; wild-type DNA will be able to bind RNA polymerase. If the repressor cannot bind to DNA, RNA polymerase will be able to bind, and the operon will always be on. A good repressor, regardless of source, will be able to bind both DNAs. If the repressor cannot bind tRNA, it cannot be removed from the operator, and the operon will always be off. In some of the partial diploids, this repressor will be able to bind to both operators, and thus will appear dominant to wild-type.

32. Lysis will result. Since *cI* controls transcription of O_L and O_R, it will not be able to bind to O_L and O_R, and transcription will occur, leading to lysis.

34. a. Operator and repressor. b. make a partial diploid with wild-type; the operator mutant will make β-galactosidase constitutively, and the repressor mutant will make it only in the presence of lactose. Four mutations are possible in the *lac* operon: mutations in the *z* gene or the promoter will never make the enzyme. Mutations in the repressor will always make the enzyme because the repressor cannot bind DNA. In operator mutations, a good repressor can never bind DNA. In a partial diploid, i^-o^+/i^+o^+, the wild-type repressor is *trans* acting and can bind to both operators, creating an inducible situation. In i^+o^-/i^+o^+, repressor cannot bind to o^-, and this DNA is always on, even though the wild-type DNA is off in the absence of lactose.

36. The operator is probably between S and T, and the promoter is probably to the left of Q. Deletions of the operator sequence will prevent binding by the repressor, and the operon will always be on. This result is seen in 5 and 4. Both strains 3 and 4 are missing part of the region between S and T. A promoter deletion will never make RNA, and this is seen in deletion 1. The DNA of the normal cell must bend so that the operator lies in front of Q.

Chapter 14
The Eukaryotic Chromosome

1. In general, prokaryotes are small, have a relatively small circular chromosome, and have little internal cellular structure as compared with eukaryotes. Most prokaryotic mRNAs are polycistronic, under operon control; eukaryotic mRNAs are highly processed, monocistronic, and not under operon control. Prokaryotes are mostly single-celled organisms, whereas eukaryotes are mostly multicellular. Eukaryotes have repetitive DNA, absent for the most part in prokaryotes. Prokaryotic chromosomes are not complexed with protein to anywhere near the same extent that eukaryotic chromosomes are.

2. The evidence for the uninemic nature of the eukaryotic chromosome is summarized in figures 14.1, 14.2, and 14.3.

3. Assume that each chromosome contained two complete copies of the same DNA. Following the protocol of figure 14.1, the results would be chromosomes, before separation, that consisted of either two labeled chromatids or only one labeled chromatid, in a 1 : 1 ratio (barring sister chromatid exchanges). The labeled chromatid in chromosomes with just one chromatid labeled will have twice the label of each chromatid in the chromosomes in which both chromatids are labeled. (See illustration at top of next column.)

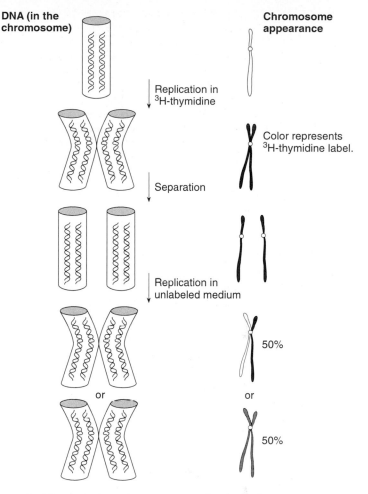

4. The major protein components of eukaryotic chromosomes are the histones, which form nucleosomes, compacting the DNA, and nonhistone proteins, which form the scaffold and have other functions.

5. The length of DNA associated with nucleosomes was determined by footprinting, in which free DNA was digested, leaving only those segments protected by nucleosomes. Nucleosome hypersensitive sites are sites not in a nucleosomal state; they seem to be sites involved in replication, transcription, and other DNA activities.

6. The nucleosome is composed of two each of histones H2A, H2B, H3, and H4. Histone H1 seems to be associated with DNA as it enters and exits the nucleosome.

7. See figures 14.8 and 14.9 for the relationship of the 110-, 300-, and 2,000-Å chromosome fibers.

8. G-bands and C-bands are illustrated in figures 14.17 and 14.18. R-bands are the reverse of G-bands. Their structures are summarized in table 14.3.

9. See figure 14.17.

10. See figure 14.12.

11. Polytene chromosomes are chromosomes that underwent endomitosis: They consist of numerous synapsed copies of the same chromatid (e.g., in the *Drosophila* salivary glands). Regions of active transcription in polytene chromosomes form diffuse areas called puffs or Balbiani rings (see figure 14.12). Lampbrush chromosomes occur in amphibian oocytes (see figure 14.16).

12. A chromosomal puff can be stage-specific, tissue-specific, constitutive, or environmentally induced. It is an area of active transcription in a polytene chromosome.

13. Satellite DNA differs in its base sequence from the main quantity of DNA and thus forms a satellite band during buoyant density analysis. It is usually centromeric heterochromatin, composed of a highly repetitive DNA.

14. A kinetochore is a protein-RNA structure that connects the chromosome to the microtubules of the spindle. It occurs in a visible constriction of the chromosome called a centromere. Yeast centromeres are about 220 bp long with three consensus regions (see figure 14.21).

15. Telomeres are at chromosome ends. They protect the ends of chromosomes, preventing them from being degraded by exonucleases, and allow them to be properly replicated. Telomeres are repetitions of a five- to eight-base sequence.

16. Cot is a renaturation scale of DNA (concentration × time); see figure 14.27 for various cot curves. $Cot_{1/2}$ values have been used to construct a scale of unique DNA length in DNA samples. The repetitiveness of mouse DNA can be determined from the cot curve of figure 14.28, which indicates three different fractions.

17. Highly repetitive DNA usually makes up the centromeric and telomeric regions of the chromosome. Unique DNA, making up the bulk of structural genes, has a large component that is transcribed. Repetitive DNA is composed of dispersed DNA (e.g., Alu family), multiple copies of transcribed DNA (e.g., rRNA, histones), and diverged copies of ancestral genes (e.g., globin family genes).

18. Construct a radioactive Alu probe. Then use this probe to locate Alu sequences on the chromosomes.

19. The most direct method of determining the direction of transcription of the histone genes would be to sequence the region, from which transcriptional information can be ascertained.

20. See figure 14.30.

21. Cloning and then sequencing the region would provide the answer. Analysis would show genes of similar sequence to the active genes, but lacking the sequences for transcription.

22. a. Both measurements should yield approximately the same size molecule.
    b. Method 1 will yield a DNA size larger than that determined by method 2.

    If the chromosome is composed of only one molecule, both methods give total DNA per chromosome. If the chromosome has more than one molecule, each molecule (method 2) will be less than the total amount of DNA per chromosome (method 1). Assume for example that the chromosome has two identically sized molecules. The molecular size determined by method 2 will be one half the size calculated by method 1.

24. Nucleases will not digest DNA to which proteins, especially histones, are bound. Approximately 200 bp must be wrapped around proteins, and there must be some unprotected regions between these 200-bp regions. Multiples of 200 bp appear because the nuclease does not cut at each unprotected sequence.

26. No. The average coding part of a gene has about 1,000 bp. A 200-bp region would give maximally a protein of 66 amino acids. Most proteins have much more than 66 amino acids.

28. Repetitive regions must be separated by nonrepetitive regions. If the repetitive sequences are about 400-bp long, the 400-bp renaturation mix will contain

(a)

in about equal amounts. The 4,000-bp mix will contain mostly molecules that look like this:

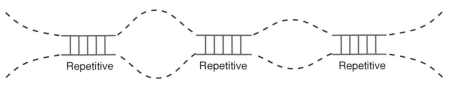

(b)

In separating single-stranded and double-stranded molecules, this type of molecule will behave like a double-stranded DNA. We get an unusually high amount of "apparent" renaturation.

30.

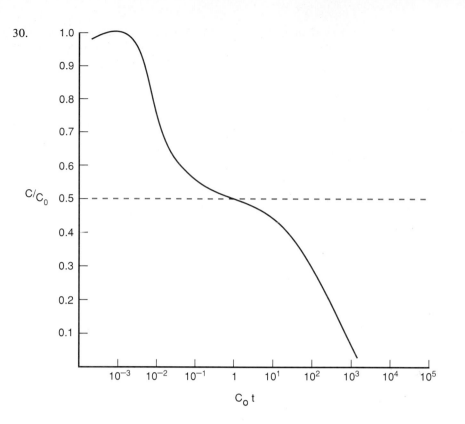

32.

cot	c/c₀
0.01	$\frac{1}{1.01} = 0.99$
0.05	$\frac{1}{1.05} = 0.95$
0.1	$\frac{1}{1.1} = 0.90$
0.5	$\frac{1}{1.5} = 0.67$
1.0	$\frac{1}{2} = 0.5$
5.0	$\frac{1}{6} = 0.167$
10	$\frac{1}{11} = 0.099$
50	$\frac{1}{51} = 0.02$
100	$\frac{1}{101} = 0.01$

Chapter 15
Gene Expression: Control in Eukaryotes

1. A fate map can be constructed by marking cells in the developing embryo with vital dyes and determining where these dyes end up in later developmental stages, thus tracing the fate of the marked cells.

2. Zygote $\rightarrow P_1 \rightarrow E \rightarrow p \rightarrow 1 \rightarrow p \rightarrow p \rightarrow a$

3. A homeotic mutant is one in which cells follow a different developmental pathway; they are "switched" from one pathway to another. Since a single mutation results in such drastic changes of the phenotype, homeotic genes are viewed as binary developmental switches from one state to another.

4. *Nasobemia* causes legs to grow at antennae sites; *bithorax* causes two thoraxes to be produced; *fushi tarazu* results in too few segments in the developing fruit fly.

5. The homeo box is a sequence of about 180 bp found in homeotic and segmentation genes. It was discovered with recombinant DNA techniques (Southern blot). Its significance is that it may define a common mechanism of action of these genes. A homeo domain is the amino acid translation of the homeo box, which may use a helix-turn-helix motif to interact with DNA.

6. Symmetry can arise either by symmetrical cell divisions or by nonsymmetrical movements, divisions, or deaths of cells.

7. Aberrant development can give information about normal development. Historically, the major tool of genetics has been the analysis of mutant organisms.

8. The helix-turn-helix motif (see fig. 15.13) is two alpha helices separated by a short turn within the protein, providing the structure to interact with DNA. Three other motifs are the zinc finger, the leucine zipper, and the copper fist.

9. Assuming that development will take place at all, very early development should reflect the cytoplasm (frog) but soon be taken over by nuclear control (*Xenopus*). A *Xenopus* adult with *Xenopus* germ cells should result.

10. Amphibians have very large eggs, development is external to the female, a ready supply of zygotes is available, and they are easily manipulated experimentally.

11. See figure 15.15.

12. An antibody recognizes an antigen by "lock and key" complementary structure; see figure 15.18.

13. The components of an Ig light chain are V, J, and C regions; the components of an Ig heavy chain are V, D, J, and C regions.

14. Any mature B lymphocyte should produce only one antibody; it has the potential to produce in excess of 10^8 different kinds.

15. The V-J joining recognition signal is a heptamer and nonamer separated by 23 and 12 bp; see figure 15.20.

16. V-J joining is a looping-out type of site-specific recombination, equivalent to an integrase-excisionase system found in phage λ. The rest is transcription, splicing, and translation.

17. B and T lymphocytes are produced in the bone marrow and mature either in the bone marrow (B cells) or in the thymus gland (T cells). B cells produce immunoglobulins; cytotoxic T cells kill infected host cells.

18. A T-cell receptor is an immunoglobulin-like molecule located on the surface of T cells enabling them to identify infected host cells.

19. The major histocompatibility complex is a gene complex whose protein products are involved in immunity, including antigen presentation, intercell communication, and complement production.

20. The gross chromosomal abnormalities associated with cancers are primarily deletions and translocations.

21. Assuming that each cancer might be controlled by a single locus and assuming that breast cancer appears only in women and prostate cancer appears only in men (sex limited), pancreatic and prostate cancer are probably controlled by autosomal recessive genes; colon cancer is probably controlled by a dominant gene (autosomal or sex linked); and breast cancer by a recessive gene, either autosomal or sex linked.

22. Retinoblastoma is frequently associated with a deletion of chromosomal band 13q14. Wilm's tumor is associated with the deletion of chromosomal band 11p13.

23. The protein product of the retinoblastoma gene, p105, binds with oncogene proteins. Thus the protein may somehow suppress transformation; when bound by oncogene proteins, p105 may be rendered ineffective. Hence p105 seems to act to suppress transformation and thus the gene is called an anti-oncogene.

24. The disease occurs only in homozygotes indicating that the normal condition is dominant (functional) and the retinoblastoma condition (nonfunctional) results from the failure of an enzymatic step.

25. Animal viruses can have DNA or RNA, either single- or double-stranded. They can be enveloped or nonenveloped. They can have simple or complex protein coats.

26. Minimally, a retrovirus must have the *gag*, *pol*, and *env* genes. The protein products of *gag* are core viral proteins; of *pol* are reverse transcriptase, protease, and integrase; and of *env* are envelope glycoproteins.

27. The following are translation mechanisms: normal translation; readthrough translation; and splice and then translation.

28. See figure 15.32 for the mechanism by which long terminal repeats are generated at the ends of retroviral DNA; they encode an enhancer and promoter, among other signals.

29. The v and c refer to viral and cellular, respectively. A proto-oncogene is a cellular oncogene within a nontransformed cell.

30. Cellular oncogenes can be activated by mutation or enhanced transcription resulting from new placement of a gene by either deletion of nearby material or translocation.

31. The v-*src* gene has no introns and the virus can function without it.

32. Burkitt's lymphoma can arise when c-*myc* is translocated adjacent to the IgM constant gene promoter, which is very active in B lymphocytes. Hence c-*myc* is transcribed at a much greater rate than normal.

34. Since azacytidine prevents methylation, the observed increase in transcription suggests that the presence of methyl groups inhibits transcription.

36. One explanation is that all promoters for antibody genes are nonfunctional, but this is unlikely. Similarly, we can imagine a defect in a homeobox responsible for all antibody production, but again this possibility is unlikely. More likely is a defect in a transcriptional factor or a defect in the process of joining the variable and constant regions.

38. In order to have more than one gene, we must be able to identify more than one different protein. If the mRNA is first translated to yield a polyprotein that is then cleaved to yield individual, different proteins, a cell could avoid the problem of not being able to initiate translation in the middle of the mRNA. Thus a "typical" eukaryotic mRNA might contain only one translated unit that is then cleaved to form several proteins.

40. In clone 1, the virus has inserted next to the oncogene, and in clone 2, the insertion occurred within the oncogene. We see one band that is common to all three cell lines; this band must represent the normal oncogene. The fact that this band is present in all three clones indicates that the insertion of a virus has occurred in only one of the two copies of the gene present in the cell. If it had inserted within both genes, we should not have seen the normal band. We see a larger fragment in clone 1, indicating that a restriction site has been lost or altered.

Clone 1 also indicates that the DNA of the virus does not contain a site for the restriction enzyme used. In clone 2, the two new bands suggest that a new site has been created. If this site were outside the gene, we would expect to see only one new band.

42. a. The virus could increase the amount of mRNA for the growth factor. b. Use northern blots to quantify the amount of oncogene mRNA present in infected and uninfected cells. If a growth factor is overproduced, the excess molecules could cause cells to divide more rapidly and become cancerous. Many retroviruses contain regions that behave as enhancers of transcription of cellular oncogenes. If the virus inserts near an oncogene, the enhancer could stimulate transcription. An increase in transcription of a specific gene can be followed by quantifying the amount of specific RNA present in infected and uninfected cells. The intensity of the hybridized band should be higher in cells that are making more RNA.

Chapter 16
DNA: Its Mutation, Repair, and Recombination

1. For example, if the twenty individual cultures of table 16.1 had values of 15, 13, 15, 20, 17, 14, 21, 19, 16, 13, 27, 14, 15, 26, 12, 21, 14, 17, 12, 14, then the mutation theory would not have been supported because the variation between the individual and bulk cultures would not have been different.

2. Temperature-sensitive mutations are most valuable to study critical functions of a cell because they display the mutant phenotype only at the restrictive temperature. At the permissive temperature, the cells can be kept alive. For example, temperature-sensitive mutations of enzymes of DNA replication can be studied, whereas many of the mutants, if not temperature sensitive, would die.

3. Reading across each row, we gather more and more information.

   Row 1: 1, 6, and 7 are part of one complementation group.
   Row 2: 2 and 5 are part of one complementation group.
   Row 3: 3 and 4 are part of one complementation group.
   Row 4: No new information.
   Row 5: No new information.
   Row 6: Reinforces that 6 and 7 are part of the same complementation group.
   Row 7: No new information.

   Thus we conclude that there are three complementation groups present: 1, 6, and 7 are mutually noncomplementing as are 2 and 5, and 3 and 4. The half table missing is a mirror image because a cross of 1 and 3 is the same as a cross of 3 and 1 (reciprocity). The diagonal always contains negative elements because every mutant is a structural allele of itself.

4. The overlap patterns of the deletions define six regions: overlap of 1 and 2; 2 only; overlap of 2 and 3; 3 only; overlap of 3 and 4; 4 only. We can thus localize mutations M to S according to these six areas:
   Mutation M: region of 3 only.
   Mutation N: region of 2 and 3 overlap.
   Mutation O: region of 4 only.
   Mutation P: region of 3 and 4 overlap.
   Mutation Q: region of 2 only.
   Mutation R: region of 1 and 2 overlap.
   Mutation S: region of all but unique and overlap areas of 4.

The figure shows where each of these mutations are in the A cistron.

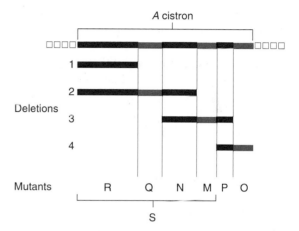

5. All mutants should be crossed in pairwise combinations yielding heterozygote daughters. (Presumably, earlier crosses indicated that these are X-linked loci.) All F_1 daughters will be wild-type: The mutations complement and therefore are not alleles. When eosin flies are crossed in a similar fashion, daughters will be wild-type except when the parents were eosin and white. In that case, daughters will be mutant showing the lack of complementation and hence that the mutations are alleles.

6. When organisms that have phenotypes controlled by functional alleles are mated, their progeny, for the most part, are mutant; the alleles do not complement each other. Structural alleles are functional alleles that have changes at identical nucleotides. In the cis configuration, one cistron has two mutant sites.

7. Prokaryotic and phage genes appear not to have intervening sequences. Benzer and Yanofsky worked at a time when introns were unknown and it was assumed that the length of a gene was transcribed and then translated.

8. In intra-allelic complementation, mutant polypeptides form a functional enzyme or protein. This gives the impression of genic complementation, which would lead the researcher to believe the mutations were in nonallelic genes.

9. The rex gene of phage λ represses growth of phage T4 rII mutants.

10. See figure 16.14 for the tautomeric forms of the bases. Figure 16.15 shows one base pair in which adenine is in the imino form and therefore can pair with cytosine; we diagram the remaining base pairs here. In a similar fashion, table 16.4 presents all tautomeric base pairings. These result in transition mutations: AT becomes GC; GC becomes AT; TA becomes CG; and CG becomes TA.

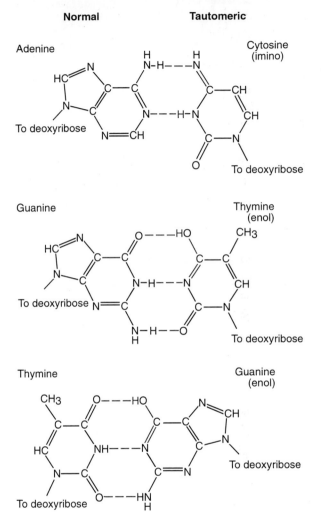

Normal	**Tautomeric**
Adenine	Cytosine (imino)
Guanine	Thymine (enol)
Thymine	Guanine (enol)

11. In replicating DNA, a transition mutation can occur by tautomerization of a base in the template strand (template transition) or entering the progeny strand (substrate transition).

12. Transversion can occur by a combination of tautomeric and *anti-syn* shifts (see fig. 16.18) or by replication opposite a missing base in the template (see fig. 16.22).

13. 5BU (pyrimidine analogue) and 2AP (purine analogue) are incorporated into DNA as thymine and adenine, respectively. However, each undergoes tautomeric shifts more frequently than the normal base. Both cause transitions. Nitrous acid also promotes transitions by converting cytosine into uracil, which acts like thymine, and adenine into hypoxanthine, which acts like guanine. Proflavin induces insertions and deletions by intercalating and buckling DNA. Ethyl ethane sulfonate removes purine rings and thus promotes transitions and transversions.

14. A point mutation can be either a missense (new amino acid), nonsense (stop codon created), or frameshift mutation (addition or deletion of a number of bases not divisible by three). These can all be corrected by simple reversion (restoration). Intragenic suppression can rectify the frameshift. Intergenic suppression (tRNA change) can rectify all three.

15. See figure 16.24.

16. Intergenic suppression is restoration of function of a mutated gene by a mutation at another locus. Intragenic suppression is restoration by a second mutation at the same locus. Both involve a second mutation event at a different position from the first. Neither usually restores the wild-type completely.

17. Thymine dimerization can be restored by photoreactivation (DNA photolyase required), excision repair (see fig. 16.28), or postreplicative repair (see fig. 16.34).

18. Excision repair endonucleases can recognize dimerizations, mismatched bases, and apurinic-apyrimidinic sites.

19. RecA causes single-stranded DNA to invade double-stranded DNA. It also acts as a protease, cleaving the λ cI repressor protein (and is involved in the catalysis of LexA). Cleaving the cI repressor causes λ to enter the lytic cycle.

20. See figure 16.36 for a diagram of recombination. The RecA and RecBCD proteins are required as is DNA ligase.

21. The same enzymes are needed for reciprocal and nonreciprocal recombination: RecA, RecBCD, and DNA ligase.

22. Three genes. Gene *A*: mutants 1, 4, 8; gene *B*: mutants 2, 5; gene *C*: mutants 6, 7. Mutant 3 probably contains a deletion that spans genes *A* and *C*. Begin by finding mutants that do not complement. These should be mutations in the same gene. Mutants 2 and 5 are in the same gene. Initially, we suspect that mutants 1, 3, 4, and 8 are in the same gene, which is different from the gene that contains 6 and 7. If mutant 3 is in gene *A*, it should complement 6 and 7, and it does not. One explanation is that 3 is a deletion spanning genes *A* and *C*. Alternatively, mutant 3 could be in gene *A* but be a polar mutation. Either possibility implies that the order is *B A C*.

24.

Begin with deletions that yield mostly "−"s. These must be large deletions that cover most of the other deletions. Mutations 1 and 5 are such mutations. Since they give no wild-type, they must overlap:

<pre>
_____ 1 _____

        ___ 5 ___
</pre>

Now look at mutant 2. It gives wild-type recombinants with 1 but not 5. Therefore it must overlap the region deleted in 5. Mutant 3, by similar logic, must cover part of deletion 1. We can draw these results as follows. Broken lines indicate we do not yet know how long the deletion is:

<pre>
_____ 1 _____ _ _ _ _ 2 _ _ _ _

_ _ _ _ 3 _ _ _ _____ 5 _____
</pre>

Now look at 4. It gives "−" (no wild-type recombinants) with both 1 and 5, and therefore must be in the region that 1 and 5 overlap. If 4 extended to and overlapped either 2 or 3, we expect to see "−" with them. Since this prediction is not met, 4 must be a small deletion spanning at least part of the overlap of 1 and 5. Since mutant 6 gives no wild-type with 1, 4, or 5, it must be within the common region deleted in all three strains.

26. We know that the anticodon pairs with the codon, and we expect nonsense suppressors to contain an altered anticodon. The fact that the nonsense codon can be read by a tRNA with a normal anticodon but altered dihydrouridine loop suggests that this region of the tRNA also interacts somehow with the mRNA. Or, the way in which this loop interacts with the ribosome causes the anticodon sequence to be misread.

28. $x^+ \rightarrow x$: AT → GC; $y^+ \rightarrow y$: GC → AT. This problem requires logic and a knowledge of how mutagens work. For x, the key is the response to HA, which only causes GC → AT transitions. Mutant x is reverted by HA, therefore x must be GC, and the normal x^+ was AT. AP-induced mutations can also be reverted by AP. Since y is not reverted by HA, y must be AT. Therefore y^+ must be GC.

30. A deletion that spans regions of more than one gene. Consider the following two genes: ____1_|_2____. By definition, mutations in gene 1 will complement mutations in gene 2, but not other mutations in gene 1. If we have a deletion, x, that covers part of both 1 and 2,

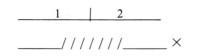

presumably gene 1 will be nonfunctional because it is missing the last part of the protein. Gene 2 will be nonfunctional because the beginning portion of the gene is missing. The following genotypes will give no complementation:

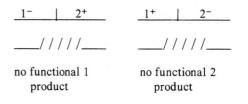

no functional 1   no functional 2
product   product

We could also get no complementation when mutants are in two genes if we have a bacterial operon in which one of the genes contains a polar nonsense mutation. Such a mutation eliminates all distal functions. Thus, if the operon is A B C D, and we construct the following partial diploid $\dfrac{A^-(\text{nonsense}) \; B^+ \; C^+ \; D^+}{A^+ \qquad\qquad B^- \; C^+ \; D^+}$, we will get no complementation because the top DNA is effectively $A^- \; B^- \; C^- \; D^-$.

32. The auxotroph probably contains a deletion. If a few bases are missing, nothing is available to cause transitions or transversions. It is highly unlikely that the correct number of missing bases could be spontaneously and correctly inserted.

Chapter 17
Non-Mendelian Inheritance

1. Persistence of an environmentally induced trait into later generations is known as dauermodification and does not imply genetic control. After a suitable number of generations, the phenotype returns to normal, indicating an environmental rather than genetic response.

2. Sex linkage shows patterns of inheritance (e.g., zig-zag) that differ among sexes in reciprocal crosses. These patterns are not seen in snails. Also, snails are hermaphroditic: each individual can be both male and female.

3. Although the genetic scheme predicts shell coiling perfectly, one could do experiments involving the injection of cytoplasm into eggs to test the viral hypothesis.

4. A sinistral female must have had a dd mother and therefore must have at least one d allele. If she produces dextral offspring, she must have a D allele. Therefore, her genotype must be Dd; her father provided her D allele and therefore could have been DD or Dd.

5. By looking at different species, it is clear that no one gene for oxidative phosphorylation has to be in the mitochondrial genome.

6. In matings, the petite phenotype follows Mendelian ratios without exception.

7. The rule of thumb is that suppressive petite mitochondria will dominate a cell, whereas neutral petite mitochondria will be lost in a competitive situation. Therefore

segregational petite $\times$ segregational petite $\rightarrow$ segregational petites

segregational petite $\times$ neutral petite $\rightarrow$ segregational petites

segregational petite $\times$ suppressive petite $\rightarrow$ suppressive petites

neutral petite $\times$ neutral petite $\rightarrow$ neutral petites

neutral petite $\times$ suppressive petite $\rightarrow$ suppressive petites

suppressive petite $\times$ suppressive petite $\rightarrow$ suppressive petites

8. Cytoplasmic inheritance is determined by the different results seen in reciprocal crosses (see fig. 17.9). An *iojap*-type chromosomal gene will have a phenotype of inducing variegation (in Mendelian ratios dependent on the exact crosses) that is then inherited cytoplasmically.

9. In 0.02% of the offspring cells, the mt^- allele of the streptomycin locus is inherited. In essence, these cells seem to have inherited chloroplast genes from the mt^- parent. These cells thus provide us with a window on the possibility of chloroplast genotypes from both parents being viable within the same cell. Thus recombination among other chloroplast genes can be looked for in this class of offspring (0.02% of total). If recombination occurs, map distances can be calculated by the usual methods, keeping in mind that we are taking data only from within this 0.02% of offspring.

10. Both mitochondria and plastids contain DNA and ribosomes with prokaryotic affinities. Defects in either can produce similar phenotypic effects of reduced growth.

11. Mitochondria and chloroplasts have prokaryotic affinities. They both have circular chromosomes and their metabolism is affected by prokaryotic inhibitors (e.g., antibiotics). Certain prokaryotic mRNAs will hybridize with the organelle's DNA. There are numerous other aspects of biochemistry, morphology, and physiology that help to demonstrate affinities.

12. Both strains have the Kk genotype; of sensitive exconjugants, all will produce sensitive (kk) offspring by autogamy. Of killer exconjugants, 1/3 (KK) will produce only killer offspring while 2/3 (Kk) will segregate both killers (KK) and sensitives (kk) through autogamy.

13. One way to determine that two loci are involved is to look at the proportion of offspring that lose mu particles after autogamy. In some strains, 1/2 the offspring will lose mu particles, indicating one locus was initially segregating (autogamy of $M_1m_1m_2m_2$ yields $M_1M_1m_2m_2$ or $m_1m_1m_2m_2$ in a 1:1 ratio). In other strains, 1/4 of the offspring will lose mu after autogamy, indicating that two unlinked loci were segregating (autogamy in $M_1m_1M_2m_2$ yields $M_1M_1M_2M_2$, $M_1M_1m_2m_2$, $m_1m_1M_2M_2$, or $m_1m_1m_2m_2$ in a 1:1:1:1 ratio).

14. To begin with, controlled breeding experiments would be carried out over several generations to ascertain that normal Mendelian ratios did not occur. Once non-Mendelian inheritance is confirmed, the crosses could be examined for maternal effects. Given that these exist, an infectious agent could be looked for by injecting cytoplasm from infected (CO_2-sensitive) organisms into uninfected (CO_2-resistant) eggs. The CO_2-resistant organisms would then become infected. The infected cytoplasm could be searched for a bacterium of virus. The causative agent, known as the sigma virus, would be discovered. When isolated and purified, it would prove to be the infective agent.

15. Several bits of evidence lead to the conclusion that a trait is controlled by a plasmid gene: Existence of a plasmid that can be seen by electron microscopy or through density-gradient centrifugation; a change in multiple aspects of the phenotype; infectious nature of the change; lack of linkage to host chromosomal loci; and disappearance of the trait with disappearance of the plasmid.

16. Human mitochondrial DNA does not have introns. Finding an intron would suggest that the mitochondria had acquired a nuclear gene.

18. a. All type 1, 1.5 and 3.7 kb. b. all type 2, 2.5 and 6.0 kb. Recall that the chloroplast DNA from mt^- cells does not appear in progeny.

20. Conjugation will produce exconjugants with the same genotype, but which haploid micronucleus survives in each cell will be random. Therefore, 1/4 of the time the genotypes of the exconjugants are expected to be KK, 1/2 of the time Kk, and 1/4 of the time kk. The sensitive exconjugant will remain sensitive, regardless of which of the above genotypes is present. Autogamy will not affect the genotype of homozygous exconjugants, but it will affect the heterozygote. Among the killer exconjugants we expect:

Killer exconjugants	Autogamous products	Phenotypic ratio
1/4 KK	all KK	1/4 killer
1/2 Kk → 1/2 KK		1/4 killer
↘ 1/2 kk		1/4 sensitive (Kappa are lost)
1/4 kk	all kk	1/4 sensitive (Kappa are lost)

22. Female parent: D/d; male parent: $?/d$; (A): dd. Since selfing produces only sinistral snails, individual (A) must not produce dextral cytoplasm; therefore it must be homozygous for sinistral coiling, dd. The individual must get one d allele from each parent, so each parent must be at least heterozygous. Since (A) has dextral coiling, this information must be in the cytoplasm of the egg, and the mother must be heterozygous. The father could be either D/d or d/d.

24. One mate-killer:3 non–mate-killers. The maintenance of mate-killer requires the presence of two dominant nuclear genes (M_1 and M_2). The double heterozygote is $M_1m_1M_2m_2$. Autogamy produces four different homozygous lines in equal frequency:

$M_1M_1M_2M_2$	mate-killer maintained
$M_1M_1m_2m_2$	no mate-killer
$m_1m_1M_2M_2$	no mate-killer
$m_1m_1m_2m_2$	no mate-killer

26. Maroon-like must effect the cytoplasm of the egg. The first cross is unusual and alerts us to maternal inheritance. A true Mendelian factor should produce 1/2 maroon-like males and 1/2 females. The genotypes of the F_1 females are $ma\text{-}1^+/ma\text{-}1$ and $ma\text{-}1/ma\text{-}1$; half of the females should be of each genotype. If the wild-type allele produces wild-type cytoplasm for the progeny, regardless of the genotype of the progeny, any female that is $ma\text{-}1^+/ma\text{-}1$ will produce all wild-type progeny.

Chapter 18
Quantitative Inheritance

1. The y variable is follicles; the x variable is eggs laid. The x mean is 38.4; the y mean is 37.5. $\text{Cov}(x, y) = 113.2$, $s_x^2 = 116.6$, $s_y^2 = 163.0$, $r = 0.82$, $b = 0.97$, $a = 0.3$.

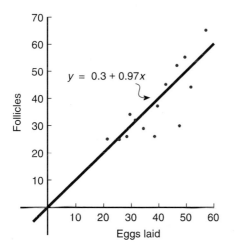

2. Three in two hundred is approximately 1 in $64 = 1/(4)^3$; therefore three loci (see table 18.1). Each effective allele contributes about 1/2 pound over the 2-pound base (3-pound difference divided by six effective alleles: $AABBCC = 5$ pounds, $aabbcc = 2$ pounds).

3. Two in five hundred is approximately 1 in $256 = 1/(4)^4$; therefore four loci. Each effective allele contributes about 1/4 inch above 2 inches. For an approximate graph, see the right side of figure 18.1.

4. Independent assortment: for example, $AaBbCcDd$ parents can have $AABBCCDD$ offspring.

5. Polygenes are different from traditional Mendelian genes only in that several loci have the same phenotypic effect. Through an additive model and environmental variability, a quantitative trait arises.

6. First, outstanding athletic ability must be defined. Then genetic effects must be assessed through heritability analyses such as twin studies and correlations among relatives.

7. Improved nutrition, medicine, and sanitation affect almost all people, allowing genetic potential to be universally affected. Thus variation in a population can be genetic, whereas the mean value can be under the control of the environment. In other words, impoverished people have genetic variation about a low mean, whereas well-fed people have genetic variation about a high mean.

8. Good nutrition should increase the heritability of height by eliminating some of the environmental variance; it affects only the denominator in a heritability equation.

9. Grow the plants under uniform environmental conditions to see the environmental role in height. By crossing the plants and looking at the extremes of the F_2 generation, it is possible to estimate the number of genes involved in this trait.

10. Individuals of intermediate color can produce both lighter and darker offspring by independent assortment. That is, $AaBbCcDd$ parents can produce $AABBCCDD$ and $aabbccdd$ offspring. However, if each effective allele adds color, then individuals with the base color (white) who marry each other ($aabbccdd$) presumably cannot have children with darker skin.

11. Begin by graphing the data. The general form of the data indicates that additive genes predominate. Excessive skew is caused by nonadditive effects. The number of loci involved is difficult to determine because the long-eared parent was so variable; more information is needed. (Probably four to eight loci are involved.)

12. For example:

P_1 Cy/Pm H/S $Ce/M(4) \times +$ (wild-type)
e.g.,

F_1 (male) $Cy/+$ $H/+$ $Ce/+ \times$ (backcross) $+$ (wild-type female)
F_2 $Cy/+$ $+/+$ $+/+$; $+/+$ $H/+$ $+/+$, etc.
or,

F_1 (male) $Cy/+$ $H/+$ $Ce/+ \times$ (backcross) Cy/Pm H/S $Ce/M(4)$ (female)
F_2 Cy/Pm $H/+$ $Ce/+$; $+/Pm$ $+/S$ $+/M(4)$; etc.

13. Thorax length: $V_D + V_I = (100 - [43 + 51]) = 6$; $H_N = 43/100 = 0.43$; $H_B = 49/100 = 0.49$. Eggs laid: $V_D + V_I = 44$; $H_N = 0.18$; $H_B = 0.62$.

14. $H = (4 - 0.9)/(5 + 5) = 0.31$; H (high line) $= (4 - 3)/5 = 0.2$; H (low line) $= (3 - 0.9)/5 = 0.42$. Part of the difference may be due to the number of alleles available for selection in each direction and nonadditive factors.

15. $r = 0.43$; $H_N = 0.43/0.50 = 0.86$ (narrow-sense heritability); environmental variance, a component of the total phenotypic variance appears in the denominator of the heritability equations (18.11 and 18.12). Thus environmental factors that lower environmental variance (more uniform environments or environmental effects) increase heritability. And, factors that raise the environmental variance (less uniform environments or environmental effects) decrease heritability.

16. H = final difference in means divided by the summed cumulative selection differentials = $(23 - 18)/(13 + 14) = 0.19$. (Hint: Y_O becomes Y with each new generation. Thus selection differentials for the high line are 3, 2, 3, 3, 2, summing to 13. The selection differentials for the low line are 3, 2, 2, 4, 3, summing to 14.)

18. 50 cm. The difference between the two heights is 20 cm. This difference must result from the presence of dominant alleles. The tall plant has four dominant alleles, so each allele contributes an average of 5 cm to the height of the plant. The heterozygote has two dominant alleles; $2 \times 5 = 10$ cm.

20. $H_B = 0.72$; $H_N = 0.45$. First calculate V_{ph} as $4.0 + 1.8 + 0.5 + 2.5 = 8.8$. Broad sense heritability (H_B) is total genetic variance divided by total variance.
$$\frac{V_A + V_D + V_I}{V_{ph}} = \frac{4.0 + 1.8 + 0.5}{8.8} = \frac{6.3}{8.8} = 0.72.$$
Narrow sense heritability (H_N) is additive genetic variance divided by total variance.
$$\frac{V_A}{V_{ph}} = \frac{4.0}{8.8} = 0.45$$

22. 4.5 g
$$H_N = \frac{\text{gain}}{\text{selection differential}}. \text{ Thus}$$
$(H_N)(\text{selection differential}) = \text{gain}$

$(0.5)(9.0) = 4.5$ g

24. Eight. The frequency of individuals in the F_2 that resemble one parent is $\frac{1}{4^n}$, where n = the number of genes involved. In this case, 100 of 6,200,000 were like one parent. $\frac{100}{6,200,000}$ is approximately $\frac{1}{64,000}$ which approximates $\frac{1}{4^8}$. So we probably have 8 genes involved.

26. 176. To solve this problem, realize that heritability (H) = gain/selection differential = $(Y_O - \bar{Y})/(Y_P - \bar{Y})$

$$0.4 = \frac{\text{gain}}{Y_P - \bar{Y}} = \frac{\text{gain}}{185 - 170}$$
$\text{gain} = (0.4)(15) = 6.0$
$\text{gain} = Y_O - \bar{Y}$
$6.0 = Y_O - 170$
$Y_O = 176$ lbs

Chapter 19
Population Genetics: The Hardy-Weinberg Equilibrium and Mating Systems

1. The frequencies of the three genotypes are $f(MM) = 41/100 = 0.41$; $f(MN) = 38/100 = 0.38$; $f(NN) = 21/100 = 0.21$. The frequency of M, p, is the frequency of MM homozygotes plus half the frequency of heterozygotes:

$p = f(MM) + (1/2)f(MN) = 0.41 + (1/2)(0.38)$
$= 0.41 + 0.19 = 0.60$

$q = 1 - p = 1 - 0.60 = 0.40$

Alternatively

$$p = f(MM) = \frac{2(\#MM) + \#MN}{2 \times \text{total}} = \frac{2(41) + 38}{200}$$
$$= \frac{120}{200} = 0.60$$
$q = 1 - p = 0.40$

We do the following chi-square test:

	MM	MN	NN	Total
Observed	41	38	21	100
Expected	$p^2 \times 100$	$2pq \times 100$	$q^2 \times 100$	
	36	48	16	100
Chi-Square	0.694	2.083	1.563	4.340

The critical chi-square (0.05, one degree of freedom) = 3.841. We thus reject the null hypothesis that this population is in Hardy-Weinberg proportions.

2. a. We cannot count genes to determine allelic frequencies. If we assume Hardy-Weinberg equilibrium: $p = f(A)$, $q = f(a)$. Then there are nine aa individuals, or $f(aa) = 0.09$. If the Hardy-Weinberg equilibrium is assumed, $q^2 = f(aa) = 0.09$. Therefore $q = \sqrt{0.09} = 0.3$ and $p = 0.7$. Genotype frequencies, assuming random mating, will be $f(AA) = p^2 = 0.7^2 = 0.49$; $f(Aa) = 2pq = 2(0.7)(0.3) = 0.42$; and $f(aa) = q^2 = 0.3^2 = 0.09$. Since there are zero degrees of freedom (number of phenotypes − number of alleles = 2 − 2 = 0), we cannot do a chi-square test.

   b. The frequencies of the three genotypes are as follows: $f(FF) = 137/420 = 0.326$; $f(FS) = 196/420 = 0.467$; and $f(SS) = 87/420 = 0.207$. The allelic frequencies are $p = f(F) = f(FF) + (1/2)f(FS) = 0.326 + 0.467/2 = 0.560$; and $q = f(S) = 1 - 0.560 = 0.440$. We can thus set up the following chi-square table:

	FF	*FS*	*SS*	**Total**
Observed	137	196	87	420
Expected	$p^2 \times 420$	$2pq \times 420$	$q^2 \times 420$	
	131.7	207.0	81.3	420
Chi-Square	0.213	0.585	0.400	1.198

The critical chi-square, at 0.05 with one degree of freedom, is 3.841. We thus fail to reject the null hypothesis that the population does not deviate from Hardy-Weinberg proportions.

3. Here we must assume Hardy-Weinberg equilibrium because of dominance. The $f(tt) = 65/215 = 0.302$. Thus $q = f(t) = \sqrt{f(tt)} = \sqrt{0.302} = 0.55$; and $p = f(T) = 1 - 0.55 = 0.45$. Since there are zero degrees of freedom (number of phenotypes − number of alleles = 2 − 2 = 0), we cannot do a chi-square test to determine if the population is mating at random (in Hardy-Weinberg proportions).

4. Due to dominance, we must assume Hardy-Weinberg equilibrium. Assume c is the cystic fibrosis allele and C is the normal allele. Then $f(cc) = 1/2{,}500 = 0.0004$; $q = f(c) = \sqrt{0.0004} = 0.02$; and $p = f(C) = 1 - 0.02 = 0.98$. The frequency of heterozygotes is $2pq = 2(0.02)(0.98) = 0.0392$. Or 3.92% of persons carries the cystic fibrosis allele.

5. Let $p = f(A)$; $q = f(B)$; and $r = f(O)$. Due to dominance, we cannot count alleles and therefore must assume Hardy-Weinberg equilibrium. Thus $r = \sqrt{f(OO)} = \sqrt{0.09} = 0.3$. Type A + type O include the *AA, AO,* and *OO* genotypes. Therefore, assuming random mating, $(p + r)^2 = 0.07 + 0.09 = 0.16$; and $p + r = \sqrt{0.16} = 0.4$; $p = 0.4 - r = 0.4 - 0.3 = 0.1$. And, $q = 1 - p - r = 1 - 0.3 - 0.1 = 0.6$. However, since there is one degree of freedom (number of phenotypes − number of alleles = 4 − 3 = 1), we can do a chi-square test (see next column):

	Type A	**Type B**
Observed	7	72
Expected	$(p^2 + 2pr) \times 100$	$(q^2 + 2qr) \times 100$
	$(0.01 + 0.06) \times 100$	$(0.36 + 0.36) \times 100$
	7	72

	Type AB	**Type O**
Observed	12	9
Expected	$2pq \times 100$	$r^2 \times 100$
	0.12×100	0.09×100
	12	9

	Total
Observed	100
Expected	100

Since the observed and expected values are identical, the chi-square value is zero. The population thus appears to be mating at random.

6. a. Hardy-Weinberg proportions are achieved in one generation of random mating (multiple-allele extension) if the locus is autosomal. If the locus is sex linked, with different initial frequencies in the two sexes, then approach to equilibrium is gradual. If the locus is autosomal but there are different frequencies in the two sexes, then Hardy-Weinberg proportions are achieved in two generations.

   b. If the loci are not in equilibrium to begin with (linkage disequilibrium), then equilibrium is achieved asymptotically.

7. First we determine allelic frequencies by counting alleles.

$$p_A = f(A) = (58 + 12)/100 = 0.7$$
$$q_A = f(a) = 1 - p_A = 0.3$$
$$p_B = f(B) = (58 + 22)/100 = 0.8$$
$$q_B = f(b) = 1 - p_B = 0.2$$

We calculate the expected linkage arrangement, assuming linkage equilibrium, by simple probability, to generate the following chi-square table:

	AB	*ab*	*Ab*	*aB*	**Total**
Observed	58	8	12	22	100
Expected	$p_A p_B \times 100$	$q_A q_B \times 100$	$p_A q_B \times 100$	$q_A p_B \times 100$	
	$(0.7)(0.8)(100)$	$(0.3)(0.2)(100)$	$(0.7)(0.2)(100)$	$(0.3)(0.8)(100)$	
	56	6	14	24	100
Chi-Square	0.071	0.667	0.286	0.167	1.191

The critical chi-square (0.05, one degree of freedom) = 3.841. We thus fail to reject the null hypothesis of no deviation from linkage equilibrium.

8. a. Most human populations should be in Hardy-Weinberg proportions at the taster locus regardless of what the allelic frequencies are. 3:1 is a family ratio when the parents are heterozygotes or a population ratio when $p = q$, given dominance.
   b. With no other information, it is probably safest to assume $p = q = 0.5$. At equal frequencies of alleles, the dominant phenotype (tasting) will occur in 75% of people ($p^2 + 2pq$).

9. Inbreeding is disadvantageous when it causes recessive deleterious alleles to become homozygous. This occurs in normally outbred populations of diploids that have built up these harmful alleles. In species that normally inbreed, these deleterious alleles are probably no longer present; they were either removed by selection long ago or cannot build up in the population because of the regular pattern of inbreeding.

10. There are three paths in this path diagram—one with **B** as the common ancestor (six ancestors in the path) and two with **C** as the common ancestor (one to D, with four ancestors, and one to E, with three ancestors). Although no other inbreeding coefficients are given, B is the offspring of half siblings (with A as the common ancestor) and has an inbreeding coefficient of $(1/2)^3 = 1/8$. Thus, summing the three paths:

$$F_1 = (1/2)^6(1 + 1/8) + (1/2)^4 + (1/2)^3$$
$$= 0.0176 + 0.0625 + 0.125 = 0.2051$$

11. There are four paths passing through A, the only common ancestor ($F_A = 0.01$). All have ABCI as one side of the path. The four other legs of the paths are ADEHI, ADGHI, AFGHI, and AFEHI. Since each path has six ancestors, the inbreeding coefficient is

$$F_1 = 4(1/2)^6(1.01) = 0.063$$

12. There is one path, with A as a common ancestor. The path has a total of three ancestors. Given that F_A is 0.2, the inbreeding coefficient of G is

$$F_G = (1/2)^3(1.2) = 0.15$$

Path diagram

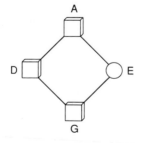

13. Using the formula $F = (2pq - H)/2pq$, we calculate that $2pq = 0.48$, and $H = 38/100 = 0.38$. Therefore, $F = (0.48 - 0.38)/0.48 = 0.208$.

14. a. In order to prove that the perpendicular to the base divides it into the ratio of p and q, we use the following diagram:
   In this diagram we need to prove $x/y = p/q$, or

$$\frac{x}{y} = \frac{D + (1/2)H}{R + (1/2)H}$$

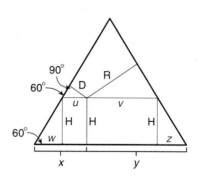

Remember that the three perpendiculars, D, R, and H, are the proportions of the three genotypes, *AA, aa,* and *Aa,* respectively. Thus $D + (1/2)H = p$, by definition as is $q = R + (1/2)H$. We do the following manipulation:

$$x = u + w$$

Since

$$\sin 60° = \sqrt{3}/2 = D/u, \ u = 2D/\sqrt{3}$$
$$\text{Tan } 60° = \sqrt{3} = H/w, \ w = H/\sqrt{3}$$

Thus

$$x = u + w$$
$$x = 2D/\sqrt{3} + H/\sqrt{3}$$

or

$$x\sqrt{3}/2 = D + (1/2)H$$

Similarly

$$y = v + z$$

and

$$y\sqrt{3}/2 = R + (1/2)H$$

Dividing

$$\frac{(x\sqrt{3}/2)}{(y\sqrt{3}/2)} = \frac{D + (1/2)H}{R + (1/2)H}$$

or

$$\frac{x}{y} = \frac{D + (1/2)H}{R + (1/2)H}$$

b. To prove that the populations in Hardy-Weinberg equilibrium are a parabola:

At equilibrium $D = p^2$
$$R = q^2$$
$$H = 2pq$$

Thus

$$(2pq)^2 = 4p^2q^2 \quad \text{or}$$
$$H^2 = 4DR \quad \text{or}$$
$$H^2 - 4DR = 0$$

This is the general form of the equation of a parabola, $y^2 - 2ax = 0$.

16. $p = 0.2$, $q = 0.8$. We need to set up an equation that satisfies the condition:
$q^2 = 2(2pq) = 4pq$
substitute $1 - q$ for p
$q^2 = 4q(1 - q) = 4q - 4q^2$
$0 = 4q - 5q^2$
$0 = q(4 - 5q)$
either $q = 0$ or $5q = 4$
q cannot equal 0, or there would be no recessive homozygotes. So, q must $= 0.8$.
Check by substituting:
$q^2 = 0.64$, $2pq = 2(0.2)(0.8) = 0.32$

18. $M = 0.43$; $N = 0.57$. No, it is not in equilibrium. If it were, we would expect $(0.57)^2$ NN individuals $= 487$, $(0.43)^2$ MM individuals $= 277$, and $2(0.43)(0.57)$ MN individuals $= 735$. The observed numbers are very different from the expected, and a chi-square analysis will confirm this. (Chi-square, one degree of freedom $= 527.0$; $p << 0.01$.)

20. BB: 0.303; Bb: 0.577; bb: 0.119. For this problem you have to calculate allelic frequencies for males and females separately. Assume $p = f(A)$, $q = f(a)$; for the breeding women, $p = \dfrac{20 + 6}{32} = 0.81$ and $q = \dfrac{6}{32} = 0.19$.

For the males, $p = \dfrac{4 + 8}{32} = 0.375$ and $q = \dfrac{20}{32} = 0.625$.

If we assume random mating among males and females, we can set up a Punnett square to calculate offspring.

♀	0.375 B ♂	0.625 b
0.81 B	0.303 BB	0.506 Bb
0.19 b	0.071 Bb	0.119 bb

22. $f(A)$: 0.792; $f(B)$: 0.208. The easiest way to calculate frequencies is to do it empirically. We have 300 people, so we have 600 alleles.
$f(A) = \dfrac{2 \times 200(AA) + 75(AB)}{600} = \dfrac{475}{600} = 0.792$
$f(B) = 1 - f(A) = \dfrac{125}{600} = 0.208$

24. 0.208. First calculate allelic frequencies.
$p = f(m) = \dfrac{2(21) + 38}{200} = 0.4$ and $q = f(n) =$
$1 - f(m) = 0.6$. Inbreeding coefficient $F = (2pq - H)/2pq$, where H is observed frequency of heterozygotes.
$2pq = 2(0.4)(0.6) = 0.48$; $H = 0.38$, so
$F = \dfrac{(0.48 - 0.38)}{0.48} = 0.208$
Heterozygotes have been reduced by 21%, presumably because of inbreeding.

Chapter 20
Population Genetics: Processes that Change Allelic Frequencies

1. a. The equilibrium frequency of a is $\hat{q} = \mu/(\mu + v)$.
   Therefore
   $\hat{q} = (6 \times 10^{-5})/(6 \times 10^{-5} + 7 \times 10^{-7})$
   $= 0.00006/0.0000607 = 0.988$
   b. If $q = 0.90$ at generation n, then

   $q_{n + 1} = q_n + \mu p_n - v q_n$
   $= 0.90 + (6 \times 10^{-5})(0.10)$
   $- (7 \times 10^{-7})(0.90)$
   $= 0.9000066$

2. We use equation 20.12 to calculate the migration rate, m:

   $m = (q_c - q_1)/(q_2 - q_1)$

   In this case $q_c = 0.45$, $q_1 = 0.62$, and $q_2 = 0.03$. Thus

   $m = (0.45 - 0.62)/(0.03 - 0.62) = -0.17/-0.59$

   $= 0.288$

3. The population had a deficiency of heterozygotes. Given that natural selection is not acting on the population, this could have resulted from inbreeding or subdivision—the Wahlund effect—or both. Further information would be needed to fully understand what is happening in this population.

4. The ultimate fate of the A allele is either loss or fixation. The probability of loss is 0.3; one hundred generations; two thousand generations.

5. When there is no back mutation, we have a simple model:

   $A \xrightarrow{\mu} a$

   Therefore

   $q_{n + 1} = q_n + \mu p_n$

   $\Delta q = q_{n + 1} - q_n = q_n + \mu p_n - q_n = \mu p_n$

   At equilibrium:

   $\Delta q = 0$, or $\hat{p} = 0$ and $\hat{q} = 1.0$

6. In stabilizing selection, extremes of a distribution are selected against. In directional selection, one extreme is favored over the other. In disruptive selection, both extremes are favored over the middle of the distribution.

7. Fitnesses could be 1, 1/2, and 1/4 for *AA*, *Aa*, and *aa*, respectively (or 1, $1 - s$, and $1 - 1.5s$, where $s = 0.5$):

	AA	*Aa*	*aa*
Before selection	p^2	$2pq$	q^2
Fitnesses (W)	1	$1 - s$	$1 - 1.5s$
Ratio after selection	p^2	$2pq(1 - s)$	$q^2(1 - 1.5s)$
Frequencies after selection	$p^2/\overline{W}$	$2pq(1 - s)/\overline{W}$	$q2(1 - 1.5s)/\overline{W}$

	Total
Before selection	1
Ratio after selection	$\overline{W} = 1 - 2pqs - 1.5sq^2$
Frequencies after selection	1

Then

$$q_{n+1} = pq(1 - s)/\overline{W} + q^2(1 - 1.5s)/\overline{W}$$

$$\Delta q = q_{n+1} - q$$

$$= pq(1 - s)/\overline{W} + q^2(1 - 1.5s)/\overline{W} - q\overline{W}/\overline{W}$$

This equation simplifies to
$$\Delta q = sq(q[1.5 - 0.5q] - 1)/\overline{W}$$
At equilibrium ($\Delta q = 0$), $\hat{q} = 0$ or 1 (setting the expression within parentheses = 0). If we substitute various values of q into the Δq equation, we see that the equilibrium is stable. At any value of q other than $\hat{q}$ (0 or 1), Δq will be negative, indicating a decrease in q in the next generation. Thus $\hat{q} = 1$ is a trivial equilibrium point. Any population of individuals segregating both alleles will eventually lose the *a* allele.

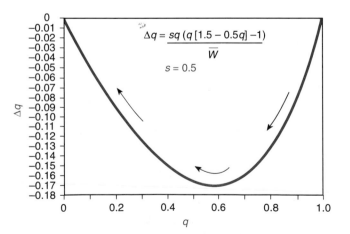

$$\Delta q = \frac{sq\,(q\,[1.5 - 0.5q] - 1)}{\overline{W}}$$

$$s = 0.5$$

8. Heterozygote disadvantage:

	AA	*Aa*
Before selection	p^2	$2pq$
Fitnesses (W)	1	$1 - s$
After selection	$p^2/\overline{W}$	$2pq(1 - s)/\overline{W}$

	aa	Total
Before selection	q^2	1
Fitness (W)	1	
After selection	$q^2/\overline{W}$	$\overline{W} = 1 - 2pqs$

Then

$$q_{n+1} = (pq[1 - s] + q^2)/\overline{W}$$

$$\Delta q = q_{n+1} - q = (pq[1 - s] + q^2)/\overline{W} - q\overline{W}/\overline{W}$$

$$= (pq[1 - s] + q^2 - q[1 - 2pqs])/\overline{W}$$

which simplifies to

$$\Delta q = spq(2q - 1)/\overline{W}$$

At $\Delta q = 0$, $\hat{q} = 0$, 1, or 0.5 ($2q - 1 = 0$, therefore $q = 0.5$). The equilibrium points of 0 and 1 are stable—if perturbed slightly (less than 0.5), the population will return to these values. The value $\hat{q} = 0.5$ is, however, unstable—if perturbed, it will continue away from the equilibrium point.

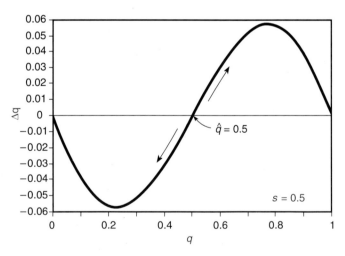

$\hat{q} = 0.5$

$s = 0.5$

9. **a.** This is heterozygous advantage. The total is divided by $\overline{W}$ because losses have occurred in the homozygous categories.

b. $q_{n+1} = (pq + q^2[1 - s_2])/\overline{W}$

c. Assume $s_2 = s_1$. Then substituting into equation 20.24:

$$\hat{q} = s_1/(s_1 + s_2) = s_1/(s_1 + 2s_1) = s_1/3s_1 = 0.33$$

$$\hat{p} = 0.67$$

If $s_1 = 0.1$ and $s_2 = 0.3$, then

$$\hat{q} = 0.1/(0.1 + 0.3) = 0.1/0.4 = 0.25; \hat{p} = 0.75$$

Heterozygotes $= 2\hat{p}\hat{q} = 0.375 = 37.5\%$

10.

	AA	Aa
Before selection	p^2	$2pq$
Fitnesses (W)	$1 - s$	$1 - s$
After selection	$p^2(1 - s)/\overline{W}$	$2pq(1 - s)/\overline{W}$

	aa	Total
Before selection	q^2	1
Fitness (W)	1	
After selection	$q^2/\overline{W}$	$\overline{W} = 1 - p^2s - 2pqs$

$$p_{n+1} = (p^2[1 - s] + pq[1 - s])/\overline{W}$$

11. $\hat{q} \approx \sqrt{\mu/s}$; see equation 20.22

$$\hat{p} \approx 1 - \sqrt{\mu/s}$$

Substituting

$$\hat{q} = 0.018; \hat{p} = 0.982$$

12. Since only heterozygotes survive, $\hat{q} = 0.5$. This can also be derived from equation 20.24:

$$\hat{q} = s_1/(s_1 + s_2)$$

If $s_1 = s_2 = 1$, then

$$\hat{q} = 1/2$$

13. By inspection, you can see that there has been a relative decrease in sepia-eyed flies and a relative increase in wild-type homozygotes and heterozygotes. We need to convert these changes to fitnesses. We do so as follows:

$$s^+/s^+ = 30/25 = 1.2$$

$$s^+/s = 60/50 = 1.2$$

$$s/s = 10/25 = 0.4$$

These "fitness" values can be changed so that the highest fitness is 1.0 by dividing each by 1.2, yielding fitnesses of 1, 1, and 0.33. This is then an example of selection against the homozygous recessive genotype (see table 20.2) in which $s = 0.67$ ($1 - 0.33$).

14. The *ST* inversion seems to do best at lower elevations, the *AR* at higher elevations. *CH* (and others) do not appear to be affected by altitude. To test this hypothesis, we would grow caged populations of flies with different initial frequencies of the various inversions at different elevations. We predict that, regardless of initial conditions, they would eventually equilibrate at the values in the table for the given altitudes. From there we could begin to investigate what attribute of altitude is affecting the flies (temperature, pressure, oxygen content, or other).

16. $p = 0.56; q = 0.44$. We can use many formulas, including setting up a selection model (table 20.3). Here we will calculate frequencies empirically by calculating the number of alleles contributed by each genotype. In a population, 0.25 of the gametes will result from *AA* individuals (fitness $\times$ frequency $=$ gamete frequency), 0.4 from *Aa* (0.8×0.5), and 0.15 from *aa* (0.6×0.25). Since all gametes from *AA* and *aa* will have only one allele, we have $0.25A$ and $0.15a$. Half of the gametes from *Aa* will be of each allele, so *Aa* gives $0.2A$ and $0.2a$ to the gamete pool. This gives total alleles as $0.25 + 0.4 + 0.15 = 0.8$ of the original. Therefore

$$p = \frac{0.25 + 0.20}{0.8} = \frac{0.45}{0.80} = 0.56; q = 1 - p = 0.44.$$

18. 0.714. $\hat{q} = \frac{u}{u + v} = \frac{5 \times 10^{-5}}{5 \times 10^{-5} + 2 \times 10^{-5}} = \frac{5}{7}$
$= 0.714$

20. 0.575. We use the formula in equation 20.15, $q_n = \frac{q}{1 - sq^2} - \frac{sq^2}{1 - sq^2}$. Since the relative fitness, $W = 0.4$, $s = 1 - W = 0.6$

$$q_n = \frac{0.7 - (0.6)(0.49)}{1 - (0.6)(0.49)} = \frac{0.406}{0.706} = 0.575$$

22. 0.47. We again use equation 20.12: $m = \frac{q_c - q_1}{q_2 - q_1}$ where $m = 0.1$, $q_c = 0.45$, and $q_2 = 0.25$.

$$0.1 = \frac{0.45 - q_1}{0.25 - q_1}$$

$$0.1 (0.25 - q_1) = 0.45 - q_1$$

$$0.025 - 0.1q = 0.45 - q_1$$

$$0.9q_1 = 0.425$$

$$q_1 = \frac{0.425}{0.9} = 0.47$$

Chapter 21
Genetics of the Evolutionary Process

1. In this example, $p = 0.7$ and $q = 0.3$. The expected distribution is *FF*, 49; *FS*, 42; and *SS*, 9. The chi-square value is 8.163, indicating that this population is not in Hardy-Weinberg proportions. There are too many heterozygotes and too few homozygotes indicating heterozygote advantage. Frequency-dependent selection cannot be determined from one sample; one would have to see a change in selection coefficients as allelic frequencies

changed. However, frequency-dependent selection would not look like heterozygote advantage. Rather, it would look like directional selection favoring the rare allele. A transient polymorphism is also an example of directional selection and would not look like heterozygote advantage.

2. Genetic variability can be maintained by heterozygote advantage (e.g., sickle-cell anemia in people); frequency-dependent selection (rare-male mating advantage in *Drosophila*); transient polymorphism (industrial melanism in moths during an increase or decrease in industrialization); life-stage selection; differential selection in heterogeneous environments; and neutrality.

3. Neo-Darwinism is the application of population genetics to Darwinian evolution. Darwinian evolution works by natural selection favoring the most fit organisms in competition among the overproduced young of any species.

4. By definition, a population increase is a period during which more young are surviving per adult than normal. Whatever the circumstances are (e.g., new food supply, favorable weather, reduction in predators), competition is relaxed and therefore offspring that normally would not survive (under stable numbers) do survive.

5. Each process lets reproductive isolating mechanisms evolve while some barrier to breeding arises (see fig. 21.3).

6. In the neutral gene hypothesis, it is suggested that polymorphisms occur in populations due to a balance of mutation and random genetic drift of neutral or nearly neutral alleles. The alternative view is that all variation is maintained by selective mechanisms. To distinguish these views, it is necessary to either look for selection in a large number of loci or determine some global pattern in distribution of allelic frequencies. Clearly both views are correct some of the time.

7. Presumably, in mammals, selection could involve types of substrates acted upon by electrophoretic variants; functioning at different pHs and ionic strengths in various cellular compartments; resistance to enzyme inhibitors; interaction with other proteins and membrane components; and others.

8. The investigators have suggested that this is a form of Batesian mimicry in which the normal leaves are mimicking those that have been attacked by caterpillars. In this way other butterflies and moths (and other insects) may be deterred from laying their eggs on these leaves to avoid the competition from larvae (caterpillars) already there.

9. Evolutionary rates determined from the protein, mtDNA, or nuclear DNA clocks are calculated assuming average rates of change. Looking at the end product (contemporary species) can provide insights into the actual rate at which species diverged by comparing rates in different lineages of known time of divergence. However, fluctuations in rate may be impossible to determine from extant species. In order to settle this controversy, much more detailed fossil lineages are needed.

10. The formula for K is $K = -\ln(1 - d/n)$, in which d is the number of amino acid differences and n is the total number of amino acid sites. Thus, human being–dog: $K = -\ln(1 - 1/8) = 0.133$; human being–chicken: $K = -\ln(1 - 3/8) = 0.470$; and dog–chicken: $K = -\ln(1 - 3/8) = 0.470$.

11. Constraint refers to the limitations on changes that can take place. Some changes will result in nonfunctional proteins and enzymes and thus cannot be in a successful lineage. For example, many base changes that led to new amino acids in enzyme active sites will disrupt enzymatic activity. If these mutations take place, they will be eliminated by natural selection.

12. Natural selection leads to adapted organisms; neutralism does not necessarily change that view. Neutralism is a theory to account for the tremendous amount of molecular variation that occurs in natural populations, variation that until the 1960s was not believed to be present.

13. Individual selection is defined by the fitnesses of genotypes of individuals, along the lines of the selection models that we looked at in chapter 20. Group selection is a concept, true under only a few circumstances, in which natural selection favors actions by individuals that are beneficial to the group at the expense of the fitness of the individuals. Kin selection is natural selection favoring behavior in individuals that benefits the individuals and their relatives that carry similar alleles.

14. Eight first cousins, on average, carry the complete genome of an individual. Therefore, from an evolutionary point of view, an individual and his or her eight cousins include the same alleles.

16. Yes. Two distinct species should not yield fertile progeny. We see a great reduction in the numbers of offspring from hybrids, indicating that hybrid inviability is one isolating mechanism operating.

18. Punctuated equilibrium proposes that species remain unchanged for long periods of time, and that major changes occur only periodically. If species A existed for 10 million years and suddenly (geologically speaking) changed dramatically to species X and Y, there would be few fossils because of the relatively short time in which intermediate forms were present. Another argument is, barring a complete sequence, there will always be gaps in the fossil record.

20. 5×10^7 years. $k = K/2T$, so $T = K/2k = 0.2/2 (2 \times 10^{-9}) = 0.5 \times 10^7$.

22. No. Certain proteins, and hence certain coding sequences, are likely to be conserved. Since most introns seem to have no function, more mutations in introns would be expected to accumulate over time. Since cDNA does not contain introns, we underestimate the number of base changes, and species seem to be more closely related than they may be.

24. Individuals are driven to preserve and enhance the frequency of their own alleles in the population. We expect, and this has been confirmed in many populations, that males will kill the young of other males, but not those they fathered.

Suggestions for Further Reading

Chapter 2
Mendel's Principles

Beadle, G. W., and E. L. Tatum. 1941. Genetic control of biochemical reactions in *Neurospora. Proc. Nat. Acad. Sci., USA* 27:499–506.

Borror, D. 1960. *Dictionary of Word Roots and Combining Forms.* Palo Alto, Calif.: Mayfield.

Brink, R. E., ed. 1967. *Heritage from Mendel.* Madison: University of Wisconsin Press.

Corcos, A., and F. Monaghan. 1985. Role of de Vries in the recovery of Mendel's work. I. Was de Vries an independent discoverer of Mendel? *J. Hered.* 76:187–90.

———. 1987. Correns, an independent discoverer of Mendelism? I. An historical/critical note. *J. Hered.* 78:330.

Douglas, L., and E. Novitski. 1977. What chance did Mendel's experiments give him of noticing linkage? *Heredity* 38:253–57.

Fisher, R. A. 1936. Has Mendel's work been rediscovered? *Ann. Sci.* 1:115–37.

Garrod, A. E. 1909. *Inborn Errors of Metabolism.* London: Hodder and Stoughton.

Grant, V. 1975. *Genetics of Flowering Plants.* New York: Columbia Univ. Press.

Haring, V., J. E. Gray, B. A. McClure, M. A. Anderson, and A. E. Clarke. 1990. Self-incompatibility: A self-recognition system in plants. *Science* 250:937–41.

Hommes, F. A., ed. 1990. *Techniques in Diagnostic Human Biochemical Genetics: A Laboratory Manual.* New York: Wiley.

Horowitz, N. H. 1990. George Wells Beadle (1903–1989). *Genetics* 124:1–6.

———. 1991. Fifty years ago: The *Neurospora* revolution. *Genetics* 127:631–35.

King, R. C., and W. D. Stansfield. 1990. *Dictionary of Genetics,* 4th ed. New York: Oxford University Press.

Koch, R., and F. de la Cruz. 1991. The danger of birth defects in the children of women with phenylketonuria. *J NIH Res.* 3:61–63.

Mariani, C., M. De Beuckeleer, J. Tructtner, J. Leemans, and R. Goldberg. 1990. Induction of male sterility in plants by a chimaeric ribonuclease gene. *Nature* 347:737–41.

Perkins, D. D. 1992. Neurospora: The organism behind the molecular revolution. *Genetics* 130:687–701.

Peters, J. A. 1959. *Classical Papers in Genetics.* Englewood Cliffs, N.J.: Prentice-Hall.

Pilgrim, I. 1986. A solution to the too-good-to-be-true paradox and Gregor Mendel. *J. Hered.* 77:218–20.

Reaume, A. G., D. A. Knecht, and A. Chovnick. 1991. The *rosy* locus in *Drosophila melanogaster:* Xanthine dehydrogenase and eye pigments. *Genetics* 129:1099–1109.

Rick, C. M. 1991. Tomato paste: A concentrated review of genetic highlights from the beginnings to the advent of molecular genetics. *Genetics* 128:1–5.

Stern, C., and E. R. Sherwood, eds. 1966. *The Origin of Genetics, A Mendel Source Book.* San Francisco: Freeman.

Stubbe, H. 1972. *History of Genetics.* Cambridge, Mass.: MIT Press.

Sturtevant, A. H. 1965. *A History of Genetics.* New York: Harper & Row.

Voipio, P. 1990. When and how did Mendel become convinced of the idea of general successive evolution? *Hereditas* 113:179–81.

Weiling, F. 1971. Mendel's "too good" data in *Pisum* experiments. *Folia Mendeliana* 6:75–77.

———. 1986. What about Fisher's statement of the "too good" data of J. G. Mendel's *Pisum* paper? *J. Hered.* 77:281–83.

Wellner, D., and A. Meister. 1981. A survey of inborn errors of amino acid metabolism and transport in man. *Ann. Rev. Biochem.* 50:911–68.

Woolf, C. M., and J. R. Swafford. 1988. Evidence for eumelanin and pheomelanin producing genotypes in the Arabian horse. *J. Hered.* 79:100–106.

Yamamoto, F., H. Clausen, T. White, J. Marken, and S. Hakomori. 1990. Molecular genetic basis of the histo-blood group ABO system. *Nature* 345:229–33.

Chapter 3
Mitosis and Meiosis

Allen, R. D. 1987. The microtubule as an intracellular engine. *Sci. Amer.*, Feb., 42–48.

Broek, D., R. Bartlett, K. Crawford, and P. Nurse. 1991. Involvement of p34^{cdc2} in establishing the dependency of S phase on mitosis. *Nature* 349:388–93.

Frontiers of biology: The cell cycle. [Five articles.] 1989. *Science* 246:603–34.

Gallagher, G. L. 1990. Evolutions: The mitotic spindle. *J. NIH Res.* 2:103–04.

———. 1992. Evolutions: The cell cycle. *J. NIH Res.* 4:124–28.

Heneen, W. 1982. The centromeric region in the scanning electron microscope. *Hereditas* 97:311–14.

Hyams, J. S., and B. R. Brinkley, eds. 1989. *Mitosis: Molecules and Mechanisms.* San Diego: Academic Press.

Hyman, A. A., and T. J. Mitchison. 1991. Two different microtubule-based motor activities with opposite polarities in kinetochores. *Nature* 351:206–11.

John, B., and K. Lewis. 1975. *Chromosome Hierarchy.* New York: Oxford University Press.

Margulis, L., and D. Sagan. 1986. *Microcosmos: Four Billion Years of Evolution from Our Microbial Ancestors.* New York: Summit Books.

Marx, J. 1991. The cell cycle: spinning further afield. *Science* 252:1490–92.

McIntosh, J. R., and K. L. McDonald. 1989. The mitotic spindle. *Sci. Amer.*, Oct., 48–56.

Murray, A. W., and M. W. Kirschner. 1991. What controls the cell cycle. *Sci. Amer.*, March, 56–63.

Norbury C., and P. Nurse. 1992. Animal cell cycles and their control. *Ann. Rev. Biochem.* 61:441–70.

Rhoades, M. M. 1950. Meiosis in maize. *J. Hered.* 41:59–67.

Strange, C. 1992. Cell cycle advances. *BioScience* 42:252–56.

Stubbe, H. 1972. *History of Genetics.* Cambridge, Mass.: MIT Press.

Sutton, W. S. 1903. The chromosomes in heredity. *Biol. Bull.* 4:231–51.

Swanson, C. P. 1957. *Cytology and Cytogenetics.* Englewood Cliffs, N.J.: Prentice-Hall.

Vallee, R. B., and H. S. Shpetner. 1990. Motor proteins of cytoplasmic microtubules. *Ann. Rev. Biochem.* 59:909–932.

von Wettstein, D., S. W. Rasmussen, and P. B. Holm. 1984. The synaptonemal complex in genetic segregation. *Ann. Rev. Genet.* 18:331–413.

Whittaker, R. H. 1969. New concepts of kingdoms of organisms. *Science* 163:150–60.

Chapter 4
Probability and Statistics

Berger, J. O., and D. A. Berry. 1988. Statistical analysis and the illusion of objectivity. *Amer. Sci.* 76:159–65.

Dixon, W. J., and F. J. Massey, Jr. 1983. *Introduction to Statistical Analysis.* 4th ed. New York: McGraw-Hill.

Howson, C., and P. Urbach. 1991. Bayesian reasoning in science. *Nature* 350:371–74.

Jefferys, W. H., and J. O. Berger. 1992. Ockham's razor and Bayesian analysis. *Amer. Sci.* 80:64–72.

Maynard Smith, J. 1968. *Mathematical Ideas in Biology.* New York: Cambridge University Press.

Moore, D. S. 1985. *Statistics: Concepts and Controversies.* San Francisco: Freeman.

Ross, S. M. 1976. *A First Course in Probability.* New York: Macmillan.

Siegel, S. 1956. *Nonparametric Statistics for the Behavioral Sciences.* New York: McGraw-Hill.

Yoccoz, N. G. 1991. Use, overuse, and misuse of significance tests in evolutionary biology and ecology. *Bull. Ecol. Soc. America* 72:106–11.

Chapter 5
Sex Determination, Sex Linkage, and Pedigree Analysis

Baker, B. S., G. Hoff, T. C. Kauman, M. F. Wolfner, and T. Hazelrigg. 1991. The *doublesex* locus of *Drosophila melanogaster* and its flanking regions: A cytogenetic analysis. *Genetics* 127:125–38.

Barr, M., and E. Bertram. 1949. A morphological distinction between neurones of the male and female, and the behavior of the nucleolar satellite during accelerated nucleoprotein synthesis. *Nature* 163:676–77.

Borsanni, G., et al. 1991. Characterization of a murine gene expressed from the inactive X chromosome. *Nature* 351:325–29.

Bridges, C. 1932. The genetics of sex in *Drosophila.* In *Sex and Internal Secretions, A Survey of Recent Research,* edited by E. Allen, 53–93. Baltimore, Md.: Williams & Wilkins.

Brockdorff, N., et al. 1991. Conservation of position and exclusive expression of mouse *Xist* from the inactive X chromosome. *Nature* 351:329–31.

Brown, C., et al. 1991. A gene from the region of the human X inactivation centre is expressed exclusively from the inactive X chromosome. *Nature* 349:38–44.

Burgoyne, P. S. 1982. Genetic homology and crossing over in the X and Y chromosomes of mammals. *Hum. Genet.* 61:85–90.

Charlesworth, B. 1991. The evolution of sex chromosomes. *Science* 251:1030–33.

Cherfas, J. 1991. Sex and the single gene. *Science* 252:782.

Cline, T. W. 1988. Evidence that *sisterless-a* and *sisterless-b* are two of several discrete "numerator elements" of the X/A sex determination signal in *Drosophila* that switch *Sxl* between two alternative stable expression states. *Genetics* 119:829–62.

Corcos, A. F. 1983. Pattern baldness: Its genetics revisited. *Amer. Biol. Teacher* 45:371–75.

Davies, K. 1991. The essence of inactivity. *Nature* 349:15–16.

Eicher, E. M., and L. L. Washburn. 1986. Genetic control of primary sex determination in mice. *Ann. Rev. Genet.* 20:327–60.

Elbrecht, A., and R. G. Smith. 1992. Aromatase enzyme activity and sex determination in chickens. *Science* 255:467–70.

Erickson, J. W., and T. W. Cline. 1991. Molecular nature of the *Drosophila* sex determination signal and its link to neurogenesis. *Science* 251:1071–74.

Farrow, M., and R. Juberg. 1969. Genetics and laws prohibiting marriage in the United States. *J. Amer. Med. Assoc.* 209:534–38.

Franco, B., et al. 1991. A gene deleted in Kallmann's syndrome shares homology with neural cell adhesion and axonal path-finding molecules. *Nature* 353:529–36.

Gartler, S., and A. Riggs. 1983. Mammalian X-chromosome inactivation. *Ann. Rev. Genet.* 17:155–90.

Goodfellow, P. J., S. M. Darling, N. S. Thomas, and P. N. Goodfellow. 1986. A pseudoautosomal gene in man. *Science* 234:740–43.

Graham, J. B., and R. W. Winters. 1961. Familial hypophosphatemia: An inherited demand for increased vitamin D. *Ann. N.Y. Acad. Sci.* 91:667–73.

Grant, S. G., and V. M. Chapman. 1988. Mechanisms of X chromosome regulation. *Ann. Rev. Genet.* 22:199–233.

Hardy, R. W., et al. 1984. Cytogenetic analysis of a segment of the Y chromosome of *Drosophila melanogaster*. *Genetics* 107:591–610.

Harley, V. R., et al. 1992. DNA binding activity of recombinant SRY from normal males and XY females. *Science* 255:453–56.

Inoue, K., K. Hoshijima, H. Sakamoto, and Y. Shimura. 1990. Binding of the *Drosophila Sex-lethal* gene product to the alternative splice site of *transformer* primary transcript. *Nature* 344:461–63.

Kennison, J. 1983. Analysis of Y-linked mutations to male sterility in *Drosophila melanogaster*. *Genetics* 103:219–34.

Koopman, P., J. Gubbay, N. Vivian, P. Goodfellow, and R. Lovell-Badge. 1991. Male development of chromosomally female mice transgenic for *Sry*. *Nature* 351:117–21.

Lindsley, D. L., and G. G. Zimm. 1992. The genome of *Drosophila melanogaster*. San Diego: Academic Press.

Lucchesi, J. C. 1983. The relationship between gene dosage, gene expression, and sex in *Drosophila. Develop. Genet.* 3:275–82.

Lyon, M. F. 1962. Sex chromatin and gene action in the mammalian X chromosome. *Amer. J. Hum. Genet.* 14:135–48.

———. 1990. Evolution of the X chromosome. *Nature* 348:585–86.

Mardon, G., et al. 1989. Duplication, deletion, and polymorphism in the sex-determining region of the mouse Y chromosome. *Science* 243:78–80.

McKusick, V. A. 1990. *Mendelian Inheritance in Man: Catalogs of Autosomal Dominant, Autosomal Recessive, and X-Linked Phenotypes*, 9th ed. Baltimore, Md.: Johns Hopkins University Press.

Morgan, T. H. 1910. Sex limited inheritance in *Drosophila. Science* 32:120–22.

Roberts, L. 1988. Zeroing in on the sex switch. *Science* 239:21–23.

Shapiro, L., et al. 1979. Noninactivation of an X-chromosome locus in man. *Science* 204:1224–26.

Villeneuve, A. M., and B. J. Meyer. 1990. The role of *sdc-1* in the sex determination and dosage compensation decisions in *Caenorhabditis elegans*. *Genetics* 124:91–114.

Walker, C. L., et al. 1991. The Barr body is a looped X chromosome formed by telomere association. *Proc. Nat. Acad. Sci., USA* 88:6191–95.

Williamson, J., and M. Bentley. 1983. Dosage compensation in *Drosophila:* NADP-enzyme activities and cross-reacting material. *Genetics* 103:649–58.

Wright, W. G. 1988. Sex change in the Mollusca. *TREE* 3:137–40.

Chapter 6
Linkage and Mapping in Eukaryotes

Anonymous. 1990. The human genome. *J. NIH Res.* 2:133–60.

Anonymous. 1991. Human genetic disorders. *J. NIH Res.* 3:143–68.

Ashburner, M. 1991. *DIS69: The Genetic Maps of* Drosophila. Cambridge, Mass.: Drosophila Information Service.

Carson, S. D., W. M. Henry, and T. B. Shows. 1985. Tissue factor gene localized to human chromosome 1 (1pter → 1p21). *Science* 229:991–93.

Cavener, D. R., D. C. Otteson, and T. C. Kaufman. 1986. A rehabilitation of the genetic map of the 84B-D region in *Drosophila melanogaster. Genetics* 114:111–23.

Creighton, H. B., and B. McClintock. 1931. A correlation of cytological and genetical crossing over in *Zea mays. Proc. Nat. Acad. Sci. USA* 17:492–97.

Crow, J. F. 1983. *Genetics Notes,* 8th ed. Minneapolis, Minn.: Burgess Publishing.

———. 1988. A diamond anniversary: the first chromosome map. *Genetics* 118:1–3.

———. 1990. Mapping functions. *Genetics* 125:669–71.

Emery, A. E. H., and D. L. Rimoin, eds. 1990. *Principles and Practice of Medical Genetics*. New York: Churchill Livingstone.

Fincham, J. R. S., P. R. Day, and A. Radford. 1979. *Fungal Genetics,* 4th ed. Berkeley: University of California Press.

Glass, N. L., J. Grotelueschen, and R. L. Metzenberg. 1990. *Neurospora crassa A* mating-type region. *Proc. Nat. Acad. Sci. USA* 87:4912–16.

Hooper, C. 1991. An exciting "if" in Alzheimer's. *J. NIH Res.* 3:65–70.

Lefevre, G., and W. Watkins. 1986. The question of the total gene number in *Drosophila melanogaster. Genetics* 113:869–95.

Lindsley, D., and E. Grell. 1968. *Genetic Variations of* Drosophila melanogaster. Washington, D.C.: Carnegie Institution.

Link, A. J., and M. V. Olson. 1991. Physical map of *Saccharomyces cerevisiae* genome at 110-kilobase resolution. *Genetics* 127:681–98.

Littlefield, J. W. 1964. Selection of hybrids from matings of fibroblasts in vitro and their presumed recombinants. *Science* 145:709–10.

Lyon, M. F. 1990. L. C. Dunn and mouse genetic mapping. *Genetics* 125:231–36.

McKusick, V. A. 1992. *Mendelian Inheritance in Man: Catalogs of Autosomal Dominant, Autosomal Recessive, and X-linked Phenotypes,* 10th ed., two vols. Baltimore, Md.: Johns Hopkins University Press.

Morgan, T. H. 1919. *The Physical Basis of Heredity*. Philadelphia: Lippincott.

Morton, N. E. 1955. Sequential test for the detection of linkage. *Amer. J. Human. Genet.* 7:277–318.

Nasim, A., P. Young, and B. F. Johnson, eds. 1989. *Molecular Biology of Fission Yeast.* New York: Academic Press.

Panthier, J., J. Guénet, H. Condamine, and F. Jacob. 1990. Evidence for mitotic recombination in $W^{ei}/+$ heterozygous mice. *Genetics* 125:175–82.

Risch, N. 1992. Genetic linkage: Interpreting Lod scores. *Science* 255:803–04.

Roman, H. 1986. The early days of yeast genetics: A personal narrative. *Ann. Rev. Genet.* 20:1–12.

Ruddle, F. H., and R. S. Kucherlapati. 1974. Hybrid cells and human genes. *Sci. Amer.,* July, 36–44.

Sorsa, V. V. 1988. *Chromosome Maps of* Drosophila. Two volumes. Boca Raton, Fl.: CRC Press.

Sowers, A. E., ed. 1987. *Cell Fusion.* New York: Plenum Press.

Staben, C., and C. Yanofsky. 1990. *Neurospora crassa a* mating-type region. *Proc. Nat. Acad. Sci. USA* 87:4917–21.

Stern, C. 1936. Somatic crossing over and segregation in *Drosophila melanogaster. Genetics* 21:625–730.

Sturtevant, A. H. 1913. The linear arrangement of six sex-linked factors in *Drosophila,* as shown by their mode of association. *J. Exp. Zool.* 14:43–59.

Sugawara, O., M. Oshimura, M. Koi, L. A. Annab, and J. C. Barrett. 1990. Induction of cellular senescence in immortalized cells by human chromosome 1. *Science* 247:707–10.

Touchette, N. 1991. The link between yeast and neuronal differentiation: A bridge to FAR. *J. NIH Res.* 3:53–58.

Verma, R. S., and A. Babu. 1989. *Human Chromosomes. Manual of Basic Techniques.* Elmsford, N.Y.: Pergamon Press.

Chapter 7
Linkage and Mapping in Prokaryotes and Viruses

Bachmann, B. J. 1987. Linkage map of *Escherichia coli* K–12, ed. 7. Vol. 2 of *Escherichia coli and Salmonella typhimurium: Cellular and Molecular Biology,* 807–76. Washington, D.C.: American Society for Microbiology.

Birge, E. A. 1988. *Bacterial and Bacteriophage Genetics.* New York: Springer-Verlag.

Brock, T. D. 1990. *The Emergence of Bacterial Genetics.* Cold Spring Harbor, N.Y.: Cold Spring Harbor Laboratory Press.

Cairns, J., G. Stent, and J. Watson, eds. 1966. Phage and the origins of molecular biology. *Cold Spr. Harb. Symp. Quant. Biol.* 31.

Drlica, K., and M. Riley, eds. 1990. *The Bacterial Chromosome.* Washington, D.C.: American Society of Microbiology.

Freifelder, D. 1987. *Microbial Genetics.* Boston: Science Books International.

Garfield, E. 1990. A tribute to Joshua Lederberg: Highlights of a remarkable scientific career. *Current Contents* 33:5–11.

Hardy, K., ed. 1986. *Bacterial Plasmids.* 2d ed. Washington, D.C.: American Society of Microbiology.

Hogle, J. M., M. Chow, and D. J. Filman. 1985. Three-dimensional structure of poliovirus at 2.9 Å resolution. *Science* 229:1358–65.

Ippen-Ihler, K. A., and E. G. Minkley, Jr. 1986. The conjugation system of F, the fertility factor of *Escherichia coli. Ann. Rev. Genet.* 20:593–624.

Jacob, F., and E. Wollman. 1961. *Sexuality and the Genetics of Bacteria.* New York: Academic Press.

Kaplan, M., and R. Webster. 1977. The epidemiology of influenza. *Sci. Amer.,* Dec., 88–106.

Lamb, R., and P. Choppin. 1983. The gene structure and replication of influenza virus. *Ann. Rev. Biochem.* 52:467–506.

Lederberg, J. 1987. Genetic recombination in bacteria: A discovery account. *Ann. Rev. Genet.* 21:23–46.

———. 1989. Replica plating and indirect selection of bacterial mutants: Isolation of preadaptive mutants in bacteria by sib selection. *Genetics* 121:395–99.

Lederberg, J., et al. 1951. Recombination analysis of bacterial heredity. *Cold Spr. Harb. Symp. Quant. Biol.* 16:413–43.

Lim, D. V. 1989. *Microbiology.* St. Paul, Minn.: West Publishing Co.

Low, K. B., and D. Porter. 1978. Modes of gene transfer and recombination in bacteria. *Ann. Rev. Genet.* 12:249–87.

Mandecki, W., K. Krajewska-Grynkiewicz, and T. Klopotowski. 1986. A quantitative model for nonrandom generalized transduction, applied to the phage P22-*Salmonella typhimurium* system. *Genetics* 114:633–57.

Martin, S. J. 1978. *The Biochemistry of Viruses.* Cambridge: Cambridge University Press.

Maynard Smith, J., C. G. Dowson, and B. G. Spratt. 1991. Localized sex in bacteria. *Nature* 349:29–31.

Miller, J. H. 1992. *A short course in bacterial genetics.* Plainview, N.Y.: Cold Spring Harbor Laboratory Press.

O'Brien, S. J., ed. 1987. *Genetic Maps.* Vol. 4. Cold Spring Harbor, New York: Cold Spring Harbor Laboratory.

Prescott, L. M., J. P. Harley, and D. A. Klein. 1990. *Microbiology.* Dubuque, Iowa: Wm. C. Brown.

Sanders, F. K. 1981. *Viruses.* Burlington, N.C.: Carolina Biological Supply Company Scientific Publications.

Stahl, F. 1989. The linkage map of phage T4. *Genetics* 123:245–48.

Stent, G., and R. Calendar. 1978. *Molecular Genetics, An Introductory Narrative.* 2d ed. San Francisco: Freeman.

Tsao, J., et al. 1991. The three-dimensional structure of canine parvovirus and its functional implications. *Science* 251:1456–64.

Wood, W. B., and R. S. Edgar. 1967. Building a bacterial virus. *Sci. Amer.,* July, 60–74.

Chapter 8
Cytogenetics

Allore, R., et al. 1988. Gene encoding the β subunit of S100 protein is on chromosome 21: Implications for Down syndrome. *Science* 239:1311–13.

Ashworth, A., S. Rastan, R. Lovell-Badge, and G. Kay. 1991. X-chromosome inactivation may explain the difference in viability of XO humans and mice. *Nature* 351:406–8.

Barch, M. J., ed. 1991. *The ACT Cytogenetics Laboratory Manual.* New York: Raven Press.

Benavente, E., and J. Orellana. 1991. Chromosome differentiation and pairing behavior of polyploids: An assessment on preferential metaphase I associations in colchicine-induced autotetraploid hybrids within the genus *Secale. Genetics* 128:433–42.

Borgaonkar, D. S. 1989. *Chromosomal Variation in Man: A Catalogue of Chromosomal Variants and Anomalies.* 5th ed. New York: Alan R. Liss, Inc.

Brewster, T., and P. Gerald. 1978. Chromosome disorders associated with mental retardation. *Pediatr. Ann.* 7: 82–89.

Cochran, D. 1983. Alternate-2 disjunction in the German cockroach. *Genetics* 104:215–17.

Craymer, L. 1984. Techniques for manipulating chromosomal rearrangements and their applications to *Drosophila melanogaster.* II. Translocations. *Genetics* 108:573–87.

Cribiu, E. P., et al. 1989. Identification of chromosomes involved in a Robertsonian translocation in cattle. *Genet. Sel. Evol.* 21:555–60.

Daniel, A., ed. 1988. *The Cytogenetics of Mammalian Autosomal Rearrangements.* New York: Alan R. Liss, Inc.

Davies, K. 1991. The essence of inactivity. *Nature* 349:15–16.
————. 1991. Breaking the fragile X. *Nature* 351:439–40.

Dellarco, V., P. Voytek, and A. Hollaender, eds. 1985. *Aneuploidy. Etiology and Mechanisms.* New York: Plenum Press.

Dewald, G., M. Hammond, J. Spurbeck, and S. Moore. 1980. Origin of chi46,XX/46,XY chimerism in a human true hermaphrodite. *Science* 207:321–23.

Epstein, C. J. 1988. Mechanisms of the effects of aneuploidy in mammals. *Ann. Rev. Genet.* 22:51–75.

Futch, D. 1966. A study of speciation in South Pacific populations of *Drosophila ananassae. Univ. Texas Stud. Genet.* 6615:79–120.

Garner, J. 1988. *Backgrounds of Human Cytogenetics: A Bibliography.* Brookville, N.Y.: Confrontation Magazine Press, Long Island University.

Hassold, T. J., and P. A. Jacobs. 1984. Trisomy in man. *Ann. Rev. Genet.* 18:69–97.

Hoffman, M. 1991. Unraveling the genetics of fragile X syndrome. *Science* 252:1070.

Karpechenko, G. D. 1928. Polyploid hybrids of *Raphanus sativus* L. X *Brassica oleracea* L. *Zeit. Indukt. Abstam. Vererbung.* 48:1–85.

Khush, G. S., R. J. Singh, S. C. Sur, and A. L. Librojo. 1984. Primary trisomics of rice: Origin, morphology, cytology and use in linkage mapping. *Genetics* 107:141–63.

Lejeune, J., M. Gautier, and R. Turpin. 1959. Etude des chromosomes somatiques de neuf enfants mongoliens. *C. R. Acad. Sci. (Paris)* 248:1721–22.

Lewis, R. 1991. Genetic imprecision. *BioScience* 41:288–93.

Loidl, J., F. Ehrendorfer, and D. Schweizer. 1990. EM analysis of meiotic chromosome pairing in a pentaploid *Achillea* hybrid. *Heredity* 65:11–20.

Orr, H. A. 1990. "Why polyploidy is rarer in animals than in plants" revisited. *Amer. Nat.* 136:759–70.

Patterson, D. 1987. The causes of Down syndrome. *Sci. Amer.,* Aug., 52–60.

Penrose, L. S. 1933. The relative effects of paternal and maternal age in mongolism. *J. Genet.* 27:219–24.

Schmid, W. 1977. Cytogenetical problems in prenatal diagnosis. *Hereditas* 86:37–44.

Simpson, J. L. 1982. Abnormal sexual differentiation in humans. *Ann. Rev. Genet.* 16:193–224.

Stebbins, G. L. 1971. *Chromosomal Evolution in Higher Plants.* Reading, Mass.: Addison-Wesley.

Sturtevant, A. 1925. The effects of unequal crossing over at the *Bar* locus in *Drosophila. Genetics* 10:117–47.

Tartof, K. D. 1988. Unique crossing over then and now. *Genetics* 120:1–6.

Valentine, G. 1975. *Chromosome Disorders: An Introduction for Clinicians.* 3d ed. Philadelphia: Lippincott.

Warburton, D., J. Byrne, and N. Canki. 1990. *Chromosomal Anomalies and Prenatal Development: An Atlas.* New York: Oxford University Press.

White, M. J. D. 1973. *Animal Cytology and Evolution.* 3d ed. New York: Cambridge University Press.

Witkin, H. A., et al. 1976. Criminality in XYY and XXY men. *Science* 193:547–55.

Chapter 9
Chemistry of the Gene

Adams, R. L. P. 1991. *DNA Replication.* Oxford: IRL Press.

Alberts, B. M. 1987. Prokaryotic DNA replication mechanisms. *Phil. Trans. Roy. Soc. London, Series B* 317:395–420.

Arscott, P. G., et al. 1989. Scanning tunnelling microscopy of Z-DNA. *Nature* 339:484–86.

Avery, O., C. MacLeod, and M. McCarty. 1944. Studies on the chemical nature of the substance inducing transformation of *Pneumococcal* types. *J. Exp. Med.* 79:137–58.

Beebe, T. P., et al. 1989. Direct observation of native DNA structures with the scanning tunneling microscope. *Science* 243:370–72.

Cairns, J. 1963. The chromosomes of *E. coli. Cold Spr. Harb. Symp. Quant. Biol.* 28:43–46.

Campbell, J. L. 1986. Eukaryotic DNA replication. *Ann. Rev. Biochem.* 55:733–71.

Chargaff, E., and J. Davidson, eds. 1955. *The Nucleic Acids.* New York: Academic Press.

Chase, J. W., and K. R. Williams. 1986. Single-stranded DNA binding proteins required for DNA replication. *Ann. Rev. Biochem.* 55:103–36.

Dickerson, R. 1983. The DNA helix and how it is read. *Sci. Amer.,* Dec., 94–111.

Dickerson, R., et al. 1982. The anatomy of A-, B-, and Z-DNA. *Science* 216:475–85.

Driscoll, R. J., M. G. Youngquist, and D. D. Baldeschwieler. 1990. Atomic-scale imaging of DNA using scanning tunnelling microscopy. *Nature* 346:294–96.

Dulbecco, R. 1987. *The Design of Life.* New Haven: Yale University Press.

Flower, A. M., and C. S. McHenry. 1990. The γ subunit of DNA polymerase III holoenzyme of *Escherichia coli* is produced by ribosomal frameshifting. *Proc. Nat. Acad. Sci. USA* 87:3713–17.

Fraenkel-Conrat, H., and B. Singer. 1957. Virus reconstitution. II. Combination of protein and nucleic acid from different strains. *Biochem. Biophys. Acta.* 24: 540–48.

Gallagher, G. L. 1991. Evolutions: DNA replication. *J. NIH Res.* 3:101–04.

Griffith, F. 1928. Significance of pneumococcal types. *J. Hygiene* 27:113–59.

Hegstrom, R. A., and D. K. Kondepudi. 1990. The handedness of the universe. *Sci. Amer.,* Jan., 108–15.

Hershey, A., and M. Chase. 1952. Independent functions of viral protein and nucleic acid in growth of bacteriophage. *J. Gen. Physiol.* 36:39–56.

Hwang, D. S., and A. Kornberg. 1990. A novel protein binds a key origin sequence to block replication of an E. coli minichromosome. *Cell* 63:325–31.

Johnston, J. 1992. Cry threedom! Triple-helix research takes off. *J. NIH Res.* 4:36–42.

Kelly, T., and B. Stillman. 1988. Eukaryotic DNA replication. *Cancer Cells* 6:1–385.

Kornberg, A. 1989. *For the Love of Enzymes: The Odyssey of a Biochemist.* Cambridge, Mass.: Harvard University Press.

Kornberg, A., and T. A. Baker. 1992. *DNA Replication.* 2d ed. San Francisco: Freeman.

Lederberg, J. 1991. The gene (H. J. Muller 1947). *Genetics* 129:313–16.

Marahens, Y., and B. Stillman. 1992. A yeast chromosomal origin of DNA replication defined by multiple functional elements. *Science* 255:817–23.

Marians, K. J. 1992. Prokaryotic DNA replication. *Ann. Rev. Biochem.* 61:673–719.

Matson, S. W., and K. A. Kaiser-Rogers. 1990. DNA helicases. *Ann. Rev. Biochem.* 59:289–329.

McHenry, C. S. 1988. DNA polymerase III holoenzymes of *Escherichia coli. Ann. Rev. Biochem.* 57:519–50.

Meselson, M., and F. Stahl. 1958. The replication of DNA in *Escherichia coli. Proc. Nat. Acad. Sci., USA* 44:671–82.

Moffat, A. S. 1991. Triplex DNA finally comes of age. *Science* 252:1374–75.

Prusiner, S. B. 1991. Molecular biology of prion diseases. *Science* 252:1515–22.

Rich, A., A. Nordheim, and A. Wang. 1984. The chemistry of left-handed Z-DNA. *Ann. Rev. Biochem.* 53:791–846.

Riordan, M. L., and J. C. Martin. 1991. Oligonucleotide-based therapeutics. *Nature* 350:442–43.

Salas, M. 1991. Protein-priming of DNA replication. *Ann. Rev. Biochem.* 60:39–71.

Stillman, B. 1989. Initiation of eukaryotic DNA replication in vitro. *Ann. Rev. Cell Biol.* 5:197–245.

Strobel, S. A., and P. B. Dervan. 1990. Site-specific cleavage of a yeast chromosome by oligonucleotide-directed triple-helix formation. *Science* 249:73–75.

Stryer, L. 1988. *Biochemistry.* 3d ed. New York: Freeman.

Touchette, N. 1991. Scrapie prion: Teaching an old dogma new tricks. *J. NIH Res.* 3:49–54.

Trifonov, E. N. 1991. DNA in profile. *Trends Biochem. Sci.* 16:467–70.

Tsurimoto, T., T. Melendy, and B. Stillman. 1990. Sequential initiation of lagging and leading strand synthesis by two different polymerase complexes at the SV40 DNA replication origin. *Nature* 346:534–39.

Wang, A., et al. 1981. Left-handed double helical DNA; variations in the backbone conformation. *Science* 211:171–76.

Wang, J. 1982. DNA topoisomerases. *Sci. Amer.,* July, 94–109.

Wang, T. S.-F. 1991. Eukaryotic DNA polymerases. *Ann. Rev. Biochem.* 60:513–52.

Watson, J. D. 1968. *The Double Helix.* New York: Signet.

Watson, J. D., and F. H. C. Crick. 1953. Molecular structure of nucleic acids: A structure for deoxyribose nucleic acid. *Nature* 171:737–38.

Watson, J. D., et al. 1987. *Molecular Biology of the Gene.* 4th ed. Menlo Park, Calif.: Benjamin/Cummings.

Weissmann, C. 1991. The prion's progress. *Nature* 349: 569–71.

———. 1991. A "unified theory" of prion propagation. *Nature* 352:679–83.

Westheimer, F. H. 1987. Why nature chose phosphates. *Science* 235:1173–78.

Chapter 10
Gene Expression: Transcription

Barinaga, M. 1990. Introns pop up in new places—What does it mean? *Science* 250:1512.

Belfort, M. 1990. Phage T4 introns: Self-splicing and mobility. *Ann. Rev. Gen.* 24:363–85.

Björk, G. R., et al. 1987. Transfer RNA modification. *Ann. Rev. Biochem.* 56:263–87.

Brennan, C. A., A. J. Dombroski, and T. Platt. 1987. Transcription termination factor rho is an RNA-DNA helicase. *Cell* 48:945–52.

Brewer, B. J. 1988. When polymerases collide: Replication and the transcriptional organization of the E. coli chromosome. *Cell* 53:679–86.

Brody, E., and J. Abelson. 1985. The "spliceosome": Yeast pre-messenger RNA associates with a 40S complex in a splicing-dependent reaction. *Science* 228:963–67.

Butler, J. S., and T. Platt. 1988. RNA processing generates the mature 3′ end of yeast CYC1 messenger RNA in vitro. *Science* 242:1270–74.

Cech, T. R. 1986. RNA as an enzyme. *Sci. Amer.,* Nov., 64–75.

———. 1986. A model for the RNA-catalyzed replication of RNA. *Proc. Nat. Acad. Sci., USA* 83:4360–63.

———. 1988. *Molecular Biology of RNA.* New York: Alan R. Liss.

———. 1990. Self-splicing of group I introns. *Ann. Rev. Biochem.* 59:543–68.

Cheng, S.-W. C., et al. 1991. Functional importance of sequence in the stem-loop of a transcription terminator. *Science* 254:1205–07.

Company, M., J. Arenas, and J. Abelson. 1991. Requirement of the RNA helicase-like protein PRP22 for release of messenger RNA from spliceosomes. *Nature* 349:487–93.

Conaway, J. W., and R. C. Conaway. 1991. Initiation of eukaryotic messenger RNA synthesis. *J. Biol. Chem.* 266:17, 721–24.

Crick, F. 1970. Central dogma of molecular biology. *Nature* 227:561–63.

Darnell, J., Jr. 1985. RNA. *Sci. Amer.*, Oct., 68–78.

De Duve, C. 1984. *A Guided Tour of the Living Cell, vol. 2.* New York: Scientific American Books.

Dinter-Gottlieb, G. 1986. Viroids and virusoids are related to group I introns. *Proc. Nat. Acad. Sci., USA* 83:6250–54.

Doolittle, R. F. 1991. Counting and discounting the universe of exons. *Science* 253:677–79.

Doolittle, W. F. 1987. The origins and function of intervening sequences in DNA: A review. *Amer. Nat.* 130:915–28.

Dorit, R. L., L. Schoenbach, and W. Gilbert. 1990. How big is the universe of exons? *Science* 250:1377–82.

Edelson, E. 1990. Transcription factors: Governors for the genetic engine. *Mosaic* 21:2–9.

Fischer, U., and R. Lührmann. 1990. An essential signaling role for the m3G cap in the transport of U1 snRNP to the nucleus. *Science* 249:786–90.

Gall, J. 1991. Spliceosomes and snurposomes. *Science* 252:1499–1500.

Gallagher, G. L. 1990. Evolutions: Transcription. *J. NIH Res.* 2:102–4.

Gibbons, A. 1991. Molecular scissors: RNA enzymes go commercial. *Science* 251:521.

Grabowski, P., R. Padgett, and P. Sharp. 1984. Messenger RNA splicing in vitro: An excised intervening sequence and a potential intermediate. *Cell* 37:415–27.

Green, M. R. 1986. Pre-mRNA splicing. *Ann. Rev. Genet.* 20:671–708.

Guthrie, C. 1991. Messenger RNA splicing in yeast: Clues to why the spliceosome is a ribonucleoprotein. *Science* 253:157–63.

Guthrie, C., and B. Patterson. 1988. Spliceosomal snRNAs. *Ann. Rev. Genet.* 22:387–419.

Hall, B. D., and S. Spiegelman. 1961. Sequence complementarity of T2-DNA and T2-specific RNA. *Proc. Nat. Acad. Sci., USA* 47:137–46.

Heintz, N. 1991. Transcriptional regulation [introducing 12 articles]. *Trends Biochem. Sci.* 16:393.

Helmann, J. D., and M. J. Chamberlain. 1988. Structure and function of bacterial sigma factors. *Ann. Rev. Biochem.* 57:839–72.

Herendeen, D. R., K. P. Williams, G. A. Kassavetis, and E. P. Geiduschek. 1990. An RNA polymerase-binding protein that is required for communication between an enhancer and a promoter. *Science* 248:573–78.

Hoffman, M. 1991. RNA editing: What's in a mechanism? *Science* 253:136–38.

Holley, R. W., et al. 1965. Structure of a ribonucleic acid. *Science* 147:1462–65.

Hou, Y.-M., and P. Schimmel. 1988. A simple structural feature is a major determinant of the identity of a transfer RNA. *Nature* 333:140–45.

Jahn, M., M. J. Rogers, and D. Söll. 1991. Anticodon and acceptor stem nucleotides in tRNA[Gln] are major recognition elements for *E. coli* glutaminyl-tRNA synthetase. *Nature* 352:258–60.

Kainz, M., and J. Roberts. 1992. Structures of transcription elongation complexes in vivo. *Science* 255:838–41.

Lake, J. 1976. Ribosome structure determined by electron microscopy of *Escherichia coli* small subunits, large subunits and monomeric ribosomes. *J. Mol. Biol.* 105:131–59.

———. 1983. Evolving ribosome structure: Domains in archaebacteria, eubacteria, and eucaryotes. *Cell* 33:318–19.

Lewin, B. 1984. First true RNA catalyst found. *Science* 223:266–67.

Lewis, R. 1991. Introns in cyanobacteria stir up evolutionary debate. *J. NIH Res.* 3:54–58.

Liu, L. F., and J. C. Wang. 1987. Supercoiling of the DNA template during transcription. *Proc. Nat. Acad. Sci., USA* 84:7024–27.

Maniatis, T. 1991. Mechanisms of alternative pre-mRNA splicing. *Science* 251:33–34.

McSwiggen, J. A., and T. R. Cech. 1989. Stereochemistry of RNA cleavage by the *Tetrahymena* ribozyme and evidence that the chemical step is not rate-limiting. *Science* 244:679–83.

Meadows, J. 1990. The splice of life. *J. NIH Research* 2:54–57.

Miller, O. L., Jr., et al. 1970. Electron microscopic visualization of transcription. *Cold Spr. Harb. Symp. Quant. Biol.* 35:505–12.

Noller, H. 1984. Structure of ribosomal RNA. *Ann. Rev. Biochem.* 53:119–62.

Nomura, M. 1984. The control of ribosome synthesis. *Sci. Amer.*, Jan., 102–14.

Pabo, C. D., and R. T. Sauer. 1992. Transcription factors: Structural families and principles of DNA recognition. *Ann. Rev. Biochem.* 61:1053–95.

Pace, N. R. 1992. New horizons for RNA catalysis. *Science* 256:1402–03.

Peterson, M. G., N. Tanese, B. F. Pugh, and R. Tjian. 1990. Functional domains and upstream activation properties of cloned human TATA binding protein. *Science* 248:1625–30.

Peterson, M. G., et al. 1991. Structure and functional properties of human general transcription factor IIE. *Nature* 354:369–73.

Piken, W. A., D. B. Olsen, F. Benseler, H. Aurup, and F. Eckstein. 1991. Kinetic characterization of ribonuclease-resistant 2′-modified hammerhead ribozymes. *Science* 253:314–17.

Platt, T. 1986. Transcription termination and the regulation of gene expression. *Ann. Rev. Biochem.* 55:339–72.

Ptashne, M., and A. A. F. Gann. 1990. Activators and targets. *Nature* 346:329–31.

Rennie, J. 1991. Proofreading genes. *Sci. Amer.*, May, 28–29.

Riesner, D., and H. J. Gross. 1985. Viroids. *Ann. Rev. Biochem.* 54:531–64.

Sawadogo, M., and A. Sentenac. 1990. RNA polymerase B (II) and general transcription factors. *Ann. Rev. Biochem.* 59:711–54.

Schafer, D. A., J. Gelles, M. P. Sheetz, and R. Landick. 1991. Transcription by single molecules of RNA polymerase observed by light microscopy. *Nature* 352:444–48.

Sharp, P. A. 1987. Splicing of messenger RNA precursors. *Science* 235:766–71.

Shih, M.-C., P. Heinrich, and H. M. Goodman. 1988. Intron existence predated the divergence of eukaryotes and prokaryotes. *Science* 242:1164–66.

Simpson, L. 1990. RNA editing—a novel genetic phenomenon. *Science* 250:512–13.

Smith, C. W. J., J. G. Patton, and B. Nadal-Ginard. 1989. Alternative splicing in the control of gene expression. *Ann. Rev. Genet.* 23:527–77.

Steitz, J. A. 1988. "Snurps." *Sci. Amer.,* June, 56–63.

Symons, R. H. 1992. Small catalytic RNAs. *Ann. Rev. Biochem.* 61:641–71.

Von Hippel, P., et al. 1984. Protein-nucleic acid interactions in transcription: A molecular analysis. *Ann. Rev. Biochem.* 53:389–446.

Waldrop, M. M. 1989. Did life really start out in an RNA world? *Science* 246:1248–49.

Wassarman, D. A., and J. A. Steitz. 1991. Alive with DEAD proteins. *Nature* 349:463–64.

Weiss, R. 1990. Guide RNA: Ghostwriters in the cell. *J. NIH Res.* 2:49–52.

Wittmann, H. 1983. Architecture of prokaryotic ribosomes. *Ann. Rev. Biochem.* 52:35–65.

Wu, J., and J. L. Manley. 1991. Base pairing between U2 and U6 snRNAs is necessary for splicing of a mammalian pre-mRNA. *Nature* 352:818–21.

Young, R. A. 1991. RNA polymerase II. *Ann. Rev. Biochem.* 60:689–715.

Zaug, A. J., and T. R. Cech. 1986. The intervening sequence RNA of *Tetrahymena* is an enzyme. *Science* 231:470–75.

Zeigler, D. R., and D. H. Dean. 1990. Orientation of genes in the *Bacillus subtilis* chromosome. *Genetics* 125:703–8.

Chapter 11
Gene Expression: Translation

Baker, K. P., and G. Schatz. 1991. Mitochondrial proteins essential for viability mediate protein import into yeast mitochondria. *Nature* 349:205–8.

Bernabeu, C., and J. A. Lake. 1982. Nascent polypeptide chains emerge from the exit domain of the large ribosomal subunit: Immune mapping of the nascent chains. *Proc. Nat. Acad. Sci., USA* 79:3111–15.

Berry, M. J., et al. 1991. Recognition of UGA as a selenocysteine codon in Type I deiodinase requires sequences in the 3′ untranslated region. *Nature* 353:273–76.

Bonitz, S., et al. 1980. Codon recognition rules in yeast mitochondria. *Proc. Nat. Acad. Sci., USA* 77:3167–70.

Capel, M. S., et al. 1987. A complete mapping of the proteins in the small ribosomal subunit of *Escherichia coli*. *Science* 238:1403–6.

Caras, I. W., and G. N. Weddell. 1989. Signal peptide for protein secretion directing glycophospholipid membrane anchor attachment. *Science* 243:1196–98.

Caron, F., and E. Meyer. 1985. Does *Paramecium primaurelia* use a different genetic code in its macronucleus? *Nature* 314:185–88.

Chapeville, F., et al. 1962. On the role of soluble ribonucleic acid in coding for amino acids. *Proc. Nat. Acad. Sci., USA* 48:1086–92.

Chothia, C., and A. V. Finkelstein. 1990. The classification and origins of protein folding patterns. *Ann. Rev. Biochem.* 59:1007–39.

Cigan, A. M., L. Feng, and T. F. Donahue. 1988. $tRNA_i^{Met}$ functions in directing the scanning ribosome to the start site of translation. *Science* 242:93–97.

Crick, F. H. C. 1958. On protein synthesis. *Symp. Soc. Exp. Biol.* 12:138–63.

———. 1966. Codon-anticodon pairing: The wobble hypothesis. *J. Mol. Biol.* 19:548–55.

Crick, F. H. C., et al. 1961. General nature of the genetic code for proteins. *Nature* 192:1227–32.

De Duve, C. 1988. The second genetic code. *Nature* 333:117–18.

Ellis, R. J., and S. M. van der Vies. 1991. Molecular chaperones. *Ann. Rev. Biochem.* 60:321–47.

Eriani, G., et al. 1990. Partition of tRNA synthetases into two classes based on mutually exclusive sets of sequence motifs. *Nature* 347:203–6.

Fontaine, B., et al. 1990. Hyperkalemic periodic paralysis and the adult muscle sodium channel α-subunit gene. *Science* 250:1000–2.

Fox, T. D. 1987. Natural variation in the genetic code. *Ann. Rev. Genet.* 21:67–91.

Francke, C., et al. 1982. Electronic microscopic visualization of a discrete class of giant translation units in salivary gland cells of *Chironomus tentans*. *EMBO Jour.* 1:59–62.

Frank, J., A. Verschoor, M. Radermacher, and T. Wagenknecht. 1990. Morphologies of eubacterial and eucaryotic ribosomes as determined by three-dimensional electron microscopy. In *The Ribosome, Structure, Function, & Evolution,* edited by W. E. Hill, et al., 107–13. Washington, D.C.: American Society for Microbiology.

Gething, M.-J., and J. Sambrook. 1992. Protein folding in the cell. *Nature* 355:33–45.

Glick, B., and G. Schatz. 1992. Import of proteins into mitochondria. *Ann. Rev. Genet.* 25:21–44.

Grafstein, D. 1983. Stereochemical origins of the genetic code. *J. Theoret. Biol.* 105:157–74.

Grivell, L. 1983. Mitochondrial DNA. *Sci. Amer.,* March, 78–89.

Grunberg-Manago, M. 1963. Polynucleotide phosphorylase. *Progr. Nucleic Acid Res.* 1:93–133.

Guthrie, C. 1980. II. Folding up a transfer RNA molecule is not simple. *Quart. Rev. Biol.* 55:335–52.

Hershey, J. W. B. 1991. Translational control in mammalian cells. *Ann. Rev. Biochem.* 60:717–55.

Hill, W. E., et al., eds. 1990. *The Ribosome, Structure, Function, & Evolution.* Washington, D.C.: American Society for Microbiology.

Hooper, C. 1990. Origami enzymes: Hints of a new preoccupation for molecular chaperones. *J. NIH Research* 2:63–66.

Knighton, D. R., et al. 1991. Crystal structure of the catalytic subunit of cyclic adenosine monophosphate-dependent protein kinase. *Science* 253:407–14.

Lake, J. A. 1985. Evolving ribosome structure: Domains in archaebacteria, eubacteria, eocytes and eukaryotes. *Ann. Rev. Biochem.* 54:507–30.

Landry, S. J., and L. M. Gierasch. 1991. Recognition of nascent polypeptides for targeting and folding. *Trends in Biochem. Sci.* 16:159–63.

Leinfelder, W., et al. 1988. Gene for a novel tRNA species that accepts L-serine and cotranslationally inserts selenocysteine. *Nature* 331:723–25.

Macejak, D. G., and P. Sarnow. 1991. Internal initiation of translation mediated by the 5′ leader of a cellular mRNA. *Nature* 353:90–94.

Maitra, U., E. Stringer, and A. Chaudhuri. 1982. Initiation factors in protein biosynthesis. *Ann. Rev. Biochem.* 51:869–900.

Martin, J., et al. 1991. Chaperonin-mediated protein folding at the surface of groEL through a 'molten globule'-like intermediate. *Nature* 352:36–42.

Moazed, D., and H. F. Noller. 1989. Interaction of tRNA with 23S rRNA in the ribosomal A, P, and E sites. *Cell* 57:585–97.

Moldave, K. 1985. Eukaryotic protein synthesis. *Ann. Rev. Biochem.* 54:1109–49.

Moras, D. 1992. Structural and functional relationships between aminoacyl-tRNA synthetases. *Trends Biochem. Sci.* 17:159–64.

Nirenberg, M. W., and P. Leder. 1964. RNA code-words and protein synthesis: The effect of trinuclotides upon the binding of tRNA to ribosomes. *Science* 145:1399–1407.

Nirenberg, M. W., and J. H. Matthei. 1961. The dependence of cell-free protein synthesis in *E. coli* upon naturally occurring or synthetic polyribonucleotides. *Proc. Nat. Acad. Sci., USA* 47:1588–1602.

Noller, H. F. 1991. Ribosomal RNA and translation. *Ann. Rev. Biochem.* 60:191–227.

Noller, H. F., et al. 1990. Structure of rRNA and its functional interactions in translation. In *The Ribosome, Structure, Function, & Evolution,* edited by W. E. Hill, et al., 73–92. Washington, D.C.: American Society for Microbiology.

Normanly, J., and J. Abelson. 1989. tRNA identity. *Ann. Rev. Biochem.* 58:1029–49.

Ochoa, S. 1980. The pursuit of a hobby. *Ann. Rev. Biochem.* 49:1–30.

Pfanner, N., and W. Neupert. 1990. The mitochondrial protein import apparatus. *Ann. Rev. Biochem.* 59:331–53.

Poritz, M. A. 1990. An *E. coli* ribonucleoprotein containing 4.5S RNA resembles mammalian signal recognition particle. *Science* 250:1111–17.

Preer, J. R., et al. 1985. Deviation from the universal code shown by the gene for surface protein 51A in *Paramecium. Nature* 314:188–90.

Richards, F. M. 1991. The protein folding problem. *Sci. Amer.,* Jan., 54–63.

Rould, M. A., et al. 1989. Structure of *E. coli* glutaminyl-tRNA synthetase complexed with tRNA^Gln and ATP at 2.8 Å resolution. *Science* 246:1135–42.

Ruff, M., et al. 1991. Class II aminoacyl transfer RNA synthetases: Crystal structure of yeast aspartyl-tRNA synthetase complexed with tRNA^Asp. *Science* 252:1682–89.

Sanger, F. 1988. Sequences, sequences, and sequences. *Ann. Rev. Biochem.* 57:1–28.

Sasavage, N. L., et al. 1982. Nucleotide sequence of bovine prolactin messenger RNA: Evidence for sequence polymorphism. *J. Biol. Chem.* 257:678–81.

Schimmel, P. 1987. Aminoacyl tRNA synthetases: General scheme of structure-function relationships in the polypeptides and recognition of transfer RNAs. *Ann. Rev. Biochem.* 56:125–58.

Schulman, L. H., and J. Abelson. 1988. Recent excitement in understanding transfer RNA identity. *Science* 240:1591–92.

Schulman, L. H., and H. Pelka. 1988. Anticodon switching changes the identity of methionine and valine transfer RNAs. *Science* 242:765–68.

Scott, J., et al. 1983. Structure of a mouse submaxillary messenger RNA encoding epidermal growth factor and seven related proteins. *Science* 221:236–40.

Stern, S., et al. 1989. RNA-protein interactions in 30S ribosomal subunits: Folding and function of 16S rRNA. *Science* 244:783–90.

Stryer, L. 1988. *Biochemistry.* 3d ed. New York: Freeman.

Van Knippenberg, P. H. 1990. Aspects of translation initiation in *Escherichia coli.* In *The Ribosome, Structure, Function, & Evolution,* edited by W. E. Hill, et al., 265–74. Washington, D.C.: American Society for Microbiology.

Verner, K., and G. Schatz. 1988. Protein translocation across membranes. *Science* 241:1307–13.

Watson, J. D. 1963. The involvement of RNA in the synthesis of proteins. *Science* 140:17–26.

Wickner, W. T., and H. F. Lodish. 1985. Multiple mechanisms of protein insertion into and across membranes. *Science* 230:400–7.

Yonath, A., K. R. Leonard, and H. G. Wittmann. 1987. A tunnel in the large ribosomal subunit revealed by three-dimensional image reconstruction. *Science* 236:813–16.

Chapter 12
DNA Cloning and Sequencing

Adams, M. D., et al. 1991. Complementary DNA sequencing: Expressed sequence tags and human genome project. *Science* 252:1651–56.

Adelman, J. P., et al. 1987. Two mammalian genes transcribed from opposite strands of the same DNA locus. *Science* 235:1514–17.

American Association for the Advancement of Science. 1986. Biotechnology. *Science* 232:1305–1476.

———. 1989. The new harvest: Genetically engineered species. *Science* 244:1225–1412.

———. 1990. Frontiers in biology: Plants. *Science* 250:923–66.

Andreason, G. L., and G. A. Evans. 1988. Introduction and expression of DNA molecules in eukaryotic cells by electroporation. *BioTechniques* 6:650–60.

Arnheim, N., and H. Erlich. 1992. Polymerase chain reaction strategy. *Ann. Rev. Biochem.* 61:131–56.

Arnheim, N., H. Li., and X. Cui. 1991. Genetic mapping by single sperm typing. *Animal Genetics* 22:105–15.

Bartlett, R. J., et al. 1987. A new probe for the diagnosis of myotonic muscular dystrophy. *Science* 235:1648–50.

Baskin, Y. 1988. Genetically engineered microbes: The nation is not ready. *Amer. Sci.*, 76, 338–40.

Benfey, P. N., and N.-H. Chua. 1989. Regulated gene in transgenic plants. *Science* 244:174–81.

Booth, W. 1988. Animals of invention. *Science* 240:718.

Botstein, D., R. L. White, M. Skolnick, and R. W. Davis. 1992. Construction of a genetic linkage map in man using restriction fragment length polymorphisms. *J. NIH Res.* 4:66–74.

Camper, S. A. 1987. Research applications of transgenic mice. *BioTechniques* 5:638–50.

Cantor, C. R. 1990. Orchestrating the human genome project. *Science* 248:49–51.

Caruthers, M. H. 1985. Gene synthesis machines. DNA chemistry and its uses. *Science* 230:281–85.

Cherfas, J. 1991. Ancient DNA: Still busy after death. *Science* 253:1354–56.

Courteau, J., et al. 1991. Genome databases. *Science* 254:201–7.

Culliton, B. J. 1990. Gene therapy: Into the home stretch. *Science* 249:974–76.

DePamphilis, M. L., et al. 1988. Microinjecting DNA into mouse ova to study DNA replication and gene expression and to produce transgenic animals. *BioTechniques* 6:662–80.

Dicker, A. P., et al. 1989. Sequence analysis of a human gene responsible for drug resistance: A rapid method for manual and automated direct sequencing of products generated by the polymerase chain reaction. *BioTechniques* 7:830–38.

Dietz, H. C., et al. 1991. Marfan syndrome caused by a recurrent *de novo* missense mutation in the fibrillin gene. *Nature* 352:337–39.

Donis-Keller, H. 1990. *Human Genome Mapping Techniques.* New York: Stockton Press.

Elder, J., R. Spritz, and S. Weissman. 1981. Simian virus 40 as a eukaryotic cloning vehicle. *Ann. Rev. Genet.* 15: 295–340.

Finn, C., et al. 1984. The structural gene for tetanus neurotoxin is on a plasmid. *Science* 224:881–84.

Garza, D., J. W. Ajioka, D. T. Burke, and D. L. Hartl. 1989. Mapping the *Drosophila* genome with yeast artificial chromosomes. *Science* 246:641–46.

Gasser, C. S., and R. T. Fraley. 1992. Transgenic crops. *Sci. Amer.*, June, 62–69.

Gilbert, W. 1981. DNA sequencing and gene structure. *Science* 214:1305–12.

Gusella, J. F. 1986. DNA polymorphism and human disease. *Ann. Rev. Biochem.* 55:831–54.

Guyer, R. L., and D. E. Koshland, Jr. 1989. The molecule of the year. *Science* 246:1543–46.

Hagelberg, E., I. C. Gray, and A. J. Jeffreys. 1991. Identification of the skeletal remains of a murder victim by DNA analysis. *Nature* 352:427–29.

Hamilton, D. P. 1990. Down to the wire for the NF gene. *Science* 249:236–38.

Harvey, W. 1988. Cracking open marine algae's biological treasure chest. *Bio/Technology* 6:488–92.

Horsch, R. B., et al. 1985. A simple and general method for transferring genes into plants. *Science* 227:1229–31.

Jasny, B. R., et al. 1991. Genome maps 1991. *Science* 254:247–62.

Johnson, I. 1983. Human insulin from recombinant DNA technology. *Science* 219:632–37.

Johnston, S. A., et al. 1988. Mitochondrial transformation in yeast by bombardment with microprojectiles. *Science* 240:1538–41.

Johnston-Dow, L., et al. 1987. Optimized method for fluorescent and radio labeled DNA sequencing. *BioTechniques* 5:754–65.

Kilbane, J. J., II, and B. A. Bielaga. 1991. Instantaneous gene transfer from donor to recipient microorganisms *via* electroporation. *BioTechniques* 10:354–65.

Kirby, L. T. 1990. *DNA Fingerprinting: An Introduction.* New York: Stockton Press.

Lasic, D. 1992. Liposomes. *Amer. Sci.* 80:20–31.

Lim, K., and C. -B. Chae. 1989. A simple assay for DNA transfection by incubation of the cells in culture dishes with substrates for beta-galactosidase. *BioTechniques* 7:576–79.

Maddox, J. 1991. The case for the human genome. *Nature* 352:11–14.

Mannino, R. J., and S. Gould-Fogerite. 1988. Liposome mediated gene transfer. *BioTechniques* 6:682–90.

Martin, C., et al. 1991. Improved chemiluminescent DNA sequencing. *BioTechniques* 11:110–13.

Marx, J. L. 1982. Building bigger mice through gene transfer. *Science* 218:1298.

———. 1987. Assessing the risks of microbial release. *Science* 237:1413–17.

———. 1988. Cloning sheep and cattle embryos. *Science* 239:463–64.

———. 1988. DNA fingerprinting takes the witness stand. *Science* 240:1616–18.

———. 1988. Foreign gene transferred into maize. *Science* 240:145–46.

———. 1988. Multiplying genes by leaps and bounds. *Science* 240:1408–10.

———. 1989. The cystic fibrosis gene is found. *Science* 245:923–25.

Molecular advances in genetic disease. [Reports + 8 articles.] 1992. *Science* 256:766–813.

Mullis, K. B. 1990. The unusual origin of the polymerase chain reaction. *Sci. Amer.*, Apr., 56–65.

Nakamura, Y., et al. 1987. Variable number of tandem repeat (VNTR) markers for human gene mapping. *Science* 235:1616–22.

National Research Council. 1988. *Mapping and Sequencing the Human Genome.* Washington, D.C.: National Academy Press.

Netzer, W. 1988. Scorched fields and sizzling crops intensify pace of development of drought-tolerant plants. *Genet. Engineering News* 8:1, 32–33.

Neufeld, P. J., and N. Colman. 1990. When science takes the witness stand. *Sci. Amer.*, May, 46–53.

Normark, S., et al. 1983. Overlapping genes. *Ann. Rev. Genet.* 17:499–525.

Oliver, S. G., et al. 1992. The complete DNA sequence of yeast chromosome III. *Nature* 357:38–46.

Ow, D. W., et al. 1986. Transient and stable expression of the firefly luciferase gene in plant cells and transgenic plants. *Science* 234:856–59.

Palmiter, R. D., and R. L. Brinster. 1986. Germ-line transformation of mice. *Ann. Rev. Genet.* 20:465–99.

Pastan, I., V. Chaudhary, and D. J. Fitzgerald. 1992. Recombinant toxins as novel therapeutic agents. *Ann. Rev. Biochem.* 61:331–54.

Patrusky, B. 1991. Drosophila botanica (the fruit fly of plant biology). *Mosaic* 22:32–43.

Pestka, S. 1983. The purification and manufacture of human interferons. *Sci. Amer.*, Aug., 37–43.

Pool, R. 1989. In search of the plastic potato. *Science* 245:1187–89.

Richa, J., and C. W. Lo. 1989. Introduction of human DNA into mouse eggs by injection of dissected chromosome fragments. *Science* 245:175–77.

Roberts, L. 1989. Genome mapping goal now in reach. *Science* 243:424–25.

———. 1989. Genome project under way, at last. *Science* 243:167–68.

———. 1990. Huntington's gene: So near, yet so far. *Science* 247:624–27.

———. 1991. Fight erupts over DNA fingerprinting. *Science* 254:1721–23.

Roberts, S. S. 1990. Yeast of burden: Yoking the YAC. *J. NIH Res.* 2:77–79.

Ruano, G., and K. K. Kidd. 1991. Coupled amplification and sequencing of genomic DNA. *Proc. Nat. Acad. Sci., USA* 88:2815–19.

Sambrook, J., E. Fritsch, and T. Maniatis. 1989. *Molecular Cloning: A Laboratory Manual.* 2d ed. 3 vols. Cold Spring Harbor, N.Y.: Cold Spring Harbor Laboratory.

Sanger, F. 1988. Sequences, sequences, and sequences. *Ann. Rev. Biochem.* 57:1–28.

Sanger, F., et al. 1978. The nucleotide sequence of bacteriophage ϕX174. *J. Molec. Biol.* 125:225–46.

Schell, J. S. 1987. Transgenic plants as tools to study the molecular organization of plant genes. *Science* 237:1176–83.

Shigekawa, K., and W. J. Dower. 1988. Electroporation of eukaryotes and prokaryotes. A general approach to the introduction of macromolecules into cells. *BioTechniques* 6:742–51.

Struhl, K. 1983. The new yeast genetics. *Nature* 305:391–96.

Sun, M. 1988. Preparing the ground for biotech tests. *Science* 242:503–5.

Watkins, P. C. 1988. Restriction fragment length polymorphism (RFLP): Applications in human chromosome mapping and genetic disease research. *BioTechniques* 6:310–19.

Watson, J. D., M. Gilman, J. Witkowski, and M. Zoller. 1992. *Recombinant DNA.* 2d ed. New York: Freeman.

Weaver, R. F. 1984. Beyond supermouse: Changing life's genetic blueprint. *National Geographic,* Dec., 818–47.

White, R., and J.-M. Lalouel. 1988. Chromosome mapping with DNA markers. *Sci. Amer.*, Feb., 40–48.

Williams, R. S., et al. 1991. Introduction of foreign genes into tissues of living mice by DNA-coated microprojectiles. *Proc. Nat. Acad. Sci., USA* 88:2726–30.

Wilson, G. G., and N. E. Murray. 1991. Restriction and modification systems. *Ann. Rev. Genet.* 25:585–627.

Wong, W. K. R., et al. 1988. Wood hydrolysis by *Cellulomonas fimi* endoglucanase and exoglucanase coexpressed as secreted enzymes in *Saccharomyces cerevisiae. Bio/Technology* 6:713–19.

Wood, K. V., et al. 1989. Complementary DNA coding click beetle luciferases can elicit bioluminescence of different colors. *Science* 244:700–2.

Wu, R., L. Grossman, and K. Moldave, eds. 1989. *Recombinant DNA Methodology.* San Diego: (Harcourt Brace Jovanovich, Publishers) Academic Press.

Yuan, R. 1981. Structure and mechanism of multifunctional restriction endonucleases. *Ann. Rev. Biochem.* 50: 285–315.

Chapter 13
Gene Expression: Control in Prokaryotes and Phages

Bachmair, A., D. Finley, and A. Varshavsky. 1986. In vivo half-life of a protein is a function of its amino-terminal residue. *Science* 234:179–86.

Bass, S., V. Sorrells, and P. Youderian. 1988. Mutant trp repressors with new DNA-binding specificities. *Science* 242:240–45.

Beckwith, J., and D. Zipser. 1970. *The Lactose Operon.* Cold Spring Harbor, N.Y.: Cold Spring Harbor Laboratory.

Condit, R., F. M. Stewart, and B. R. Levin. 1988. The population biology of bacterial transposons: A priori conditions for maintenance as parasitic DNA. *Amer. Nat.* 132:129–47.

Dickson, R., et al. 1975. Genetic regulation: The *lac* control region. *Science* 187:27–35.

Eguchi, Y., T. Itoh, and J. -I. Tomazawa. 1991. Antisense RNA. *Ann. Rev. Biochem.* 60:631–52.

Geiduschek, E. P. 1991. Regulation of expression of the late genes of bacteriophage T4. *Ann. Rev. Genet.* 25:437–60.

Gilbert, W., and B. Müller-Hill. 1966. Isolation of the *Lac* repressor. *Proc. Nat. Acad. Sci., USA* 56:1891–98.

Gold, L. 1988. Posttranscriptional regulatory mechanisms in *Escherichia coli. Ann. Rev. Biochem.* 57:199–233.

Green, P. J., O. Pines, and M. Inouye. 1986. The role of antisense RNA in gene regulation. *Ann. Rev. Biochem.* 55:569–97.

Holzman, D. 1991. A "jumping gene" caught in the act. *Science* 254:1728–29.

Jacob, F. 1988. *The Statue Within. An Autobiography.* New York: Basic Books. (Translated from the French by F. Philip, Alfred P. Sloan Foundation Series.)

Jacob, F., and J. Monod. 1961. Genetic regulatory mechanisms in the synthesis of proteins. *J. Mol. Biol.* 3:318–56.

Johnson, W., C. Moran, Jr., and R. Losick. 1983. Two RNA polymerase sigma factors from *Bacillus subtilis* discriminate between overlapping promoters for a developmentally regulated gene. *Nature* 302:800–4.

Jordan, S. R., and C. O. Pabo. 1988. Structure of the lambda complex at 2.5 Å resolution: Details of the repressor-operator interactions. *Science* 242:893–99.

Kantrowitz, E. R., and W. N. Lipscomb. 1988. *Escherichia coli* aspartate transcarbamylase: The relation between structure and function. *Science* 241:669–74.

Kleckner, N. 1983. Transposon Tn*10*. In *Mobile Genetic Elements*. edited by J. Shapiro. New York: Academic Press.

———. 1990. Regulating Tn*10* and IS*10* transposition. *Genetics* 124:449–54.

Kolter, R., and C. Yanofsky. 1982. Attenuation in amino acid biosynthetic operons. *Ann. Rev. Genet.* 16:113–34.

Konigsberg, W., and G. Godson. 1983. Evidence for use of rare codons in the *dnaG* gene and other regulatory genes of *Escherichia coli*. *Proc. Nat. Acad. Sci., USA* 80: 687–91.

Kroos, L., B. Kunkel, and R. Losick. 1989. Switch protein alters specificity of RNA polymerase containing a compartment-specific sigma factor. *Science* 243:526–29.

Landy, A. 1989. Dynamic, structural, and regulatory aspects of λ site-specific recombination. *Ann. Rev. Biochem.* 58:913–49.

Lilley, D. M. J. 1991. When the CAP fits bent DNA. *Nature* 354:359–60.

Lobell, R. B., and R. F. Schleif. 1990. DNA looping and unlooping by AraC protein. *Science* 250:528–32.

Maresca, B., and S. Lindquist, eds. 1991. *Heat Shock*. Berlin: Springer Verlag.

Marx, J. L. 1985. A crystalline view of protein-DNA binding. *Science* 229:846–48.

Mizuuchi, K., and R. Craigie. 1986. Mechanism of bacteriophage mu transposition. *Ann. Rev. Genet.* 20:385–429.

Murialdo, H. 1991. Bacteriophage lambda DNA maturation and packaging. *Ann. Rev. Biochem.* 60:125–53.

Ohno, S. 1988. Codon preference is but an illusion created by the construction principle of coding sequences. *Proc. Nat. Acad. Sci., USA* 85:4378–82.

Oxender, D., G. Zurawski, and C. Yanofsky. 1979. Attenuation in the *Escherichia coli* tryptophan operon: Role of RNA secondary structure involving the tryptophan codon region. *Proc. Nat. Acad. Sci., USA* 76:5524–28.

Pabo, C., and R. Sauer. 1984. Protein-DNA recognition. *Ann. Rev. Biochem.* 53:293–321.

Ptashne, M. 1992. *A Genetic Switch: Gene Control and Phage λ*. 2d ed. Cambridge, Mass.: Cell Press and Blackwell Scientific Publications.

Ptashne, M., A. Johnson, and C. Pabo. 1982. A genetic switch in a bacterial virus. *Sci. Amer.*, Nov., 128–40.

Reidhaar-Olson, J. F., and R. T. Sauer. 1988. Combinatorial cassette mutagenesis as a probe of the informational content of protein sequences. *Science* 241:53–57.

Reznikoff, W. S., et al. 1985. The regulation of transcription initiation in bacteria. *Ann. Rev. Genet.* 19:355–87.

Rogers, S., R. Wells, and M. Rechsteiner. 1986. Amino acid sequences common to rapidly degraded proteins: The PEST hypothesis. *Science* 234:364–68.

Schultz, S. C., G. C. Shields, and T. A. Steitz. 1991. Crystal structure of a CAP-DNA complex: The DNA is bent by 90°. *Science* 253:1001–7.

Shapiro, J. 1983. *Mobile Genetic Elements*. New York: Academic Press.

Simons, R. W., and N. Kleckner. 1988. Biological regulation by antisense RNA in prokaryotes. *Ann. Rev. Genet.* 22:567–600.

Takeda, Y., et al. 1983. DNA-binding proteins. *Science* 221:1020–26.

Tanaka, K., T. Shiina, and H. Takahashi. 1988. Multiple principal sigma factor homologs in Eubacteria: Identification of the "*rpoD* box." *Science* 242:1040–42.

Weintraub, H. M. 1990. Antisense RNA and DNA. *Sci. Amer.*, Jan., 40–46.

Yanofsky, C. 1984. Comparison of regulatory and structural regions of genes of tryptophan metabolism. *Mol. Biol. Evol.* 1:143–61.

Chapter 14
The Eukaryotic Chromosome

Adolph, K. W. 1989. *Chromosomes: Eukaryotic, Prokaryotic, and Viral*. 2 vols. Boca Raton, Fla.: CRC Press.

Allen, J., et al. 1980. The structure of histone H1 and its location in chromatin. *Nature* 288:675–79.

Arndt-Jovin, D. J., et al. 1985. Fluorescence digital imaging microscopy in cell biology. *Science* 230:247–56.

Bank, A., J. G. Mears, and F. Ramirez. 1980. Disorders of human hemoglobin. *Science* 207:486–93.

Bennett, M. D., A. Gropp, and U. Wolf, eds. 1984. *Chromosomes Today*. Vol. 8. London: George Allen & Unwin.

Berendes, H. D., F. M. A. van Breugel, and T. K. H. Holt. 1965. Experimental puffs in salivary gland chromosomes of *Drosophila hydei*. *Chromosoma* (Berlin) 16:35–46.

Blackburn, E. H. 1991. Structure and function of telomeres. *Nature* 350:569–73.

———. 1992. Telomerases. *Ann. Rev. Biochem.* 61:113–29.

Blackburn, E. H., and J. W. Szostak. 1984. The molecular structure of centromeres and telomeres. *Ann. Rev. Biochem.* 53:163–94.

Bonne-Andrea, C., M. L. Wong, and B. M. Alberts. 1990. *In vitro* replication through nucleosomes without histone displacement. *Nature* 343:719–26.

Brinkley, B. R., A. Tousson, and M. M. Valdivia. 1985. The kinetochore of mammalian chromosomes: Structure and function in normal mitosis and aneuploidy. In *Aneuploidy: Etiology and Mechanisms*, edited by V. L. Dellarco, P. E. Voytek, and A. Hollaender, 243–67. New York: Plenum Press.

Britten, R., and D. Kohne. 1968. Repeated sequences in DNA. *Science* 161:529–40.

Burlingame, R. W., et al. 1985. Crystallographic structure of the octameric histone core of the nucleosome at a resolution of 3.3 Å. *Science* 228:546–53.

Callan, H. G. 1986. *Lampbrush Chromosomes*. Vol. 36, *Molecular Biology, Biochemistry, and Biophysics*. New York: Springer-Verlag.

Clarke, L., and J. Carbon. 1985. The structure and function of yeast centromeres. *Ann. Rev. Genet.* 19:29–56.

Comings, D. 1978. Mechanisms of chromosome banding and implications for chromosome structure. *Ann. Rev. Genet.* 12:25–46.

Edström, J. -E., H. Sierakowska, and K. Burvall. 1982. Dependence of Balbiani ring induction in *Chironomus* salivary glands on inorganic phosphate. *Devel. Biol.* 91:131–37.

Eissenberg, J. C., et al. 1985. Selected topics in chromatin structure. *Ann. Rev. Genet.* 19:485–536.

Gall, J. G. 1990. Tying up the loose ends. *Nature* 344:108–9.

Gall, J. G., E. H. Cohen, and D. D. Atherton. 1974. The satellite DNAs of *Drosophila virilis*. *Cold Spr. Harb. Symp. Quant. Biol.* 38:417–21.

Gray, J. W., et al. 1987. High-speed chromosome sorting. *Science* 238:323–29.

Gross, D. S., and W. T. Garrard. 1988. Nuclease hypersensitive sites in chromatin. *Ann. Rev. Biochem.* 57:159–97.

Gustafson, J. P., and R. Appels, eds. 1988. *Chromosome Structure and Function*. New York: Plenum Press.

Henderson, E., et al. 1987. Telomeric DNA oligonucleotides form novel intramolecular structures containing guanine-guanine base pairs. *Cell* 51:899–908.

Jelinek, W., and C. Schmid. 1982. Repetitive sequences in eukaryotic DNA and their expression. *Ann. Rev. Biochem.* 51:813–44.

Karlsson, S., and A. W. Nienhuis. 1985. Developmental regulation of human globin genes. *Ann. Rev. Biochem.* 54:1071–1108.

Kavenoff, R., L. C. Klotz, and B. H. Zimm. 1974. On the nature of chromosome-sized DNA molecules. *Cold Spr. Harb. Symp. Quant. Biol.* 38:1–8.

Klug, A., et al. 1980. A low resolution structure for the histone core of the nucleosome. *Nature* 287:509–16.

Konkel, D., S. Tilghman, and P. Leder. 1978. The sequence of the chromosomal mouse β-globin major gene: Homologies in capping, splicing and poly(A) sites. *Cell* 15:1125–32.

Kornberg, R., and A. Klug. 1981. The nucleosome. *Sci. Amer.* 244:52–64.

Long, E., and I. Dawid. 1980. Repeated genes in eukaryotes. *Ann. Rev. Biochem.* 49:727–64.

Maniatis, T., et al. 1980. The molecular genetics of human hemoglobins. *Ann. Rev. Genet.* 14:145–78.

Manuelidis, L. 1990. A view of interphase chromosomes. *Science* 250:1533–40.

Maxson, R., R. Cohn, and L. Kedes. 1983. Expression and organization of histone genes. *Ann. Rev. Genet.* 17: 239–77.

Melamed, M. R., T. Lindmo, and M. L. Mendelsohn, eds. 1990. *Flow Cytometry and Sorting*. 2d ed. New York: Wiley-Liss.

Moyzis, R. K. 1991. The human telomere. *Sci. Amer.*, Aug., 48–55.

Moyzis, R. K., et al. 1988. A highly conserved repetitive DNA sequence, (TTAGGG)$_n$, present at the telomeres of human chromosomes. *Proc. Nat. Acad. Sci., USA* 85:6622–26.

Orkin, S. H., and H. H. Kazazian, Jr. 1984. The mutation and polymorphism of the human β-globin gene and its surrounding DNA. *Ann. Rev. Genet.* 18:131–71.

Osheim, Y. N., and O. L. Miller, Jr. 1983. Novel amplification and transcriptional activity of chorion genes in *Drosophila melanogaster* follicle cells. *Cell* 33:543–53.

Paulson, J., and U. Laemmli. 1977. The structure of histone depleted metaphase chromosomes. *Cell* 12:817–28.

Roberts, L. 1988. Chromosomes: The ends in view. *Science* 240:982–83.

Roberts, S. S. 1990. In situ hybridization: Nowhere to hide for nucleotides. *J. NIH Res.* 2:82–86.

Shippen-Lentz, D., and E. H. Blackburn. 1990. Functional evidence for an RNA template in telomerase. *Science* 247:546–52.

Stamatoyannopoulos, G. 1991. Human hemoglobin switching. *Science* 252:383.

Stark, G., and G. Wahl. 1984. Gene amplification. *Ann. Rev. Biochem.* 53:447–91.

Sumner, A. T. 1990. *Chromosome Banding*. Cambridge, Mass.: Unwin Hyman.

Sun, J. -M., R. Wiaderkiewicz, and A. Ruiz-Carrillo. 1989. Histone H5 in the control of DNA synthesis and cell proliferation. *Science* 245:68–71.

Taylor, J. H., P. S. Woods, and W. L. Hughes. 1957. The organization and duplication of chromosomes as revealed by autoradiographic studies using tritium-labeled thymidine. *Proc. Nat. Acad. Sci., USA* 43:122–28.

Touchette, N. 1990. Evolutions: Chromatin structure. *J. NIH Res.* 2:95–96.

Uberbacher, E. C., et al. 1986. Shape analysis of the histone octamer in solution. *Science* 232:1247–49.

Verma, R., ed. 1988. *Heterochromatin: Molecular and Structural Aspects*. Cambridge: Cambridge University Press.

Verma, R., and A. Babu. 1989. *Human Chromosomes: Manual of Basic Techniques*. New York: Pergamon Press.

Vig, B. 1982. Sequence of centromere separation: Role of centromeric heterochromatin. *Genetics* 102:795–806.

Vuorio, E., and B. de Crombrugghe. 1990. The family of collagen genes. *Ann. Rev. Biochem.* 59:837–72.

Wolffe, A. P. 1991. Developmental regulation of chromatin structure and function. *Trends Cell Biol.* 1:61–66.

Yu, G. -L., J. D. Bradley, L. D. Attardi, and E. H. Blackburn. 1990. *In vivo* alteration of telomere sequences and senescence caused by mutated *Tetrahymena* telomerase RNAs. *Nature* 344:126–32.

Chapter 15
Gene Expression: Control in Eukaryotes

Akira, S., K. Okazaki, and H. Sakano. 1987. Two pairs of recombination signals are sufficient to cause immunoglobulin V-(D)-J joining. *Science* 238:1134–38.

Alt, F. W., T. K. Blackwell, and G. D. Yancopoulos. 1987. Development of the primary antibody repertoire. *Science* 238:1079–87.

Ames, B. 1979. Identifying environmental chemicals causing mutations and cancer. *Science* 204:587–93.

Amit, A. G., et al. 1986. Three-dimensional structure of an antigen-antibody complex at 2.8 Å resolution. *Science* 233:747–53.

Atkinson, M. A., and N. K. Maclaren. 1990. What causes diabetes? *Sci. Amer.*, July, 62–71.

Barnes, D. M. 1989. Regulating transcription with a copper fist. *J. NIH Res.* 1:66–67.

Beardsley, T. 1991. Smart genes. *Sci. Amer.*, Aug., 86–95.

Bellvé, A. R., and H. J. Vogel, eds. 1991. *Molecular Mechanisms in Cellular Growth and Differentiation.* San Diego: Academic Press.

Bender, A., and G. F. Sprague, Jr. 1989. Pheromones and pheromone receptors are the primary determinants of mating specificity in the yeast *Saccharomyces cerevisiae. Genetics* 121:463–76.

Benjamini, E., and S. Leskowitz. 1991. *Immunology: A Short Course.* 2d ed. New York: Wiley-Liss.

Berg, D. E., and M. M. Howe. 1989. *Mobile DNA.* Washington, D.C.: American Society for Microbiology.

Bhattacharyya, M. K., et al. 1990. The wrinkled-seed character of pea described by Mendel is caused by a transposon-like insertion in a gene encoding starch-branching enzyme. *Cell* 60:115–22.

Bishop, J. M. 1982. Oncogenes. *Sci. Amer.,* Mar., 81–92.

———. 1987. The molecular genetics of cancer. *Science* 235:305–11.

Bjorkman, P. J., and P. Parham. 1990. Structure, function, and diversity of class I major histocompatibility complex molecules. *Ann. Rev. Biochem.* 59:253–88.

Blackwell, T. K., and F. W. Alt. 1989. Mechanism and developmental program of immunoglobulin gene rearrangement in mammals. *Ann. Rev. Genet.* 23:605–36.

Briggs, R., and T. King. 1952. Transplantation of living nuclei from blastula cells into enucleated frog's eggs. *Proc. Nat. Acad. Sci., USA* 38:455–63.

Brookmeyer, R. 1991. Reconstruction and future trends of the AIDS epidemic in the United States. *Science* 253:37–42.

Brosius, J. 1991. Retroposons—seeds of evolution. *Science* 251:753.

Caplan, S., and J. Kurjan. 1991. Role of α-factor and the *MFα1* α-factor precursor in mating in yeast. *Genetics* 127:299–307.

Coen, E. S., and E. M. Meyerowitz. 1991. The war of the whorls: Genetic interactions controlling flower development. *Nature* 353:31–37.

Cohen, L. A. 1987. Diet and cancer. *Sci. Amer.,* Nov., 42–48.

Croce, C. M., and G. Klein. 1985. Chromosome translocations and human cancer. *Sci. Amer.,* Mar., 54–60.

Davies, D. R., E. A. Padlan, and S. Sheriff. 1990. Antibody-antigen complexes. *Ann. Rev. Biochem.* 59:439–73.

Davis, M. M. 1990. T cell receptor gene diversity and selection. *Ann. Rev. Biochem.* 59:475–96.

De Robertis, E. M., G. Oliver, and C. V. E. Wright. 1990. Homeobox genes and the vertebrate body plan. *Sci. Amer.,* July, 46–52.

Doe, C. Q., et al. 1988. Expression and function of the segmentation gene *fushi tarazu* during *Drosophila* neurogenesis. *Science* 239:170–75.

Donehower, L. A., et al. 1992. Mice deficient for p53 are developmentally normal but susceptible to spontaneous tumours. *Nature* 356:215–21.

Doolittle, R. F., et al. 1990. Origins and evolutionary relationships of retroviruses. *Quart. Rev. Biol.* 64:1–30.

Duncan, I. 1987. The bithorax complex. *Ann. Rev. Genet.* 21:285–319.

Dyson, N., et al. 1989. The human papilloma virus-16 E7 oncoprotein is able to bind to the retinoblastoma gene product. *Science* 243:934–37.

Edelman, G. M. 1989. Topobiology. *Sci. Amer.,* May, 76–88.

Edwards, J. W., and G. M. Coruzzi. 1990. Cell-specific gene expression in plants. *Ann. Rev. Genet.* 24:275–303.

Evans, R. M., and S. M. Hollenberg. 1988. Zinc fingers: Gilt by association. *Cell* 52:1–3.

Federoff, N. 1984. Transposable genetic elements in maize. *Sci. Amer.,* June, 84–98.

Feldman, M., and L. Eisenbach. 1988. What makes a tumor cell metastatic? *Sci. Amer.,* Nov., 60–85.

Ferguson, E. L., and H. R. Horvitz. 1985. Identification and characterization of 22 genes that affect the vulval cell lineages of the nematode *Caenorhabditis elegans. Genetics* 110:17–72.

Franza, B. R., Jr., B. R. Cullen, and F. Wong-Staal, eds. 1988. *The Control of Human Retrovirus Gene Expression.* Cold Spring Harbor, N.Y.: Cold Spring Harbor Laboratory.

French, D. L., R. Laskov, and M. D. Scharff. 1989. The role of somatic hypermutation in the generation of antibody diversity. *Science* 244:1152–57.

Gallo, R. C. 1987. The AIDS virus. *Sci. Amer.,* Jan., 46–56.

Gehring, W. J. 1985. The molecular basis of development. *Sci. Amer.,* Oct., 152–62.

———. 1987. Homeo boxes in the study of development. *Science* 236:1245–52.

Gierl, A., H. Saedler, and P. A. Peterson. 1989. Maize transposable elements. *Ann. Rev. Genet.* 23:71–85.

Gilmore, T. D. 1991. Malignant transformation by mutant Rel proteins. *Trends Genet.* 7:318–22.

Goldberg, R. B. 1988. Plants: Novel developmental processes. *Science* 240:1460–67.

Greenwald, I. S., P. W. Sternberg, and H. R. Horvitz. 1983. The *lin-12* locus specifies cell fates in *Caenorhabditis elegans. Cell* 34:435–44.

Grey, H. M., A. Sette, and S. Buus. 1989. How T cells see antigen. *Sci. Amer.,* Nov., 56–64.

Hake, S. 1992. Unraveling the knots in plant development. *Trends Genet.* 8:109–14.

Haluska, F. G., Y. Tsujimoto, and C. M. Croce. 1987. Oncogene activation by chromosome translocation in human malignancy. *Ann. Rev. Genet.* 21:321–45.

Harrison, S. C. 1991. A structural taxonomy of DNA-binding domains. *Nature* 353:715–19.

Harrison, S. C., and A. K. Aggarwal. 1990. DNA recognition by proteins with the helix-turn-helix motif. *Ann. Rev. Biochem.* 59:933–69.

Hill, A. V. S., et al. 1991. Common West African HLA antigens are associated with protection from severe malaria. *Nature* 352:595–600.

Hodgkin, J. 1990. Sex determination compared in *Drosophila* and *Caenorhabditis. Nature* 344:721–28.

Holliday, R. 1989. A different kind of inheritance. *Sci. Amer.,* June, 60–73.

Hollstein, M., D. Sidransky, B. Vogelstein, and C. C. Harris. 1991. p53 mutations in human cancers. *Science* 253:49–53.

Hood, L., M. Kronenberg, and T. Hunkapiller. 1985. T cell antigen receptors and the immunoglobulin supergene family. *Cell* 40:225–29.

Horvitz, H. R., and J. E. Sulston. 1990. "Joy of the worm." *Genetics* 126:287–92.

Houck, M. A., J. B. Clark, K. R. Peterson, and M. G. Kidwell. 1991. Possible horizontal transfer of *Drosophila* genes by the mite *Proctolaelaps regalis*. *Science* 253:1125–29.

Ingham, P. W. 1988. The molecular genetics of embryonic pattern formation in *Drosophila*. *Nature* 335:25–34.

Johnson, P. F., and S. L. McKnight. 1989. Eukaryotic transcriptional regulatory proteins. *Ann. Rev. Biochem.* 58:799–839.

Journal of NIH Research. 1991. Special section: AIDS. 3:61–95, 133–36.

———. 1991. Special section: Cancer in focus. 3:41–72.

Kennel, S., et al. 1984. Monoclonal antibodies in cancer detection and therapy. *BioScience* 34:150–56.

Kenyon, C. 1988. The nematode *Caenorhabditis elegans*. *Science* 240:1448–53.

Kinzler, K. W., et al. 1991. Identification of a gene located at chromosome 5q21 that is mutated in colorectal cancers. *Science* 251:1366–70.

Knudson, A. G., Jr. 1971. Mutation and cancer: Statistical study of retinoblastoma. *Proc. Nat. Acad. Sci., USA* 68:820–23.

———. 1986. Genetics of human cancer. *Ann. Rev. Genet.* 20:231–51.

Kurjan, J. 1992. Pheromone response in yeast. *Ann. Rev. Biochem.* 61:1097–1129.

Landschulz, W. H., P. F. Johnson, and S. L. McKnight. 1988. The leucine zipper: A hypothetical structure common to a new class of DNA binding proteins. *Science* 240:1759–64.

Leder, P. 1982. The genetics of antibody diversity. *Sci. Amer.,* May, 102–15.

Lewin, B. 1984. The continuing tale of a small worm. *Science* 225:153–56.

Lewis, S., A. Gifford, and D. Baltimore. 1985. DNA elements are asymmetrically joined during site-specific recombination of kappa immunoglobulin genes. *Science* 228:677–85.

Lindquist, S. 1986. The heat-shock response. *Ann. Rev. Biochem.* 55:1151–91.

Maddox, J. 1991. AIDS research turned upside down. *Nature* 353:297.

Mahowald, A. P., ed. 1990. *Genetics of Pattern Formation and Growth Control: The Forty-Eighth Annual Symposium of the Society for Developmental Biology, Berkeley, California, June 18–21, 1989.* New York: Wiley-Liss.

Malacinski, G. M., ed. 1988. *Developmental Genetics of Higher Organisms.* New York: Macmillan.

Malkin, D., et al. 1990. Germ line p53 mutations in a familial syndrome of breast cancer, sarcomas, and other neoplasms. *Science* 250:1233–38.

Marx, J. 1984. What do oncogenes do? *Science* 223:673–76.

———. 1989. How DNA viruses may cause cancer. *Science* 243:1012–13.

Max, E. E., J. G. Seidman, and P. Leder. 1979. Sequences of five potential recombination sites encoded close to an immunoglobulin κ constant region gene. *Proc. Nat. Acad. Sci., USA* 76:3450–54.

McClintock, B. 1984. The significance of responses of the genome to challenge. *Science* 226:792–801.

McKnight, S. L. 1991. Molecular zippers in gene regulation. *Sci. Amer.,* Apr., 54–64.

Melton, D. A. 1991. Pattern formation during animal development. *Science* 252:234–41.

Milstein, C. 1986. From antibody structure to immunological diversification of immune response. *Science* 231:1261–68.

Mitelman, F. 1988. *Catalog of Chromosome Aberrations in Cancer.* 3d ed. New York: Alan R. Liss.

Mlot, C. 1991. A well-rounded worm. *Science* 252:1619–20.

Moffat, A. S. 1991. Making sense of antisense. *Science* 253:510–11.

Nasmyth, K., and D. Shore. 1987. Transcriptional regulation in the yeast life cycle. *Science* 237:1162–70.

Oshima, Y., and I. Takano. 1971. Mating types in *Saccharomyces:* Their convertability and homothallism. *Genetics* 67:327–35.

Pavletich, N. P., and C. O. Pabo. 1991. Zinc finger-DNA recognition: Crystal structure of a Zif268-DNA complex at 2.1 Å. *Science* 252:809–17.

Pines, M., ed. 1992. *From Egg to Adult.* Bethesda: Howard Hughes Medical Institute.

Rubin, G. M. 1988. *Drosophila melanogaster* as an experimental organism. *Science* 240:1453–59.

Schleif, R. 1988. DNA binding by proteins. *Science* 241:1182–87.

Science. 1988. AIDS. 239:573–622 (eight articles).

———. 1990. Frontiers in biotechnology: Tolerance in the immune system. 248:1335–93 (nine articles).

———. 1991. Cancer. 254:1131–77 (seven articles).

Scientific American. 1988. What science knows about AIDS. Oct. (entire issue, 10 articles.)

———. 1990. The body against itself. Dec., 106–15.

St. Clair, M. H., et al. 1991. Resistance to ddI and sensitivity to AZT induced by a mutation in HIV-1 reverse transcriptase. *Science* 253:1557–59.

Stanbridge, E. J. 1990. Human tumor suppressor genes. *Ann. Rev. Genet.* 24:615–57.

Steward, F., M. Mapes, and K. Mears. 1958. Growth and organized development of cultured cells. II. Organization in cultures grown from freely suspended cells. *Amer. J. Bot.* 45:705–8.

Strominger, J. L. 1989. Developmental biology of T-cell receptors. *Science* 244:943–50.

Struhl, K. 1989. Molecular mechanisms of transcriptional regulation in yeast. *Ann. Rev. Biochem.* 58:1051–77.

Sulston, J., and H. Horvitz. 1977. Post-embryonic cell lineages of the nematode, *Caenorhabditis elegans*. *Develop. Biol.* 56:110–56.

Sulston, J., et al. 1983. The embryonic cell lineage of the nematode, *Caenorhabditis elegans*. *Develop. Biol.* 100:64–119.

Taylor, R. 1991. Evolutions: The HIV-1 genome. *J. NIH Res.* 3:118–20.

Tonegawa, S. 1985. The molecules of the immune system. *Sci. Amer.,* Oct., 122–31.

Travers, A. A. 1989. DNA conformation and protein binding. *Ann. Rev. Biochem.* 58:427–52.

Vaishnav, Y. N., and F. Wong-Staal. 1991. The biochemistry of AIDS. *Ann. Rev. Biochem.* 60:577–630.

Varmus, H. 1988. Retroviruses. *Science* 240:1427–35.

Vogelstein, B. 1990. A deadly inheritance. *Nature* 348:681–82.

Vollbrecht, E., B. Veit, N. Sinha, and S. Hake. 1991. The developmental gene *Knotted-1* is a member of a maize homeobox gene family. *Nature* 350:241–43.

Von Boehmer, H., and P. Kisielow. 1991. How the immune system learns about self. *Sci. Amer.,* Oct., 74–81.

Waldman, T. A. 1991. Monoclonal antibodies in diagnosis and therapy. *Science* 252:1657–62.

Weinberg, R. A. 1988. Finding the anti-oncogene. *Sci. Amer.,* Sep., 44–51.

Weissman, B. E., et al. 1987. Introduction of a normal human chromosome 11 into a Wilm's tumor cell line controls its tumorigenic expression. *Science* 236:175–80.

Yunis, J. 1983. The chromosomal basis of human neoplasia. *Science* 221:227–36.

———. 1986. Chromosomal rearrangements, genes, and fragile sites in cancer: Clinical and biologic implications. In *Important Advances in Oncology 1986,* edited by V. T. DeVita, Jr., S. Hellman, and S. A. Rosenberg, 93–128. Philadelphia: Lippincott.

Chapter 16
DNA: Its Mutation, Repair, and Recombination

Alber, T. 1989. Mutational effects on protein stability. *Ann. Rev. Biochem.* 58:765–98.

Albertini, R. J., J. A. Nicklas, J. P. O'Neill, and S. H. Robison. 1990. In vivo somatic mutations in humans: Measurement and analysis. *Ann. Rev. Genet.* 24:305–26.

Ames, B. N., R. Magaw, and L. S. Gold. 1987. Ranking possible carcinogenic hazards. *Science* 236:271–80.

Amrein, M., et al. 1988. Scanning tunneling microscopy of recA-DNA complexes coated with a conducting film. *Science* 240:514–16.

Amundsen, S. K., A. M. Neiman, S. M. Thibodeaux, and G. R. Smith. 1990. Genetic dissection of the biochemical activities of RecBCD enzyme. *Genetics* 126:25–40.

Atkins, J. F., R. B. Weiss, S. Thompson, and R. F. Gesteland. 1991. Towards a genetic dissection of the basis of triplet decoding, and its natural subversion: Programmed reading frame shifts and hops. *Ann. Rev. Genet.* 25:201–28.

Benzer, S. 1961. On the topography of the genetic fine structure. *Proc. Nat. Acad. Sci., USA* 47:403–15.

Chu, G., and E. Chang. 1988. Xeroderma pigmentosum group E cells lack a nuclear factor that binds to damaged DNA. *Science* 242:564–67.

Coverley, D., et al. 1991. Requirement for the replication protein SSB in human DNA excision repair. *Nature* 349:538–41.

Cox, M. M. 1987. Enzymes of general recombination. *Ann. Rev. Biochem.* 56:229–62.

Craig, N. L. 1988. The mechanism of conservative site-specific recombination. *Ann. Rev. Genet.* 22:77–105.

Denniston, C. 1982. Low level radiation and genetic risk estimation in man. *Ann. Rev. Genet.* 16:329–55.

Drake, J. W. 1991. Spontaneous mutation. *Ann. Rev. Genet.* 25:125–46.

Dunderdale, H. J., et al. 1991. Formation and resolution of recombination intermediates by *E. coli* RecA and RuvC proteins. *Nature* 354:506–10.

Echols, H., and M. F. Goodman. 1991. Fidelity mechanisms in DNA replication. *Ann. Rev. Biochem.* 60:477–511.

Egelman, E. H., and X. Yu. 1989. The location of DNA in RecA-DNA helical filaments. *Science* 245:404–7.

Epstein, R. H., et al. 1963. Physiological studies of conditional lethal mutations of bacteriophage T4D. *Cold Spr. Harb. Symp. Quant. Biol.* 28:375–92.

Gallagher, G. L. 1991. Evolutions: DNA repair. *J. NIH Res.* 3:94–96.

Gartler, S., and D. Stadler. 1990. Herschel L. Roman (1914–1989). *Genetics* 126:1–3.

Green, M. M. 1990. The foundations of genetic fine structure: A retrospective from memory. *Genetics* 124:793–96.

Holliday, R. 1974. Molecular aspects of genetic exchange and gene conversion. *Genetics* 78:273–87.

Howard-Flanders, P., S. C. West, and A. Stasiak. 1984. Role of RecA protein spiral filaments in genetic recombination. *Nature* 309:215–20.

Huang, J.-C., J. Svoboda, T. Reardon, and A. Sancar. 1992. Human nucleotide excision nuclease removes thymine dimers from DNA by incising the 22nd phosphodiester bond 5′ and the 6th phosphodiester bond 3′ to the photodimer. *Proc. Nat. Acad. Sci., USA* 89:3664–68.

Kucherlapati, R. S., and G. R. Smith, eds. 1988. *Genetic Recombination.* Washington, D.C.: American Society for Microbiology.

Lahue, R. S., K. G. Au, and P. Modrich. 1989. DNA mismatch correction in a defined system. *Science* 245:160–64.

Lee, C., and M. Levitt. 1991. Accurate prediction of the stability and activity effects of site-directed mutagenesis on a protein core. *Nature* 352:448–51.

Loeb, L. A., and B. D. Preston. 1986. Mutagenesis by apurinic/apyrimidinic sites. *Ann. Rev. Genet.* 20:201–30.

Loechler, E. L., C. L. Green, and J. M. Essigmann. 1984. *In vivo* mutagenesis by O^6-methylguanine built into a unique site in a viral genome. *Proc. Nat. Acad. Sci., USA* 81:6271–75.

Luria, S., and M. Delbrück. 1943. Mutations of bacteria from virus sensitivity to virus resistance. *Genetics* 28:491–511.

Marx, J. 1990. Animal carcinogen testing challenged. *Science* 250:743–45.

Modrich, P. 1987. DNA mismatch correction. *Ann. Rev. Biochem.* 56:435–66.

———. 1991. Mechanisms and biological effects of mismatch repair. *Ann. Rev. Genet.* 25:229–53.

Mossman, K. L. 1992. Low-dose ionizing radiation: The question of cancer. *J. NIH Res.* 4:51–53.

Muller, H. J. 1927. Artificial transmutation of the gene. *Science* 66:84–87.

Neel, J. V., and S. E. Lewis. 1990. The comparative radiation genetics of humans and mice. *Ann. Rev. Genet.* 24:327–62.

Pearlman, D. A., et al. 1985. Molecular models for DNA damaged by photoreaction. *Science* 227:1304–08.

Potter, H., and D. Dressler. 1976. On the mechanism of genetic recombination: Electron microscopic observation of recombination intermediates. *Proc. Nat. Acad. Sci., USA* 73:3000–04.

Radman, M., and R. Wagner. 1988. The high fidelity of DNA duplication. *Sci. Amer.,* Aug., 40–46.

Ripley, L. S. 1990. Frameshift mutation: Determinants of specificity. *Ann. Rev. Genet.* 24:189–213.

Roman, H., and M. M. Ruzinski. 1990. Mechanisms of gene conversion in *Saccharomyces cerevisiae. Genetics* 124:7–25.

Sancar, A., and G. B. Sancar. 1988. DNA repair enzymes. *Ann. Rev. Biochem.* 57:29–67.

Siciliano, M. J. 1987. Chromosomal assignment of human genes coding for DNA repair functions. In *Isozymes: Current Topics in Biological and Medical Research.* Vol. 15, *Genetics, Development, and Evolution,* edited by M. C. Rattazzi, J. G. Scandalios, and G. S. Whitt, 217–23. New York: Alan R. Liss.

Singer, B., and J. T. Kusmierak. 1982. Chemical mutagenesis. *Ann. Rev. Biochem.* 52:655–93.

Slilaty, S. N., and J. W. Little. 1987. Lysine-156 and serine-119 are required for LexA repressor cleavage: A possible mechanism. *Proc. Nat. Acad. Sci., USA* 84:3987–91.

Smith, G. R. 1987. Mechanism and control of homologous recombination in *Escherichia coli. Ann. Rev. Genet.* 21:179–201.

Smith, M. 1985. In vitro mutagenesis. *Ann. Rev. Genet.* 19:423–62.

Stadler, L. J. 1928. Mutations in barley induced by X rays and radium. *Science* 68:186–87.

Stahl, F. W. 1987. Genetic recombination. *Sci. Amer.,* Feb., 91–101.

Story, R. M., I. T. Weber, and T. A. Steitz. 1992. The structure of *E. coli recA* protein monomer and polymer. *Nature* 355:318–25 [Erratum, p. 567].

Streisinger, G., and J. Owen. 1985. Mechanisms of spontaneous and induced frameshift mutation in bacteriophage T4. *Genetics* 109:633–59.

Tanaka, K., et al. 1990. Analysis of a human DNA excision repair gene involved in a group A xeroderma pigmentosum and containing a zinc-finger domain. *Nature* 348:73–76.

Taylor, A., and G. R. Smith. 1980. Unwinding and rewinding of DNA by the RecBC enzyme. *Cell* 22:447–57.

———. 1992. RecBCD enzyme is altered upon cutting DNA at a Chi recombination hotspot. *Proc. Nat. Acad. Sci., USA* 89:5226–30.

Tennant, R. C., et al. 1987. Prediction of chemical carcinogenicity in rodents from in vitro genetic toxicity assays. *Science* 236:933–41.

Watson, J. D. 1991. Salvador E. Luria (1912–1991). *Nature* 350:113.

West, S. C. 1992. Enzymes and molecular mechanisms of genetic recombination. *Ann. Rev. Biochem.* 61:603–40.

Yanofsky, C., et al. 1967. The complete amino-acid sequence of the tryptophan synthetase *A* protein (α subunit) and its colinear relationship with the genetic map of the *A* gene. *Proc. Nat. Acad. Sci., USA* 57:296–98.

Chapter 17
Non-Mendelian Inheritance

Beeman, R. W., K. S. Friesen, and R. E. Denell. 1992. Maternal-effect selfish genes in flour beetles. *Science* 256:89–92.

Birky, C., Jr. 1983. Relaxed cellular controls and organelle heredity. *Science* 222:468–75.

Blackburn, E. H., and K. M. Karrer. 1986. Genomic reorganization in ciliated protozoans. *Ann. Rev. Genet.* 20:501–21.

Broda, P. 1979. *Plasmids.* San Francisco: Freeman.

Cann, R. L., M. Stoneking, and A. C. Wilson. 1987. Mitochondrial DNA and human evolution. *Nature* 325:31–36.

Chang, D. D., and D. A. Clayton. 1987. A mammalian mitochondrial RNA processing activity contains nucleus-encoded RNA. *Science* 235:1178–84.

Chomyn, A., et al. 1986. URF6, last unidentified reading frame of human mtDNA, codes for an NADH dehydrogenase subunit. *Science* 234:614–18.

Christian, J., and C. Lemunyan. 1958. Adverse effects of crowding on lactation and reproduction of mice and two generations of their progeny. *Endocrinology* 63:517–29.

Ciferri, O., and L. Dure, III, eds. 1983. *Structure and Function of Plant Genomes.* New York: Plenum.

Costanzo, M. C., and T. D. Fox. 1990. Control of mitochondrial gene expression in *Saccharomyces cerevisiae. Ann. Rev. Genet.* 24:91–113.

Dilts, J. A., and R. L. Quackenbush. 1986. A mutation in the R body-coding sequence destroys expression of the killer trait in *P. tetraurelia. Science* 232:641–43.

Evans, R. J., K. M. Oakley, and G. D. Clark-Walker. 1985. Elevated levels of petite formation in strains of *Saccharomyces cerevisiae* restored to respiratory competence. I. Association of both high and moderate frequencies of petite mutant formation with the presence of aberrant mitochondrial DNA. *Genetics* 111:389–402.

Felsenstein, J. 1991. Allan Charles Wilson (1934–1991). *Nature* 353:19.

Gallagher, G. L. 1989. Evolutions: Mitochondria. *J. NIH Res.* 1:135–36.

Garesse, R. 1988. *Drosophila melanogaster* mitochondrial DNA: Gene organization and evolutionary considerations. *Genetics* 118:649–63.

Grivell, L. 1983. Mitochondrial DNA. *Sci. Amer.,* Mar., 78–89.

Grun, P. 1976. *Cytoplasmic Genetics and Evolution.* New York: Columbia University Press.

Gyllensten, U., D. Wharton, A. Josefsson, and A. C. Wilson. 1991. Paternal inheritance of mitochondrial DNA in mice. *Nature* 352:255–57.

Hoeh, W. R., K. H. Blakley, and W. M. Brown. 1991. Heteroplasmy suggests limited biparental inheritance of *Mytilus* mitochondrial DNA. *Science* 251:1488–90.

Hoffman, M. 1991. How parents make their mark on genes. *Science* 252:1250–51.

Hurt, E. C., and G. Schatz. 1987. A cytosolic protein contains a cryptic mitochondrial targeting signal. *Nature* 325:499–503.

Ippen-Ihler, K. A., and E. G. Minkley, Jr. 1986. The conjugation system of F, the fertility factor of *Escherichia coli. Ann. Rev. Genet.* 20:593–624.

Landis, W. 1987. Factors determining the frequency of the killer trait within populations of the *Paramecium aurelia* complex. *Genetics* 115:197–205.

Little, M., V. Van Heyningen, and N. Hastie. 1991. Dads and disomy and disease. *Nature* 351:609–10.

Margulis, L. 1981. *Symbiosis in Cell Evolution.* San Francisco: Freeman.

McKusick, V. 1992. *Mendelian Inheritance in Man.* 10th ed. Baltimore: Johns Hopkins University Press.

Neale, D. B., K. A. Marshall, and R. R. Sederoff. 1989. Chloroplast and mitochondrial DNA are paternally inherited in *Sequoia sempervirens* D. Don Endl. *Proc. Nat. Acad. Sci., USA* 86:9347–49.

Palca, J. 1990. The other human genome. *Science* 249:1104–5.

Palmer, J. D. 1985. Comparative organization of chloroplast genomes. *Ann. Rev. Genet.* 19:325–54.

Palmer, J. D., R. A. Jorgensen, and W. F. Thompson. 1985. Chloroplast DNA variation and evolution in *Pisum:* Patterns of change and phylogenetic analysis. *Genetics* 109:195–213.

Perrimon, N., L. Engstrom, and A. P. Mahowald. 1989. Zygotic lethals with specific maternal effect phenotypes in *Drosophila melanogaster.* I. Loci on the X chromosome. *Genetics* 121:333–52.

Pfanner, N., and W. Neupert. 1990. The mitochondrial protein import apparatus. *Ann. Rev. Biochem.* 59:331–53.

Preer, J., L. Preer, and A. Jurand. 1974. *Kappa* and other endosymbionts in *Paramecium aurelia. Bact. Rev.* 38:113–63.

Rapacz, J., et al. 1991. Identification of the ancestral haplotype for apolipoprotein B suggests an African origin of *Homo sapiens sapiens* and traces their subsequent migration to Europe and the Pacific. *Proc. Nat. Acad. Sci., USA* 88:1403–6.

Rhoades, M. 1946. Plastid mutations. *Cold Spr. Harb. Symp. Quant. Biol.* 11:202–7.

Rodgers, J. 1991. Mechanisms Mendel never knew. *Mosaic* 22:2–11. (Entire issue on nontraditional inheritance.)

Sager, R. 1972. *Cytoplasmic Genes and Organelles.* New York: Academic Press.

Sapienza, C. 1990. Parental imprinting of genes. *Sci. Amer.,* Oct., 52–60.

Schon, E. A., et al. 1989. A direct repeat is a hotspot for large-scale deletion of human mitochondrial DNA. *Science* 244:346–49.

Sonneborn, T. 1950. Methods in the general biology and genetics of *Paramecium aurelia. J. Exp. Zool.* 113:87–148.

Stephanou, G., and S. Alahiotis. 1983. Non-Mendelian inheritance of "heat-sensitivity" in *Drosophila melanogaster. Genetics* 103:93–107.

Stringer, C. B. 1990. The emergence of modern humans. *Sci. Amer.,* Dec., 98–104.

Stuttard, C., and K. Rozee, eds. 1980. *Plasmids and Transposons.* New York: Academic Press.

Tzagoloff, A., and A. M. Myers. 1986. Genetics of mitochondrial biogenesis. *Ann. Rev. Biochem.* 55:249–85.

Walbot, V., and E. Coe, Jr. 1979. Nuclear gene *iojap* conditions a programmed change to ribosome-less plastids in *Zea mays. Proc. Nat. Acad. Sci., USA* 76:2760–64.

Wallace, D. C. 1992. Diseases of the mitochondrial DNA. *Ann. Rev. Biochem.* 61:1175–1212.

Wallace, D. C., et al. 1988. Mitochondrial DNA mutation associated with Leber's hereditary optic neuropathy. *Science* 242:1427–30.

Wickner, R. B., 1986. Double-stranded RNA replication in yeast: The killer system. *Ann. Rev. Biochem.* 55:373–95.

Williamson, D., and D. Poulson. 1979. Sex ratio organisms (spiroplasmas) of *Drosophila. The Mycoplasmas* 3:175–208.

Yang, D., et al. 1985. Mitochondrial origins. *Proc. Nat. Acad. Sci., USA* 82:4443–47.

Chapter 18
Quantitative Inheritance

Allen, G. 1965. Twin research: Problems and prospects. *Prog. Med. Genet.* 4:242–69.

Barton, N. H., and M. Turelli. 1989. Evolutionary quantitative genetics: How little do we know? *Ann. Rev. Genet.* 23:337–70.

Bouchard, T. J., Jr., and M. McGue. 1981. Familial studies of intelligence: A review. *Science* 212:1055–59.

Bouchard, T. J., et al. 1990. Sources of human psychological differences: The Minnesota study of twins reared apart. *Science* 250:223–28.

Chapman, A. B., ed. 1985. *General and Quantitative Genetics.* Amsterdam: Elsevier Science Publishers.

Crow, J. 1957. Genetics of insect resistance to chemicals. *Ann. Rev. Entomol.* 2:227–46.

Cunningham, P. 1991. The genetics of thoroughbred horses. *Sci. Amer.,* May, 92–98.

Falconer, D. S. 1989. *Introduction to Quantitative Genetics.* 3d ed. London: Longman.

Feldman, M., and R. Lewontin. 1975. The heritability hang-up. *Science* 190:1163–68.

Fletcher, R. 1991. *Science, Ideology, and the Media: The Cyril Burt Scandal.* Transaction Publishers. [Reviewed by L. Loevinger. 1991. *Nature* 352:120.]

Harrison, G. A., and J. J. T. Owen. 1964. Studies on the inheritance of human skin color. *Ann. Human Genet.* 28:27–37.

Hay, D. A. 1985. *Essentials of Behaviour Genetics.* Melbourne: Blackwell Scientific Publications.

Holden, C. 1987. The genetics of personality. *Science* 237:598–601.

Holt, S. B. 1961. Inheritance of dermal ridge patterns. In *Recent Advances in Human Genetics,* edited by L. S. Penrose, 101–19. London: J. and A. Churchill.

Hunt, E. 1983. On the nature of intelligence. *Science* 219:141–46.

Kagan, J., J. Reznick, and N. Snidman. 1988. Biological bases of childhood shyness. *Science* 240:167–71.

Kolata, G. 1986. Manic-depression: Is it inherited? *Science* 232:575–76.

Paterson, A. H., et al. 1991. Mendelian factors underlying quantitative traits in tomato: Comparison across species, generations, and environments. *Genetics* 127:181–97.

Plomin, R. 1990. The role of inheritance in behavior. *Science* 248:183–88.

Prochazka, M., et al. 1987. Three recessive loci required for insulin-dependent diabetes in nonobese diabetic mice. *Science* 237:286–89.

Rose, R. J., et al. 1990. Social contact and sibling similarity: Facts, issues, and red herrings. *Behav. Genetics* 20:763–78.

Stern, C. 1970. Model estimates of the number of gene pairs involved in pigmentation variability of the Negro-American. *Human Hered.* 20:165–68.

Weber, K. E. 1990. Selection on wing allometry in *Drosophila melanogaster*. *Genetics* 126:975–89.

Weir, B. S. 1987. Quantitative genetics in 1987. *Genetics* 117:601–2.

Woolf, C. M. 1990. Multifactorial inheritance of common white markings in the Arabian horse. *J. Heredity* 81:250–56.

Chapter 19
Population Genetics: The Hardy-Weinberg Equilibrium and Mating Systems

Begley, S. 1991. A question of breeding. *Natl. Wildlife,* Feb.–March, 12–16.

Bittles, A. H., W. M. Mason, J. Greene, and N. A. Rao. 1991. Reproductive behavior and health in consanguineous marriages. *Science* 252:789–94.

Buss, D. M. 1985. Human mate selection. *Amer. Sci.* 73:47–51.

Cavalli-Sforza, L. L., and W. F. Bodmer. 1971. *The Genetics of Human Populations.* San Francisco: Freeman.

Crow, J. F. 1986. *Basic Concepts in Population, Quantitative, and Evolutionary Genetics.* New York: W. H. Freeman.

———. 1987. Population genetics history: A personal view. *Ann. Rev. Genet.* 21:1–22.

———. 1988. Eighty years ago: The beginnings of population genetics. *Genetics* 119:473–76.

———. 1988. Sewall Wright (1889–1988). *Genetics* 119:1–4.

Crow, J. F., and M. Kimura. 1970. *An Introduction to Population Genetics Theory.* New York: Harper & Row.

De Finetti, B. 1926. Considerazioni matematiche sul l'ereditarieta mendeliana. *Metron* 6:1–41.

Doolittle, D. P. 1987. *Population Genetics: Basic Principles.* Berlin: Springer-Verlag.

Fisher, R. A. 1930. *The Genetical Theory of Natural Selection.* Oxford: Clarendon Press.

Haldane, J. B. S. 1932. *The Causes of Evolution.* New York: Harper & Row.

Hardy, G. H. 1908. Mendelian proportions in a mixed population. *Science* 28:49–50.

Kimura, M. 1988. Thirty years of population genetics with Dr. Crow. *Jap. J. Genet.* 63:1–10.

Maynard Smith, J. 1989. *Evolutionary Genetics.* Oxford: Oxford University Press.

Mettler, L. E., T. G. Gregg, and H. E. Schaffer. 1988. *Population Genetics and Evolution.* 2d ed. Englewood Cliffs, N.J.: Prentice-Hall.

Ralls, K., J. D. Ballou, and A. Templeton. 1988. Estimates of lethal equivalents and the cost of inbreeding in mammals. *Conservation Biol.* 2:185–93.

Russell, E. S. 1989. Sewall Wright's contributions to physiological genetics and to inbreeding theory and practice. *Ann. Rev. Genet.* 23:1–18.

Wallace, B. 1981. *Basic Population Genetics.* New York: Columbia University Press.

Weinberg, W. 1908. Über den Nachweis der Vererbung beim Menschen. *Jahresh. Ver. Vater. Naturk. Wuerttemb.* 64:368–82.

Wright, S. 1931. Evolution in Mendelian populations. *Genetics* 16:97–159.

———. 1968–78. *Evolution and the Genetics of Populations: A Treatise.* 4 vols. Chicago: University of Chicago Press.

Chapter 20
Population Genetics: Processes that Change Allelic Frequencies

Avigad, S., et al. 1990. A single origin of phenylketonuria in Yemenite Jews. *Nature* 344:168–70.

Baskin, Y. 1991. Rhino biology: Keeping tabs on an endangered species. *Science* 252:1256–57.

Cavalli-Sforza, L., and W. Bodmer. 1971. *The Genetics of Human Populations.* San Francisco: Freeman.

Cebra-Thomas, J. A., et al. 1991. Allele- and haploid-specific product generated by alternative splicing from a mouse *t complex responder* locus candidate. *Nature* 349:239–41.

Crow, J. F. 1986. *Basic Concepts in Population, Quantitative, and Evolutionary Genetics.* New York: Freeman.

———. 1987. Population genetics history: A personal view. *Ann. Rev. Genet.* 21:1–22.

———. 1990. Fisher's contributions to genetics and evolution. *Theoret. Pop. Biol.* 38:263–75.

——— 1992. Centennial: J. B. S. Haldane, 1892–1964. *Genetics* 130:1–6.

———. 1992. Erwin Schrödinger and the hornless cattle problem. *Genetics* 130:237–39.

Crow, J., and M. Kimura. 1970. *An Introduction to Population Genetics Theory.* New York: Harper & Row.

Diamond, J. 1988. Founding fathers and mothers. *Nat. Hist.,* June, 10–15.

Dobzhansky, T. 1958. Genetics of natural populations. XXVII. The genetic changes in populations of *Drosophila pseudoobscura* in the American Southwest. *Evolution* 12:385–401.

Doolittle, D. P. 1987. *Population Genetics: Basic Principles.* Berlin: Springer-Verlag.

Edwards, A. W. F. 1990. Fisher, $\overline{W}$, and the fundamental theorem. *Theoret. Pop. Biol.* 38:276–84.

Fivush, B., R. Parker, and R. Tamarin. 1975. Karyotype of the beach vole, *Microtus breweri,* an endemic island species. *J. Mammal.* 56:272–73.

Graham, J. W., J. W. Welch, and K. I. McPhee. 1983. *Waterloo BASIC.* Waterloo, Ontario: WATCOM Publications.

Hill, G. E. 1991. Plumage coloration is a sexually selected indicator of male quality. *Nature* 350:337–39.

Holland, J. H. 1992. Genetic algorithms. *Sci. Amer.,* July, 66–72.

Kimura, M. 1955. Solution of a process of random genetic drift with a continuous model. *Proc. Nat. Acad. Sci., USA* 41:144–50.

———. 1988. Thirty years of population genetics with Dr. Crow. *Jap. J. Genet.* 63:1–10.

Koenig, W. D. 1989. Sex-biased dispersal in the contemporary United States. *Ethology and Sociobiology* 10:263–78.

Lewontin, R. C. 1985. Population genetics. *Ann. Rev. Genet.* 19:81–102.

Lyttle, T. W. 1991. Segregation distorters. *Ann. Rev. Genet.* 25:511–57.

Mallett, J., et al. 1990. Estimates of selection and gene flow from measures of cline width and linkage disequilibrium in Heliconius hybrid zones. *Genetics* 124:921–36.

Maynard Smith, J. 1991. The population genetics of bacteria. *Proc. Royal Soc. London, series B* 245:37–41.

McCauley, D. E. 1991. Genetic consequences of local population extinction and recolonization. *TREE* 6:5–8.

Mettler, L., T. Gregg, and H. E. Schaffer. 1988. *Population Genetics and Evolution.* 2d ed. Englewood Cliffs, N.J.: Prentice-Hall.

O'Brien, S. J., D. E. Wildt, and M. Bush. 1986. The cheetah in genetic peril. *Sci. Amer.,* May, 84–92.

Powers, D. A., T. Lauerman, D. Crawford, and L. DiMichele. 1991. Genetic mechanisms for adapting to a changing environment. *Ann. Rev. Genet.* 25:629–59.

Pyke, G. H. 1991. What does it cost a plant to produce floral nectar? *Nature* 350:58–59.

Rood, S. B., et al. 1988. Gibberellins: A phytohormonal basis for heterosis in maize. *Science* 241:1216–18.

Thoday, J. M., and J. B. Gibson. 1962. Isolation by disruptive selection. *Nature* 193:1164–66.

Wallace, B. 1981. *Basic Population Genetics.* New York: Columbia University Press.

———. 1991. *Fifty Years of Genetic Load. An Odyssey.* Ithaca, N.Y.: Cornell University Press.

Weber, K. E. 1992. How small are the smallest selectable domains of form? *Genetics* 130:345–53.

Chapter 21
Genetics of the Evolutionary Process

Barlow, G. W. 1991. Nature-nurture and the debates surrounding ethology and sociobiology. *Amer. Zool.* 31:286–96.

Bethell, T. 1976. Darwin's mistake. *Harper's,* Feb., 70–75.

Bowcock, A. M., et al. 1991. Drift, admixture, and selection in human evolution: A study with DNA polymorphisms. *Proc. Nat. Acad. Sci., USA* 88:839–43.

Brakefield, P. M. 1987. Industrial melanism: Do we have all the answers? *TREE* 2:117–22.

Breeuwer, J. A. J., and J. H. Werren. 1990. Microorganisms associated with chromosome destruction and reproductive isolation between two insect species. *Nature* 346:558–60.

Bush, G. L. 1975. Modes of animal speciation. *Ann. Rev. Ecol. Syst.* 6:339–64.

Carson, H. L. 1987. The genetic system, the deme, and the origin of species. *Ann. Rev. Genet.* 21:405–23.

Clark, B. 1975. The causes of biological diversity. *Sci. Amer.,* Aug., 50–60.

Clutton-Brock, T. H., and P. H. Harvey, eds. 1978. *Readings in Sociobiology.* San Francisco: Freeman.

Cohn, J. P. 1992. Naked mole-rats. *BioScience* 42:86–89.

Conover, D. O., and D. A. Van Voorhees. 1990. Evolution of a balanced sex ratio by frequency-dependent selection in a fish. *Science* 250:1556–58.

Coyne, J. A. 1992. Genetics and speciation. *Nature* 355:511–15.

Crow, J. F. 1986. *Basic Concepts in Population, Quantitative, and Evolutionary Genetics.* New York: Freeman.

———. 1987. Population genetics history: A personal view. *Ann. Rev. Genet.* 21:1–22.

———. 1992. Twenty-five years ago in GENETICS: Identical triples. *Genetics* 130:395–98.

Darwin, C. 1845. *The Voyage of the Beagle.* 1962 ed. Garden City, N.Y.: Doubleday.

———. 1859. *The Origin of Species by Means of Natural Selection or the Preservation of Favored Races in the Struggle for Life.* London: John Murray.

Drickamer, L. C., and S. H. Vessey. 1991. *Animal Behavior: Mechanism, Ecology and Evolution.* 3rd rev. Dubuque, Iowa: Wm. C. Brown.

Eldredge, N., and S. J. Gould. 1972. Punctuated equilibria: An alternative to phyletic gradualism. In *Models in Paleobiology,* edited by T. Schopf, 85–115. San Francisco: Freeman.

Fisher, A. 1991. Sociobiology, a special report: A new synthesis comes of age. *Mosaic* 22:2–17.

Ford, E. B. 1975. *Ecological Genetics.* 4th ed. London: Chapman and Hall.

Futuyma, D. J. 1986. *Evolutionary Biology.* 2d ed. Sunderland, Mass.: Sinauer Associates.

Gillespie, J. H. 1988. More on the overdispersed molecular clock. *Genetics* 118:385–86.

———. 1991. *The Causes of Molecular Evolution.* Oxford: Oxford University Press.

Gojobori, T., E. N. Moriyama, and M. Kimura. 1990. Molecular clock of viral evolution, and the neutral theory. *Proc. Nat. Acad. Sci., USA* 87:10015–18.

Grant, P. R. 1986. *Ecology and Evolution of Darwin's Finches.* Princeton: Princeton University Press.

———. 1991. Natural selection and Darwin's finches. *Sci. Amer.,* Oct., 82–87.

Hamilton, W. D. 1964. The genetical evolution of social behavior, I, II. *J. Theoret. Biol.* 7:1–52.

Honeycutt, R. L. 1992. Naked mole-rats. *Amer. Sci.* 80:43–53.

Hutter, P., J. Roote, and M. Ashburner. 1990. A genetic basis for the inviability of hybrids between sibling species of *Drosophila. Genetics* 124:909–20.

Kaneshiro, K. Y. 1988. Speciation in the Hawaiian *Drosophila. BioScience* 38:258–63.

Kimura, M. 1985. *The Neutral Theory of Molecular Evolution.* New York: Cambridge University Press.

———. 1987. Molecular evolutionary clock and the neutral theory. *J. Mol. Evol.* 26:24–33.

Koehn, R. 1970. Functional and evolutionary dynamics of polymorphic esterases in catostomid fishes. *Amer. Fish. Soc., Trans.* 99:219–28.

Kuhn, T. S. 1970. *The Structure of Scientific Revolutions.* 2d ed. Chicago: University of Chicago Press.

Landman, O. E. 1991. The inheritance of acquired characteristics. *Ann. Rev. Genet.* 25:1–20.

Lewin, R. 1988. Molecular clocks turn a quarter century. *Science* 239:561–63.

Lewontin, R. 1985. Population genetics. *Ann. Rev. Genet.* 19:81–102.

———. 1991. Twenty-five years ago in GENETICS: Electrophoresis in the development of evolutionary genetics: Milestone or millstone? *Genetics* 128:657–62.

Lewontin, R., and J. Hubby. 1966. A molecular approach to the study of genic heterozygosity in natural populations. II. *Genetics* 54:595–609.

Li, W., and D. Graur. 1990. *Fundamentals of Molecular Evolution.* Sunderland, Mass.: Sinauer Associates.

Maynard Smith, J. 1989. *Evolutionary Genetics.* New York: Oxford University Press.

Mettler, L. E., T. G. Gregg, and H. E. Schaffer. 1988. *Population Genetics and Evolution.* 2d ed. Englewood Cliffs, N.J.: Prentice-Hall.

Niemalä, P., and J. Tuomi. 1987. Does the leaf morphology of some plants mimic caterpillar damage? *Oikos* 50:256–57.

Otte, D., and J. A. Endler. 1989. *Speciation and Its Consequences.* Sunderland, Mass.: Sinauer Associates.

Packer, C., D. A. Gilbert, A. E. Pusey, and S. J. O'Brien. 1991. A molecular genetic analysis of kinship and cooperation in African lions. *Nature* 351:562–65.

Queller, D. C., J. E. Strassmann, and C. R. Hughes. 1988. Genetic relatedness in colonies of tropical wasps with multiple queens. *Science* 242:1155–57.

Rinderer, T. E., et al. 1991. Hybridization between European and Africanized honey bees in the Neotropical Yucatan peninsula. *Science* 253:309–11.

Ritland, D. B., and L. P. Brower. 1991. The viceroy butterfly is not a batesian mimic. *Nature* 350:497–98.

Sarkar, S. 1992. Sex, disease, and evolution—variations on a theme from J. B. S. Haldane. *BioScience* 42:448–54.

Scherer, S. 1990. The protein molecular clock: Time for a reevaluation. Edited by M. K. Hecht, B. Wallace, and R. J. MacIntyre. *Evolutionary Biology* 24:83–106.

Selander, R. K., A. G. Clark, and T. S. Whittam, eds. 1991. *Evolution at the Molecular Level.* Sunderland, Mass.: Sinauer Associates.

Stebbins, G. 1977. *Processes of Organic Evolutionary Biology.* 3d ed. Englewood Cliffs, N.J.: Prentice-Hall.

Strickberger, M. W. 1990. *Evolution.* Boston: Jones and Bartlett.

Tamarin, R., and M. Sheridan. 1987. Behavior-genetic mechanisms of population regulation in microtine rodents. *Amer. Zool.* 27:921–27.

Templeton, A. R. 1981. Mechanisms of speciation—a population genetic approach. *Ann. Rev. Ecol. Syst.* 12:23–48.

Wilson, A. C. 1985. The molecular basis of evolution. *Sci. Amer.,* Oct., 164–73.

Wilson, E. O. 1975. *Sociobiology: The New Synthesis.* Cambridge, Mass.: Harvard University Press.

Wood, B. 1992. Origin and evolution of the genus *Homo. Nature* 355:783–90.

Young, W. 1985. *Fallacies of Creationism.* Calgary, Alberta: Detselig Enterprises, Ltd.

Zuckerkandl, E., and L. Pauling. 1965. Evolutionary divergence and convergence in proteins. In *Evolving Genes and Proteins,* edited by V. Bryson and H. Vogel, 97–166. New York: Academic Press.

Glossary

A

A (aminoacyl) site The site on the ribosome occupied by an aminoacyl-tRNA just prior to peptide bond formation.

acentric fragment A chromosomal piece without a centromere.

acrocentric chromosome A chromosome whose centromere lies very near one end.

activation energy (ΔG) Energy needed to initiate a chemical reaction.

active site The part of an enzyme where the actual enzymatic function is performed.

adaptive value *See* fitness.

additive model A mechanism of quantitative inheritance in which alleles at different loci either add a fixed amount to the phenotype or add nothing.

adenine *See* purines.

adjacent-1 segregation A separation of centromeres during meiosis in a reciprocal translocation heterozygote such that unbalanced zygotes are produced.

adjacent-2 segregation Separation of centromeres during meiosis in a translocation heterozygote such that homologous centromeres are pulled to the same pole.

A DNA The form of DNA at high humidity; it has tilted base pairs and more base pairs per turn than does B DNA.

affected Individuals in a pedigree that exhibit the specific phenotype under study.

allele Alternative form of a gene.

allelic exclusion A process whereby only one immunoglobulin light chain and one heavy chain gene are transcribed in any one cell; the other genes are repressed.

allopatric speciation Speciation in which the evolution of reproductive isolating mechanisms occurs during physical separation of the populations.

allopolyploidy Polyploidy produced by the hybridization of two species.

allosteric protein A protein whose shape is changed when it binds a particular molecule. In the new shape the protein's ability to react to a second molecule is altered.

allotype Mutant of a nonvariant part of an immunoglobulin gene that follows the rules of simple Mendelian inheritance.

allozygosity Homozygosity in which the two alleles are alike but unrelated. *See* autozygosity.

allozymes Forms of an enzyme, controlled by alleles of the same locus, that differ in electrophoretic mobility. *See* isozymes.

alternate segregation A separation of centromeres during meiosis in a reciprocal translocation heterozygote such that balanced gametes are produced.

alternative splicing Various ways of splicing out introns in eukaryotic pre-mRNAs resulting in one gene producing several different mRNA and protein products.

altruism A form of behavior in which an individual risks lowering its fitness for the benefit of another.

Alu family A dispersed, intermediately repetitive DNA sequence found in the human genome about three hundred thousand times. The sequence is about 300 bp long. The name Alu comes from the restriction endonuclease that cleaves it.

aminoacyl-tRNA synthetases Enzymes that attach amino acids to their proper tRNAs.

amphidiploid An organism produced by hybridization of two species followed by somatic doubling. It is an allotetraploid that appears to be a normal diploid.

anagenesis The evolutionary process whereby one species evolves into another without any splitting of the phylogenetic tree. *See* cladogenesis.

anaphase The stage of mitosis and meiosis in which sister chromatids or homologous chromosomes are separated by spindle fibers.

aneuploids Individuals or cells exhibiting aneuploidy.

aneuploidy The condition of the cell or of an organism that has additions or deletions of whole chromosomes from the expected, balanced number of chromosomes.

angiosperms Plants whose seeds are enclosed within an ovary. Flowering plants.

antibody A protein produced by a B lymphocyte that protects the organism against antigens.

anticoding strand The DNA strand that forms the template for both the transcribed mRNA and the coding strand.

anticodon The three-base sequence on tRNA complementary to a codon on mRNA.

antigen A foreign substance capable of triggering an immune response in an organism.

antimutator mutations Mutations of DNA polymerase that decrease the overall mutation rate of a cell or of an organism.

anti-oncogene A gene that represses malignant growth and whose absence results in malignancy (e.g., retinoblastoma).

antiparallel strands Strands, as in DNA, that run in opposite directions with respect to their 3' and 5' ends.

antisense RNA RNA product of *mic* (*mRNA-interfering complementary* RNA) genes that regulates another gene by base pairing, and thus blocking, its mRNA.

antisense strand *See* anticoding strand.

antiterminator protein A protein that, when bound at its normal attachment sites, lets RNA polymerase read through normal terminator sequences (e.g., the *N*- and *Q*-gene products of phage λ).

AP endonucleases Endonucleases that initiate excision repair at apurinic and apyrimidinic sites on DNA.

ascospores Haploid spores found in the asci of Ascomycete fungi.

ascus The sac in Ascomycete fungi that holds the ascospores.

assignment test A test that determines whether a locus is on a specific chromosome by observation of the concordance of the locus and the specific chromosome in hybrid cell lines.

assortative mating The mating of individuals with similar phenotypes.

ataxia-telangiectasia A disease in human beings caused by a defect in X-ray–induced repair mechanisms.

attenuator region A control region at the promoter end of repressible amino acid operons that exerts transcriptional control based on the translation of a small leader peptide gene.

attenuator stem *See* terminator stem.

autogamy Nuclear reorganization in a single *Paramecium* cell similar to the changes that occur during conjugation.

autopolyploidy Polyploidy in which all the chromosomes come from the same species.

autoradiography A technique in which radioactive molecules make their location known by exposing photographic plates.

autosomal set A combination of nonsex chromosomes consisting of one from each homologous pair in a diploid species.

autosomes The nonsex chromosomes.

autotrophs Organisms that can utilize carbon dioxide as a carbon source.

autozygosity Homozygosity in which the two alleles are identical by descent (i.e., they are copies of an ancestral gene).

auxotrophs Organisms that have specific nutritional requirements.

B

bacillus A rod-shaped bacterium.

backcross The cross of an individual with one of its parents or an organism with the same genotype as a parent.

back mutation The process that causes reversion. A change in a nucleotide pair in a mutant gene that restores the original sequence and hence the original phenotype.

bacterial lawn A continuous cover of bacteria on the surface of a growth medium.

bacteriophages Bacterial viruses.

balanced lethal system An arrangement of recessive lethal alleles that maintains a heterozygous chromosome combination. Homozygotes for any lethal-bearing chromosome perish.

Balbiani rings The larger polytene chromosomal puffs. Generally synonymous with *puffs*. *See* chromosome puffs.

Barr body Heterochromatic body found in the nuclei of normal females but absent in the nuclei of normal males.

Batesian mimicry Form of mimicry in which an innocuous model gains protection by resembling a noxious or dangerous host.

B DNA The right-handed, double-helical form of DNA described by Watson and Crick.

β-galactosidase The enzyme that splits lactose into glucose and galactose (coded by a gene in the *lac* operon).

β-galactoside acetyltransferase An enzyme that is involved in lactose metabolism and encoded by a gene in the *lac* operon.

β-galactoside permease An enzyme involved in concentrating lactose in the cell (coded by a gene in the *lac* operon).

binary fission Simple cell division in single-celled organisms.

binomial expansion The terms generated when a binomial expression is raised to a particular power.

binomial theorem The theorem that gives the terms of the expansion of a binomial expression raised to a particular power.

biochemical genetics The study of the relationships between genes and enzymes, specifically the role of genes in controlling the steps in biochemical pathways.

biolistic A method (*biological* bal*listic*) of transfecting cells by bombarding them with microprojectiles coated with DNA.

biological species concept Organisms are classified in the same species if they are potentially capable of interbreeding and producing fertile offspring.

bivalents Structures, formed during prophase of meiosis I, consisting of the synapsed homologous chromosomes. Equivalent to a tetrad of chromatids.

blastomeres Cells making up the blastula.

blastopore The embryonic opening of the future gut.

blastula The first developmental stage of a developing embryo.

blunt-end ligation The ligating or attaching of blunt-ended pieces of DNA by T4 DNA ligase. Used in creating hybrid vectors.

bottleneck A brief reduction in size of a population, which usually leads to random genetic drift.

branch migration The process in which a crossover point between two duplexes slides along the duplexes.

breakage and reunion The general mode by which recombination occurs. DNA duplexes are broken and reunited in a crosswise fashion according to the Holliday model.

breakage-fusion-bridge cycle Damage that a dicentric chromosome goes through each cell cycle.

bubbles The nucleic acid configuration during replication in eukaryotic chromosomes, or the shape of heteroduplex DNA at the site of a deletion or insertion.

buoyant density of DNA A measure of the density or size of DNA determined by the equilibrium point reached by DNA after density gradient centrifugation.

C

CAAT box An invariant DNA sequence at about -70 in many eukaryotic promoters.

calculus of the genes Apparent calculation by the genes to determine when a particular altruistic behavior is beneficial to inclusive fitness and hence worth doing.

cancer An informal term for a diverse class of diseases marked by abnormal cell proliferation.

cancer-family syndromes Pedigree patterns in which unusually large numbers of blood relatives develop certain kinds of cancers.

cap A sequence of methyl groups added to the 5′ end of eukaryotic mRNA.

capsid The protein shell of a virus.

capsomere Protein clusters making up discrete subunits of a viral protein shell.

carcinoma Tumor arising from epithelial tissue (e.g., glands, breast, skin, linings of the urogenital and respiratory systems).

cassette mechanism The mechanism by which homothallic yeast cells alternate mating types. The mechanism involves two silent transposons (cassettes) and a region where these cassettes can be expressed (cassette player).

catabolite activator protein (CAP) A protein that when bound with cyclic AMP can attach to sites on sugar-metabolizing operons to enhance transcription of these operons.

catabolite repression Repression of certain sugar-metabolizing operons in favor of glucose utilization when glucose is present in the environment of the cell.

catalyst A substance that increases the rate of a chemical reaction without itself being permanently changed.

cDNA *See* complementary DNA.

cell cycle The cycle of cell growth, replication of the genetic material, and nuclear and cytoplasmic division.

cell-free system A mixture of cytoplasmic components from cells, lacking nucleic acids and membranes. Used for *in vitro* protein synthesis and other purposes.

centimorgan A chromosome mapping unit. One centimorgan equals 1% recombinant offspring.

central dogma The original postulate that information can be transferred only from DNA to DNA, from DNA to RNA, and from RNA to protein.

centric fragment A piece of chromosome containing a centromere.

centrioles Cylindrical organelles, found in eukaryotes (except in higher plants), that organize the formation of the spindle.

centromere markers Loci located near their centromeres.

centromeres Constrictions in eukaryotic chromosomes in which the kinetochores lie.

centromeric fission Creation of two chromosomes from one by splitting the centromere.

chaperone *See* molecular chaperone.

Chargaff's rule Chargaff's observation that in the base composition of DNA the quantity of adenine equaled the quantity of thymine and the quantity of guanine equaled the quantity of cytosine (equal purine and pyrimidine content).

Charon phages Phage lambda derivatives used as vehicles in DNA cloning.

chiasmata X-shaped configurations seen in tetrads during the latter stages of prophase I of meiosis. They represent physical crossovers. (Singular: *chiasma*.)

chimeras *See* mosaics.

chimeric plasmid Hybrid, or genetically mixed, plasmid used in DNA cloning.

chi site Sequence of DNA at which the RecBCD protein cleaves one of the strands during recombination.

chi-square distribution The sampling distribution of the chi-square statistic. A family of curves whose shapes depend on degrees of freedom.

chloroplast The organelle that carries out photosynthesis and starch grain formation.

chromatids The subunits of a chromosome prior to anaphase of meiosis or mitosis. At anaphase of meiosis II or mitosis, when the centromeres divide and the sister chromatids separate, each chromatid is then a chromosome.

chromatin The nucleoprotein material of the eukaryotic chromosome.

chromomeres Dark regions of chromatin condensation in eukaryotic chromosomes at meiosis or mitosis.

chromosomal theory of inheritance The theory that chromosomes are linear sequences of genes.

chromosome The form of the genetic material in viruses and cells. A circle of DNA in prokaryotes; a DNA or an RNA molecule in viruses; a linear nucleoprotein complex in eukaryotes.

chromosome jumping A technique of isolating clones from a genomic library that are not contiguous but skip a region between known points on the chromosome. This is done usually to bypass regions that are difficult or impossible to "walk" through or regions known not to be of interest.

chromosome puffs Diffuse, uncoiled regions in polytene chromosomes where transcription is actively taking place.

chromosome walking A technique for studying segments of DNA, larger than can be individually cloned, by using overlapping probes.

cis Meaning "on the near side of"; refers to geometric configurations of atoms or mutants on the same chromosome.

cis-dominant Mutants (e.g., of an operator) that control the functioning of genes on that same piece of DNA.

cis-trans complementation test A mating test to determine whether two mutants on opposite chromosomes will complement each other: a test for allelism.

cistron Term coined by Benzer for the smallest genetic unit that exhibits the *cis-trans* position effect; synonymous with *gene*.

cladogenesis The evolutionary process whereby one species splits into two or more species. *See* anagenesis.

***ClB* method** A technique devised by Muller to rapidly screen fruit flies for recessive X chromosome lethal mutations. The *ClB* chromosome carries a recessive lethal (*l*), a dominant marker (*B*), and an inversion (crossover suppressor, *C*).

clinal selection Selection that changes gradually along a geographic gradient.

clone A group of cells arising from a single ancestor.

coccus A spherical bacterium.

coding strand The DNA strand with the same sequence as the transcribed mRNA (given U in RNA and T in DNA). Compare with "anticoding strand."

codominance The relationship of alleles such that the phenotype of the heterozygote shows the individual expression of each allele.

codon preference The idea that for amino acids with several codons, one or a few are preferred and are used disproportionately. They would correspond with tRNAs that are abundant.

codons The sequences of three RNA or DNA nucleotides that specify either an amino acid or termination of translation.

coefficient of coincidence The number of observed double crossovers divided by the number expected based on the independent occurrence of crossovers.

coefficient of relationship, *r* The proportion of alleles held in common by two related individuals.

cointegrate A fusion of two elements. An intermediate structure in transposition.

colicinogenic factors *See* col plasmids.

Col plasmids Plasmids that produce antibiotics (colicinogens) used by the host to kill other strains of bacteria.

common ancestry The state of two individuals when they are blood relatives. When two parents have a common ancestor, their offspring will be inbred.

competence factor A surface protein that binds extracellular DNA and enables the cell to be transformed.

complementarity The correspondence of DNA bases in the double helix such that adenine in one strand is opposite thymine in the other strand and cytosine in one strand is opposite guanine in the other. This relationship explains Chargaff's rule.

complementary DNA (cDNA) DNA synthesized by reverse transcriptase using RNA as a template.

complementation The production of the wild-type phenotype by a cell or an organism that contains two mutant genes. If complementation occurs, the mutants are almost certainly nonallelic.

complementation group Cistron (determined by the *cis-trans* complementation test).

complete linkage The state in which two loci are so close together that alleles of these loci are virtually never separated by crossing over.

complete medium A culture medium that is enriched to contain all of the growth requirements of a strain of organisms.

component of fitness A particular aspect in the life cycle of an organism upon which natural selection acts.

composite transposon A transposon constructed of two IS elements flanking a control region that frequently contains host genes.

concordance The amount of similarity in phenotype between individuals.

conditional-lethal mutant A mutant that is lethal under one condition but not lethal under another condition.

confidence limits A statistical term for a pair of numbers that predict the range of values within which a particular parameter lies.

conjugation A process whereby two cells come in contact and exchange genetic material. In prokaryotes the transfer is a one-way process.

consanguineous Meaning "between blood relatives"; usually refers to inbreeding or incestuous matings.

consensus sequence A sequence of the common nucleotides found in many different DNA or RNA samples of homologous regions (e.g., promoters).

conservative replication A postulated mode of DNA replication in which an intact double helix acts as a template for a new double helix; known to be incorrect.

conserved sequence A sequence found in many different DNA or RNA samples (e.g., promoters) that is invariant in the sample.

constitutive heterochromatin Heterochromatin that surrounds the centromere. *See* satellite DNA.

constitutive mutant A mutant whose transcription is no longer under regulatory control.

continuous replication In DNA, uninterrupted replication in the 5' to 3' direction using a 3' to 5' template.

continuous variation Variation measured on a continuum rather than in discrete units or categories (e.g., height in human beings).

copper fist Configuration of a DNA-binding protein that resembles a fist closed around a penny. In this case, the penny is copper ions; the knuckles of the fist of the yeast ACE1 protein interact with the promoter of the metallothionein gene, enhancing its transcription.

copy-choice hypothesis An incorrect hypothesis that stated that recombination resulted from the switching of the DNA-replicating enzyme from one homologue to the other.

corepressor The metabolite that when bound to the repressor (of a repressible operon) forms a functional unit that can bind to its operator and block transcription.

correlation coefficient A statistic that gives a measure of how closely two variables are related.

cosmid A hybrid plasmid that contains *cos* sites at each end. *Cos* sites are recognized during head filling of lambda phages. Cosmids are useful for cloning large segments of foreign DNA (up to 50 kb).

cotransduction The simultaneous transduction of two or more genes.

cot values (cot$_{1/2}$) The product of C_o, the original concentration of denatured, single-stranded DNA and *t*, time in seconds, giving a useful index of renaturation. Cot$_{1/2}$ values are the midpoint values on cot curves—cot values plotted against concentration of remaining single-stranded DNA—which estimate the length of unique DNA in a sample.

coupling Allele arrangement in which mutants are on the same chromosome and wild-type alleles on the homologue.

covariance A statistical value measuring the simultaneous deviations of x and y variables from their means.

criss-cross pattern of inheritance The phenotypic pattern of inheritance controlled by X-linked recessive alleles.

critical chi-square A chi-square value for a given degrees of freedom and probability level to which an experimental chi-square is to be compared.

crossbreed Fertilization between separate individuals.

cross-fertilization *See* crossbreed.

crossing over A process in which homologous chromosomes exchange parts by a breakage-and-reunion process.

crossover suppression The apparent lack of crossing over within an inversion loop in heterozygotes. Due to mortality of zygotes carrying defective crossover chromosomes rather than actual suppression.

crossovers *See* chiasmata.

cryptic coloration Coloration that allows an organism to match its background and hence become less vulnerable to predation or recognition by prey.

cyclic AMP A form of AMP (adenosine monophosphate) used frequently as a second messenger in eukaryotic hormone nets and in catabolite repression in prokaryotes.

cytokinesis The division of the cytoplasm of a cell into two daughter cells. *See* karyokinesis.

cytoplasmic inheritance Extra-chromosomal inheritance controlled by nonnuclear genomes.

cytosine *See* pyrimidines.

cytotoxic T lymphocytes T cells that are responsible for attacking host cells that have been infected with an invading bacterium or virus.

D

dauermodification The persistence for several generations of an environmentally induced trait.

degenerate code A code in which several code words have the same meaning. The genetic code is degenerate because there are many instances in which different codons specify the same amino acid.

degrees of freedom An estimate of the number of independent categories in a particular statistical test or experiment.

deletion chromosome A chromosome with part deleted.

deme A locally interbreeding population.

denatured Loss of natural configuration (of a molecule) through heat or other treatment. Denatured DNA is single-stranded.

denominator elements Genes on the autosomes of *Drosophila* that regulate the sex switch (*sxl*) to the off condition (maleness). Refers to the denominator of the X/A genic balance equation.

density-gradient centrifugation A method of separating molecular entities by their differential sedimentation in a centrifugal gradient.

depauperate fauna A fauna, especially common on islands, lacking many species found in similar habitats.

derepressed The condition of an operon that is transcribing because repressor control has been lifted.

deterministic Referring to events that have no random or probabilistic aspects but proceed in a fixed, predictable fashion.

development The process of orderly change that an individual goes through in the formation of structure.

diakinesis The final stage of prophase I of meiosis when chiasmata terminalize.

dicentric chromosome A chromosome with two centromeres.

dictyotene A prolonged diplonema of primary oocytes that can last many years.

dideoxy method A method of DNA sequencing that uses chain-terminating (dideoxy) nucleotides.

dihybrid An organism heterozygous at two loci.

dimerization The chemical union of two identical molecules.

diploid The state of having each chromosome in two copies per nucleus or cell.

diplonema (diplotene stage) The stage of prophase of meiosis I in which chromatids appear to repel each other.

directional selection A type of selection that removes individuals from one end of a phenotypic distribution and thus causes a shift in the distribution.

disassortative mating The mating of two individuals with dissimilar phenotypes.

discontinuous replication In DNA, the replication in short 5' to 3' segments using the 5' to 3' strand as a template while going backward, away from the replication fork.

discontinuous variation Variation that falls into discrete categories (e.g., the color of garden peas).

discrete generations Generations that have no overlapping reproduction. All reproduction takes place between individuals in the same generation.

dispersive replication A postulated mode of DNA replication combining aspects of conservative and semiconservative replication; known to be incorrect.

disruptive selection A type of selection that removes individuals from the center of a phenotypic distribution and thus causes the distribution to become bimodal.

D-loop Configuration found during DNA replication of chloroplast and mitochondrial chromosomes wherein the origin of replication is different on the two strands. The first structure formed is a displacement loop, or D-loop.

DNA cloning *See* gene cloning.

DNA-DNA hybridization When DNA from the same or different sources is heated and then cooled, double helices will re-form at homologous regions. This technique is useful for determining sequence similarities and degrees of repetitiveness among DNAs.

DNA glycosylases Endonucleases that initiate excision repair at the sites of various damaged or improper bases in DNA.

DNA gyrase A topoisomerase that relieves supercoiling in DNA by creating a transient break in the double helix.

DNA ligase An enzyme that closes nicks or discontinuities in one strand of double-stranded DNA by creating an ester bond between adjacent 3'–OH and 5'–PO_4 ends on the same strand.

DNA polymerase One of several classes of enzymes that polymerize DNA nucleotides using single-stranded DNA as a template.

DNA-RNA hybridization When a mixture of DNA and RNA is heated and then cooled, RNA can hybridize (form a double helix) with DNA that has a complementary nucleotide sequence.

docking protein Responsible for attaching (docking) a ribosome to a membrane by interacting with a signal particle attached to a ribosome destined to be membrane bound.

dominant An allele that expresses itself even when heterozygous. Also the trait controlled by that allele.

dosage compensation A mechanism by which species with sex chromosomes ensure that one sex does not have a differential activity of alleles on the sex chromosomes.

dot blotting A blotting technique of DNA already cloned that eliminates the electrophoretic separation step. Autoradiographs reveal dots rather than bands on a gel, indicative of a cloned gene.

double digest The product formed when two different restriction endonucleases act on the same segment of DNA.

double helix The normal structural configuration of DNA consisting of two helices rotating about the same axis.

double reduction The condition in polyploids in which a heterozygous individual produces homozygous gametes.

doublesex An allele that converts fruit fly males and females into developmental intersexes.

downstream A convention on DNA related to the position and direction of transcription by RNA polymerase $(5' \rightarrow 3')$. Downstream (or 3' to) is in the direction of transcription whereas upstream (5' to) is in the direction from which the polymerase has come.

dyad Two sister chromatids attached to the same centromere.

E

E (exit site) Site on the ribosome through which depleted tRNAs pass during ejection.

electrophoresis The separation of molecular entities by electric current.

electroporation A technique for transfecting cells by the application of a high-voltage electric pulse.

elongation factors (EF-Ts, EF-Tu, EF-G) Proteins necessary for the proper elongation and translocation processes during translation at the ribosome in prokaryotes. Replaced by eEF1 and eEF2 in eukaryotes.

endogenote Bacterial host chromosome.

endomitosis Chromosomal replication without nuclear or cellular division that results in cells with many copies of the same chromosome.

endonucleases Enzymes that make nicks internally in the backbone of a polynucleotide. They hydrolyze internal phosphodiester bonds.

enhancer A eukaryotic DNA sequence that increases transcription of a region even if the enhancer is distant from the region being transcribed.

enriched medium *See* complete medium.

enzyme Protein catalyst.

episomes Term of Jacob and Wollman for genetic particles that can either exist independently in a cell or become integrated into the host chromosome.

epistasis The masking of the action of alleles of one gene by allele combinations of another gene.

equational division The second meiotic division is an equational division because it does not reduce chromosome numbers.

euchromatin Regions of eukaryotic chromosomes that are diffuse during interphase. Presumably the actively transcribing DNA of the chromosomes.

eukaryotes Organisms with true nuclei.

euploidy The condition of a cell or organism that has one or more complete sets of chromosomes.

evolution In Darwinian terms, a gradual change in phenotypic frequencies in a population that results in individuals with improved reproductive success.

evolutionary rates The rate of divergence between taxonomic groups, measurable as amino acid substitutions per million years.

excision repair A process whereby cells remove part of a damaged DNA strand and replace it through DNA synthesis using the undamaged strand as a template.

exconjugant Each of the two cells that separates after conjugation has taken place.

exogenote DNA that a bacterial cell has taken up through one of its sexual processes.

exon A region of a gene that has intervening sequences (introns) that is actually translated or *ex*pressed.

exon shuffling The hypothesis put forward by Walter Gilbert that exons code for functional units of a protein and that evolution of new genes proceeded by recombination or exclusion of exons.

exonucleases Enzymes that digest nucleotides from the ends of polynucleotide molecules. They hydrolyze phosphodiester bonds of terminal nucleotides.

experimental design A branch of statistics that attempts to outline the way in which experiments should be carried out so the data gathered will have statistical value.

exploitation competition A form of competition that revolves around the superior ability to gather resources, rather than an active interaction among organisms for these resources.

expression vector A hybrid vector (plasmid) that expresses its cloned genes.

expressivity The degree of expression of a genetically controlled trait.

eyes Referring to the configuration of replicating DNA in eukaryotic chromosomes.

F

F₁ *See* filial generation.

factorial The product of all integers from the specified number down to one (unity).

Fanconi's anemia A disease in human beings with a syndrome of congenital malformations; associated with various cancers.

fate map A map of the developmental fate of a zygote or early embryo showing the adult organs that will develop from material at a given position on the zygote or early embryo.

F-duction *See* sexduction.

fecundity selection The forces acting to cause one genotype to be more fertile than another genotype.

feedback inhibition A posttranslational control mechanism in which the end product of a biochemical pathway inhibits the activity of the first enzyme of this pathway.

fertility factor The plasmid that allows a prokaryote to engage in conjugation with, and pass DNA into, an F⁻ cell.

F factor *See* fertility factor.

filial generation Offspring generation. F_1 is the first offspring, or filial, generation; F_2 is the second; and so on.

fimbriae *See* pili.

first division segregation (FDS) The allele arrangement of ordered spores that indicates the lack of recombination between a locus and its centromere.

fitness, *W* The relative reproductive success of a genotype as measured by survival, fecundity, or other life history parameters.

fluctuation test An experiment by Luria and Delbrück that compared the variance in number of mutations among small cultures with subsamples of a large culture to determine the mechanism of inherited change in bacteria.

Fokker-Planck equation An equation that describes diffusion processes. It is used by population geneticists to describe random genetic drift.

footprinting A technique to determine the length of nucleic acid in contact with a protein. While in contact the free DNA is digested. The remaining DNA is then isolated and characterized.

founder effect Genetic drift observed in a population founded by a small, nonrepresentative sample of a larger population.

F-pili Sex pili. Hairlike projections on an F⁺ or Hfr bacterium involved in anchorage during conjugation and presumably through which DNA passes.

fragile site A chromosomal region that has a tendency to break.

fragile-X syndrome The most common form of inherited mental retardation. Named for its association with an X chromosome with a tip that breaks or appears uncondensed. Inheritance involves imprinting.

frameshift A mutation in which there is an addition or deletion of nucleotides that causes the codon reading frame to shift.

frequency-dependent selection A selection whereby a genotype is at an advantage when rare and at a disadvantage when common.

functional alleles Mutants that fail to complement each other in a *cis-trans* complementation test.

fundamental number The number of chromosome arms in a somatic cell of a particular species.

G

gamete A germ cell having a haploid chromosome complement. Gametes from parents of opposite sexes fuse to form zygotes.

gametic selection The forces acting to cause differential reproductive success of one allele over another in a heterozygote.

gametophyte The haploid stage of a plant life cycle that produces gametes (by mitosis). It alternates with a diploid, sporophyte generation.

G-bands Eukaryotic chromosomal bands produced by treatment with Giemsa stain.

gene Inherited determinant of the phenotype. *See* cistron; locus.

gene amplification A process or processes by which the cell increases the number of a particular gene within the genome.

gene cloning Production of large numbers of a piece of DNA after that piece of DNA is inserted into a vector and taken up by a cell. Cloning occurs as the vector replicates.

gene conversion In Ascomycete fungi a 2:2 ratio of alleles is expected after meiosis, yet a 3:1 ratio is sometimes observed. The mechanism of gene conversion is explained by repair of heteroduplex DNA produced by the Holliday model of recombination.

gene family A group of genes that has arisen by duplication of an ancestral gene. The genes in the family may or may not have diverged from each other.

gene flow The movement of genes from one population to another by way of interbreeding of individuals in the two populations.

gene pool All of the alleles available among the reproductive members of a population from which gametes can be drawn.

generalized transduction Form of transduction in which any region of the host genome can be transduced. *See* specialized transduction.

genetic code The linear sequences of nucleotides that specify the amino acids during the process of translation at the ribosome.

genetic engineering Popular term for recombinant DNA technology. *See* recombinant DNA technology.

genetic fine structure The structure of the gene analyzed at the level of the smallest units of recombination and mutation (nucleotides).

genetic load The relative decrease in the mean fitness of a population due to the presence of genotypes that have less than the highest fitness.

genetic polymorphism The occurrence together in the same population of more than one allele at the same locus, with the least frequent allele occurring more frequently than can be accounted for by mutation.

genic balance theory The theory of Bridges that the sex of a fruit fly is determined by the relative number of X chromosomes and autosomal sets.

genome The entire genetic complement of a prokaryote or virus or the haploid genetic complement of a eukaryotic species.

genomic library A set of cloned fragments making up the entire genome of an organism or species.

genophore The chromosome (genetic material) of prokaryotes and viruses.

genotype The genes that an organism possesses.

germ-line theory A theory to account for the high degree of antibody variability found. The germ-line theory suggests that every B lymphocyte has all the genes for every type of immunoglobulin but transcribes only one. *See* somatic-mutation theory.

Giemsa stain A complex of stains specific for the phosphate groups of DNA.

Goldstein-Hogness box *See* TATA box.

gray crescent A cortical region of the egg of frogs and some salamanders that forms just after fertilization on the side opposite sperm penetration.

group I introns Self-splicing introns that require a guanine-containing nucleotide for splicing; they release the intron in a linear form.

group II introns Self-splicing introns that do not require an external nucleotide for splicing. The intron is released in a lariat form.

group selection Selection for traits that would be beneficial to a population at the expense of the individual possessing the trait.

guanine *See* purines.

guide RNA (gRNA) RNA that guides the insertion of uridines (RNA editing) into mRNAs in trypanosomes. Found in transcripts from minicircles and maxicircles of DNA in kinetoplasts.

gynandromorphs Mosaic individuals having simultaneous aspects of both the male and the female phenotype.

H

haplodiploidy The sex-determining mechanism found in some insect groups among which males are haploid and females are diploid.

haploid The state of having one copy of each chromosome per nucleus or cell.

HAT medium A selection medium for hybrid cell lines; contains hypoxanthine, aminopterin, and thymidine. HPRT$^+$ TK$^+$ cell lines can survive in this medium.

heat-shock proteins Proteins that appear in a cell after the cell has been subjected to elevated temperatures.

helicase A protein that unwinds DNA at replicating Y-junctions.

helix-turn-helix motif Configuration found in DNA-binding proteins consisting of a recognition helix and a stabilizing helix separated by a short turn.

hemizygous The condition of loci on the X chromosome of the heterogametic sex of a diploid species.

heritability A measure of the degree to which the variance in the distribution of a phenotype is due to genetic causes. In the broad sense it is measured by the total genetic variance divided by the total phenotypic variance. In the narrow sense it is measured by the genetic variance due to additive genes divided by the total phenotypic variance.

hermaphrodite An individual with both male and female genitalia.

heterochromatin Chromatin that remains tightly coiled (and darkly staining) throughout the cell cycle.

heteroduplex analysis Duplex DNA formed by strands from different sources will have loops and bubbles in regions where the two DNAs differ. This heterogeneous DNA is referred to as a *heteroduplex.* Electron microscopic observation (analysis) of this DNA is a useful tool in recombinant DNA work.

heteroduplex DNA *See* hybrid DNA.

heteroduplex mapping The use of heteroduplex analysis to determine the location of various inserts, deletions, or heterogeneities in a piece of DNA.

heterogametic The sex with heteromorphic sex chromosomes; during meiosis, it produces different kinds of gametes in regard to these sex chromosomes.

heterogeneous nuclear mRNA (hnRNA) The original RNA transcripts found in eukaryotic nuclei before posttranscriptional modifications.

heterokaryon A cell that contains two or more nuclei from different origins.

heteromorphic chromosomes Chromosomes of which the members of a homologous pair are not morphologically identical (e.g., the sex chromosomes).

heteroplasmy The existence within an organism of genetic heterogeneity within the populations of mitochondria or chloroplasts.

heterothallic A botanical term used for organisms in which the two sexes reside in different individuals.

heterotrophs Organisms that require an organic form of carbon as a carbon source.

heterozygote A diploid or polyploid with different alleles at a particular locus.

heterozygote advantage A selection model in which heterozygotes have the highest fitness.

heterozygous DNA *See* hybrid DNA.

Hfr High frequency of recombination. A strain of bacteria that has incorporated an F factor into its chromosome and can then transfer the chromosome during conjugation.

histones Arginine- and lysine-rich basic proteins making up a substantial portion of eukaryotic nucleoprotein.

hnRNA *See* heterogeneous nuclear mRNA.

Hogness box *See* TATA box.

holandric trait Trait controlled by a locus found only on the Y chromosome. Involves father to son transmission.

holoenzyme The complete enzyme including all subunits. Often used in reference to RNA and DNA polymerases.

homeo box A consensus sequence of about 180 base pairs discovered in homeotic genes in *Drosophila.* Also found in other developmentally important genes from yeast to human beings.

homeo domain The sixty-amino acid protein translated from the homeo box.

homeotic mutants Mutants in which a given cell develops along a pathway normally followed by a different cell type.

homogametic The sex with homomorphic sex chromosomes; it produces only one kind of gamete in regard to the sex chromosomes.

homologous chromosomes Members of a pair of essentially identical chromosomes that synapse during meiosis.

homologous recombination Breakage and reunion between homologous lengths of DNA mediated by RecA and RecBCD.

homomorphic chromosomes Morphologically identical members of a homologous pair of chromosomes.

homoplasmy The existence within an organism of only one type of plastid, usually referring to genetic identity of mitochondria or chloroplasts.

homothallic A botanical term used for groups whose individuals are not of different sexes.

homozygote A diploid or a polyploid with identical alleles at a locus.

hormone net The integration and control of the various interacting hormones in an individual.

hormones Chemicals that are secreted by one type of cell and act on a second type of cell.

H-Y antigen The histocompatibility Y-antigen, a protein found on the cell surfaces of male mammals.

hybrid Offspring of unlike parents.

hybrid DNA DNA whose two strands have different origins.

hybridoma A cell resulting from the fusion of a spleen cell and multiple myeloma cell. These cells can be maintained indefinitely in cell culture in which they produce monoclonal antibodies.

hybrid plasmid A plasmid that contains an inserted piece of foreign DNA.

hybrid screening Radioisotope technique used to determine whether a hybrid plasmid contains a particular gene or DNA region.

hybrid vector *See* hybrid vehicle.

hybrid vehicle An episome or plasmid containing an inserted piece of foreign DNA.

hybrid zone Geographical region in which previously isolated populations that have evolved differences come into contact and form hybrids.

hypervariable loci Loci with many alleles; especially those whose variation is due to variable members of tandem repeats.

hypostatic gene A gene whose expression is masked by an epistatic gene.

I

identity by descent The state of two alleles when they are identical copies of the same ancestral allele (autozygous).

idiogram A photograph or diagram of the chromosomes of a cell arranged in an orderly fashion. *See* karyotype.

idiotypic variation Variation in the variable parts of immunoglobulin genes.

idling reaction The production of guanosine tetraphosphate ($3'$–ppGpp–$5'$) by the stringent factor when a ribosome encounters an uncharged tRNA in the A site.

immunity The ability of an organism to resist infection.

immunoglobulins (Igs) Specific proteins produced by derivatives of B lymphocytes that protect an organism from antigens.

imprinting *See* molecular imprinting.

inbreeding The mating of genetically related individuals.

inbreeding coefficient, *F* The probability of autozygosity.

inbreeding depression A depression of vigor or yield due to inbreeding.

incestuous A mating between blood relatives who are more closely related than the law of the land allows.

inclusive fitness The expansion of the concept of the fitness of a genotype to include benefits accrued to relatives of an individual since relatives share parts of their genomes. Hence an apparently altruistic act toward a relative may in fact enhance the fitness of the individual performing the act.

incomplete dominance The situation in which both alleles of the heterozygote influence the phenotype. The phenotype is usually intermediate between the two homozygous forms.

independent assortment, rule of Mendel's second rule, describing the independent segregation of alleles of different loci.

inducible system A system (a coordinated group of enzymes involved in a catabolic pathway) is inducible if the metabolite upon which it works causes transcription of the genes controlling these enzymes. These systems are primarily prokaryotic operons.

induction Regarding temperate phages, the process of causing a prophage to become virulent.

industrial melanism The darkening of moths during the recent period of industrialization in many countries.

initiation codon The mRNA sequence AUG, which specifies methionine, the first amino acid used in the translation process. (Occasionally GUG is recognized as an initiation codon.)

initiation complex The complex formed for initiation of translation. It consists of the 30S ribosomal subunit, mRNA, N-formyl-methionine tRNA, and three initiation factors.

initiation factors (IF1, IF2, IF3) Proteins (prokaryotic with eukaryotic analogues) required for the proper initiation of translation.

initiator proteins Proteins that recognize the origin of replication on a replicon and take part in primosome construction.

insertion mutagenesis Change in gene action due to an insertion event that either changes a gene directly or disrupts control mechanisms.

insertion sequences (IS) Small, simple transposons. *See* transposable element.

inside marker The middle locus of three linked loci.

intercalary heterochromatin Heterochromatin, other than centromeric heterochromatin, dispersed throughout eukaryotic chromosomes.

interference competition A form of competition that involves a fight or other active interaction among organisms.

intergenic suppression A mutation at a second locus that apparently restores the wild-type phenotype to a mutant at a first locus.

interkinesis The abbreviated interphase that occurs between meiosis I and II. No DNA replication occurs here.

interphase The metabolically active, nondividing stage of the cell cycle.

interrupted mating A mapping technique in which bacterial conjugation is disrupted after specified time intervals.

intersex An organism with external sexual characteristics that have attributes of both sexes.

intervening sequences (introns) Sequences of DNA within a gene that are transcribed but later removed prior to translation.

intra-allelic complementation The restoration of activity of an enzyme made of subunits in a heterozygote of two mutants that, when homozygous, do not have activity. Caused by interaction of the subunits in the protein.

intragenic suppression A second change within a mutant gene that results in an apparent restoration of the original phenotype.

intron *See* intervening sequences.

inversion The replacement of a section of a chromosome in the reverse orientation.

inverted repeat sequence A nucleotide sequence read in opposite orientation on the same double helix.

in vitro Biological or chemical work done in the test tube (literally, "in glass") rather than in living systems.

iojap A locus in corn that produces variegation.

ionizing radiation Radiation, such as X rays, that causes atoms to release electrons and become ions.

IS elements *See* insertion sequences.

isochromosome A chromosome with two genetically and morphologically identical arms.

isoschizomers Restriction endonucleases that recognize the same target sequence and cleave it the same way.

isozymes Different electrophoretic forms of the same enzyme. Unlike allozymes, isozymes are due to differing subunit configurations rather than allelic differences.

J

junctional diversity Variability in immunoglobulins caused by variation in the exact crossover point during V-J, V-D, and D-J joining.

K

kappa particles The bacteria-like particles that give a *Paramecium* the killer phenotype.

karyokinesis The process of nuclear division. *See* cytokinesis.

karyotype The chromosome complement of a cell. *See* idiogram.

kinetochores The chromosomal attachment points for the spindle fibers, located within the centromeres.

kin selection The mode of natural selection that acts on an individual's inclusive fitness.

L

***lac* operon** The inducible operon including three loci involved in the uptake and breakdown of lactose.

ladder gel *See* stepladder gel.

lagging strand Strand of DNA being replicated discontinuously.

lampbrush chromosomes Chromosomes of amphibian oocytes having loops suggestive of a lampbrush.

leader The length of mRNA from the 5′ end to the initiation codon, AUG.

leader peptide gene A small gene within the attenuator control region of repressible amino acid operons. Translation of the gene tests the concentration of amino acids in the cell.

leader transcript The mRNA transcribed by the attenuator region of repressible amino acid operons. The transcript is capable of several alternative stem-loop structures dependent on the translation of a short leader peptide gene.

leading strand Strand of DNA being replicated continuously.

leptonema (leptotene stage) The first stage of prophase I of meiosis in which chromosomes become distinct.

lethal-equivalent alleles Alleles whose summed effect is that of lethality—for example, four alleles, each of which would be lethal 25% of the time (or to 25% of their bearers), are equivalent to one lethal allele.

leucine zipper Configuration of a DNA-binding protein in which leucine residues on two helices interdigitate, in zipper fashion, to stabilize the protein. Discovered in a protein in the rat.

leukemia Cancer of the bone marrow resulting in excess production of leukocytes.

level of significance The probability value in statistics used to reject the null hypothesis.

linkage The association of loci on the same chromosome.

linkage disequilibrium The condition among alleles at different loci such that allelic combinations in a gamete do not occur according to the product rule of probability.

linkage equilibrium The condition among alleles at different loci such that any allelic combination in a gamete occurs as the product of the frequencies of each allele at its own locus.

linkage groups Associations of loci on the same chromosome. In a species there are as many linkage groups as there are homologous pairs of chromosomes.

linkage number The number of times one strand of a helix coils about the other.

locus The position of a gene on a chromosome. (Plural: *loci*.)

***Lod* score method** A technique (*l*ogarithmic *od*ds) for determining the most likely recombination frequency between two loci from pedigree data.

lymphoma Cancer of the lymph nodes and spleen that causes excessive production of lymphocytes.

Lyon hypothesis The hypothesis that suggests that the Barr body is an inactivated X chromosome.

lysate The contents released from a lysed cell.

lysis The breaking open of a cell by the destruction of its wall or membrane.

lysogenic The state of a bacterial cell that has an integrated phage (prophage) in its chromosome.

M

major histocompatiblity complex A group of highly polymorphic genes whose products appear on the surface of cells imparting to them the property of "self" (belonging to that organism). Some other functions are also involved.

mapping The study of the position of genes on chromosomes.

mapping function The mathematical relationship between measured map distance and actual recombination frequency.

map unit The distance equal to 1% recombination between two loci.

marker A locus or allele whose phenotype provides information about a chromosome or chromosomal segment during genetic analysis. *See* centromere markers; inside marker; outside marker.

mate-killer A phenotype of *Paramecium* induced by intracellular bacteria-like mu particles.

maternal effect The effect of the maternal parent's genotype on the phenotype of her offspring.

mating type In many species of microorganisms, individuals can be divided into two types. Mating can take place only between individuals of opposite mating types due to the interaction of cell surface components.

mean The arithmetic average; the sum of the data divided by the sample size.

mean fitness of the population, $\overline{W}$ The sum of the fitnesses of the genotypes of a population weighted by their proportions; hence a weighted mean fitness.

meiosis The nuclear process in diploid eukaryotes that results in gametes or spores with only one member of each original homologous pair of chromosomes per nucleus.

merozygote A bacterial cell having a second copy of a particular chromosomal region in the form of an exogenote.

metacentric chromosome A chromosome whose centromere is located in the middle.

metafemale A fruit fly with an X/A ratio greater than unity.

metagon An RNA necessary for the maintenance of mu particles in *Paramecium*.

metamale A fruit fly with X/A ratio below 0.5.

metaphase The stage of mitosis or meiosis in which spindle fibers are attached to kinetochores and the chromosomes are positioned in the equatorial plane of the cell.

metaphase plate The plane of the equator of the spindle into which chromosomes are positioned during metaphase.

metastasis The migration of cancerous cells to other parts of the body.

metrical variation *See* continuous variation.

microtubules Hollow cylinders made of the protein tubulin that make up, among other things, the spindle fibers.

mimicry A phenomenon in which an individual gains an advantage by looking like the individuals of a different species.

minimal medium A culture medium for microorganisms that contains the minimal necessities for growth of the wild-type.

mismatch repair A form of excision repair initiated at the sites of mismatched bases in DNA.

missense mutation Mutations that change a codon for one amino acid into a codon for a different amino acid.

mitochondrion The eukaryotic cellular organelle in which the Krebs cycle and electron transport reactions take place.

mitosis The nuclear division producing two daughter nuclei identical to the original nucleus.

mitotic apparatus *See* spindle.

mixed families Groups of four codons sharing their first two bases and coding for more than one amino acid.

molecular chaperone A protein that aids in the folding of a second protein. The chaperone prevents proteins from forming structures that would be inactive.

molecular evolutionary clock A measurement of evolutionary time in nucleotide substitutions per year.

molecular imprinting The phenomenon in which there is differential expression of a gene depending on whether it was maternally or paternally inherited.

monoclonal antibody The antibody from a clone of cells producing the same antibody. An individual with multiple myeloma usually produces monoclonal antibodies.

monohybrids Offspring of parents that differ in only one genetic characteristic. Usually implies heterozygosity at a single locus under study.

monosomic A diploid cell missing a single chromosome.

morphological species concept Organisms are classified in the same species if they appear similar.

mosaicism The condition of being a mosaic. *See* mosaics.

mosaics Individuals made up of two or more cell lines.

mRNA Messenger RNA; the basic function of the nucleotide sequence of mRNA is to determine the amino acid sequence in proteins.

Müllerian mimicry A form of mimicry in which noxious species evolve to resemble each other.

multihybrid An organism heterozygous at numerous loci.

multinomial expansion The terms generated when a multinomial is raised to a power.

mu particles Bacteria-like particles found in the cytoplasm of *Paramecium* that have the mate-killer phenotype.

mutability The ability to change.

mutants Alternative phenotypes to the wild-type; the phenotypes produced by alternative alleles.

mutation The process by which a gene or chromosome changes structurally; the end result of this process.

mutational load Genetic load caused by mutation.

mutation rate The proportion of mutants per cell division in bacteria or single-celled organisms or the proportion of mutations per gamete in higher organisms.

mutator mutations Mutations of DNA polymerase that increase the overall mutation rate of a cell or of an organism.

muton A term coined by Benzer for the smallest mutable site within a cistron.

N

natural selection The process in nature whereby one genotype leaves more offspring than another genotype because of superior life history attributes such as survival or fecundity.

nearest-neighbor analysis A technique of transferring radioactive atoms between adjacent nucleotides in DNA used to demonstrate that the two strands of DNA run in opposite directions.

negative interference The phenomenon whereby a crossover in a particular region enhances the occurrence of other apparent crossovers in the same region of the chromosome.

N-end rule The life span of a protein is determined by its amino-terminal (N-terminal) amino acid.

neo-Darwinism The merger of classical Darwinian evolution with population genetics.

neoplasm New growth of abnormal tissue.

neutral gene hypothesis The hypothesis that most genetic variation in natural populations is not maintained by selection.

NF *See* fundamental number.

nickase *See* DNA gyrase.

noncoding strand *See* anticoding strand.

nondisjunction The failure of a pair of homologous chromosomes to separate properly during meiosis.

nonhistone proteins The proteins remaining in chromatin after the histones are removed. The scaffold structure is made of nonhistone proteins.

nonparental ditype (NPD) A spore arrangement in Ascomycetes that contains only the two recombinant-type ascospores (assuming two segregating loci).

nonparentals *See* recombinants.

nonrecombinants In mapping studies, offspring that have alleles arranged as in the original parents.

nonsense codon One of the mRNA sequences (UAA, UAG, UGA) that signals the termination of translation.

nonsense mutations Mutations that change a codon for an amino acid into a nonsense codon.

normal distribution Any of a family of bell-shaped frequency curves whose relative position and shape are defined on the basis of the mean and standard deviation.

northern blotting A gel transfer technique used for RNA. *See* Southern blotting.

N segments Sequences of nucleotides, added in a template-free fashion, at the joining junctions of heavy-chain antibody genes.

nuclear transplantation The technique of placing a nucleus from another source into an enucleated cell.

nuclease One of the several classes of enzymes that degrade nucleic acid. *See* endonucleases; exonucleases.

nuclease hypersensitive site Region of eukaryotic chromosome that is specifically vulnerable to nuclease attack because it is not wrapped as nucleosomes.

nucleolar organizer The chromosomal region around which the nucleolus forms; site of tandem repeats of the major rRNA gene.

nucleolus The globular, nuclear organelle formed at the nucleolar organizer. Site of ribosome construction.

nucleoprotein The substance of eukaryotic chromosomes consisting of proteins and nucleic acids.

nucleoside A sugar-base compound that is a nucleotide precursor. Nucleotides are nucleoside phosphates.

nucleosomes Arrangements of DNA and histones forming regular spherical structures in eukaryotic chromatin.

nucleotide Subunits that polymerize into nucleic acids (DNA or RNA). Each nucleotide consists of a nitrogenous base, a sugar, and one or more phosphate groups.

null hypothesis The statistical hypothesis that states that there are no differences between observed and expected data.

nullisomic A diploid cell missing both copies of the same chromosome.

numerator elements Genes on the X chromosome in *Drosophila* that regulate the sex switch (*sxl*) to the on condition (femaleness). Refers to the numerator of the X/A genic balance equation.

nutritional-requirement mutants *See* auxotrophs.

O

Okazaki fragments Segments of newly replicated DNA produced during discontinuous DNA replication.

oncogene Genes capable of transforming a cell. They are found in the active state in retroviruses and transformed cells and in the inactive state in nontransformed cells in which they are called proto-oncogenes.

one-gene-one-enzyme hypothesis Hypothesis of Beadle and Tatum that one gene controls the production of one enzyme. Later modified to the concept that one cistron controls the production of one polypeptide.

oogenesis The process of ovum formation in female animals.

oogonia Cells in females that produce primary oocytes by mitosis.

operator A DNA sequence that is recognized by a repressor protein or repressor-corepressor complex. When the operator is complexed with the repressor, transcription is prevented.

operon A sequence of adjacent genes all under the transcriptional control of the same operator.

outbreeding The mating of genetically unrelated individuals.

outside markers Loci on either side of another locus or specified region.

ovum Egg. The one functional product of each meiosis in female animals.

P

P (peptidyl) site The site on the ribosome occupied by the peptidyl-tRNA just before peptide bond formation.

P_1 Parental generation.

pachynema (pachytene stage) The stage of prophase I of meiosis in which chromatids are first distinctly visible.

palindrome A sequence of words, phrases, or nucleotides that reads the same regardless of which direction one starts from; the sites of recognition of type II restriction endonucleases.

panmictic Referring to unstructured (random-mating) populations.

paracentric inversion A chromosomal inversion that does not include the centromere.

paramecin A toxin liberated by a "killer" *Paramecium*.

parameters Measurements of attributes of a population; denoted by Greek letters.

parapatric speciation Speciation in which the evolution of reproductive isolating mechanisms occurs when a population enters a new niche or habitat within the range of the parent species.

parental ditype (PD) A spore arrangement in Ascomycetes that contains only the two nonrecombinant-type ascospores.

parental imprinting *See* molecular imprinting.

parentals *See* nonrecombinants.

parthenogenesis The development of an individual from an unfertilized egg that did not arise by meiotic chromosome reduction.

partial digest A restriction digest that has not been allowed to go to completion and thus contains pieces of DNA that have restriction endonuclease sites that have not been cleaved.

partial dominance *See* incomplete dominance.

Pascal's triangle A triangular array of numbers made up of the coefficients of the binomial expansion.

passenger DNA Foreign DNA incorporated into a plasmid.

path diagram A modified pedigree showing only the direct line of descent from common ancestors.

pedigree A representation of the ancestry of an individual or family; a family tree.

penetrance The normal appearance in the phenotype of genetically controlled traits.

peptidyl transferase The enzymatic center responsible for peptide bond formation during translation at the ribosome.

pericentric inversion A chromosomal inversion that includes the centromere.

permissive temperature A temperature at which temperature-sensitive mutants are normal.

PEST hypothesis Degradation of a protein in under two hours is signaled by a region within the protein rich in proline (P), glutamic acid (E), serine (S), or threonine (T).

petite mutations Mutations of yeast that produce small, anaerobic-like colonies.

phages *See* bacteriophages.

phenocopy A phenotype that is not genetically controlled but looks like a genetically controlled phenotype.

phenotype The observable attributes of an organism.

pheromone A chemical signal, analogous to a hormone, that passes information between individuals.

phosphodiester bond A diester bond linking two nucleotides together (between phosphoric acid and sugars) to form the nucleotide polymers DNA and RNA.

photoreactivation The process whereby dimerized pyrimidines (usually thymines) in DNA are restored by an enzyme (deoxyribodipyrimidine photolyase) that requires light energy.

phyletic evolution *See* anagenesis.

phyletic gradualism The process of gradual evolutionary change over time.

phylogenetic tree A diagram showing evolutionary lineages of organisms.

pili (fimbriae) Hairlike projections on the surface of bacteria; Latin for "hair."

plaques Clear areas on a bacterial lawn caused by cell lysis due to viral attack.

plasmid An autonomous, self-replicating genetic particle, usually of double-stranded DNA.

plastid A chloroplast prior to the development of chlorophyll.

pleiotropy The phenomenon whereby a single mutant affects several apparently unrelated aspects of the phenotype.

point mutations Small mutations that consist of a replacement, addition, or deletion of one or a few bases.

poky mutations Mutations of *Neurospora* that produce a petitelike phenotype.

polar bodies The small cells (which eventually disintegrate) that are the by-products of meiosis in female animals. One functional ovum and potentially three polar bodies result from meiosis of each primary oocyte.

polarity Meaning "directionality" and referring either to an effect seen in only one direction from a point of origin or to the fact that linear entities (such as a single strand of DNA) have ends that differ from each other.

polarity gene A mitochondrial gene with alleles that are preferentially found in daughter mitochondria after a recombinational event between mitochondria.

polar mutant An organism with a mutation, usually within an operon, that prevents the expression of genes distal to itself.

pollen grain The male gametophyte in higher plants.

poly-A tail A sequence of adenosine nucleotides added to the 3′ end of eukaryotic mRNAs.

polycistronic Referring to prokaryotic messenger RNAs that contain several genes within the same mRNA transcript.

poly-dA/poly-dT technique A method of inserting foreign DNA into a vehicle by making 5′ poly-dA and 5′ poly-dT tails on the vehicle and foreign DNAs. Also feasible with poly-dG and poly-dC.

polygenic inheritance *See* quantitative inheritance.

polymerase chain reaction (PCR) A method to amplify rapidly DNA segments in cycles of denaturation, primer addition, and replication.

polymerize To form a complex compound by linking together many smaller elements.

polynucleotide phosphorylase An enzyme that can polymerize diphosphate nucleotides without the need for a primer. The function of this enzyme *in vivo* is probably in its reverse role as an RNA exonuclease.

polyploids Organisms with greater than two chromosome sets.

polyribosome *See* polysome.

polysome The configuration of several ribosomes simultaneously translating the same mRNA. Shortened form of the term *polyribosome*.

polytene chromosome Large chromosome consisting of many chromatids formed by rounds of endomitosis followed by synapsis.

population A group of organisms of the same species relatively isolated from other groups of the same species. *See* deme.

position effect An alteration of phenotype caused by a change in the relative arrangement of the genetic material.

positive interference When the occurrence of one crossover reduces the probability that a second will occur in the same region.

postreplicative repair A DNA repair process initiated when DNA polymerase bypasses a damaged area. Enzymes in the *rec* system are used.

posttranscriptional modifications Changes in eukaryotic mRNA made after transcription has been completed. These changes include additions of caps and tails and removal of introns.

preemptor stem A configuration of leader transcript mRNA that does not terminate transcription in the attenuator-controlled amino acid operons.

Pribnow box Relatively invariant sequence of six nucleotides in prokaryotic promoters centered at the position −10 with the consensus sequence of TATAAT.

primary oocytes The cells that undergo meiosis in female animals.

primary spermatocytes The cells that undergo meiosis in male animals.

primary structure The sequence of polymerized amino acids in a protein.

primary transcript The product of eukaryotic transcription before posttranscriptional modification takes place.

primase An enzyme that creates an mRNA primer for Okazaki fragment initiation.

primer In DNA replication, a length of double-stranded DNA that continues as a single-stranded template in the 3′ to 5′ direction.

primosome A complex of two proteins, a primase and helicase, that initiates RNA primers on the lagging DNA strand during DNA replication.

prion Infectious agent responsible for several neurological diseases (scrapie, kuru, Creutzfeld-Jakob syndrome). It appears to be a protein.

probability The expectation of the occurrence of a particular event.

probability theory The conceptual framework concerned with quantification of probabilities. *See* probability.

proband *See* propositus.

probe In recombinant DNA work, a radioactive nucleic acid complementary to a region being searched for in a restriction digest or genome library.

processivity The ability of an enzyme to repetitively continue its catalytic function without dissociating from its substrate.

product rule The rule that states that the probability of the occurrence of independent events is the product of their separate probabilities.

progeny testing Breeding of offspring to determine their genotypes and that of their parents.

prokaryotes Organisms that lack true nuclei.

promoter A region of DNA to which RNA polymerase binds in order to initiate transcription.

proofread Technically, to read for the purpose of detecting errors for later correction. DNA polymerase has 3′ to 5′ exonuclease activity, which it uses during polymerization to remove nucleotides it has recently added. This is a correcting ability to remove errors in replication.

prophage A temperate phage integrated into the host chromosome.

prophase The initial stage of mitosis or meiosis in which chromosomes become visible and the spindle apparatus forms.

proplastid Mutant plastids that do not grow and develop into chloroplasts.

propositus (proposita) The person through whom a pedigree was discovered.

proto-oncogene The nonactivated form of a cellular oncogene in an untransformed cell.

prototrophs Strains of organisms that can survive on minimal medium.

pseudoalleles Alleles that are functionally but not structurally allelic. Within gene families, pseudoalleles are alleles that are not expressed.

pseudoautosomal gene A gene that occurs on both sex-determining heteromorphic chromosomes.

pseudodominance The phenomenon in which a recessive allele shows itself in the phenotype when only one copy of the allele is present, as in hemizygous alleles or in deletion heterozygotes.

punctuated equilibrium The evolutionary process involving long periods without change (stasis) punctuated by short periods of rapid speciation.

Punnett square A diagrammatic representation of a particular cross used to predict the progeny of the cross.

purines Nitrogenous bases of which guanine and adenine are found in DNA and RNA.

pyrimidines Nitrogenous bases of which thymine is found in DNA, uracil in RNA, and cytosine in both.

Q

quantitative inheritance The mechanism of genetic control of traits showing continuous variation.

quantitative variation *See* continuous variation.

quaternary structure The association of polypeptide subunits to form the final structure of a protein.

R

RAM mutants *R*ibosomal *am*biguity *m*utants that allow incorrect tRNAs to be incorporated into the translation process.

random genetic drift Changes in allelic frequency due to sampling error.

random mating The mating of individuals in a population such that the union of individuals with the trait under study occurs according to the product rule of probability.

random strand analysis Mapping studies in organisms that do not keep together all the products of meiosis.

readthrough Transcription or translation beyond the normal termination signals in DNA or RNA, respectively.

realized heritability Heritability measured by a response to selection.

recessive An allele that does not express itself in the heterozygous condition.

reciprocal altruism An apparently altruistic behavior done with the understanding that the recipient will reciprocate at some future date.

reciprocal cross A cross with the phenotype of each sex reversed as compared with the original cross. Made to test the role of parental sex on inheritance pattern.

reciprocal translocation A chromosomal configuration in which the ends of two nonhomologous chromosomes are broken off and become attached to the nonhomologues.

recombinant DNA technology Techniques of gene cloning. Recombinant DNA refers to the hybrid of foreign and vector DNA. *See* gene cloning.

recombinants In mapping studies, offspring with allelic arrangements made up of a combination of the original parental alleles.

recombination The nonparental arrangement of alleles in progeny that can result from either independent assortment or crossing over.

recon A term coined by Benzer for the smallest recombinable unit within a cistron.

***rec* system** Several loci controlling genes (*recA*, *recB*, *recC*, and others) involved in postreplicative DNA repair.

reductional division The first meiotic division. It reduces the number of chromosomes and centromeres to half that of the original cell.

regulator gene A gene primarily involved in control of the production of another gene's product.

relative Darwinian fitness *See* fitness.

relaxed mutant A mutant that does not exhibit the stringent response under amino acid starvation.

release factors (RF1 and RF2) Proteins in prokaryotes responsible for termination of translation and release of the newly synthesized polypeptide when a nonsense codon appears in the A site of the ribosome. Replaced by eRF in eukaryotes.

Renner complexes Specific gametic chromosome combinations in *Oenothera*.

repetitive DNA DNA made up of copies of the same nucleotide sequence.

replica plating A technique to rapidly transfer microorganism colonies to numerous petri plates.

replication The process of copying.

replicons A replicating genetic unit including a length of DNA and its site for the initiation of replication.

replisome The DNA-replicating structure at the Y-junction consisting of two DNA polymerase III enzymes and a primosome (primase and DNA helicase).

repressible system A coordinated group of enzymes, involved in a synthetic pathway (anabolic), is repressible if excess quantities of the end product of the pathway lead to the termination of transcription of the genes for the enzymes. These systems are primarily prokaryotic operons.

repressor The protein product of a regulator gene that acts to control transcription of inducible and repressible operons.

reproductive isolating mechanisms Environmental, behavioral, mechanical, and physiological barriers that prevent two individuals of different populations from producing viable progeny.

reproductive success The relative production of offspring by a particular genotype.

repulsion Allelic arrangement in which each homologous chromosome has mutant and wild-type alleles.

resistance transfer factor Infectious transfer part of R plasmids.

restricted transduction *See* specialized transduction.

restriction digest The results of the action of a restriction endonuclease on a DNA sample.

restriction endonucleases Endonucleases that recognize certain DNA sequences, which they cleave. They are thought to protect cells from viral infection; useful in recombinant DNA work.

restriction fragment length polymorphism (RFLP) Variations in banding patterns of electrophoresed restriction digests.

restriction map A physical map of a piece of DNA showing recognition sites of specific restriction endonucleases separated by lengths marked in numbers of bases.

restrictive temperature A temperature at which temperature-sensitive mutants display the mutant phenotype.

retinoblastoma A childhood cancer of retinoblast cells caused by the inactivation of an anti-oncogene.

retrotransposon (retroposon) A class of genetic elements that includes retroviruses and transposons that have an intermediate RNA stage.

reverse transcriptase An enzyme that can use RNA as a template to synthesize DNA.

reversion The return of a mutant to the wild-type phenotype by way of a second mutational event.

R factors *See* R plasmids.

rho-dependent terminator A DNA sequence signaling the termination of transcription; termination requires the presence of the rho protein.

rho-independent terminator A DNA sequence signaling the termination of transcription; the rho protein is not required for termination.

rho protein A protein that is involved in the termination of transcription.

ribosomes Organelles at which translation takes place. They are made up of two subunits consisting of RNA and proteins.

ribozyme Catalytic or autocatalytic RNA.

RNA editing The insertion of uridines into mRNAs after transcription is completed; controlled by guide RNA. May also involve insertion of cytidines in some organisms, or possible deletions of bases.

RNA phages Phages whose genetic material is RNA. They are the simplest phages known.

RNA polymerase The enzyme that polymerizes RNA by using DNA as a template. It can also act as a primase, initiating Okazaki fragments during DNA replication. (Also known as *transcriptase* or *RNA transcriptase*.)

RNA replicase A polymerase enzyme that catalyzes the self-replication of single-stranded RNA.

Robertsonian fusion Fusion of two acrocentric chromosomes at the centromere.

rolling-circle replication A model of DNA replication that accounts for a circular DNA molecule producing linear daughter double helices.

R plasmids Plasmids that carry genes that control resistance to various drugs.

rRNA Ribosomal RNA. RNA components of the subunits of the ribosomes.

rule of independent assortment *See* independent assortment, rule of.

rule of segregation *See* segregation, rule of.

S

sampling distribution The distribution of frequencies with which various possible events could occur or a probability distribution defined by a particular mathematical expression.

sarcoma Tumor of tissue of mesodermal origin (e.g., muscle, bone, cartilage).

satellite DNA Highly repetitive eukaryotic DNA primarily located around centromeres. Satellite DNA usually has a different buoyant density than the rest of the cell's DNA.

scaffold The eukaryotic chromosome structure remaining when DNA and histones have been removed; made from nonhistone proteins.

scanning hypothesis Proposed mechanism by which the eukaryotic ribosome recognizes the initiation region of an mRNA after binding the 5' capped end of it. The ribosome scans the mRNA for the initiation codon.

scientific method A procedure used by scientists to test hypotheses by making predictions about the outcome of an experiment before the experiment is performed. The results provide support or refutation of the hypothesis.

screening technique A technique to determine the genotype or phenotype of an organism.

secondary oocytes The cells formed by meiosis I in female animals.

secondary spermatocytes The products of the first meiotic division in male animals.

secondary structure The flat or helical configuration of the polypeptide backbone of a protein.

second division segregation (SDS) The allele arrangement in the spores of Ascomycetes with ordered spores that indicates a crossover between a locus and its centromere.

segregation, rule of Mendel's first principle describing how genes are passed from one generation to the next.

segregational load Genetic load caused when a population is segregating less fit homozygotes because of heterozygote advantage.

selection *See* natural selection.

selection coefficients, s, t The sum of forces acting to lower the relative reproductive success of a genotype.

selection-mutation equilibrium An equilibrium allelic frequency resulting from the balance between selection against an allele and mutation recreating this allele.

selective medium A culture medium that is enriched with a particular substance to allow the growth of particular strains of organisms.

selfed *See* self-fertilization.

self-fertilization Fertilization in which the two gametes are from the same individual.

selfish DNA A segment of the genome with no apparent function although it can control its own copy number.

semiconservative replication The mode by which DNA replicates. Each strand acts as a template for a new double helix. *See* template.

semisterility Nonviability of a proportion of gametes or zygotes.

sense strand *See* coding strand.

sex chromosomes Heteromorphic chromosomes whose distribution in a zygote determines the sex of the organism.

sex-controlled traits Traits that appear more often in one sex than in another.

sex-determining region Y (*SRY*) The sex switch, or testis-determining factor, in human beings, located on the Y chromosome. (*Sry* in mice.)

sexduction A process whereby a bacterium gains access to and incorporates foreign DNA brought in by a modified F factor during conjugation.

sex-influenced traits *See* sex-controlled traits.

Sex-lethal A gene in *Drosophila,* located on the X chromosome, that is a sex switch, directing development toward femaleness when it is in the "on" state. It is regulated by numerator and denominator elements that act to influence the genic balance equation (X/A).

sex-limited traits Traits expressed in only one sex. They may be controlled by sex-linked or autosomal loci.

sex linked The inheritance pattern of loci located on the sex chromosomes (usually the X chromosome in XY species); also refers to the loci themselves.

sex-ratio phenotype A trait in *Drosophila* whereby females produce mostly, if not only, daughters.

sex switch A gene in mammals, normally found on the Y chromosome, that directs the indeterminate gonads toward development as testes.

sexual selection The forces, determined by mate choice, acting to cause one genotype to mate more frequently than another genotype.

Shine-Dalgarno hypothesis A proposal that prokaryotic mRNA is aligned at the ribosome by complementarity between the mRNA upstream from the initiation codon and the 3′ end of the 16S rRNA.

siblings (sibs) Brothers and sisters.

sigma factor The protein that gives promoter-recognition specificity to the RNA polymerase core enzyme of bacteria.

signal hypothesis The major mechanism whereby proteins that must insert into or across a membrane are synthesized by a membrane-bound ribosome. The first thirteen to thirty-six amino acids synthesized, termed a *signal peptide,* are recognized by a *signal recognition particle* that draws the ribosome to the membrane surface. The signal peptide may be removed later from the protein.

signal peptide *See* signal hypothesis.

signal recognition particle *See* signal hypothesis.

single-strand binding proteins Proteins that attach to single-stranded DNA, usually near the replicating Y-junction, to stabilize the single strands.

sister chromatids *See* chromatids.

site-specific recombination A crossover event, such as the integration of phage lambda, that requires homology of only a very short region and uses an enzyme specific for that recombination.

skew A distortion of the shape of the normal distribution toward one side or the other.

snRNPs Small nuclear ribonucleoproteins; components of the spliceosome, the intron-removing apparatus in eukaryotic nuclei.

sociobiology The study of the evolution of social behavior in animals.

somatic doubling A disruption of the mitotic process that produces a cell with twice the normal chromosome number.

somatic hypermutation The occurrence of a high level of mutation in the variable regions of immunoglobulin genes.

somatic-mutation theory A theory to account for the high degree of antibody variability. It suggests that mutation of a basic immunoglobulin gene accounts for all the different types of immunoglobulins produced by B lymphocytes. *See* germ-line theory.

SOS box The region of the promoters of various genes that is recognized by the LexA repressor. Release of repression results in the induction of the SOS response.

SOS response Repair systems (recA, uvr) induced by the presence of single-stranded DNA that usually occurs from postreplicative gaps caused by various types of DNA damage. The RecA protein, stimulated by single-stranded DNA, is involved in the inactivation of the LexA repressor, thereby inducing the response.

Southern blotting A method, first devised by E. M. Southern, used to transfer DNA fragments from an agarose gel to a nitrocellulose gel for the purpose of DNA-DNA or DNA-RNA hybridization during recombinant DNA work.

spacer DNA Regions of nontranscribed DNA between transcribed segments.

specialized transduction Form of transduction based on faulty looping out by a temperate phage. Only neighboring loci to the attachment site can be transduced. *See* generalized transduction.

speciation A process whereby, over time, one species evolves into a different species (anagenesis) or whereby one species diverges to become two or more species (cladogenesis).

species A group of organisms belong to the same species if they are capable of interbreeding to produce fertile offspring.

sperm cells The gametes of males.

spermatids The four products of meiosis in males that develop into sperm.

spermatogenesis The process of sperm production.

spermatogonium A cell type in the testes of male vertebrates that gives rise to primary spermatocytes by mitosis.

spermiogenesis The process by which spermatids mature into sperm cells.

spindle The microtubule apparatus that controls chromosome movement during mitosis and meiosis.

spiral cleavage The cleavage process in mollusks and some invertebrates whereby the spindle at mitosis is tipped in relation to the original egg axis.

spirillum A spiral bacterium.

spliceosome Protein-RNA complex that removes introns in eukaryotic nuclear RNAs.

sporophyte The stage of a plant life cycle that produces spores by meiosis and alternates with the gametophyte stage.

stabilizing selection A type of selection that removes individuals from both ends of a phenotypic distribution thus maintaining the same distribution mean.

standard deviation The square root of the variance.

standard error of the mean The standard deviation divided by the square root of the sample size. It is the standard deviation of a sample of means.

stasipatric speciation Instantaneous speciation caused by polyploidy.

statistics Measurements of attributes of a sample from a population; denoted by Roman letters. *See* parameters.

stem-loop structure A lollipop-shaped structure formed when a single-stranded nucleic acid molecule loops back on itself to form a complementary double helix (stem), topped by a loop.

stepladder gel A DNA-sequencing gel. The numerous bands in each lane give the appearance of a stepladder.

stochastic A process with an indeterminate or random element as opposed to a deterministic process that has no random element.

stringent factor A protein that catalyzes the formation of an unusual nucleotide (guanosine tetraphosphate) during the stringent response under amino acid starvation.

stringent response A translational control mechanism of prokaryotes that represses tRNA and rRNA synthesis during amino acid starvation.

structural alleles Mutant alleles that are altered at identical base pairs.

structural genes Nonregulatory genes.

submetacentric chromosome A chromosome whose centromere lies between its middle and its end but closer to the middle.

subtelocentric chromosome A chromosome whose centromere lies between its middle and its end but closer to the end.

sum rule The rule that states that the probability of the occurrence of mutually exclusive events is the sum of the probabilities of the individual events.

supercoiling Negative or positive coiling of double-stranded DNA that differs from the relaxed state.

supergenes Several loci, which usually control related aspects of the phenotype, in close physical association.

suppressor gene A gene that, when mutated, apparently restores the wild-type phenotype to a mutant of another locus.

survival of the fittest In evolutionary theory, survival of only those organisms best able to obtain and utilize resources (fittest). This phenomenon is the cornerstone of Darwin's theory.

Svedberg unit A unit of sedimentation during centrifugation. Abbreviation is S, as in 50S.

swivelase *See* DNA gyrase.

sympatric speciation Speciation in which the evolution of reproductive isolating mechanisms occurs within the range and habitat of the parent species. This speciation may be common in parasites.

synapsis The point-by-point pairing of homologous chromosomes during zygotene or in certain dipteran tissues that undergo endomitosis.

synaptonemal complex A proteinaceous complex that apparently mediates synapsis during zygotene stage and then disintegrates.

synteny test A test that determines whether two loci belong to the same linkage group by observing concordance in hybrid cell lines.

synthetic medium A chemically defined substrate upon which microorganisms are grown.

T

TACTAAC box A consensus sequence surrounding the lariat branch point of eukaryotic mRNA introns.

target theory A theory that predicts response curves based on the number of events required to cause the phenomenon. Used to determine whether point mutations are single events.

TATA box An invariant DNA sequence at about −25 in the promoter region of eukaryotic genes; analogous to the Pribnow box in prokaryotes.

tautomeric shift Reversible shifts of proton position in a molecule. Bases in nucleic acids shift between keto and enol forms or between amino and imino forms.

T-cell receptors Surface proteins of T cells that allow the T cells to recognize host cells that have been infected.

telocentric chromosome A chromosome whose centromere lies at one of its ends.

telomerase An enzyme that adds telomeric sequences to the ends of eukaryotic chromosomes. No template is necessary.

telomere The ends of linear chromosomes that are required for replication and stability.

telophase The terminal stage of mitosis or meiosis in which chromosomes uncoil, the spindle breaks down, and cytokinesis usually occurs.

temperate phage A phage that can enter into lysogeny with its host.

temperature-sensitive mutant An organism with an allele that is normal at a permissive temperature but mutant at a restrictive temperature.

template A pattern serving as a mechanical guide. In DNA replication, each strand of the duplex acts as a template for the synthesis of a new double helix.

template strand *See* anticoding strand.

terminator sequence A sequence in DNA that signals the termination of transcription to RNA polymerase.

terminator stem A configuration of the leader transcript that signals transcription termination in attenuator-controlled amino acid operons.

tertiary structure The further folding of a protein, bringing α-helices and β-sheets into three-dimensional arrangements.

testcross The cross of an organism with a homozygous recessive organism.

testing of hypotheses The determination of whether to accept or reject a proposed hypothesis based on the likelihood of the experimental results.

testis-determining factor (*TDF*) General term for the gene determining maleness in human beings (*Tdf* in mice).

tetrads The meiotic configuration of four chromatids first seen in pachytene. There is one tetrad (bivalent) per homologous pair of chromosomes.

tetranucleotide hypothesis The hypothesis, based on incorrect information, that DNA could not be the genetic material because its structure was too simple—that is, that repeating subunits contain one copy each of the four DNA nucleotides.

tetraploids Organisms with four whole sets of chromosomes.

tetratype (TT) A spore arrangement in Ascomycetes that consists of two parental and two recombinant spores, indicating a single crossover between two linked loci.

theta structure An intermediate structure formed during the replication of a circular DNA molecule.

three-point cross A cross involving three loci.

thymine *See* pyrimidines.

topoisomerase An enzyme that can relieve (or create) supercoiling in DNA by creating transitory breaks in one (type I) or both (type II) strands of the helical backbone.

topoisomers Forms of DNA with the same sequence but differing in their linkage number (coiling).

totipotent The state of a cell that can give rise to any and all adult cell types, as compared with a differentiated cell whose fate is determined.

trailer The length of mRNA from the nonsense codon to the 3′ end or, in polycistronic mRNAs, from a nonsense codon to the next gene's leader.

trans Meaning "across" and referring usually to the geometric configuration of mutant alleles across from each other on a homologous pair of chromosomes.

trans-acting Referring to mutations of, for example, a repressor gene, that act through a diffusable protein product; the normal mode of action of most recessive mutations.

transcription The process whereby RNA is synthesized from a DNA template.

transcription factors Eukaryotic proteins that aid RNA polymerase to recognize promoters. Analogous to prokaryotic sigma factors.

transducing particle A defective phage, carrying part of the host's genome.

transduction A process whereby a cell can gain access to and incorporate foreign DNA brought in by a viral particle.

transfection The introduction of foreign DNA into eukaryotic cells.

transfer operon (*tra*) Sequence of loci that impart the male (F-pili–producing) phenotype on a bacterium. The male cell can transfer the F plasmid to an F⁻ cell.

transform In oncology, the conversion of a normal cell into a cancerous one.

transformation A process whereby prokaryotes take up DNA from the environment and incorporate it into their genomes, or the conversion of a eukaryotic cell into a cancerous one.

transformer An allele in fruit flies that converts chromosomal females into sterile males.

transgenic Eukaryotic organisms that have taken up foreign DNA.

transition mutation A mutation in which a purine/pyrimidine base pair is replaced with a base pair in the same purine/pyrimidine relationship.

translation The process of protein synthesis wherein the primary structure of the protein is determined by the nucleotide sequence in mRNA.

translocase (EF-G) Elongation factor in prokaryotes necessary for proper translocation at the ribosome during the translation process. Replaced by eEF2 in eukaryotes.

translocation A chromosomal configuration in which part of a chromosome becomes attached to a different chromosome. Also a part of the translation process in which the mRNA is shifted one codon in relation to the ribosome.

transposable genetic element A region of the genome, flanked by inverted repeats, a copy of which can be inserted at another place; also called a transposon or a jumping gene.

transposon *See* transposable genetic element.

transversion mutation A mutation in which a purine replaces a pyrimidine or vice versa.

trihybrid An organism heterozygous at three loci.

triploids Organisms with three whole sets of chromosomes.

trisomic A diploid cell with an extra chromosome.

tRNA Transfer RNA. Small RNA molecules that carry amino acids to the ribosome for polymerization.

true heritability *See* heritability.

true speciation *See* cladogenesis.

tumor Abnormal growth of tissue.

tumor suppressor genes Genes that normally control unlimited cellular growth. When both copies of the gene are mutated, cellular transformation follows. Examples are the *p53* gene and the genes for retinoblastoma and Wilm's tumor.

two-point cross A cross involving two loci.

two-strand double crossover A double crossover that involves only two of the four chromatids of a tetrad.

type I error In statistics, the rejection of a true hypothesis.

type II error In statistics, the accepting of a false hypothesis.

typological thinking The concept that organisms of a species conform to a specific norm. In this view, variation is considered abnormal.

U

unequal crossing over Nonreciprocal crossing over caused by mismatching of homologous chromosomes. Usually occurs in regions of tandem repeats.

uninemic chromosome A chromosome consisting of one double helix of DNA.

unique DNA A length of DNA with no repetitive nucleotide sequences.

unmixed families Groups of four codons sharing their first two bases and coding for the same amino acid.

unusual bases Other bases, in addition to adenine, cytosine, guanine, and uracil, found primarily in tRNAs.

upstream A convention on DNA related to the position and direction of transcription by RNA polymerase ($5' \rightarrow 3'$). Downstream (or $3'$ to) is in the direction of transcription whereas upstream ($5'$ to) is in the direction from which the polymerase has come.

uracil *See* pyrimidines.

V

variable-number-of-tandem-repeats (VNTR) loci Loci that are hypervariable because of tandem repeats. Presumably, variability is generated by unequal crossing over.

variance The average squared deviation about the mean of a set of data.

variegation Patchiness; a type of position effect that results when particular loci are contiguous with heterochromatin.

vehicle plasmid A plasmid containing a piece of passenger DNA; used in recombinant DNA work.

virion A virus particle.

viroids Bare RNA particles that are plant pathogens.

V-J joining The joining of a variable gene and a joining gene in the first step of the formation of a functioning immunoglobulin gene.

W

Wahlund effect A subdivided population contains fewer heterozygotes than predicted despite the fact that all subdivisions are in Hardy-Weinberg proportions.

western blotting A technique for probing for a particular protein using antibodies. *See* Southern blotting.

wild-type The phenotype of a particular organism when first seen in nature.

Wilm's tumor A childhood kidney cancer caused by the inactivation of an anti-oncogene.

wobble Referring to the reduced constraint of the third base of an anticodon as compared with the other bases thus allowing additional complementary base pairings.

X

X-inactivation center (*XIC*) Locus on the X chromosome in mammals at which inactivation is initiated.

X linked *See* sex linked.

X-ray crystallography A technique, using X rays, to determine the atomic structure of molecules that have been crystallized.

xeroderma pigmentosum A disease in human beings caused by a defect in the UV mutation repair system.

Y

yeast artificial chromosome (YAC) Originating from a bacterial plasmid, a YAC contains additionally a yeast centromeric region (CEN) and a yeast origin of DNA replication (ARS). YACs are capable of cloning very large pieces of DNA.

Y-junction The point of active DNA replication where the double helix opens up so that each strand can serve as a template.

Y linked Inheritance pattern of loci located on the Y chromosome only. Also refers to the loci themselves.

Z

Z DNA A left-handed form of DNA found under physiological conditions in short GC segments that are methylated. It may be important in regulating gene expression in eukaryotes.

***ZFY* gene** Originally believed to be the human male sex-switch gene, located on the short arm of the Y chromosome. *ZFY* stands for zinc finger on the Y chromosome.

zinc finger Configuration of a DNA-binding protein that resembles a finger with a base, usually cysteines and histidines, binding a zinc ion. Discovered in a transcription factor in *Xenopus*.

zygonema (zygotene stage) The stage of prophase I of meiosis in which synapsis occurs.

zygotic induction When a prophage is passed into an F⁻ cell during conjugation it may begin vegetative growth.

zygotic selection The forces acting to cause differential mortality of an organism at any stage (other than gametes) in its life cycle.

Index

Panmictic population, 544
Paper chromatography, of peptides, 285–86
Papillary cystadenocarcinoma of ovary, 426
Papovavirus, 429
Paracentric inversion, 171
Paracodon, 275
Paramecin, 494
Paramecium
    autogamy in, 492–94
    conjugation in, 492
    kappa particles in, 493–95
    killer trait in, 492–95
    life cycle of, 492
    mate-killer infection in, 494–95
    mu particles in, 494–95
Parameter, 507
Paramyxovirus, 429
Parapatric speciation, 561–63
Parental, 100
Parental ditype (PD) ascus, 118–19
Parental imprinting. *See* Imprinting
Parthenogenesis, 58
Partial diploid. *See* Merozygote
Partial dominance, 22
Partial restriction digest, 326–27
Parvovirus, 429
Pascal's triangle, 66
Passenger DNA, 306
Patau syndrome. *See* Trisomy 13
Patch formation, 472
Patent, on genetically engineered organism, 321, 323
Path diagram, 534–36
Pattern baldness, 82
Pauling, Linus, 205, 570
Pauling (unit), 570
PBS. *See* Primer-binding site
PCR. *See* Polymerase chain reaction
PD ascus. *See* Parental ditype ascus
Pea. *See* Garden pea
Pedigree analysis, 82–87, 121
    of dominant trait, 83–84
    with inbreeding, 534–36
    of recessive trait, 84–85
    symbols used in, 83–84
    of X-linked trait, 85–87
Peer review, 5
Penetrance, 82–83
Penicillin
    microbial origin of, 143
    mode of action of, 143
    resistance to, 141
    selection technique using, 145–46
Penicillium, antibiotics of, 143
Peppered moth, 575
Pepsin, 285
Peptic ulcer, 515
Peptidase gene, 123, 127
Peptide
    electrophoresis of, 285–86
    paper chromatography of, 285–86
Peptide bond, 269
    formation of, 279, 281–82
Peptide map, 285
Peptidyl transferase, 279, 281, 287, 291–92
Peptidyl-tRNA, 277
Pericentric inversion, 171–72
Permissive temperature, 454
PEST hypothesis, 375–77
Petite mutant, yeast, 487–88
Petri plate, 140
Petunia, genetically engineered, 373
p53 gene, 428, 438
Phage. *See also* Virus
    chromosomes of, 136
    control of gene expression in, 348–77
    Hershey and Chase experiment with, 201–2
    life cycle of, 152–56
    linkage and mapping in, 135–61

mRNA of, 235
    phenotypes of, 141–42
    recombination in, 152–55
    replication in, 227
    as research organism, 6, 136–37
    sexual processes in, 143–52
    temperate, 155–56
    transducing. *See* Transduction
    as vector, 304, 306–7
Phage f2, 261
Phage lambda, 140, 152, 154–57, 240–41, 306–9, 470
    early transcription in, 361–62
    generation interval of, 60
    late transcription in, 361–62
    linear forms of, 361
    lysogenic cycle in, 360–65
    lytic cycle in, 360–65
    map of, 360
    operons of, 360–65
Phage lambda repressor, 354, 360–66
Phage M13, 337, 341
Phage MS2, 261, 374, 399
Phage P22, 156
Phage φX174, 240, 337–40
    map of, 338
Phage Qβ, 261, 299
Phage R17, 261
Phage receptor, 161
Phage resistance, 141, 154, 161, 442–44
Phage SP8, 236
Phage T1, 152, 161, 442–44
Phage T2, 136, 141, 144, 152, 154, 201–2, 235, 348
Phage T3, 152
Phage T4, 143, 152–53, 161, 236, 290–91, 363, 371, 399, 446–50, 465
Phage T5, 152
Phage T6, 152, 161
Phage T7, 152, 240
Phaseolus vulgaris. See Bean
Phenocopy, 83, 482
Phenotype, 5
    definition of, 20
    environmental effects in distribution of, 504–5
    mate choice and, 531
Phenotypic ratio, 28–29, 36
Phenylalanine, 268
Phenylalanine tRNA, 250
Phenylisothiocyanate (PITC), 285–86
Phenylketonuria (PKU), 35, 64, 529, 552
pho genes, 161
Phosphate group, 208
Phosphodiester bond, 205–8
3-Phosphoglycerate dehydrogenase, 161
Phosphomannose isomerase, 160
Phosphoribosyl anthranilate transferase, 161
Phosphorylation, of proteins, 48, 433
Phosphoserine phosphatase, 161
Photoreactivation, 465
phr gene, 465, 471
Phyletic evolution. *See* Anagenesis
Phyletic gradualism, 563–64
Phylogenetic tree, 570–71
Physical containment, 322–23
Physical isolation mechanisms, 561
Picornavirus, 429
Pigeon, chromosome number in, 47
Pigmentation, 32–35
    in moths, 483–84
pil gene, 161
Pili, 148
Pisum sativum. See Garden pea
PITC. *See* Phenylisothiocyanate
Pitcairn Island population, 546
PKU. *See* Phenylketonuria

Plant
    cloning vectors for, 319–21
    crop. *See* Crop plant
    evolution of, 187–88
    genetic engineering in, 319–21
    life cycle of, 45
    meiosis in, 44, 54, 56
    mitochondrial inheritance in, 485
    mitosis in, 51, 74
    polyploidy in, 187–88
    transgenic, 319–25, 342, 373
    variegation in, 489–90
Plant breeding, 511–12, 514
Plant virus, 138
Plaque, viral, 140–42, 144, 154
Plasmid, 10, 45, 495–97. *See also* F factor
    chimeric, 306
    copy number of, 372
    uncovering of, 496–97
    as vector, 304, 306–10, 315, 317, 322
    in yeast, 317–19
Plasmid pBR313, 315–16
Plasmid pBR322, 307, 309, 319
Plasmid pCL1, 317–18
Plasmid pCL2, 317–18
Plasmid R222, 496
Plasmid Ti, 319, 321, 323–25
Plastid, 488–89
Platelet-derived growth factor, 433
Pleiotropy, 38
Ploidy, sex determination by, 75
pls genes, 161
Pneumocystis pneumonia, 434
Point mutation, 452–54
Polar body, 57–58
Polar mutant, 365
Polar nucleus, 59
pol genes, 161, 228, 430–32, 436
Poliovirus, 127, 139
Pollen grain, 59, 490
Poly-A polymerase, 253
Poly-A tail, on mRNA, 253, 255, 312
Polycistronic mRNA, 283–84, 350, 382
Polydactyly, 83–84
Poly-dA/poly-dT cloning technique, 310–12
Polygalacturonase, 342
Polygenic inheritance, 502–6
    in beans, 510–11
    in humans, 506, 515–17
    location of genes, 505–6
Polymerase chain reaction (PCR), 333, 339
Polymorphism, 11, 561–62
    genetic, 564
    maintenance of, 564–68
    rate of, 564
    survey of genic heterozygosity, 568
    transient, 567
Polynucleotide kinase, 326
Polynucleotide phosphorylase, 294
Polyploidy, 76, 167, 186
    in animals, 187–88
    in plants, 187–88
    speciation and, 562
Polyribosome, 282–84
Polysome. *See* Polyribosome
Polytene chromosome, 108, 381, 389–91, 395
Poly-T primer, 312
POP' site, 361–62
Population, 523
    mean fitness of, 549–50
    size of, 524–25
    small-sized, 544–47
Population analysis, in inbreeding, 535–36
Population genetics, 11
    computer modeling in, 553
    Hardy-Weinberg equilibrium, 522–36

    mating systems, 522–36
    models for, 541
    processes that change allelic frequencies, 540–54
Population statistics, 506–11
Porin, 372–73
Position effect, 169, 172, 176, 444–45
Positive interference, 106
Postreplicative repair, 467–71
Posttranscriptional modification, of mRNA, 250, 253, 255, 262–63
Posttranslational control, in prokaryotes, 374–77
Postzygotic isolation mechanisms, 560
Potato spindle tuber viroid, 257–58
Poultry
    body weight in, 513
    egg production in, 513
Poxvirus, 429
p105 protein, in retinoblastoma, 428
Prader-Willi syndrome, 484
Predation, 573, 575
Prediction, 4, 64
Preemptor stem, 359
Preer, J., 493
Prezygotic isolation mechanisms, 560
Pribnow box, 238, 240–42, 372
Primase, 218, 220, 223–25, 228, 247
Primer
    for dideoxy method of DNA sequencing, 337–39
    DNA, 218
    replication, 223
Primer-binding site (PBS), 430
Primosome, 223–25
Prion, 203, 263
Prion protein, 203
Probability, 63–65, 122
Probability theory, 64
Proband, 84
Probe. *See* Gene probe
Processivity, 218
Product rule, 65–66
Proflavin, 290–91, 461, 463
pro genes, 161
Progeny testing, 21
Prokaryote, 45, 136
    chromosomes of, 136
    control of gene expression in, 348–77
    DNA of, 144
    eukaryote vs., 46, 382
    linkage in, 135–61
    mapping in, 135–61
    mRNA in, 283–84
    plasmids in, 495–97
    ribosomes of, 245, 291
    RNA polymerase of, 247
    transcriptional control in, 348–72
    transcription in, 250–51
    translation in, 10, 250–51, 282–83
Prolactin, 290
Proline, 268–69
Promoter, 236–41, 244
    consensus sequences in, 238, 240–41, 251–52
    efficiency of, 372
    Escherichia coli, 241
    in eukaryote, 251–52, 416
    lac operon, 350–51, 372
    in nuclease hypersensitive sites, 386–87
    phage lambda, 241, 361–64, 366–67
    phage T4, 371
Proofreading, in replication, 218, 220–21, 465
Prophage, 155–56, 360–61, 470
Prophase
    meiosis I, 53–54
    meiosis II, 54, 56
    mitotic, 48–49, 51
Proplastid, 489–90

RNA polymerase II, 247, 251–52
RNA polymerase III, 246–47
RNA primer, 220, 223–24
RNA replicase, 261
RNA transcript, 244–45
  leader region, 245
  primary, 253
  trailer region, 245
RNA virus, 10, 137–38, 144, 202–3, 261, 263, 428–29
Robertson, W., 175
Robertsonian fusion, 175–76
Rolling-circle model, of replication, 227–30
Root tip, mitosis in, 51
ros oncogene, 432
Rounding to a ratio, 64
Roundworm. *See Ascaris*
Rous, Peyton, 105, 429–30
Rous sarcoma virus, 261, 429, 433
R plasmid, 495–96
rpo genes, 161, 228, 372
RPS4X gene, 184
RPS4Y gene, 184
rRNA. *See* Ribosomal RNA
RTF. *See* Resistance transfer factor

S

Saccharomyces cerevisiae. See Yeast
Sager, Ruth, 490–91
Saintpaulia. See African violet
Salmonella typhimurium
  in Ames test, 456
  flagella of, 370–71
Sample variance, 507–8
Sampling distribution, 67–69
Sampling error, 544–45
Sanger, Frederick, 105, 334
Sarcoma, 425, 428
  Kaposi's, 434
Satellite DNA, 392–94, 399
Scaffold protein, 387, 389
Scanning hypothesis, 277
Scatter plot, 509–10
Schizophrenia, 515, 517
Schulman, L., 275
Scientific journals, 5
Scientific method, 4–5, 64
Screening technique, 141
Seasonal isolation mechanisms, 560
Sea urchin, histone genes of, 400
Second division segregation, 115–18
Second filial generation. *See* F₂ generation
Second genetic code, 275
Second messenger, 354
Segmentation genes, 411–13
Segregation, 7, 17, 55
  adjacent-1, 174–75, 177
  adjacent-2, 175, 177
  alternate, 174, 177
  first division, 115–18
  in *Neurospora*, 115–18
  rule of, 19–22
  second division, 115–18
  after translocation, 174–77
Segregational load, 566
Segregational petite, 487
Selection
  artificial, 547, 559
  natural. *See* Natural selection
  potential response to, 512. *See also* Heritability
Selection coefficient, 547, 550
Selection experiment, 511–12
Selective medium, 140
Selenocysteine, 299
Self-fertilization, 16, 21–22, 512, 532
Selfing. *See* Self-fertilization
Selfish DNA, 261, 371

Self-splicing, 254–57
Semiconservative replication, 212–13, 382–83
Semisterility, 171
Sequenator, 286
Sequence, -10. *See* Pribnow box
Sequence, -35, 240–41, 372
ser genes, 161
Serine, 268
Serine deaminase, 160
Serine hydroxymethyl transferase, 160
Serum proteins, 89
Sex chromatin. *See* Barr body
Sex chromosome, 46
  abnormal number of, 183–84
  compound systems, 79
  nondisjunction of, 179–80
  XO system, 75, 78–79
  XY system, 75–78
  ZW system, 79
Sex-controlled trait. *See* Sex-influenced trait
Sex determination
  in birds, 79
  in *Drosophila*, 76–77
  in fish, 79
  genic balance theory of, 76–77
  in humans, 75–78
  in insects, 78–79
  in mice, 78
  patterns of, 75
Sex-determining region Y, 78
Sexduction, 143, 145, 151–52
Sex-influenced trait, 82
Sex-lethal gene. *See Sxl* gene
Sex-limited trait, 82
Sex-linked trait, 79–82
  in *Drosophila*, 79–81
  recessive, 85
Sex pilus. *See* F-pilus
Sex ratio, 65–66
Sex-ratio phenotype, in *Drosophila*, 495
Sex-reversed individual, 77–78
Sex-switch gene, 77
Sexual processes, in bacteria, 143–52
Sexual selection, 547
Shapiro, J. A., 367
Sharp, Philip A., 254
Sheep
  generation interval of, 60
  horns in, 82
  wool and fleece length in, 513
Shepherd's purse, seed capsule shape in, 36
Sheppard, P. M., 573
Sherman, P., 577
Shigella, R plasmids in, 496
Shikimic acid, 160
Shine-Dalgarno sequence, 277, 279, 283–84, 374
Shub, D., 261
Sibling, 83
Sickle-cell anemia, 38, 400, 564–66
Sigma factor, 236, 241–42, 244, 371–72
Signal hypothesis, 287–90
Signal peptidase, 288–89
Signal peptide, 287–89
Signal recognition particle (SRP), 288–89
Signal sequence, 486
Signal-sequence receptor, 289
Significance, level of, 71
Singer, B., 203
Single-strand binding protein (ssb protein), 224–25, 228, 467, 474
Single transformant, 146
sis oncogene, 432–33
Sister chromatid exchange, 383
Sister chromatids, 46–49, 54
Site-directed mutagenesis, 458–60

Site-specific recombination, 361, 420–24
Site-specific variation, in genetic code, 299
Skin color, 515–16
ski oncogene, 432
Slipper limpet, sex determination in, 75
Small nuclear ribonucleoprotein (snRNP), 259–60
Smith, Hamilton, 105, 304
Snail, shell attributes in, 172, 482–83
Snapdragon, flower color in, 32–36
Snell, George, 105
snRNP. *See* Small nuclear ribonucleoprotein
Sociobiology, 11, 572–77
Sodium channel, 290
Soft selection, 566
Solenoid model, for chromatin, 387–89
Somatic-cell hybridization, 123–28
Somatic crossing over, 119–21
Somatic doubling, 187
Somatic hypermutation, 421
Sonneborn, Tracy, 492
SOS box, 470–71
SOS response, 364–65, 470–71
Southern, E. M., 315
Southern blotting, 314–15
Spacer DNA, 247–48, 400
Spacer RNA, 246
Specialized transduction, 156–57
Speciation
  allopatric, 561–63
  evolution and, 558–64
  forms of, 559–60
  parapatric, 561–63
  sympatric, 561–63
  true. *See* Cladogenesis
Species, definition of, 523, 560
Species concept, 560
Spermatid, 57–58
Spermatocyte, 57–58
Spermatogenesis, 57–58
Spermatogonium, 57–58
Sperm cells, 57–58
Spermiogenesis, 57–58
Sperm nucleus, 59
S phase, 48
Spina bifida, 517
Spindle apparatus, 45, 48–50, 52–53, 187
Spiral cleavage, in invertebrates, 483
Spirillum, 136
Spirochete, and sex-ratio trait in *Drosophila*, 495
Spliceosome, 257–60
Splicing, 254–59, 401, 421, 423–24, 432
  alternative, 259, 421
Spontaneous mutagenesis, 454–55
Spontaneous mutation, 452
Spore
  formation of, 47, 50–58
  fungal, 112–19
  isolation in *Neurospora*, 113
  ordered, 113
  unordered, 118–19
Sporophyte, 45, 52, 58–59
Sporulation genes, 238
src oncogene, 432–33
SRP. *See* Signal recognition particle
SRY gene, 78
ssb gene, 228
ssb protein. *See* Single-strand binding protein
Stabilizing selection, 547–48
Stable mutational equilibrium, 542
Stadler, Lewis J., 452
Stage-specific chromosome puffs, 390–91
Stahl, Franklin W., 212–13
Stanbridge, E., 427
Standard deviation, 67, 507–8

Standard error of mean, 508
Starvation, amino acid, 359, 374
Stasis, 563
Statistic, 507
Statistics, 64, 67–71
  population, 506–11
Stature, 127
Stebbins, G. L., 560
Steitz, Joan A., 259
Stem-loop structure, 242–44, 357–59, 373, 422
Stepladder gel, 337, 339
Sterile caste system, in insects, 576–77
Stern, Curt, 119–20
Steroid sulphatase gene, 79, 91
Steward, F., 406
Stochastic event, 64
Strawberry, 167
Streptococcus pneumoniae, transformation in, 200–201
Streptomyces, antibiotics of, 143
Streptomycin, 143, 291, 486, 491
str gene, 161
Stringent factor, 374–75
Stringent response, 374–75
Strong inference, 4
Structural allele, 444–47
Structural protein, 198
Sturtevant, Alfred H., 110, 176
Submetacentric chromosome, 45–46
Substrate transition, 455
Subtelocentric chromosome, 45
N-Succinyl-diaminopimelic acid deacylase, 160
suc gene, 161
Sucrose density-gradient centrifugation, 245
sul genes, 470–71
Summer squash, shape of, 36
Sum rule, 64–65
supB gene, 161
Supercoiled DNA, 224–27, 241–42
Supergene, 172
Superinfection, 155, 363
Suppression
  crossover, 171
  intergenic, 461–65
  intragenic, 452–53
Suppressive petite, 487–88
Suppressor gene, 461–65
Surface antigen 1 gene, 127
Survival of fittest, 558–59
Sutton, Walter, 60
SV40, 240, 319–20, 339, 428
Svedberg unit, 245
Sweet pea, flower color in, 36
Swine
  birth weight in, 513
  growth rate of, 513
  litter size in, 513
Sxl gene, 77
Symmetry, bilateral, 408, 410
Sympatric speciation, 561–63
Synapsis, 53
Synaptonemal complex, 52–53
Synteny test, 123
Synthetic medium, 139

T

TACTAAC box, 259
Takano, I., 417
TAR sequence, 436
TATA box, 251–52
tat gene, 435–36
Tatum, Edward L., 7, 36–38, 105, 147–48, 151
Tautomeric shift, 454–55
Taylor, J., 382
Tay-Sachs disease, 23, 552
T-cell receptor, 419–21, 424–25